ECOLOGICAL STRESS AND THE NEW YORK BIGHT: SCIENCE AND MANAGEMENT

Garry F. Mayer
Editor
National Oceanic and Atmospheric Administration
Office of Marine Pollution Assessment
Stony Brook, New York

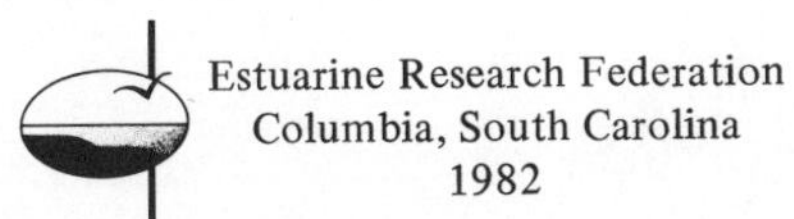

Estuarine Research Federation
Columbia, South Carolina
1982

Proceedings of a symposium on the Ecological Effects of Environmental Stress, New York, New York, June 10-15, 1979.

Symposium Sponsors

National Oceanic and Atmospheric Administration
Office of Marine Pollution Assessment
Estuarine Research Federation
New York Sea Grant Institute
New Jersey Marine Sciences Consortium

First published in 1982 by the Estuarine Research Federation, Columbia, South Carolina.
Printed in the United States of America.

Distributed by the Estuarine Research Federation
Belle W. Baruch Institute for Marine Biology and Coastal Research
University of South Carolina
Columbia, South Carolina 29208.

Library of Congress Catalog Card Number 82-71795
ISBN 0-9608990-0-6

Publication of this volume was supported by grants from the NOAA Office of Marine Pollution Assessment.

Cover illustration: Oil slick surrounds Statue of Liberty, New York harbor. (Courtesy EPA Documerica, Chester Higgins.)

Dedicated to

Bostwick H. Ketchum
1912 - 1982

Premier oceanographer and
pioneer in the study
of the New York Bight

CONTENTS

Contributed Papers on the Effects of Pollutants on Fishes

Contributed Papers on the Effects of Pollutants on Plankton and Neuston

PREFACE

An often-voiced criticism of the environmental regulatory process is the failure to maintain adequate lines of communication among legislators and managers charged with developing and implementing environmental policy and scientists engaged in environmental research. The former two groups have a need for definitive information and unequivocal pronouncements on the environmental effects or ecological implications of specific actions. Members of the latter group, on the other hand, often are unable to respond precisely. They are trained to report on observations, to avoid speculation, and to enumerate the strengths and weaknesses of conclusions. Such opposing approaches have estranged these groups and have limited their interactions.

In 1979, the National Oceanic and Atmospheric Administration's (NOAA) Marine EcoSystems Analysis (MESA) New York Bight Project organized and funded a symposium that attempted to bridge the gap between scientists and policymakers. Its purpose was to summarize evidence on the effects of human wastes on the biota and ecosystems of the New York Bight and to explore whether this information could be used to improve the management of major sources of marine pollution to the region. Entitled "Ecological Effects of Environmental Stress," the symposium brought together members of government, academia, industry, environmental groups, and the citizenry for five days of discussion and debate. Engineers, lawyers, administrators, public officials, and scientists served on panels that met in New York City between June 11 and June 15, 1979. Findings of these panels, together with background papers prepared by individual panel members, are presented in this volume.

The organization of the volume reflects that of the symposium. In all, nine panels were convened: five on science and four on management. Three science panels reviewed organismic effects of pollutants on fishes, bottom-dwelling organisms, and plankton/neuston (*i.e.*, effects on cells, tissues, and individual plants and animals). The remaining two science panels explored implications of toxicant additions and nutrient and carbon loadings to communities and ecosystems of the New York Bight. Reports of the science panels are presented in the first section of the volume.

Management panels focused on problems associated with domestic wastes (*i.e.*, sewage sludge, raw sewage, and sewer effluent), dredged materials, industrial wastes, and non-point sources of wastes. On the basis of the science panels' assessments of environmental effects in the Bight and their causes, management panels were asked to review major pollution sources in the New

York - New Jersey metropolitan area and to explore ways of reducing or eliminating the input of materials identified as being of concern. Reports of the management panels are found in the second section of this volume.

The final five sections of the volume are composed of technical papers originally prepared as source documents for the June 1979 symposium. These papers were contributed by members of the science and management panels and are arranged in accordance with the composition of the science panels. Manuscripts by management panelists are grouped with those of the two science panels on community and ecosystem effects.

Panel reports contain extensive cross references to contributed papers. This was done to provide uninitiated readers with an easy entree to the vast literature on waste disposal problems in the New York Bight and to facilitate the task of more knowledgeable readers in locating specific details of issues covered in technical papers. Contributed papers may be of special interest to those in search of syntheses and reviews of pollution research pertinent to the New York Bight. Extensive bibliographies are presented at the close of most contributed papers.

Panel reports are aimed at a wider readership than are the contributed papers. The former represent broad-ranging summaries written for readers who may lack scientific or engineering backgrounds. Efforts were made in editing panel reports to reduce jargon and to define technical terms that may be unfamiliar to the layperson. Contributed papers, on the other hand, are detailed technical reports on specific research efforts. They are similar in tone to papers published in scientific journals. By adopting a two-part format, the Editorial Board sought to produce a volume on the New York Bight of use to a range of readers from technical specialists to governmental leaders, decision-makers, members of the public, and students who may lack extensive technical backgrounds.

ACKNOWLEDGEMENTS

Many individuals and organizations contributed to the organization of the symposium and production of this book. Although NOAA funded both activities in entirety, invaluable assistance and support were provided by three co-sponsors: the New York Sea Grant Institute, the Estuarine Research Federation, and the New Jersey Marine Sciences Consortium. The New York Sea Grant Institute was instrumental in arranging the details of the 1979 symposium; the Estuarine Research Federation shouldered much of the burden for publishing and distributing this volume. I am particularly indebted to the support and advice of J.S. O'Connor, R.L. Swanson, D.F. Squires, F.J. Vernberg, and R.B. Abel of these organizations.

Many people have given generously of their time and have significantly influenced the form and content of the volume. Chief among these are the members of the Editorial Board, who conceived of the book's format, provided guidelines for assembling the manuscripts, and implemented the peer review process. Editorial assistance was furnished by S. Stoeppelwerth and S. Risoli. Special thanks are extended to R.K. Monahan for her patient and tireless attention to the myriad details associated with this volume. Ms. Monahan assisted in all aspects of the preparation of this book and is largely responsible for developing the index. N. Porthouse assisted in the latter task, entering index items into a computer for storage and manipulation.

Typographical and graphic arts services were furnshed by Stan Hitt Graphics, Inc., Washington, DC.

Garry F. Mayer
April 1982

TAKING CLAMS OR OTHER
SHELLFISH FROM THIS UNCERTIFIED
AREA IS PROHIBITED.
CLOSED AREA
VIOLATIONS ARE PUNISHABLE BY
FINE, JAIL SENTENCE OR BOTH
N.Y.S. DEPT. OF
ENVIRONMENTAL CONSERVATION

REPORTS OF SCIENCE PANELS

Opposite: Harvesting valuable fin- and shellfish resources from areas of the New York Bight, portions of the Hudson-Raritan estuary, and segments of adjacent coastal waters is restricted because of contamination by pathogens or toxicants. Scientific investigations have attempted to characterize the effects of pollution on organisms and ecosystems within the region and to examine the implications of such effects to human populations. (Courtesy Garry F. Mayer, NOAA Office of Marine Pollution Assessment.)

EFFECTS OF POLLUTANTS ON BENTHOS

M.R. Carriker[1]
J.W. Anderson
W.P. Davis
D.R. Franz[2]
G.F. Mayer[2,3]
J.B. Pearce
T.K. Sawyer
J.H. Tietjen[2]
J.F. Timoney
D.R. Young

Abstract. Intensive quantitative ecological studies of the benthos of the New York Bight were not begun until the early 1960's, by which time significant alteration of benthic communities had already occurred. This paper summarizes investigations on the effects of organic loading, petroleum hydrocarbons, synthetic organics (*e.g.*, pesticides and industrial compounds), and toxic metals on the structure and function of benthic communities of the New York Bight.

INTRODUCTION

The sea bottom, especially in shallow waters, is an important site for organic decomposition and nutrient regeneration (Rowe and Smith, 1977; Rowe, 1978). These and related functions of the benthos have taken on added significance as coastal waters have become the repository for society's toxic and non-toxic wastes. In heavily impacted areas such as the New York Bight, the use of benthic ecosystems to process and store wastes competes with more traditional uses as a direct source of food for humans, and as feeding and breeding areas for commercially and recreationally important species of finfish and shellfish. Unfortunately, these rival uses are often incompatible and may lead to the destruction of existing assemblages of bottom-dwelling organisms, loss of valuable fisheries resources, reduction in the ability of the benthos to cope with pollutants, and generation of human health hazards. The purpose of this document is to summarize current knowledge of the effects of human wastes on the benthos of the New York Bight and adjacent waters, and to explore the consequences of alterations in the diversity and abundance of the organisms that compose these ecosystems.

BENTHIC ENVIRONMENT OF THE NEW YORK BIGHT AREA

The distribution and abundance of benthos in the New York Bight area are determined by both natural and human-related factors. Among the most important natural factors are sediment composition and salinity. Sandy bottoms

[1]Chairman, Panel on the Effects of Pollutants on Benthos.
[2]Compiler of revised panel report.
[3]Not original member of panel.

predominate over most of the inner continental shelf, but within the estuary and adjacent areas of the shelf directly influenced by the estuary, significant amounts of fine sediments (silts and clays) are deposited. Species adapted to life in fine sediments such as silts and muds generally are not able to tolerate shifting sandy bottoms, where organic matter fails to accumulate and where animals are in danger of being buried by unstable sediments. Likewise, animals living on hard bottoms such as gravel or boulders often lack physiological and behavioral mechanisms to withstand the accumulation of particulate material, or to tolerate low dissolved oxygen conditions that may develop in softer bottom sediments.

Although each species responds independently to its physical and chemical environment, groups of species with similar requirements co-occur and form assemblages or communities. In the more saline portions of the lower estuary and inner New York Bight, where bottom salinities remain relatively constant, benthic assemblages occur as a patchwork related to differences in sediment composition. Moving into the harbor and up-river, the nature of assemblages associated with particular substrates gradually changes with the disappearance of species less tolerant of freshwater.

Deposition of fine sediments occurs in all estuaries and in many coastal areas including the New York Bight. However, the Bight is unusual in that silts and clays originating in the Hudson-Raritan estuary are highly contaminated by a variety of carbon-rich materials as well as toxic metals and toxic organic compounds. Contaminated particulates are washed out of the estuary and redeposited in the New York Bight. In addition, portions of the inner Bight receive massive doses of contaminated particulates through the ocean disposal of dredged materials and sewage sludge. Particulates from each of these sources are responsible for a shift in the composition of bottom sediments in substantial portions of the inner Bight from essentially "clean," sandy substrates to highly polluted, muddy substrates (see Steimle *et al.*, this volume, for further details).

IDENTIFYING POLLUTION EFFECTS

Coastal benthic organisms have received considerable attention in pollution studies. Most bottom-dwelling organisms are sedentary or sessile (fixed in place) as adults and can neither readily avoid nor escape large areas of contamination. Toxicants or carbon-rich wastes may become integrated into the substratum, producing chronic stress. The poor mobility of benthic assemblages also permits repeated observations over time, theoretically making possible detailed examinations of the nature and rates of pollution-induced changes. Comparisons of benthic assemblages in the field usually are based on some aspect of the numerical composition of the faunas. The total number of species in an assemblage (species richness) and degree to which an assemblage is numerically dominated by individuals of one or a few species (dominance) are measures of diversity. Each of these parameters may be measured separately or combined into a single coefficient of diversity.

Most marine biologists agree that species diversity generally declines as pollution effects become more severe. However, the interpretation of diversity measures is made difficult by the variety of factors that may influence diversity. For example, the composition of bottom assemblages at a given location commonly varies over time because of natural periodic and aperiodic environmental changes. Such variations (*e.g.*, fluctuations in temperature, salinity, quantity and quality of food items, predation, and disease) may be independent of pollution but may produce changes in diversity indices and related environmental measures that parallel, enhance, or counteract those brought about by pollution. The importance of natural variables, singly or in combination, will differ among species and within species under different environmental conditions. Responses of benthos to interactive effects of these variables can be complex (Vernberg and Coull, 1975; Livingston, this volume; Young and Young, this volume). An appreciation of how these natural variables affect benthic biota, or at least those organisms that play important roles in given benthic systems, is necessary before the added effect(s) of human-introduced perturbations can be evaluated.

Elucidation of contaminant effects on particular benthic organisms or assemblages requires some knowledge of feeding habits. Trophic relationships may determine the types, quantities, and sources of pollutants likely to reach an organism, as well as the likely fates of pollutants. Unfortunately, the understanding of trophic relationships that exist among benthic organisms and between the benthos and the overlying water column is imperfect. While attempts have been made to erect food webs for benthos based on buccal (mouth) morphologies and gut contents (Perkins, 1958; Odum and Heald, 1975), conclusions drawn from such studies may be misleading because not all ingested materials are digested and assimilated (Edmondson, 1957), and because the full range of dietary items may not be revealed without extensive gut content analyses.

Animal populations may be affected indirectly by the action of toxicants, particularly action on food stocks. For example, Alongi and Tietjen (1980) demonstrated that growth and reproduction of marine nematodes can be influenced by the types of bacteria and algae ingested. Considerable evidence suggests that the benthic bacterial flora of the New York area has been affected by the presence of toxicants. Timoney and Port (this volume) observed the presence of mercury-resistant bacteria in New York Bight sediments, and suggested that resistance to metals may permit bacteria to survive and serve as food for nematodes and deposit-feeders living in metal-enriched sediments. These consumers, however, may prefer to feed on bacterial species that are not resistant to metals contamination. If deposit-feeders are not able to alter their feeding patterns to utilize non-preferred foods (*i.e.*, resistant species), their growth and reproduction could be affected adversely by food limitation.

A major problem in assessing benthic contaminant effects is that only within the past twenty years have accurate, quantitative sampling techniques been applied to a spectrum of substrates and locations. While this has yielded information on populations of benthic organisms under a variety of environmental conditions, most estuarine and near-coastal habitats have been subjected to human influences for more than four decades. Consequently, there are few quantitative data on the "before impact" distributions of benthic species. The problem is exacerbated because many methods have been employed to study the benthos. This limits the scope of many "comparative" studies. In addition, there are few locales for which simultaneous, comprehensive observations of all size classes of benthic organisms have been made. The smallest organisms (microbenthos–*e.g.,* bacteria and protozoa) are difficult to quantify and intermediate-sized organisms (meiobenthos–usually less than 2 mm in length) present some degree of taxonomic difficulty to the non-expert. Even large forms such as lobsters and crabs are difficult to quantify accurately because their size and mobility enable them to escape capture by most benthic sampling devices.

Study of toxicant effects in the New York Bight is confounded because impacted areas rarely are subjected to only one class of contaminants. For example, dredged materials and sewage sludge that are disposed in the New York Bight apex are enriched by trace metals, polynuclear aromatic hydrocarbons (PAH's), polychlorinated biphenyls (PCB's), and other contaminants (Anderson, this volume; O'Connor *et al.,* this volume). To partition the effects of these materials on natural benthic populations would be a herculean task. As a result, much of the data on effects of specific toxicants on individual species or species assemblages have come from laboratory studies. The classical approach has been to expose an organism to a range of toxicant concentrations for a finite period and to count the number of dead animals at the end of each exposure. Typically, the concentration at which 50% of the test organisms die is noted (LC_{50}). The primary advantage of this type of experiment is that it may yield reproducible results. However, because experiments are usually of short duration (hours or days) and are conducted under artificial laboratory conditions, the results may have little ecological significance and may be at variance with what occurs under natural situations. It is not uncommon for bioassay studies to indicate that a benthic species is killed or affected by nannogram per gram concentrations of a toxicant in the laboratory and yet to find viable populations of that organism exposed to milligram per gram concentrations of the same toxicant under natural situations (Gray, 1974). The reverse also has been noted, where effects in the field are associated with seemingly "harmless" concentrations of toxicants in the laboratory.

There are several possible explanations for differences between field and laboratory observations. First, benthic species that thrive in sediments containing high levels of toxicants may have adapted physiologically and genetically to toxicants, whereas bioassay individuals, perhaps taken from clean sediments, may not be resistant. Bryan and Hummerstone (1971) showed evidence of such adaptation for the polychaete, *Nereis diversicolor,* in metal-rich sediments in Cornish estuaries. Metal-resistant bacteria present in the New York Bight also appear to have adapted to metal stress (Litchfield *et al.,* this volume). Second, the chemical and physical state (sorption, species) of a given toxicant may differ in laboratory tests from the natural form of the pollutant. In laboratory situations, toxicants usually are present in solution; whereas in the field, chelation, adsorption, and other physico-chemical factors may bind or remove toxicants from solution, rendering them relatively harmless. Other problems may include synergistic interactions between two or more contaminants, delayed responses to continuing exposures to subacute toxicant concentrations, and sublethal responses (*e.g.*, reduced disease resistance) that occur in the field but are not readily detectable in simple laboratory experiments (Waldichuk, 1973).

One example of an apparent pollution-associated disease with no known specific cause is the black gill condition of rock crabs *(Cancer irroratus)*. An average of 4.5% of the rock crabs collected from coastal dumpsites near New York, New Jersey, Delaware, and Maryland have blackened gill filaments, compared to less than 1% for crabs from control areas ranging from Cape Hatteras to Georges Bank. Black gills are often fouled by sediment particles, bacteria, protozoa, and copepods. Incidence of this condition appears to be associated with the presence of sewage-laden silt; however, the biological significance of black gill is unknown at this time (Sawyer, this volume).

HISTORICAL PERSPECTIVES OF THE NEW YORK BIGHT AND LOWER HUDSON ESTUARY

Although the lower Hudson estuary and adjacent New York Bight (Figures 1 and 2) might be expected to be among the best-studied estuarine systems, surprisingly few data are available on the benthos from the period predating the middle decades of this century. Only during the past ten years have data accumulated on substances such as heavy metals, PCB's, petroleum hydrocarbons, and other contaminants in sediments, water, and animal tissues. Intensive quantitative studies of the benthos of the New York Bight were not begun until approximately 1960, by which time significant alterations of the Bight had already occurred (Franz, this volume). Since then, the quantitative and qualitative ecology of the benthos have been studied by several workers (Pearce *et al.*, 1976, 1977; Walker *et al.*, 1979; Tietjen, 1980; Boesch, this volume; Steimle *et al.*, this volume).

Records of New York's oyster industry provide the only well-documented account of changes in estuarine conditions over the past 150 years (Franz, this volume). Populations of American oysters *(Crassostrea virginica)* at one time occurred throughout much of the lower Hudson estuary from Sandy Hook, New Jersey to Ossining, New York (Ingersoll, 1881). In the early 1800's, major concentrations of oysters occurred along the southern and eastern shores of Staten Island, the Staten Island Kills, Newark Bay, Jamaica Bay, and Gowanus and Gravesend Bays in Brooklyn (Ingersoll, 1881). Productive oyster grounds existed in the upper harbor, along the east and west shores of Manhattan, on the New Jersey shore in the area of Bergen, and in the area of City Island, New York.

Beginning in the 1870's, reports of polluted shellfish suggested that waste disposal practices had a negative impact on the estuary (Blackford, 1885, 1887). Solid wastes (*e.g.*, street sweepings, cellar dirt, ballast, and materials dredged

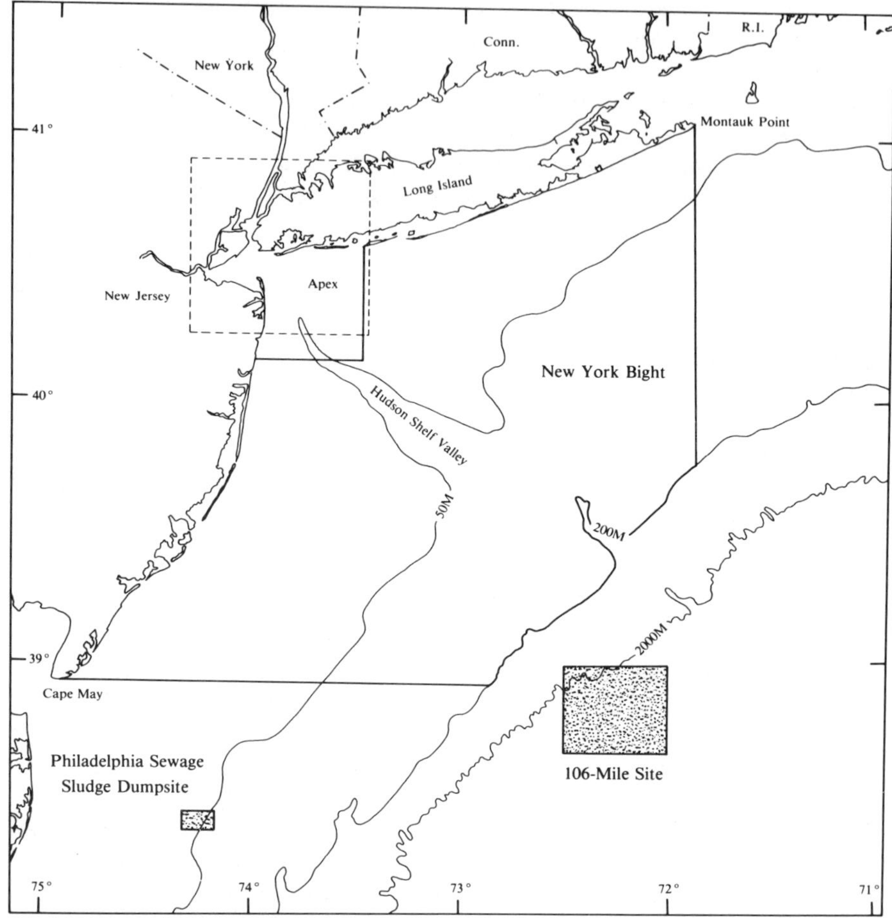

Figure 1: New York Bight and adjacent waters. Dashes outline area depicted in Figure 2.

from the harbor) created major problems for local oyster grounds, as did sludge acids from sugar refineries. Oysters from Staten Island, Newark Bay, and the Hudson River were unmarketable because of contamination by coal oil; other accounts reported that heavy metals, especially copper, were affecting the survival and marketability of oysters in Raritan Bay. By 1905, eutrophication associated with estuarine discharges of untreated sewage was of concern, particularly as the link between typhoid outbreaks and sewage-contaminated shellfish became evident (Anonymous, 1904). By 1921, all commercial shellfish beds in the lower harbor were condemned for public health reasons (U.S. Works Progress Administration, 1939).

Thus, the decade beginning in 1870 is a bench mark in the degradation of the lower harbor; it may be considered the major chronological datum for evaluating subsequent pollutional impacts in the New York Bight and surrounding waters.

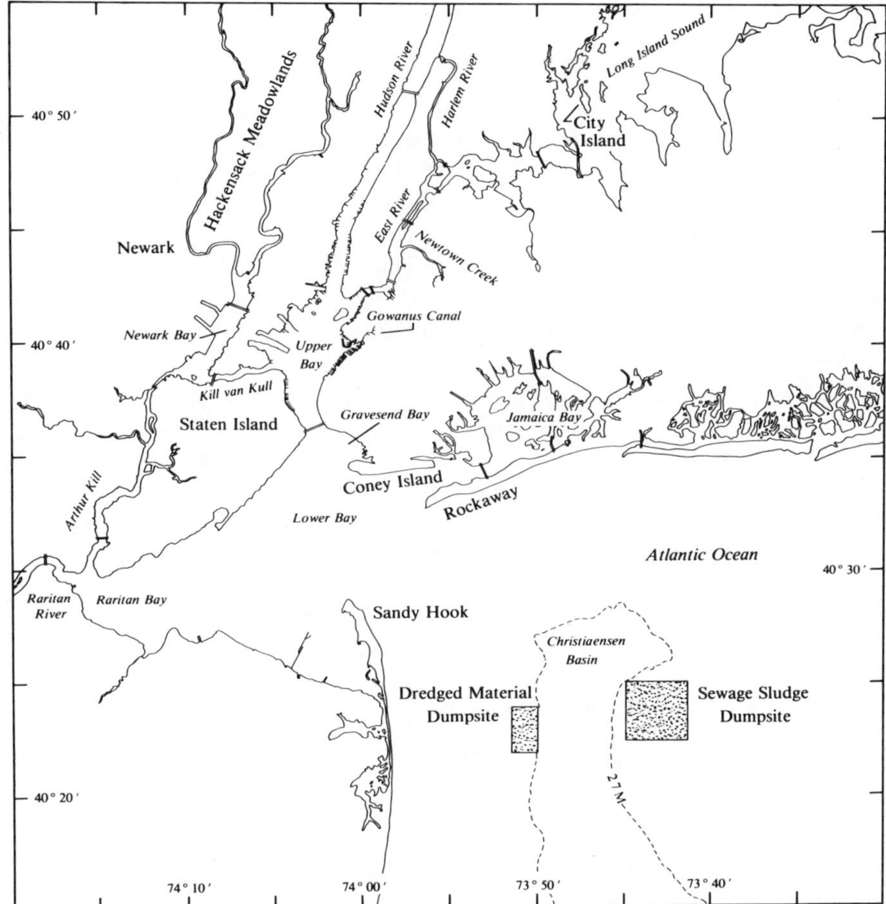

Figure 2: Lower Hudson-Raritan estuary and inner New York Bight.

Since that time, addition of contaminants to areas of the inner Bight has led to elevated toxicant concentrations and measurable changes in population density, species composition, species diversity, physiology, behavior, and physical appearance of benthic animals.

ORGANIC LOADING

The relative effect of nutrient and carbon inputs (*i.e.*, organic loading) on benthic community structure and function in the New York Bight is difficult to assess because elevated levels of other contaminants usually accompany high organic loads. In addition, increased organic loading tends to be associated with a shift in the type of substratum from sands to silts and clays, and with a change in faunal composition. As the substratum becomes finer, suspension feeders are replaced by deposit-feeding animals (Boesch, this volume). Significant faunal changes also occur in the meiobenthos (Tietjen, 1977, 1980).

Sediments at the dredged materials and sewage sludge disposal sites have organic loads that far exceed those of most neighboring coarse sediments (Gross, 1976; Mueller *et al.,* 1976a). This also is true of sediments of the Christiaensen Basin, a topographic depression that lies between the two dumpsites at the head of the Hudson Shelf Valley. The Basin is a natural depositional site for particulates flushed from the Hudson-Raritan estuary or resuspended from adjacent dumpsites. Other organically enriched sediments occur in patches off the coasts of the inner Bight, particularly off Brooklyn, Queens, and southwestern Long Island. These sediments include materials of sewage origin (Swanson, 1979).

As was suggested above, it is difficult to assess the impact of man-generated carbon loadings on the benthos of the inner Bight because the carbon-rich, fine-grained sediments of this area also contain high levels of toxicants. According to Boesch (this volume), the most meaningful comparisons of Christiaensen Basin/dumpsite faunas are with those along a gradient extending from the Basin down the Hudson Shelf Valley. It is along this gradient that sediment granulometry remains relatively constant but organic and toxicant loadings decrease.

The work of Steimle *et al.* (this volume) and Boesch (this volume) indicate that faunas of the dumpsites and the Christiaensen Basin differ considerably from those that might be expected in comparable, uncontaminated portions of the continental shelf. In all, more than 240 km^2 of bottom are affected (Boesch, this volume). Boesch's analysis reveals that dense populations of the polychaete, *Capitella capitata,* occur in a 10-15 km^2 area of the Basin east of the sludge disposal site. Beyond this zone, most of the Basin supports dense populations of the polychaetes, *Nephthys incisa, Pherusa affinis, Mediomastus ambiseta, Tharyx* sp., and *Asabellides* sp., the burrowing anemone, *Ceriantheopsis americanus,* and the bivalve, *Nucula proxima.* Populations of these species decline with increasing distance from the Basin, which suggests that either organic enrichment, the presence of toxicants, or predator exclusion (brought about by periodic low oxygen levels or elevated toxicant levels) may be responsible for the dense populations.

Tube-dwelling amphipods of the families Ampeliscidae and Corophiidae and the crab, *Cancer irroratus,* which are abundant elsewhere in the Hudson Shelf Valley, are rare or absent in the Christiaensen Basin. Such faunal changes are probably of ecological and economic significance because the bottom fauna that currently dominates the Christiaensen Basin does not include the preferred prey of commercially important Bight finfish (Boesch, this volume). The implication is that areas impacted by high carbon loadings (and associated toxicants) probably reduce the production of Bight fisheries. At this time, it is impossible to judge the significance of such a reduction.

The most probable direct effect of increased organic loading on benthos is a reduction in dissolved oxygen (DO), which results from biological and chemical oxidation of the added organic material. During the summer, DO levels in the bottom waters of the inner Bight generally are low and occasionally fall below 4 mg/liter (the level generally indicated by EPA as being stressful to benthic organisms) (Pearce, 1972; Segar and Berberian, 1976; Swanson *et al.*, 1979). DO concentrations of less than 1 mg/liter have been reported at the dredged materials and sewage sludge disposal sites (Pearce, 1972; Thomas *et al.*, 1976). Segar and Berberian (1976) suggested that low oxygen levels may be partially responsible for degradation of apex benthic communities.

When DO levels in well-oxygenated sediments decline, resident benthos may be replaced by species tolerant of low oxygen levels. Frequent or prolonged episodes of low DO, such as are associated with continued high discharges of organic matter, may cause less efficient decomposition of organic carbon by benthic communities, which in turn, may compound accumulation of organic material (Tietjen, this volume). This is likely the case in the Christiaensen Basin, where measurements of total organic carbon and the fecal steroid, coprostanol, suggest that sewage-related carbon has accumulated since at least the early 1950's (Hatcher and McGillivary, 1979).

TOXIC METALS

Metals are introduced into the New York Bight primarily from dredged material disposal, municipal and industrial wastewater, atmospheric fall-out, and surface runoff (Mueller *et al.*, 1976a). Metals either are dissolved in seawater or become adsorbed onto organic particles, which subsequently are deposited into the bottom sediments or are flushed from the Bight. Deposition of particulates associated with dumping represents a prime avenue for introducing toxicants into sediments of the inner Bight. On a per weight basis, dredged materials provide the single largest input of cadmium, chromium, copper, lead and zinc to the Bight (24-80%) (Swanson, 1977). Only mercury deviates from this pattern, with approximately 4.5% of the mass loadings provided by dredge spoil dumping and 70% from wastewater discharge. Sewage sludge dumping generally adds small quantities of metals to the apex (approximately 2-6% of cadmium, copper, lead and zinc mass loadings, and approximately 15% of chromium mass loadings; Mueller *et al.*, 1976b). However, Furr *et al.* (1976) report mercury concentrations of 15 ppm in New York City sewage sludge; these are among the highest mercury concentrations reported for sewage sludge. As the result of dumping and the settling of contaminant-laden particulates flushed from the estuary, concentrations of metals at the dumpsites and in the Christiaensen Basin are one to two orders of magnitude greater than at locations in the outer portions of the Bight (Steimle *et al.*, this volume).

Levels of metals in the tissues of deposit- and filter-feeding invertebrates commonly are ten times greater in animals living near sources of contamination (such as dumpsites) than in organisms from other regions of the Bight. While numerous surveys have examined metal levels in organisms of the Bight and adjacent waters, O'Connor and Rachlin (this volume) question the utility of such data for assessing environmental impact. Many factors combine to determine the metal concentrations in a given organism at any time. As a result, body burdens may not be related directly to metals levels in the surrounding media or substrates. For example, biochemical mechanisms regulating uptake, storage, and discharge of metals in animals may vary from element to element, within elements depending upon the species of metal involved, and between groups of organisms. In addition, tolerance for and internal distribution of specific metals within a given organism may vary with time and according to the past history of that organism (O'Connor and Rachlin, this volume).

Proteins such as metallothioneins may bind metals and thereby regulate their concentrations in benthic organisms. Levels of these proteins increase with exposure to toxic metals. Thus protein complexing is one manner by which benthic animals may "adapt" to increased metal loads. Ecological implications of metal body burdens depend in part upon the chemical form of the metal. Elevated levels of a biochemically inactive metal bound to large organic molecules and stored in organs such as the hepatopancreas (liver) or kidney, may represent a considerably smaller risk to an organism than low but active levels of the same metal (see O'Connor and Rachlin, this volume, for a more detailed discussion of bioaccumulation).

The manner in which metals are taken up by the benthos from the environment also affect body burdens (O'Connor and Rachlin, this volume). Although contaminated particulates may play a role in the uptake of toxic metals by the benthos, generally it is thought that interstitial waters (water between sediment particles) and waters immediately overlying the bottom are the major sources of metals for benthic organisms. Surface flocculents may represent an additional source (Boesch, this volume; Young, this volume a).

Despite abnormally high levels of toxic metals in sediments of the New York Bight apex (Steimle *et al.*, this volume) and at selected locations within the Hudson-Raritan estuary (*e.g.*, Raritan Bay: Greig and McGrath, 1977; Newark Bay and the Hackensack Meadowlands: Selikoff *et al.*, 1980), there are few, if any, instances in the New York area where benthic mortalities have been linked solely to chronic exposures to toxic metals. This does not mean that metals cause no effects but that effects are difficult to separate from those induced by other pollutants. At sublethal concentrations, toxic metals, acting alone or synergistically with other contaminants, may reduce reproductive output, adversely affect the survival of larvae and adults, and reduce tolerance to other forms of stress. These possibilities cannot easily be demonstrated to occur under natural conditions. Metals have been associated with changes in the species composition and diversity of the macrobenthos of Raritan Bay (Greig and McGrath, 1977), the inner Bight (Steimle *et al.*, this volume; Boesch, this volume), and the Philadelphia dumpsite (Lear *et al.*, this volume). Metals also are associated with changes in the nematode assemblages (meiofauna) of the Bight apex (Tietjen, 1980). In each of these cases, however, cause and effect relationships have not been demonstrated.

Metals, especially mercury and cadmium, have been shown to be toxic to sedimentary bacteria (Timoney and Port, this volume; Litchfield, *et al.*, this volume). Metal-resistant strains of bacteria occur where concentrations of metals in sediments are elevated; such bacteria are rare in non-contaminated areas. The genetic mechanism conferring resistance in these bacteria may be plasmid-mediated (*i.e.*, associated with genetic material outside the chromosome that may be exchanged between bacterial cells) (Litchfield *et al.*, this volume). Associated with this resistance to metals by natural populations of bacteria in the Bight is resistance to various classes of antibiotics. It is unlikely at present that antibiotic/metal resistant strains of bacteria pose any threat to human health (Timoney and Port, this volume); however, more intensive uses of inner Bight and estuarine shellfish resources might warrant a reexamination of this problem.

Resistance to metals enables microorganisms in the sediment to function in the presence of otherwise debilitating concentrations of toxic metals. Even in heavily impacted areas, bacteria retain their ability to detoxify contaminated sediments, regenerate nutrients, and provide food in the form of bacterial biomass to the meiofaunal component of the benthic community (Timoney and Port, this volume). However, the fraction of active (resistant) bacteria in extremely polluted areas may be small (Litchfield *et al.*, this volume).

PETROLEUM HYDROCARBONS

Petroleum hydrocarbons are a heterogeneous assemblage of compounds that includes the constituents of crude oil as well as the many substances resulting from petroleum refining and fuel combustion. This broad group represents the largest class of toxic contaminants (by volume) introduced to the Bight (Michael, this volume). Toxicity varies with molecular size, complexity, and solubility in water (Anderson, this volume; Michael, this volume). Within the petroleum hydrocarbons, the fraction known as the polynuclear aromatic hydrocarbons (PAH's) have been singled out for specific attention because of their carcinogenicity in mammals and persistence in marine sediments.

Despite their importance, few data are available on the quantity and distribution of petroleum hydrocarbons in the New York Bight and adjacent estuaries. Past estimates of petroleum hydrocarbon inputs to the Bight have been included in calculations of "oil and grease" loadings (*e.g.*, Mueller *et al.*, 1976a, b). It is extremely difficult to assess from such data the volumes of ecologically significant petroleum hydrocarbons remaining in the environment (Michael, this volume). More recently, attempts have been made to assess the concentrations of a broad range of toxic organic substances in living and non-living compartments of the Bight. Preliminary results, including those for PAH's, are reported in this volume by Anderson and O'Connor *et al.*

Most of the petroleum hydrocarbons in the waters and sediments of the New York Bight originate in the Hudson-Raritan estuary and are contributed by urban runoff, barge dumping (dredged materials and sewage sludge), and municipal wastewaters (O'Connor *et al.*, this volume; Michael, this volume).

Combustion products deposited on city streets that are subsequently washed into sewer systems form a major constituent of the runoff component.

Most recent analyses reported low levels of PAH's in the sea surface microlayer of the inner and outer New York Bight, and from sediments of the outer Bight (Anderson, this volume). Sediments from the Christiaensen Basin, Lower Bay, and Newark Bay were contaminated with PAH's at levels generally 10-100 times greater than those found in the outer Bight. The most contaminated sediments were found in Gowanus Canal (off Upper Bay) and in Newtown Creek (off the East River). In these two locations, PAH's were present in concentrations of parts per million, rather than parts per billion—two to six orders of magnitude above those in the outer Bight.

The significance of elevated sediment PAH concentrations to the benthos is unclear. Experimental evidence suggests that oil-contaminated sediment must contain high concentrations of hydrocarbons (several hundred parts per million) to be lethal; however, significant sublethal effects have been observed at moderate (1-50 ppm) levels of pollution. Included are impaired burrowing activities, reduced feeding rates, altered energy budgets, weight loss, behavioral modifications, and alterations of internal biochemistry (Anderson, this volume).

The severity of effects is also dependent upon the nature of the petroleum hydrocarbons involved, their availability to the benthos, (*e.g.*, dissolved in water overlying the bottom, in interstitial waters, adsorbed to particles, or contained in food items), the duration of exposure, and the organisms impacted. Two- and 3-ring PAH's, such as naphthalenes and phenanthrenes, are rapidly accumulated from solution. Short-term exposures from both water and sediment are likely to produce relatively high levels (1-10 ppm) of tissue contamination. If organisms are exposed to a constant supply of these compounds, as in the case of sessile organisms, high levels may be accumulated and death may occur. Mobile benthos, on the other hand, may avoid continuous exposures, and tissues can be depurated in relatively short periods (days to weeks) in clean water. Higher molecular weight PAH's such as benzo(a)pyrene and chrysene (4- and 5-ring compounds) may be taken up slowly by deposit-feeding benthos and retained for significant periods. Short-term exposure from either water or sediments probably will produce relatively little contamination; long-term exposure, however, can result in significant accummulation of these PAH's (Anderson, this volume).

The most recent analyses indicate that PAH's are present in many species from the inner New York Bight and lower Hudson-Raritan estuary (Anderson, this volume). Napthalenes, phenanthrenes, biphenyl, and other 2- and 3-ring aromatic compounds appear to have been taken up from the water. Biphenyls, in particular, were common in biotic and abiotic samples. PAH compounds of higher molecular weight also were found in the organisms sampled. As might be expected, elevated levels of petroleum hydrocarbons generally were found in detoxifying organs (*e.g.*, fish liver, crustacean hepatopancreas) rather than in muscle (Anderson, this volume).

The implications of these findings are difficult to assess. Data gathered from studies of the effects of oil spills on relatively pristine environments document a variety of community effects ranging from mass mortalities (Sanders *et al.*, 1980) to small changes in community structure, such as the disappearance of rare or sensitive species, or changes in the proportion of certain species (Michael, this volume). Although it is impossible to partition the specific effects of petroleum hydrocarbons relative to those of other contaminants, it appears that in localized areas of the Bight and estuary, hydrocarbon levels are sufficiently high to be solely responsible, in theory, for benthic mortalities (Anderson, this volume), or changes in benthic community structure (Michael, this volume).

Limited available data have not revealed instances of unacceptable PAH levels in edible flesh of finfishes or shellfishes taken in the Bight. The occurrence of higher molecular weight PAH's, particularly in the detoxifying organs of these animals, suggests that tainting may be a potential problem for deposit- and detritus-feeders from the vicinity of the dumpsites and estuary. High (7 ppm) levels of PAH's occur in the hepatopancreas of lobsters *(Homarus americanus)* taken from Raritan Bay (O'Connor *et al.*, this volume). Human consumption of this organ from contaminated lobsters could contribute a burden of PAH's (Anderson, this volume).

A range of studies clearly have demonstrated deleterious effects of petroleum hydrocarbons on individuals, populations, and communities. The continued disposal of petroleum hydrocarbons in the New York Bight will increase the toxic effects on benthic animals as these materials continue to accumulate in sediments. A major effort is needed to reduce as many sources of petroleum hydrocarbon contamination as possible. Better information urgently is required on the long-term sublethal effects of these compounds on populations and communities.

CHLORINATED HYDROCARBONS

Chlorinated hydrocarbons, including chlorinated pesticides, such as DDT and dieldrin, and polychlorinated biphenyls (PCB's), are among the better known contaminants of the New York Bight and Hudson-Raritan estuary ecosystems. These materials have been discharged into the environment over a period of years. Despite recent bans and limitations on the domestic manufacture or use of some of these substances, inputs to marine and estuarine environments continue.

Chlorinated hydrocarbons are of particular concern because of their persistence and toxicity. Stable compounds such as PCB's accumulate in sediments, which, in turn, may act as continual sources of ecosystem contamination. Chlorinated hydrocarbons have been found in the tissues of estuarine and near-shore animals. If present in sufficient quantity, they could pose a threat to humans eating contaminated organisms. The hazard arises from the potential carcinogenicity and mutagenicity of these substances (O'Connor and Stanford, 1979).

A number of chlorinated pesticides have been reported from the inner Bight and adjacent waters; however, DDT (including its metabolites, DDE and DDD) and dieldrin are the dominant pesticides in the waters, sediments and biota of this region (O'Connor *et al.,* this volume). These compounds originate from agricultural applications upstream of the metropolitan area and from urban domestic use in household pest control.

Significant concentrations of dieldrin have been measured in bottom sediments in the Christiaensen Basin and at the sewage sludge dumpsite. DDT is more widespread, occurring in bottom sediments throughout much of the lower harbor, the Christiaensen Basin, and the sewage sludge and dredged materials dumpsites in the Bight apex. As sewage sludge contains high concentrations (1,000 ppm) of DDT, this material probably reaches the Bight not only from sewage sludge but also from raw and untreated sewage discharges throughout the region (O'Connor *et al.,* this volume).

The impact of high sediment DDT levels is unclear. Available data (O'Connor *et al.,* this volume) indicate relatively low DDT levels in the flesh of winter flounder (*Pseudopleuronectes americanus*) from the Christianensen Basin (9-20 ppm), lobster from Raritan Bay (26 ppm), and blue mussels (*Mytilus edulis*) from Sandy Hook, New Jersey or Coney Island, New York (32-65 ppm).

For similar situations in the Southern California Bight, Young (this volume b) reports that DDT concentrations in both bottom particulates and the tissues of select benthic invertebrates and fishes declined along a gradient with increasing distance from offshore municipal sewer outfalls. Young also indicates that a ten-fold reduction in DDT inputs from Los Angeles County outfalls between 1971 and 1975 was reflected by only a 1.6-fold reduction in sediment DDT concentrations and a 1.5-fold reduction in the DDT levels of Dover sole flesh *(Microstomus pacificus).* (Similar data were obtained for PCB's). Thus, sediments fouled with chlorinated hydrocarbons may cause elevated contaminant levels in bottom-feeding fishes (and perhaps other benthic organisms) long after the major inputs have been reduced.

The largest point sources of PCB's to the Bight/estuary region were discharges from capacitor manufacturing plants located on the Hudson, approximately 400 km north of Manhattan. While these discharges have ceased, PCB's bound to river sediments are being transported downstream to New York harbor and the Bight. Within the greater New York area, major inputs of PCB's stem from sewage treatment plant effluent, dredged materials and sewage sludge disposal, urban runoff, and atmospheric fallout (O'Connor *et al.,* this volume).

Although elevated PCB levels occur in the sediments of New York harbor, Raritan Bay, and the inner Bight, no consistent relationship has been established between PCB concentrations in the flesh of benthic species and PCB levels at the site of capture. However, elevated concentrations have been demonstrated in the liver of finfishes and hepatopancreas of lobster (O'Connor *et al.,* this volume). Both dietary uptake (from contaminated particles or contaminated prey) as well as direct uptake from water represent routes by which PCB's may enter benthic organisms.

CONCLUSIONS

1. Temporal and spatial trends exist in the distribution and abundance of benthic organisms in lower Raritan Bay and the New York Bight apex as a result of pollution. The benthic macrofauna in the western third of Raritan Bay and the Christiaensen Basin has been severely affected by pollution. The panel agrees with Atwood *et al.* (1979) that the assimilation capacities for certain contaminants in these areas may have been exceeded. Dumping, as currently practiced, is highly destructive in localized areas; significant changes in fauna and micro-flora have taken place. Effects are usually synergistic, and it has not been possible to isolate effects of individual contaminants.

2. Field survey data from studies such as those on black gills in crustaceans have provided substantial evidence that pollution-associated disease syndromes are significantly more prevalent in and adjacent to the Bight apex than in fringe or control areas.

3. In sandy sediments impacted by dredged materials and sewage sludge, the normal nematode species complex has been replaced by species characteristic of muds. Several of the replacement species have been shown (in laboratory studies) to possess lower metabolic rates than species of the original sand fauna. Nematodes with lower metabolic rates may contribute less to the organic decomposition process than normal sand fauna. Continued dumping may result in the establishment of meiofaunal assemblages that do not function efficiently in organic decomposition; this may result in a continued buildup of organics in sediments.

4. Bacterial populations highly resistant to metals are correlated with high concentrations of mercury and other metals in Bight sediments. Many mercury-resistant bacteria reduce toxic methyl mercury to less toxic, more volatile forms that dissipate into the sediments and overlying water. Selection of a heavy metal resistant flora simultaneously has resulted in the persistence of a gene pool for antibiotic resistance in sedimentary bacteria because of a genetic linkage phenomenon.

5. Reported values for concentrations of certain dissolved metals (*e.g.*, copper) in waters of Raritan Bay and in subsurface water above the dump site in the Bight exceed toxicity values for the benthos obtained from laboratory tests. Such tests, however, are generally conducted with the use of simple ionic forms of the dissolved metals and not the complex forms of these metals that naturally occur.

6. Petroleum hydrocarbons with two to five aromatic rings (PAH's) represent the most toxic and least degradable components of oil. Inputs to the New York estuarine system and Bight come from multiple sources and are at a dangerously high level. Analyses of PAH's show that concentrations decrease from very high in sewage sludge (1,000 ppm) and estuarine (2,000 ppm) sediments to low (0.03 ppm) in samples from the outer Bight. Organs of fish, lobster, and mussels collected from the estuary and the Bight contained high levels of PAH's (up to 4.4 ppm); flesh contained only about 0.1 ppm. These few data show that PAH's

from the contaminated environment are reaching the biota from either water or particulates.

7. Higher molecular weight chlorinated hydrocarbons such as polychlorinated biphenyls (PCB's) are significant because they are resistant to degradation, concentrated by marine organisms, and accumulate in bottom sediments where they are available to the benthos and fishes that prey on benthos. High loading of Hudson River sediments by PCB wastes poses a serious long-term threat to the ecosystems of the Hudson-Raritan estuaries and the New York Bight.

8. Detailed characterization and quantification of chemical contaminants in the New York Bight and estuaries at this writing are inadequate to permit conclusions regarding ecological impacts. Available data suggest, however, that contamination should be of interest to agencies with public health responsibilities, particularly in the Hudson-Raritan estuary.

REFERENCES

Alongi, D.M. and J.H. Tietjen. 1980. Population growth and trophic interactions among free-living marine nematodes. *In:* Marine Benthic Dynamics, K.R. Tenore and B.C. Coull (eds.), University of South Carolina Press, Columbia, SC. pp. 151-166.

Anderson, J.W. This volume. The transport of petroleum hydrocarbons from sediments to benthos and the potential effects. pp. 165-179.

Anonymous. 1904. Sewage and polluted oysters as a cause of typhoid and other gastro-intestinal disturbances. U.S. Chem. Bur. Bull. 156: 28-44.

Atwood, D., D.W. Brown, V. Cabelli, J. Farrington, C. Garside, G. Han, D.V. Hansen, G. Harvey, K.S. Kamlet, J. O'Connor, L. Swanson, D. Swift, J. Thomas, J. Walsh, and T. Whitledge. 1979. The New York Bight. *In:* Assimilative Capacity of United States Coastal Waters for Pollutants, E.D. Goldberg (ed.), Proceedings of a Workshop, Working Paper No. 1, Federal Plan for Ocean Pollution Research Development and Monitoring, NOAA/ERL, Boulder, CO. pp. 148-178.

Blackford, E.G. 1885. Report of the commissioner of fisheries of the State of New York in charge of oyster investigations. Assembly of the State of New York, Vol. 6, Document No. 85. 137 pp.

Blackford, E.G. 1887. Second report of the oyster investigation and of the survey of oyster territory for the years 1885 and 1886. Assembly of the State of New York, Vol. 6, Document No. 28-36. 387 pp.

Boesch, D.F. This volume. Ecosystem consequences of alternations of benthic community structure and function in the New York Bight region. pp. 543-568.

Bryan, G.W. and L.G. Hummerstone. 1971. Adaptation of the polychaete *Nereis diversicolor* to sediments containing high concentrations of heavy metals. I. General observations and adaptation to copper. J. Mar. Biol. Assoc. U.K. 51: 845-863.

Edmonson, W.T. 1957. Trophic relations of the zooplankton. Trans. Am. Micros. Soc. 76: 205-245.

Franz, D.R. This volume. An historical perspective on mollusks in lower New York harbor, with emphasis on oysters. pp.181-197.

Furr, A.K., A.W. Lawrence, S.S. Tong, M.Z. Grandolfo, R.A. Hofstader, C.A.

Bache, W.H. Guterman, and B.J. Lisk. 1976. Multielements and chlorinated hydrocarbon analysis of municipal sewage sludges of American cities. Environ. Sci. Tech. 10: 683-687.

Gray, J.S. 1974. Synergistic effects of three heavy metals on growth rates of a marine ciliate protozoan. *In:* Pollution and Physiology of Marine Organisms, F.J. Vernberg and W.B. Vernberg (eds.), Academic Press, New York. pp. 465-485.

Greig, R.A. and R.A. McGrath. 1977. Trace metals in sediments of Raritan Bay. Mar. Pollut. Bull. 8: 188-192.

Gross, M.G. 1976. Sources of urban wastes. Am. Soc. Limnol. Oceanogr. Spec. Symp. 2: 150-161.

Hatcher, P.G. and P.A. McGillivary. 1979. Sewage contamination in the New York Bight: Coprostanol as an indicator. Environ. Sci. Technol. 13: 1225-1229.

Ingersoll, E. 1881. The oyster industry. *In:* The History and Present Conditions of the Fisheries Industry, S. Baird and G. Goode (eds.), U.S. Government Printing Office, Washington, DC. 251 pp.

Lear, D.W., M.L. O'Malley, W.C. Muir, and G. Pence. This volume. Environmental effects of sewage sludge at the Philadelphia dumping site. pp. 481-493.

Litchfield, C.D., M.A. Devanas, J. Zindulis, M. Meskill, J. Freedman, and C. McClean. This volume. Influence of cadmium on the microbial populations and processes in sediments from the New York Bight apex. pp. 587-604.

Livingston, R.L. This volume. Long-term variability in coastal systems: Background noise and environmental stress. pp. 605-620.

Michael, A.D. This volume. The potential contribution of petroleum hydrocarbons to changes in benthic communities of the New York Bight. pp. 621-630.

Mueller, J.A., J.S. Jeris, A.R. Anderson, and C.F. Hughes. 1976a. Contaminant Inputs to the New York Bight. NOAA Tech. Memo. ERL MESA-6. 347 pp.

Mueller, J., A.R. Anderson, and J.S. Jeris. 1976b. Contaminants entering the New York Bight: Sources, mass loads, significance. Am. Soc. Limnol. Oceanogr. Spec. Symp. 2: 162-170.

O'Connor, J.M., J.B. Klotz, and Theo. J. Kneip. This volume. Sources, sinks and distribution of organic contaminants in the New York Bight ecosystem. pp. 631-653.

O'Connor, J.M. and J.W. Rachlin. This volume. Perspectives on metals in New York Bight organisms: factors controlling accumulation and body burdens. pp. 655-673.

O'Connor J.S. and H.M. Stanford (eds.). 1979. Chemical Pollutants of the New York Bight; Priorities for Research. NOAA Special Report. 217 pp.

Odum, W.E. and E.J. Heald. 1975. The detritus-based food web of an estuarine mangrove community. *In:* Estuarine Research, Part I, L.E. Cronin (ed.), Academic Press, New York. pp. 265-286.

Pearce, J.B. 1972. The effects of solid waste disposal on benthic communities in the New York Bight. *In:* Marine Pollution and Sea Life, M. Ruivo (ed.), Fishing News (books) Ltd., London. pp. 404-411.

Pearce, J.B., J.V. Carracciolo, M.B. Halsey, and L.H. Rogers. 1976. Temporal and spatial distributions of benthic macroinvertebrates in the New York Bight. Am. Soc. Limnol. Oceanogr. Spec. Symp. 2: 394-403.

Pearce, J.B., L.H. Rogers, J.V. Carracciolo, and M.B. Halsey. 1977. Distribution and abundance of benthic organisms in the New York Bight apex, five seasonal cruises, August 1973 through September 1974. NOAA Tech. Memo ERL MESA-32. 803 pp.

Perkins, E.J. 1958. The food relationships of the microbenthos, with particular reference to that found at Whitstable, Kent. Ann. Mag. Nat. Hist. 13: 64-77.

Rowe, G.T. 1978. Benthic nutrient regeneration and rate of primary production in continental shelf water—a reply. Nature 274: 189-190.

Rowe, G.T. and K.L. Smith. 1977. Benthic-pelagic coupling in the mid-Atlantic Bight. *In:* Ecology of Marine Benthos, B.C. Coull (ed.), University of South Carolina Press, Columbia, SC. pp. 55-56.

Sanders, H.L., J.F. Grassle, G.R. Hampson. L.S. Morse, S. Garner-Price, and C. Jones. 1980. Anatomy of an oil spill: long-term effects from the grounding of the barge *Florida* off West Falmouth, Massachusetts. J. Mar. Res. 38: 265-380.

Sawyer, T.K. This volume. Distribution and seasonal incidence of "black gill" in the rock crab, *Cancer irroratus.* pp. 199-211.

Segar, D.A. and G.A. Berberian. 1976. Oxygen depletion in the New York Bight apex: causes and consequences. Am. Soc. Limnol. Oceanogr. Spec. Symp. 2: 220-239.

Selikoff, I.J., T.W. Clarkson, T.J. Kneip, N. Nelson, W.J. Nicholson, J. Sanders, R.H. Wilson, and P. Grandjean. 1980. Mercury in the Hackensack Meadowlands. Report to Hon. Brendan Byrne, Governor of New Jersey, by the Advisory Committee on Mercury. Environmental Sciences Laboratory, Mount Sinai School of Medicine of CUNY, New York. 37 pp.

Steimle, F., J. Carracciolo, and J.B. Pearce. This volume. Impacts of dumping on New York Bight apex benthos. pp. 213-223.

Swanson, R.L. 1977. Status of ocean dumping research in New York Bight. J. Waterway, Port, Coastal and Oceanic Division, Am. Soc. Civ. Eng. 103: 9-24.

Swanson, R.L. 1979. Testimony presented at the public hearing on the issuing of permits for the continued ocean dumping of municipal sewage sludge. U.S. Environmental Protection Agency, Region II. October 2, 1979, Newark, NJ. 32 pp.

Swanson, R.L., C.J. Sindermann, and G. Han. 1979. Oxygen depletion and the future: an evaluation. *In:* Oxygen Depletion and Associated Benthic Mortalities in New York Bight, 1976, R.L. Swanson and C.J. Sindermann (eds.), NOAA Prof. Pap. 11: 335-345.

Thomas, J.P., W.C. Phoel, F.W. Steimle, J.E. O'Reilly, and C.A. Evans. 1976. Seabed oxygen consumption—New York Bight apex. Am. Soc. Limnol. Oceanogr. Spec. Symp. 2: 354-369.

Tietjen, J.H. 1977. Population distribution and structure of the free-living nematodes of Long Island Sound. Mar. Biol. 43: 123-136.

Tietjen. J.H. 1980. Population structure and species composition of the free-living nematodes inhabiting sands of the New York Bight apex. Estuarine Coastal Mar. Sci. 10: 61-73.

Tietjen, J.H. This volume. Potential roles of meiofauna in polluted ecosystems and the impact of pollution on meiofauna. pp. 225-233.

Timoney, J.F. and J.G. Port. This volume. Heavy metal and antibiotic resistance in *Bacillus* and *Vibrio* from sediments of New York Bight. pp. 235-248.

U.S. Works Progress Administration, New York City. 1939. Report relative to shellfish industry in New York Harbor. US-WPA Rep. No. 465-97-3-131. 264 pp.

Vernberg, W.B. and B.C. Coull. 1975. Multiple factor effects of environmental parameters on the physiology, ecology and distribution of some marine meiofauna. Cah. Biol. Mar. 16: 721-732.

Waldichuk, M. 1973. Trends in methodology for evaluation of effects of pollutants on marine organisms and ecosystems. CRC Critical Reviews in Environmental Control 3: 167-211.

Walker, H.A., S.B. Saila, and E.L. Anderson. 1979. Exploring the data structure of New York Bight benthic data using post-collection stratification of samples and linear discrimate analysis for species comparison. Estuarine Coastal Mar. Sci. 9: 101-120.

Young, D.K. and M. Young. This volume. Marine macrobenthos as indicators of environmental stress. pp.527-539.

Young, D.R. This volume a. A comparative study of trace metal contamination in the Southern California and New York Bights. pp. 249-262.

Young, D.R. This volume b. Chlorinated hydrocarbon contaminants in the Southern California and New York Bights. pp. 263-276.

EFFECTS OF POLLUTANTS ON FISHES

C.J. Sindermann[1]
S.C. Esser
E. Gould
B.B. McCain
J.L. McHugh
R.P. Morgan II
R.A. Murchelano
M.J. Sherwood
P.R. Spitzer

Abstract. As human populations in the New York and New Jersey coastal areas have increased, there have been concomitant increases in the pollution of adjacent marine environments and in the stresses placed on local fish and shellfish populations. A variety of chemical pollutants has been found in elevated concentrations in organisms, sediments, and waters of the New York Bight. These include toxic metals, chlorinated hydrocarbons, and petroleum and its components. Response to pollutants is species-specific, but early life history stages and gonadal tissues of adults are particularly sensitive. Contaminants of particular importance to fishes and shellfishes are mercury, cadmium, silver, PCB's, DDT and its metabolites, and petroleum hydrocarbons.

Only in the most severely degraded waters of the Northeast are there localized disappearances of fish and shellfish. Catch statistics for commercially valuable species in the New York Bight reveal few changes in abundance that are directly attributable to pollution. In part, this may reflect the difficulty of distinguishing between the effects of natural environmental variations, overfishing, and pollution. It would appear that effects on individual organisms and on localized population segments are more promising indicators of pollution-related environmental stress than effects on total populations. Since toxicants may have severe and varied effects on individual fish and shellfish, management alternatives to continued marine pollution should be explored.

INTRODUCTION

During the past decade there has been increasing concern about the well-being of coastal and estuarine organisms and about the real and potential impacts of human activities on ocean resources and ecosystems. This concern has led to a demand for more and better scientific information on the effects of natural and human-induced environmental changes on fisheries. There has been a gradual increase in polluted coastal and estuarine areas—an increase that generally is proportional to the density and level of industrialization of the adjacent human population. Pesticides and other synthetic chemicals that have been added to coastal waters, even in low concentrations, can drastically affect the physiology of fishes and shellfishes. Most organisms have had little evolutionary experience with these compounds and are metabolically ill-equipped to deal with them. Large volumes of organic material in the form of sewage sludge and effluents have been added that can produce anaerobic or low-oxygen environments. These loadings often are accompanied by other contaminants, such as toxic metals and petroleum hydrocarbons, which can alter

[1] Chairman, Panel on the Effects of Pollutants on Fishes.

the normal functions of enzymes in fishes and in the food organisms they consume.

That chemical pollutants cause stress in marine animals has been demonstrated repeatedly. Descriptions of lethal and sublethal effects of toxic materials abound in the scientific literature. It is much more difficult, however, to demonstrate that stress from pollutants can have significant quantifiable effects on species abundance. Apart from localized effects in severely contaminated coastal and estuarine zones, this has not been demonstrated satisfactorily at the present time.

This report summarizes the present status of knowledge concerning pollutant effects on fishes and shellfishes of the New York Bight. Principal classes of contaminants reviewed are toxic metals, chlorinated hydrocarbons, and petroleum and its components. Technical papers presented in the second half of this book (*e.g.*, Calabrese *et al.*, this volume; Esser, this volume; McCain and Malins, this volume; Murchelano, this volume; Murchelano and Ziskowski, this volume; Sherwood, this volume) document in greater detail fisheries-related work on these categories of pollutants.

EXISTING POLLUTANT LEVELS AND POLLUTANT INPUTS

Toxic metals

Toxic metals are significant contaminants of the New York Bight and include elements that are either "major perceived threats" or "potentially significant threats" to the human users or living marine resources of the area (O'Connor and Stanford, 1979). Wastewater supplies 70% of the mercury input to the Bight and dredged material supplies the major portion (24-80%) of other toxic metals. The daily mass load of toxic metals (metric tons/day) moving into the Bight varies from 0.3 for mercury, to 2.4 for cadmium, to 13 for lead and copper, to 230 for iron (Mueller *et al.*, 1976: Table 38).

Metal levels in Bight sediments are generally higher in the apex than in outer portions of the Bight and vary with topography and sediment grain size. Elevated trace metal concentrations have been observed in association with fine-grained sediments (*e.g.*, silts and clays) and with topographic lows (Carmody *et al.*, 1973; Swanson, 1977). Within the New York Bight apex, especially high concentrations of zinc, chromium, copper, lead, nickel, iron, mercury, and cadmium occur at the dredged materials disposal area and in the Christiaensen Basin and Hudson Shelf Valley (Krom *et al.*, 1981). Table 1 lists the ranges of metals in sediments from the ocean dumpsite area off New York City. Carmody *et al.*, (1973) estimated that approximately 60 km^2 of the inner Bight had copper, chromium, lead, and zinc levels ten times above normal and that approximately 170 km^2 had concentrations five times above normal.

Toxic metals in the water column are less well sampled than those in the sediments. Values for the Bight apex vary widely with sampling location and depth, season, tidal stage (especially in the region of the Sandy Hook-Rockaway

Table 1. Metal levels in sediments from the ocean dumpsite area of New York City. From Greig and McGrath (1977).

	Range (ppm, dry wt)
Cadmium	0.01 - 12
Chromium	0.02 - 53
Copper	0.1 - 720
Nickel	0.5 - 240
Lead	0.02 - 400
Zinc	0.02 - 730

transect) (Alexander and Alexander, 1977), and ocean dumping activities. Metals such as copper, cadmium, and zinc are present predominantly in the dissolved phase; significant quantities of other metals, including iron, occur as particulates (Segar and Cantillo, 1976). Segar and Cantillo (1976) and Alexander *et al.* (1978) respectively reported the following average dissolved metal levels (ppb) from the inner Bight: cadmium, 0.8 and <0.1; copper, 4.5 and 5.6; iron, 12 and 4.6; and zinc, 32 and 19. Dissolved lead, nickel, and mercury levels averaged <3, 11.3, and <0.2 ppb, respectively (Alexander *et al.*, 1978). Average particulate metal levels in the apex were less than 1 ppb for most metals (cadmium, copper, iron, lead, and nickel), but reached 74.1 ppb for iron (Alexander *et al.*, 1978).

A variety of toxic metals have been reported from the fishes and shellfishes of the Bight. For example, Greig and Wenzloff (1977) reported < 1 ppm cadmium, chromium, mercury, and silver, 1 ppm nickel and lead, 13 ppm copper, and 30 ppm zinc (wet wt) in the livers of Bight finfishes. Metal levels in zooplankton are proportionately higher, ranging from 1 and 2 ppm (dry wt) for silver and cadmium, respectively, to highs of 54, 69, and 625 ppm (dry wt) for copper, lead, and zinc, respectively (Greig *et al.*, 1977). A comprehensive discussion of metal levels in the biota of the Bight and adjacent waters is presented by O'Connor and Rachlin (this volume).

Chlorine and Chlorinated Hydrocarbons

Chlorine

The primary input of chlorine to the New York Bight and Hudson-Raritan estuary results from sewage treatment. Additional sources include the treatment of cooling water used in heat exchangers and the treatment of process effluent. Although chlorine may be considered a non-persistent pollutant, a variety of chlorinated organic compounds are produced as a consequence of chlorine use. These represent a potential threat to living marine resources. For example, chlorine readily reacts with nitrogenous organic materials such as are found in sewage to form chloramines. The latter are toxic to fishes (U.S. Environmental Protection Agency, 1976). The chlorination of sewage effluents and power plant cooling water also has been reported to result in the generation of 2-chlorophenol and related compounds. Chlorophenols may cause tainting of fish flesh at low concentrations (U.S. Environmental Protection Agency, 1980a). The significance of chlorine and its derivatives has not been examined in the New York Bight region.

Chlorinated hydrocarbons

The use and dispersal of pesticides and industrial chemicals constitute primary pathways for the chlorinated hydrocarbon contamination of the New York Bight. The New York Bight drainage is not a major site for the production of halogenated hydrocarbons with the possible exception of toxaphene (MESA, 1978). However, pesticides are used extensively on a daily basis in the New York Bight region and in upstream agricultural areas. DDT and dieldrin are the dominant chlorinated pesticides in the sediments and biota of the Bight (O'Connor *et al.*, this volume).

The New York-New Jersey region annually utilizes large quantities of halogenated hydrocarbons and synthetic organics for industrial and manufacturing processes. This is reflected in the presence of various industrial halorganics including chlorobenzene, chlorophenols, and halogenated diphenyl ethers in the Bight ecosystem (MESA, 1978). Perhaps the best documented of the industrial halorganics are the polychlorinated biphenyls (PCB's). The tragedy of PCB input into the Hudson River (over 300,000 kg between 1950 and 1973) is recorded in several papers (*e.g.*, Walker, 1976, Horn *et al.*, 1979). The resultant environmental concentrations of PCB's remain a subject for study by federal and state agencies.

Recent work by MacLeod *et al.* (1981) and by O'Connor *et al.* (this volume) suggests that significant sources of both PCB's and DDT exist in the lower Hudson-Raritan estuary. Sewage sludge is estimated to contribute 106-231 kg DDT and 600-2,000 kg PCB per year to the Bight apex. Dredged materials from the New York harbor area are estimated to contribute 65 kg DDT and 3,500 kg PCB to the inner Bight per year (O'Connor *et al.*, this volume). The latter may increase significantly as contaminated sediments from the upper Hudson River move downstream (Horn *et al.*, 1979).

Chemical analyses of chlorinated hydrocarbons in the sediments and tissues of winter flounder (*Pseudopleuronectes americanus*) from Sandy Hook/Raritan Bay and the apex dumping area have revealed elevated levels of PCB's. PCB concentrations ranged from 1.4 mg/kg to 9.9 mg/kg, wet wt in the livers of winter flounder from Sandy Hook/Raritan Bay and from 0.48 mg/kg to 10 mg/kg, wet wt in the livers of flounders from the apex dumping area. These levels are comparable to those in fishes from contaminated areas in southern California and Puget Sound. Levels of total DDT in fish tissues from the two New York-New Jersey areas mentioned above, however, were several orders of magnitude lower than those in fish tissues from a contaminated site off southern California (Sherwood, this volume).

Little work has been done on the contaminants of technical grade pesticides and industrial grade organics. These impurities may be more important as pollutants than is believed presently. Additionally, little is known about ambient levels of chlorinated aliphatics from industrial sources in the New York Bight. As society's use of synthetic organic compounds grows, such halogenated organics may be expected to take on added significance as potential environmental hazards.

Petroleum hydrocarbons

"As much as 80% of the [ship] traffic in the Bight is petroleum-related [Purrett-Carroll, 1979]"; consequently, petroleum-associated aromatic hydrocarbons are, and will continue to be, likely causes of degraded ecosystems in the New York Bight. Chemical analyses for polynuclear aromatic hydrocarbons (PAH's) in sediments and tissues of animals from the New York Bight and adjacent waters were performed recently by MacLeod *et al.* (1981). PAH concentrations in sediments from the Christiaensen Basin were greatly elevated, generally ranging from 40 ppb to 1,000 ppb, dry wt; highest values were observed for pyrene, benz(a)anthracene, fluoranthene, and phenanthrene. Comparable values were observed in highly contaminated sediments from the Raritan Bay-Lower Bay area of the Hudson Raritan estuary.

Liver tissue from winter flounder taken from Raritan Bay exhibited elevated PAH levels. (Comparable analyses were not made on specimens from the Bight.) Naphthalene, 1- and 2-methylnaphthalenes, and biphenyl were present in concentrations of 500, 1,000, 700, and 200 ppb, dry wt, respectively. These levels are similar to those observed in the livers of experimental English sole (*Parophrys vetulus*) maintained on sediment contaminated with approximately 400 ppm crude oil for one month in laboratory studies (McCain *et al.*, 1978). Winter flounder levels also are of the same order of magnitude as was reported for the livers of starry flounder (*Platichthys stellatus*) exposed in the laboratory to 0.9 ± 0.1 ppm seawater-soluble fractions of crude oil for one week (Roubal *et al.*, 1978).

Elevated concentrations of several high molecular weight PAH's were detected in digestive glands of lobsters (*Homarus americanus*) from Raritan Bay. PAH's found in high concentrations (ppm, dry wt) included pyrene (1.5), benz(a)-anthracene (0.6), chrysene (0.8), and benzo(a)pyrene (0.3) (MacLeod *et al.*, 1981). Many of the aromatic hydrocarbons mentioned above and found in sediments and animal tissues are known or suspected to cause pathological changes in a variety of mammals.

CONTAMINANT EFFECTS

All life history stages of fishes and shellfishes are susceptible to the effects of pollutants. On the basis of the vulnerability of these organisms, it would seem that populations in contaminated waters should dwindle and disappear, yet this has not happened on the North American east coast. Localized disappearances of species have been recorded only from severely degraded waters, and even in such areas, other species remain and, in some instances, are abundant. This is true particularly of coastal fish species, many of which spend much of their early lives in waters that are to some extent contaminated.

Death and disappearance may be criteria that are too extreme for determining whether pollutants have had harmful effects on marine biota. Sublethal effects may represent more promising indicators of long-term pollution impacts. Sublethal effects are difficult to quantify and difficult to sort out from the effects of the other environmental stressors. They include physiological and biochemical alterations, behavioral modifications, and pathological changes. Lethal and sublethal effects of each major class of contaminants are considered below.

Toxic Metal Effects

Metal toxicity varies not only with such expected parameters as salinity and metal salt form, but also with life stage and species. Bioassay research with early-life stages of marine animals has shown the particular sensitivity of these life forms to toxic metals, notably mercury and silver. Lead is twice as toxic as cadmium to early-life stages, but is two hundredfold less toxic than mercury or silver (Calabrese *et al.,* this volume).

The order of metal toxicity changes in adults from that which is observed in juveniles. Cadmium produces more severe metabolic effects in adult winter flounder and lobster than either mercury or silver, despite far lower rates of cadmium uptake by the tissues of these animals. Similar exposure of flounder to five times as much lead produced less metabolic disturbance. Low salinity enhanced or altered some effects and, in combination with suboptimal temperatures, increased metal toxicity to early-life forms (Calabrese *et al.,* this volume).

In considering the potential hazard of trace metals to marine life, mercury has been ranked first, with cadmium and silver next, if toxicity and abundance in the environment are used as criteria (Calabrese *et al.,* this volume). Recent research (*e.g.,* Sunda and Guillard, 1976; Sunda and Gillespie, 1979) has indicated that copper can be toxic to diatoms, dinoflagellates, and fish eggs and larvae at concentrations only slightly above natural levels. Although the toxicity of this metal to bivalve larvae and adult animals is less than that of the three metals mentioned above, its potential effect on important food chain organisms calls for more intensive study.

High tissue concentrations of a metal are not always quantitatively related to metabolic disorders. Metals can be immobilized biochemically by an animal's detoxifying mechanisms and thus rendered metabolically inert. For example, exposure of lobsters for one month to low levels of mercury (6 ppb) produced significant tissue uptake of that metal and some metabolic abnormalities, whereas a similar exposure to cadmium produced no detectable tissue uptake of metal but metabolic abnormalities as great as, or greater than, those seen in the mercury-exposed animals. Similar results were observed for winter flounder. The adults of these species apparently possess a tolerance or sequestering mechanism for mercury and silver that is lacking in juvenile forms. This enables normal tissue metabolism to occur in the presence of large body burdens of these metals (Calabrese *et al.*, this volume).

Detrimental effects of chronic exposure to trace metal concentrations, such as those occurring in the most polluted portions of the inner Bight and adjacent estuarine waters, are demonstrable in laboratory experiments with adult marine organisms. Chronic exposure to low levels of toxic metals reveals two general effects: disruption of respiratory and glycolytic (metabolic) patterns, and lowered sensitivity to metabolic regulators. The former produces abnormal expenditures of energy reserves; the latter reduces an organism's capacity to adapt and survive in a naturally changing environment. Exposure to mercury and cadmium at concentrations as low as 6 ppb impaired physiological control mechanisms (Calabrese *et al.*, this volume). Stress of this nature drains energy reserves and renders marine life more susceptible to debilitation and death by predation, disease, and environmental fluctuations.

Toxic metal pollution is greatest in the inshore waters of the Bight and adjacent estuary. These highly polluted areas are the breeding and nursery grounds for many commercially and recreationally important marine species. The sensitivity of early-life stages to toxic metals has been observed repeatedly in bioassays. High contaminant-induced mortality in the early-life stages of fishes could lower recruitment to fisheries stocks; however, such effects have not been demonstrated in the New York Bight area (Longwell and Hughes, this volume).

Chlorinated Hydrocarbon Effects

Only a few chlorinated hydrocarbons (*e.g.*, toxaphene) are designed to be piscicides. Acute and sublethal effects of chlorinated hydrocarbons in estuarine and marine fishes generally result from the use of pesticides in other applications, and from the use of industrial compounds whose chemical structures are similar to those of pesticides. Effects of pesticides on fishes have been observed for almost all life stages. Fish deaths have been reported in areas exposed to field application of pesticides. Major pesticide impacts are seen in early embryos, late larvae, and gonadal tissues of adults.

Industrial chlorinated organics have effects that are similar to those of pesticides. Although acute toxicity is not usually observed in fishes exposed to industrial chemicals, sublethal and chronic effects may occur. Laboratory investigations suggest that chronic PCB exposure may cause a variety of effects among fishes including reduced survival (embryos, juveniles, adults), lesions, hemorrhaging of skin, liver abnormalities, and behavioral modifications (U.S. Environmental Protection Agency, 1980b). Freeman and Idler (1975) observed PCB interference with the male sex hormone metabolism and reproduction. Smith *et al.* (1979) reported high prevalence of enlarged livers with tumors and other abnormalities in Atlantic tomcod (*Microgadus tomcod*) from the Hudson River estuary. Although the exact causes of the tomcod abnormalities are unknown, PCB's are suspected of having a role.

Sherwood (this volume) postulates a link between fin erosion and elevated sediment PCB levels for Dover sole and starry flounder from the Southern California Bight and Puget Sound area, respectively. A similar relationship has not been demonstrated for fin erosion in the flatfishes of the New York Bight (Murchelano and Ziskowski, this volume; Sherwood, this volume). Because of the large volume of PCB's present in the sediments of the upper Hudson River, these compounds are likely to continue as important contaminants of estuarine and migratory fishes in the New York Bight.

Petroleum Hydrocarbon Effects

Recent reviews of the short-term (acute and subacute) effects of aromatic petroleum hydrocarbons on aquatic animals have been published by Craddock (1977), Hodgins *et al.* (1977), Johnson (1977), and Neff (1979). Laboratory investigations have shown that exposure to petroleum hydrocarbon concentrations of 0.5-3 ppm may cause behavioral changes and death in a variety of fish and shellfish species. Although seawater concentrations of aromatic hydrocarbons are generally much lower than those utilized in experimental situations, concentrations of 1-3 ppm have been reported for the surface waters of the New York Bight (McCain and Malins, this volume).

The effects of long-term exposures of marine animals to aromatic hydrocarbons are not as well-documented. Examples of recent research applicable to the New York Bight include two studies in which flatfishes were exposed to contaminated sediments. In one study, Sherwood (1976) observed early signs of fin erosion in Dover sole exposed for 13 months to marine sediments from a highly polluted area near Los Angeles. In the other study (McCain *et al.*, 1978), English sole were exposed to sediment-associated crude oil for four months. Compared to control sole residing on clean sediment, the oil-exposed sole took up substantial amounts of petroleum hydrocarbons, developed liver abnormalities, lost weight, and had higher mortalities. Although the species used in these studies are not present in the New York Bight, representatives of the same family are found in or near the Bight area (Smith, 1976).

Other Ecological Effects

Disease

In addition to effects of specific contaminants that can be demonstrated experimentally, there are ecologic effects in the New York Bight that are visible but not fully understood. Among these are several fish and shellfish diseases (*i.e.*, fin erosion, shell disease, and "black gill"), which are numerically more prevalent in the Bight (Murchelano, this volume).

Fin erosion affects many species of Bight fishes but is most prevalent at present in bottom-dwelling flatfishes. The disease is common in fishes from the most polluted parts of the Bight (Ziskowski and Murchelano, 1975; Murchelano and Ziskowski, 1976) and is probably related to depressed environmental quality (Murchelano and Ziskowski, this volume). No specific cause for fin rot in New York Bight flatfishes has been determined; however, it appears to be associated with the exposure of susceptible species to contaminated sediments. Fin rot is a chronic disease that may cause mortality or decreased survival.

Shell disease affects Bight crabs, lobsters, and shrimp (Gopalan and Young, 1975; Young and Pearce, 1975). The disease usually is not fatal because the infected outer skeleton is shed periodically by the growing animal. High prevalence of the disease in specific areas of the Bight suggests that the microorganisms that cause shell disease are more abundant in the affected areas.

"Black gill" aptly describes a disease of crabs and lobsters in which the gills of the affected animals are blackened through melanization and accumulation of sedimentary materials (Sawyer *et al.,* 1979; Sawyer, this volume). The disease is most prevalent in the area of the Bight contaminated with sewage sludge. Severely fouled gills may impair respiration and reduce survival of affected animals.

Effects on fish-eating birds

During the 19th and first half of the 20th Centuries, record nesting densities were reported for osprey (*Pandion haliaetus*) populations in the northeastern United States. Extensive estuaries and shallow bays supported rich anadromous and near-shore fish populations, which were essential to the wellbeing of this fish-eating bird. Beginning in the 1950's, however, osprey populations in the northeast experienced extensive reproductive failures and population declines. By the mid-1970's, the number of active nests in the region had declined to one-tenth that of the mid-1940's (P.R. Spitzer, personal communication). Subsequent investigations suggested a link between reduced eggshell thickness and contaminated fishes in the diets of these birds. Spitzer *et al.* (1977) observed negative correlations between shell thickness and DDE, PCB, and mercury concentrations in osprey eggs; the strongest correlation was that with DDE (a DDT metabolite). Recovery of osprey populations in the New Jersey-New York-New England region was first noted in the mid- to late-1970's and followed the 1972 federal ban on DDT use. A similar recovery has been seen in the brown pelican (*Pelecanus occidentalis*) of the Pacific coast. Recovery of local osprey populations in Connecticut may have been enhanced by restrictions in the use of dieldrin during the early 1970's (Spitzer *et al.*, 1978).

Catch statistics in the New York Bight

If the landings of major species in the New York Bight area are examined, few changes in abundance can be found that are directly attributable to pollution. Surf clams (*Spisula solidissima*) suffered a major mortality in 1976 off the New Jersey coast from anoxia caused by a combination of large-scale oceanographic and meteorological events. Nitrogenous wastes from the Hudson River and adjacent areas may have been contributory, in that had nitrogen enrichment from human sources not been present, the zone of anoxia might have been smaller and less persistent. Adult surf clams cannot move to escape low oxygen, and because they cannot remain closed for very long, they always are threatened by such anoxic events. The 1976 loss was large (approximately 163,000 metric tons of clam meats), and it affected commercial catches thereafter. It must be pointed out, however, that the surf clams were killed by anoxia, not by pollution directly. Other than the surf clam, the ocean quahog (*Arctica islandica*) was the only species whose stock was affected by anoxia in 1976 (Swanson and Sindermann, 1979).

Judging from data on commercial landings of other species caught in the New York Bight, no signs of adverse effects on abundance can be seen. Menhaden (*Brevoortia tyrannus*) landings are much lower than in the late 1950's and early 1960's, but higher than the late 1960's. Similarly, oyster (*Crassostrea virginica*) landings are well below the levels of the early 1950's and earlier, but considerably above those of the mid-1960's. Both resources clearly are overfished; major fluctuations are fishery- or climate-related. Scup (*Stenotomus chrysops*), weakfish (*Cynoscion regalis*), and Atlantic mackerel (*Scomber scombrus*) landings are lower than previously, but considerably higher at present than during the 1960's or early 1970's. Landings of shad (*Alosa sapidissima*) and alewives (*A. aestivalis*) are down, but the shad catch has been low since the late 1940's, a trend caused by events in the river, not outside. Bluefish (*Pomatomus saltatrix*) landings were almost as high in 1978 as in the 1930's, and certainly would be larger if the recreational catch were known. Silver hake (*Merluccius bilinearis*) catches have been higher recently than for almost 50 years. Hard clam (quahog, *Mercenaria mercenaria*) landings are somewhat lower, but there is no evidence that pollution in the New York Bight has been a cause. Yellowtail flounder (*Limanda ferruginea*) landings are low at present, but the cause is overfishing, not pollution. Butterfish (*Peprilus triacanthus*) landings also are down, but also as a result of overfishing. Sea bass (Serranidae) catches are considerably lower than they were in the early 1950's, but higher than in the early 1970's. There is no evidence that pollution was the cause of the changes seen. Landings of the other major species (Atlantic cod, *Gadus morhua;* summer flounder, *Paralichthys dentatus;* Atlantic croaker, *Micropogon undulatus;* winter flounder; red hake, *Urophycis chuss;* sea scallop, *Placopecten magellanicus;* blue crab, *Callinectes sapidus;* ocean quahog; soft clam, *Mya arenaria;* American lobster; tilefish, *Lopholatilus chamaeleonticeps;* and striped bass, *Morone saxatilis*) either are down for reasons presently not known or are up considerably from previous levels. The effects of natural environmental variations on fisheries are large, and create a background "noise" against which it is difficult to measure human effects.

Many species that migrate from salt to fresh water to spawn are now at low abundance levels. American shad have been declining for 100 years or more. Alewives have been seriously reduced by a combination of domestic and foreign fishing and should recover at least partically now that foreign fishing is under control. Striped bass also are presently low, but this is a species that is dominated by strong year classes with intervening years of poor production. For approximately the past 40 years (until 1970) striped bass abundance had been generally increasing.

CONCLUSIONS

After discussions and reviews of manuscripts submitted for this symposium (see second portion of the volume), the fishes working group reached the following conclusions:

1. Each of the major classes of pollutants (toxic metals, chlorinated hydrocarbons, petroleum hydrocarbons) is capable of lowering the capacity of fish and shellfish species to adjust to changing environmental conditions, whether natural or human-induced. Concentrations of a number of pollutants found in Bight waters are in the range of those causing effects in experimental studies. Although sensitivity to pollutants is species-specific, survival of stressed animals is in part a function of competition and predation.

2. Early embryo and late larval stages appear to be the most sensitive to all pollutants. Gonadal tissue is also very sensitive. A large number of fish species spawn within the heavily impacted New York Bight and adjacent estuarine areas. Consequently, the potential impact of pollutants on the reproduction of local stocks is a major concern.

3. Fishes in the New York Bight area may have the ability to cope with at least some pollution stresses through a variety of mechanisms. Up to a finite point, organisms ranging from bacteria to vertebrates have an innate capacity to deal with stress through either genetic, biochemical, physiological, or behavioral mechanisms. Fishes in the New York Bight generally have a high reproductive capacity. For a stress to have a truly significant effect on these fishes, the stress would have to affect a large proportion of the eggs or larvae.

Since fishes are motile, the impact of pollution may be lessened through the dispersal of early life stages and the mobility of juveniles and adults. Effects may be present in localized populations, but recruitment from contiguous areas may result statistically in a lower overall impact.

Fish species vary in sensitivity to pollutants. Pollution may result in the reduction of more sensitive species, with the possible increase in the more resistant species. Such selective action can alter community structure. Other effects in the form of increasing occurrence of abnormalities also may occur.

4. Exposure to low levels of toxic metals increases mortality at early life stages, may render adults more susceptible to predation and disease, and may reduce reproductive capability.

High tissue concentrations of toxic metals are not always quantitatively related to metabolic abnormalities. Trace metal pollution, in concentrations found in the waters of the New York Bight, is seldom directly lethal. One exception is the especially sensitive early-life forms (embryos and larvae), where mortalities from exposure to toxic metals could represent potential and possibly significant reduction in fishery-stock recruitment. The major impact of toxic metals on adult marine animals, which strongly bioaccumulate metals, is a reduced capacity to adapt to normal environmental changes (*e.g.*, salinity, temperature) and a greater susceptibility to predation and disease. Such stresses may well determine the relative success of one species in competition with other more tolerant species.

5. Toxic effects of chlorinated hydrocarbons include mortality (particularly of early life stages), lowered resistance to disease, and decreased survival of offspring. The presence of elevated levels of PCB's and other chlorinated hydrocarbons in the New York Bight ecosystem is of concern because these compounds have been shown in laboratory work and in field studies conducted outside of the greater New York-New Jersey area to be capable of interfering with ovarian and testicular function.

6. Chemical analyses of bottom sediments, plankton, and tissues from fishes and lobsters from the New York Bight have established the existence of toxic levels of polynuclear aromatic hydrocarbons. Adult and early life forms of fishes and shellfishes have been killed or have developed pathological conditions when experimentally exposed to equivalent concentrations of these hydrocarbons.

7. High prevalences of fin erosion, shell disease, and black gill disease in New York Bight fishes and shellfishes indicate that the area is a stressful environment for some bottom-dwelling species. These diseases are related to depressed environmental quality, especially of bottom sediments.

8. Cytotoxic and mutagenic effects of pollutants have been demonstrated in the New York Bight. Mortality of eggs and reduced egg viability (as measured by abnormal chromosomes and cell divisions in embryos) have a direct relationship with extent of pollution (Longwell and Hughes, this volume).

9. The New York Bight apex and associated topographic lows are most severely impacted by pollutants; beyond this area there is an uneven and seasonally variable gradient of decreasing impacts. Many fish species utilizing the Bight are migratory and may encounter the high impact zone in their movements.

RECOMMENDATIONS

As estuarine and coastal waters become increasingly polluted, greater stress is put upon fish and shellfish stocks, threatening their health and recruitment. If a marine environment is to be maintained that is capable of supporting healthy, renewable resource populations, it is imperative that the effects of pollution first be determined and that the results of remedial actions be monitored. The linking of specifically designed laboratory research to broader, more general field activities is one move in this direction.

Based on discussions and reviews of manuscripts submitted for this symposium, the panel recommends the following:

1. Adverse effects of pollution in the New York Bight, and in the apex in particular, have been found to occur in several species of fish and shellfish of recreational and commercial value. Management must consider alternatives to further pollutant inputs to alleviate these impacts.

2. Effects on individual organisms and on localized population segments, rather than on total populations, should be used as criteria for action by regulatory and management groups.

3. Measurements of abundance presently are not sufficiently precise to identify pollution effects. Programs should be developed that can lead to better understanding of large-scale changes in abundance and of natural factors causing changes in abundance. In particular, environmental and fisheries statistics must be improved.

4. Continuous monitoring of physical, chemical, biological, and geological factors should be undertaken in impacted and unimpacted areas. Monitoring should be capable of detecting changes in the environment and living resources of the New York Bight. Such a program should be built upon the information and insights gathered during the MESA New York Bight Project and concurrent fisheries research conducted by the National Marine Fisheries Service. Monitoring should be balanced carefully with laboratory studies.

5. In order to better understand the present and/or future impacts of hydrocarbon pollutants and metals on the ecosystems of the Bight, additional laboratory and field studies are recommended. The two types of studies should be complementary and closely coordinated. Laboratory investigations should involve exposure of adults and early life stages (eggs and larvae) of commercially or ecologically important fish and invertebrate species to pollutant levels found in the Bight. Chronic and acutely toxic effects should be evaluated using bioassays and a variety of gross morphological, behavioral, physiological, and histopathological criteria. All laboratory studies should be performed in an interdisciplinary manner involving the fields of biology, chemistry, and pathology.

6. Lipid-soluble chlorinated organics should not be released into waters of the New York Bight, and their use should be restricted. Research should focus on the effects of chlorinated hydrocarbons and impurities in technical grade industrial organic chemicals.

7. The by-products resulting from chlorination of sewage or other effluents should be analyzed. It is recommended that determinations be made as to whether the by-products are accumulating in fishes.

8. Most research on contaminants of fishes and shellfishes has focused on toxic metals, petroleum hydrocarbons, and chlorinated hydrocarbons. Other contaminants in the New York Bight also may be present and are detectable by state-of-the-art techniques. Research and monitoring are recommended to assess the prevalence of all contaminants in fishes and shellfishes from the New York Bight.

REFERENCES

Alexander, J.E. and E.C. Alexander. 1977. Chemical Properties. MESA New York Bight Atlas Monograph 2. New York Sea Grant Institute, Albany, NY. 47 pp.

Alexander, J.E., R. Hollman, and T. White. 1978. Heavy metal concentrations at the apex of the New York Bight. NOAA Tech. Memo. ERL MESA-34. 34 pp.

Calabrese, A., E.Gould, and F.P. Thurberg. This volume. Effects of toxic metals in marine animals of the New York Bight: some laboratory observations. pp. 281-297.

Carmody, D.J., J.B. Pearce, and W.E. Yasso. 1973. Trace metals in sediments of New York Bight. Mar. Pollut. Bull. 4: 132-135.

Craddock, D.R. 1977. Acute toxic effects of petroleum on arctic and sub-arctic marine organisms. *In:* Effects of Petroleum on Arctic and Sub-arctic Marine Environments and Organisms, Vol. II, D.C. Malins (ed.), Academic Press, New York. pp. 1-93.

Esser, S.C. This volume. Long-term changes in some finfishes of the Hudson-Raritan estuary. pp. 299-314.

Freeman, H.C. and D.R. Idler. 1975. The effect of polychlorinated biphenyl on steroidogenesis and reproduction in the brook trout (*Salvelinus fontinalis*). Can. J. Biochem. 53: 666-670.

Gopalan, U.K. and J.S. Young. 1975. Incidence of shell disease in shrimp in the New York Bight. Mar. Pollut. Bull. 6: 149-153.

Greig, R.A., A. Adams, and D.R. Wenzloff. 1977. Trace metal content of plankton and zooplankton collected from the New York Bight and Long Island Sound. Bull. Environ. Contam. Toxicol. 18: 3-8.

Greig, R.A. and R.A. McGrath. 1977. Trace metals in sediments from Raritan Bay. Mar. Pollut. Bull. 8: 188-192.

Greig, R.A. and D.R. Wenzloff. 1977. Trace metals in finfish from the New York Bight and Long Island Sound. Mar. Pollut. Bull. 8: 198-200.

Hodgins, H.O., B.B. McCain, and J.W. Hawkes. 1977. Marine fish and invertebrate diseases, host disease resistance, and pathological effects of petroleum. *In:* Effects of Petroleum on Arctic and Sub-arctic Marine Environments and Organisms, Vol. II, D.C. Malins (ed.), Academic Press, New York. pp. 95-173.

Horn, E.G., L.J. Hetling, and J. Tofflemire. 1979. The problem of PCBs in the Hudson River system. Ann. N.Y. Acad. Sci. 320: 591-609.

Johnson, F.G. 1977. Sublethal biological effects of petroleum hydrocarbon exposures: Bacteria, algae, and and invertebrates. *In:* Effects of Petroleum on Arctic and Subarctic Marine Environments and Organisms, Vol. II, D.C. Malins (ed.), Academic Press, New York. pp. 271-318.

Krom, M.D., K.K. Turekian, and N.H. Cutshall. 1981. The fate of trace metals in the sediments of the New York Bight. Program and Abstracts. Third International Ocean Disposal Symposium, Woods Hole, MA. pp. 50-51.

Longwell, A.C. and J.B. Hughes. This volume. Cytologic, cytogenetic, and embryologic state of Atlantic mackerel eggs from surface waters of the New York Bight in relation to pollution. pp. 381-388.

McCain, B.B., H.O. Hodgins, W.E. Gronlund, J.W. Hawkes, D.W. Brown, M.S. Myers, and J.H. Vandermeulen. 1978. Bioavailability of crude oil from experimentally oiled sediments to English sole (*Parophrys vetulus*), and pathological consequences. J. Fish. Res. Board Can. 35: 657-664.

McCain, B.B. and D.C. Malins. This volume. Effects of petroleum hydrocarbons on selected demersal fishes and crustaceans. pp. 315-326.

MacLeod, W.D., Jr., L.S. Ramos, A.J. Friedman, D.G. Burrows, P.G. Prohaska, D.L. Fisher, and D.W. Brown. 1981. Analyses of residual chlorinated hydrocarbons, aromatic hydrocarbons and related compounds in selected sources, sinks, and biota of New York Bight. NOAA Tech. Memo. OMPA-6. 128 pp.

MESA. 1978. MESA New York Bight Project Technical Development Plan for Fiscal Year 1979. NOAA Special Report. 159 pp.

Mueller, J.A., J.S. Jeris, A.R. Anderson, and C.F. Hughes. 1976. Contaminant inputs to the New York Bight. NOAA Tech. Memo. ERL MESA-6. 347 pp.

Murchelano, R.A. This volume. Some pollution-associated diseases and abnormalities of marine fish and shellfish: a perspective for the New York Bight. pp. 327-346.

Murchelano, R.A. and J. Ziskowski. 1976. Fin rot disease studies in the New York Bight. Am. Soc. Limnol. Oceanogr. Spec. Symp. 2: 329-336.

Murchelano, R.A. and J. Ziskowski. This volume. Fin rot disease in the New York Bight (1973-1977). pp. 347-358.

Neff, J.S. 1979. Polycyclic Aromatic Hydrocarbons in the Aquatic Environment: Sources, Fates and Biological Effects. Applied Science Publishers Ltd., London. 262 pp.

O'Connor, J.M., J.B. Klotz, and T.J. Kneip. This volume. Sources, sinks, and distribution of organic contaminants in the New York Bight ecosystem. pp. 631-653.

O'Connor J.M. and J. Rachlin. This volume. Perspectives on metals in New York Bight organisms: factors controlling accumulation and body burden. pp. 655-673.

O'Connor J.S. and H.M. Stanford (eds.). 1979. Chemical Pollutants of the New York Bight; Priorities for Research. NOAA Special Report. 217 pp.

Purrett-Carroll, L. 1979. New light on the New York Bight. NOAA 9: 44-47.

Roubal, W.T., S.I. Stranahan, and D.C. Malins. 1978. The accumulation of low molecular weight aromatic hydrocarbons of crude oil by coho salmon (*Oncorhynchus kisutch*) and starry flounder (*Platichthys stellatus*). Arch. Environ. Contam. Toxicol. 7: 237-249.

Sawyer, T.K. This volume. Distribution and seasonal incidence of "black gill" in the rock crab, *Cancer irroratus*. pp. 199-211.

Sawyer, T.K., S.A. MacLean, J.E. Bodammer, and B.A. Harke. 1979. Gross and microscopical observations on gills of rock crabs (*Cancer irroratus*) and lobsters (*Homarus americanus*) from nearshore waters of the eastern United States. *In:* Proceedings of the Second Biennial Crustacean Health Workshop, Texas A & M Univ. Sea Grants Publ. TAMSG 79-114. pp. 68-91.

Segar, D.A. and A.Y. Cantillo. 1976. Trace metals in the New York Bight. Am. Soc. Limnol. Oceanogr. Spec. Symp. 2: 171-198.

Sherwood, M. 1976. Fin erosion disease induced in the laboratory. *In:* Annual Report of the Southern California Coastal Water Research Project, Southern California Coastal Water Research Project, El Segundo, CA. pp. 143-148.

Sherwood, M.J. This volume. Fin erosion, liver condition, and trace contaminant exposure in fishes from three coastal regions. pp. 359-377.

Smith, C.E., T.H. Peck, R.J. Klauda, and J.B. McLaren. 1979. Hepatomas in Atlantic tomcod *Microgadus tomcod* (Waldbaum) collected in the Hudson River estuary in New York. J. Fish Diseases 2: 313-319.

Smith, C.L. 1976. The Hudson River fish fauna. *In:* Hudson River Ecology, Hudson River Environmental Society. 11 pp.

Spitzer, P.R., R.W. Risebrough, J.W. Grier, and C.R. Sindelar, Jr. 1977. Eggshell thickness-pollutant relationships among North American ospreys. *In:* Transactions of the North American Osprey Research Conference, J.C. Ogden (ed.), U.S. National Park Service. pp. 115-119.

Spitzer, P.R., R.W. Risebrough, W. Walker II, R. Hernandez, A. Poole, D. Puleston, and I.C.T. Nisbet. 1978. Productivity of ospreys in Connecticut-Long Island increases as DDE residues decline. Science 202: 333-335.

Sunda, W.G. and P.A. Gillespie. 1979. The response of a marine bacterium to cupric ion and its use to estimate cupric ion activity in seawater. J. Mar. Res. 37: 761-777.

Sunda, W.G. and R.R.L. Guillard. 1976. The relationship between cupric ion activity and the toxicity of copper to phytoplankton. J. Mar. Res. 34: 511-529.

Swanson, R.L. 1977. Status of ocean dumping research in New York Bight. J. Waterway Port Coast. Ocean Div., Amer. Soc. Civil Eng. 103 (WWI): 9-24.

Swanson, R.L. and C.J. Sindermann (eds.). 1979. Oxygen Depletion and Associated Benthic Mortalities in New York Bight, 1976. NOAA Prof. Pap. 11: 345 pp.

U.S. Environmental Protection Agency. 1976. Quality Criteria for Water. U.S. Environmental Protection Agency, Washington, DC. 256 pp.

U.S. Environmental Protection Agency. 1980a. Ambient Water Quality Criteria for Phenol. EPA 440/5-80-066. Office of Water Regulations and Standards, U.S. Environmental Protection Agency, Washington, DC. 93 pp.

U.S. Environmental Protection Agency. 1980b. Ambient Water Quality Criteria for Polychlorinated Biphenyls. EPA 440/5-80-068. Office of Water Regulations and Standards, U.S. Environmental Protection Agency, Washington, DC. 191 pp.

Walker, C.R. 1976. Polychlorinated biphenyl compounds (PCBs) and fishery resources. Fisheries 1: 19-25.

Young, J.S. and J.B. Pearce. 1975. Shell disease in crabs and lobsters from New York Bight. Mar. Pollut. Bull. 6: 101-105.

Ziskowski, J. and R. Murchelano. 1975. Fin erosion in winter flounder. Mar. Pollut. Bull. 6: 26-29.

EFFECTS OF POLLUTANTS ON PLANKTON AND NEUSTON

R. Lee
A.C. Longwell
T.C. Malone
L.S. Murphy
D.R. Nimmo
H.B. O'Connors, Jr.[1]
L.S. Peters
K.D. Wyman[1,2]

Abstract. A non-technical summary of the known and potential effects of environmental stress on plankton/neuston populations in the New York Bight is presented. This summary was prepared by the plankton/neuston panel. The panel concluded that field investigations have detected no significant pollution-induced changes in the growth and distribution of New York Bight plankton populations. The potential for chronic and acute impacts on plankton/neuston populations, however, has been demonstrated in controlled laboratory situations investigating chemical pollutants that occur in the Bight and adjacent waters. The problems associated with interpreting field studies conducted in the Bight and extrapolating laboratory studies to plankton populations in the Bight are discussed. Recommendations for future research and regulatory efforts are provided.

INTRODUCTION

Pollution-related stresses on plankton populations have been associated with effects such as reduced yields of food species, elevated contamination levels in seafood, and diminished aesthetic and recreational appeal of marine environments. Currently, there is no consensus that all of these effects have occurred in nature, or that they pertain equally to the New York Bight; nevertheless, the potential clearly exists. The state of plankton populations in the New York Bight and adjacent waters must be considered with the actual and potential impacts of domestic and chemical wastes, if prudent pollution management strategies are to be developed for the region. This paper presents a non-technical summary of these subjects prepared by the symposium panel on the effects of pollutants on plankton/neuston populations of the New York Bight.

The panel first undertook to characterize the current plankton/neuston populations of the area and the natural environmental factors that affect their structure (*e.g.*, species composition) and function (*e.g.*, productivity). It then considered how the plankton/neuston populations differ from those of similar but less impacted areas, and how they may have changed with time. Major chemical pollutants of the Bight (O'Connor and Stanford, 1979) and sewage-derived nutrient enrichment were assessed for their known or possible effects on plankton/neuston. This assessment was based on laboratory and field observations reported in the scientific literature and in technical papers

[1] Compiler of revised panel report.
[2] Not original member of panel.

presented at this symposium. The panel, however, acknowledged the difficulty of doing research leading to clearly defined cause and effect relationships. Finally, the panel provided recommendations for future research to identify causes of change in plankton/neuston populations and regulatory efforts to reduce the potential for undesirable effects resulting from pollution.

PLANKTON AND NEUSTON

Marine organisms that float and drift with the flow of ocean currents and tides are termed plankton. The importance of these organisms is reflected in the ability of some plankton to convert solar energy to chemical energy and the ability of other plankton to transfer this energy as food for larger species inhabiting the Bight and adjacent waters. Plankton also have shown a capacity to assimilate or accumulate a variety of inorganic and organic chemicals and subsequently to transfer these chemicals throughout food chains that may include humans. Some of the chemicals are suspected carcinogens, mutagens, and teratogens.

Planktonic organisms associated with the surface film of the ocean are termed neuston. Bacterial counts may be ten to one hundred times higher in surface films than just below the surface. Species of plants (phytoplankton) and invertebrate animals (zooplankton) are also quantitatively and qualitatively different in surface films than in subsurface samples (Sieburth, 1971; Taguchi and Nakajima, 1971). In addition, buoyant fish eggs are closely associated with the sea surface. Yet the role of the neuston in accumulating, distributing, and transforming pollutants is neither studied nor understood.

Typical representatives of planktonic organisms occurring in the New York Bight and adjacent waters are presented in Figure 1. Those familiar with plankton will recognize that only a few representatives of a diverse group of organisms are illustrated.

Phytoplankton

In nearshore (neritic) waters, nanophytoplankton, diatoms, and dinoflagellates (Figure 1A, B, and C, respectively) comprise the bulk of the phytoplankton. These organisms are single-celled plants that occur as solitary cells or in chains, and are responsible for generating organic material known as primary production. They require sunlight for photosynthesis; thus, growing populations are restricted to the lighted surface layer, or euphotic zone. The extent of the euphotic zone is dependent upon the quantity and quality of the total suspended particulates in the seawater. In the clearest ocean water, the euphotic zone may extend deeper than 200 meters; in the turbid New York Bight apex and adjacent waters, it may be less than 10 meters deep.

In addition to sunlight, phytoplankton require nutrients for growth. These include carbon dioxide, nitrogen, (*e.g.*, ammonium, nitrate, nitrite, and urea), phosphate, sulfate, and other inorganic compounds. Some species (*i.e.*, diatoms) require large amounts of silicon (as silicate) for the formation of cell coverings.

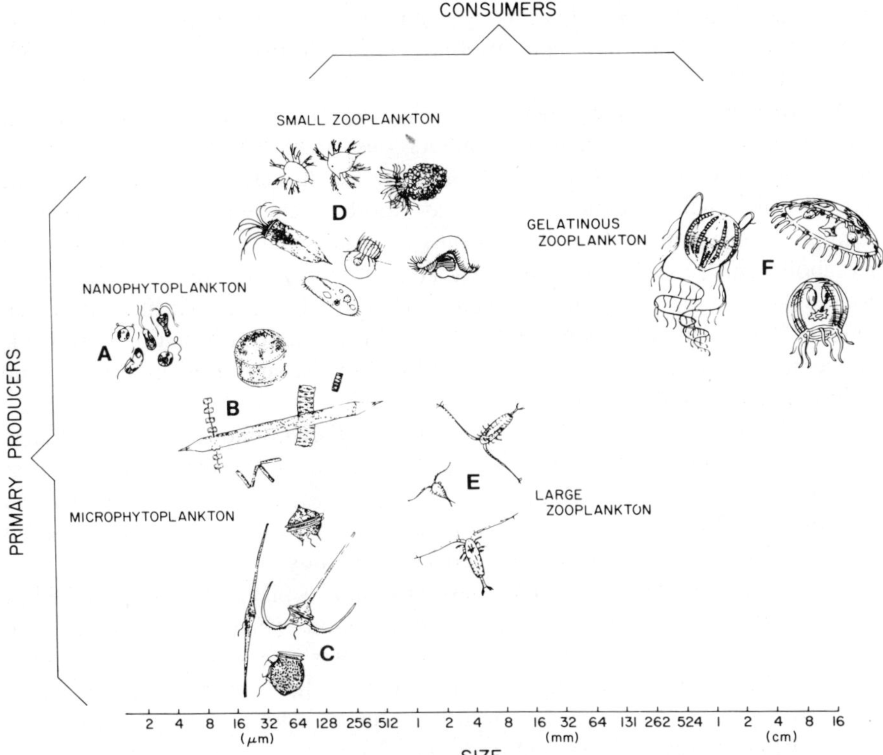

Figure 1. Representatives of plankton occurring in the New York Bight and adjacent waters. (A) nanophytoplankton; (B) diatoms; (C) dinoflagellates; (D) microzooplankton; (E) calanoid copepods; (F) gelatinous zooplankton.

Many require vitamins and other organic growth factors. Most of these nutrients occur in excess of that required for continued phytoplankton growth. However, depending upon the season and locale, the quantity and rate of supply of usable forms of nitrogen and silicate may limit phytoplankton growth and abundance (Ryther and Dunstan, 1971; Malone *et al.,* 1980a).

Vertical mixing of the water column, resulting from storm activity, convective cooling, and tidal currents, is an important process controlling the availability of light and the abundance of essential nutrients for phytoplankton populations. Phytoplankton growth and abundance are also influenced by the sinking of phytoplankton cells from the euphotic zone and by zooplankton grazing. These two factors are especially important in the inner portions of the New York Bight (Malone and Chervin, 1979).

Zooplankton

The zooplankton include members of practically every major taxonomic group of animals. Many zooplankton species remain planktonic throughout their life cycle. Others (*e.g.*, floating eggs and weakly swimming larvae of fish, shellfish, and bottom-dwelling invertebrates) occur in the zooplankton for only a short, but critical, part of their cycles.

Protozoans and larval stages of larger invertebrates contribute to the microzooplankton, the smaller members (less than 500 μm in size) of the zooplankton community (Figure 1D). These organisms efficiently capture a wide variety of small organic particles as food, including bacteria, detritus, nanophytoplankton, and smaller diatoms (Capriulo and Carpenter, 1980). High metabolic and reproductive rates suggest that protozoans are important in the food chain dynamics of nearshore waters. Their small size and fragility, however, renders them difficult to study and collect. Insufficient information is available concerning their seasonal abundance, distribution, and role in pristine and polluted coastal waters.

Gelatinous zooplankton, including jellyfish and ctenophores (Figure 1F), may be very abundant in spring, summer, and fall zooplankton communities. These organisms reach a size of several centimeters and yet efficiently feed on smaller organisms. As was the case for protozoans, the delicate nature of gelatinous zooplankton makes them difficult to collect and study.

Crustaceans compose by far the largest group of planktonic particle-feeding animals. Within this group, the calanoid copepods (Figure 1E) are of great importance, often making up 70-90% of the mass of zooplankton collections. The success of copepod populations and all other zooplankton organisms depends upon factors such as hydrographic conditions, food availability, competition between species, and intensity of predation.

FIELD STUDIES IN THE NEW YORK BIGHT REGION

The plankton of the New York Bight and adjacent waters have been the subject of few investigations, all of which were restricted to small geographic areas, limited to short periods of time, and focused on selected taxonomic groups. Malone (1977a), in reviewing earlier studies of plankton in the Bight, found the data too sketchy to describe plankton distributions comprehensively, let alone to provide useful predictions of the impacts of human activities.

Numerous problems hamper attempts to make comparisons among historical studies for the purpose of identifying changes in plankton/neuston populations. The variety of collecting gear and sampling techniques used by previous investigators makes data assessment difficult. The relatively large mesh sizes used in plankton nets have grossly undersampled the microzooplankton. In addition, historical data often reflect less sophisticated sampling technology than is presently employed. Comparability of plankton data also depends upon standardized sampling depths, sampling times (*i.e.*, time of day and year), and station locations. These problems, as well as large natural population fluctuations and inherent spatial variability, have precluded observation of clearly defined effects of waste discharges on plankton communities. Indeed, Judkins *et al.* (1980), having given proper consideration to the above problems, concluded that the zooplankton in the New York Bight had not substantially changed in the 15 years between 1960 and 1975.

Comparisons of the Bight with less impacted regions are subject to similar problems and confounded by differences in environmental and hydrographic properties. Although plankton communities in the heavily impacted Hudson-Raritan estuary are different from those encountered in the moderately polluted apex and relatively pristine outer Bight (Malone, 1977a; Judkins *et al.*, 1980), these differences are seen in less impacted areas and may be explained by the differential ability of plankton species to tolerate environmental factors such as temperature, salinity, and particulate loads.

EFFECTS OF POLLUTANTS ON PLANKTON

Although there is no evidence of long-term changes in the plankton/neuston populations of the New York Bight, it is still possible that chemical and wastewater discharges have impacted these populations. Laboratory studies have demonstrated that many pollutants have the potential to influence processes governing plankton growth and to affect the plankton themselves in an acute or sublethal fashion. The detrimental impact of some pollutants (*e.g.*, nutrients) is small, and remedial efforts to reduce environmental concentrations of these substances may produce undesirable effects. The toxicities of other contaminants, however, are well documented by laboratory studies. The presence of such materials in high concentrations in New York Bight waters may have serious consequences for plankton/neuston populations. Finally, some substances have not been adequately assessed for their impact upon plankton/neuston populations and deserve attention by scientists and policy-makers.

Sewage-Derived Nutrient Enrichment

Sewage-derived nutrient enrichment probably has little undesirable impact upon plankton/neuston populations of the New York Bight and adjacent waters. Despite discharges of approximately 300-600 metric tons of nitrogen per day into the Hudson-Raritan estuary, phytoplankton biomass and productivity in this body of water remain relatively low compared to those in other areas having comparable nutrient concentrations (Malone, this volume). Phytoplankton consume less than 10% of the available nitrogen (Garside *et al.*, 1976) and account for only 5% to 50% of the total organic carbon input into the estuary (Malone, this volume). These data reflect the effects of high concentrations of river-borne suspended sediments, which limit the extent of the euphotic zone, and high freshwater dilution rates, which transport phytoplankton biomass and nutrients out of the estuary and into the apex (Malone, 1977b; Malone *et al.*, 1980b).

Phytoplankton growth in the apex appears to be limited by light and temperature under most conditions (Malone, this volume), although silicon limitation may occur in some instances (Malone *et al.*, 1980a). Primary production, sustained by a constant nutrient flux into the apex from the estuary, accounts for 70% to 80% of the particulate organic carbon input within 20 km of the estuary mouth (Segar and Berberian, 1976). The fate of primary production in the apex varies according to season. Diatom production (Figure 1B), which dominates the phytoplankton community during winter and early spring, sinks to the bottom in the absence of density stratification and grazing pressure, where it contributes to the degree of oxygen demand in the apex (Malone and Chervin, 1979). In contrast, nanophytoplankton production (Figure 1A) predominates during summer months, but is prevented from sinking to the bottom by density stratification and zooplankton grazing. Zooplankton fecal material also may contribute to low oxygen stress in the Bight, but probably to a smaller extent than phytoplankton remains reaching the bottom.

Severe oxygen depletion has occurred in the Bight when phytoplankton production was unchecked and was not efficiently washed from the Bight or incorporated into the food web (Segar and Berberian, 1976; Malone, 1978). Such events were recorded in 1968, 1971, 1974, 1976, and 1977. Blame for the most severe of these events—the 1976 anoxia—has been placed primarily on atypical wind and current patterns that greatly reduced flushing in the Bight and imported into the apex phytoplankton that had bloomed over the outer portions of the continental shelf (Swanson *et al.*, 1979; Falkowski *et al.*, 1980). The role of human wastes in 1976 is uncertain, but they may have increased the probability of the episode (O'Connor, 1979). The remaining four events are considerably less well studied; however, nutrient and carbon inputs through the Sandy Hook/Rockaway Point transect and from the New Jersey coast probably acted in combination with anomalies in the hydrographic regime to produce local areas of low dissolved oxygen (O'Connor, 1979). In all of these episodes, it is unlikely that ocean dumping was a major factor. Ocean dumping of sewage sludge and dredged materials contributes to only 10% of the summer phytoplankton nitrogen demand in the apex and has no significant effect on phytoplankton growth rates (Malone, 1976).

Productivity in the Bight beyond the apex occurs in the absence of human-related nutrient enrichment. Vertical mixing, sinking, and light limitation are important processes that control phytoplankton growth during winter and early spring, while zooplankton grazing, local mixing events, and nutrient limitation are important in the summer and fall seasons.

Malone (this volume) concluded that secondary treatment is having a beneficial effect on water quality in the estuary and apex. Removal of organic inputs improves dissolved oxygen levels and reduces bacterial densities, but does not alter phytoplankton production. Tertiary (advanced wastewater) treatment, however, does not appear necessary and may have a detrimental impact on local fisheries by reducing primary production in the estuary and apex.

Chlorinated Hydrocarbons

Substantial levels of several chlorinated hydrocarbons presently are found in the Hudson River and New York Bight (O'Connor *et al.*, this volume).

Polychlorinated biphenyls (PCB's) and chlorinated pesticides are of particular concern because of their high concentrations in upper river sediments, inputs from the lower Hudson-Raritan estuary, chemical stability, and ability to bioaccumulate throughout the food web.

Recently, factors determining chlorinated hydrocarbon accumulation by plankton and its biological effect on plankton populations have been discussed (Peters and O'Connor, this volume; O'Connors *et al.,* this volume). Under experimental conditions, phytoplankton growth rates and photosynthetic activity have been significantly reduced by exposure to low concentrations of DDE (the major metabolite of DDT), chlordane, dieldrin, and PCB (Aroclor 1254). The size and species compositions of phytoplankton assemblages characteristic of the New York Bight area were altered following experimental exposure to PCB's. Large diatoms were found to be more sensitive to PCB's than were nanophytoplankton.

In laboratory situations, zooplankton accumulated lethal amounts of PCB's directly from the water and from grazing on PCB-contaminated phytoplankton. In addition, some zooplankton were observed to feed less efficiently on smaller particles than on larger particles. It has been hypothesized that contaminants that reduce the size composition of phytoplankton assemblages may favor zooplankton that feed most efficiently on smaller particles. O'Connors *et al.* (1978) speculated that such a change could alter the flow of biomass in marine food webs and possibly affect harvestable fish yields. It is debatable, however, whether this mechanism applies to the New York Bight, since diatom production and peak zooplankton abundances do not co-occur in this system. Indeed, zooplankton in the Bight reach peak abundance while grazing on summer nanophytoplankton blooms.

Toxic chemicals may have a more direct impact on harvestable fish yields by affecting reproductive success. Longwell and Hughes (this volume) demonstrated statistically significant associations among measures of mackerel (*Scomber scombrus*) egg health (*e.g.,* embryo moribundity, genetic abnormalities), toxic metal concentrations in surface waters and zooplankton, and toxic hydrocarbon concentrations in zooplankton. Elevated levels of egg mortality generally were observed in the more polluted areas of the Bight. Whether this has affected fish abundances cannot be determined since declines in fish populations also are attributable to overfishing, climatic and hydrographic variations, and changes in spawning grounds (see also Sindermann *et al.,* this volume).

Petroleum Hydrocarbons

The Hudson-Raritan estuary receives over 870 metric tons of oil and grease each day from municipal runoff, shipping, and barge activities (Mueller *et al.,* 1976), yet few field studies have identified impacts on plankton that are directly associated with oil pollution. Cod (*Gadus morhua*) and pollack (*Pollachius virens*) eggs sampled in the vicinity of the ARGO MERCHANT oil spill showed signs of external contamination and decreased egg health (Longwell, 1977).

Copepods also were observed with fouled feeding appendages and ingested oil. Changes in the plankton community in response to the oil spill, however, could not be demonstrated.

Several changes have been observed in plankton populations exposed to petroleum and its derivatives under controlled experimental conditions (*e.g.*, Menzel, 1977). Large increases in numbers of bacteria were detected following addition of napthalenes (a major group of water-soluble petroleum compounds) to experimental enclosures (Lee *et al.*, 1978). A portion of the increase in bacterial density may have resulted from the growth of hydrocarbon-degrading bacteria, since napthalenes were rapidly broken down in treated enclosures. Microbial growth also may have been stimulated by organic compounds secreted by phytoplankton in response to the presence of hydrocarbons. In contrast, however, neither the degradation of hydrocarbons nor bacterial density was enhanced following exposure of plankton communities to some crude oils and No. 2 fuel oil (Hodson *et al.*, 1977). This suggests that the nature and concentration of hydrocarbons, as well as the species of plankton being investigated, are important factors influencing the potential impact of petroleum hydrocarbons on plankton.

The species composition of phytoplankton populations shifted from diatoms to microflagellates (nanophytoplankton) in hydrocarbon-treated enclosures. This shift reflects the increased sensitivity of some diatom species (Lee *et al.*, 1978) and increased photosynthetic activity of microflagellates (Parsons *et al.*, 1976) following exposure to hydrocarbons.

Increases in the number of protozoans and other microzooplankton accompanied increases in the bacteria and nanoplankton that these organisms feed upon. No significant differences were noted in copepod abundance. Copepods appear to be affected only at high concentrations of petroleum hydrocarbons (Lee and Anderson, 1977; Burdugo *et al.*, 1977; Berman and Heinle, 1980). On the other hand, ctenophores suffered heavy mortality in enclosed plankton communities exposed to petroleum hydrocarbons, which relaxed predation pressure on copepods and microzooplankton (Lee *et al.*, 1978).

Extrapolation of experimental results to events occurring in the New York Bight must be made with caution. Particular concern must be given to potential artificialities in the results which arise from confining plankton populations to enclosures (Menzel, 1977). Furthermore, concentrations of hydrocarbons in underlying waters following large oil spills are low and dissipate rapidly. Population changes that were observed in experimental situations may not occur under these circumstances.

Toxic Metals

The New York Bight receives significantly greater quantities of toxic metals (particularly zinc, copper, and lead) than do surrounding areas (Mueller *et al.*,

1976). Zooplankton collected from the New York Bight reflect these greater inputs by having higher metal concentrations than do plankton from surrounding areas (Grieg *et al.*, 1977). As with other pollutants, however, changes in plankton populations of the New York Bight as a result of elevated metal concentrations have not been demonstrated and must be inferred from laboratory and controlled ecosystem studies.

O'Connor and Stanford (1979) reviewed numerous laboratory studies demonstrating the inhibition of phytoplankton growth and photosynthesis following exposure to toxic metals. However, the results of these studies must be used with caution in predicting changes in plankton/neuston populations. In the laboratory most of the variability encountered in the real world is controlled by testing the effect of one substance on a single species, under defined conditions. Metals toxicity may be affected by chelating agents (*i.e.*, compounds that reduce the availability of metals in the environment) released during spring freshets and by the presence of other metals. In addition, interactions with non-metal contaminants may enhance or reduce the toxic impact of metals (Swedmark and Granmo, in press). The potential impact of a single metal on plankton/neuston populations in the Bight is certainly complicated by the multitude of contaminants present in the region.

The size and species composition of phytoplankton populations enclosed in controlled ecosystem containers shifted from diatoms to smaller microflagellates following exposure to copper and mercury (Thomas and Seibert, 1977; Thomas *et al.*, 1977). Smaller zooplankton and ctenophores also were sensitive to heavy metals (Beers *et al.*, 1977; Reeve *et al.*, 1977). The success of copepods, however, was primarily influenced by the presence or absence of predators. Despite changes in the structure of the plankton community within the enclosures, factors such as photosynthetic and nitrogen uptake rates were only temporarily affected.

Ocean Dumping of Industrial Wastes and Dredged Materials

Industrial wastes currently are disposed at a dumpsite located 196 km (106 n. mi.) southeast of the Bight apex (DWD 106). These wastes may be acidic or alkaline and contain a variety of metals and organic compounds (see O'Connor and Park, this volume). Investigations examining the effects of these wastes on plankton populations have been reviewed by the U.S. Environmental Protection Agency (1980) and National Oceanic and Atmospheric Administration (1981). Laboratory studies suggest that phytoplankton are affected by DWD 106 wastes only at very high concentrations (Murphy *et al.*, this volume), yet short-term changes in phytoplankton structure have occurred at another dumpsite off Puerto Rico (Murphy *et al.*, in press). Feeding rates were lower for copepods exposed to realistic concentrations of DWD 106 waste than for those unexposed to the waste (Capuzzo and Lancaster, 1981). Relatively little ocean water is affected by dumping, however, and it is unlikely that ocean dumping of these wastes has a significant, large-scale impact on plankton populations. Similar conclusions have been reached regarding the impact of acid waste disposal in the Bight apex (Wiebe *et al.*, 1973; Grice *et al.*, 1973).

Dredged materials may represent a large source of toxic metals and hydrocarbons to the New York Bight. The potential toxicities of these substances on plankton populations already have been discussed. In addition, dredging operations and dumping of spoils may have an impact on plankton communities which is independent of chemical toxicity. Nimmo *et al.* (this volume) reported significant mortality and reproductive failure among zooplankton exposed to highly turbid conditions, such as may exist during dredging and dumping operations.

Other Pollutants

Many chemical substances have not been assessed adequately for their impact on plankton/neuston populations in the New York Bight. Among these are organic compounds (including N-nitroso compounds and complexes formed in chlorinated sewage) and radionuclides. Some of these are suspected carcinogens and mutagens, and, therefore, may cause abnormalities in eggs and larvae of valuable seafood species. Little is known about their potential for incorporation into food species.

RESISTANCE TO POLLUTANTS

Several studies indicate that individual plankton populations may have evolved resistance to inorganic and organic pollutants. Fisher *et al.* (1973) reported that oceanic phytoplankton showed greater sensitivity to PCB's than did estuarine clones of the same species. Murphy *et al.* (this volume) demonstrated that phytoplankton isolated from polluted bays and estuaries were able to tolerate pollutants better than clones of the same species isolated from unpolluted estuaries. The nature of this resistance is not clearly understood but may reflect the development of structural adaptations at the cellular level or physiological responses which permit phytoplankton to exclude, inactivate, or metabolize toxic substances.

The significance of pollution resistance in plankton is under debate. Some have questioned the validity of pollution toxicity studies that used plankton as bioassay organisms. These individuals suggest that such investigations may have overestimated or underestimated the impact of particular pollutants depending upon whether the plankton used in the tests were isolated from pristine or polluted waters. Others have speculated that cellular modifications which enable phytoplankton cells to limit pollution uptake may also hamper the accumulation of essential nutrients. Were this the case, plankton productivity in polluted waters could be below normal. Finally, the argument has been advanced that because resistant strains may concentrate toxicants, there is a greater likelihood of transferring these substances up the food web and ultimately to humans. At the present time, none of the above hypotheses have been proven. All represent areas that must be explored further before conclusions can be drawn.

CONCLUSIONS AND RECOMMENDATIONS

1. Field investigations in the New York Bight and adjacent waters indicate there has been no detectable impact of pollutants on the growth and distribution of plankton populations. These investigations, however, are recent and have not been carried out over a sufficient length of time to detect chronic effects. Such effects require extended periods of exposure before expression and may be more serious and long-lasting than more easily observed, short-term impacts. An adequate long-term, multidisciplinary program to monitor plankton/neuston populations in the Bight must be developed. The analysis of data collected from such a program would be greatly assisted if the spatial and temporal variability of the Bight's plankton regime could be better defined. The panel, however, recognizes that the time and costs of such a research effort are formidable.

2. Controlled (*e.g.*, laboratory) studies have provided circumstantial evidence of pollution-related changes in the structure and function of New York Bight plankton/neuston populations. Continued experimental effort is required to supplement field observations. Specifically, research should be directed at better understanding (a) the rates and transport mechanisms of toxic substances through planktonic food webs, (b) the interactions between pollutants and the effects of such interaction on marine plankton, (c) the nature of resistance to toxicants evolved by marine plankton, and (d) the mutagenic and carcinogenic potential of pollutants. Data collected from such research would be valuable to managers wishing to develop risk assessment models and to identify other substances that present the greatest hazards to the productivity of the Bight.

3. If pollution-induced changes in plankton/neuston populations have occurred in the New York Bight, recovery is likely to be slow following remedial actions such as the abatement of toxic substances. Recovery will be dependent on rates of elimination of toxicants from the Bight system and rates of repopulation.

4. Experimental studies have demonstrated that many pollutants have similar potential impacts on plankton populations (*e.g.*, reduction in size and species diversity of phytoplankton communities). Regulation of individual pollutants may not be sufficient to cause a noticeable "recovery" of plankton populations.

5. Efforts should be continued to upgrade sewage treatment to the secondary level. Such efforts need to be done in conjunction with research designed to increase understanding of the freshwater flow in the estuary. Tertiary treatment is unnecessary as it applies to nutrient loading.

6. Restrictions specifying uses and "safe" levels of toxic substances should be reassessed to reduce the potential for impact on plankton populations and to lower levels of pollutants in valuable seafood species. Better enforcement of existing regulations would help reach these ends.

REFERENCES

Beers, J.R., M.R. Reeve, and G.D. Grice. 1977. Controlled Ecosystem Pollution Experiment: effect of mercury on enclosed water columns. IV. Zooplankton population dynamics and production. Mar. Sci. Comm. 3: 355-394.

Berdugo, V., R.P. Harris, and S.C. O'Hara. 1977. The effects of petroleum hydrocarbons on reproduction of an estuarine planktonic copepod in laboratory cultures. Mar. Pollut. Bull. 8: 138-143.

Berman, M.S. and D.R. Heinle. 1980. Modification of the feeding behavior of marine copepods by sub-lethal concentrations of water-accommodated fuel oil. Mar. Biol. 56: 59-64.

Capriulo, G.C. and E.J. Carpenter. 1980. Grazing by 35 μm to 202 μm microzooplankton in Long Island Sound. Mar. Biol. 56: 319-326.

Capuzzo, J.M. and B.A. Lancaster. 1981. The effects of pollutants on marine zooplankton at Deep Water Dumpsite 106: preliminary findings. *In:* Ocean Dumping of Industrial Wastes, B.H. Ketchum, D.R. Kester, and P.K. Park (eds.), Plenum Press, New York. pp. 411-419.

Falkowski, P.G., T.S. Hopkins, and J.J. Walsh. 1980. An analysis of factors affecting oxygen depletion in the New York Bight. J. Mar. Res. 38: 479-506.

Fisher, N.S., L.B. Graham, E.S. Carpenter, and C.F. Wurster. 1973. Geographic differences in phytoplankton sensitivity to PCBs. Nature. 241: 548-549.

Garside, C., T.C. Malone, O.A. Roels, and B.A. Sharfstein. 1976. An evaluation of sewage-derived nutrients and their influence on the Hudson estuary and New York Bight. Estuarine Coastal Mar. Sci. 4: 281-289.

Greig, R.A., A. Adams, and D.R. Wenzloff. 1977. Trace metal content of plankton and zooplankton collected from the New York Bight and Long Island Sound. Bull. Environ. Contam. Toxicol. 18: 3-8.

Grice, G.D., P.H. Wiebe, and E. Hoagland. 1973. Acid-iron waste as a factor affecting the distribution and abundance of zooplankton in the New York Bight. I. Laboratory studies on the effects of acid waste on copepods. Estuarine Coastal Mar. Sci. 1: 45-50.

Hodson, R.E., F. Azam, and R.F. Lee. 1977. Effects of four oils on marine bacterial populations: Controlled Ecosystem Pollution Experiment. Bull. Mar. Sci. 27: 119-126.

Judkins, D.C., C.D. Wirick, and W.E. Esaias. 1980. Composition, abundance, and distribution of zooplankton in the New York Bight, September 1974 - September 1975. Fish. Bull. 77: 669-683.

Lee, R.F. and J.W. Anderson. 1977. Fate and effect of napthalenes: Controlled Ecosystem Pollution Experiment. Bull. Mar. Sci. 27: 127-134.

Lee, R.F., M. Takahashi, and J. Beers. 1978. Short term effects of oil on plankton in controlled ecosystems. *In:* Proceedings of the Conference on Assessment of Ecological Impacts of Oil Spills, American Institute of Biological Sciences, Arlington, VA. pp. 634-650.

Longwell, A.C. 1977. A genetic look at fish eggs and oil. Oceanus 20: 46-58.

Longwell, A.C. and J.B. Hughes. This volume. Cytologic, cytogenetic, and embryologic state of Atlantic mackeral eggs from surface waters of the New York Bight in relation to pollution. pp. 381-388.

Malone, T.C. 1976. Phytoplankton productivity in the apex of the New York Bight: Environmental regulation of productivity/chlorophyll *a*. Am. Soc. Limnol. Oceanogr., Spec. Symp. 2: 260-272.

Malone, T.C. 1977a. Plankton Systematics and Distributions. MESA New York Bight Atlas Monograph 13, New York Sea Grant Institute, Albany, NY. 45 pp.

Malone, T.C. 1977b. Environmental regulation of phytoplankton productivity in the lower Hudson estuary. Estuarine Coastal Mar. Sci. 5: 157-171.

Malone, T.C. (ed.) 1978. The 1976 *Ceratium tripos* bloom in the New York Bight: causes and consequences. NOAA Tech. Rep. NMFS Circ. 410. 14 pp.

Malone, T.C. This volume. Factors influencing the fate sewage-derived nutrients in the lower Hudson estuary and New York Bight. pp. 389-400.

Malone, T.C. and M.B. Chervin. 1979. The production and fate of phytoplankton size fractions in the plume of the Hudson River, New York Bight. Limnol. Oceanogr. 24: 683-697.

Malone, T.C., C. Garside, and P.J. Neale. 1980a. Effects of silicate depletion on photosynthesis by diatoms in the plume of the Hudson River. Mar. Biol. 58: 197-204.

Malone, T.C., P.J. Neale, and D. Boardman. 1980b. Influences of estuarine circulation on the distribution and biomass of phytoplankton size fractions. *In:* Estuarine Perspectives, V.S. Kennedy (ed.), Academic Press, New York. pp. 249-262.

Menzel, D.W. 1977. Summary of experimental results: Controlled Ecosystem Pollution Experiment. Bull. Mar. Sci. 27: 142-145.

Mueller, J.A., J.S. Jeris, A.R. Anderson, and G.F. Hughes. 1976. Contaminant inputs to the New York Bight. NOAA Tech. Memo. ERL MESA-6. 347 pp.

Murphy, L.S., R.L. Guillard, and J. Gavis. This volume. Evolution of resistant phytoplankton strains through exposure to marine pollutants. pp. 401-412.

Murphy, L.S., E.M. Huagen, and J.F. Brown. In press. Phytoplankton: comparison of laboratory bioassay and field measurements. *In:* Industrial and Sewage Wastes in the Ocean, I.W. Duedall, B.H. Ketchum, P.K. Park, and D.R. Kester (eds.), J. Wiley and Sons, New York.

National Oceanic and Atmospheric Administration. 1981. Assessment report on the effects of waste dumping in 106-mile ocean waste disposal site: Dumpsite Evaluation Report 81-1. NOAA Special Report. 319 pp.

Nimmo, D.R., T.L. Hamaker, E. Mathews, and W.T. Young. This volume. The long-term effects of suspended sediment on survival and reproduction of the mysid shrimp, *Mysidopsis bahia,* in the laboratory. pp. 413-422.

O'Connor, J.M., J.B. Klotz, and T.J. Kneip. This volume. Sources, sinks, and distribution of organic contaminants in the New York Bight ecosystem. pp. 631-653.

O'Connor, J.S. 1979. A perspective on natural and human factors. *In:* Oxygen Depletion and Associated Benthic Mortalities in New York Bight, 1976, R.L. Swanson and C.J. Sindermann (eds.), NOAA Prof. Pap. 11: 323-333.

O'Connor, J.S. and H.M. Stanford (eds.). 1979. Chemical Pollutants of the New York Bight; Priorities for Research. NOAA Special Report. 217 pp.

O'Connor, T.P. and P.K. Park. This volume. Consequences of industrial waste disposal at the 106-mile ocean waste disposal site. pp. 675-697.

O'Connors, H.B., Jr., C.F. Wurster, C.D. Powers, D.C. Biggs, and R.G. Rowland. 1978. Polychlorinated biphenyls may alter trophic pathways by reducing phytoplankton size and production. Science 201: 737-739.

O'Connors, H.B., Jr., C.D. Powers, C.F. Wurster, K.D. Wyman, G.M. Nau-Ritter, and R.G. Rowland. This volume. Fates and biological effects on plankton of particle-sorbed PCB's in coastal waters. pp. 423-449.

Parsons, T.R., W.K.W. Li, and R. Waters. 1976. Some preliminary observations on the enhancement of phytoplankton growth by low levels of mineral hydrocarbons. Hydrobiol. 51: 85-89.

Peters, L.S. and J.M. O'Connor. This volume. Factors affecting PCB and DDT uptake by zooplankton and fish from the Hudson estuary. pp. 451-465.

Reeve, M.R., J.C. Gamble, and M.A. Walter. 1977. Experimental observations on the effects of copper on copepods and other zooplankton: Controlled Ecosystem Pollution Experiment. Bull. Mar. Sci. 27: 92-104.

Ryther, J.H. and W.M. Dunstan. 1971. Nitrogen, phosphorous, and eutrophication in the coastal marine environment. Science 171: 1008-1013.

Segar, D.A. and G.A. Berberian. 1976. Oxygen depletion in the New York Bight apex: causes and consequences. Am. Soc. Limnol. Oceanogr., Spec. Symp. 2: 220-239.

Sieburth, J.McN. 1971. Distribution and activity of oceanic bacteria. Deep-Sea Res. 18: 1111-1121.

Sindermann, C.J., S.C. Esser, E. Gould, B.B. McCain, J.L. McHugh, R.P. Morgan II, R.A. Murchelano, M.J. Sherwood, and P.R. Spitzer. This volume. Effects of pollutants on fishes. pp. 23-38.

Swanson, R.L., C.J. Sindermann, and G. Han. 1979. Oxygen depletion and the future: an evaluation. *In:* Oxygen Depletion and Associated Benthic Mortalities in New York Bight, 1976, R.L. Swanson and C.J. Sindermann (eds.), NOAA Prof. Pap. 11: 335-345.

Swedmark, M. and A. Granmo. In press. Effects of mixtures of heavy metals and a surfactant on the development of cod (*Gadus morhua* L.). *In:* The Early Life History of Fish: Recent Studies, R. Lasker and K. Sherman (eds.), Rapp. P.-V. Reun. Cons. Int. Explor. Mer 178: 95-103.

Taguchi, S. and K. Nakajima. 1971. Plankton and seston in the sea surface of three inlets of Japan. Bull. Plankton Soc. Japan 18: 20-36.

Thomas, W.H. and D.L.R. Seibert. 1977. Effects of copper on the dominance and diversity of algae: Controlled Ecosystem Pollution Experiment. Bull. Mar. Sci. 27: 23-33.

Thomas, W.H., D.L.R. Seibert, and M. Takahashi. 1977. Controlled Ecosystem Pollution Experiment: effect of mercury on enclosed water columns. III. Phytoplankton population dynamics and production. Mar. Sci. Comm. 3: 331-345.

U.S. Environmental Protection Agency. 1980. Environmental Impact Statement (EIS) for 106-Mile Ocean Waste Disposal Site Designation. U.S. Environmental Protection Agency, Oil and Special Materials Control Division, Marine Protection Branch, Washington, DC.

Wiebe, P.H., G.D. Grice, and E. Hoagland. 1973. Acid iron waste as a factor affecting the distribution and abundance of zooplankton in the New York Bight. II. Spatial variations in the field and implications for monitoring studies. Estuarine Coastal Mar. Sci. 1: 51-64.

EFFECTS OF NUTRIENTS AND CARBON LOADINGS ON COMMUNITIES AND ECOSYSTEMS

A.J. Mearns[1]
E. Haines
G.S. Kleppel
R.A. McGrath
J.J.A. McLaughlin
D.A. Segar
J.H. Sharp
J.J. Walsh
J.Q. Word
D.K. Young
M.W. Young

Abstract. Waters of the Hudson River estuary and the New York Bight apex experience unpleasant and costly ecological problems that are caused, in part, by excessive inputs of nitrogen and carbon from shore-based sewage discharges. A principal problem is summertime depression and localized depletion of dissolved oxygen in the estuary and bottom waters of the apex. In the estuary, decay of organic matter (carbon) from sewage and runoff, and from phytoplankton blooms stimulated by nitrogen from sewage, reduce dissolved oxygen to levels insufficient for diverse and abundant populations of typical estuarine organisms.

Most of the nitrogen discharged into the estuary is flushed into surface waters of the apex, where it is assimilated by coastal phytoplankton in an area of up to 1,350 km^2 (late winter and early spring). Organic matter produced by this excess growth, coupled with appropriate oceanographic conditions, may contribute to episodes of oxygen depletion in limited portions of apex bottom waters. Oxygen-demanding materials from sludge dumping do not appear to contribute greatly to oxygen depletion. The large-scale anoxic episode of 1976 was not caused by these inputs.

A large-scale reduction in nitrogen input (*i.e.*, >90%), especially in the summer, might alleviate low oxygen episodes in the estuary and apex, but would not prevent the kind of episode that occurred in 1976. Nutrients other than nitrogen and toxic materials may encourage the growth of some species and inhibit others, depending on location and time of year. It is still unclear what changes might result from controlling these substances.

INTRODUCTION

About 200,000 metric tons of nitrogen and 950,000 metric tons of organic carbon enter waters of the New York Bight each year from clearly identifiable human sources. Most of these nutrient and carbon inputs are derived together with toxic materials from waste discharges into the Hudson-Raritan estuary and New York Bight apex. Scientists and laymen alike are concerned that the human inputs (which are in addition to natural loadings) may have had an adverse effect on the diversity and productivity of local marine ecosystems. Many believe that reductions of inputs will increase the richness and abundance of marine life in coastal waters; others, however, are not certain that a large problem exists.

The purpose of this report is to summarize the findings of a panel inquiring

[1] Chairman, Panel on the Effects of Nutrients and Carbon Loadings on Communities and Ecosystems.

into the effects of nutrients and carbon from waste discharges on the abundance, diversity, and production of marine life in the New York Bight and adjacent estuarine waters. The objectives of the inquiry were to: (1) identify the most serious ecological problems associated with present nutrient and carbon loadings; (2) determine to what extent oxygen depletions are caused by these loadings; (3) identify uncertainties that remain in the assessment of effects, and how such uncertainties might be resolved; and (4) suggest what can be done to alleviate identified ecological problems without causing new or larger difficulties.

Approach

To accomplish the objectives of the panel, recent published and unpublished data on nutrient and carbon inputs and on nutrient cycling were reviewed. The panel examined data from the New York-New Jersey area, as well as from other coastal areas. Attention was focused on the phytoplankton and benthic invertebrate communities (benthic macrofauna) because these elements of the ecosystem are the basis of food webs that support harvestable fish and shellfish resources. The panel reviewed data on phytoplankton production and biomass, factors limiting phytoplankton growth, dissolved oxygen levels and annual variations in oxygen, spatial and temporal variations in benthic invertebrate densities, and variations in diversity and community structure of benthic invertebrates. Three regions were examined in detail: the Hudson River estuary, including Upper and Lower New York Bays and Raritan Bay; the New York Bight apex; and the New York Bight beyond the apex.

The analysis presented below begins with a review of why there is concern about nutrient and carbon inputs, and what local coastal waters might be like in the absence of inputs introduced by waste discharges. Next, the magnitude and fate of recent inputs are summarized, together with a region-by-region account of the known or suspected effects of these inputs on phytoplankton, dissolved oxygen, and benthic communities. Finally, the answers, uncertainties, and recommendations of the panel are presented.

BACKGROUND

Expected Conditions

The waters off New York and New Jersey may be divided into: (1) offshore areas of strong circulation where overall biological production is low and is controlled by the ocean; (2) shallow inshore areas of variable circulation (*e.g.*, the apex) where overall production is somewhat higher than in offshore areas and partially is under the influence of natural inputs (*e.g.*, rivers); and (3) estuarine areas (*e.g.*, New York harbor and the Hudson River) where production is high and is stimulated by nutrient inputs from rivers and runoff.

Within these areas, diverse phytoplankton communities, dominated in the spring by diatoms (large, siliceous phytoplankton) and in the summer by chlorophytes (small green algae), should form the basis of the food web. Under normal circumstances, the benthos should be abundant (thousands per square meter) and should include a variety of molluscs, polychaetes, crustaceans, and echinoderms. These organisms should serve as food for abundant species of bottomfish and shellfish. In addition, anadromous fishes such as shad and striped bass should make annual, generally unimpeded migrations through the area and into local river systems.

Inputs of nutrients and carbon from land are important in maintaining rich and productive coastal plant and animal communities. However, excessive inputs, such as from sewage or urban runoff, cause biological changes that eventually alter the abundance and variety of marine fish and shellfish populations. This is especially true in areas of poor circulation, such as New York harbor, where oxygen-depleted waters are not replaced effectively by oxygenated water.

Factors Limiting Phytoplankton Production

In coastal marine waters, phytoplankton productivity (*i.e.*, primary productivity) usually is limited by nitrogen, but rarely by other nutrients such as phosphorus (Ryther and Dunstan, 1971). Therefore, increased nitrogen inputs could stimulate the growth of phytoplankton. In situations where nitrogen loading is extremely high, massive phytoplankton blooms could result. During such blooms, oxygen is consumed by all living organisms associated with the bloom and by the decay of their excretory products. Following the bloom, plankton die and undergo microbial decomposition. The products of decomposition may enhance the growth of bottom-dwelling organisms. The decomposition of excess phytoplankton may create an abnormal demand on the dissolved oxygen supply and, in extreme situations, could completely exhaust bottom water oxygen, thus reducing the abundance and diversity of benthic organisms, including fishes and shellfishes.

Nutrients other than nitrogen may become limiting under certain conditions. For example, silica is required for the growth of diatoms, which are considered a significant link in some marine food webs and important to some commercial fisheries (Ryther, 1969). If nitrogen inputs exceed available silica inputs, silica may limit the growth of diatoms and allow the growth of other phytoplankters not requiring silica (Officer and Ryther, 1980). In such situations, the animals that use diatom-based food chains (*e.g.*, clams) could become food-limited relative to those that do not (Officer and Ryther, 1980).

Other factors can limit primary production in coastal waters. These include inert or living suspended solids that scatter and absorb light required for photosynthesis, and unusual concentrations of specific toxicants or trace elements (Dunstan, 1975; McLaughlin *et al.*, 1977; Kleppel, 1979; McIsaac *et al.*, 1979). Runoff and wastewaters contribute to the suspended solids load and may limit phytoplankton production. Self-shading by phytoplankton can limit productivity. Toxic materials can act directly on phytoplankton to decrease production (O'Connors *et al.*, 1978). Finally, certain minerals required by phytoplankton (*e.g.*, iron) can occur in low or fluctuating concentrations. When nutrients such as nitrogen are available at saturating concentrations, essential minerals may become limiting factors (Dunstan, 1975).

Some phytoplankton blooms and declines in dissolved oxygen levels are natural events, even in unimpacted waters. Variations in runoff and natural nutrient input, and occasional but unusual wind conditions that bring to the surface cold water containing low oxygen and high nutrient concentrations, may cause local plankton blooms and reduced levels of dissolved oxygen in shallow inshore waters.

Factors Limiting the Benthos

While several factors can limit benthic production, one of the most common is the availability of appropriate food. In many inshore coastal areas, bottom fauna biomass is determined primarily by the fallout of organic matter produced by plankton (Parsons *et al.,* 1977). In addition, bottom life can be altered by increased inputs of both nutrients and carbon. For example, Raymont (1949) and Gross (1949) experimentally enriched Scottish sea lochs with inorganic fertilizers, and over a period of two to three years observed a doubling of benthic production and accompanying increases in the growth rates of bottom-feeding fishes.

Potential Effects of Sewage

Sewage contains nitrogen and other nutrients in forms essential to many kinds of phytoplankton. It also contains carbon in forms that can undergo decay by marine microorganisms. In the absence of toxic materials, sewage discharged into coastal waters serves to stimulate primary production and the growth of benthic populations, including fishes. Field experiments in which processed sewage sludge was added to subtidal benthic populations resulted in increased biomass and enhanced the densities of certain benthic species (Young and Young, 1978, this volume). In areas affected by sewage outfalls in California, growth rates of Dover sole, a benthic-feeding flatfish, were greater than in control areas (Mearns and Harris, 1975) and the abundance of many invertebrates increased (Bascom *et al.,* 1979). However, as has been mentioned, large-scale nutrient loading also may lead to oxygen depletion, benthic mortalities, and less diverse bottom communities. Semi-enclosed and poorly flushed areas such as estuaries are especially sensitive to such effects.

Sewage and other wastewaters may contain toxic materials such as trace metals, synthetic hydrocarbons, and petroleum hydrocarbons. The production expected from nutrient and carbon inputs may be reduced by the presence of toxic materials.

SITUATION IN THE NEW YORK BIGHT AND ADJACENT WATERS

Inputs of Nutrients and Carbon

Each year waters of the New York Bight and adjacent estuaries receive nearly 9 million metric tons of suspended solids, 1 million metric tons of organic carbon, 0.2 million metric tons of nitrogen, and 50 thousand tons of phosphorus from the adjacent coastlines and human populations of New York and New Jersey (Mueller *et al.*, 1976) (Tables 1 and 2). Recycling, advection, upwelling, and mixing contribute to the loading and maintenance of nutrients in the Bight (Walsh *et al.*, 1978).

Of the land-based sources, barge dumping is the largest source of suspended solids and total phosphorus (63% and 50% of all land-based inputs, respectively). Barge dumping of sewage sludge and dredged materials also contributes approximately one-third of the oxygen demand in the apex (Falkowski *et al.*, 1980). However, the Hudson-Raritan estuary represents the largest source of total organic carbon and total nitrogen (including ammonia, nitrate, and nitrite), and is a major source of phosphorus (Table 1) (Segar and Berberian, 1976; Falkowski *et al.*, 1980). Inputs of nitrogen and carbon from the New Jersey coastline and from the atmosphere are relatively small, but locally important.

Considering only coastal sources, runoff (urban and rural) is the largest source of suspended solids and is a competitive source with municipal wastewater of total organic carbon (Table 2). Industrial sources of nutrients and carbon are trivial compared to municipal wastewaters and other sources. As noted above, the Hudson-Raritan estuary is the dominant geographical nutrient source because it receives most of the liquid wastewater generated by the adjacent urban population of 16 million (Mueller *et al.*, 1976). The Hudson River is also a large source of silica (40-340 metric tons per day). The quantity supplied by the river varies with river flow and includes a small amount of silica that is of sewage origin (20 metric tons per day or 6-50%) (McLaughlin *et al.*, this volume).

Hudson-Raritan Estuary

Phytoplankton and Red Tides

Most authors agree that only small amounts (10%) of the nitrogen entering the estuary are actually used in primary production within the estuary (Garside, *et al.*, 1976; Ingram, 1979; McLaughlin *et al.*, this volume). Depending on time of year and locality, factors limiting production include low light penetration, increased flushing (Malone, 1977; Kleppel, 1979), and low concentrations of silica (caused by diatom utilization) (Malone, 1977). Flushing may be particularly important in the late spring and summer when high river flow coincides with effective grazing of blooms and their wide dispersal in upper stratified waters.

Episodic blooms (*i.e.*, 100,000 cells/ml) of several "red tide" flagellates have occurred in the nearshore waters of New Jersey and Lower New York Bay since the early 1960's. Although the factors causing these blooms are still poorly understood, laboratory investigations suggest (1) that urea and nitrate from sewage effluents and river runoff may be the prefered nitrogen forms of major New Jersey red tide algae; and (2) that certain strongly chelated metals from industrial effluents in the Raritan Bay area influence initiation of the blooms

Table 1. Mass emissions of suspended solids, carbon, nitrogen, and phosphorus into the New York Bight and adjacent waters summarized by location of input. After Mueller *et al.* (1976).

	Total mass input (thousand metric tons/year)	Percent contribution by location			
		Direct	Coastal		
			Hudson NY harbor	New Jersey	Long Island
Suspended solids	8,760	68	31	0.6	0.1
Total organic carbon	949	37	58	4.0	0.6
Total nitrogen	190	29	65	4.0	2.0
Ammonia nitrogen	77	28	67	3.0	2.0
Nitrate and nitrite nitrogen	44	33	55	10.0	2.0
Total phosphorus	50	51	45	2.0	2.0

Table 2. Mass emissions of suspended solids, carbon, nitrogen, and phosphorus into the New York Bight and adjacent waters summarized by type of input. After Mueller *et al.* (1976).

	Total mass input (thousand metric tons/year)	Percent contribution by source				
		Direct		Coastal		
				Wastewater		
		Barge	Atmosphere	Municipal	Industrial	Runoff
Suspended solids	8,760	63	5	4	0.2	28
Total organic carbon	949	25	12	29	1	33
Total nitrogen	190	16	13	40	2	29
Ammonia nitrogen	77	24	4	55	3	14
Nitrate and nitrite nitrogen	44	0.1	33	6	0.3	61
Total phosphorus	50	50	0.7	35	1	13

(Mahoney and McLaughlin, 1977). Formerly, all algae were thought to prefer ammonia over urea or nitrate as a nitrogen source.

Dissolved Oxygen

As previously noted, wastewaters and runoff are both important sources of carbon to the estuary. The contribution of carbon to the benthic environment from these sources relative to that from phytoplankton production is poorly understood. In either case, low dissolved oxygen levels ($\leqslant$10% saturation) are common in many parts of the estuary, presumably as a result of high carbon loading (Suszkowski, 1973). Clearly, the dissolved oxygen levels are too low to support diverse populations of fish and shellfish.

Benthos

A recent study of the Lower Bay Complex (*i.e.,* Raritan Bay, Sandy Hook Bay, and Lower Bay) found abnormally low densities and numbers of benthic macrofaunal species (McGrath, 1974). As recently as 1957, Dean (1975) showed that a species of the amphipod, *Ampelisca,* was a dominant member of the benthic community in the Lower Bay Complex. This organism was sampled in densities of up to 13,000 individuals/m^2 and was present at 125 out of 193 stations. In contrast, 16 years later, McGrath (1974) collected only one individual of this genus in samples from 88 stations. Total abundance of all macrobenthic species averaged less than 100 individuals/m^2 in 1973, a noticeable decrease from that recorded by Dean (1975). Thus, there have been recent and dramatic changes in the abundance and diversity of the benthos of the Lower Bay Complex.

The silty sediments of the Lower Bay Complex and New York harbor are a sink for river-borne solids, organic materials, and various associated contaminants. Raritan Bay sediments contain approximately 5% organic matter and are comparable to sewage-enriched areas in southern California (Word and Mearns, 1979) and the Baltic Sea (Anger, 1975) that support benthic densities two orders of magnitude greater than those observed in New York harbor.

High levels of toxic metals have been reported from sediments adjacent to sewage outfalls in California, but benthic densities in these affected regions reach 21,000 individuals/m^2, or four times higher than background. In contrast, Sandy Hook Bay has significantly lower metal concentrations, yet the sediments of this area are associated with lower densities (100 individuals/m^2) of benthic infauna (McGrath, 1974). Minor differences between sediment types and habitats in the Hudson-Raritan estuary and southern California do not explain the drastic differences in benthic invertebrate densities.

Michael (this volume) noted that the patterns of distribution and abundance of macrobenthos in Raritan Bay are related to petroleum hydrocarbon contaminants, while McDonough (1976) concluded that low densities of macrobenthic organisms were related to high toxic metal levels (despite the fact that these levels are lower than those reported from southern California sites).

The residence time of such contaminants in sediments is a function of physical dispersal processes and bioturbation (*i.e.*, biological reworking of sediments). The effects of physical dispersal processes may be seen by comparing Lower New York Bay with the Ambrose Channel. The Channel is a physically active area whose sediments exhibit low concentrations of organic matter and a dense benthic population dominated by haustoriid amphipods. The converse is characteristic of Lower New York Bay (McGrath, unpublished data).

It is probable that benthic production, particularly of suspension feeders, is partly dependent on net phytoplankton production, and that the nanoplankton, which dominate the lower estuary in summer, are not utilized as effectively. Although further investigation is required to determine whether this hypothesis is correct, it appears that macrofaunal abundance and diversity in the Lower Bay Complex are not limited by carbon or nitrogen inputs.

New York Bight Apex

Inputs and Recycling

As noted above, river runoff, wastewaters, and direct barge dumping are major sources of carbon, nitrogen, and other nutrients for the apex; however, coastal and oceanic processes also contribute to inputs and modify concentrations. Although oceanographic inputs of nitrogen by advection and diffusion generally are believed to be small, Walsh *et al.* (1978) have suggested that variations in storm frequencies and upwelling may be important in delivering deep water nitrogen to localized areas.

In the summer, a thermocline develops that effectively separates the upper (20 m) mixed layer from bottom waters. During this period, wastewaters and estuarine inputs become extremely important contributors of nitrogen and carbon to upper waters, while the importance of bottom deposits as a source of nutrients is reduced. As a result, dissolved nitrogen concentrations in apex waters are higher than in adjacent coastal areas at equivalent depths (Malone, this volume).

Phytoplankton

Most of the nitrogen discharged from the estuary is assimilated by phytoplankton in the apex; 70-80% is used within 20 km of the mouth of the estuary (Malone, this volume). Malone also estimated that the area required to assimilate the dissolved inorganic nitrogen input (ammonia, nitrate, and nitrite) varies seasonally from 670 km^2 to 1,350 km^2 (*i.e.*, from 54% to over 100% of the 1,250 km^2 area of the apex).

The production of phytoplankton in the part of the apex most influenced by Hudson River water is limited by factors other than nitrogen (Garside *et al.*, 1976; Malone, 1977, this volume). Consequently, Garside *et al.* (1976) concluded that changes in nitrogen input into the apex would not change primary production in the affected area (up to 1,300 km^2 in the winter) but would change the size of the area where nitrogen is not limiting.

Dissolved Oxygen

The apex has suffered several episodes of nearly anaerobic water and less severe depressions occur almost yearly. In terms of biological effects, midsummer is a critical period. As noted above, the thermocline (at approximately 20 m) reduces mixing between the surface mixed layer and bottom water, creating an isolated bottom water pool. Phytoplankton grow rapidly in the surface mixed layer and may concentrate at the thermocline. On the average, dissolved oxygen levels in the bottom water drop to 10-40% saturation (2-4 ml/l) (Malone, this volume), compared to an average of 50-90% saturation in bottom waters south of the apex (Sharp, this volume). There is a considerable year-to-year variation in oxygen levels. For example, during July 1976 a region-wide "anoxic" episode was triggered by unusual wind conditions and a bloom of *Ceratium tripos* (Walsh *et al.,* 1978). Dissolved oxygen concentrations dropped to below 10% saturation over much of the coastal shelf south of the apex (Swanson and Sindermann, 1979; Falkowski *et al.,* 1980). In contrast, conditions in 1978 produced only slight dissolved oxygen depressions (*i.e.,* 80-90% saturation) over most of the shelf (Sharp, this volume).

Despite a number of comprehensive surveys, there is no agreement concerning the degree to which wastewaters contribute to dissolved oxygen depressions. Depressions to 50% saturation are not uncommon. Although further reductions (to 25% saturation or less) occur with greater frequency, they remain unusual. Increased primary production stimulated by excess nitrogen inputs to the estuary probably contribute to the problem (Segar and Berberian, 1976); however, the degree of wastewater contribution remains at issue (Falkowski *et al.,* 1980).

Benthos

The condition of the benthos in the New York Bight apex is markedly different from that in the estuary. As a result of direct dumping, organic carbon concentrations in sediments are higher at the apex dumpsites than in adjacent areas at comparable depths (Steimle *et al.,* this volume). Benthic population densities are an average of at least two orders of magnitude greater than in the estuary (Steimle *et al.,* this volume). However, there is a noticeable absence or paucity of amphipods in both areas (Pearce, 1972; McGrath, 1974), as also is the case in sediments surrounding a southern California sewage outfall (Word *et al.,* 1977). Ampeliscid amphipods are thought to be sensitive indicators of chronic petrochemical pollution (Blumer *et al.,* 1970). There is also an absence of echinoderms, as in a similar area off the coast of California (Word and Mearns, 1979).

The impact of sludge dumping in the apex centers around the western portion of the sewage sludge disposal site. The most depauperate portion of the dumpsite occupies an area of 10-15 km^2, while enhanced (in terms of abundance) transitional areas occupy 240 km^2 (Boesch, this volume). The areas of impact compare favorably with those observed at an equivalent point source discharge in southern California (Bascom *et al.*, 1979). While the two locations are not directly comparable, their relative sizes and faunal similarities suggest a need to more thoughtfully compare outfalls and dumpsites in various coastal areas around the United States (Mearns and Word, this volume).

Waters Beyond the Apex

Studies at coastal sites distant from the apex indicate that primary production is nitrogen limited and that variations in dissolved oxygen concentrations are largely consistent with expected physical oceanographic variations (Sharp, this volume). Of interest is the extent to which distant waters will be affected if nitrogen inputs continue, are reduced, or increase. If the relationships between primary production, dissolved oxygen depression, and waste inputs were linear, slight changes in nutrient inputs would produce little change offshore. However, it is more likely that effects of nutrient inputs are geometric or exponential, as in the case of benthic communities affected by solids emissions at southern California outfalls (Mearns and Word, this volume). If such a relationship were true for the New York Bight, a slight increase in nitrogen input could trigger a much larger increase in primary production (Segar and Berberian, 1976). Alternatively, a modest reduction in input could greatly reduce the potential for effects at distant sites as well as in the apex.

CONCLUSIONS AND RECOMMENDATIONS

The most serious ecological consequences of nutrient and carbon loading are the effects on dissolved oxygen, especially in the lower Hudson estuary and in the apex (with allowance for the panel's uncertainties about the latter area). The principal causes of reduced benthic diversity and abundance appear to be toxicants, rather than nitrogen or carbon inputs, except where bottom water dissolved oxygen levels are nearly zero. Biological oxygen demand, hydrogen sulfide, and toxic metals are less likely causes of these effects than chlorinated hydrocarbons, petrochemicals, or some as yet unidentified anthropogenic material.

The occurrence of low dissolved oxygen levels in the Hudson-Raritan estuary probably is caused by local carbon loading. Outside the estuary, in the apex and adjacent coastal shelf waters, low dissolved oxygen episodes are exaggerated by nutrient input primarily from estuarine wastewater discharges, but there remains uncertainty about the extent of contribution to these episodes. Low dissolved oxygen episodes in coastal shelf waters are not initiated by these inputs but by oceanographic and meterological conditions.

The panel on nutrients and carbon loadings concludes that:

1. Control of nitrogen and carbon inputs from wastewaters will reduce dissolved oxygen problems in the apex and beyond; however, zero discharge will not eliminate them. Episodes of low dissolved oxygen will occur in the Bight as a

result of natural causes; however, such events may be less severe after nitrogen and carbon inputs are reduced.

2. Unless source control can be made so effective that nitrogen becomes a limiting factor (*i.e.*, removal of more than 90% supplied to the estuary), managers will have little control over primary production and species composition in the estuary, *per se.* The control of other potentially limiting nutrients represents an alternate but insufficiently explored approach for reaching these ends.

3. The most critical area for benthos is the Lower Bay Complex of the Hudson-Raritan estuary, rather than the offshore dumpsites. In the estuary, benthic production is limited by factor(s) other than organic carbon. Benthic production probably is not limited by sediment-bound metals or hydrogen sulfide, but rather by some as yet unidentified anthropogenic material. A management strategy specific to this part of the ecosystem will be required if benthic productivity is to be improved. Benthic environments in other parts of the Hudson-Raritan estuary should be examined in greater detail.

4. The factors limiting biological production and species composition in the New York Bight and adjacent waters are sufficiently complex to merit caution in implementing any management plan. Reductions in waste inputs will assist in improving the quality and biological diversity of local waters; however, the panel is unable to predict the exact nature or location of such changes.

REFERENCES

Anger, K.F. 1975. On the influence of sewage pollution on inshore benthic communities in the South of Kiel Bay. Helgol. Wiss. Meeresunters. 27: 408-438.

Bascom, W., A.J. Mearns, and J.Q. Word. 1979. Establishing boundaries between normal, changed and degraded areas. *In:* Annual Report of the Southern California Coastal Water Research Project, Southern California Coastal Water Research Project, El Segundo, CA. pp. 81-94.

Blumer, M., J. Sass, G. Souza, H. Sanders, F. Grassle, and G. Hampson. 1970. The West Falmouth oil spill; persistance of pollution eight months after the accident. Woods Hole Oceanogr. Inst. Tech. Rep. WHOI-70-44. 32 pp.

Boesch, D.F. This volume. Ecosystem consequences of alterations of benthic community structure and function in the New York Bight region. pp. 543-568.

Dean, D. 1975. Raritan Bay macrobenthos survey, 1957-60. NOAA NMFS Data Rep. 99. 51 pp.

Dunstan, W.M. 1975. Problems of measuring and predicting influence of effluents on marine phytoplankton. Environ. Sci. Tech. 9: 635-638.

Falkowski, P.G., T.S. Hopkins, and J.J. Walsh. 1980. An analysis of factors affecting oxygen depletion in the New York Bight. J. Mar. Res. 38: 479-506.

Garside, C., T.C. Malone, O.A. Roels, and B.A. Sharfstein. 1976. An evaluation of sewage-derived nutrients and their influence on the Hudson River Estuary and New York Bight. Estuarine Coastal Mar. Sci. 4: 281-289.

Gross, F. 1949. Further observations on fish growth in fertilized sea loch (Loch Craiglin). J. Mar. Biol. Assoc. U.K. 28: 1-8.

Ingram, R.J. 1979. Selected phytoplankton nutrients in the Lower Hudson

Estuary. Ph.D. Thesis, Fordham University. 156 pp.
Kleppel, G.S. 1979. Aspects of primary productivity in the East River of New York Harbor. Ph.D. Thesis, Fordham University. 153 pp.
Mahoney, J.B. and J.J.A. McLaughlin. 1977. The association of phytoflagellate blooms in Lower New York Bay with hypertrophication. J. Exp. Mar. Biol. Ecol. 8: 53-65.
Malone, T.C. 1977. Environmental regulation of phytoplankton productivity in the Lower Hudson Estuary. Estuarine Coastal Mar. Sci. 5: 157-171.
Malone, T.C. This volume. Factors influencing the fate of sewage-derived nutrients in the Lower Hudson estuary and New York Bight. pp. 389-400.
McDonough, K.B. 1976. A benthic index of environmental quality for the New York Bight Apex and Raritan Bay. M.S. Thesis, State University of New York at Stony Brook. 89 pp.
McGrath, R.A. 1974. Benthic macrofaunal census of Raritan Bay—preliminary results, Pap. 24. *In:* Hudson River Ecology, Third Symposium on Hudson River Ecology, March 22-23, 1973, Hudson River Environmental Society, Inc. 40 pp.
McIsaac, J.J., R.C. Dugdale, S.A. Huntsman, and H.L. Conway. 1979. The effect of sewage on uptake of inorganic nitrogen. J. Mar. Res. 37: 51-66.
McLaughlin, J.J.A., M.P. Brown, R.J. Ingram, and G.S. Kleppel. 1977. Nitrogen, phosphorus, and light considerations for phytoplankton modeling in the New York harbor estuarine system. Fifth International Congr. Protozool. Abstr. 364.
McLaughlin, J.J.A., G.S. Kleppel, M.P. Brown, R.J. Ingram, and W.B. Samuels. This volume. The importance of nutrients to phytoplankton production in New York Harbor. pp. 469-479.
Mearns, A.J. and L. Harris. 1975. Age, length, and weight relationships in southern California populations of Dover sole. Southern California Coastal Water Research Project Rep. TM 219. 17 pp.
Mearns, A.J. and J.Q. Word. This volume. Forecasting effects of sewage solids on marine benthic communities. pp. 495-512.
Michael, A.D. This volume. The potential contribution of petroleum hydrocarbons to changes in benthic communities of the New York Bight. pp. 621-630.
Mueller, J.A., A.R. Anderson, and J.S. Jeris. 1976. Contaminants in the New York Bight. J. Water Pollut. Control Fed. 48(10): 2309-2326.
O'Connors, H.B., Jr., C.F. Wurster, C.D. Powers, D.C. Biggs, and R.G. Rowland. 1978. Polychlorinated biphenyls may alter marine trophic pathways by reducing phytoplankton size and production. Science 20: 737-739.
Officer, C.B. and J.H. Ryther. 1980. The possible importance of silicon in marine eutrophication. Mar. Ecol. Progr. Ser. 3: 83-91.
Parsons, T.R., M. Takahashi, and B. Hargrave. 1977. Biological Oceanographic Processes. Pergamon Press, New York. 332 pp.
Pearce, J.B. 1972. The effects of waste disposal on benthic communities in the New York Bight. *In:* Marine Pollution and Sea Life, M. Ruivo (ed.), Fishing News (books) Ltd., London. pp. 404-411.
Raymont, J.E.C. 1949. Further observations on changes in the bottom fauna of a fertilized sea-loch. J. Mar. Biol. Assoc. U.K. 28: 9-19.
Ryther, J. 1969. Photosynthesis and fish production in the sea. Science. 166: 72-76.
Ryther, J.H. and W.M. Dunstan. 1971. Nitrogen, phosphorus, and eutrophication in the coastal marine environment. Science 171: 1008-1013.
Segar, D.A. and G.A. Berberian. 1976. Oxygen depletion in the New York Bight

apex: causes and consequences. Am. Soc. Limnol. Oceanogr. Spec. Symp. 2: 220-239.

Sharp, J.H. This volume. *In situ* phosphate regeneration in Middle Atlantic coastal waters. pp. 513-525.

Steimle, F., J. Caracciolo, and J.B. Pearce. This volume. Impacts of dumping on New York Bight apex benthos. pp. 213-223.

Suszkowski, D.J. 1973. Sewage pollution in New York Harbor: a historical perspective. M.S. Thesis, State University of New York at Stony Brook. 68 pp.

Swanson, R.L. and C.J. Sindermann (eds.). 1979. Oxygen Depletion and Associated Benthic Mortalities in New York Bight, 1976. NOAA Prof. Pap. 11. 345 pp.

Walsh, J.J., T.E. Whitledge, F.W. Barvenik, C.D. Wirick, and S.D. Howe. 1978. Wind events and food chain dynamics within the New York Bight. Limnol. Oceanogr. 23: 659-683.

Word, J.Q., B.L. Myers, and A.J. Mearns. 1977. Animals that are indicators of marine pollution. *In:* Annual Report of the Southern California Coastal Water Research Project, Southern California Coastal Water Research Project, El Segundo, CA. pp. 199-206.

Word, J.Q. and A.J. Mearns. 1979. The 60-meter control survey. *In:* Annual Report of the Southern California Coastal Water Research Project, Southern California Coastal Water Research Project, El Segundo, CA. pp. 41-56.

Young, M.W. and D.K. Young. This volume. Marine macrobenthos as indicators of environmental stress. pp. 527-538.

Young, D.K. and M.W. Young. 1978. Regulation of species densities of seagrass-associated macrobenthos: evidence from field experiments in the Indian River Estuary, Florida. J. Mar. Res. 36: 569-593.

EFFECTS OF TOXIC SUBSTANCES ON COMMUNITIES AND ECOSYSTEMS

D.A. Wolfe[1]
D.F. Boesch
A. Calabrese
J.J. Lee
C.D. Litchfield
R.J. Livingston
A.D. Michael
J.M. O'Connor
M. Pilson
L.V. Sick

Abstract. The New York Bight receives a wide variety of wastes dumped into the Hudson River and the Hudson-Raritan estuary, as well as direct barged inputs of dredged materials, sewage sludge, and chemical wastes. These pollutant additions have caused widespread degradation of the marine and estuarine environments, with effects being most severe in the western portions of Raritan Bay and the New York Bight apex. Beyond the inner apex, there is an uneven, seasonably variable gradient of decreasing impacts.

The scientific evidence that the changes are caused by pollutant additions is overwhelming and conclusive; it is supported by extensive field observations of ecological degradation and measurements of occurrences of pollutants, laboratory experiments on organism responses to toxicant exposures, and comparisons with observations on disturbed and undisturbed systems in other coastal areas. The evidence shows a progressive deterioration of the marine environment caused by the discharge of pollutants into the area. The trend of these changes is away from the traditional aesthetic and resource values associated with the marine environment.

Unfortunately, the New York Bight has been subjected to a wide variety of pollutants for such a long time that it is not possible at this time to ascribe specific effects to specific pollutants. Observed changes represent cumulative effects of combinations of toxic materials, nutrient and organic loading, and habitat modifications. The selection of approaches for stopping or reversing the environmental degradation, therefore, is exceedingly difficult.

Despite the current shortcomings of ecological science, environmental managers should develop management alternatives for alleviating the impacts of pollution in the New York Bight. Environmental managers and environmental scientists must approach environmental problem-solving jointly and cooperatively. Managers and scientists should develop mutually agreeable plans that consider the range of options available for environmental management, and the specific risks and costs associated with each option. Proposed courses of action should be field-tested on a limited, experimental scale to assess their probable ecological consequences prior to full-scale implementation.

INTRODUCTION

The New York Bight receives an unknown proportion of the wastes discharged into the Hudson River and Hudson-Raritan estuary, as well as direct barged inputs of dredged materials, sewage sludge, and chemical wastes. Major research efforts have been mounted as the result of concern at local, state, and

[1]Chairman, Panel on the Effects of Toxic Substances on Communities and Ecosystems.

federal levels about the potential ecological effects of pollutants in the Bight. Pollutants may render seafood unfit for human consumption (with consequent closing of shellfish areas), become stranded as solid wastes on recreational beaches, or change the composition, structure, or productivity of marine communities in the Bight. This paper addresses changes in community or ecosystem-level characteristics (*i.e.*, changes that transcend those on individual organisms), and discusses why these changes seldom can be ascribed to specific toxicants added to the marine environment.

Identifying the causes of ecological change is desirable for environmental management, because it permits the regulation of pollutants most centrally involved in environmental degradation. However, because of the scale, complexity, and long history of multiple contaminant additions in the New York Bight, separating the effects of individual pollutants from each other, from the effects of natural environmental fluctuations, and from the effects of other human activities such as sport and commercial fishing or habitat modification (*e.g.*, dredging) is a monumental scientific task. Many years of field observations coupled closely with a program of field and laboratory experimental research are required to generate reliable evidence for specific causes of environmental change.

In general, research efforts in the New York Bight have not established definite cause and effect relationships between specific pollutants and changes at the ecosystem level. This is a result of the spatial heterogeneity and temporal variability of the environment, the simultaneous and interacting effects of multiple contaminants on the biotic system, and the ability of ecosystems to adapt to physical and chemical alterations without undergoing major changes in fundamental structure or function. Perhaps the greatest point at issue is the word "major", for the ecosystems of the Bight have changed with contaminant additions. Determining whether the costs of ecosystem changes are great enough to warrant acceptance of alternate costs associated with reducing wastes is a policy judgement.

In subsequent sections of this paper, the general problems of identifying toxicant effects at the ecosystem level of organization are discussed. Where possible, ecological changes observed in the New York Bight are related to pollutant effects, and the types of research required to establish the causality of such effects are identified. Only questions of ecological change at the ecosystem or community level of organization are addressed. Omitted are specific biochemical, behavioral, and physiological effects of contaminants on individual organisms.

PROBLEMS IN IDENTIFYING TOXICANT EFFECTS IN ECOSYSTEMS

Spatial Heterogeneity and Temporal Variability

Natural spatial and temporal variability of marine systems and lack of understanding of the sources of such variability make it very difficult to distinguish toxicant effects (and to design monitoring programs to detect toxicant effects). Even with prior knowledge of the natural ecosystems of the Bight and its estuaries, the significance of human interventions through toxicant discharges cannot be readily identified, quantified, or assessed.

Spatial heterogeneity of aquatic systems and the sampling approaches required to answer specific questions have been reviewed elsewhere (Michael *et al.,* 1977; Green, 1979). Statistical tools currently are available for the design of sampling approaches to overcome problems of spatial heterogeneity.

Temporal variability is less tractable analytically than spatial heterogeneity. At any instant, a natural system reflects temporal sequences of past biological events as well as ongoing processes. These processes are superimposed upon one another, and occur at varying rates and periodicities. In addition, these processes differ along poorly understood habitat or microhabitat gradients. These and other problems give rise to "background noise" in the statistics of ecosystem descriptions (Livingston, this volume).

Natural catastrophic changes and long-term system responses complicate detection of human-induced effects. Storms and hurricanes have variable impacts on coastal areas. Biological changes may be associated with heavy rainfall and runoff, as noted for soft-bottom communities in the lower Chesapeake Bay (Boesch *et al.,* 1976) and for the benthic fauna of northern Gulf of Mexico beaches (Saloman and Naughton, 1977). Temperature changes, especially winter lows, often affect coastal systems (Snelson and Bradley, 1978). Few such effects have been documented for longer than two years. There is ample (though indirect) evidence from commercial fisheries data that long-term population changes are affected by climatic features (Sutcliffe *et al.,* 1977; Meeter *et al.,* 1979; Van Winkle *et al.,* 1979), but fisheries data also are affected by factors such as fishing effort and socioeconomic conditions (Clark and Brown, 1977; Meeter *et al.,* 1979). If effects of human intervention are to be discerned, long-term multidisciplinary analyses must be undertaken on all measurable sources of variation to determine the significance of individual factors. Specific cycles of natural variability must be defined on the same time frame, and the sources of that variability must be understood.

Delayed Expression of Contaminant Effects

Toxicant effects at the ecosystem level of organization are most likely to be expressed as gradual shifts in the proportion of individuals representing each species in the community. This is because contaminants in the ecosystem generally affect species composition and either species diversity or the evenness of distribution of species in a community (Pielou, 1967; Patrick, 1967; Gray and Mirya, 1979). The time course for expression of toxic effects at the community level may be on the order of years (Slobodkin, 1961; Levins, 1968) provided that toxicant effects are sublethal and do not include massive reproductive failure. Sublethal effects, however, may be translated into rapidly expressed community effects if the stressed system is unstable at the outset (Slobodkin, 1961; Rozensweig and MacArthur, 1963; Levins, 1968).

Human activities in the New York Bight have imposed effects on many components of the ecosystem, raising contaminant levels in living and non-living parts of the environment, reducing or eliminating populations in certain areas, and altering the characteristics of a variety of habitats (Calabrese *et al.*, this volume; O'Connor and Rachlin, this volume; Litchfield *et al.*, this volume). Because effects generally preceded the initiation of local research, most available data do not provide a perspective on the response period for contaminant effects on Bight ecosystems.

If the effects of toxicants on coastal aquatic ecosystems are mediated through processes similar to other sources of stress, changes in community composition are likely to be expressed on a time scale of one to several years. Thus, even if specific toxicants in the Bight impose direct measurable effects on aquatic communities, the effects of increasing or decreasing contaminant levels are likely to occur in time frames longer than those provided by short-term studies or monitoring efforts of one to two years duration.

Recovery

Once the source of contaminants in the Bight is removed, recovery of the ecosystem will proceed with a period of delay that depends upon (1) the rate of elimination of accumulated contaminants and (2) the rate of repopulation by the indigenous community.

The resilience of communities following non-toxic, physical disturbance varies widely. In general, communities in dynamic environments such as estuaries and coastal beaches are more resilient than those that normally do not experience wide environmental fluctuations and frequent disturbance (Boesch and Rosenberg, 1981). Disturbed sediments are recolonized within weeks by indigenous benthic macrofauna in shallow coastal habitats; in the deep sea, however, recovery is much slower (more than one year). In frequently disturbed habitats, the early colonizers may be indigenous species, whereas in more constant habitats, opportunistic species that otherwise are rare in the natural community may colonize initially and be replaced subsequently by indigenous species.

The rate of recovery is further influenced by the persistence of contaminants in the environment. For example, Rosenberg (1976) found that recovery of macrobenthos to pre-disturbance conditions required eight years after abatement of pulp mill waste discharge. The presence of toxicants less readily degradable than those of pulp mill wastes probably necessitates even longer recovery periods.

The time needed for recovery of benthic deposits is not readily predictable. Mechanisms for reducing biologically active contaminant levels include biological, physical, and chemical processes (*e.g.*, scouring, leaching, transport by currents, sediment reworking, and burial). Accumulation of contaminants in organisms for subsequent removal by wide-ranging predators is an ineffective mechanism for recovery, because the biomass is small in proportion to the mass of sediments. Bioturbation (*i.e.*, reworking of sediments by organisms) may affect recovery either by burying of surficial materials or by relocating deeper contaminants to a surface that is subject to scour. Thus, the time for recovery of benthic deposits will be controlled by the rate of sediment transport away from the site, the rate of sediment exchange with "clean" areas, or the rate of degradation of the contaminant in question.

Relative to the benthic system, recovery in the water column is rapid after removal of contaminant inputs. Most organic contaminants have a high affinity for finely divided solids and tend to be scavenged from the water column by particulates. When metals are present in high concentrations, they also tend to sorb to particulates and become deposited in the sediments. Those contaminants that remain in the water column are mixed and dispersed by currents and decline rapidly to "background" concentrations. In semi-enclosed systems, such as the Hudson-Raritan estuary or the inner Bight, recovery in the water column may be slowed by continual resuspension of contaminated bottom sediments.

Unavailability of Historical Descriptive Data

Historical data on community composition, ecosystem function, and population dynamics in the Bight region are incomplete and inadequate for many purposes. Redfield and Walford (1951), in their discussion of the disposal of wastes at sea, stressed the need for an adequate background of chemical, physical, and biological data from the Bight region prior to making any decisions to increase the rate of ocean dumping. Comprehensive studies on the effects of dumping were not begun for more than 15 years, at which time dumping activities had already been proceeding for over 60 years. Present research and monitoring efforts, therefore, are forced to assess the ecological impacts of dumping on populations that have been influenced by dumping for decades. No historical data exist for many classes of contaminants. Despite the fact that petroleum hydrocarbons have been major pollutants in the Bight for over a century, no measurements were made before the last two decades. Few synthetic organics were measured in the Bight before 1978, although many were in common usage in the 1950's.

Improvements in analytical technology have made historical data comparisons questionable, even where old data exist. In many instances, the development of new analytical techniques has made the detection of previously unmeasurable or unknown compounds possible. These problems make it impossible for scientists to establish an adequate historical perspective on contamination (especially by toxic organic compounds) in the New York Bight.

CHANGES IN THE NEW YORK BIGHT RELATED TO TOXIC MATERIALS

Biological communities living in the New York Bight have undergone changes in structure and function as a result of anthropogenic inputs. Inshore areas are more seriously impacted; effects are harder to detect in the offshore areas. In the absence of detailed comparisons with older data, current information about the

Bight must be compared with expectations derived from an understanding of the ecology of comparable areas, relatively unpolluted situations, and discrete pollution incidents. In the case of petroleum hydrocarbons, for example, concentrations of 300-2,000 ppm have been found in the sediments of the Bight apex (Farrington *et al.*, 1977; Michael, this volume). There is good evidence from studies of the 1969 oil spill at West Falmouth, Massachusetts, that petroleum hydrocarbons at these levels affect the structure of benthic communities. The West Falmouth example typifies the inferential processes that are required to deduce causal relationships in the New York Bight.

The West Falmouth oil spill clearly demonstrates the effect of a specific group of toxicants on the structure and function of a marine community. A spill of No. 2 fuel oil occurred in a fairly pristine area where no other contaminants had occurred in significant quantity. Effects ranged from almost total mortality of all benthic species in heavily impacted areas to alteration of community structure with the loss of selected species in areas exposed to weathered components of fuel oil at levels as low as 20 ppm in sediments (Michael *et al.*, 1975).

Susceptibilities varied among faunal components. In the broadest terms, crustacea were most severely affected, followed by polychaetes and molluscs, in that order. Long-term sampling by Michael *et al.* (1975), Grassle and Grassle (1976), and Sanders (1978) documented a successional recovery process. A similar successional sequence was observed by McCall (1977) after animals were removed from sediment plots in Long Island Sound, where no contaminants were present to interfere.

At locations where fuel oil residues remained after several years, community structure reflected changes seen at the time of the spill. Low amounts of petroleum hydrocarbons in the sediments were associated with the losses of various crustacean species, including the amphipod, *Ampelisca.* There was an increase in the proportion of opportunistic species such as the polychaetes, *Capitella* or *Mediomastus,* which were the first to colonize sediments in areas where the fauna was eliminated. In locations where heavy oiling occurred, the fauna consisted of a monoculture of *Capitella.*

Wormald (1976) documented the collapse and recovery of a meiofaunal community following an oil spill on a sandy beach in Hong Kong. Initially, the meiofauna was almost destroyed. Nematodes reappeared within one month, but the harpacticoid copepods did not return for eight months, by which time the amount of oil had decreased by 50%. Normal populations returned to the area within 10 to 15 months of the spill.

Concentrations of petroleum hydrocarbons in the New York Bight apex exceed those found to have affected the West Falmouth biota. Similar alterations in benthic and other communities of the Bight could have been caused by petroleum hydrocarbons (Michael, this volume). However, supporting experimental evidence is required to substantiate the suggested causal relationship.

Effects on Ecosystem Structure and Function

Pollutant effects can change the composition of ecosystems (structure), as well as the ways in which ecosystem activities occur (function). Alterations in species composition and abundance may produce changes in trophic structure (food web) which, in turn, may modify productivity, energy flows, and the movement (*i.e.*, rates and pathways) of materials such as nutrients, minerals, and toxicants through ecosystems. Evidence for toxicant-related changes in community structure and function in the New York Bight and adjacent waters is presented below. This evidence is based on observations made in the vicinity of the Bight, inferences from laboratory studies, and examinations of similar environments outside the Bight.

Microbiota

Atlas and Bartha (1973) examined the persistence and distributions of oil-degrading bacteria in Raritan Bay, an area which is heavily polluted with petroleum hydrocarbons. The predominant genera were typical of marine systems, but a majority of the bacteria from the Bay were capable of growth on media containing petroleum mixtures. Litchfield *et al.* (1976) obtained similar data for this system. The abundance and taxonomic composition of sediment microbes fluctuated widely among stations in Raritan Bay and within the same station during a 30-month period. The degree of fluctuation obscured single pollutant-induced effects in either the numbers or the apparent generic structure of the sedimentary microflora.

In the water column and sediments of the Bight apex, microbial densities are dependent on seasonal factors (nutrients and temperature) and on sediment type. In general, the total abundance of active metabolizing forms approaches that of a more eutrophic system than is typical of offshore zones (C.D. Litchfield, personal communication).

Recent studies have shown that bacterial populations from metal-contaminated sediments of New York Bight exhibit increased resistance to the effects of toxic metal exposure. Increased resistance has been demonstrated for mercury and zinc (Timoney *et al.*, 1978; Timoney and Port, this volume), and cadmium (Litchifield *et al.*, this volume). Metals-resistant strains also exhibit elevated resistance to antibiotics. Thus, mechanisms exist within natural microbial populations to adapt to the stresses posed by a variety of toxic metals.

Aspects that have yet to be considered are the cumulative effects of shifts in the bacterial community on bacterivorous predators and the effects of bacterial resistance factors on their hosts. Until aspects such as these are understood, the importance of toxicant stress to the structure of the microbial community and to the relationship between this community and the proper functioning of the marine ecosystem will remain unknown.

Phytoplankton

The species composition and standing crop (abundance) of phytoplankton in the Hudson-Raritan estuary and in the New York Bight apex are substantially different than in offshore waters of the Bight (Malone, 1977). Standing crop, measured either in terms of chlorophyll *a* or cell density, is very high in lower New York Bay and in the Bight apex, where primary productivity is considerably greater than in offshore waters (Malone, 1976, 1977).

Patterns of phytoplankton community structure can be related to nutrient stimulation, salinity, temperature, and zooplankton grazing (Malone, 1977). The role of toxicants emanating from the estuary or dumped into the Bight apex is not documented.

Experiments have been conducted to observe the effects of toxicants on pelagic communities using large plastic enclosures in Saanich Inlet, British Columbia (Grice *et al.*, 1977; Menzel and Case, 1977) and Loch Ewe, Scotland (Gamble *et al.*, 1977). Addition of either copper (10 μg/1) or mercury (5 μg/1) resulted in rapid decline of dominant centric diatoms and eventual dominance of smaller microflagellates (nanoplankton) and dinoflagellates (Gamble *et al.*, 1977; Thomas and Seibert, 1977; Thomas *et al.*, 1977). Phytoplankton standing crop and productivity were depressed initially but recovered quickly to control levels. Thus, the structure of phytoplankton communities, and their resultant trophic availability, may be affected by a toxic pollutant with only ephemeral effects on rate processes (*e.g.*, photosynthesis, growth, and reproduction).

O'Connors *et al.* (1978) found that polychlorinated biphenyls (PCB's) at concentrations of 1-10 μg/1 reduced the average size of phytoplankton cells. Greve and Parsons (1977) postulated that either nutrient stimulation or toxicants may reduce average cell size. The significance to the ecosystem of a shift from diatoms and other relatively large phytoplankton toward microflagellates is unclear. O'Connors *et al.* suggested that such a shift could increase the number of trophic levels and divert the flow of energy from harvestable fish, which rely on short food chains, to jellyfish and other gelatinous predators, which are favored by long food chains. Others, however, question the likelihood of a trophic shift (see Lee *et al.*, this volume, for a discussion of this subject). In any case, the hypothesized relationship between toxicants and phytoplankton cell size is undoubtedly oversimplified, as other conditions may also favor small motile phytoplankters (Eppley and Weiler, 1979).

Zooplankton

The zooplankton of the New York Bight and inshore waters is characterized by large population fluctuations and seasonal changes in species composition (Malone, 1977). The variable composition reflects alternating intrusions of estuarine and coastal or oceanic waters into the Bight. These highly dynamic conditions make it difficult to observe the effects of toxicants in the field. Thus far, no field study of Bight zooplankton has been able to detect effects of waste discharges (Wiebe *et al.*, 1973; Malone, 1977).

Effects of toxicants on zooplankton community structure have been investigated in the Saanich Inlet and Loch Ewe experiments discussed previously. Addition of copper ($<$ 5 μg/1) or mercury (5 μg/1) resulted in effects on the zooplankton community attributable to direct toxicity, changes in the quality or quantity of food, and intensity of predation. Small zooplankters appeared to be most sensitive (Beers *et al.*, 1977a, b), but some large predators (*e.g.*, ctenophores) also were relatively sensitive (Reeve *et al.*, 1977).

Differential sensitivity of zooplankters can alter plankton structure. For example, if predators are more sensitive to a toxicant than grazers, grazers might proliferate and locally reduce phytoplankton crops. If populations of phytoplankton or zooplanktonic grazers are reduced, or if their species composition is altered, large consumers such as fish larvae may be affected, even though the toxicant was not directly lethal to the large consumers (Koeller and Parsons, 1977; Koeller and Wallace, 1977). If a toxicant results in a shift in the phytoplankton community to small flagellates, adjustment in the zooplankton may favor small grazers. Reductions in populations of grazing crustaceans that constitute the prey of juvenile fishes may affect growth and survival of the fish (Koeller and Parsons, 1977).

Meiobenthos

The meiobenthos of the New York Bight and adjacent waters are poorly known (see Carriker *et al.*, this volume for discussion). Nematodes and harpacticoid copepods generally are dominant, but members of other phyla and the young of macrobenthos occasionally are important. Tietjen (1980) investigated the nematode component of the meiobenthic community in the New York Bight apex. In fine sand habitats, species diversity was inversely correlated with the concentration of trace metals. This was not the case in muddier sediments. Tietjen concluded that the trophic structure of the nematode community may be simplified by contaminants in sandy sediments

Macrobenthos

Alterations resulting from human activities have been demonstrated most clearly for the macrobenthos, particularly the infauna (organisms living within sediments). The macrobenthos of Raritan and Sandy Hook Bays is depauperate, exhibiting low faunal abundance and low diversity (McGrath, 1974; Franz, this volume). Comparison of McGrath's 1973 survey (McGrath, 1974) with Dean's 1957-1960 survey (Dean, 1975) suggests that considerable degradation of the macrobenthos has occurred recently, although the differences could have arisen in part from normal annual variability. For example, amphipods of the genus *Ampelisca,* which were abundant in the earlier survey, were scarce in 1973. Franz (this volume) demonstrated a dramatic reduction in the species richness of molluscs and in the importance of the estuary's oyster industry by the 1920's.

The macrobenthos of the Bight apex has been studied intensively (Rowe, 1971; National Marine Fisheries Service, 1972; Pearce, 1972; Pearce *et al.,* 1976; Botton, 1979; Boesch, this volume). No effects of human activities have been demonstrated for the macrobenthos of sandy sediments south of Long Island, east of New Jersey, or on Cholera Bank (east of the sewage sludge dumpsite). The silt and clay sediments of the Christiaensen Basin and upper Hudson Shelf Valley (see Boesch, this volume: Figure 1) are heavily contaminated with trace metals and organic material resulting in part from dumping of sewage sludge and dredge spoil (Carmody *et al.*, 1973; Gross, 1976; Hatcher and Keister, 1976; Freeland *et al.*, 1979). Boesch (this volume) compared macrobenthos from the Christiaensen Basin, near the sewage sludge disposal site, with macrobenthos from other stations in the Hudson Shelf Valley. Organic content and metals concentrations were lower in silt and clay sediments at stations distant from the dump site. Several species of benthic invertebrates were unusually abundant in the vicinity of the Basin, possibly as a result of the enrichment of sediments with organic materials derived from the Hudson River plume, dumping, or nutrient stimulation of phytoplankton productivity. The large populations in the Basin also could reflect decreased grazing by predators, which may be excluded from the area by low dissolved oxygen levels or toxicants (Boesch, this volume). On the other hand, several species of benthic crustaceans were abundant elsewhere in the Shelf Valley but were rare or absent in the Basin.

It is impossible to determine the degree to which toxicants in bottom sediments are responsible for the exclusion of species from contaminated sediments in the Bight apex. Exclusion also could result from the indirect effects of elevated levels of organic materials in the sediments. The organic enrichment of sediments may result in depletion of interstitial and near-bottom dissolved oxygen and release of toxic hydrogen sulfide. Species susceptible to toxic effects of trace metals and synthetic organic compounds (*e.g.*, crustaceans in particular) also generally are intolerant of low oxygen levels and sulfide.

Alterations of macrobenthic community structure resulting from organic enrichment and/or from toxicants may have deleterious consequences for bottom-living fishes that prey on the benthos. Benthic invertebrates that tend to be disadvantaged (*e.g.*, amphipods) are important prey for many fishes, whereas most of the species that become more abundant are not important prey (Boesch, this volume). Thus, although the total biomass and productivity of the benthos may be increased by organic enrichment, the yield to higher trophic levels may, in fact, be reduced. The benthic macrofauna of the Hudson-Raritan estuary is even more depauperate than that of the "enriched zone" of the apex. It constitutes a very degraded food source for resident and transient fishes (McGrath, 1974; Boesch, this volume).

The recovery of the benthos in affected sections of the estuary and Bight apex after cessation of the discharge of toxic and organic materials would probably take many years. Rosenberg (1976) found that the macrobenthos of sediments heavily contaminated with pulp mill wastes in a Swedish fjord required about eight years to return to natural conditions. These sediments were not heavily contaminated with toxicants. The silt and clay sediments of the Christiaensen Basin and upper Hudson Shelf Valley are subject to resuspension, and the rate of sedimentation is low (Freeland *et al.*, 1979). In addition, the activities of benthic organisms can be expected to expose shallowly buried toxicants. Thus, the concentration of many persistent toxicants (*e.g.*, trace metals, organochlorides, and polynuclear aromatic hydrocarbons) in surface sediments is not likely to decline rapidly in the most impacted portion of the Bight.

Fishes

Declines or fluctuations in landings (McHugh, 1977) and stocks (Edwards, 1976; Grosslein and Azarovitz, in press) of coastal fishes in the New York Bight region have been attributed primarily to fishing pressure, climatic variations, and modifications of critical coastal habitats. The pervasiveness of these effects has made it impossible to ascribe reduction in fishery resources to toxicant additions *per se,* or to toxic effects on food chains supporting the fisheries. In the Hudson-Raritan estuary, however, toxicants and organic loadings have almost certainly had effects on fishes. In certain of the most heavily contaminated and oxygen-stressed areas, populations of fishes are greatly reduced.

Toxicants in sediments of the Bight apex have been linked with certain sublethal effects on fishes, particularly fin erosion of demersal fishes (Murchelano and Ziskowski, 1976, this volume; Sherwood, this volume). Although the consequences of these effects to the species populations cannot be measured, some increase in mortality is a likely result. In addition, as discussed above, the possibility exists that both the benthic and planktonic communities have been altered by nutrient and organic enrichment and/or by toxicant additions in ways that reduce the efficiency of the trophic coupling to fishes. Thus, while it is likely that the direct effects of toxicants on fishes (in the Bight region) are small compared to the effects of fishing, natural factors, and habitat modifications, toxic effects cannot be dismissed completely.

Toxicant Effects on Nutrient Cycling

Physical, chemical, and biological processes, singly or in combination, can mediate potential effects of toxic substances. For example, enhanced concentrations of trace metals (Szekielda *et al.,* 1972; Sick *et al.,* 1978) and petroleum hydrocarbons (Klemas and Polis, 1977) have been found in convergence zones associated with fronts. Filter-feeding zooplankton, which may congregate in such zones, are known to process and concentrate significant amounts of trace metals in fecal pellets that are potentially available both to water column and benthic organisms (Boothe and Knauer, 1972; Wolfe and Rice, 1972; Small *et al.,* 1973; Honjo, 1978; Sick and Johnson, 1978).

Chemical form and degree of bonding to other materials can have significant effects on the bioavailability and potential toxicity of pollutants (Sunda and Guillard, 1976; Sunda *et al.,* 1978; Luoma, 1979; Rueter *et al.,* 1979). The rate and magnitude of biological productivity also will have an impact on the extent of dispersion and toxicity of potential pollutants (Morris, 1971). The sources and quantities of trace metal additions to the New York Bight ecosystem are generally known. Enhanced trace metals in the New York Bight system can be attributed to the relatively high discharge rates of the Hudson River, the several dump sites in the Bight, and the proximity of the Bight to a large population center with highly contaminated estuaries. Dredged material contributes the major portion (24%-80%) of the toxic metal input to the Bight, except for mercury, 70% of which is attributable to wastewater. Sewage sludge dumping contributes less than 6% of the toxic metal load (Swanson, 1977). Large fractions of the cadmium, chromium, copper, iron, mercury, lead, and zinc inputs to the Bight are attributable to the Hudson River discharge.

Within the water column, there is ample evidence that trace metal enhancement has occured in plankton inhabiting the New York Bight. Greig *et al.* (1977a) reported significantly higher concentrations of copper, nickel, lead, and zinc in zooplankton collected in the New York Bight apex relative to plankton from either Long Island Sound or coastal New Jersey to the south. Benthic organisms from the New York Bight, especially from the apex, contain enhanced tissue concentrations of several trace metals. Greig *et al.* (1977b) reported that tissues of the rock crab (*Cancer irroratus*) from the New York Bight had silver concentrations approximately twice as high as tissue from crabs collected at a control site. Similarly, both cadmium and silver were more highly concentrated in digestive gland tissue of the channeled whelk (*Busycon canaliculatum*) taken from the Bight than in the tissue of whelks from a control coastal site. The average concentrations of silver, arsenic, copper, and zinc in clams (*Spisula solidissima* and *Arctica islandica*) were generally higher in samples from the Bight than in samples from similar depths off the coasts of Delaware and Virginia (Wenzloff *et al.*, 1979). For certain metals, concentrations in clam tissues decreased consistently with latitude in a southerly direction from the center of the Bight.

Potentially toxic organic materials have accumulated to higher levels in the estuary and inner Bight than in offshore areas (O'Connor *et al.*, this volume). Polychlorinated biphenyls enter New York harbor from the Hudson River, where residues remain from earlier industrial discharges, from sewage treatment plant discharges, and from atmospheric inputs. Chlorinated pesticides enter the region in runoff from agricultural areas in the Hudson-Raritan watershed and on Long Island. Elevated concentrations of polynuclear aromatic hydrocarbons occur in sedimentary deposits of waterways, in dredged materials, and in the Christiaensen Basin, suggesting that runoff and the combined sewer outfalls carry spilled petroleum products and incompletely combusted hydrocarbons to the harbor system. The sediments of the River and harbor are the major sinks for these organic materials, and thus elevated concentrations also occur in sediments at the dredged materials dumpsite. Recent analyses of organic contaminants in biota of the Bight region are summarized by O'Connor *et al.* (this volume).

Although enhanced concentrations of potentially toxic metals and organic compounds have been found in selected living and non-living components of the New York Bight apex ecosystem, the effects of such enhancement on overall nutrient and mineral cycles have not been investigated. Studies of nutrient regeneration by the benthic community of the Bight (Rowe *et al.*, 1976) have suggested that the nitrogen cycle may be altered from that of neighboring, less polluted areas. On the other hand, Walsh *et al.* (1978) observed that large-scale seasonal and annual variability in nutrient cycles and primary productivity was greater than the differences between the New York Bight and surrounding areas. Thus, the significance of human-induced modifications will be exceedingly difficult to ascertain.

RELEVANCE OF ECOSYSTEM CHANGES TO MANAGEMENT DECISIONS

The New York Bight receives a wide variety of wastes that are dumped into the Hudson River and the Hudson-Raritan estuary, as well as direct inputs of dredged materials, sewage sludge, and chemical wastes. These pollutant additions have caused widespread degradation of the marine and estuarine environments. The known effects are most severe in the Hudson-Raritan estuary and the New York Bight apex. Beyond the inner apex, there is an uneven, seasonably variable gradient of decreasing impacts.

The scientific evidence that the changes are caused by pollutant additions is overwhelming and conclusive; it includes extensive field observations of ecological degradation and measurements of pollutant occurrences, laboratory experiments on organism responses to toxicant exposures, and comparisons with observations on disturbed and undisturbed systems in other coastal areas. Yet, as discussed in the foregoing sections of this paper, the intrinsic complexities of ecosystem structure and function, and of toxicant interactions within the ecosystem, render very difficult the judgements of when a significant change has occurred, and whether the change was the result of a particular contaminant. The spatial heterogeneity and temporal variability within an area the size of the New York Bight are so great that in most cases, intensive sampling programs with carefully stratified designs are required to quantify statistically significant deviations from normal ecosystem characteristics. Even so, one can neither ascribe with certainty the causes for those changes to particular toxicants introduced into the system, nor readily judge from the analyses alone whether ecosystem function has been significantly and deleteriously impacted. Despite the difficulties of establishing rigorous scientific evidence, however, the ecological changes that have occurred are readily visible in many areas of the Hudson-Raritan estuary, and for many people these changes are significant indicators of deleterious impact.

Conclusions regarding magnitude and causes of ecological change are most readily apparent in sudden catastrophic events, such as the AMOCO CADIZ or the Ixtoc 1 oil spills. Changes are so dramatic that very few scientific data are required to support the conclusions of significance and causality (although scientific determination of when the disturbed system has recovered to a normal state may be very tenuous, indeed). However, even in acute pollution situations like oil spills, it is easy to document only relatively short-term and local ecological effects, such as obliteration of an amphipod population or destruction of a normal salt marsh community. It may be impossible, except through intuitive logic, to relate those changes to subsequent effects on a population of transient predators upon amphipods, or on a fish population whose juvenile forms depend upon salt marshes for habitat and food. Furthermore, people are most likely to base their response to a spill situation upon potential threats to economic resources, such as commercial fishes, crabs, or oysters, or upon aesthetic values totally apart from ecologic considerations. Thus, indistinct perceptions of ecosystem stress do not influence, either to any great extent or in any consistent manner, the selection of environmentally optimal management alternatives in pollution situations.

In situations like the New York Bight, where multiple insults have resulted in gradual and cumulative changes and where people have gradually acclimated to different standards of environmental quality, the problem is much more difficult than in instances of acute pollution. Potential approaches to improved environmental management are numerous, and all approaches have measurable social, economic, or environmental costs associated with their implementation. Yet society still is unable to quantify confidently the relative rates or magnitudes of ecosystem improvement that may result from the various options. Thus, ecosystem characteristics alone tend not to yield information directly applicable to decisions about the control or regulation of potentially deleterious toxicant additions to marine systems such as the New York Bight. Nonetheless, general "health" of the ecosystem must be a major concern to decision-makers with regulatory authority over potentially deleterious discharges.

The ecological studies cited in this paper were for the most part not designed and conducted to test and select specific management options. They were designed instead to characterize the present status of the New York Bight ecosystem. The study results provide a useful context for identifying and selecting potential management options that will protect the Bight from further deterioration or promote its recovery. Ecological characterization to date also provides a point of reference against which future changes may be compared.

As management options are identified and proposed for the New York Bight (*e.g.*, reduce or eliminate effluents and dumping; regulate the timing and frequency of discharges; relocate the dump sites; treat discharges to remove toxicants), specific studies and analyses will be required to determine the environmental advantages and economic costs of each proposal. Environmental studies instituted in response to such management options should be designed to meet specific information requirements for decision points and alternatives that are identified in advance of the study. Some of these studies should be large-scale and long-term field experiments in which specific management alternatives are instituted on a trial basis in order to test comparative merits of different approaches. Thus, the design of environmental studies becomes an integral part of an overall decision analysis framework that outlines the options in sequence of importance, identifies the information needs at each step, evaluates the benefits, costs, and disadvantages, and specifies the decision strategy for each option under different possible circumstances. Environmental managers must work in concert with environmental scientists to define the needed program of studies.

Continued long-term, comprehensive ecological study of systems receiving anthropogenic toxicant additions is necessary in representative areas to provide improved understanding of how such systems respond over the long term to pollutant stresses and to mitigative measures. The New York Bight should be reexamined periodically to determine whether the general health of the system is deteriorating or improving in conjunction with environmental regulatory measures applied to pollutant discharges (water quality criteria and best-available-technology standards, location and timing of ocean dumping) or to resource utilization (public health standards for swimming beaches or for food organisms harvested from the area). Such long-term spot-checking or monitoring will provide a basis for the more or less rigid regulation of pollutant discharges in this and similar systems.

CONCLUSIONS AND RECOMMENDATIONS

1. Although varying degrees of ecological change can be documented readily throughout the New York Bight region, and although the degradation can be ascribed to pollutant inputs of human origin, the evidence that changes are caused by toxicants (as opposed to nutrient and carbon loading) is only circumstantial. No ecological or community-level change can be ascribed to any specific toxicant. If such cause-effect relationships are to be established, a balanced program of field and laboratory research must be undertaken. Greater emphasis must be placed on experimental approaches to ecosystem analysis in order to deduce causal relationships and mechanisms.

2. Ecosystem and community-level characteristics such as species composition and abundance, productivity, and diversity are highly variable in both time and space and should not be proposed as a direct basis for regulating pollutant discharges.

3. For effective management-directed planning of environmental research, environmental managers and environmental scientists must jointly and cooperatively develop an action plan that encompasses the potential options available for environmental management and the specific information requirements for assessing the relative overall risks associated with each option.

4. Specific management alternatives should be instituted, on a trial basis, in selected areas and on a limited scale, in order to test the relative ecological merits of different approaches. Tests should be accompanied by extensive programs of environmental sampling and measurement that are carefully designed to meet previously specified decision options regarding the tests. Such tests may take several years to complete.

5. Community characteristics, especially of the benthos, should be monitored at selected stations as an integrative, long-term measure of the effectiveness of whatever management practices or regulations are eventually adopted. A long-term commitment of at least 5-10 years may be required before the ecological results of changed management approaches become evident.

REFERENCES

Atlas, R.M. and R. Bartha. 1973. Abundance, distribution and oil biodegradation potential of microorganisms in Raritan Bay. Environ. Pollut. 4: 291-300.

Beers, J.R., G.L. Stewart, and K.D. Hoskins. 1977a. Dynamics of microzooplankton populations treated with copper: Controlled Ecosystem Pollution Experiment. Bull. Mar. Sci. 27: 66-79.

Beers, J.R., M.R. Reeve, and G.D. Grice. 1977b. Controlled Ecosystem Pollution Experiment: effect of mercury on enclosed water columns. IV. Zooplankton population dynamics and production. Mar. Sci. Comm. 3: 355-394.

Boesch, D.F. This volume. Ecosystem consequences of alterations of benthic community structure and function in the New York Bight region. pp. 543-568.

Boesch, D.F., R.J. Diaz, and R.K. Virnstein. 1976. Effects of tropical storm Agnes on soft-bottom macrobenthic communities of the James and York estuaries and the lower Chesapeake Bay. Chesapeake Sci. 17: 246-259.

Boesch, D.F. and R. Rosenberg. 1981. Response to stress in marine benthic communities. *In:* Stress Effects on Natural Ecosystems, G.M. Barrett and R. Rosenberg (eds.), John Wiley and Sons, New York. pp. 179-200.

Boothe, P.N. and G.A. Knauer. 1972. The possible importance of fecal material in the biological amplification of trace and heavy metals. Limnol. Oceanogr. 17: 270-274.

Botton, M.L. 1979. Effects of sewage sludge on the benthic invertebrate community of the inshore New York Bight. Estuarine Coastal Mar. Sci. 8: 169-180.

Calabrese, A., E. Gould, and F.P. Thurberg. This volume. Effects of toxic metals in marine animals of the New York Bight: some laboratory observations. pp. 281-297.

Carmody, D.J., J.B. Pearce, and W.E. Yasso. 1973. Trace metals in sediments of New York Bight. Mar. Pollut. Bull. 4: 132-135.

Carriker, M.R., J.W. Anderson, W.P. Davis, D.R. Franz, G.F. Mayer, J.B. Pearce, T.K. Sawyer, J.H. Tietjen, J.F. Timoney, and D.R. Young. This volume. Effects of pollutants on benthos. pp. 3-21.

Clark, S.H. and B.E. Brown. 1977. Changes in biomass of finfishes and squids from the Gulf of Maine to Cape Hatteras, 1963-74, as determined from research vessel survey data. Fish. Bull. 75: 1-22.

Dean, D. 1975. Raritan Bay macrobenthos survey, 1957-60. NOAA NMFS Data Rep. 99. 51 pp.

Edwards, R.L. 1976. Middle Atlantic fisheries: recent changes in populations and outlook. Am. Soc. Limnol. Oceanogr. Spec. Symp. 2: 302-311.

Eppley, R.W. and C.S. Weiler. 1979. The dominance of nanoplankton as an indicator of marine pollution: a critique. Oceanol. Acta. 2: 241-245.

Farrington, J.W., N.M. Frew, P.M. Gschwend, and B.W. Tripp. 1977. Hydrocarbons in cores of Northwestern Atlantic coastal and continental margin sediments. Estuarine Coastal Mar. Sci. 5: 793-808.

Franz, D.R. This volume. An historical perspective on molluscs in lower New York harbor, with emphasis on oysters. pp. 181-197.

Freeland, G.L., D.J.P. Swift, and R.A. Young. 1979. Mud deposits near the New York Bight dumpsite: origin and behavior. *In:* Ocean Dumping and Marine Pollution, H.O. Palmer and M.G. Gross (eds.), Dowden, Hutchinson and Ross, Stroudsburg, PA. pp. 73-95.

Gamble, J.C., J.M. Davies, and J.H. Steele. 1977. Loch Ewe bag experiment, 1974. Bull. Mar. Sci. 27: 146-175.

Grassle, J.F. and J.P. Grassle. 1976. Opportunistic life histories and genetic systems in marine benthic polychaetes. J. Mar. Res. 32: 253-284.

Gray, J.S. and F.B. Mirya. 1979. A possible method for the detection of pollution-induced disturbances on marine benthic communities. Mar. Pollut. Bull. 10: 142-146.

Green, R.H. 1979. Sampling Design and Statistical Methods for Environmental Scientists. John Wiley and Sons, New York. 257 pp.

Greig, R.A., A. Adams, and D.R. Wenzloff. 1977a. Trace metal content of plankton and zooplankton collected from the New York Bight and Long Island Sound. Bull. Environ. Contam. Toxicol. 18: 3-8.

Greig, R.A., D.R. Wenzloff, A. Adams, B. Nelson, and C. Shelpuk. 1977b. Trace metals in organisms from disposal sites of the middle eastern United States. Arch. Environ. Contam. Toxicol. 6: 395-409.

Greve, W. and T.R. Parsons. 1977. Photosynthesis and fish production: Hypothetical effects of climatic change and pollution. Helgol. Wiss. Meeresunters. 30: 666-672.

Grice, C.D., M.R. Reeve, P. Koeller, and D.W. Menzel. 1977. The use of large volume, transparent, enclosed sea-surface water columns in the study of stress on plankton ecology. Helgol. Wiss. Meeresunters. 30: 118-133.

Gross, M.G. 1976. Sources of urban wastes. Am. Soc. Limnol. Oceanogr. Spec. Symp. 2: 150-161.

Grosslein, M.D. and T. Azarovitz. In press. Fish Distribution. MESA New York Bight Atlas Monograph 15, New York Sea Grant Institute, Albany, NY.

Hatcher, P.G. and L.E. Keister. 1976. Carbohydrates and organic carbon in New York Bight sediments as possible indicators of sewage contamination. Am. Soc. Limnol. Oceanogr. Spec. Symp. 2: 240-248.

Honjo, S. 1978. Sedimentation of materials in the Sargasso Sea at 5,367 m deep station. J. Mar. Res. 36: 464-491.

Klemas, V. and D.F. Polis. 1977. A study of density fronts and their effects on coastal pollutants. Remote Sensing Environ. 6: 95-126.

Koeller, P. and T.R. Parsons. 1977. The growth of young salmonids (*Onchorhynchus keta*): Controlled Ecosystem Pollution Experiment. Bull. Mar. Sci. 27: 114-118.

Koeller, P.A. and G.T. Wallace. 1977. Controlled Ecosystem Pollution Experiment: effects of mercury on enclosed water columns. V. Growth of juvenile chum salmon (*Onchorhynchus keta*). Mar. Sci. Comm. 3: 395-406.

Lee, R., A.C. Longwell, T.C. Malone, L.S. Murphy, D.R. Nimmo, H.B. O'Connors, Jr., L.S. Peters, and K.D. Wyman. This volume. Effects of pollutants on plankton and neuston. pp. 39-52.

Levins, R. 1968. Toward an evolutionary theory of the niche. *In:* Evolution and Environment, E.T. Drake (ed.), Yale Univ. Press, New Haven, CT. pp. 325-340.

Litchfield, C.D., J.P. Nakas, and R.H. Vreeland. 1976. Bacterial flux in some New Jersey estuarine sediments. Am. Soc. Limnol. Oceanogr. Spec. Symp. 2: 340-353.

Litchfield, C.D., M.A. Devanas, J. Zindulis, M. Meskill, J. Freedman, and C. McClean. This volume. Influence of cadmium on the microbial populations and processes in sediments from the New York Bight apex. pp. 587-604.

Livingston, R.J. This volume. Long-term variability in coastal systems: background noise and environmental stress. pp. 605-620.

Luoma, S.N. and G.W. Bryan. 1979. Trace metal bioavailability: modeling chemical and biological interactions of sediment-bound zinc. *In:* Chemical Modeling in Aqueous Systems, E.A. Jenne (ed.), American Chemical Society Symposium Series No. 93. pp. 577-609.

Malone, T.C. 1976. Phytoplankton productivity in the apex of the New York Bight: environmental regulation of productivity/chlorophyll *a*. Am. Soc.

Limnol. Oceanogr. Spec. Symp. 2: 260-272.

Malone, T.C. 1977. Plankton Systematics and Distribution. MESA New York Bight Atlas Monograph 13, New York Sea Grant Institute, Albany, NY. 45 pp.

McCall, P.L. 1977. Community patterns and adaptive strategies of the infaunal benthos of Long Island Sound. J. Mar. Res. 35: 211-266.

McGrath, R. 1974. Benthic macrofaunal census of Raritan Bay—preliminary results, Pap. 24. *In:* Hudson River Ecology, Third Symposium on Hudson River Ecology, March 22-23, 1973, Hudson River Environmental Society, Inc., Tuxedo, NY. 40 pp.

McHugh, J.L. 1977. Fisheries and fishery resources of New York Bight. NOAA Tech. Rep. NMFS Circ. 401. 50 pp.

Meeter, D., R.J. Livingston, and G.C. Woodsum. 1979. Short and long-term hydrological cycles of the Apalachiocola drainage system with application to Gulf coastal populations. *In:* Ecological Processes in Coastal and Marine Systems, R.J. Livingston (ed.), Plenum Press, New York. pp. 315-338.

Menzel, D.W. and J. Case. 1977. Concept and design: Controlled Ecosystem Pollution Experiment. Bull. Mar. Sci. 27: 1-7.

Michael, A.D. This volume. The potential contribution of petroleum hydrocarbons to changes in benthic communities of the New York Bight. pp. 621-630.

Michael, A.D., C. Van Raalte, and L.S. Brown. 1975. Long term effects of an oil spill at West Falmouth. *In:* Proceedings of Joint Conference on the Prevention and Control of Oil Spills, San Francisco, American Petroleum Institute, Washington, DC. pp. 573-581.

Michael, A.D., D. Boesch, C. Hershner, R.J. Livingston, K. Roos, and D. Straughan. 1977. Benthos. *In:* Oil Spill Studies: Strategies and Techniques, A.P.I. Publication No. 4286, American Petroleum Institute, Washington, DC. pp. 57-75.

Morris, A.W. 1971. Trace metal variations in sea water of the Menai Straits caused by a bloom of *Phaeocystis.* Nature 233: 427-428.

Murchelano, R.A. and J. Ziskowski. 1976. Fin rot disease studies in the New York Bight. Am. Soc. Limnol. Oceanogr. Spec. Symp. 2: 329-339.

Murchelano, R.A. and J. Ziskowski. This volume. Fin rot disease in the New York Bight (1973-1977). pp. 347-358.

National Marine Fisheries Service. 1972. The Effects of Waste Disposal in the New York Bight. National Marine Fisheries Service, Sandy Hook Laboratory, Highlands, NJ. 749 pp.

O'Connor, J.M., J.B. Klotz, and T.J. Kneip. This volume. Sources, sinks, and distribution of organic contaminants in the New York Bight ecosystem. pp. 631-653.

O'Connor, J.M. and J. Rachlin. This volume. Perspectives on metals in New York Bight organisms: factors controlling accumulation and body burdens. pp. 655-673.

O'Connors, H.B., Jr., C.F. Wurster, C.D. Powers, D.C. Biggs, and R.G. Rowland. 1978. Polychlorinated biphenyls may alter marine trophic pathways by reducing phytoplankton size and production. Science 201: 737-739.

Patrick, R. 1967. Diatom communities in estuaries. *In:* Estuaries, G.H. Lauff (ed.), American Association for the Advancement of Science, Washington, DC. pp. 311-315.

Pearce, J. 1972. The effects of waste disposal on benthic communities in the New York Bight. *In:* Marine Pollution and Sea Life, M. Ruvio (ed.), Fishing News (books) Ltd., London. pp. 404-411.

Pearce, J.B., J.V. Caracciolo, M.B. Halsey, and L.H. Rogers. 1976. Temporal and spatial distributions of benthic macroinvertebrates in the New York Bight. Am. Soc. Limnol. Oceanogr. Spec. Symp. 2: 394-403.

Pielou, E.C. 1967. The use of information theory in the study of the diversity of biological populations. Proc. 5th Berkeley Symp. Math. Stat. and Prob. 4: 163-177.

Redfield, A.C. and L.A. Walford. 1951. A Study of the Disposal of Chemical Waste at Sea. Publication 201. National Academy of Sciences/National Research Council, Washington, DC. 249 pp.

Reeve, M.R., J.C. Gamble, and M.A. Walter. 1977. Experimental observations on the effects of copper on copepods and other zooplankton: Controlled Ecosystem Pollution Experiment. Bull. Mar. Sci. 27: 92-104.

Rosenberg, R. 1976. Benthic faunal dynamics during succession following pollution abatement in a Swedish estuary. Oikos 27: 414-427.

Rowe, G.T. 1971. The effects of pollution on the dynamics of the benthos of New York Bight. Thalassia Jugosl. 7: 353-359.

Rowe, G.T., R.L. Smith, and C.H. Clifford. 1976. Benthic-pelagic coupling in the New York Bight. Am. Soc. Limnol. Oceanogr. Spec. Symp. 2: 370-376.

Rozensweig, M. and R.H. MacArthur. 1963. Graphical representation and stability conditions of predator-prey interactions. Am. Nat. 97: 209-223.

Rueter, J.G., J.J. McCarthy, and E.J. Carpenter. 1979. The toxic effect of copper on *Oscillatoria theibauti.* Limnol. Oceanogr. 24: 558-561.

Saloman, C.H. and S.P. Naughton. 1977. Effect of Hurricane Eloise on the benthic fauna of Panama City Beach, Florida, U.S.A. Mar. Biol. 42: 357-363.

Sanders, H.L. 1978. *Florida* oil spill impact on the Buzzards Bay benthic fauna: West Falmouth. J. Fish. Res. Board Can. 35: 717-730.

Sherwood, M.J. This volume. Fin erosion, liver condition, and trace contaminant exposure in fishes from three coastal regions. pp. 359-377.

Sick, L.V., C.C. Johnson, and R. Engel. 1978. Trace metal enhancement in the biotic and abiotic components of an estuarine tidal front. J. Geophys. Res. 83: 4659-4667.

Sick, L.V. and C.C. Johnson. 1978. Seasonal and geographic distribution of trace metal concentrations in zooplankton and macrobenthic invertebrates from Georges Bank. Sea Grant Rep., CMS-1-78, College of Marine Studies, University of Delaware. 57 pp.

Slobodkin, L.B. 1961. Growth and Regulation of Animal Populations. Holt, Reinhart, and Winston, New York. 184 pp.

Small, L.F., S.W. Fowler, and S. Keckes. 1973. Flux of zinc through a macroplanktonic crustacean. *In:* Radioactive Contamination of the Marine Environment, International Atomic Energy Agency, Vienna, Austria. pp. 437-452.

Snelson, F.F. and W.K. Bradley, Jr. 1978. Mortality of fishes due to cold on the East Coast of Florida, January, 1977. Fla. Sci. 41: 1-12.

Sunda, W. and R.R.L. Guillard. 1976. The relationship between cupric ion activity and the toxicity of copper to phytoplankton. J. Mar. Res. 34: 511-529.

Sunda, W.G., D.W. Engel, and R.M. Thuolte. 1978. Effect of chemical speciation on toxicity of cadmium to grass shrimp, *Palaemonetes pugio*: importance of free cadmium ion. Environ. Sci. Technol. 12: 409-413.

Sutcliffe, W.H., Jr., K. Drinkwater, and B.S. Muir. 1977. Correlations of fish catch and environmental factors in the Gulf of Maine. J. Fish. Res. Board Can. 34: 19-30.

Swanson, R.L. 1977. Status of ocean dumping research in New York Bight. J.

Waterway, Port, Coastal, and Ocean Division, Am. Soc. Civ. Eng. 103: 9-24.

Szekielda, K.H., S.L. Kupferman, V. Klemas, and D.V. Polis. 1972. Element enrichment in organic films and foam associated with aquatic frontal systems. J. Geophys. Res. 77: 5278-5282.

Thomas, W.H. and D.L.R. Seibert. 1977. Effects of copper on the dominance and diversity of algae: Controlled Ecosystem Pollution Experiment. Bull. Mar. Sci. 27: 23-33.

Thomas, W.H., D.L.R. Seibert, and M. Takahashi. 1977. Controlled Ecosystem Pollution Experiment: effect of mercury on enclosed water columns. III. Phytoplankton population dynamics and production. Mar. Sci. Comm. 3: 331-354.

Tietjen, J.H. 1980. Population structure and species composition of the free-living nematodes inhabiting sands of the New York Bight Apex. Estuarine Coastal Mar. Sci. 10: 61-73.

Timoney, J.F., J. Port, J. Giles, and J. Spanier. 1978. Heavy-metal and antibiotic resistance in the bacterial flora of sediments of New York Bight. Appl. Environ. Microbiol. 36: 465-472.

Timoney, J.F. and J.G. Port. This volume. Heavy metal and antibiotic resistance in *Bacillus* and *Vibro* from sediments of New York Bight. pp. 235-248.

Van Winkle, W., B.L. Kirk, and B.W. Rust. 1979. Periodicities in Atlantic Coast striped bass (*Morone saxatilis*) commercial fisheries data. J. Fish. Res. Board Can. 36: 54-62.

Walsh, J.J., T.E. Whitledge, F.W. Barnevick, S.O. Howe, C.D. Wirick, W.E. Esaias, and J.T. Scott. 1978. Wind events and food chain dynamics within the New York Bight. Limnol. Oceanogr. 23: 659-683.

Wenzloff, D.R., R.A. Grieg, A.S. Merrill, and J.W. Ropes. 1979. A survey of heavy metals in the surf clam, *Spisula solidissima,* and the ocean quahog, *Arctica islandica,* of the Mid-Atlantic coast of the United States. Fish. Bull. 77: 280-285.

Wiebe, P.H., G.D. Grice, and E. Hoagland. 1973. Acid iron waste as a factor affecting the distribution and abundance of zooplankton in the New York Bight. II. Spatial variations in the field and implications for monitoring studies. Estuarine Coastal Mar. Sci. 1: 51-64.

Wolfe, D.A. and T.R. Rice. 1972. Cycling of elements in estuaries. Fish. Bull. 70: 959-972.

Wormald, A.P. 1976. Effects of a spill of marine diesel oil on the meiofauna of a sandy beach at Picnic Bay, Hong Kong. Environ. Pollut. 11: 117-130.

REPORTS OF MANAGEMENT PANELS

Opposite: Fresh Kills landfill on Staten Island, New York. This facility contains a variety of wastes. Its importance as a source of contamination to area waters has not been adequately quantified. (Courtesy EPA Documerica.)

MANAGEMENT OF DOMESTIC WASTES[1]

C.G. Gunnerson[2]
R. Adamski
P.W. Anderson
R.S. Dewling
D.A. Julius
G.S. Kleppel
D.W. Lear
A.J. Mearns
J.T. Miller
E.P. Myers
J.J. Roswell
S. Schroeder
D.A. Segar
C.J. Sindermann
C.M. Stern

Abstract. Studies of municipal waste discharges and sewage sludge dumping reveal that these sources account for more than half of the loadings of the most serious pollutants in the New York Bight region. Liquid municipal discharges account for 35-70% and municipal sludge dumping for 5-15% of total input, depending upon the contaminant under consideration.

Increasing costs of conventional waste disposal practices and regulation have not been accompanied by corresponding benefits. Consequently, there is increasing need to determine willingness to pay for alternative policy or institutional and technological options for waste generation and disposal practices in the New York metropolitan area.

INTRODUCTION

The objectives of the panel report on domestic wastes are to identify major ecological problems in the New York Bight and its estuaries, to determine the extent to which domestic wastes are responsible for these problems, and to suggest remedial measures for reducing current and future impacts. Included on the panel were representatives of the National Oceanic and Atmospheric Administration (NOAA), the U.S. Environmental Protection Agency (EPA), the City of New York, environmental action groups, academia, and the public. A matrix approach was used to identify priority items for consideration in the management of domestic and municipal wastes. A list of the issues discussed and a ranking of their relative importance are provided in Table 1. Only medium- and high-priority items for which institutional or technological responses are at hand were considered in detail by the panel.

The report is divided into three sections: an introduction to the issues, a survey of perceived pollution problems, and an examination of the management of major municipal and domestic waste sources. Background to the problems is provided by reviews of legal, political, and economic issues relevant to domestic

[1]Contributions and discussions presented in this report have been summarized by the Chairman.

[2]Chairman, Panel on the Management of Domestic Wastes.

Table 1. Municipal waste discharge and effects.
Relative importance: L = low, M = medium, H = high.

	Nitrogen/BOD[1]	Trace organics	Pathogens	Petroleum hydrocarbons	Metals	Trace nutrients	Chlorine
EFFECTS							
Anoxic events	H	L	L	L	L	M	L
Red tides	H	L	L	L	M	M	L
Fin erosion	L	H	M	M	H	L	L
Damaged benthos	M	H	L	H	H	L	L
Damaged fisheries	M	H	H	M	M	L	L
Damaged beaches[2]	M	L	H	L	L	L	L
SOURCES							
Raw sewage discharges	M	L	H	L	L	M	L
Combined sewer overflows	M	L	H	M	M	M	L
Sewage treatment plant effluents	H	L	M	L	H	H	H
Sewage sludge dumping	L	H	M	M	H	L	L

[1] Biochemical oxygen demand.

[2] Including aesthetic and health considerations (*e.g.*, the presence of floatables and suspended matter).

waste management. This is followed by discussions of effects of waste management practices including enrichment, anoxia, and fish kills; red tides; damaged benthos; impacted fisheries; and fouled beaches. The final section analyzes the major sources of municipal and domestic wastes in the region (*i.e.*, raw sewage, combined sewer overflows, sewage effluent, and sewage sludge).

Where possible within each section, recommendations are provided for the formulation of management policies. Problems associated with the ocean disposal of sewage sludge were discussed extensively by the panel. Also provided are probable responses of agencies to suggestions for revising the December 31, 1981 deadline for eliminating sludge dumping. A general summary of technology and management options relating to municipal and domestic wastes is provided at the conclusion of this report.

LEGAL BACKGROUND

If management options for the disposal of municipal and domestic wastes conflict with federal law, either the options must be discarded or the law must be changed, if municipal dischargers are not to face the risk of enforcement penalties. Pertinent laws include the Clean Water Act, the Clean Air Act, the Marine Protection, Research and Sanctuaries Act, the Coastal Zone Management Act, the Toxic Substances Control Act, the Resource Conservation and Recovery Act, the Marine Pollution Research and Development and Monitoring Act, the Fishery Conservation and Management Act, and the National Environmental Policy Act. Compliance schedules and regulations established in association with these acts play a large role in determining management practices. In addition, the National Environmental Policy Act applies where federal construction grant money is involved, and requires a complete environmental assessment of projected activities and an in-depth consideration of alternatives.

Under the Clean Water Act, publicly owned treatment works (POTW's) must achieve at least secondary treatment unless a Section 301(h) variance is obtained. The original July 1, 1977 deadline for secondary treatment has been extended for most non-complying POTW's. A Section 301(h) variance allows a POTW with an existing discharge to deep marine or well-mixed estuarine waters to install less than secondary treatment if a "balanced indigenous population" of marine life would be protected. A number of other conditions also must be met, including adequate control of toxic pollutants. Whether or not a variance is obtained, POTW's must comply with state water quality standards. The content of POTW effluent and sludge may be affected by EPA pretreatment regulations under the Clean Water Act, which requires some removal of toxic pollutants.

Sewage sludge disposal is governed by different laws, depending on the method of disposal. Under the Marine Protection, Research and Sanctuaries Act, the ocean dumping of sewage sludge which "may unreasonably degrade or endanger human health, welfare, communities, or the marine environment, ecological systems, or economic potentialities" is scheduled to be terminated by

December 31, 1981. Land disposal of sewage sludge is governed principally by Section 405 of the Clean Water Act and by the Resource Conservation and Recovery Act. Incineration of sewage sludge is governed by the Clean Air Act. Applicable requirements of state implementation plans must be met for incineration, as well as new source performance standards in the case of new facilities.

POLITICAL FACTORS

Political constraints on the management of domestic wastes are imposed by the limits of public awareness and by the limits of public participation in remedial program planning and implementation. Changes in public interest and priorities are revealed periodically by replacement of elected officials. These changes are followed later by changes in law and, eventually, by changes in administrative, programmatic, and fiscal response. Agents of change include the media, religious and secular organizations, and the public schools. Major actors include the Congress (which responds to pressures from organized environmental, recreational, industrial, commercial, and taxpayer constituencies); federal, state, and local regulatory and operating agencies; environmental leaders; the news media; and the general public. The general public is impacted by decisions on waste management; however, it usually reacts to such decisions rather than initiates them.

Many types of information may be considered in making legislative and administrative decisions. Included are: values generally held by society, some of which may be in conflict (*e.g.*, interest in environmental quality versus concern over rising costs); factual information that may be easily understood by laymen; results of directed technical research on management-related problems; and knowledge of the condition, capacity, and operation of infrastructure systems, such as those for water supply, waste disposal, and transportation. Institutions likely to contribute interpretations of such data include professional, academic religious, fraternal, and political action groups.

Problems in environmental management arise from three factors: (1) inadequate communication between the unorganized elements of the public and the political decision-makers; (2) increased responsibilities assigned by legislation to regulatory and operating agencies without corresponding increases in staffing and funding; and (3) inappropriate legislative responses to environmental problems. The first two items are self-explanatory and require little discussion. With regard to item (2), the operation of over-taxed agencies often is crisis-controlled, which precludes timely sharing of information with the affected public.

The legislative response to environmental crises during the past decade has been to mandate minimum (or zero) environmental and public health risk. The approach generally taken assumes that ecological systems have no capacity to deal with wastes (*i.e.*, to assimilate or disperse them) and that individuals are not prepared to take recognizable risks. Section 301(h) of the Clean Water Act (which provides for waivers of secondary treatment requirements) and the "bubble" concept in air pollution control (which provides for allocating or substituting pollutant loadings within an air basin or airshed) represent a break in the tradition. Nevertheless, the uncertainty of risks to future generations remains a major factor in the public's assessment of pollution-related problems.

Possible short- and long-term approaches to environmental management problems include: (1) use by appropriate agencies of direct and media communications to gain public recognition of the costs of attaining environmental goals; (2) application of findings that the public is willing to pay for environmental benefits; and (3) augmentation of knowledge gained by the public through citizen advisory committees, which were required in over one hundred federal programs for consumer and environmental projects.

ECONOMIC FACTORS

The use of an economic framework to compare wastes management options is often helpful in clarifying the underlying environmental and political tradeoffs. In addition, it can be a useful tool for demonstrating to the general public the link between tax requirements and the standards chosen for water and air quality.

Each management option has associated costs and benefits. In order to select among options, the first step is to gather reliable information on the cost of each alternative over time (*i.e.,* both capital and recurrent costs) and the benefits that are likely to result. The latter often are difficult to quantify, since attempts must be made both to link specific actions with adverse effects (damaged fisheries, polluted beaches, and so forth) and to attach monetary values to healthy fish populations and clean beaches. The second step is to present the cost information and information on benefits expected to those who will be asked to bear the financial burden, so that a final decision can be made. In the case of the New York Bight, this will involve federal, state, and municipal parties. Federal and state contributions to each alternative can be estimated readily since they take the form of a specified percentage grant toward certain categories of capital costs. Political and economic realities suggest that additional funding for construction expenditures and all recurrent expenses will be the responsibility of the municipal communities. In order for voters and their representatives to make an informed choice among alternatives, the local share of costs should be expressed in terms of a recurrent (annual or monthly) user charge.

Financing methods will depend upon existing administrative arrangements. However, both financial accountability and user choice could be enhanced if the present system of financing water and sewage services through the general revenue fund is gradually replaced by one based on direct user charges. As a first step toward this end, New York City water and sewage accounts should be segregated from the general city budget to clarify the real cost of producing those services.

POLLUTION EFFECTS

Enrichment and Anoxia

Enrichment of the lower Hudson River, New York harbor, and the New York Bight is a consequence of discharging nitrogen, phosphorus, and carbon compounds in the drainage basin. The results occasionally are spectacular. Nanoplankton (microalgae) grow faster than the river can dilute them (Malone, 1977). Phytoplankton blooms annually in the harbor in response to "urban hypertrophication" (Mahoney and McLaughlin, 1977). Massive plankton blooms and occasional red tides are found off the Long Island and New Jersey coasts (Swanson and Sindermann, 1979).

During 1976, bottom water in a large region of the inner shelf of the New York Bight became depleted in oxygen. Zero oxygen concentrations and some sulfide were present in some areas. Oxygen concentrations remained low in some parts of the region from June to September. The anoxia and/or the toxic effects of sulfide killed 92% of the surf clam population over 3,100 km^2 of the ocean floor and severely affected other benthic species. Most investigators attributed the low oxygen conditions to biodegradation of phytoplankton blooms and benthic organisms. Although excessive nutrient levels (nitrogen, carbon, trace metals, vitamins) from raw and treated sewage discharges in the estuary are major factors in stimulating typical summer algal blooms in these waters, human nutrient contributions were not responsible for the phytoplankton that contributed most of the oxygen demand in the 1976 anoxia event (Swanson and Sindermann, 1979).

Less severe events have occurred in recent years on the inner shelf of the New York Bight, mostly in waters off the northern New Jersey coast. During years when anoxia has not occurred, summer oxygen concentrations in bottom waters off northern New Jersey (particularly in the New York Bight apex) normally drop below 50% of saturation. It is in this region that the discharge from the Hudson-Raritan River estuary is dispersed and the ocean dumping of sewage sludge, dredged materials, and industrial wastes takes place. To the south, off the coast of southern New Jersey, summer oxygen concentrations in an average year drop to about 50-70% of saturation (Swanson and Sindermann, 1979).

The lower average summer oxygen concentrations in the apex region are caused by a combination of factors: poor flushing of the region, high oxygen demand resulting from decomposition of biological material formed by high primary production, and additional oxygen demand from organic material in ocean dumped solids. Spring and summer primary production in the apex region is higher than elsewhere on the shelf, and is supported by the influx of river water that contains high concentrations of dissolved nitrogen. The latter originates predominantly in sewage and sewage treatment plant discharges. Lower concentrations of dissolved nitrogen limit summer primary production in more distant parts of the Bight.

Annual flushing of excess nitrogen in the apex prevents long-term accumulation problems, as with phosphorus accumulation in lakes. Nevertheless, prudent reduction of the total nitrogen discharge may be a sound management goal. Nitrogen removal in the sewage treatment process can be achieved by any one of several operationally proven treatment technologies. At most, such removal would be necessary during a period of six months in any given year. A monitoring program for relevant water quality parameters could provide information that would allow the use of nitrogen removal only during those years and time periods when it is necessary.

Red Tides

"Red tide" and "black surf" are caused by blooms of phytoplankton normally present in low numbers in the Bight. At best, they are aesthetically unpleasant; at worst, they are pathogenic. Bathing beaches may be closed or avoided during such blooms, and economic loss may result. Nutrient enrichment of coastal waters enhances the blooms. The frequency, intensity, and longevity of red tides may be symptomatic of ecological degradation. They have been recorded in New Jersey coastal waters and in Lower New York Bay since the early 1960's (Mahoney and McLaughlin, 1977).

Though the causes of "red tides" are poorly understood, it is believed that stimulation of "seed" populations by a specific suite of environmental conditions results in blooms (Murphy *et al.,* 1975). Stimulation of seed populations in Raritan Bay occurs near outfalls that discharge large quantities of nitrogenous wastes. Several New Jersey red tide species appear to use urea and nitrate in preference to ammonia when all other nutrients are abundant (Mahoney and McLaughlin, 1977). Laboratory studies reveal that many dinoflagellates are sensitive to the presence of microconstituents and chelated trace metals. Some species will not grow if the minimum quantities of microconstituents are absent. Though little is known about the biochemistry of metal-chelator systems relative to red tide phytoplankters, it is significant that both metals and chelators are contributed by industrial and municipal wastes. Since urea is also an occasional industrial waste component, removal of organic nitrogen and (possibly) metals though wastewater treatment may reduce the frequency of red tides.

Damaged Benthos

The most persistent effects of ecological stress in the estuarine-marine habitat are revealed in benthic (bottom) organisms and communities. Geographical areas of concern are the lower Hudson-Raritan estuary, the Bight apex, and the Bight. Pollutant inputs amenable to partial management through domestic sewage systems include toxic metals, petroleum hydrocarbons, and halogenated hydrocarbons. Nutrient enrichment, organic carbon and associated oxygen demand, and bacteria and viruses, which may be pathogenic to the benthic fauna, are also of concern.

The immediate consequence of organic and chemical oxygen demands is the reduction of late summer dissolved oxygen values to averages of 0-2 mg/liter in the Arthur Kill, approximately 2 mg/liter in Kill van Kull and the East River, 2-4 mg/liter at the mouth of the Raritan River, and 3-4 mg/liter at the Narrows (Interstate Sanitation Commission, 1979). Impacts of other contaminants extend over greater distances (Gunnerson, 1981).

Observations of benthic ecosystems in the lower Hudson River and Bight apex and, to a lesser extent, in the outer Bight reveal damage to lobsters, oysters, clams, crabs, scallops, worms, and the minute life forms on which they feed. Toxic metals, petroleum hydrocarbons, halogenated hydrocarbons (especially PCB's), nutrient enrichment, organic carbon, and enteric bacteria all contribute to this degradation (Gunnerson, 1981). While additional research is needed to identify optimal control measures, presently available technologies such as source control, and reduction of contaminated waste solids entering the Bight, should be undertaken as essential short- and long-term control measures.

Damaged Fisheries

Damage to fisheries from pollutants results from reductions in the abundance of exploited populations and from reductions in the availability of products to consumers because of community health concerns. Experimental studies have shown that three principal classes of contaminants affect survival and well-being of fishes and shellfishes. These three classes are chlorinated hydrocarbons, toxic metals, and petroleum components. These same substances, with the addition of pathogenic microorganisms, can cause public health problems for fishery products. Additionally, anoxia related to organic inputs can have measurable impacts on fish and shellfish stocks. It is also true that a number of natural environmental factors affect abundance of fish; careful assessment of fisheries resources is necessary to separate effects of natural and fishing pressures from those of pollution. (See Sinderman *et al.,* this volume, for a detailed discussion of effects on fishes.)

Damaged Beaches

Beach pollution incidents occur frequently in recreational areas in the Hudson estuary and New York Bight apex. Most beaches in the harbor and many in the estuary are closed to public bathing. Floating paper, plastics, tar or grease balls, oil, wood, and dead or dying marine life are carried by the winds, currents, and waves, and are deposited daily on the beaches of New York City, Long Island, and New Jersey. Major beach pollution incidents of stranded floatables occurred in 1976. In June, significant quantities of plastic, rubber, grease balls and other materials of obvious sewage origin, and of trash and wood polluted the beaches of Long Island. Most beaches were closed for periods of a few days to allow for cleanup and, as a precaution, to allow health officials to determine risks to beach users. Bacteriological data generally are considered the most important elements of beach monitoring programs. These data indicated no risk to bathers during the incident. A similar incident occurred in September on New Jersey beaches (Figure 1). In both cases, the primary sources of the floatables

Figure 1. A Sandy Hook, New Jersey beach littered with floatables. (Courtesy J.B. Pearce, NOAA National Marine Fisheries Service.)

were thought to be raw sewage in combined sewer overflows and inadequately treated sewage discharges to the New York harbor (Swanson *et al.*, 1977). Screening and skimming will remove these floatables, which then can be disposed of in land-based systems to prevent recurrence of such incidents.

MUNICIPAL POLLUTION SOURCES AND PROBLEMS

Sanitary disposal of human wastes is the major consideration in design and operation of sewers. Conventional primary (sedimentation) and secondary (biological) treatment remove visible evidence, putrescible organic materials, and (less efficiently) pathogens from sewage. Tertiary treatment, which can be physical, chemical, or biological, is ordinarily required to remove nitrogen and phosphorus. These materials otherwise would fertilize and enhance plankton blooms in receiving waters. Sludge from primary, secondary, and tertiary treatment presents corollary problems of treatment and disposal. Problems with disposal or reclamation of either effluent or sludge from domestic wastes are minimal. Both treatment and disposal become increasingly complicated and expensive when hazardous or toxic organic or inorganic materials are discharged by either manufacturers or users. Mercury, for instance, can enter the system from industrial wastes, from household use of mercurochrome, or from dental amalgams.

Sources of wastes from the New York metropolitan area include raw sewage discharged to the Hudson and East Rivers, combined sewer overflows, sewage treatment plant effluents, urban runoff, industries (mostly in northern New Jersey), and sewage sludge. These are discussed in turn.

Raw Sewage

Approximately 6.6 m^3/s (150 mgd) of raw sewage is discharged into the Hudson River from western Manhattan. Another 2.2 m^3/sec (50 mgd) is discharged from Brooklyn into Upper Bay. These discharges pose obvious aesthetic insults (from recognizable fecal and plastic materials) and pose possible public health risks. Sharply lowered dissolved oxygen levels near discharge points quickly merge with generally reduced levels in the estuary. These levels are associated with dispersed organic loadings from the drainage basin, urban runoff, and waste treatment plants.

Combined Sewer Overflows

Raw sewage is discharged from approximately 200 discharge points whenever total sewage plus stormwater flows exceeds two times peak dry weather flow. Floating materials of obvious sewage origin, as well as oil and grease from household garbage grinders and other sources, create visible pollution. Pathogens, organic loadings, and toxic materials contribute to pollution of the estuary and Bight. Overflows occur mostly during the spring months when river flows are high. Technological approaches to minimizing the impacts of combined sewer overflows include screening and the separate disposal of floatables, chlorination of pathogens (which is generally ineffective because of the high chlorine demand of suspended organic solids), storage and subsequent return to sewers of flows over two times the dry weather flows, and source separation of domestic and industrial waste components.

Sewage Treatment Plant Effluents

Figure 2 shows locations of 28 major municipal treatment plant discharges in the New York metropolitan area. Table 2 summarizes discharge sources, quantities, treatment, and year of initial construction. Estimated effects of present discharges on dissolved oxygen are shown in Figure 3. Also shown are estimated improvements expected when presently authorized construction is completed ("baseline"), when the Manhattan and Brooklyn raw sewage discharges are treated and bypass regulators receive adequate maintenance ("secondary"), and when all sewage and 90% of storm water overflows receive secondary plus single-stage lime treatment followed by filtration, carbon adsorption, and disinfection ("zero discharge"). Estimated costs in 1980 dollars of the three levels of treatment are listed in Table 3.

Seventy-five percent of conventional construction costs (82% for innovative technology) are borne by the federal government, 12.5% by the State of New York, and the balance by the City. The City has been unable to raise its share of construction costs on schedule, so there have been construction delays. It also

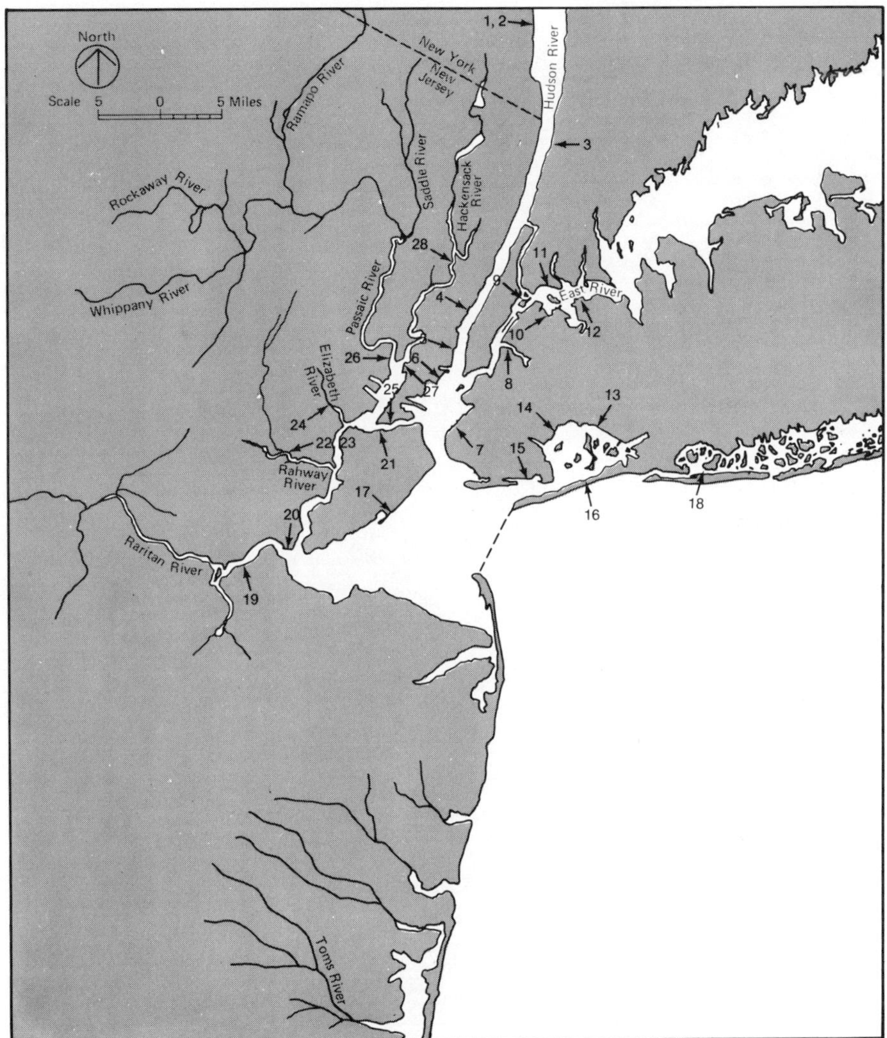

Figure 2. Major municipal discharges in the New York Bight area. From Mueller and Anderson (1978).

has been unable to raise adequate operation and maintenance funds to maintain existing sewers and treatment facilities. This has resulted in recommendations for federal assistance in operation and maintenance funding. Such funding for sewerage facilities is particularly vulnerable during periods of declining or steady-state economic conditions, and the funding may be diverted to more visible activities. The resulting lack of maintenance activities, staffing, supervision, and quality control is revealed in effluent quality that falls short of design assumptions (U.S. Environmental Protection Agency, 1977). The above costs do not include those to either government or industry for control of toxic or hazardous substances now discharged into the sewerage system. Although not

Table 2. Major municipal discharges. Locations of discharges shown on Figure 2. Treatment type: P = primary, MA = modified aeration, S = secondary. After Mueller and Anderson (1978), Hydroscience (1979), and Interstate Sanitation Commission (1979).

Location	Source	Year	Flow[1] m^3/s	Flow[1] mgd	Estimated percent industrial	Treatment[2]	Tributary population, 1975
1	Orangetown SD #2	–	0.66	15	–	S	–
2	Rockland Co. Dist. #1	–	0.96	22	0.7	S	–
3	Yonkers Joint Treatment	–	4.1	92	6.6	S	–
4	West New York	1953	0.44	10	–	P	–
5	Hoboken	1958	0.74	17	40	P	–
6	Jersey City–East	–	1.6	37	–	P	–
7	New York City–Owls Head	1952	4.4	100	1.3	MA	720,800
8	–Newtown Creek	1967	13	300	6.8	MA	968,300
9	–Wards Island	1936	11	245	1.0	S	1,278,400
10	–Bowery Bay	1942	5.6	128	6.8	S	698,500
11	–Hunts Point	1952	7.0	160	0.9	S	717,200
12	–Tallmans Island	1938	3.1	70	2.6	S	405,000
13	–Jamaica	1943	4.4	100	2.1	S	573,500
14	–26th Ward	1945	3.5	80	1.6	S	338,200
15	–Coney Island	1935	4.4	100	0.3	MA	692,900
16	–Rockaway	1952	0.88	20	0.1	S	118,700
17	–Oakwood Beach	1954	0.88	20	0.01	S	156,600
18	Bay Park (Pearsall's Hassock)	–	3.0	68	9.8	S	–
19	Middlesex Co. Sanitary Auth.	1958	5.3	120	50	S	–
20	Perth Amboy	–	0.31	7	46	P	–
21	NYC–Port Richmond	1953	1.6	36	14	P	171,600
22	Linden–Roselle Joint Meeting	–	1.1	26	30	S	–
23	Rahway Valley	–	1.5	35	26	S	–
24	Essex–Union Joint Meeting	–	3.3	75	17	S	–
25	Bayonne	1954	0.74	17	22	P	–
26	Passaic Valley	1924	13	300	39	P	–
27	Jersey City–West	–	1.1	24	24	P	–
28	Bergen Co. Sanitary Auth.	–	2.8	64	20	S	–
29	Atlantic City	–	0.70	16	1.2	P	–

[1] Not including 6.0 m^3/s (137 mgd) primary and secondary effluent from 138 smaller communities or 8.8 m^3/s (200 mgd) raw sewage from Manhattan and Brooklyn.

[2] 1980 baseline treatment for New York City and most New Jersey discharges.

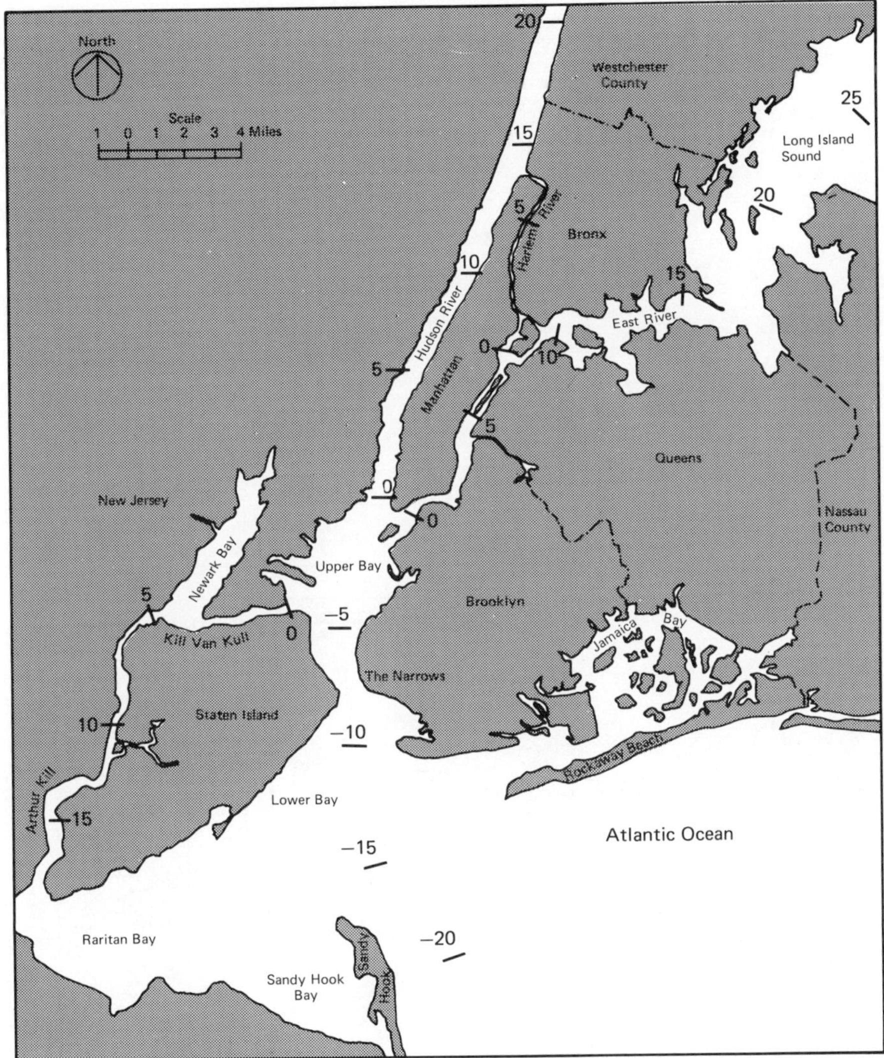

Figure 3a. Location of transects shown in Figure 3b. Distance shown in miles; 1 mi = 1.6 km. After Hazen and Sawyer (1978).

yet achieved, mandated source control measures eventually will reduce discharges of these materials to the New York Bight.

In spite of large investments made in conventional sewerage systems since 1950 and scheduled through the year 2000, total waste loadings to the Bight as measured by biological oxygen demand (BOD) and suspended solids have remained essentially constant and are likely to continue to do so (J.A. Mueller, Manhattan College, personal communication). Estimated annualized capital and operation and maintenance costs listed in Table 3 are for comparison of present treatment options. Although federal and state portions of construction costs are

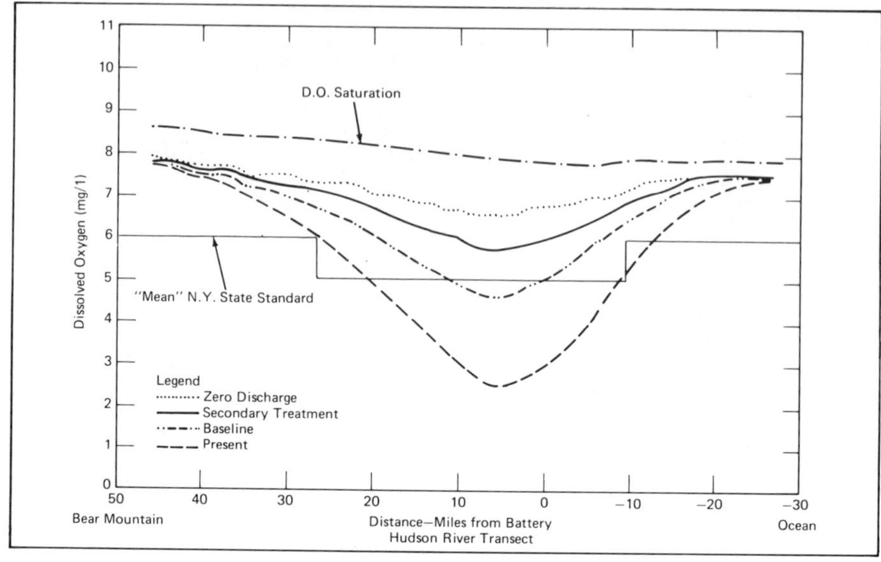

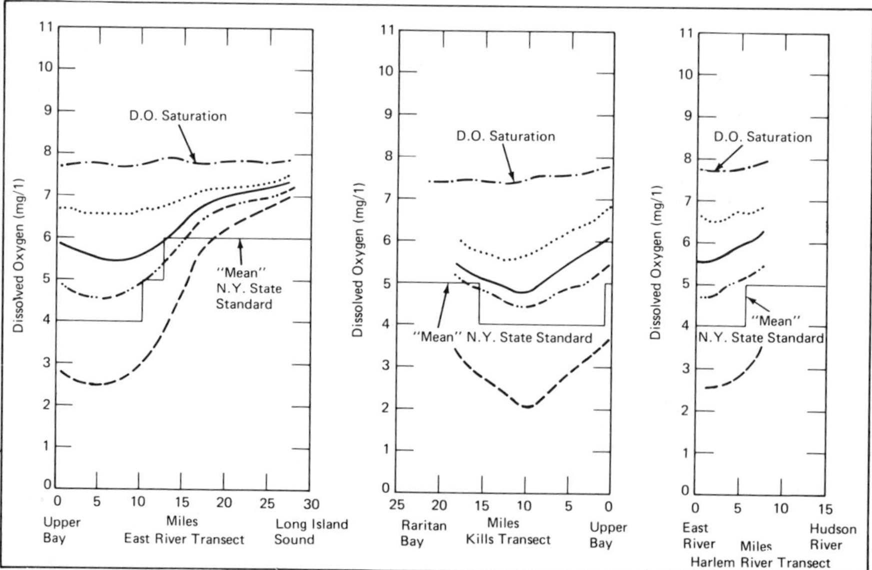

Figure 3b. Predicted dissolved oxygen concentrations under zero discharge, secondary treatment, baseline, and present conditions. Data are from the lower layer of Hudson River waters and from vertically integrated stations on other transects. All computations are time-averaged to include both wet and dry weather loadings. After Hazen and Sawyer (1978).

welcomed as a source of local employment, they represent tradeoffs with other regional and local uses of these funds and carry with them a mortgage on the future because of their requirement for operation and maintenance funds.

Table 3. Estimated capital and operating costs for alternative treatment plans. Costs estimated in millions of 1980 dollars. After Hazen and Sawyer (1978), adjusted by ratio of Engineering News Record 1980/1975 indices of 3200/2209 = 1.45.

	Baseline	Secondary	Zero discharge
New York City			
Initial construction	1,885	3,770	11,600
Annual capital cost[1]	190	380	1,160
Annual operation and maintenance	90	100	300
New Jersey and other			
Initial construction	1,160	1,305	3,625
Annual capital cost[1]	120	130	360
Annual operation and maintenance	40	60	140
Total annual costs			
Capital plus operation and maintenance	440	670	1,960

[1] Assuming 10% annual capital recovery factor.

A minimum estimate of average annual costs in 1980 dollars of the conventional approach (Table 3, "baseline") to New York metropolitan area sewerage is $1,050 million (Gunnerson, in press). This is based on an estimated $2 billion in construction costs since 1950 and another $3 billion required by the year 2000, an annual capital recovery factor of 10%, annual operation and maintenance costs at 30% of capital costs, and $400 million per year for the cost of 95-115 liters of water per capita per day (lcd) for flushing the sanitary portion of the wastes (U.S. Army Corps of Engineers, 1977). Continuing degradation of receiving water quality in spite of these expenditures, and increasing competition for fiscal and other resources clearly indicate a need to examine alternative institutions and technologies for New York metropolitan area waste management.

Sewage Sludge

In 1979, 5.4 million metric tons of sewage sludge with about 5% solids were dumped into the New York Bight. Sludge production will increase to 8 million metric tons by 1981, when the mandated ban on further dumping is scheduled to go into effect. Sludge dumping contributes to marked environmental and ecological effects over 75 km^2 of the 28,000 km^2 New York Bight. It accounts for only 5-15% of the total pollution load to the Bight, depending upon the parameter being measured. Unless a cessation of sludge dumping is accompanied by significant improvements in dredge spoil and other waste disposal practices, recovery or improvement of the 75 km^2 almost certainly will be masked by the effects of other waste sources. Beaches will be unaffected. While elimination of sludge dumping is a readily attainable step, uncertainties as to its cost-effectiveness relative to those of alternatives have led to proposals for extending the timing of rate of compliance. The reported costs of sludge dumping vary greatly. Table 4 contrasts ocean disposal costs at three potential sites in the New York Bight region.

Probable responses to judicial or legislative changes permitting alternative dumping locations and/or phased reductions in quantities of sewage sludge dumped into the New York Bight vary widely. Regulatory agency interests, supported by judicial findings, take a negative view of proposed changes because (1) non-enforcement would penalize those who previously had complied with the regulations; (2) feasible environmentally acceptable land-based alternatives exist; (3) the marine environment would continue to be impacted; and (4) valuable resources (*e.g.*, nutrients and energy) would continue to be wasted.

Conservationist and resource management interests also respond negatively. They insist that ecological impacts of sludge dumping are too elusive and too important to take chances with. These groups further maintain that feasible alternatives that utilize or recycle the resources are available.

The scientific view of phased reductions notes (1) that the contribution of sludge to Bight pollution is relatively small compared to other inputs; (2) that significant recovery of other damaged ecosystems has occurred in one to five years, but that recovery of the Bight from sludge dumping may take longer because of other inputs; and (3) that further research and application of Southern California Coastal Water Research Project findings should lead to better prediction of benefits and of changes in biological impacts of ocean dumping (*e.g.*, Mearns and Word, this volume).

The technological view of graduated reductions is positive, since such a course would permit application of a multi-media approach to sludge management that considers the assimilative capacity of air, land, and water. This approach offers considerable operational flexibility and potential cost-savings. Design and specifications preparation, equipment delivery, and construction would be more efficient. Land use and allocation of space to waste disposal would be improved, and the risk of universal application of a wrong technology would be minimized. A multifaceted approach also would be likely to produce a least-cost combination of technologies whose capacity would most closely match demand.

Political response to the possibility of graduated reductions is positive because this course would allow for the development of acceptable compromises among constituencies.

TECHNOLOGY AND POLICY OPTIONS

Inflation and other economic pressures have increased competition for public resources and decreased the margins for error in their allocation. Needs for the water supply and wastewater disposal sectors may be projected from planned and current operation and construction costs. These are summarized for the New York area in Table 3. Benefits to the area's receiving waters from expenditures projected for the period 1950-2000 are likely to be limited because waste loadings to the New York Bight essentially have remained at 1950 levels. This

Table 4. Cost of sewage sludge dumping at three possible sites in the New York Bight region. Costs presented per metric ton (per short ton), dry wt.

	"12-mile" site (apex)	"65-mile" site (mid-shelf)	"106-mile" site (beyond shelf)
EPA[1]	\$39 (\$35)	\$154 (\$140)	\$308 (\$280)
Middlesex County, NJ[2]	\$55 (\$50)	–	\$132 (\$120)

[1] U.S. Environmental Protection Agency (1979).
[2] Based on the 1980 contract for disposal of approximately 907,000 metric tons (1,000,000 short tons), wet wt., of Middlesex County sludge.

emphasizes the need to reappraise the technological, institutional, and policy alternatives for dealing with domestic wastes.

Water supply service levels vary from 2 lcd, needed to sustain life, to 20 lcd, at which all health benefits of a safe domestic water supply are realized (Kalbermatten *et al.*, in press), to 337 lcd, which provides for a variety of amenities and conveniences in New York City (U.S. Army Corps of Engineers, 1977). Waste disposal service level options include kitchen garbage grinders, separate systems for disposal of garbage, mandatory flow-limiting devices, and dual waste disposal systems for new construction. Dual systems would permit minimum quantities of water from toilets (black water) to be transported and treated at minimum cost, possibly with methane recovery, in appropriately scaled facilities. Water from kitchens, baths, and laundries (gray water) can be discharged to surface drainage systems and thus minimize the present sanitary problems of storm water overflows. Alternatively, grey water could be reused either routinely or during drought without hazard (Bollman, 1979).

Related technological choices that control the amounts of toxic or hazardous substances discharged to the Bight include remedial dredging to remove contaminated bottom sediments from the harbor, estuaries, and Hudson River. Recent experience in Japan has shown potential cost-effectiveness of remedial dredging (Peterson and Randolph, 1979). Studies in the United States reveal costs of \$60 to \$200 million for the previously mentioned removal of PCB-contaminated sediments in the upper Hudson River, to \$1 billion for dredging kepone from the James River, or for dredging highly contaminated sediments from New York harbor and the lower Hudson River (Carcich and Tofflemire, 1979).

In any evaluation of alternatives, questions of scale, redundancy, availability of adequately trained operating staff, and resulting system reliability and resilience should be considered. For example, control of toxic components of industrial wastes can take advantage of specialized professional competence available at the source but not at municipal treatment plants. Limitations of the latter are revealed clearly in a recent EPA document which shows that, in practice, treatment plant performance does not conform to design criteria (U.S.

Environmental Protection Agency, 1977). Examination of operating data from 52 municipal primary treatment plants and 130 secondary treatment plants from around the country are summarized in Table 5. The reported ranges of 0% to 99% BOD and suspended solids removal indicate problems. Low values reflect operating and supervision problems; high values suggest analytical and supervision problems. Even greater variance was reported for metals and other trace constituents. Certainly, operation and maintenance problems are not peculiar to New York. The solution lies in allocating greater resources to plant operation and maintenance, and to recruiting and retaining qualified staff.

Policy options include rejection or acceptance of industrial wastes in the municipal sewerage system, mandatory industrial recirculation, dual water supply systems, increasing block tariffs of metered water supplies to individual households, and charging full marginal costs for services rendered. Preliminary assessments for New York City have been made of some of these options by the Temporary State Commission (1975), U.S. Army Corps of Engineers (1977), Hazen and Sawyer (1978), and U.S. Environmental Protection Agency (1979). Their findings were generally cautious with respect to residential water metering, water use restrictions, pricing, distribution system leakage detection and control, and educational programs. They were optimistic with respect to industrial recirculation, mostly because it was an effective way to comply with the mandates of the Clean Water Act (PL 92-500).

The preceding discussion of options is illustrative rather than exhaustive and is intended to show the need and possible approaches to monitoring and reassessment of the conventional engineering and fiscal solutions to municipal waste problems in the New York metropolitan area. It is recommended that appropriate studies be undertaken to determine economic costs and benefits of alternative technological, institutional, and fiscal options (such as matching capacity to demand), and that corresponding policy options be clearly identified (such as maximizing short-term employment, maximizing resource conservation, choosing between equity and efficiency, and minimizing long-term operation and maintenance costs). In making these assessments, it is necessary to remember that it costs less to supply water than it does to dispose of it, that there is a variety of options whose costs and benefits can be compared, and that consideration and development of public willingness to select and pay for these benefits is essential.

AFTERWORD[3]

On November 2, 1981 Judge Abraham D. Sofaer of the U.S. District Court, Southern District of New York issued his Final Judgment in *The City of New York vs. the United States Environmental Protection Agency* (80 Civ. 1677). This decision will play an important role in determining the future of ocean

[3] Prepared by Peter W. Anderson, U.S. Environmental Protection Agency, Region II.

Table 5. Percentage removals of waste constituents by primary, trickling filter, and activated sludge plants. BOD–biological oxygen demand; COD–chemical oxygen demand; P-total–total phosphorus; SS–suspended solids; TKN–total Kjeldahl nitrogen; TOC–total organic carbon. After U.S. Environmental Protection Agency (1977).

Parameter	Primary plants				Trickling filter plants				Activated sludge plants			
	Mean	Standard deviation	Max. min.	No. of plants	Mean	Standard deviation	Max. min.	No. of plants	Mean	Standard deviation	Max. min.	No. of plants
SS	51	18	92/17	47	75	19	97/20	66	75	22	99/9	62
BOD	30	22	89/0	52	77	18	96/5	60	84	15	99/18	65
TOC	24	19	56/0	30	64	18	84/8	23	73	12	89/42	13
COD	26	–	82/0	18	71	–	95/34	36	75	–	94/24	40
P-total	13	8	24/0	7	26	22	99/0	24	42	25	92/0	36
TKN	22	20	60/0	7	50	27	94/7	20	34	26	92/5	11
NH_3	20	16	64/0	42	41	30	99/0	48	49	31	99/4	47
Phenol	38	–	50/25	2	50	28	85/0	12	69	31	98/0	16
Metals												
Cd	8	17	76/0	31	20	25	75/0	35	17	27	88/0	44
Cr	26	26	80/0	36	37	30	99/0	48	46	34	98/0	54
Pb	24	26	88/0	34	37	31	93/0	41	39	32	95/0	49
Hg	27	29	75/0	21	30	23	67/0	20	39	32	99/0	34
Cu	26	24	77/0	44	54	24	95/0	49	57	24	95/0	63
Ni	6	18	92/0	28	21	23	86/0	32	20	21	80/0	44
Zn	31	22	88/0	38	46	22	89/0	52	58	25	99/0	58
Fe	40	22	89/0	27	50	26	90/0	30	63	27	98/8	35
Mn	15	20	81/0	16	31	23	72/0	21	38	32	93/0	19

dumping of sewage sludge. The suit stemmed from New York City's contention that EPA's Ocean Dumping Criteria were more restrictive than required by the Marine Protection, Research, and Sanctuaries Act (MPRSA). This Act prohibits ocean dumping of wastes that will "unreasonably degrade" the marine environment and specifies nine factors for evaluating materials proposed for such disposal (Section 102). A 1977 amendment to MPRSA mandates an end to ocean dumping of harmful sewage sludge by December 31, 1981.

The Ocean Dumping Criteria challenged in the suit were promulgated in accordance with MPRSA Section 102 to assess the suitability of permit applications for the dumping of specific wastes. Under the permitting procedure, if a waste (municipal sludge, industrial waste, or dredged material) does not violate a series of "environmental impact prohibitions, limits, and conditions" (including bioassay tests) (40 CFR 227, Subpart B), it is presumed not to "unreasonably degrade" the marine environment. The permit application then may be evaluated with regard to the need for ocean dumping, the availability of alternatives to this form of disposal, and the impact on other uses of the ocean.

Prior to the end of 1981, if a waste did not comply with environmental impact criteria, an applicant could be granted an interim permit if it previously had dumped its wastes into the ocean, if alternatives immediately available were less environmentally acceptable than ocean dumping, and if the applicant provided for the implementation of land-based disposal or compliance with the environmental criteria by December 31, 1981. All sewage sludge dumpers in the New York - New Jersey metropolitan area, including New York City, were required to phase out ocean dumping by the end of 1981, because EPA had determined in 1976 that environmentally acceptable land-based alternatives exist for the region. New York City, however, contended that the adverse consequences and costs of land-based disposal would greatly exceed the effects of continued dumping in the heavily polluted New York Bight. On this basis, the City urged EPA to renew its interim dumping permit past the 1981 date. When EPA refused to hear the City's contentions, the City brought suit to compel EPA to consider its evidence.

Judge Sofaer's decision stated that the 1981 deadline shall remain in effect, but only for ocean dumping that may "unreasonably degrade" the marine environment at a particular dumpsite, in view of relevant statutory criteria. The decision further indicated that the environmental risks of ocean dumping must be balanced against those associated with land-based alternatives. In its Final Judgment, the Court ordered EPA to:

(1) Revise its regulations to allow review of "all relevant statutory criteria" (*i.e.*, balancing). EPA may use existing regulations in evaluating applications for dumping permits, as long as such reviews are consistent with applicable law. EPA may not establish a conclusive presumption of "unreasonable degradation" with regard to a waste solely upon the waste's non-compliance with environmental impact criteria.

(2) Evaluate the City's pending application in accordance with item 1 above.

(3) Take no enforcement action against the City with respect to its ocean

dumping activities, as long as such dumping is in accordance with the conditions of the City's last permit.

(4) Allow the City to continue dumping at the 12-mile sewage sludge site, pending action on the City's petition to redesignate the site (site designation expired December 31, 1981). If EPA declines to redesignate the site, the City (unless a court intervenes) will take all steps necessary for a rapid and orderly removal of dumping to a location designated by EPA.

The Final Judgment pertains only to New York City and will not be appealed by EPA. Other cases involving two sewerage authorities in New York currently are being litigated. Since these cases are before Judge Sofaer, it is assumed that final judgments will be similar to that for the New York City suit. In May 1982, judgments were issued in cases involving six sewerage authorities in New Jersey. Negotiated Consent Judgments for these cases, issued by Judge H. Lee Sarokin of the District of New Jersey, contained provisions similar to those in the New York City judgment. Of particular interest is that intervention in these cases was sought but denied to two Congressmen, the National Wildlife Federation, and other environmental groups. These groups had raised the issue that Judge Sofaer's decision did not take into account provisions of the London Ocean Dumping Convention that prohibit the dumping of wastes (including sewage sludges) containing mercury and cadmium (as elements or compounds), organohalogens, and oil as other than trace contaminants.

REFERENCES

Bollman, F.H. 1979. Impact of Severe Drought in Marin County, California. Bulletin 208, California Department of Water Resources, Sacramento, CA. 46 pp.

Carcich, I.G. and T.J. Tofflemire. 1979. Distribution and concentration of PCB in the Hudson River and associated management problems. *In:* Management of Bottom Sediments Containing Toxic Substances, S.A. Peterson and K.K. Randolph (eds.), EPA 600/3-79-102, U.S. Environmental Protection Agency, Environmental Research Laboratory, Corvallis, OR. pp. 1-23.

Gunnerson, C.G. 1981. The New York Bight ecosystem. *In:* Marine Environmental Pollution, Vol. 2, Dumping and Mining, R.A. Geyer (ed.), Elsevier Scientific Publishing Co., Amsterdam. pp. 313-378.

Gunnerson, C.G. In press. Waste disposal in the New York metropolitan area. *In:* The Ocean Disposal of Municipal Wastes: the Impact on Estuarine and Coastal Waters, E.P. Myers (ed.), MIT Press, Cambridge, MA.

Hazen and Sawyer. 1978. Section 208 Areawide Waste Treatment Management Planning Program, Draft Final Report to City of New York. Hazen and Sawyer, Engineers, New York.

Hydroscience. 1979. Tasks 512 and 522. Baseline and Alternatives. *In:* The City of New York Section 208 Areawide Waste Treatment Management Plan. New York City Department of Environmental Protection, New York.

Interstate Sanitation Commission. 1979. Report on the Water Pollution Control Activities. Interstate Sanitation Commission, New York. 112 pp.

Kalbermatten, J.M., D.S. Julius, and C.G. Gunnerson. In press. Appropriate Sanitation Alternatives: A Technical and Economic Appraisal. Johns Hopkins University Press, Baltimore, MD.

Mahoney, J.B. and J.J.A. McLaughlin. 1977. The association of phytoplankton blooms in lower New York Bay with hypertrophication. J. Exp. Mar. Biol. Ecol. 28: 53-65.

Malone, T.C. 1977. Environmental regulation of phytoplankton primary productivity in the lower Hudson Estuary. Estuarine Coastal Mar. Sci. 5: 157-171.

Mearns, A.J. and J.Q. Word. This volume. Forecasting effects of sewage solids on marine benthic communities. pp. 495-512.

Mueller, J.A. and A.R. Anderson. 1978. Industrial Wastes. MESA New York Bight Atlas Monograph 30, New York Sea Grant Institute, Albany, NY. 39 pp.

Murphy, E.B., K.A. Sterdinger, B.S. Roberts, J. Williams, and J.W. Jolley, Jr. 1975. An explanation for the Florida east coast *Gymnodinium breve* red tide of November 1972. Limnol. Oceanogr. 20: 481-486.

Peterson, S.A. and K.K. Randolph (eds.). 1979. Management of bottom sediments containing toxic substances. EPA 600/3-79-102. U.S. Environmental Protection Agency, Environmental Research Laboratory, Corvallis, OR. 391 pp.

Sindermann, C.J., S.C. Esser, E. Gould, B.B. McCain, J.L. McHugh, R.P. Morgan II, R.A. Murchelano, M.J. Sherwood, and P.R. Spitzer. This volume. Effects of pollutants on fishes. pp. 23-38.

Swanson, R.L., G.M. Hansler, and J. Marotta. 1977. Long Island Beach Pollution: June 1976. NOAA/MESA Special Report. 75 pp.

Swanson, R.L. and C.J. Sindermann (eds.). 1979. Oxygen Depletion and Associated Benthic Mortalities in the New York Bight, 1976. NOAA Prof. Pap. 11. 345 pp.

Temporary State Commission on the Water Supply Needs of Southeastern New York. 1975. Water for Tomorrow. State of New York, Albany, NY. 55 pp.

U.S. Army Corps of Engineers. 1977. Northeastern United States Water Supply Study, Summary Report, North Atlantic Division, U.S. Army Corps of Engineers, New York. 94 pp.

U.S. Environmental Protection Agency. 1977. Federal Guidelines, State and Local Pretreatment Programs, Vol. 1. EPA 430/9-76-017a, U.S. Environmental Protection Agency, Section F, Washington, DC. 27 pp.

U.S. Environmental Protection Agency. 1979. Computer Summary of Construction Grant Programs to 1979. U.S. Environmental Protection Agency, Region II, New York.

MANAGEMENT OF DREDGED MATERIAL

R.B. Gordon[1]
W.F. Bohlen
H.J. Bokuniewicz
M. de Picciotto
J. Johnson
K.S. Kamlet
T.F. McKinney
J.R. Schubel
D.J. Suszkowski
T.D. Wright

Abstract. New York harbor is in a naturally shallow estuary. Without dredging, maximum water depth in the harbor would be approximately 6 m, too shallow for most vessels engaged in international trade. Sediment reaches the harbor from both the sea and the rivers. Over a period of many years, appropriate engineering works could reduce the amount of sediment that accumulates in the harbor. However, periodic maintenance dredging will be required as long as New York harbor remains a major seaport.

Much of the muddy sediment that is deposited in the harbor is contaminated with a range of undesirable and toxic substances; more contamination is added while the sediment remains in the harbor. The most serious problem associated with maintenance dredging is disposal of contaminated sediments. For many years, the accepted practice has been to discharge such material at an open-water site, the Mud Dump, off the entrance to New York harbor. As dredging continues, the size of the Mud Dump will increase. Concern about the ecological consequences of continued use of the Mud Dump has lead to exploration of alternative disposal methods. Most are more expensive; some, such as containerized ocean disposal, are enormously more expensive. No one scheme will solve the problem of disposal of contaminated dredged material. Nevertheless, this paper outlines the main elements of a regional dredging plan that will provide for continued economic operation of the Port while minimizing environmental damage.

INTRODUCTION

The waters of New York harbor are naturally shallow because of the accumulation of sediment from rivers, storm drains, and shore erosion; biological production of particulates and solid debris; and sediment carried into the harbor from the New York Bight by estuarine circulation. Dredging is required to maintain channels deep enough for safe navigation of ships. The need for dredging may be reduced somewhat in the future by measures designed to control siltation, but it will be necessary to dispose of large quantities of dredged material for many years. As long as the Port continues operation with seagoing vessels, dredged channels will be required. Therefore, an important environmental goal is to reduce the impact of required dredging and disposal on the harbor, Bight, and surrounding land.

Dredging-related activities in the Port of New York produce physical disturbances at the site where sediment is excavated by the dredge and at the site

[1] Chairman, Panel on the Management of Dredged Material.

where the dredged material is placed. The environmental impacts of these changes generally are well documented and understood; most agree that the physical impacts are not large. However, much of the material dredged has acquired a burden of undesirable or toxic substances prior to entering the harbor or during its stay in the waters of the harbor. Disposal of contaminated dredged material may disperse pollutants to areas that would not have been affected had there been no dredging.

Those who are familiar with the issue will agree that some of the material dredged from the Port of New York is clean, and that only physical effects arising from its disposal need be considered. They also will agree that another fraction may be so contaminated that it must be disposed in sites where it is contained. The greatest part (perhaps as much as 80%) of the material dredged from the Port of New York falls neither in the clean nor obviously contaminated categories. The consequences of different modes of disposal of this material are subject to some dispute, and the amount of material that falls into this classification will likely change as criteria for evaluating the degree of contamination are adjusted.

This report discusses the problem of establishing criteria for the contamination of dredged material. In addition, suggestions are offered for reducing undesirable impacts of the disposal of this material. In cases where measures can be implemented at modest cost, their use may be desirable, even in the absence of clearly documented deleterious effects of disposal operations. However, other measures will be very costly and may have important effects on the competitive position of the Port. Use of such options will be considered only after a clearly defined need has been established.

Economic Importance of Dredging

In 1978, the Port of New York handled approximately 62 million metric tons (61 million long tons) of import/export ocean-borne general and bulk cargoes. This volume is exclusive of all coastwise and domestic waterborne traffic, which would add approximately 54 million metric tons (53 million long tons) to the total. Inter-harbor traffic volumes are excluded from these statistics. Import/export cargoes moving through the Port in 1978 were valued at $37 million. The value of import/export cargoes handled by the Port of New York in this year was the largest of all American ports of entry. A total of 7,620 oceangoing vessels called at the Port of New York in 1978 (Port Authority of New York and New Jersey, 1979).

The import/export industry in the Port of New York supported approximately 174,000 jobs in 1978, with an estimated annual payroll of $3.6 billion. This segment of Port industry, including intercoastal, domestic, and inter-harbor traffic, represents a regional input to the economy of $7.25 billion. This is approximately 10% of the metropolitan area's contribution to the gross national product (Port Authority of New York and New Jersey, 1979).

Dredging Methods

Dredging in New York harbor is accomplished by mechanical and hydraulic means. The most widely used mechanical dredge is the bucket or clamshell dredge. While in operation, these dredges are immobile and anchored to the harbor floor with piles. Material is lifted off the bottom in buckets (Figure 1) and placed either in dump scows or behind bulkheads.

Two types of hydraulic dredges are used within New York harbor (Suszkowski, 1977). Fixed hydraulic dredges vacuum sediments off the bottom and pump them in slurry form through a pipeline to a dump scow or an adjacent upland area (Figure 2). Hopper dredges are mobile devices (Figure 3a); arms located on the sides of the dredge are dragged along the harbor floor and bottom (Figure 3b). Dredged material is stored in on-board hoppers for subsequent disposal. Hopper dredges have the advantage of being able to work in areas with heavy traffic and high current speeds.

Disposal Sites and Volumes Dredged

Since 1888, open-water disposal sites near the entrance to New York harbor have received materials dredged from access channels and ship berths. The disposal sites have been moved four times during this period. The present site

Figure 1. Mechanical (bucket) dredge in operation. (Courtesy U.S. Army Engineer Waterways Experiment Station.)

Figure 2. Hydraulic cutterhead dredge with dredging apparatus exposed. (Courtesy U.S. Army Corps of Engineers.)

(Mud Dump, Figure 4) is located approximately 10 km (6 mi) east of Highlands, New Jersey, and 16 km (10 mi) south of Rockaway Beach, New York. The average volume of dredged materials excavated in the Port has risen from approximately 4.6 million m^3/yr (6 million yd^3/yr) in the 1930's to approximately 6-7.6 million m^3/yr (8-10 million yd^3/yr) in the 1970's (Suszkowski, 1977). Future annual volumes are expected to remain in this range

Figure 3a. Small hopper dredge. (Courtesy U.S. Army Corps of Engineers.)

Figure 3b. Vacuum arm of a hopper dredge. (Courtesy U.S. Army Corps of Engineers.)

unless major deepening and widening of existing channels or the construction of substantial new waterfront facilities are undertaken. The latter is not envisioned at present.

Of the volume currently dredged, approximately 4-5 million m^3 (5-7 million yd^3) result from channel maintenance dredging by the U.S. Army Corps of Engineers (COE). The remainder results from maintenance dredging of ship berths by owners and users. Port Authority dredging, which is second in volume to that of the COE, averages 600,000 m^3/yr (800,000 yd^3/yr) (New York District, COE, unpublished data).

CHARACTERIZING DREDGED MATERIALS

Proper management of dredging and dredged material disposal requires adequate characterization of the material, particularly its physical and chemical properties, and its polluting potential. Characterization of physical properties is relatively straightforward and non-controversial. Diagnostic characterization of chemical properties, and particularly of the polluting potential of dredged material, is much more difficult and controversial.

Public Laws 92-532 (Marine Protection, Research, and Sanctuaries Act of 1972) and 92-500 (Federal Water Pollution Control Act Amendments of 1972)

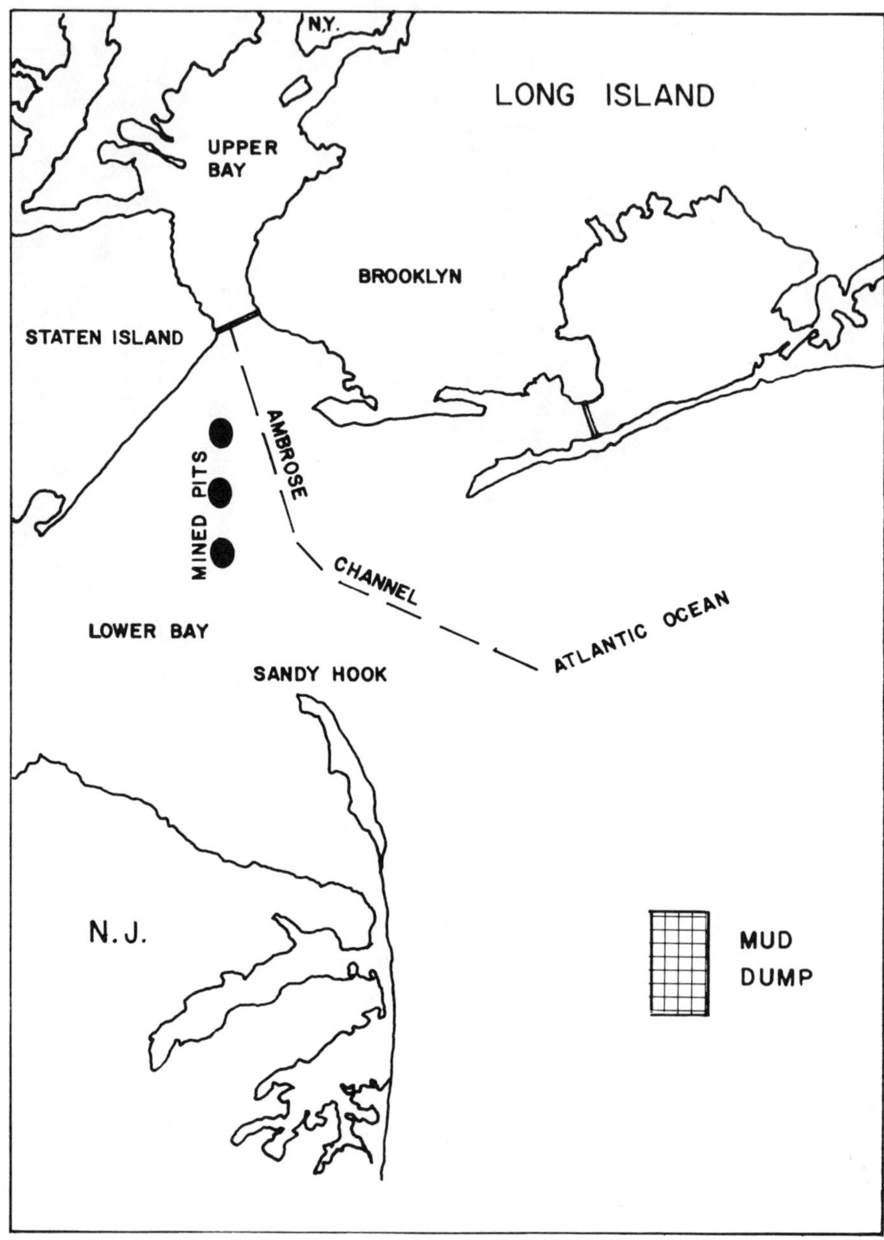

Figure 4. New York harbor and present dredged material disposal site (Mud Dump).

require the U.S. Environmental Protection Agency (EPA) to consult with the COE in developing guidelines and criteria for regulating discharges of dredged and fill material. Based upon Section 103 of the Marine Protection, Research, and Sanctuaries Act of 1972, the New York District, COE, has administered its ocean dumping program in accordance with EPA ocean dumping criteria (Environmental Protection Agency/Corps of Engineers Technical Committee on Criteria for Dredged and Fill Material, 1977). These criteria specify a testing program intended to ensure that dumping will not have unacceptable, adverse impacts on the marine environment and its living resources. Tests subject several species of marine animals to proposed dredged material to determine whether mortality or toxicant bioaccumulation will occur as a result of exposure.

Bioassay testing requires that the dredged material be separated into three phases: liquid, suspended particulate, and solid. Liquid and suspended particulate phase bioassays are used to evaluate water column impacts; solid phase bioassays are used to evaluate impacts associated with material deposited at the proposed dump site. All three tests are designed to evaluate short-term, unacceptable adverse impacts. Bioaccumulation testing, used primarily in association with the solid phase, is conducted to assess the potential for long-term accumulation of contaminants in organisms residing in the vicinity of the proposed dump site.

Bioassay and bioaccumulation testing of dredged material has been carried out only since 1977. When first implemented, the introduction of bioassay test results was a subject of considerable debate. However, the results of over 100 sets of bioassays submitted to the New York District, COE, have shown that the majority of New York harbor sediments are generally non-toxic (Suszkowski and Mansky, in press). The results of laboratory bioaccumulation testing have been used to predict the potential for chronic impacts to organisms residing in the vicinity of the Mud Dump (Engler, 1980). Although bioaccumulation studies routinely are being used in the permitting process, shortcomings remain, and further refinement of this technique will be required. How to extrapolate long-term, environmentally significant effects from statistically significant laboratory results remains unclear. Additional research will be required to develop the combination of laboratory and field studies to accomplish this end. (For further discussion of issues relating to the evaluation of dredged material and a description of the recently utilized "matrix approach", see Kamlet, in press.)

The COE estimates that approximately 10% of the material dredged from greater New York harbor is clean sand, and that less than 10% is so highly contaminated that it is not suitable for open-water disposal. If regulatory criteria are made more stringent, the percentage of material not acceptable for open-water ocean disposal will increase. Most material dredged from New York harbor comes from repeated maintenance dredging of authorized projects. The characteristics of such material, including pollutant levels, generally will not change substantially over a period of at least several years. A program should be initiated to characterize sediments accumulating in the approved dredging projects in the Port. It is important to determine whether contaminant concentrations in channel deposits tend to be spatially uniform (horizontally and vertically), or whether there are scattered hot spots of contamination.

Special attention should be given to shoals that require frequent dredging. Distinguishing channels dominated by contaminated sediments from those dominated by uncontaminated sediments is a tractable problem; delineating areas of contaminated and uncontaminated sediments within channels also may be tractable. Ideally, disposal methods and locations should be matched to the properties and contamination status of dredged materials.

Alternative disposal sites should be identified; in addition, assessments should be made of how materials of differing quality would behave if placed at alternate sites, and what environmental effects would result. Economic analyses also should be made. Without these assessments, disposal options cannot be evaluated appropriately, and management will continue to be in response to crises. Proposed assessments should utilize the considerable body of information that exists. In addition, these assessments will require a sampling program to characterize potential dredged materials and their distribution, and also will require a series of pilot studies of disposal operations. The cost of such programs will be small compared to the cost of dredging and the cost of losses, should prolonged delays in dredging occur.

DISPOSAL OF HIGHLY CONTAMINATED MATERIALS

Alternatives to conventional disposal are needed for material not suitable for placement at open-water sites. For toxic materials, several containment alternatives are possible, including land containment, placement in subaqueous borrow pits, and construction of islands. Some materials also might be processed to reduce adverse properties before disposal. This section discusses each of these options as they might be applied to materials unsuitable for open-water disposal.

Sub-bottom Containment

Open-water disposal of contaminated dredged material might be accomplished by placing the material in depressions on the sea floor and covering it with clean sediment. Subaqueous borrow pits are one type of depression that might be used. In New York harbor several large holes from previous mining operations already exist. West of Ambrose Channel there are two small pits and one large pit that have a combined volume of about 19 million m^3 (25 million yd^3) (Figure 4). Over the next three years, it has been estimated that about 21 million m^3 (27 million yd^3) of sand may be needed for large construction projects in the New York metropolitan area (J. Marrotta, New York Office of General Services, personal communication). As a result, if borrow pits could be demonstrated to be acceptable disposal sites, the volumes of existing or potential subaqueous pits would be sufficiently large to contain much of the contaminated dredged material now being removed from the Port of New York. The technology to carry out disposal operations in a mined pit is available (Johanson *et al.*, 1976); dredged sediment can be placed accurately on the pit floor. Borrow pits also could be used as temporary repositories for sandy dredged material held for subsequent use, such as beach nourishment.

Although some questions remain to be addressed, several submarine pits in the floor of New York harbor appear suitable for the containment of dredged sediment. The slopes of pit sides and the roughness of pit floors should restrict the spread of dredged sediment along the harbor bottom during disposal. In addition, the pits act as natural traps for fine-grained sediment. Such conditions would enhance the probability of containing dredged material.

If a deposit of dredged mud were to be covered or capped successfully with clean sediment, the underlying mud would be protected from disturbances by storms and bioturbation (reworking by benthic organisms). The cover material could be either uncontaminated mud or sand. A sand cap would have the advantage of restoring the area as closely as possible to its original condition. Several technical problems require solution before a full-scale capping operation can be undertaken. For example, little is known about the possibility of erosion during storms. Toward this end, a pilot study has been initiated in two borrow pits in Lower Bay (Bokuniewicz, in press).

Engineering measures are available to enhance the success of covering operations. It may be beneficial to use small pits that can be filled and capped rapidly, to minimize the period during which the mud is exposed. In addition, placement of cover material through a pipeline that discharges a few feet above the contaminated deposit could reduce the likelihood of contaminant resuspension. Equipment capable of submerged discharge is not readily available, but alterations to existing equipment to add this capability should not be prohibitive.

A more expensive but geographically more flexible approach to sub-bottom containment is disposal of dredged material in areas confined by submerged dikes. However, the cost and environmental problems caused by dike construction probably would make this approach cost-effective only for very contaminated materials. To the Panel's knowledge, it has never been attempted. The hazards of such dike construction under heavy sea conditions have yet to be fully evaluated, but a study concerning the construction of a large diked containment island in Lower Bay was done by Howard, Needles, Tammen, and Bergendoff (1976).

Underwater disposal of contaminated dredged material in New York harbor could be carried out without modification of existing laws. Sediments that could be placed in pits or submerged dikes in the harbor would be subject to the guidelines of Section 404(b)(1) of the Clean Water Act, but would not come under the Ocean Dumping Act.

Island Creation

Island creation involves the containment of dredged material within confined areas in the protected waters of harbors. With proper controls (design, implementation, maintenance, and monitoring), this alternative could provide a disposal method for dredged material not acceptable for open-water disposal. Several sites have been studied in the New York harbor area for protected water containment. Some of these sites have been studied for the combined disposal of dredged material and garbage; others were considered for ultimate use as shoreline recreational facilities. The West Bank Containment area, once proposed for the west bank of Ambrose Channel in the vicinity of Hoffman and Swinburne Islands, would have a site capacity of 61 million m^3 (80 million yd^3) (Conner *et al.,* 1979). All dredged material removed from New York harbor for the next 20 years could be accommodated in this area (D. Suszkowski, New York District, COE, personal communication). If only contaminated materials were disposed of at this location, the life of the site would be extended further. Other potential sites within the New York harbor are likely to have significantly less capacity.

Possible negative impacts of such islands that require evaluation on a site-specific basis include loss of bottomlands, changes in existing circulation patterns, and temporary construction impacts such as water quality effects. In addition, if island creation were to be considered for sediments demonstrated to cause severe water quality impacts, the implications of releases of effluent waters from the containment structure would have to be addressed. The effectiveness of presently available wastewater treatment technology for treating such effluent is unclear.

Land Containment

Upland disposal of dredged material affects hydrologic and biological systems that are quite different from those affected by ocean disposal. The differences confound comparisons of alternative disposal methods. How, for example, does one choose between the likelihood and consequences of polluted groundwater and sea food contamination? This section discusses technical information relating to the formation of a policy on upland disposal of contaminated dredged materials from the New York harbor area.

Inland disposal of dredged materials in wetlands or surface waters presently is regulated by the COE and EPA under Section 404 of the Clean Water Act of 1977. Effluent discharged from an upland site also is regulated by this law. The Resource Conservation and Recovery Act of 1976 provides EPA with separate authority for regulating land disposal; however, final regulations have yet to be promulgated.

The New York District, COE, identified several hundred potential upland sites for the containment of dredged material (Conner *et al.,* 1979; Corps of Engineers, 1980; Leslie *et al.,* 1980). Many of these were in agricultural or wetland areas, and the states of New York and New Jersey objected to this potential use for such properties. As a result, the New York District, COE, no longer is considering agricultural and wetland sites (D. Suszkowski, New York District, COE, personal communication). Numerous other potential upland containment sites remain, but their availability is uncertain (Corps of Engineers, 1980).

Considerable information exists on the requirements for and operation of a terrestrial containment facility. The COE Dredged Material Research Program has studied on-land containment (*e.g.*, Chen *et al.*, 1978; Gambrell *et al.*, 1978; Palermo *et al.*, 1978; Spaine *et al.*, 1978), and much research has been sponsored by EPA on the land disposal of many types of solid waste. A terrestrial containment site obviously should not be in an unsuitable area such as atop an aquifer, in a wetland, or in an area of high runoff. The soil underlying the site should have the chemical capacity to adsorb and bind contaminants to particles. If a suitable site is not found, perhaps one could be created by lining an area with an impermeable clay layer. Precautions to prevent infiltration and to maintain satisfactory effluent quality probably are more costly (but perhaps more reliable) than for subaqueous disposal. Rehandling facilities may have to be constructed, and material may have to be transferred several times to place it on the site.

Any comparison of upland and ocean disposal must consider public health. If leaching were to occur at a land disposal site, and if monitoring/restoration efforts failed, polluted groundwater could result. A possible consequence could be contaminated wells. With regard to effects on animal life, many believe that bioconcentration of contaminants to high levels is more likely in open water than on land because aquatic organisms have a well-developed capacity to bioconcentrate nutrients and other chemicals in minute amounts from the surrounding medium. It also is argued that soil conservation, vegetation management, and other terrestrial site management techniques may encourage adsorption of contaminants to sediment and organic particles in soil and discourage bioaccumulation by food-chain species. Although extensive research has shown this to be possible for sewage sludge, it has not been demonstrated fully for dredged material. Adequate evaluation of the leaching potential of contaminated dredged materials in upland sites requires further knowledge of the mechanisms of leaching under alternate cycles of drying and freezing. This knowledge should be developed locally for the specific materials in question, through experiments simulating land disposal.

Upland sites are more accessible for subsequent monitoring than either open ocean sites or harbor borrow pits. Monitoring wells can be established on land to permit rapid detection of unanticipated groundwater pollution. Systematic monitoring is far more difficult at an ocean disposal site. Capture and treatment of contaminated leachates is performed at many solid waste landfills. Aside from problems of leachate, subsequent wind erosion of upland disposal sites remains to be fully understood. Propagation of pollutants away from the disposal site could produce a variety of poorly understood but possibly significant effects.

Processing Contaminated Sediments.

Release of contaminants at disposal sites may be reduced by addition of materials that chemically or physically bind contaminants within deposited dredged material. Unfortunately, little information is available concerning the long-term effectiveness of fixative procedures. Costs of these procedures are relatively high, ranging from \$10/m^3 to \$100/m^3 of dredged material (New York District, COE, unpublished data). (Dredging in the Port of New York now costs about \$1.74/m^3; Conner *et al.*, 1979.) In some cases these techniques might be supplemented by direct chemical processing to reduce contaminant levels sufficiently to permit open-water disposal.

To date, the use of direct processing has been confined to relatively small-scale, special purpose projects such as the removal of oils from sediments entrained by industrial cooling water intakes. More recently, direct processing has been discussed as a means of removing quantities of polychlorinated biphenyls (PCB's) associated with sediments in the upper Hudson River (W.F. Bohlen, University of Connecticut, personal communication). The results of these evaluations indicate that when applied to large-volume operations, available direct processing methods are inefficient and add significantly to total project costs. For the upper Hudson, a specific application of direct processing increased the cost per cubic meter of sediment removed by at least a factor or two. Similar processing expenses would increase removal costs for New York harbor sediments by a factor of five. As a result, the present utility of direct processing within continuing large-scale operations such as those conducted in New York harbor is limited. Its potential, however, should be evaluated periodically.

Dredging Methods for Contaminated Sediment

The removal of highly contaminated sediments may require special techniques designed to minimize dredge-induced resuspension. Resuspension is minimized with high efficiency dredges such as the Pneuma dredge used extensively in Japan and employed recently by the COE for cleanup of a PCB spill in the Duwamish River, Washington. The unique suction head in this hydraulic system minimizes agitation at the dredging point and significantly reduces the volume of effluent water by providing a discharge containing more than 60% solids. The configuration also can be used to place spoil within upland disposal sites and subaqueous borrow pits.

REDUCTION OF REQUIRED DREDGING

Reduction or Control of Sedimentation

The principal cause of shoaling in dredged navigation channels in the Port of New York is deposition of fine-grained material from rivers and sewage discharges. Table 1 lists the weights and volumes of materials deposited by these sources. Littoral drift northeast along northern New Jersey and west along Long Island beaches contributes an additional 540,000 metric tons and 450,000 metric tons of primarily sandy material per year, respectively, and accounts for about 10% of New York harbor maintenance dredging volume. In July 1959, the Public Works Committee of the U.S. House of Representatives authorized a study to determine how to reduce shoaling in the federal channel and pier slips

Table 1. Fine-grain sediment deposited in the Port of NewYork. After Gross (1974).

Source	Amount deposited (10^6 kg/yr)	(10^6 m^3/yr)[1]
Hudson River	726	1.31
Raritan River	63	0.11
Passaic River	86	0.16
Subtotal	875	1.58
Combined sewer overflows	69	0.12
Untreated sewage	127	0.23
Solids in sewage plant effluents	75	0.14
Subtotal	271	0.49

[1] Average dredged material density = 554 kg/m^3.

along a 18 km (11 mi) reach of the lower Hudson River between the George Washington Bridge and the Battery. This study (Panuzio, 1965) utilized hydraulic models of the lower Hudson at the Waterways Experiment Station, Vicksburg, Mississippi. Two proposals were described that were projected to reduce shoaling significantly. The first involved dredging the Hudson in the vicinity of the George Washington Bridge to restore the pre-World War II configuration of the riverbed. Disposal of dredged material in this area produced a constriction that remains today. It was projected that if this plan were put into effect, a 40% reduction in the rate of navigation channel shoaling would occur. The second plan involved periodically closing the Harlem River by means of a movable gate at Spuyten Duyvil. The gate was to be open on the ebb to allow flow from the Hudson into the Harlem but was to be closed during flood. This proposal was projected to reduce the rates of channel shoaling by approximately 40% and slip shoaling by 20%. Neither plan would reduce harbor-wide shoaling, but both would transfer the problem to another location. Increased shoaling would occur along with salinity intrusion into upstream locations in the Hudson River.

Cost-benefit analyses for the two plans should be reassessed in light of current economic and environmental conditions. Increased attention should be given to sedimentation control procedures at upland sources to reduce, as much as possible, future sources of siltation (see also Kneip *et al.*, this volume). At present, the extent of the problem is not known because adequate sediment yield data are unavailable. A better stream monitoring program is needed before appropriate planning strategies can be developed.

Limitation of Dredging to Active Channels

Changes in the pattern of vessel movements through New York harbor may reduce the need to maintain some established navigation channels to authorized depths. While this could reduce the amount of dredged material produced in the harbor, any such reduction would likely be quite small.

ALTERNATIVE OPEN-WATER DISPOSAL SITES

As was suggested earlier, a large portion of dredged material is acceptable for disposal in open water. In general, open-water disposal is the most economical and convenient alternative. One element of the COE's Dredged Material Research Program examined the effect of this practice on a site in Long Island Sound off Eaton's Neck, New York. Although planned disposal at this site was terminated prematurely, a significant body of information was obtained. This information indicated that, rather than undergoing degradation, the site became a highly productive habitat for lobsters and other organisms of commercial and recreational importance (Bokuniewicz *et al.*, 1977).

The New York District, COE, has monitored dredging and disposal programs in the vicinity of the New York Bight. Major COE findings include:

(1) Dredged material settles quickly to the bottom when dumped.

(2) Material remains on the bottom at the Mud Dump and is not subject to significant erosion or movement.

(3) Contaminants are not released readily to the water column during dredging operations or after subsequent disposal.

(4) For those contaminants that have been monitored, releases are quickly dispersed and do not present short-term hazards.

(5) Bioaccumulation at the Mud Dump appears to be no higher than bioaccumulation at other locations in the New York Bight apex.

(6) Water quality in the New York Bight appears to be influenced most by the Hudson River plume.

Not all observers agree with the above statements. Freeland and Merrill (1977) and Conner *et al.* (1979) speculated that material from the Mud Dump may be dispersed through storm-related activity. The latter authors estimated that as much as 10% of the dredged material disposed annually may be removed from the site in this fashion. Dayal *et al.* (1981) examined the composition of sediments at the Mud Dump and found that most solids dumped between 1973 and 1978 are present at the site, but that some metals may be mobilized from the dredged material, "in varying degrees either during the dumping process or following deposition of the dumped materials [p. 163]." In a separate study, Boehm (1981) reported resuspension of particulates from the dumpsite, based on evidence from toxic organics analyses.

The short-term impacts of open-water disposal are well defined and generally are of minor significance. Long-term impacts are more difficult to define, but physical change in substrate composition is the most apparent impact. Such a change need not be adverse. The placement of different substrate in an area may create an environment which previously did not exist. If such an environment is desirable, the change could be beneficial rather than adverse.

Undocumented long-term impacts of ocean disposal of dredged material may exist. For this reason many organizations are conducting research on this subject. For example, the COE is continuing long-term research at the Duwamish Waterway Disposal Site (Washington), which received material contaminated by PCB's. Monitoring efforts by the New York District, COE, likewise will continue.

Site Capacity

The retention of a high percentage of dumped materials in depositional mounds on the seafloor has necessitated the relocation of the Mud Dump a number of times since 1890 to avoid interference with navigation (Gross, 1976). The Panel estimated that the rate of shoaling of the center of the present site is 0.16 m per million cubic meters deposited. For the estimated annual dredging of New York harbor, the shoaling rate would be 1.4 m/yr. When new work dredging is included, this rate increase to about 1.5 m/yr. The present site has an available depositional depth of 15 m before shoaling presents a problem. A 10-year minimum life span is indicated based on estimated rates. When the site's capacity is exhausted, another site will be needed.

Future site selection, long-term management, and assessment of cumulative impacts must take site capacity considerations into account. Other considerations being equal, sites with greater capacity (greater depth) should be selected to diminish the total bottom area affected by disposal operations. Within the New York Bight, the headward portions of the Hudson Shelf Valley and the adjacent Christiaensen Basin offer the deepest practical bottom areas. It may be necessary to abandon the present Mud Dump before it is filled to capacity because of changing environmental conditions, a change in disposal strategy, new discoveries, or new laws. However, this Panel has not seen compelling evidence to indicate that the present site should not continue to be used.

Selection of Alternate Sites and Procedures

In 1977, a COE-sponsored workshop considered alternative sites for the open-water disposal of uncontaminated dredged mud (Mitre Corporation, 1977). Three general options were considered: increasing the area of the present site, choosing a new site, or revising the disposal strategy. The workshop concluded that the two most suitable sites within the Bight are the existing Mud Dump and the Christiaensen Basin. If an additional disposal area is needed, serious consideration should be given to increasing the size of the present site, presuming that conditions near the site are similar to those within the designated disposal area. This alternative would not involve additional economic costs and may be the most practical solution.

The Christiaensen Basin site would be in a large, natural basin that is slightly deeper than the present site (mean depth 31.3 m vs. 24.0 m) (U.S. Environmental Protection Agency, 1982). Portions of the Basin's sea floor bear impoverished biological communities, but the majority of the Basin supports an unusually dense aggregation of species characteristic of muddy fine sands. The presence of these species may be a function of organic enrichment (Boesch, this volume). In evaluating this alternative, managers must consider that the Christiaensen Basin has been degraded for over 100 years by the effluent of the Hudson estuary, and for shorter periods by sewage sludge and dredged material disposal. The potential for rehabilitating this area should also be considered before it is committed to receive dredged sediment. The proposed Christiaensen Basin site has been estimated to have a 60-year capacity (Mitre Corporation, 1977). It is about the same distance from the harbor as the present Mud Dump, and the economics of its use would be similar.

A Basin site poses other dilemmas. The most obvious is that the Basin lies in the path of shipping lanes to the Port of New York. It also has been suggested that because the Basin is close to Long Island beaches, there may be political opposition to its use as a disposal area.

A remaining alternative involves shifting from a strategy that attempts to contain dredged material to one that disperses the material over a wide area. Techniques for dispersing dredged sediment are available; while equipment in the harbor is not ideal for this approach, discharges could be altered to enhance dispersion. Because of the uncertain ecological effects of sustained, low-level contamination, this approach was not considered an appropriate option for the Bight by a majority of the Panel members.

MANAGEMENT OF THE DISPOSAL AREA

Dredging and Disposal Methods

To maximize the probability of dredged material containment at the present open-water disposal site, dredging operations should favor the use of large-volume clam-shell dredges and associated large-volume hopper barges. Single, large-volume disposal operations are preferred over numerous small discharges, because placement of disposed material on the bottom is more effective and monitoring of disposal operations is easier. The use of hopper dredges or similar hydraulic techniques should be confined to projects where materials will be deposited alongshore rather than at open-water sites.

Project Timing

At several locations in the northeastern United States, dredging operations are not scheduled for the summer season. The primary intent is to minimize possible interference with animal populations during the period of most sensitive biological activity (*i.e.*, spawning and early larval growth), and to reduce water quality impacts that may result from limited vertical mixing and stratification in the water column. Review of the resource characteristics and local hydrographic and water quality data suggest that such schedules, when applied within the New York harbor and Bight area, might yield minimal benefits. Nutrient and carbon loadings appear to be dominated by Hudson River outflow. Oxygen depression resulting from dredged material disposal tends to be short-lived, and associated nutrient loading from dredged material is negligible when compared to river-borne supply. These factors and the higher cost of dredging in the winter, because of poor weather, suggest that there is little to be gained by seasonal restrictions on dredging and disposal in the New York Bight. Longer-term sub-lethal effects cannot be ruled out, however.

MONITORING OF OPEN WATER DISPOSAL

The U.S. Coast Guard (USCG) has responsibility for surveillance of dumping at sea. Surveillance responsibilities are set forth in Commandant Instruction 5992.9A, which is summarized briefly below:

(1) Spot checks for valid permits will be conducted on dumping vessels.

(2) Surveillance will concentrate on the dumping of contaminated material, on EPA requested vessel/dump monitoring, and on 10% of remaining disposal activities, including those at the dredged material disposal site.

(3) Surveillance will be accomplished by USCG vessels to intercept and/or escort permittee vessels, by aircraft overflight of the dump site during dumping activity, by shipriders (USCG personnel) on permittee vessels, and by radar coverage.

(4) Random surveillance missions will be conducted in designated and non-designated areas to discourage illegal dumping.

Only materials dumped at the 106-site (chemical waste site) are considered toxic by the USCG. Material dumped anywhere else is regarded, by definition, as non-toxic.

Methods should be devised to allow more extensive Coast Guard surveillance. More than 10% of the material disposed at the Mud Dump should be monitored. The USCG is conducting research on a positive location recording system that is based on electronic navigation. Regulations implementing such an electronic device were withdrawn by the USCG after being published in proposed form. Such regulations would provide a much more effective program than is available presently. The Panel feels that such a program should be implemented as soon as possible.

COSTS OF ALTERNATIVE DISPOSAL SCHEMES

Table 2 summarizes the costs of a number of disposal alternatives for dredged material in the New York Bight area. Reliable cost estimates can be provided for established disposal practices; however, the cost of novel methods can only be approximated. Unforeseen contingencies in the implementation of new techniques tend to raise costs beyond those originally projected. Consequently, estimates for untested or little-tested methods in Table 2 should be viewed as minimum values.

Several options have been omitted from Table 2. In some instances insufficient information is available to make a meaningful estimate (*e.g.*, offshore island containment). In other instances (*e.g.*, incineration), costs were

Table 2. Costs of alternative methods for dredged material disposal. Data adapted from Conner *et al.* (1979).

Method	Cost per m^3
Shallow water disposal (*e.g.*, Mud Dump or proposed Christiaensen Basin site)	$ 1.74
Subaqueous borrow pits	1.61
Confined upland disposal (based on 15 representative upland sites)	3.30 - 10.20
Protected water containment (*e.g.*, Hoffman-Swinburne Island plan)	6.10
Filling mines (other than abandoned quarries or strip mines)	6.60 - 13.15
Open ocean disposal (based on 160 km [100 mi] haul)	8.50
Disposal in abandoned piers	14.50
Containerized ocean disposal	737.00

judged to be extremely high and likely to make the practice an unreasonable alternative (Conner *et al.*, 1979).

CONCLUSIONS AND RECOMMENDATIONS

(1) The Port of New York should not be allowed to shoal to its natural depth 5.5 m (18 ft). This would exclude all ocean-going vessels, cause severe economic damage to the people who are associated with the Port's activity, and result in loss of revenue to the region, states, and municipalities.

(2) Present dredging and disposal practices are limited by the types of equipment available for use in the Port. Since dredging must continue, changes can be made only gradually, except where very small amounts of material are involved. It is important to start now to formulate plans for long-term improvements in dredging and disposal practices.

(3) Dredged material produced in the Port of New York ranges from clean to highly contaminated. It is important to distinguish among the classes of dredged material and to use available resources to improve disposal practices for materials that pose the most severe environmental hazards.

(4) Reducing contaminant discharges into the estuary represents the single most important management strategy for decreasing the proportion of highly polluted dredged materials and minimizing concerns over the ocean dumping of such sediments. However, significant improvements can be obtained by this approach only over a period of many years.

(5) The economic and technical feasibility of hydraulic engineering works that offer the potential for reduced shoaling in critical channel areas should be re-examined.

(6) A large fraction of the material dredged from the Port of New York is

suitable for open-water disposal under present criteria. However, the criteria for classifying dredged material require continued re-evaluation as scientific understanding of environmental problems develops.

(7) Uncontaminated dredge materials having appropriate engineering properties should be used as fill in development projects when economically feasible.

(8) At present, sediment not suitable for open-water disposal is being allowed to accumulate in the Port. Alternative disposal methods that may be used for this material are containment in subaqueous borrow pits and construction of islands composed of dredged materials. The use of on-land containment sites and the processing of dredged materials to remove or to immobilize contaminants are less desirable alternatives on the basis of environmental and economic factors, respectively.

(9) Open-water dredged material disposal is the major source of contaminant-bearing solids to the New York Bight.

(10) Localized effects of dredged materials on bottom-dwelling organisms are apparent in the disposal area, but the relative importance of dredged material disposal to the general degradation of the Bight and its living resources has not been established in full.

(11) Changes in dredged material disposal policy should be based on rigorous evaluation of alternatives, including assessments of associated environmental effects and economics. The effects on uses and user groups of existing and proposed sites must also be considered.

(12) A regional dredging plan should be developed for greater New York harbor. The plan should:

(a) Identify approved navigation channels and their characteristic dimensions.

(b) Establish a policy for the frequency of maintenance dredging.

(c) Establish criteria for characterizing the physical and chemical properties and polluting potential of dredged materials, and carry out necessary characterization programs.

(d) Establish criteria for assessing the suitability of various kinds (qualities) of material for disposal at individual sites (*e.g.*, overboard, subaqueous, upland).

(e) Designate and rank available disposal sites for different kinds (qualities) of dredged materials.

(f) Plan the use of disposal sites for maintenance projects on the basis of site capacity and oceanographic characteristics.

(g) Recommend dredging and disposal methods for individual projects.

(h) Provide mechanisms for amending the plan to take into account changes in utilization of regional waters and lands, improvements in dredging and disposal technology, increased scientific knowledge, and economic and social factors.

(i) Have a built-in feedback mechanism to assess the effectiveness of the plan in a recurrent fashion.

REFERENCES

Boehm, P.D. 1981. Investigations on Pollutant Organic Chemical Fluxes in the Hudson-Raritan Estuarine and New York Bight Coastal Systems. Final report to NOAA Office of Marine Pollution Assessment, NOAA Grant No. NA80AA-D-00062. Energy Resources Co., Inc., Cambridge, MA. 59 pp.

Boesch, D.F. This volume. Ecosystem consequences of alterations of benthic community structure and function in the New York Bight region. pp. 543-568.

Bokuniewicz, H.J. In press. Burial of dredged sediment under the sea floor: Can you do it? *In:* Proceedings of the 14th Annual Dredging Seminar, Texas A&M University, Center for Dredging Studies. College Station, TX.

Bokuniewicz, H., C. Bultman, M. Dowling, J. Gebert, R. Gordon, P. Kaminsky, and C. Pilbeam. 1977. Aquatic disposal field investigations, Eatons Neck disposal site, Long Island Sound; Appendix A: Investigation of the hydraulic regime and the physical characteristics of bottom sedimentation. U.S. Army Engineer Waterways Experiment Station, Dredged Material Research Program, Tech. Rep. D-77-6. 108 pp.

Chen, K.Y., B. Eichenberger, J.L. Mang, and R.E. Hoeppel. 1978. Confined disposal area effluent and leachate control (laboratory and field investigations). U.S. Army Engineer Waterways Experiment Station, Dredged Material Research Program, Tech. Rep. DS-78-7. 94 pp.

Conner, W.G., D. Aurand, M. Leslie, J. Slaughter, A. Amr, and F.I. Ravenscroft. 1979. Disposal of Dredged Material Within the New York District: Vol. I, Present Practices and Candidate Alternatives. Mitre Corp., Metrek Division, McLean, VA.

Corps of Engineers. 1980. Dredged Material Disposal Management Program for the Port of New York and New Jersey: Incremental Implementation Plan. U.S. Army Corps of Engineers, New York District, New York. 46 pp.

Dayal, R., M.G. Heaton, M. Fuhrmann, and I.W. Duedall. 1981. A geochemical and sedimentological study of the dredged material deposit in the New York Bight. NOAA Tech. Memo. OMPA-3. 174 pp.

Engler, R.M. 1980. Prediction of pollution potential through geochemical and biological procedures: Development of regulation guidelines and criteria for the discharge of dredged and fill material. *In:* Contaminants and Sediments, Vol. I., R.A. Baker (ed.), Ann Arbor Science Publishers, Inc., Ann Arbor, MI. pp. 143-169.

Environmental Protection Agency/Corps of Engineers Technical Committee on Criteria for Dredged and Fill Material. 1977. Ecological Evaluation of Proposed Discharge of Dredged Material into Ocean Waters. Environmental Effects Laboratory, U.S. Army Engineer Waterways Experiment Station, Vicksburg, MS. 24 pp.

Freeland, G.L. and G.F. Merrill. 1977. The 1973 bathymetric survey in the New York Bight apex: maps and geological implications. NOAA Tech. Memo. ERL MESA-19. 20 pp.

Gambrell, R.P., R.A. Khalid, and W.H. Patrick, Jr. 1978. Disposal alternatives for contaminated dredged material as a management tool to minimize adverse environmental effects. U.S. Army Engineer Waterways Experiment Station, Dredged Material Research Program, Tech. Rep. DS-78-8. 148 pp.

Gross, M.G. 1974. Sedimentation and waste deposition in New York harbor. Ann. N.Y. Acad. Sci. 250: 112-128.

Gross, M.G. 1976. Waste Disposal. MESA New York Bight Atlas Monograph 26, New York Sea Grant Institute, Albany, NY. 32 pp.

Howard, Needles, Tammen, and Bergendoff. 1976. Dredged Material Disposal in New York Harbor and the New York Bight. Phase One, Feasibility Study for the U.S. Army Corps of Engineers, New York District. Howard, Needles, Tammen, and Bergendoff, New York.

Johanson, E.E., S.P. Bowen, and G. Henry. 1976. State-of-the-art survey and evaluation of open-water dredged material placement methodology. U.S. Army Engineer Waterways Experiment Station, Dredged Material Research Program, Contract Rep. D-76-3. 145 pp.

Kamlet, K.S. In press. Dredge material ocean dumping: perspectives on legal and environmental impacts. *In:* Wastes in the Ocean, Vol. 2: Dredged Material Disposal in the Ocean, D.R. Kester, B.H. Ketchum, I.W. Duedall, and P.K. Park (eds.), Wiley-Interscience, New York.

Kneip, T.J., N.H. Cutshall, R. Field, F.C. Hart, P.J. Lioy, J. Mancini, J.A. Mueller, C. Sobotowski, and J. Szeligowski. This volume. Management of non-point sources. pp. 145-161.

Leslie, M., D. Aurand, D. Schultz, and R. Holman. 1980. Disposal of Dredged Material Within the New York District: Vol. II, Preliminary Evaluation of Upland Disposal. Mitre Corp., Metrek Division, McLean, VA.

Mitre Corporation. 1977. The Proceedings of the New York Dredged Material Disposal Alternatives Workshop, October 11-13, 1977. Mitre Corp., Metrek Division, McLean, VA. 107 pp.

Palermo, M.R., R.L. Montgomery, and M.E. Poindexter. 1978. Guidelines for designating, operating, and managing dredged material containment areas. U.S. Army Engineer Waterways Experiment Station, Dredged Material Research Program, Tech. Rep. DS-78-10. 89 pp.

Panuzio, F.L. 1965. Lower Hudson River siltation. *In:* Proceedings of the Inter-Agency Sedimentation Conference, U.S. Department of Agriculture, Agricultural Research Service, Misc. Publ. 970. pp. 512-550.

Port Authority of New York and New Jersey. 1979. Foreign Trade 1978: The Port of New York and New Jersey. Port Authority of New York and New Jersey, Planning and Development Department, New York. 48 pp.

Spaine, P.A., J.L. Llopis, and E.R. Perrier. 1978. Guidance for land improvement using dredged material. U.S. Army Engineer Waterways Experiment Station, Dredged Material Research Program, Tech. Rep. DS-78-21. 85 pp.

Suszkowski, D. 1977. History of dredging and disposal operations within the Port of New York. *In:* The Proceedings of the New York Dredged Material Disposal Alternatives Workshop, October 11-13, 1977. Mitre Corp., Metrek Division, McLean, VA. pp. 11-37.

Suszkowski, D.J. and J.M. Mansky. In press. The disposal of sediments dredged from New York harbor. *In:* Sixth Annual U.S./Japan Experts Meeting on the Disposal of Toxic Substances. U.S. Army Corps of Engineers.

U.S. Environmental Protection Agency. 1982. Draft Environmental Impact Statement (EIS) for New York Dredged Material Disposal Site Designation. U.S. Environmental Protection Agency, Criteria and Standards Division, Washington, DC.

MANAGEMENT OF INDUSTRIAL WASTES[1]

A.I. Mytelka[2]
J. Ciancia
R.R. Delgado
A.M. Levinson
T.P. O'Connor
J.C. Prager
W.B. Pressman
E.A. Regna
M.M. Sadat
J.T.B. Tripp
M. Wendell

Abstract. A management system for the New York Bight must be capable of achieving and protecting water quality by preventing pollutants that adversely affect marine life and the ecosystem from entering Bight waters. Although the continental shelf beyond the New York and New Jersey coasts has not yet been damaged severely by waste disposal, the Bight apex is degraded. Improved environmental conditions may permit greater industrial and recreational use of those portions of the Bight closest to urban population centers. Interdisciplinary research is needed to understand and control the effects of various pollutants on the Bight. Resulting information must be generally available. National effluent limitations and industrial pretreatment regulations must be developed and enforced. Public and private treatment facilities should be required to monitor their discharges for designated toxic materials. New and innovative approaches to attain industrial pretreatment and direct discharge standards must also be encouraged.

INTRODUCTION

One purpose of industrial wastes management is to obtain and protect proper water quality. The propriety of the quality of water should be measured by the extent to which it provides a medium hospitable to the propagation and survival of healthy marine life and suitable for lawful and intended uses. Put negatively, appropriate water quality guards against the occurrence of contaminants at levels that adversely affect marine life and the ecosystem as a whole.

Toxic materials and other pollutants damage or destroy organisms and alter the character of waters. Unfortunately, present scientific knowledge appears inadequate to define or predict with certainty the effects of individual pollutants on the Bight and its marine life (Prager, 1974). Nor is the expertise available to predict the extent to which various levels of control will improve the health of the New York Bight. Additional predictive scientific information is needed by environmental managers. However, before such capabilities can be achieved, the relationship among diverse waste sources, the many reactions of the ecological system, and the system's wide natural variations must be understood. Scientists acknowledge the complexity of interactions between natural systems and pollutants and hedge their conclusions accordingly. With

[1] This paper is based on the report of the Panel on the Management of Industrial Wastes.
[2] Chairman, Panel on the Management of Industrial Wastes.

adequate resources, the efforts of scientists eventually should provide quantitative descriptions of ecosystem workings and the ability to predict (to some extent) how given contaminants may be reflected in biological changes.

One management approach to the imperfect state of knowledge is to advocate no restrictions on pollutant discharges until it can be proven that the demonstrable harm of such discharges outweighs the costs and burdens of their regulation. This approach would be inappropriate because the detrimental effects could be very extensive and long-lasting. Such effects include the destruction or contamination by hazardous substances of commercial and recreational marine resources. In the face of such dangers, it is necessary to act on the basis of the best knowledge available while continuing to strengthen the data base, improve analytical methods, and obtain more substantial and predictive scientific knowledge for fashioning additional or improved management tools.

When more specific knowledge concerning environmental dangers of individual contaminants becomes available, some waste dischargers voluntarily may limit or abate certain pollutant discharges. However, experience shows that formulation of appropriate public policy and regulation based on law is necessary. Accordingly, the management requirements considered in this paper are those of federal, state, and local governments, including intergovernmental cooperation for a region that is interstate in extent.

Mechanisms and possible measures available to protect the Bight and its tributary waters from industrial sources of pollution are numerous. Several agencies of the federal government, including the U.S. Environmental Protection Agency (EPA), the National Oceanic and Atmospheric Administration, the Corps of Engineers, and the Coast Guard now have important, relevant responsibilities. Each governmental level and agency operates under existing and constantly evolving laws, regulations, and guidelines.

ENVIRONMENTAL STATUS OF THE NEW YORK BIGHT

Degradation is worst in the Bight apex. A recurring theme in most scientific papers of this symposium is that coastal biota beyond the apex (*i.e.*, organisms on the continental shelf beyond the New York and New Jersey coasts) is not severely damaged by waste disposal. Comparisons of observations made in Raritan Bay, Sandy Hook Bay, the Christiaensen Basin, or the upper Hudson Shelf Valley, with similar observations from elsewhere in the Bight, provide examples of impacted plankton populations, mutagenic effects on fish eggs, damaged benthic communities, diseased fishes, and elevated contaminant levels in water, sediments, or organisms.

Within those regions of the Bight and adjacent waters where ecological damage is obvious, it is severe. The worst waste disposal-related effects are along the sea floor. The bottoms of Raritan Bay and Sandy Hook Bay support only a few hardy bottom-dwelling species (Mearns *et al.*, this volume). Muddy sediments in the Christiaensen Basin and elsewhere in the Bight apex contain high concentrations of inorganic and organic contaminants and biotic populations devoid of organisms normally used as food by fishes (Boesch, this volume). Abnormal concentrations of toxic metals, chlorinated organic compounds, and petroleum-derived compounds contaminate tissues of bottom organisms sampled from the Bight (O'Connor and Rachlin, this volume; O'Connor *et al.*, this volume; Anderson, this volume).

Within the water column, near-shore and estuarine plankton communities include individuals that exhibit a resistance to various toxicants (Lee *et al.*, this volume). The occurrence of red tide organisms appears enhanced by nutrients derived from sewage effluent, and the initiation of such blooms may be associated with the chelation of metals derived from industrial effluent (Mearns *et al.*, this volume). Developing mackerel eggs from surface waters of the New York Bight apex show high incidence of mutagenic damage (Longwell and Hughes, this volume).

The Bight apex used to support a shellfish industry but does not now. Fishing, if it existed nearer the urban centers of New York and New Jersey under conditions as attractive as they once were, would be a great commercial and recreational resource. Shoreside resorts of economic significance could exist again close to urban centers if water quality improvements were to invite more sportfishing and other recreational pursuits.

There is potential for economic resurgence if marine and estuarine environmental quality were to improve. The despoiled area is limited but includes the area most attractive for both recreational and industrial use. Alleviation of existing polluted conditions in the Bight apex promises extensive social rewards. The recreational fishing industry could move closer to populated urban areas on the shores of the apex (Figure 1). Sites that presently are unacceptable for a broad range of uses because of pollution could become acceptable.

INDUSTRIAL PRETREATMENT

The industrial wastes that reach the waters of the Bight consist of a variety of substances including biodegradable organics (measured by conventional parameters such as biochemical oxygen demand and pH), toxic metals, and synthetic compounds. The conventional pollutants are more or less well handled by established treatment methods; however, toxic metals and toxic organics present more difficult problems because they are largely uncontrolled and have not yet been effectively regulated or eliminated.

Industrial wastes find their way into the waters of the Bight in several ways. Some are directly discharged from industrial facility outfalls. By far, the greater part is released into public combined sewers. These wastes either flow through public treatment plants and then into estuarine and coastal waters, or spill directly from the combined sewers into these waters. An unknown but probably small amount of contamination from industrial sources also reaches the waters of the region as surface runoff.

Figure 1. Shore-based recreational fishing in the Hudson-Raritan estuary often occurs in ecologically degraded areas. Man shown fishing off the Dykman Street (Manhattan) raw sewage outfall in the Hudson River. (Courtesy M. Kawka, NOAA Office of Marine Pollution Assessment.)

It is also important to note that the effect of organic and inorganic contamination is not a one-time affair. Very significant but unknown percentages of all such discharges are entrapped in bottom muds. Contaminants accumulate over long periods of time and may be an increasing, permanent source of further pollution to coastal waters. Some become mixed again with the waters through resuspension caused by natural turbulence, some are resuspended through turbulence that accompanies dredging operations, and some are made available through the redeposition of dredge spoils.

The toxic wastes that spill from combined sewers never have an opportunity to be removed by public treatment plants. Available processes remove relatively little of the toxic content of the sanitary sewage-industrial waste mixture that does reach treatment plants. In any case, all except the readily degradable organics reach estuarine and Bight waters through treatment plant effluent discharges, or through sludge and sludge residues.

Except for the small percentage of industrial wastes directly discharged through private outfalls under federal and state permits (NPDES/SPDES), most of the toxic metals (about 90%) and the organic toxic waste loadings to receiving waters could be decreased substantially by pretreatment. If intercepted or rendered of acceptable quality before they reached the public sewers, they would not be found in the waters of the Bight.

For those industrial wastes discharged into public sewer systems, the municipalities become responsible for the quality of the ultimate discharges to the waters of the region. For this reason, effective implementation of pretreatment regulations under Section 307 (b) of the Clean Water Act (33 U.S.C. 1317 [b]) and other applicable laws for industry discharging into municipal treatment plants is crucial.

Pursuant to EPA pretreatment regulations (40 CFR 403), public authorities responsible for treatment plants in the region that receive significant industrial wastes are supposed to implement pretreatment programs and to enforce standards for at least 129 high priority toxic industrial pollutants discharged by industries into their systems. These standards are to be based on "best available technology" (BAT), or more stringent effluent limitations as necessary to protect water quality. This list of 129 toxic pollutants includes toxic metals and most of the toxic organic compounds that have been identified to be of concern in the New York Bight to date. The long-delayed deadline for implementation of these effluent guidelines and for approval of pretreatment programs in treatment plant NPDES permits is now June 30, 1984.

Primary enforcement responsibility for pretreatment standards rests with local agencies (40 CFR 403.8). The EPA and the states of New York and New Jersey have the ultimate enforcement role through the management of the NPDES/SPDES programs for municipal treatment works.

In view of the dominant role of the Clean Water Act and regulations under it, until pretreatment standards are issued and pretreatment programs are implemented, the New York-New Jersey region will lack an enforceable, comprehensive industrial pretreatment policy. Although enforcement of these programs in the region may and should encourage recycling and reuse of certain toxic pollutants (*e.g.*, toxic metals), it can be expected that pretreatment will lead to production of larger amounts of industrial sludges containing toxic metals and toxic organics. If these sludges are disposed of in landfills, they are supposed to be regulated under Title C of the Resource Conservation and Recovery Act. Existing landfills would be similarly regulated.

IMPACT OF THE PRETREATMENT PROGRAM ON THE BIGHT

It is the consensus of the panel that the single most effective measure to reduce the toxic metals input from industrial sources to the New York Bight is an industrial pretreatment program. The probable impact of such a program can be assessed using studies of metals loadings for portions of the Hudson-Raritan estuary. For example, the total cadmium input to the estuary is estimated to be 360 kg/day (Mueller *et al.*, 1976; M.M. Sadat, personal communication). The portion of this load from municipal sources is 226 kg/day which could be reduced to 51 kg/day by pretreatment (M.M. Sadat, personal communication). In general, an overall reduction of 80% in the toxic metals loading to the estuary from all sources is achievable (M.M. Sadat, personal communication).

Areawide Section 208 water quality management plans could provide tools for management use, both as a data base for current decisions and for future projections. For example, the New York City 208 Plan (New York City, 1978) points out that toxic metals in the harbor are dangerous to marine life and contaminate sludges. It also indicates that direct metal loads (from sewage plants and industrial sources) and loads from combined sewer outfalls make up most of the total, with the remainder attributable to combined sewer runoff and atmospheric deposition.

In the panel's estimation, the control of industrial discharges into municipal sewer systems is of great importance not only for the preservation and enhancement of estuary and Bight ecosystems, but for equally important social and economic reasons. The urban waterfront zone of the estuary represents extremely valuable real estate which, if rehabilitated, can contribute significantly to the rejuvenation of urban centers (New Jersey Department of Environmental Protection, 1980). Improvement of water quality either would result in enhancement of benefits that otherwise are realized at a low level from development occurring in the face of undesirable conditions, or would encourage uses that are precluded by the existence of a poor physical environment.

Effective implementation of a pretreatment program would increase the management options for sewage sludge disposal, including consideration of ocean disposal past 1981 and a wider range of land application programs. It is known that there are toxic organic contaminants in the Bight, some portion of which is of industrial origin. An assessment of the impact of a pretreatment program on the loading of toxic organic compounds on the Bight is not underway currently. This should be done to allow an estimate of the quantities of these compounds entering the Bight.

WASTEWATER TREATMENT TECHNOLOGY

Wastewater treatment technology currently is available to remove toxic metals and toxic organics from industrial sources that have been identified as the most probable causes of stress to the New York Bight. In general, presently available practical technology is capable of removing 85-90% of the load of these pollutants, if it is applied before the industrial wastes become mixed with the municipal waste stream. In addition, in-plant control techniques can be implemented in many cases, this will result in greater total removals. The pretreatment for many industrial plants discharging to municipal systems would be similar to that employed by some industrial establishments for direct discharges to water bodies.

In general, the conventional approach to removing toxic metals (including copper, cadmium, and lead) from wastewaters involves hydroxide precipitation followed by settling or filtration (Ciancia, 1973). For metal finishing wastewaters, reduction of chromates and oxidative destruction of cyanides precedes precipitation. In the case of mercury, the most commonly used process is sulfide precipitation, but borohydride reduction to the metallic form is also used. The common methods for removing silver from wastewaters are precipitation (*e.g.*, sulfide, chloride), ion exchange, and reductive exchange (cementation). In addition to the methods mentioned above, some of the more important processes used to treat soluble toxic metal and inorganic wastewaters are evaporation, reverse osmosis, electrodialysis, and electrolytic methods. Unless metals are separated at the source (which is practiced in many cases and should be encouraged), no practical means for recovering metals from mixed sludges has been developed.

Organic wastes include a variety of compounds with widely varying properties (*e.g.*, volatility, degradability, and toxicity). No single wastewater treatment technology is applicable to all organic waste streams; however, the activated sludge process is applicable to a wide variety of biodegradable organic compounds. The activated carbon process potentially can remove many non-biodegradable toxic organics, including chlorinated hydrocarbons, phenols, and petroleum oils. Better manufacturing practices represent an additional approach that has great potential for reducing the introduction of industrial toxicants into receiving waters and sewer systems. Improved practices also would provide for the recovery of valuable materials, which should be encouraged.

COSTS FOR POLLUTANT CONTROL

A number of pollutants have been identified by this symposium as having a collective impact on the New York Bight and adjacent waters. The concern is that unless the continued discharge of toxic metals and synthetic organic compounds (such as chlorinated organics and polynuclear aromatic hydrocarbons) is limited, serious effects in the estuary and Bight will result. Consequently, treatment technology should be applied to all of these pollutants.

Estimates of metal inputs to the Bight are based on data from governmental sources and the literature (Mueller and Anderson, 1977). Capital costs were projected for removal of metals from effluents prior to their discharge to municipal waste treatment plants (U.S. Environmental Protection Agency, 1979). The pretreatment technology applied is designed to meet "best practicable technology" (BPT) requirements and will require a capital investment of approximately $60 million. This total includes the cost of providing treatment for approximately 400 small job shop platers at an average installed equipment cost of $100,000 each. Great savings could be accomplished by the establishment of electroplating parks to take advantage of economies of scale in pretreatment processes and to facilitate the disposal of toxic metals sludge. For example, if instead of 400 separate treatment facilities with batch treatment, it would be possible to set up 20 electroplating parks, each with a centrally located continuous treatment plant, the cost of treatment would drop from $40 million to around $9 million. The cost for sludge handling, which is not included in the above estimates, similarly would be reduced from about $10 million for 400 separate job shops to about $5 million for 20 parks. The feasibility of establishing electroplating parks to take advantage of these treatment economics remains to be explored.

Capital costs for dealing with non-biodegradable organics such as the chlorinated organics and polynuclear aromatic hydrocarbons have not been estimated because quantitative loadings to the Bight are not known. Neither is it known which industrial sources are the main contributors of synthetic organics. Nevertheless, it is possible to get an appreciation for the magnitude of the costs necessary to abate non-biodegradable organics. If done in the municipal treatment plants to the extent possible through carbon absorption, the additional capital costs are estimated to be on the order of at least $300-400 million (U.S. Environmental Protection Agency, 1973). Source control and pretreatment could cut these costs at least in half.

EFFLUENT CHARGES

In addition to methods already in use in the Bight area, an option to be considered for industrial waste discharges is the use of effluent charges. The imposition of effluent charges would cause industrial users of the environment to bear costs that they impose on the rest of society. To be effective as a regulatory device, charges would have to be sufficiently high that the cost of discharging wastes into water bodies would become more expensive, on average, than the cost of treating wastes or reducing them at their source. Economists generally have taken the position that effluent charges can provide powerful incentives for pollution abatement (Freeman and Haverman, 1972; Baumol and Oates, 1975, 1976).

RECOMMENDATIONS

1. To improve the quality of management decisions affecting the Bight, the panel recommends interdisciplinary research on the ecosystem and its relation to pollutant nutrients and toxicants. This research should include chemical and biological analyses, modeling, field measurements, and cost-benefit studies. The panel also recommends improved data correlation, access, and dissemination.
2. Further development and enforcement of national effluent limitations and industry pretreatment limitations are critical. No further delays should be allowed.
3. Consideration should be given to innovative approaches to attain industrial pretreatment and direct discharge standards, such as a system of effluent charges.
4. Industries and publically owned treatment works receiving significant amounts of industrial wastes should be required to monitor their effluents for all toxic materials present.

REFERENCES

Anderson, J.W. This volume. The transport of petroleum hydrocarbons from sediments to benthos and the potential effects. pp. 165-179.

Baumol, W.J. and W.E. Oates. 1975. The instruments of environmental policy. *In:* Economic Analysis of Environmental Problems, E.S. Mills (ed.), National Bureau of Economic Research, Inc., Cambridge, MA. pp. 95-128.

Baumol, W.J. and W.E. Oates. 1976. Theory of Environmental Policy. Prentice Hall, Englewood Cliffs, NJ. 272 pp.

Boesch, D.F. This volume. Ecosystem consequences of alterations of benthic community structure and function in the New York Bight region. pp. 543-568.

Ciancia, J. 1973. New waste treatment technology in the metal finishing industry. Plating 60: 1037-1042.

Freeman, A. and R. Haverman. 1972. Residuals charges for pollution control: a policy evaluation. Science 177: 322-329.

Lee, R., A.C. Longwell, T.C. Malone, L.S. Murphy, D.R. Nimmo, H.B. O'Connors, Jr., L.S. Peters, and K.D. Wyman. This volume. Effects of pollutants on plankton and neuston. pp. 39-52.

Longwell, A.C. and J.B. Hughes. This volume. Cytologic, cytogenetic, and embryologic state of Atlantic mackerel eggs from surface waters of the New York Bight in relation to pollution. pp. 381-388.

Mearns, A.J., E. Haines, G.S. Kleppel, R.A. McGrath, J.J.A. McLaughlin, D.A. Segar, J.H. Sharp, J.J. Walsh, J.Q. Word, D.K. Young, and M.W. Young. This volume. Effects of nutrients and carbon loadings on communities and ecosystems. pp. 53-65.

Mueller, J.A. and A.R. Anderson. 1977. Contaminant inputs to New York Bight. Phase II. Urban Runoff. Report to the MESA New York Bight Project, NOAA Grant No. 04-6-002-44027, Manhattan College, Bronx, NY. 110 pp.

Mueller, J.A., J.S. Jeris, A.R. Anderson, and C.F. Hughes. 1976. Contaminant inputs to the New York Bight. NOAA Tech. Memo. ERL MESA-6. 347 pp.

New Jersey Department of Environmental Protection. 1980. New Jersey Coastal Management Program and Final Environmental Impact Statement. New Jersey Department of Environmental Protection, Division of Coastal Resources, Trenton, NJ. 533 pp.

New York City. 1978. Task 431.04. Sources of heavy metals. *In:* The City of New York Section 208 Areawide Waste Treatment Management Plan, New York City Department of Environmental Protection, New York.

O'Connor, J.M., J.B. Klotz, and T.J. Kneip. This volume. Sources, sinks, and distribution of organic contaminants in the New York Bight ecosystem. pp. 631-653.

O'Connor, J.M. and J. Rachlin. This volume. Perspectives on metals in New York Bight organisms: factors controlling accumulation and body burdens. pp. 655-673.

Prager, J.C. 1974. Factors influencing estuarine microbiota. *In:* Industrial Pollution, N.I. Sax (ed.), Van Nostrand-Reinhold, New York. pp. 148-181.

U.S. Environmental Protection Agency. 1973. Design Manual for Carbon Absorption-Process. EPA 625/1-71-002A. U.S. Environmental Protection Agency, Technology Transfer Office, Washington, DC.

U.S. Environmental Protection Agency. 1979. Development Document for Existing Source Pretreatment Standards for the Electroplating Point Source Category. EPA 440/1-79-003. U.S. Environmental Protection Agency, Effluent Guidelines Division, Office of Water and Hazardous Materials, Washington, DC. 526 pp.

MANAGEMENT OF NON-POINT SOURCES

T.J. Kneip[1]
N.H. Cutshall
R. Field
F.C. Hart
P.J. Lioy
J. Mancini
J.A. Mueller
C. Sobotowski
J. Szeligowski

Abstract. Three classes of materials have major non-point source contributions to the New York Bight: polychlorinated biphenyls, petroleum hydrocarbons, and nutrients. Concern that these materials may be causing potentially significant but non-specific effects led the Panel on the Management of Non-Point Sources to recommend the continuation or initiation of programs to control inputs of these materials to the New York Bight. Recommended steps include: (1) removal of Hudson River sediments heavily contaminated with PCB's, with either incineration or permanent regulated landfills as the ultimate control; (2) collection of used crankcase oils and control of oil tanker discharges; and (3) continuation of programs aimed at reducing soil and nutrient losses from farmlands. Concern was expressed that controls on point sources may shift pollutants to non-point source discharges. This, in turn, may require modification of non-point source control programs.

Panel recommendations were based, in part, on reports or proposals whose benefits have not been tested fully. Consequently, non-point source control programs such as dredging require simultaneous monitoring and process modification to assure achievement of expected results.

INTRODUCTION

The term "non-point sources" includes all sources that cannot be accounted for as specific isolated inputs to an aquatic system. Typically, but not exclusively, these include atmospheric inputs, runoff water from land areas, riverine and estuarine inputs, and stormwater overflows from sewage systems. Although some materials such as crankcase oils and bilge washings technically are related to point sources, discharges of this type are considered "non-point" sources because of the difficulty of monitoring and controlling multiple mobile sources.

The relative importance of non-point sources varies with the substances under consideration. Problems may be caused by a range of materials from soils, whose inputs from street washing and land erosion are measured in terms of metric tons, to polynuclear hydrocarbon compounds, which are present in nanogram per cubic meter concentrations in the atmosphere and in nanogram per gram to microgram per gram levels in aquatic sediments. No easy generalizations are possible on the relative importance of individual non-point sources of pollution.

[1]Chairman, Panel on the Management of Non-Point Sources.

Papers presented in association with this symposium and in the literature provide evidence of ecosystem damage or the potential for such damage from exposure to pollutants contributed by non-point sources. Bottom-dwelling organisms that are the predominant food items for demersal fishes are less abundant in benthic communities of the New York Bight apex (Boesch, this volume). These changes are most distinct near sewage sludge and dredged material deposits (Steimle *et al.,* this volume). In 1974, mackerel eggs exhibited developmental abnormalities and greater than normal mortality; corresponding data for 1977 associated this mortality with levels of toxic metals, aromatic hydrocarbons, and chlorinated hydrocarbons. The relationship strongly suggests impact by one or a combination of toxic materials (Longwell and Hughes, this volume).

Laboratory studies indicate that polychlorinated biphenyls (PCB's) may affect phytoplankton populations by inhibiting growth and shifting species distributions toward organisms of smaller sizes (O'Connors *et al.,* 1978). Field and laboratory studies show that aromatic hydrocarbons present a risk of toxic effects in the ecosystem based on measured concentrations in sediments and organisms (McCain and Malins, this volume). Fin erosion disease is prevalent in fishes found in contact with contaminated sediments, although no clear cause and effect relationship has been demonstrated (Murchelano and Ziskowski, this volume; Sherwood, this volume).

Both chlorinated hydrocarbons and aromatic hydrocarbons are known to induce cellular enzyme systems in many aquatic organisms as well as in man. Toxicant discharges to the Bight are believed sufficient to cause enzyme inductions in exposed organisms (Malins, 1977) and may be related to the presence of liver abnormalities observed in fishes, although the significance of these pathological observations is unknown.

Methods of observing population dynamics of plankton and fish species do not afford sufficient precision as yet to separate clearly the effects of pollutants from those of fishing pressures and natural population cycles. Thus, there is no evidence of pollutant impacts on the stock size of any commercial finfish in the Bight (Sindermann *et al.,* this volume). Commercial catches have increased in the last few years, primarily as a result of reduced fishing pressure. While the data have not been analyzed, many sport fishermen believe that sport fishing for local inshore fish stocks via contracted boats has gradually shifted east along Long Island's south shore. This could reflect declining profitability of operations from Brooklyn docks caused by economic factors or shifts in inshore species distributions. If the latter hypothesis were documented, it could be an early warning comparable to the loss of fisheries resources first noted for the East River in 1878 (Carpenter *et al.,* in press) and recently repeated by the 1977 closing of the striped bass fishery in the Hudson River estuary as a result of PCB contamination (Hetling *et al.,* 1978).

The potential for pollution-related ecological effects has led several scientific panels in this symposium to call for the control or evaluation of control technology for designated toxic substances. Controls have been recommended for PCB's and similar chlorinated materials, and considerable concern has been expressed about discharges of aromatic hydrocarbons, toxic metals, nutrient nitrogen, and carbon. The possibility of combined effects causing general ecosystem degradation also requires consideration and the development of possible management options with regard to such degradation.

The non-point sources panel has reviewed the findings of the five scientific workshops (part I, this volume) to identify which pollutants contributed by non-point sources are most in need of management attention. The panel's findings and recommendations for control are presented below.

POLYCHLORINATED BIPHENYLS

Inputs

Upper Hudson River

General Electric capacitor manufacturing facilities at Fort Edward and Hudson Falls, New York were the major sources of PCB's to the upper Hudson River (*i.e.*, above Troy, New York) until the mid-1970's (Hetling *et al.*, 1978). The history of these discharges is summarized in Table 1. The use of PCB's began at Fort Edward in 1947 and at Hudson Falls in 1952. No data exist for discharges prior to 1966; however, in the early 1970's, discharges averaged 15 kg/day. By May 1977, PCB discharges from the facilities virtually had ceased.

Much of the PCB's released into the upper Hudson was contained by the Fort Edward Dam. In October 1973, this structure was removed, and PCB-laden sediments were resuspended and transported downstream. Currently, non-point sources of PCB's to the Hudson River include deposits in sediments that continue to erode and move downstream, leaching from landfills, atmospheric deposition, and land surface runoff (Hetling *et al.*, 1978).

The distribution of PCB's in the Hudson River is summarized in Table 2. Descriptive terminology and locations of various types of deposits were adopted from Hetling *et al.* (1978). "Remnant deposits" are sediments located downstream of former General Electric discharges but upstream of the previous Fort Edward Dam. A portion of the remnant area has been stabilized and the remainder of the high PCB concentrations in the area (25-1,000 ppm, dry wt) can be stabilized or removed (Hetling *et al.*, 1978).

Table 1. Historical PCB discharge to the Hudson River from General Electric facilities at Fort Edward and Hudson Falls, New York. From Hetling *et al.* (1978).

Date	Quantity (kg/day)
Before 1966	No data
Early 1970's	15
1976	0.25
May 1977	0.001

Table 2. Summary of PCB distribution in Hudson River. ID = insufficient data. Derived from Hetling *et al.* (1978).

Area	Total PCB's (kg)	Scourable PCB's (kg)	Loss rate (kg/yr)
River sediments	(288,000)		
Remnant deposits	64,000	20,000	3,900
Upper Hudson			
Hot spots	77,000	44,000	2,600
Cold spots	57,000	ID	ID
Lower Hudson	90,000	ID	ID
Land deposits	(313,000)		
Dredge spoil areas	73,000	ID	80
Landfills and dumps	240,000	–	360
Biota			
Lower Hudson	90-900	–	–
Total	600,000	64,000 +	6,900 +

"Hot spot" areas (> 50 ppm, dry wt) contain over half the PCB's in the upper Hudson below the remnant deposits but comprise only 8% of the total area (Figure 1). Their removal is amenable to economic dredging. Core samples indicate that some of the material has been buried; dredging to depths of 0.5-1.0 m would be required to achieve 95% removal of contaminated sediments. Average PCB concentrations in hot spot areas vary from 70 ppm to 160 ppm, dry wt (Hetling *et al.*, 1978).

PCB concentrations in "cold spot" areas (<50 ppm) vary from 9 ppm to 22 ppm, dry wt. Concentrations in the lower Hudson vary from 60 ppm at the Albany Turning Basin to 6 ppm in New York harbor (Hetling *et al.*, 1978).

Atmospheric Fallout

One concern of PCB control is the ability of the material to be transported in the atmosphere. It has been reported that atmospheric input is the largest source of PCB's to Lake Superior (Eisenreich *et al.*, 1979). The direct atmospheric input of PCB's to the New York Bight is not well defined. However, loading estimates of PCB's to the entire Hudson River drainage area indicate that 3.2 kg/day can be attributed to atmospheric deposition, of which only 0.45 kg/day are estimated to be in urban runoff (Mancini *et al.*, in press). Considerably less than 3.2 kg/day may reach the New York Bight because a significant portion of atmospheric fallout is retained on the land surface.

Direct atmospheric fallout to the Bight also must be considered. Assuming the mean continental fallout rate of PCB's to be 80 $\mu g/m^2/yr$ (Hetling *et al.*,

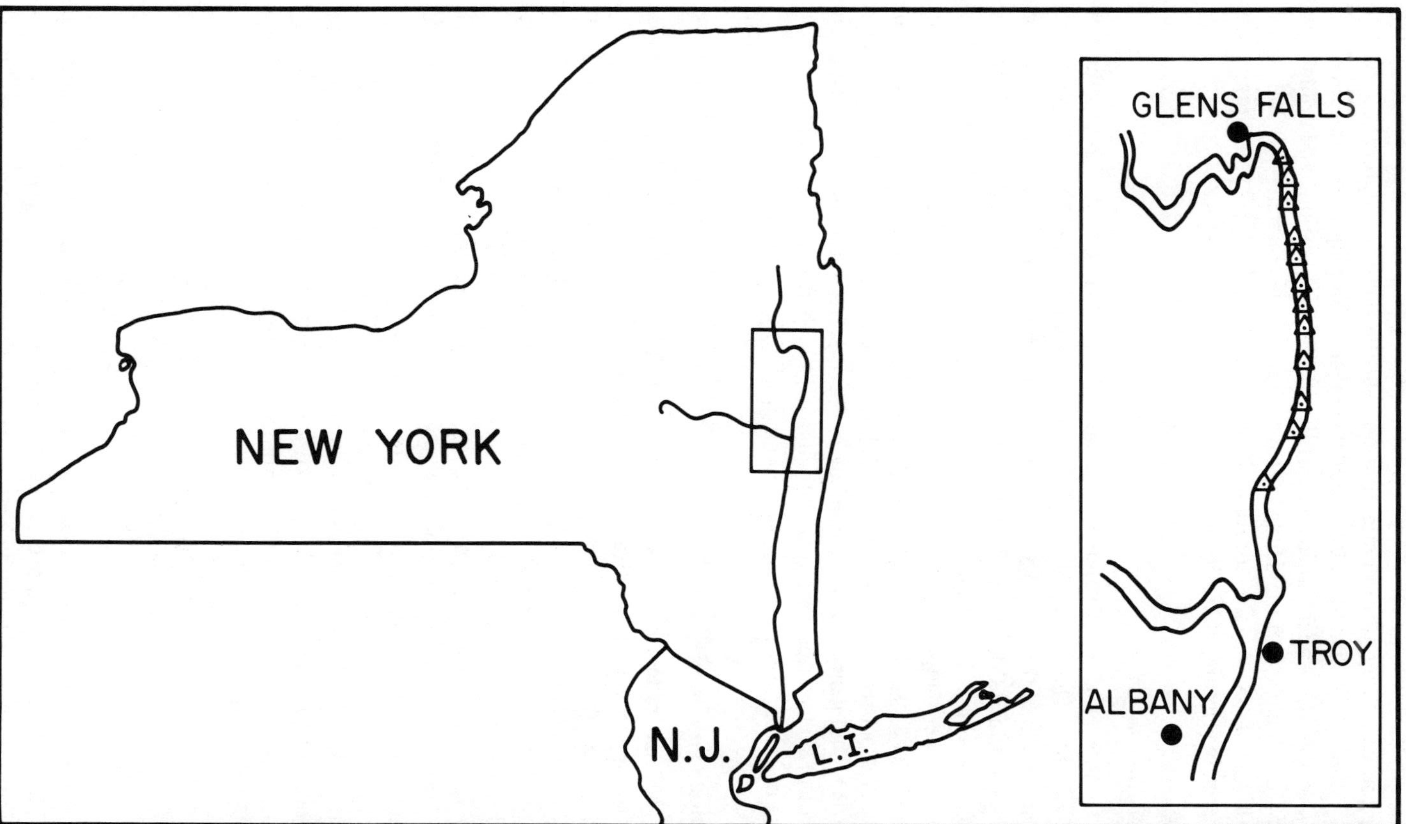

Figure 1. PCB hot spot areas in the upper Hudson River. (Courtesy New York Sea Grant Institute.)

1978), and adjusting this value to 40 $\mu g/m^2/yr$ as an estimate of average reduction caused by atmospheric dilution over the Bight, the deposition over the Bight could be estimated as 0.85 kg/day (310 kg/yr). Major emission sources of PCB's to the atmosphere include landfill and dump operations and dredge spoil dumps. Estimates for the Hudson River basin indicate that 240,000 kg and 73,000 kg of PCB's are located at landfill and dredge spoil dump sites, respectively (Hetling *et al.*, 1978).

Lower Hudson River

O'Connor *et al.* (this volume) summarized mass loadings of PCB's to the New York Bight. The disposal of dredged materials from New York harbor contributes 3,500 kg PCB/yr, while sewage sludge dumping adds between 600 kg/yr and 2,000 kg/yr. These values compare to estimates of 3,100 kg/yr passing the Green Island Dam at Troy, and a total of 2,500 kg/yr from sources south of the dam to and including the harbor area (O'Connor *et al.*, this volume). The estimates suggest that dredged materials are the most significant immediate source of PCB's to the Bight. Control of inputs that contaminate estuarine sediments includes control of inputs from the upper Hudson and control of sewage treatment plant discharges, which are the next most important source of PCB's to the estuary (O'Connor *et al.*, this volume).

Disposal Options

Significant reduction of the quantity of PCB's entering the environment is a goal of any national or local control program. The removal of PCB's from current uses (*e.g.*, transformers) or from inappropriate current disposal sites (*e.g.*, the Hudson River) will be useful only if the final disposal or destruction of the material does not create additional public health or environmental damage.

The disposal options for PCB's in lieu of dumping in the New York Bight are decomposition through incineration or deposition in controlled landfills. The U.S. Environmental Protection Agency (EPA) has facility criteria for PCB incineration which protect the environment (U.S. Environmental Protection Agency, 1979). The criteria require combustion with a 2.0 second dwell time at 1,200 ± 100°C and 3% excess oxygen in the stack gas, or a 1.5 second dwell time at 1,600 ± 100°C with 2% excess oxygen in the stack gas (U.S. Code of Federal Regulations, 40 CFR Part F61.4, Annex 1). Combustion efficiency shall be at 99.9%, as determined by the ratio:

$$\frac{[CO_2]}{[CO_2 + CO]} \times 100.$$

Monitoring requirements are specified in detail by the criteria. In the New York Bight area there are no approved PCB incinerators, but at least two applications for approval exist. The limited effort to establish PCB incineration in this area where PCB's have received so much attention suggests that land disposal rather than incineration will continue to be sought as the prime means of disposal.

The use of landfills to hold PCB's has been problematic to date. Efforts by General Electric and the New York Department of Environmental Conservation to dispose of sludge dredged from waterways around General Electric plants have raised as many concerns for environmental contamination as did the original deposits. Concern over PCB's in groundwater exists, and some monitors have shown measurable groundwater PCB levels in the area around landfill sites containing dredge spoils. Effective disposal of PCB's in landfill sites requires that spoils be encapsulated to prevent erosion, and that the cover be effective in reducing volatilization and subsequent emission of PCB's to the atmosphere.

Research is underway by the EPA and the Electric Power Research Institute to study disposal of PCB's in secure landfills. This must be encouraged. While research continues, PCB's can be found in barrels in warehouses throughout the New York Bight area. Under such storage conditions, PCB spills are a constant risk and have the potential for creating additional PCB non-point sources (Figure 2).

One alternative to current storage practices is to create a hazardous waste disposal site that meets the requirements of Section 3004 of the Resource

Figure 2. Coast Guardsmen and New Jersey fireman hose down toxic chemical dump site which exploded and burned in Elizabeth City, New Jersey. Chemicals including PCB's leaked from the site into New York Harbor. (Courtesy U.S. Coast Guard, 3rd District, Public Affairs Office.)

Conservation and Recovery Act. There can be no assurance that storage of PCB's in barrels followed by disposal in hazardous waste facilities will prevent all contamination of groundwater or surface water. However, control in these facilities can be close to absolute and is the best available short of total destruction by incineration. In 1979, the cost of disposal in hazardous waste disposal sites was projected to reach $110 per metric ton by the 1980's (F.C. Hart, personal communication).

Management Issues

River Sediments

Estimates of sedimentation in the 8 km south of the former site of the Fort Edward Dam (the area containing the highest concentration of PCB's in hot spots) suggests that less than 0.3 m of deposition is expected to occur over the next 20 years (Hetling *et al.*, 1978). This does not appear adequate to seal the sediment and prevent future transfer of PCB's to the water by leaching. Continued maintenance dredging in the upper Hudson (*i.e.*, dredging to maintain shipping channels) would remove 23,000 kg sediment over a ten-year span, with removal of 8% of the PCB's currently in the Hudson.

For the PCB's stored currently in Hudson River sediments, the following management options have been considered (Hetling *et al.*, 1978): (1) stabilize or remove remnant deposits–remove 44,000-74,000 kg PCB over a two-year period at a cost of $4.5-6.3 x 10^6 ($84-100/kg PCB); (2) remove remnant deposits and dredge hot spots–remove 140,000 kg PCB over a three-year period at a cost of $28.7 x 10^6 ($202/kg PCB); and (3) remove all sediments contaminated to over 50 μg/g–remove 185,000 kg PCB over a ten-year period at a cost of $204 x 10^6 ($1,100/kg PCB).

The second management alternative of removing (or stabilizing) remnant deposits and hot spots will remove 64% of the PCB's from the Hudson River (Hetling *et al.*, 1978). This strategy should be preferable with concurrent monitoring to determine the short-term effects of dredging operations and the long-term effects of PCB-laden sediment removal. The strategy should be modified if a need is indicated by monitoring results. Costs of the necessary monitoring are included in the above estimates at 10% of total project costs.

Disposal of contaminated sediments in encapsulated landfills should be viewed as a temporary measure, and research should be conducted on economical methods to remove and/or destroy PCB's in the sediments. PCB concentrations in dredge spoils inevitably will be lower than those of *in situ* sediments. Sediment incineration must provide for the handling, drying, and heating to over 1,200°C of large masses of inert soils; consequently, it does not appear to be an attractive alternative.

Control of Direct Sewage Discharges

To some extent, the PCB content of sewage discharges is dependent on current uses of these compounds and is not representative of atmospheric deposition and runoff inputs. Continuing long-term programs that require elimination of PCB's will reduce the PCB levels of such discharges. Control by removal at the treatment plants is unlikely to be economically feasible because of the low concentration of PCB's in the wastes and the large volumes of effluents to be handled (see also Mytelka *et al.*, this volume).

Control of Atmospheric Sources

Work completed by Amy *et al.* (1974) has indicated that PCB's are present in street dust, which is available for runoff. Concentrations on the order of 0.7-0.8 μg/g of dry solids have been measured; these concentrations are, by comparison, approximately an order of magnitude higher than those observed for mercury and DDT. It would appear that street dust could be controlled through the implementation of effective street cleaning programs. Such programs would have the added benefit of reducing pesticides, hydrocarbons, and toxic metals in runoff. Runoff provides 0.45 kg PCB/day (164 kg/yr) to the Bight, as noted previously, at a concentration about 100 times lower than that in river sediments of concern ($>$ 50 μg/g). In terms of both total PCB's and PCB concentrations in solids, runoff control should not receive as high a priority as control of the major sources.

Monitoring

Atmospheric monitoring of PCB's should be considered over the Bight and surrounding land areas to determine if concentrations are changing once control programs are implemented. Specifically, more measurements should be conducted over the Bight to get a firmer understanding of PCB distributions and to allow, in the near future, more quantitative estimates of the direct input. Reductions by a factor of two or three in major inputs, or increases of the same order in atmospheric inputs, would promote the atmosphere to a major source.

PETROLEUM AND ITS COMPONENTS

Inputs

Petroleum hydrocarbons and associated aromatic hydrocarbons are, and will continue to be, causes of ecosystem degradation in the New York Bight (Sindermann *et al.*, this volume). Accumulation of polynuclear aromatics in sediments eventually may prove harmful to benthic communities in the Bight. MacKenzie and Hunter (1979) showed preferential degradation of the less harmful compounds, with high boiling point aromatic sulfur and hydrocarbon compounds retained in sediments.

Tanacredi (1977) showed the major source of petroleum contamination in Jamaica Bay to be waste crankcase oil. This is in agreement with studies of Delaware Bay by MacKenzie and Hunter (1979). Sources have been estimated for Delaware Bay as shown in Table 3 (R. Field, personal communication). It is clear from Tanacredi's work and from the data in Table 3 that urban runoff is the major factor to be considered once existing regulations for point sources are adequately enforced.

Petroleum hydrocarbons may enter urban runoff from many dispersed sources. For example, as discussed above, aromatics in stormwater runoff may be attributed primarily to crankcase oil. Based on an average stormwater runoff concentration of 3.7 mg petroleum hydrocarbons per liter, contributions of petroleum-related substances in Philadelphia were calculated to equal 2.6 $g/m^2/yr$. In the New York area, inputs of hexane extractable material (HEM) were estimated to be 0.18-3.9 g $HEM/m^2/cm$ overflow for combined sewer overflows (CSO's), and 0.03 g $HEM/m^2/cm$ overflow for separate stormwater discharges in Jamaica Bay (Feuerstein and Maddaus, 1976). An input of 81.8 metric tons of oil and grease was reported for a CSO at the Newton Creek treatment plant in New York City (Mytelka *et al.*, 1973). The Jamaica Bay study indicated that 51% of the HEM is contributed by sewage treatment plant effluents, 41% by CSO's, and only 3% by separate stormwater discharges; however, total contributions are dependent on the size of the drainage area (or land surface). Both non-sewered and sewered areas contribute significant fractions to HEM inputs from the Borough of Staten Island and the contributing areas of New Jersey, upper New York State, and Connecticut.

Over 86% of the total hydrocarbons in Philadelphia storm runoff was associated with particulates, a distribution that probably is typical of most large urban areas. Therefore, instream solids separations being designed and considered for CSO's and separate stormwater systems will result in substantial lowering of non-point petroleum hydrocarbon inputs to the Bight, provided the solids are disposed of elsewhere.

Management Issues

Assessment of sources of petroleum hydrocarbons to New York waters is needed. Particular attention should be given to the impact of CSO's because of the large "initial flush" effect, which can result in major discharges of hydrocarbons absorbed on solids.

Table 3. Estimated sources of petroleum hydrocarbons in Delaware Bay.

	Without efficient controls (kg/day)	With efficient controls (kg/day)
Spills	2,700	2,700
Municipal	3,500-7,100	900
Refineries	11,000	900
Other industrial	4,000	2,800
Urban runoff	4,800	4,800

Water quality would benefit from controls on the dumping of crankcase oil by individuals. Elements of an effective control effort could include penalties on dumping, incentives at the retail level (*i.e.*, refundable deposits for the return of used oil), and motor vehicle inspections oriented to oil loss.

Other strategies to reduce petroleum hydrocarbon discharges should await completion of a study of petroleum hydrocarbon sources to the Bight. Remedial measures could include more stringent regulation of waterborne shipping (particularly oil tankers), better enforcement of existing regulations for wastes from motor vehicle repair shops and gasoline stations, and solids separation by holdup and settling for CSO and stormwater discharges. Currently, the latter are being considered in programs for control.

NUTRIENTS AND CARBON

The scientific panel on nutrients and carbon (Mearns *et al.*, this volume) voiced serious concern over the effects of excesses of these materials on the diversity and productivity of local marine waters. It also noted that most nutrients result from human waste discharge. A significant quantity of nutrients also originates from non-point sources.

Inputs

Importance of Non-Point Sources

Although significant, non-point sources are not major contributors of nutrients to the New York Bight. Table 4 shows that urban runoff contributes 15% of the total organic carbon, 4% of the total nitrogen, and 4% of the total phosphorus. Gaged runoff specifies river pollutant loads at a sampling site above tidal influences. Since the upstream segments of rivers draining into the New York Bight flow through major urban and industrial areas, one may safely assume that a large portion of the gaged nutrient load originates from point sources. Several studies verify this supposition. The "208" study for northeast New Jersey (New Jersey Department of Environmental Protection, 1980a) calculates that in the study area, point sources generate more than 20 times the carbon loading, more than 15 times the nitrogen, and more than 27 times the phosphate contributed

Table 4. Percent nutrient contribution to the New York Bight. From Mueller *et al.* (1976).

	Direct		Wastewater		Runoff	
	Barge	Atmospheric	Municipal	Industrial	Gaged	Urban
Total organic carbon	25	12	29	1	18	15
Total nitrogen	16	13	40	2	25	4
Total phosphorus	50	0.7	35	1	9	4

by non-point sources. The "208" plan for the upper Raritan River (New Jersey Department of Environmental Protection, 1980b) also suggests the dominance of point source nutrient loading in urbanized areas. Calculations in the plan show that in the Raritan River at South Bound Brook, point sources account for 44% of the total organic carbon load, 95% of the nitrogen, and 91% of the phosphorus.

Runoff

Nutrients from non-point sources originate from the air, natural sources, and dispersion on land by human activities, and enter surface water directly in runoff or indirectly through groundwater. All land, even undisturbed forest, contributes nutrients to surface waters. Human activities, however, greatly increase the rate of nutrient entry. Table 5 presents a comparison of average nutrient contributions from forested land, agricultural land, and urban land. Nutrients from agricultural land originate from erosion and leaching of disturbed soil, fertilizers, and animal wastes. Major sources of nutrients in urban runoff are refuse awaiting collection, dog wastes, lawn fertilizers, and septic tanks. Landfill leachates can have high concentrations of nutrients and can cause severe local problems. Nevertheless, since they comprise a small proportion of total land area, landfills do not have much effect on New York Bight nutrient levels.

Atmospheric Inputs

Mueller *et al.'s* 1976 study of contaminant sources to the New York Bight suggests that 10-12% of the total carbon and 13% of the nitrogen are derived from the atmosphere. There are two mechanisms by which these materials might enter the New York Bight. The first is localized deposition from the atmosphere and subsequent transport to waterways. The second is the movement of these materials in the atmosphere, often in the vapor phase, and their rainout into the New York Bight through precipitation. As other sources of nutrients and carbon are reduced, air pollution sources could contribute 15-20% of these materials to the New York Bight.

Carbon sources include fuel transportation and transfer, fuel storage terminals, refineries, and combustion of fuels by trucks and automobiles.

Table 5. Contributions of nutrients from forested, agricultural, and urban areas ($kg/km^2/day$). From U.S. Environmental Protection Agency (1976).

	Total nitrogen	Total phosphorus	Biological oxygen demand[1]
Forested land	7.0	0.29	9.4
Agricultural land	17.6	1.2	71
Urban land	9.4	1.5	82

[1] BOD_5

Nitrogen sources include combustion of fuel by power plants, industry, residential and commercial buildings, and vehicles.

Management Options

Although non-point source nutrient and carbon controls cannot be justified solely on the basis of their effect on the New York Bight, they are being instituted for the protection of inland waters. Because the measures noted in the following sections will have some effect on the Bight and possibly a greater effect on adjacent estuaries, the further development of these measures is encouraged. Furthermore, some of these measures have other benefits–*i.e.*, soil protection in agriculture, flood protection in controlling urban stormwater, and air quality improvement for air pollution controls.

Atmospheric

The air pollution controls that EPA has set to meet the public health and environmental goals of the Clean Air Act are the best options for dealing with related New York Bight problems. There exist pressures to avoid implementation of some of these controls because of conflicts with economic and energy goals. Ongoing control programs should lead to reductions in atmospheric carbon, nitrogen, and toxic element inputs to the Bight. The plight of the New York Bight suggests that there is value to the Bight in pursuing these air pollution programs, in addition to the values related to public health goals.

Agricultural

Various programs under the U.S. Department of Agriculture, the Soil Conservation Service, and local Soil Conservation Districts encourage farmers to institute "best management practices" for protection of water quality. This encouragement ranges from advice by local Soil Conservation District personnel to programs that provide cost sharing for instituting these practices on a farmer's land. At this time, these programs depend on voluntary participation by farmers. Furthermore, some of these practices may benefit farmers by protecting their soil or lowering their rate of fertilizer usage, in addition to protecting water quality. The New Jersey "208" program has noted 77 water quality protection "best management practices" (New Jersey Department of Environmental Protection, 1980c) that may be applied according to the unique characteristics of individual sites.

Urban/Suburban

Ameliorating the effects of nutrient pollution from urban runoff from developed land may be difficult, because modification of existing stormwater drainage systems is probably too expensive to be practical. In new developments, on the other hand, stormwater control measures may be instituted at construction. A number of local governments already require retention basins in new developments in order to control stormwater for flood protection purposes. These basins can be designed to diminish the quantity of nutrients entering surface waters through runoff. The Delaware Valley Regional Planning Commission, with funds from the "208" program in New Jersey, is investigating the efficacy of various stormwater controls, as well as legal and management considerations for implementing them (Delaware Valley Regional Planning Commission, unpublished data). Certain nonstructural measures such as street sweeping or dog waste control also may be feasible, and should receive further consideration.

CONCLUSIONS

The scientific position papers (part I, this volume) have documented effects of toxic substances on individual organisms, on local fisheries, and on biological communities. These changes may be related to the general influence of toxic substances, although other factors also may be significant. Non-point sources contribute to the total burden of toxic substances in the Bight, and logical resource management should consider strategies that would tend to reduce the quantity of toxic substances entering the system. Since the tie between particular toxic discharges and effects has not been fully defined, it would appear prudent to commit available resources in stages appropriate to the perception of the degree and probability of the various threats.

Existing programs for point source control in the region should result in reductions in the quantities of toxic materials entering the Bight. Planned pretreatment requirements for industrial wastes discharged to publicly-owned treatment plants and treatment of selected combined sewer overflows should provide further, perhaps substantial, reductions.

The point sources and combined sewer overflows currently account for well over half of the toxic metals discharged to the harbor and other inland waters. The remaining discharges of these metals can be classified as non-point source loads and are associated with upstream inputs from rivers, urban runoff, and atmospheric fallout.

Projected control activities generally address point sources where control actions historically have been more effective from the standpoint of the cost per kilogram of pollutant removed. Non-point source control measures are usually less cost-effective because of the lower concentrations and larger volumes of liquids or other wastes that must be handled. It may be concluded, therefore, that existing programs that emphasize the control of pollutant discharges are consistent with the data available, and that they appear to be effective management strategies with respect to the New York Bight.

Recommendations

Several activities related to non-point source reduction in toxic loads should be considered.

Polychlorinated Biphenyls (PCB's)

The proposed removal of contaminated sediments in the upper Hudson River represents an appropriate program for control of this non-point source, with implementation to be based on continual monitoring (option 2 of the section on PCB's).

Disposal of concentrated PCB wastes can be carried out in properly regulated landfills as an interim procedure. Destruction of concentrated wastes may be possible by proper incineration methods. If incineration proves unfeasible, regulated landfills should be upgraded to permanent repositories.

Petroleum Hydrocarbons

Development of economic incentives for the collection of used oil would have benefits in terms of reducing non-point loads of hydrocarbons from runoff to the Bight. Control of discharges from tankers should be effectively enforced.

Nutrients and Carbon

Information and education programs exist to reduce nutrients and soil lost from agricultural lands. Continuance of these programs would have the benefit of reducing costs to farmers while also reducing the loads of nutrients and toxic materials to the system.

General

Ongoing programs for control of point sources should proceed. Attention must be given to possible effects of these programs on the amounts or significance of non-point source toxic material inputs.

In summary, greatly improved programs to control point sources, together with some relatively low-cost non-point source control measures, provide realistic initial steps toward reducing the input of nutrients, carbon, and toxic substances to the Bight. The effectiveness of these steps should be monitored, with the objective of providing data for a resource management program that sequentially commits funds based on increasing knowledge of cause and effect relationships, costs, and environmental benefits.

REFERENCES

Amy, G., R. Pitt, R. Singh, W. Bradford, and M. LaGraff. 1974. Water Quality Management Planning for Urban Runoff. EPA Report 440/9-75-004. U.S. Environmental Protection Agency, Office of Planning and Standards, Washington, DC. 247 pp.

Boesch, D.C. This volume. Ecosystem consequences of alterations of benthic community structure and function in the New York Bight region. pp. 543-568.

Carpenter, J.H., P.W. Anderson, A. Calabrese, K.S. Kamlet, T.J. Kneip, and A. Michael. In press. Implications of Current and Possible Future Contaminant Levels in the Apex of the New York Bight. NOAA Special Report.

Eisenreich, S.J., G.J. Hollod, and T.C. Johnson. 1979. Accumulation of polychlorinated biphenyls (PCBs) in surficial Lake Superior sediments: Atmospheric deposition. Environ. Sci. Technol. 13: 569-573.

Feuerstein, D.L. and W.O. Maddaus. 1976. Wastewater Management Program, Jamaica Bay, N.Y. Vol. I. Summary Report. EPA 600/2-76-222a. U.S. Environmental Protection Agency, Municipal Environmental Research Laboratory, Cincinnatti, OH. 183 pp.

Griffen, P.M., C.M. McFarland, and A.R. Sears. 1978. Research on Removal or Treatment of PCB in Liquid or Sediments Dredged from the Hudson River: Semi-Annual Progress Report. General Electric Co., Corporate Research and Development, Schnectady, NY.

Hetling, L., E. Horn, and J. Tofflemire. 1978. Summary of Hudson River PCB Study Results. New York State Department of Environmental Conservation Tech. Pap. No. 51. 88 pp.

Longwell, A.C. and J.B. Hughes. This volume. Cytologic, cytogenetic, and embryologic state of Atlantic mackerel eggs from surface waters of the New York Bight in relation to pollution. pp. 381-388.

MacKenzie, M.J. and J.V. Hunter. 1979. Sources and fates of aromatic compounds in urban stormwater runoff. Environ. Sci. Technol. 13: 179-183.

Malins, D.C. 1977. Metabolism of aromatic hydrocarbons in marine organisms. Ann. N.Y. Acad. Sci. 298: 482-496.

Mancini, J., J.A. Mueller, J.J. Roswell, J. Goldman, A.I. Mytelka, T. Bober, E. Myers. In press. Urban Loads to the New York Bight. NOAA Special Report.

McCain, B.B. and D.C. Malins. This volume. Effects of petroleum hydrocarbons on selected demersal fishes and crustaceans. pp. 315-326.

Mearns, A.J., E. Haines, G.S. Kleppel, R.A. McGrath, J.J.A. McLaughlin, D.A. Segar, J.H. Sharp, J.J. Walsh, J.Q. Word, D.K. Young, and M.W. Young. This volume. Effects of nutrients and carbon loadings on communities and ecosystems. pp. 53-65.

Mueller, J.A., J.S. Jeris, A.R. Anderson, and C.F. Hughes. 1976. Contaminant inputs to the New York Bight. NOAA Tech. Memo. ERL MESA-6. 347 pp.

Murchelano, R.A. and J. Ziskowski. This volume. Fin rot disease in the New York Bight (1973-1977). pp. 347-358.

Mytelka, A.I., L.P. Cagliostro, D.J. Deutsch, and C.A. Haupt. 1973. Combined Sewer Overflow Study for the Hudson River Conference. EPA R2-73-152. U.S. Environmental Protection Agency, Office of Research and Monitoring, Washington, DC. 287 pp.

Mytelka, A.I., J. Ciancia, R.R. Delgado, A.M. Levinson, T.P. O'Connor, J.C. Prager, W.B. Pressman, E.A. Regna, M.M. Sadat, J.T.B. Tripp, and M. Wendell. This volume. Management of industrial wastes. pp. 135-143.

New Jersey Department of Environmental Protection. 1980a. Northeast New Jersey Water Quality Management Plan. Division of Water Resources, New Jersey Department of Environmental Protection, Trenton, NJ. 643 pp.

New Jersey Department of Environmental Protection. 1980b. Upper Raritan Water Quality Management Plan. Division of Water Resources, New Jersey Department of Environmental Protection, Trenton, NJ. 524 pp.

New Jersey Department of Environmental Protection. 1980c. Monmouth County Water Quality Management Plan. Division of Water Resources, New Jersey Department of Environmental Protection, Trenton, NJ. 571 pp.

O'Connor, J.M., J.B. Klotz, and T.J. Kneip. This volume. Sources, sinks, and distribution of organic contaminants in the New York Bight ecosystem. pp. 631-653.

O'Connors, H.B., Jr., C.F. Wurster, C.D. Powers, D.C. Biggs, and R.G. Rowland. 1978. Polychlorinated biphenyls may alter marine trophic pathways by reducing phytoplankton size and production. Science 201: 737-739.

Sherwood, M.J. This volume. Fin erosion, liver condition, and trace contaminant exposure in fishes from three coastal regions. pp. 359-377.

Sindermann, C.J., S.C. Esser, E. Gould, B.B. McCain, J.L. McHugh, R.P. Morgan II, R.A. Murchelano, M.J. Sherwood, and P.R. Spitzer. This volume. Effects of pollutants on fishes. pp. 23-38.

Steimle, F., J. Caracciolo, and J.B. Pearce. This volume. Impacts of dumping on New York Bight apex benthos. pp. 213-223.

Tanacredi, J.T. 1977. Petroleum hydrocarbons from effluents: detection in marine environment. J. Water Poll. Control Fed. 49: 216-226.

U.S. Environmental Protection Agency. 1976. Areawide Assessment Procedures Manual. Vols. I, II, III. EPA 600/9-76-014. U.S. Environmental Protection Agency, Municipal Environmental Research Laboratory, Cincinnati, OH. 2129 pp.

U.S. Environmental Protection Agency. 1979. Polychlorinated biphenyl criteria modification hearings. Fed. Reg. 44(106, pt. 6): 31514-31568.

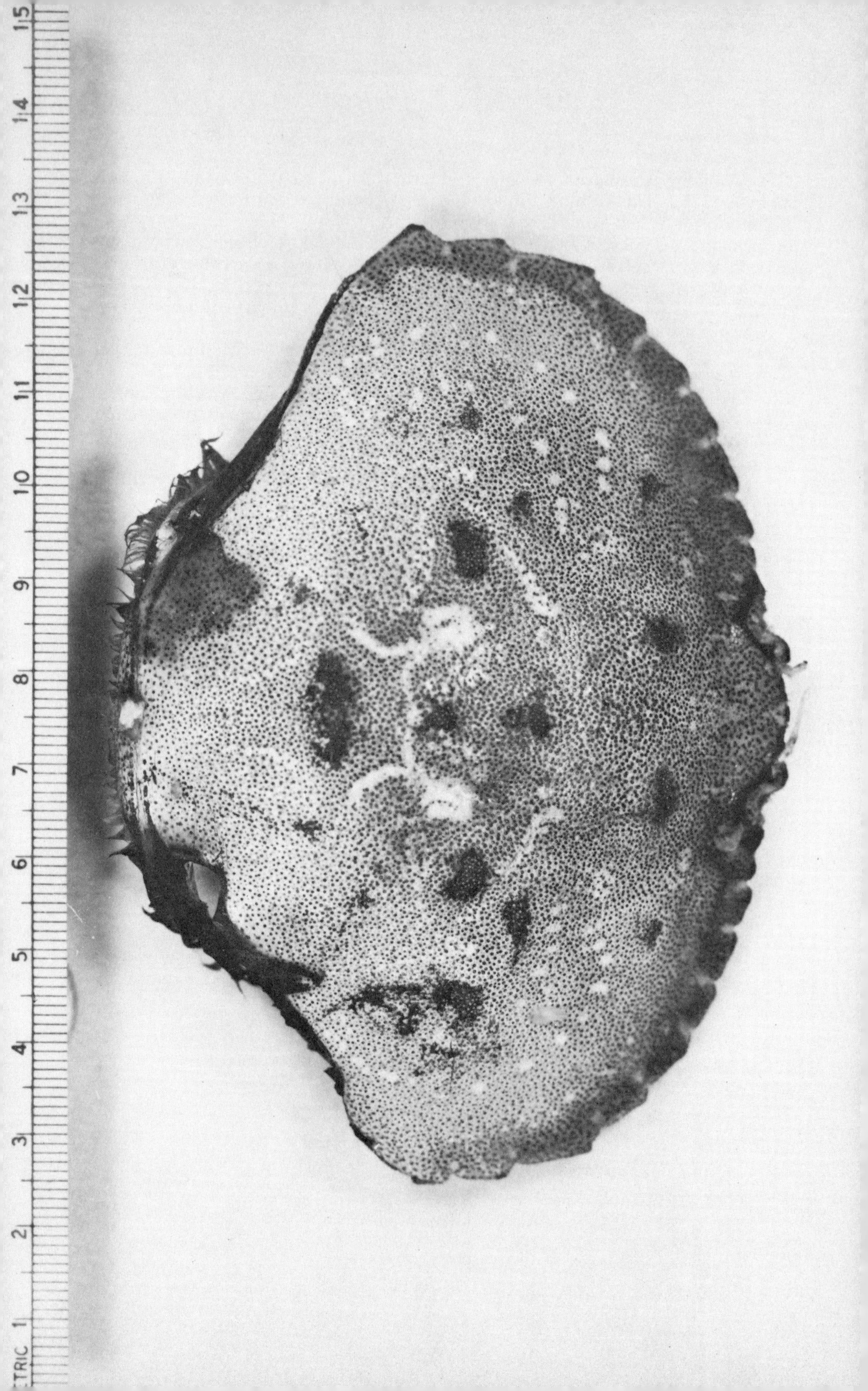

CONTRIBUTED PAPERS ON THE EFFECTS OF POLLUTANTS ON BENTHOS

Opposite: The carapace of a rock crab (*Cancer irroratus*) from Sandy Hook Bay, New Jersey. Blackened "fuzz" around the posterior margin of the shell, blackened ulcerations in the center of the carapace, and shell erosion evident opposite the 5-6 cm marks on the ruler may reflect the poor environmental quality of the inner Bight and lower Hudson-Raritan estuary. (Courtesy T.K. Sawyer, NOAA National Marine Fisheries Service.)

THE TRANSPORT OF PETROLEUM HYDROCARBONS FROM SEDIMENTS TO BENTHOS AND THE POTENTIAL EFFECTS

J.W. Anderson

Battelle
Pacific Northwest Laboratories
Marine Research Laboratory
Sequim, Washington 98382

Abstract. Polynuclear aromatic hydrocarbons (PAH) enter the New York Bight region from multiple sources. Sewage sludge and bottom sediments collected from the Bight contain significant amounts (50-500 ppb) of various PAH components, while sediments from the area slated for dredging may contain very high levels (1-120 ppm). Whereas the surface microlayer in the Bight was relatively free of petroleum hydrocarbons, plankton contained 30-200 ppb of 2- and 3-ring aromatics. Fish and lobster flesh and mackerel liver contained relatively low levels of PAH. The detoxification organs of flounder and lobster contained the highest amounts of PAH, and lobster digestive glands possessed contaminants in the full range of those detected (2- to 5-ring aromatics). Blue mussels exhibited contamination intermediate to the flesh and livers of fish and lobster. Possible routes of entry into marine species by PAH components are discussed. The implications of these data regarding tainting of food species and impacts on benthic biota from PAH and other pollutants are considered.

INTRODUCTION

There are numerous petroleum hydrocarbon inputs to the New York Bight area, including shipping, refining, municipal wastes, dumping, and aerial fallout. The rates of introduction and volumes involved are difficult to quantitate. Although it is sometimes possible to obtain data on the amounts of "oil and grease" introduced, this information is of little use in determining the contribution of effluents to the total load of more significant polynuclear (polycyclic) aromatic hydrocarbons (PAH) in the coastal environment. PAH are derived from various combustion sources and petroleum, and are most likely to remain in marine sediments for long periods. These high molecular weight compounds are present in crude and refined oils, and spills of these materials have produced sediment contamination. In the worst case, concentrations of PAH have been retained in oil-contaminated anaerobic mud for up to seven years (Teal *et al.*, 1978; Burns and Teal, 1979). Nevertheless, water concentrations of these relatively insoluble compounds, even over contaminated sediments, are three to six orders of magnitude lower than concentrations in the sediments (Burns and Teal, 1979).

The benthic biota of the New York Bight region often must tolerate multiple environmental insults over extended periods; consequently, the most significant pollutants will be those that exhibit slow degradation. The following discussion will examine the presence of these compounds in various compartments of the New York Bight marine environment, the possible routes of transfer from the sediments to animals, and the implications for the benthos and man.

DISCUSSION

Environmental Concentrations

Review of Literature

Only one case comes to mind where pollution from a known point source of petroleum hydrocarbons was studied (Armstrong *et al.*, 1977, 1979). Included in that report were detailed chemical analyses of the effluent, the receiving water, the sediments, and the distribution of benthic species. The pollutant input represented a "worst case" situation for an oil production platform, which was located in 3 m of water in a bay that contained a heavy sediment load. Alkyl-naphthalenes represented a majority of the aromatic hydrocarbons detected in the sediments, but compounds larger than trimethylphenanthrene were not included in the analyses. There was a definite correlation between the content of naphthalenes in the sediment and the number of species and individuals living in the substrate (Neff and Anderson, 1975). Concentrations of hydrocarbons in the bottom water were approximately four orders of magnitude lower than those in the sediments. Both numbers of species and individuals increased beyond 150 m from the platform along two transects, and reached a peak at stations just medial to the controls (the most distal stations), which showed somewhat lower abundances.

In a similar study, Addy *et al.* (1978) described various effects of oil on the macrobenthic community structure around a production platform in the Ekofisk field. Changes in the community correlated with the content of hydrocarbons in the sediments, but it was not clear whether a blowout in April 1977 contributed significantly to this response of the benthos. Benthic pollution by sewage discharge and petroleum effluents could have been responsible for the alterations in community structure, which extended approximately 3 km away from the central platform complex.

Calder *et al.* (1978) provided data from the spill of the AMOCO CADIZ on the coast of Brittany in March 1978, and showed significant contamination of sediments by naphthalenes and phenanthrenes. A paper by Calder and Boehm (1981) contains a discussion of the alterations in petroleum component composition within oiled sediments. Shoreline sediments retained elevated levels of alkyl-phenanthrenes and alkyl-dibenzothiophenes, but gradually lost n-alkanes and naphthalenes.

Burns and Teal (1979) reported on the most recent analyses of sediments oiled during the 1969 West Falmouth spill. They noted that degradation of sorbed oil resulted in losses of n- and branched alkanes and a relative enrichment of aromatics. The most persistent fractions were high molecular weight aromatics and naphthenes. The longest-term biotic effects were related to alterations of the behavior and reproduction of the fiddler crab, *Uca* (Krebs and Burns, 1977).

Farrington (1977) discussed the cycling of petroleum in marine sediments of the Atlantic coast of the United States. He noted the predominance of pyrolytic sources for PAH found in the sediments of Buzzards Bay, Massachusetts. The fallout of urban air hydrocarbons leads to high sediment PAH levels directly through the sinking and mixing of materials deposited on the surface of the water column, and indirectly by the washing of contaminated soils into coastal waters. Farrington cited evidence that the majority of PAH in New York Bight dumpsites are derived from combustion processes that deposit petroleum hydrocarbons on the streets; from there they are washed into storm and sanitary sewers. Farrington also noted that although it is unknown whether high PAH concentrations in the coastal zone are stressful to benthic biota, caution should be exercised in increasing such inputs.

In a recent paper, Winsor and Hites (1979) substantiated the pyrolytic origin of PAH in marine sediments near urbanized areas. A strong correlation between the distance from Boston and the concentration of PAH in Massachusetts Bay sediment samples was presented. An order of magnitude decrease in PAH was observed for each 40-km distance from Boston. Some basins in the Gulf of Maine had unusually high sediment PAH accumulations, which corresponded to the high PAH levels in fine-grained material scoured from surrounding areas.

New York Bight

The importance of petroleum hydrocarbon input to the New York Bight area from the discharge of treated wastewater was reported by Tanacredi (1977). He noted the degradation of the aquatic environment in Jamaica Bay, and a correlation between contamination patterns in water and clams and the components in waste oil and refined products. Farrington and Tripp (1977) determined the levels and distribution patterns of n-alkanes in surface sediment samples from the western North Atlantic, including areas within the New York Bight. They found strong indications of fossil fuel hydrocarbons at levels of 500-3,000 μg/g dry wt in and near the dumpsite areas. Koons and Thomas (1979) also found in abundance of C_{15} and greater hydrocarbons in samples taken from New York harbor and Hudson River stations. The range of concentrations reported was from 3,000-6,000 ppm, in areas highly impacted by man (harbors, dredge spoil, and sewage sludge disposal sites), to 40 ppm and 80 ppm on the continental shelf and rise, respectively.

The only additional information on the PAH contamination of the New York Bight marine environment is shown in Table 1. These data resulted from research sponsored by the Marine EcoSystems Analysis (MESA) New York Bight Project, in which samples were collected by the National Marine Fisheries Service, Northeast Fisheries Center and analyzed by the Northwest and Alaska Fisheries Center, National Analytical Facility. Table 1 shows the relatively low levels of PAH contamination in the water surface microlayer and bottom sediments of the outer New York Bight. Reasonably high levels of naphthalenes, biphenyl, dibenzothiophene, phenanthrene, and anthracene were found in two of the three phytoplankton samples. Sewage sludge contained 30-200 ppb concentrations of a wide range of PAH, but lower concentrations of benzo(a)pyrene, benzo(e)-pyrene, and perylene. By far, the most contaminated material was the sediment

Table 1. Concentrations of aromatic hydrocarbons in samples from New York Bight; units for the surface microlayer and sewage sludge are ng/ml, while those for all other samples are ng/g dry wt. Analyses were conducted by the Northwest and Alaska Fisheries Center on samples collected by the Northeast Fisheries Center.

	Surface microlayer			Plankton			Sewage sludge		Sediments slated for dredging			Bottom sediments		
							Composite, 3 stations		Newton Creek	Gowanus Canal	Pierhead Channel	Lower Bay	Christiaensen Basin	Outer Bight
Latitude (°N)	39°41′	40°30′	40°12′	39°41′	40°30′	40°12′	(Ward's Island, Hunt's Point, Newtown Creek)		40°44′	40°40′	40°40′	40°28′	40°25′	39°00′
Longitude (°W)	74°00′	73°56′	73°50′	74°00′	73°56′	73°50′			73°56′	74°00′	74°09′	74°02′	73°48′	72°55′
Sample size, wet	1 liter	1 liter	1 liter	10 g	10 g	10 g	0.1 liter	0.1 liter	100 g	100 g	100 g	100 g	100 g	100 g
% dry weight				9	10	12			47	43	54	46	54	83
Naphthalene	0.05	< 0.01	0.03	200	90	20	100	100	120,000	100	200	100	100	< 0.5
2-Methylnaphthalene	< 0.01	< 0.01	< 0.01	100	200	20	100	200	2,300	500	200	20	100	< 0.5
1-Methylnaphthalene	< 0.01	< 0.01	< 0.01	100	200	< 10	50	50	1,800	200	100	5	40	< 0.5
Biphenyl	< 0.01	< 0.01	< 0.01	200	200	60	50	100	700	300	200	40	50	20
Dibenzothiophene	< 0.01	< 0.02	< 0.02	200	< 10	< 20	50	100	2,700	1,000	100	40	50	< 0.5
Phenanthrene	< 0.01	< 0.01	< 0.01	200	100	< 10	200	200	14,000	1,000	300	300	300	2
Anthracene	< 0.01	0.06	0.05	100	30	< 10	50	30	9,600	500	200	100	50	5
1-Methylphenanthrene	< 0.01	< 0.01	< 0.01	< 10	< 10	< 10	50	100	1,500	1,500	100	50	40	< 0.5
Fluoranthene	0.04	0.06	< 0.01	100	< 10	< 10	100	50	10,000	2,000	50	500	400	4
Pyrene	< 0.01	< 0.01	< 0.01	< 10	< 10	< 10	100	200	7,200	3,000	500	500	400	3
Benz(a)anthracene	< 0.01	< 0.04	< 0.03	< 30	< 30	< 30	50	50	5,600	3,000	500	500	500	H 0.5
Chrysene	< 0.01	0.09	< 0.02	< 20	< 10	20	30	50	3,000	2,000	400	400	300	< 0.5
Benzo(e)pyrene	< 0.01	< 0.03	< 0.03	< 20	< 20	< 20	< 0.5	20	1,000	1,000	200	200	200	< 0.5
Benzo(a)pyrene	< 0.01	0.03	< 0.03	< 20	< 10	< 10	< 0.5	10	1,300	500	100	200	200	< 0.5
Perylene	< 0.01	2.00	1.00	< 20	< 10	< 20	< 0.5	< 0.5	400	300	50	100	50	< 0.5

designated for dredging from waterways within the estuarine system. Such material from both Newtown Creek and Gowanus Canal contained very high levels of all the PAH identified in these analyses. Many of the quantities were in the parts per million rather than in the parts per billion range. The highest value was 120 ppm for naphthalene in Newton Creek material, which also contained 10 ppm fluoranthene and 15 ppm phenanthrene. Levels of all other components in the material from the Creek and Canal were from two to five orders of magnitude above background levels for sediments of the outer Bight. Sediments from Pierhead Channel showed PAH concentrations approximately equal to those of bottom sediments from the Lower Bay and the Christiaensen Basin. These three locations were relatively consistent in the total load and distribution of PAH; concentrations were intermediate to those of the very highly contaminated waterways and the relatively clean sample from offshore. The levels for the Bay, Basin, and Channel were one to two orders of magnitude greater than background concentrations for all the compounds measured.

Perhaps more important than levels of contamination in sediments are levels of PAH in marine organisms inhabiting the vicinity. Samples of organisms collected and analyzed in the same MESA study are shown in Table 2. On the basis of numerous studies on the metabolism of petroleum hydrocarbons by marine fishes and invertebrates, one would expect to see higher levels of contamination in the digestive glands and liver, where metabolic conversions take place (Malins, 1977 a,b). Bivalves are known to be efficient accumulators of various pollutants, including petroleum hydrocarbons. The mussels analyzed in this study exhibited contamination levels intermediate to the relatively low levels detected in the flesh of fishes and lobsters, and to the higher levels in the liver and digestive glands of these species. Analyses of mackerel (Table 2) indicate that this pelagic species has low or non-detectable levels of PAH accumulation, with the exception of biphenyl. Biphenyl was found at relatively high levels in other organisms analyzed (Table 2) and also was detected in plankton (Table 1). These data suggest there must be high environmental levels of biphenyl in either water, sediments, or both. The source of biphenyl probably is industrial effluents. Table 1 reveals that biphenyl was found in sewage sludge, sediments slated for dredging, and other bottom sediments. Although no data on water samples were available, biphenyl could have been present. The presence of biphenyl in the water at parts per billion levels would explain its content in all animal samples, since this source would result in much greater uptake and accumulation in tissues. The three- to six-fold difference between the biphenyl content of lobster flesh and digestive glands (Table 2) is consistent with data presented in the literature. Anderson *et al.* (1974) reported a difference of greater than two orders of magnitude between the naphthalene content of the digestive gland and abdomen of shrimp.

It is instructive to compare naphthalene levels reported for fishes in Table 2. Of two samples of winter flounder, one showed little contamination; the other contained high amounts of naphthalenes. This difference between two samples of the same fish species points out the problems in relating levels of

Table 2. Concentrations of aromatic hydrocarbons in tissue samples from the New York Bight (ng/g dry wt). Each sample is a composite from five individual specimens. Analyses were conducted by the Northwest and Alaska Fisheries Center on samples collected by the Northeast Fisheries Center.

	Mackerel *Scomber scombrus*		Winter flounder *Pseudopleuronectes americanus*		Blue mussels *Mytilus edulis*		Lobster *Homarus americanus*				Striped bass *Morone saxatilis*					
	Liver	Roe	Liver	Liver			Digestive glands		Flesh		Flesh					
	Sheepshead Bay		Raritan Bay		Sandy Hook	Shark River	Raritan Bay				Montauk Point			Hudson River		
Latitide (°N)	40°30′	40°30′	40°28′	40°28′	40°26′	40°11′	40°33′	40°33′	40°33′		41°09′			41°00′		
Longitude (°N)	70°50′	73°50′	74°03′	74°03′	73°59′	74°01′	74°05′	74°05′	74°05′		72°13′			73°53′		
Sample size, wet	10 g	10 g	10 g	10 g	10 g	10 g	10 g	10 g	10 g	10 g	10 g	10 g	10 g	10 g	10 g	10 g
% dry weight	24	23	26	29	17	19	18	24	16	17	24	24	24	22	22	22
Naphthalene	40	< 4	500	40	10	100	70	60	24	24	< 4	30	< 4	20	< 4	< 4
2-Methylnaphthalene	< 3	< 4	1,000	< 6	60	30	60	100	< 4	< 4	< 4	< 4	< 4	3	< 3	< 2
1-Methylnaphthelene	< 3	< 4	700	< 5	< 5	10	50	70	< 50	< 40	< 3	< 4	< 3	3	< 3	< 2
Biphenyl	80	200	200	30	300	200	90	200	< 4	30	6	10	< 4	20	< 20	< 3
Dibenzothiphene	< 3	< 4	< 4	20	< 5	< 2	< 10	< 20	< 8	< 6	< 8	< 8	< 7	< 6	< 6	< 5
Phenanthrene	< 3	< 4	< 4	< 6	30	50	200	50	30	30	< 5	< 6	< 4	< 3	< 3	< 3
Anthracene	< 3	< 4	< 4	< 6	50	5	100	< 10	< 4	< 3	< 4	< 5	< 4	< 3	< 4	< 3
1-Methylphenanthrene	< 3	< 4	< 4	< 7	< 5	20	60	< 15	< 5	< 4	< 4	< 5	< 4	< 3	< 4	< 3
Fluoranthene	< 3	< 4	< 4	< 7	200	100	200	400	20	20	< 4	< 5	< 4	< 3	< 4	< 3
Pyrene	< 3	< 4	< 4	< 7	100	40	1,400	1,500	30	< 9	< 5	< 5	< 4	< 3	< 4	< 3
Benz(a)anthracene	< 3	< 4	< 4	< 10	< 5	< 2	500	600	< 10	< 8	< 10	< 10	< 9	< 7	< 8	< 7
Chrysene	< 3	< 4	< 4	< 8	< 5	30	900	800	< 6	< 5	< 5	< 5	< 5	< 4	< 4	< 3
Benzo(e)pyrene	< 3	< 4	< 4	< 9	< 5	< 2	400	300	< 6	< 6	< 6	< 6	< 5	< 5	< 6	< 5
Benzo(a)pyrene	< 3	< 4	< 4	< 8	< 5	< 2	300	300	< 7	< 6	< 5	< 6	< 5	< 5	< 6	< 4
Perylene	< 3	< 4	< 4	< 10	< 5	< 2	100	100	< 8	< 7	< 7	< 7	< 5	< 6	< 7	< 5

environmental contamination to hydrocarbon analysis of mobile fish species. However, based on comparisons of the flounder and mackerel samples, it is still possible to say that fishes living on and near the bottom contained higher concentrations of naphthalenes than those living higher in the water column. Studies by Roesijadi (1978 a,b) have shown that naphthalenes are released more rapidly from oiled sediments than are the larger PAH. This may mean that the uptake exhibited by the flounder could have been from water-borne material that was leaching from contaminated sediments. To provide an accurate assessment, one would need to know the levels of naphthalenes in the sediments, interstitial water, near-bottom water, and surface water. It is also possible that the material could have been recently ingested while the fish were feeding on contaminated benthic invertebrates. It is noteworthy that hydrocarbons of molecular weight greater than mono-methylnaphthalenes and biphenyl were not present in either mackerel (liver and roe), flounder (livers), or striped bass (flesh). High molecular weight compounds of low solubility were prominent in contaminated sediments and when accumulated are retained longer by organisms. This may be an indication that uptake by fish species was via water.

The distribution of PAH components in both blue mussels and lobsters is patchy, indicating possible multiple routes of entry. Naphthalenes and biphenyl were again well-represented. The most likely source of these soluble components is water, particularly bottom water. Mussels also contain PAH of intermediate molecular weight, but not the largest compounds (benzanthracene through perylene). That filtering activity will bring in both soluble components and those sorbed to fine particles may explain the differences between lobster and mussels in the components present.

By far, the most contaminated tissue samples were lobster digestive glands, which contained the full range of PAH, except for dibenzothiophene (Table 2). This is a point of interest, since dibenzothiophene has been associated with tissues exposed to oil spills (Grahl-Nielsen *et al.*, 1978). Grahl-Nielsen *et al.* (1979) discussed the use of dibenzothiophene in determining the contamination of areas of the North Sea by petroleum. The primary source of PAH in the New York estuarine system and Bight is probably not oil spills, but sewage effluents and combustion products of oil (discussed later)—hence the low levels of dibenzothiophene.

The presence of PAH of high molecular weight (fluoranthene and pyrene) in both the digestive glands and flesh of lobsters indicates that the habitat and/or food of this species are different from those of other organisms analyzed. Lobsters are both scavengers and predators (Cobb, 1976) and have been found to feed on crabs, gastropods, bivalves, polychaetes, fishes, and plant matter. Either by direct ingestion of contaminated detritus or by short-term transfer from contaminated food organisms, the lobsters could have consumed materials from the dumpsites. However, it is more likely that the PAH were derived from the ingestion of contaminated detritus or contaminated food organisms in the estuary. Although the PAH content of the flesh is not alarming, levels in the digestive glands indicate that ingestion of this organ by man would contribute substantially to one's total PAH burden. As in the other species, the pathway of uptake cannot be definitively described, but it is likely that several routes are involved in the transport of PAH to lobsters.

Effects of Oiled Sediment

The toxicity of oil-contaminated sediments has not been thoroughly investigated, but indications are that very high concentrations in sediment are required to produce mortality. Several studies have noted the lack of significant mortality during intermediate-length exposures to concentrations in excess of 1,000 ppm (Wells and Sprague, 1976; Anderson *et al.*, 1977; Anderson *et al.*, 1979b). However, Krebs and Burns (1977) noted the mortality and adverse responses of *Uca* living in an environment (long-term exposure) where the sediment contained 1,000-7,000 ppm. Shaw *et al.* (1976) reported mortality of the clam, *Macoma balthica,* from sediment containing 640-3,890 ppm (dry wt). The peak in mortality occurred at 29 days, and at 44 and 60 days, mortality was reduced, perhaps indicating a loss of toxic components from the substrate. By 60 days, hydrocarbon concentrations in the sediment had decreased to levels approximately equal to background. It is not known whether toxicity of contaminated sediment is the result of contact, ingestion of particles, or the uptake of interstitial water containing compounds leached from particles or freed by metabolic conversions of microorganisms.

Oil-contaminated sediment or food has been used in studies (Percy, 1976; Shaw *et al.*, 1976; Taylor and Karinen, 1977) to evaluate the survival and behavior of benthic invertebrates. The arctic amphipod, *Onisimus affinis,* avoided oil-tainted food as well as oil masses, but this response diminished with pre-exposure to oil and when weathered oil was used. The isopod, *Mesiodotea entomon,* was more resistant to oil and exhibited no preference for clean food (Percy, 1976). Taylor and Karinen (1977) studied the behavior of the detritivore, *Macoma balthica,* in response to oil-contaminated substrate and in response to oil extracts flowing over clean substrate. They reported that at least 95% of the clams burrowed to the surface before dying; this response was elicited within three days by 50% of the animals exposed to 0.37 ppm naphthalene equivalents. At a calculated dose of 0.2 ppm naphthalene equivalents, 50% of the clams would not burrow into sediment within 60 minutes. A concentration of 0.67 g oil/cm^2 stimulated 50% of the clams to move to the surface within 24 hours. Death was not observed in these four- to six-day experiments. In field studies, Shaw *et al.* (1976) showed that test doses of 5 μ liters oil/cm^2 applied to surface sediment daily for five days produced significant mortality of clams. The behavioral response of clams can be readily quantitated, and it appears that clams may be sensitive to relatively low levels of petroleum hydrocarbons. Oil sorbed to sediment seems to be tolerated to a greater extent than hydrocarbons dissolved or suspended in the water column. Krebs and Burns (1977) noted behavioral impairment in *Uca pugnax* exposed to the West Falmouth spill and in laboratory exposures to 100 ppm fuel oil in sediment. Abnormal responses were most apparent at colder exposure temperatures (13°-15°C), and mortalities were greatest during over-wintering. Alteration in the behavior of organisms may be a sensitive measure of their response to sublethal levels of toxicants in water or sediments. It is possible that the effects observed on exposure to oiled substrate result from hydrocarbons released by various means from the sediment to the interstitial water, where they are readily available for tissue accumulation.

Effects on the flatfish, *Parophrys vetulus* (English sole), from exposure to oiled sediment were reported by McCain *et al.* (1978). There was enhanced weight loss and severe hepatocellular lipid vacuolization. Roesijadi and Anderson (1979) studied one type of biochemical alteration that responded to hydrocarbon exposure in sediments. The detritivore, *Macoma inquinata,* exposed in the natural environment to oil contaminated sediment, exhibited decreases in condition index and free amino acid content, as compared to control animals in the field. That oiled sediment affected both of these parameters may mean that either the mechanical or chemical alteration of the substrate reduced the feeding response of this detritivore. Augenfeld (1980) used similar exposures in the laboratory to demonstrate reductions in the ingestion of a deposit-feeding polychaete exposed to oiled mud. Reduced sediment working rate for a closely related polychaete (*Arenicola marina*) was reported by Gordon *et al.* (1978) when the animals were exposed to oiled substrate. Therefore, it is possible that the energy budget of detritus- or deposit-feeding benthic organisms may be altered by exposures to oiled sediments, as has been reported for filter-feeding bivalves (Gilfillan, 1975; Gilfillan *et al.*, 1977; Gilfillan and Vandermeulen, 1978).

Availability of Hydrocarbons from Sediments

Until recently, little information was published regarding the bioavailability of sediment-bound petroleum hydrocarbons. Rossi (1977) reported on the comparative aspects of naphthalene uptake from water and sediment by *Neanthes arenaceodentata.* ^{14}C-naphthalene from water (150 ppb) was concentrated by a factor of approximately 40 in 24 hours, but after 28 days in sediment containing 9 ppm naphthalenes, there was no significant uptake. When ^{14}C-methylnaphthalene was used to contaminate artificial detritus on which the worms fed for 16 days, no accumulation was observed. A period of 24 hours in clean water was sufficient for the organisms to egest the apparent tissue contaminants via fecal pellets. The sediment studies of Rossi (1977) involved only a fine layer of substrate with no interstitial water. Anderson *et al.* (1977) utilized an exposure system with a 10 cm depth of sediment containing sipunculid worms (*Phascolosoma agassizii*). When exposed to high concentrations of hydrocarbons in sediment, the sipunculids exhibited low levels of tissue contamination. Two days of depuration resulted in a substantial release of naphthalenes. One- and two-week depuration periods were sufficient for the organisms to release naphthalenes completely.

It was not possible to separate uptake via ingested sediment and possible uptake from interstitial water with use of the above exposure system. Anderson *et al.* (1979b) designed a system to estimate the contribution of interstitial water to the contamination of amphipods burrowing in oiled substrate. Amphipods (*Anonyx laticoxae*) living in oiled sediment accumulated 8 ppm total naphthalenes after four days, while those in clean substrate suspended above the bottom contained 4 ppm. After 18 days of exposure, neither group contained a significant amount of contamination, while the sediment still possessed 80% of the initial total hydrocarbons and 30% of the naphthalenes. Similar findings were reported by Roesijadi *et al.* (1978b) in a study of the bioavailability of sediment-bound naphthalenes for the detritivore, *Macoma inquinata.* McCain *et al.* (1978) showed that English sole exposed to oiled sediment (700 μg/g dry wt) accumulated monomethylnaphthalenes and 1, 2, 3, 4-tetramethylbenzene to the greatest extent. Tissue content of these components decreased with time, and after 27 days of exposure only the liver contained detectable amounts.

Utilizing ^{14}C-labelled polyaromatics, Roesijadi *et al.* (1978a) studied the comparative uptake of 3- through 5-ring aromatic hydrocarbons by *M. inquinata.* They found that after an initially rapid uptake, phenanthrene (3 rings) concentrations in tissue constantly decreased during the 56-day exposure to contaminated sediment. Exposure to chrysene (4 rings) showed initial uptake for the first 20 days, followed by a decrease during the next 30 days. Benzo(a)pyrene tested in the same system was constantly accumulated during the entire 42-day exposure period. These results have recently been verified using a modified exposure system and the same organisms (Anderson *et al.*, 1979a; Augenfeld *et al.*, in press).

SUMMARY

It appears that higher molecular weight polynuclear aromatics, such as benzo(a)pyrene and chrysene in sediments, may be taken up slowly by deposit-feeding benthic organisms and retained for a significant period. Naphthalenes and phenanthrenes, which are rapidly accumulated from solution, apparently do not exhibit net uptake since at later phases of the exposures the tissues do not show contamination. Important factors to consider in this evaluation are the routes by which the hydrocarbons contact organisms and the duration of exposure. Short-term exposures from both water and sediment are likely to produce relatively high levels of tissue contamination from 2- and 3-ring aromatics (naphthalenes and phenanthrenes). If organisms are exposed to a constant supply of these compounds, high levels will be accumulated, and death probably will occur. Mobile organisms probably will not be exposed continuously, and the tissues will be depurated in a few days to a few weeks in clean water. Short-term exposure from either water or sediments to 4- and 5-ring compounds (chrysene and benzo(a)pyrene) probably will not produce significant contamination (considering very low water solubility) or effects. However, long-term exposure to these compounds in sediments is quite likely to result in significant tissue accumulation of these PAH. The effects of such an exposure will be dependent on the total concentration of PAH in the sediments, the components present, the duration of exposure, and the species involved. There is evidence that oiled substrates cause mortality of some species and alter community composition. Sublethal effects such as histopathology in flatfishes and reduced energy budget, condition index, and free amino acid content in bivalves have been observed. The conditions in these studies will have to be compared to those of the New York Bight before estimates of the impacts on the benthos of this area can be made.

IMPLICATIONS

The toxicity range for petroleum hydrocarbons in the water column is approximately 1-20 ppm for total oil and 0.3-1.0 ppm for the more toxic components (Neff *et al.*, 1976; Rice *et al.*, 1977). Under chronic exposure conditions, the latter compounds may produce sublethal effects from 0.1 ppm to 0.3 ppm (Anderson, 1977). Therefore, for hydrocarbons in the water, the lowest value that would be expected to produce adverse effects on New York Bight species would be about 100 ppb. It is not likely that surface waters contain much more than 10 ppb, but not enough specific information is available on surface or bottom water to make an assessment. The implications from tissue analyses (Table 2) are that at least some of the animals are receiving biphenyls, naphthalenes, phenanthrenes, and other compounds via the water. Sources may be site-specific and relate to effluents, or there may be larger areas in which contaminated sediments are leaching compounds to the water. It is not likely that animals are receiving naphthalenes and biphenyl from contact with sediments (including ingestion).

The presence of higher molecular weight PAH compounds in tissues of several New York Bight species indicates that the contaminants are derived from airborne particulates or estuarine and dumpsite sediments. There is less chance that these compounds will be acutely toxic to the organisms, but tainting and chronic effects are possible. Seafood tainting is a potential problem since the species analyzed show contamination, particularly in the detoxifying organs. Clams and oysters were not analyzed, but they surely must follow the pattern of blue mussels. Deposit- and detritus-feeding species from areas near the dumpsites or in the bays may show greater contamination. Fish flesh is a lesser problem from the standpoint of tainting. The recent ingestion of contaminated benthic invertebrates by flounder could explain the high hydrocarbon content of flounder liver. The most serious tainting problem at present seems to be the very high levels of PAH in organs of lobsters used as food by at least a portion of the human population.

Chronic effects on the biota and particularly the benthos are more difficult to assess. Pollutants in addition to PAH are involved, and it is not certain which class of contaminants may be the most harmful. Potentially hazardous levels of PAH in water have been defined, but such data are not available for sediments.

Certainly, life in undiluted dredged material would be impossible, but the effects of less contaminated sewage sludge and estuarine sediments on a range of marine species are difficult to predict. There is a complex distribution of clean and dirty sediments in the inner Bight. As disposal activities in the Bight continue, one would expect the buildup of contaminated sediments in some regions, which would, on the basis of PAH alone, prove to be toxic for benthic organisms.

Finally, the New York Bight receives many classes of pollutants, any one of which could produce mortality in marine species and tainting of seafood. The hazards of these classes of pollutants should be compared in relation to both problems. Further sampling will be required to answer several of the questions posed in this paper and those of others. Since the region is receiving a much higher input of PAH from multiple sources than is advisable, efforts should be made to reduce as many sources as are practical. The expertise of many individuals will probably be required to solve this complex problem and to lessen the hazards to this region of the coastal marine environment.

ACKNOWLEDGEMENTS

The author wishes to thank Dr. Joel O'Connor of the MESA New York Bight Project for furnishing the valuable analyses conducted by the Northwest and Alaska Fisheries Center on samples collected by the Northeast Fisheries Center. Funding for the preparation of this manuscript was provided by the MESA New York Bight Project, the NOAA/BLM Outer Continental Shelf Environmental Assessment Program, and the EPA Pass-thru Program to the Department of Energy.

REFERENCES

Addy, J.M., P. Levell, and J.P. Harley. 1978. Biological monitoring of sediments in Ekofisk oilfield. *In:* Proceedings of the Conference on Assessment of Ecological Impacts of Oil Spills, American Institute of Biological Sciences, Arlington, VA. pp. 514-539.

Anderson, J.W. 1977. Responses to sublethal levels of petroleum hydrocarbons: Are they sensitive indicators and do they correlate with tissue contamination? *In:* Fate and Effects of Petroleum in Marine Ecosystems and Organisms, D.A. Wolfe (ed.), Pergamon Press, Oxford, England. pp. 95-114.

Anderson, J.W., J.M. Augenfeld, E.A. Crecelius, and R.G. Riley. 1979a. Research to determine the accumulation of organic constituents and heavy metals from petroleum-impacted sediments by marine detritivores of the Alaskan Outer Continental Shelf. NOAA/OCSEAP 1979 Annual Report. Vol. 6. pp. 172-234.

Anderson, J.W., R.C. Clark, Jr., and J.J. Stegeman. 1974. Petroleum hydrocarbons. *In:* Marine Bioassays Workshop Proceedings, G.V. Cox (ed.), Marine Technology Society, Washington, DC. pp. 36-75.

Anderson, J.W., S.L. Kiesser, and J.W. Blaylock. 1979b. Comparative uptake of naphthalenes from water and oiled sediments by benthic amphipods. *In:* Proceedings of the 1979 Oil Spill Conference (Prevention, Behavior, Control, Cleanup), American Petroleum Institute, U.S. Environmental Protection Agency, and U.S. Coast Guard, Washington, DC. pp. 579-584.

Anderson, J.W., L.J. Moore, J.W. Blaylock, D.L. Woodruff, and S.L. Kiesser. 1977. Bioavailability of sediment-sorbed naphthalenes to the sipunculid worm, *Phascolosoma agassizii. In:* Fate and Effects of Petroleum in Marine Ecosystems and Organisms, D.W. Wolfe (ed.), Pergamon Press, Oxford, England. pp. 276-285.

Armstrong, H.W., K. Fucik, J.W. Anderson, and J.M. Neff. 1977. Effects of Oilfield Brine Effluent on Benthic Organisms in Trinity Bay, Texas. Publication No. 4291, American Petroleum Institute, Washington, DC.

Armstrong, H.W., K. Fucik, J.W. Anderson, and J.M. Neff. 1979. Effects of oilfield brine effluent on benthic organisms in Trinity Bay, Texas. Mar. Environ. Res. 2: 55-70.

Augenfeld, J.M. 1980. Effects of Prudhoe Bay crude oil contamination on sediment working rates of *Abarenicola pacifica.* Mar. Environ. Res. 3:307-314.

Augenfeld, J.M., R.G. Riley, B.L. Thomas, and J.W. Anderson. In press. The fate of polyaromatic hydrocarbons in an intertidal sediment exposure system. I. Bioavailability to *Macoma inquinata* (Mollusca: Pelecypoda) and *Abarenicola pacifica* (Annelida: Polychaeta). Mar. Environ. Res.

Burns, K.A. and J.M. Teal. 1979. The West Falmouth oil spill: hydrocarbons in the salt marsh ecosystem. Estuarine Coastal Mar. Sci. 8: 349-360.

Calder, J.A. and P.D. Boehm. 1981. The chemistry of AMOCO CADIZ oil in the Aber Wrach. *In:* AMOCO CADIZ: Fate and Effects of the Oil Spill, Centre Oceanologique de Bretagne, France. pp. 149-158.

Calder, J.A., J. Lake, and J. Laseter. 1978. Chemical composition of selected environmental and petroleum samples from the AMOCO CADIZ oil spill. *In:* The AMOCO CADIZ Oil Spill, A Preliminary Scientific Report, W.N. Hess (ed.), NOAA/EPA Special Report. pp. 21-84.

Cobb, S.J. 1976. The American lobster: the biology of *Homarus americanus.* Univ. Rhode Island Mar. Tech. Rep. 49. 32 pp.

Farrington, J.W. 1977 The biochemistry of oil in the ocean. Oceanus 20: 4-14.

Farrington, J.W. and B.W. Tripp. 1977. Hydrocarbons in western north Atlantic surface sediments. Geochim. Cosmochim. Acta 41: 1627-1641.

Gilfillan, E.S. 1975. Decrease of net carbon flux in two species of mussels caused by extracts of crude oil. Mar. Biol. 29: 53-57.

Gilfillan, E.S., D.W. Mayo, D.S. Page, D. Donovan, and S. Hanson. 1977. Effects of varying concentrations of petroleum hydrocarbons in sediments on carbon flux in *Mya arenaria. In:* Physiological Responses of Marine Biota to Pollutants, F.J. Vernberg, W.B. Vernberg, A. Calabrese, and F.P. Thurberg (eds.), Academic Press, New York. pp. 299-314.

Gilfillan, E.S. and J.H. Vandermeulen. 1978. Alterations in growth and physiology of soft-shell clams, *Mya arenaria,* chronically oiled with Bunker C from Chedabucto Bay, Nova Scotia, 1970-76. J. Fish. Res. Board Can. 35: 630-636.

Gordon, D.C., Jr., J. Dale, and P.D. Keizer. 1978. Importance of sediment working by the deposit-feeding polychaete *Arenicola marina* on the weathering rate of sediment-bound oil. J. Fish. Res. Board Can. 35: 591-603.

Grahl-Nielsen, O., J.T. Straveland, and S. Wilhelmsen. 1978. Aromatic hydrocarbons in benthic organisms from coastal areas polluted by Iranian crude oil. J. Fish. Res. Board Can. 35: 615-623.

Grahl-Nielsen, O., K. Westrheim, and S. Wilhelmsen. 1979. Petroleum hydrocarbons in the North Sea. *In:* Proceedings of the 1979 Oil Spill Conference (Prevention, Behavior, Control, Cleanup), American Petroleum Institute, U.S. Environmental Protection Agency, and U.S. Coast Guard, Washington, DC. pp. 629-632.

Koons, C.B. and J.P. Thomas. 1979. C^{15} + hydrocarbons in the sediments of the New York Bight. *In:* Proceedings of the 1979 Oil Spill Conference (Prevention, Behavior, Control, Cleanup), American Petroleum Institute, U.S. Environmental Protection Agency, and U.S. Coast Guard, Washington, DC. pp. 625-628.

Krebs, C.T. and K.A. Burns. 1977. Long-term effects of an oil spill on populations of the salt-marsh crab *Uca pugnax.* Science 197: 484-487.

McCain, B.B., H.O. Hodgins, W.D. Grondund, J.W. Hawkes, D.W. Brown, and M.S. Myers. 1978. Bioavailability of crude oil from experimentally oiled sediments to English sole (*Parophrys vetulus*) and pathological consequences. J. Fish. Res. Board Can. 35: 657-664.

Malins, D.C. (ed.). 1977a. Effects of Petroleum on Arctic and Subarctic Marine Environments and Organisms, Vol. 1, Academic Press, New York. 321 pp.

Malins, D.C. (ed.). 1977b. Effects of Petroleum on Arctic and Subarctic Marine Environments and Organisms, Vol. 2, Academic Press, New York. 500 pp.

Neff, J.M. and J.W. Anderson. 1975. An ultraviolet spectrophotometric method for the determination of naphthalene and alkylnaphthalenes in the tissues of oil-contaminated marine animals. Bull. Environ. Contam. Toxicol. 14: 122-128.

Neff. J.M., J.W. Anderson, B.A. Cox, R.B. Laughlin, Jr., S.S. Rossi, and H.E. Tatem. 1976. Effects of petroleum on survival, respiration and growth of marine animals. *In:* Sources, Effects, and Sinks of Hydrocarbons in the Aquatic Environment, American Institute of Biological Sciences, Arlington, VA. pp. 515-539.

Percy, J.A. 1976. Responses of arctic marine crustaceans to crude oil and oil-tainted food. Environ. Pollut. 10: 155-162.

Rice, S.P., J.W. Short, and J.F. Karinen. 1977. Comparative oil toxicity and comparative animal sensitivity. *In:* Fate and Effects of Petroleum Hydrocarbons in Marine Ecosystems and Organisms, D.A. Wolfe (ed.), Pergamon Press, Oxford, England. pp. 78-94.

Roesijadi, G. and J.W. Anderson. 1979. Condition index and free amino acid content of *Macoma inquinata* exposed to oil-contaminated marine sediments. *In:* Marine Pollution: Functional Responses, W.B. Vernberg, F.P. Thurberg, A. Calabrese and F.J. Vernberg (eds.), Academic Press, New York. pp. 69-83.

Roesijadi, G., J.W. Anderson, and J.W. Blaylock. 1978a. Uptake of hydrocarbons from marine sediments contaminated with Prudhoe Bay crude oil: influence of feeding type of test species and availability of polycyclic aromatic hydrocarbons. J. Fish. Res. Board Can. 35: 608-614.

Roesijadi, G., D.L. Woodruff, and J.W. Anderson. 1978b. Bioavailability of naphthalenes from marine sediments artificially contaminated with Prudhoe Bay crude oil. Environ. Pollut. 15: 223-229.

Rossi, S.S. 1977. Bioavailability of petroleum hydrocarbons from water, sediments and detritus to the marine annelid, *Neanthes arenaceodentata. In:* Proceedings of the 1977 Oil Spill Conference, American Petroleum Institute, Washington, DC. pp. 621-625.

Shaw, D.G., A.J. Paul, L.M. Cheek, and H.M. Feder. 1976. *Macoma balthica:* an indicator of oil pollution. Mar. Pollut. Bull. 7: 29-31.

Tanacredi, J.T. 1977. Petroleum hydrocarbons from effluents: detection in marine environments. J. Water Pollut. Control Fed. 49: 216-226.

Taylor, T.L. and J.F. Karinen. 1977. Response of the clam, *Macoma balthica* (Linnaeus), exposed to Prudhoe Bay crude oil as unmixed oil, water-soluble fraction, and oil-contaminated sediment in the laboratory. *In:* Fate and Effects of Petroleum Hydrocarbons in Marine Ecosystems and Organisms,

D.A. Wolfe (ed.), Pergamon Press, Oxford, England. pp. 229-237.

Teal, J.M., K. Burns, and J. Farrington, 1978. Analyses of aromatic hydrocarbons in sediments resulting from two spills of No. 2 fuel oil in Buzzards Bay, Massachusetts. J. Fish. Res. Board Can. 35: 510-520.

Wells, P.G. and J.B. Sprague. 1976. Effects of crude oil on American lobster (*Homarus americanus*) larvae in the laboratory. J. Fish. Res. Board Can. 33: 1604-1614.

Winsor, J.G., Jr. and R.A. Hites. 1979. Polycyclic aromatic hydrocarbons in Gulf of Maine sediments and Nova Scotia soils. Geochim. Cosmochim. Acta 43: 27-33.

AN HISTORICAL PERSPECTIVE ON MOLLUSCS IN LOWER NEW YORK HARBOR, WITH EMPHASIS ON OYSTERS

David R. Franz

Biology Department
Brooklyn College
City University of New York
Brooklyn, New York 11210

Abstract. Information on benthic organisms and communities in the lower Hudson estuary prior to the 1950's is scarce. Records from the late 1800's indicate that species richness in the shallow bay habitats along Staten Island was comparable to that presently characteristic of Great South Bay and northeastern Long Island Sound. By the 1920's many species had disappeared from Staten Island, and the number of species remaining approached that which presently exists. These data indicate that major environmental deterioration occurred during the years 1890-1920. Supporting evidence is provided by historical accounts of New York's oyster industry. Before 1900 oysters were found throughout much of the lower estuary and north to Ossining, New York, including Newark Bay, Arthur Kill, Kill Van Kull, Jamaica Bay, Raritan Bay, and the New Jersey shore of the Hudson. By the turn of the century, the shellfish industry was limited to waters south of the Narrows; by 1920 it largely had disappeared from harbor waters.

Destruction of shellfish beds was caused by a combination of factors including sewage pollution, harbor dredging, and industrial pollution. Although it is difficult to determine the relative importance of individual factors, sewage-related pollution appears to have played a major role in bringing about the general degradation of the estuarine environment. This is suggested by records of dissolved oxygen (DO) levels in the harbor that were collected beginning in 1909. These data clearly indicate that by the 1920's, DO levels over much of the harbor had declined to critical levels (0-20% saturation in summer).

While the lower Hudson estuary is seriously impacted, communities of benthic organisms exist throughout. The degree of environmental deterioration and the recovery of the estuary probably can be measured best by comparing these benthic communities with those of nearby, less polluted estuarine systems.

INTRODUCTION

The Hudson River estuary, including the complex of embayments and estuaries comprising the lower New York harbor, logically might be expected to be the most studied and the best understood major estuarine complex in the United States. For more than three hundred years the lower estuary has been a hub of human enterprise and commerce, and the New York metropolitan area has been the nucleus for the intellectual and cultural development of the nation. In addition, this estuary has supported major fisheries, notably shad (Esser, this volume) and oyster, which have provided livelihood and sustenance for significant numbers of residents. It is paradoxical that our knowledge of the biology of this estuary is remarkably lacking, particularly with regard to historical data that could be used to compare present and past conditions. From approximately 1700 to 1900 American biologists mainly were engaged in discovering and classifying the plants and animals of this continent. Local naturalists were drawn away from New York to investigate the expanding frontiers since the estuary itself excited little interest relative to the promise of distant discoveries.

By the 1870's the population and industry of New York City were expanding at a rate that soon would destroy the fabric of natural benthic communities in the lower estuary. By the early decades of the present century, the situation in New York harbor had deteriorated to critical levels. Yet, this deterioration did not elicit a careful analysis of animal and plant communities by concerned officials and agencies. Instead, research centered on engineering studies relative to pollution abatement, the potential for alleviating pollution through sewer construction and sewage treatment plants, and threats to public health generated by unlimited discharge of highly contaminated human wastes. Independent research organizations, universities, and private citizens were little inclined to collect information on the effects of pollution on natural communities or to provide "baseline" studies for subsequent research. Consequently, little historical data exist against which to compare the present biological status of the estuary or to contrast future conditions.

HISTORICAL DATA ON SPECIES RICHNESS

In the later decades of the nineteenth century, several local naturalists published lists of molluscs from the Staten Island coastal area. These lists, which appear in Hubbard and Smith (1865) and Smith (1887), can be compared with a later summary for this area by Jacot (1920) and with more recent data for this region and nearby coastal areas (*e.g.*, Dean and Haskin, 1964; McGrath, 1974).

Smith (1887) published the most comprehensive faunal list, which evidently includes shore collections as well as dredgings from Raritan Bay and perhaps the mouth of the lower harbor, as suggested by Jacot (1920). Certain of Smith's (1887) species, such as *Astarte castanea, Cyclocardia borealis, Epitonium humphreysii, Modiolus modiolus, Natica pusilla, Periploma leanum,* and *Spisula solidissima,* today occur primarily in medium to coarse sand substrates in the inner New York Bight, but not in Raritan Bay. If only those molluscan species known to occur in shallow bay and shoreline habitats are considered, Smith's data suggest that 57 species (30 gastropods, 27 bivalves) were present along the Staten Island shore during the mid-1880's. These species citations are sound and are listed in Table 1.

The fauna listed in Table 1 is comparable in species richness to the present molluscan fauna of Fishers Island Sound (off southeastern Connecticut) (Franz, 1976), where a total of 52 sublittoral species were collected, and to that of Great South Bay (Long Island), where 57 sublittoral species recently were identified (WAPORA, Inc., 1979). A measure of the similarity of these faunas can be provided by the Dice Coefficient (DC) (Hall, 1969):

$$DC = \frac{2C}{A+B};$$

Table 1. Shallow bay and shore molluscan species of Staten Island (from Smith, 1887). *Considered rare by Jacot (1920). **Considered especially rare by Jacot (1920).

Gastropoda		Bivalvia
*Mangilia cerina**		*Barnea truncata***
Busycon canaliculatum		*Cyrtopleura costata*
Busycon carica		*Mya arenaria*
Nassarius vibex		*Lyonsia hyalina*
Nassarius trivittatus		*Pandora trilineata* (probably *gouldiana*)
Ilyanassa obsoleta		*Tagelus gibbus*
Urosalpinx cinerea		*Macoma fragilis* (probably *balthica*)
Eupleura caudata		*Angulus tener* (probably *Tellina agilis*)
Nucella lapillus		*Macoma tenta*
*Anachis avara**		*Cumingia tellinoides**
Mitrella lunata		*Mulinia lateralis*
Lunatia heros		*Petricola pholadiformis*
Polinices duplicatus		*Mercenaria mercenaria*
*Cerithiopsis greeni**		*Mercenaria m. notata*
*Triphora nigrocincta**		*Gemma gemma*
*Bittium alternatum**		*Laevicardium mortoni**
Crepidula fornicata		*Mysella planulata**
Crepidula plana		*Solemya velum**
Crepidula convexa		*Yoldia limatula**
Littorina littorea		*Nucula proxima**
Lacuna vincta		*Anadara transversa*
Hydrobia minuta		*Anadara ovalis*
Odostomia bisturalis		*Modiolus demissus*
Odostomia seminuda		*Argopecten irradians*
Turbonilla interrupta		*Anomia glabra* (probably *simplex*)
Epitonium rupicola		*Crassostrea virginica*
Haminoea solitaria		*Mytilus edulis*
Cylichnella canaliculata		
*Rictaxis punctostriatus**		
*Littorina irrorata***		
Total	30	27

where A = number of species at site A, B = number of species at site B, and C = number of species shared between sites. As noted in Table 2, the Dice Coefficient of similarity between Smith's Staten Island fauna and that of Great South Bay is 86%. This suggests that the molluscan fauna presently existing in Great South Bay is qualitatively similar to that which occurred along the Staten Island shore in the late 1800's. Furthermore, the levels of species richness in Fishers Island Sound, Great South Bay, and Staten Island are high and can be interpreted as indications of the absence of significant environmental degradation.

By the 1920's, Staten Island naturalists recorded considerable environmental deterioration, which was reflected by reduced species richness. Jacot (1920) noted that development of Staten Island's north and northeast coast was accompanied by the appearance of sewage outfalls and other evidences of

Table 2. Dice Coefficient of similarity (%) for molluscs occurring in muddy/sandy habitats. Staten Island data from Smith (1887); Great South Bay data from WAPORA, Inc. (1979); Mullica River/Great Bay data from Durand and Nadeau (1972); and Raritan Bay data from McGrath (1974).

	Staten Island	Great South Bay	Mullica River/ Great Bay, New Jersey	Raritan Bay
Staten Island	–	86	57	31
Great South Bay		–	70	36
Mullica River/ Great Bay			–	40
Raritan Bay				–

environmental change, and appeared related to a decline in the number of molluscan species, compared to the fauna listed by Hubbard and Smith (1865) and Smith (1887). Jacot further observed that certain species had become "rare" or "especially rare", including: *Anachis avara, Bittium alternatum, Cerithiopsis greeni, Laevicardium mortoni, Mysella planulata, Nucula proxima, Rictaxis punctostriatus, Solemya velum, Trifora nigrocincta,* and *Yoldia limatula.* Some of these species, including *B. alternatum, S. velum,* and *L. mortoni,* frequently occur in association with eelgrass. Their disappearance may have been related to the loss of eelgrass beds.

Jacot (1920) also noted that in the Great Kills area, sediments near sewer outfalls were composed of soft, sticky clay and were inhabited only by the hardiest species. Three of the species listed above, *N. proxima, Y. limatula,* and *M. planulata,* are deposit feeders with relatively high tolerances to reducing sediments (Franz, 1973; Botton, 1979). *Nucula* has been shown to be marginally tolerant of sewage sludge and to tolerate deoxygenated conditions; *Yoldia* is less hardy and presently does not occur in areas contaminated by dredge spoils or sewage sludge (National Marine Fisheries Service, 1972). It would appear that by the 1920's, oil contamination of the Staten Island shore (Allen, 1921) and organic pollution from Staten Island sewage outfalls may have eliminated even the relatively tolerant molluscan species from nearshore habitats. It is evident from Jacot's (1920) species list that the common molluscs encountered along Staten Island (*i.e., Gemma gemma, Ilyanassa obsoleta, Modiolus demissus, Mya arenaria,* and *Nassarius trivittatus*) were essentially the same as those found today along the pollution-impacted beaches of the lower harbor and Jamaica Bay.

ENVIRONMENTAL DECLINE AS EVIDENCED BY SHELLFISH POPULATIONS

Shellfish populations, especially of the American oyster, *Crassostrea virginica,* provided a major industry and source of food to New Yorkers for more than two hundred years. The history of this industry provides a valuable measure of the changing environmental quality in the lower estuary.

Oyster Populations Before 1900

Populations of oysters occurred throughout the Hudson River estuary from Sandy Hook, New Jersey, north to Ossining, New York, a distance of about 80 km (Ingersoll, 1881). Although major portions of the estuary lacked oysters in commercial quantities because of unsuitable sediments and currents, commercial oyster grounds occurred along the southern and eastern shores of Staten Island, in Kill Van Kull, Arthur Kill, Newark Bay, and Raritan Bay, and at the mouth of the Shrewsbury River (New Jersey). Productive oyster grounds also were located in the upper harbor, including parts of the Manhattan shore, along the New Jersey shore of the Hudson River in the area of Bergen, and along the Brooklyn shore, especially in Gowanus and Gravesend Bays. South of Brooklyn, Jamaica Bay also was a major shellfish-producing area. In addition, important oyster beds were located in the upper East River, particularly in Flushing Bay, and in the western portion of Long Island Sound in the vicinity of City Island.

Beds of *Mercenaria mercenaria* (quahogs) were abundant locally over much of the lower harbor, and extensive populations occurred in Gravesend Bay, Jamaica Bay, and Raritan Bay. Clams raked in Staten Island and Jamaica Bay were marketed in New York City, but quahogs never supported an industry comparable to the oyster industry.

The over-exploitation of "wild" oyster stocks near Manhattan and Staten Island already was of concern to local officials before the Revolutionary War (Ingersoll, 1881). The early 1800's was a period of exploding growth and development in New York City. By 1815 the City's population surpassed 100,000. It reached 300,000 by 1840 and by 1850 had increased to over 500,000 (Manhattan only) (Morison, 1965). This period of growth coincided with the general depletion of natural oyster beds in and around New York City. Even the rich Raritan Bay and Staten Island beds failed early in the century, during the years preceding the War of 1812 (Kochiss, 1974).

The New York industry was rejuvenated in the mid-1800's by a technological development, the "planting" of "seed oysters" imported from Chesapeake Bay. The success of this innovation largely contributed to the remarkable revival and prosperity of New York's commercial oyster industry prior to the Civil War. During the Civil War, with supplies of seed from Virginia cut off, local seed beds were developed primarily in Newark Bay, Kill Van Kull, Raritan Bay, and western Long Island Sound near City Island. Thereafter, local seed beds played a major role in supplying seed oysters to New York's industry (Kochiss, 1974).

By the last decades of the nineteenth century, significant changes occurred in the abundance of oysters in New York harbor. Natural seed beds declined because of overharvesting and the practice of planting oysters for commercial harvest on natural seed beds (Blackford, 1885). Although providing quick profits, this practice had the effect of reducing the production of seed oysters for planting. As a result, seed beds in Newark Bay, the Kill Van Kull, and along the Staten Island shore of Raritan Bay largely were depleted by 1890. Also, refuse from oil works and sugar refineries along the New York and New Jersey shores, increasing sedimentation, and the dumping of refuse and sewage played a major part in the loss of seed beds (Blackford, 1885).

Post-1900 Decline of New York's Oyster Industry

Pollution and continued population growth became major problems for the oyster industry in the 1880's and increased in importance in the early years of the 1900's in tandem with the continued decline of natural seed beds. By 1905 oyster beds in New York harbor (excluding Jamaica Bay) were largely limited to the southeastern shore of Staten Island, the beds north of the Narrows having been destroyed by pollution (New York Bay Pollution Commission, 1905). Small oyster beds, as well as major beds of quahogs, survived in Gravesend Bay, which, however, suffered a great deal from refuse and the pollution of the harbor (Blackford, 1887). Jamaica Bay remained a major producing area for oysters although oysters were not generally "planted" there. Growing conditions were so favorable that small oysters could be brought to marketable size over one season. At its peak, Jamaica Bay is said to have produced up to 700,000 bushels of oysters per year (Interstate Sanitation Commission, 1938). In the period around the turn of the century, Jamaica Bay also supported major soft clam (*Mya arenaria*) populations that were the basis for a local fishery (Kellogg, 1901).

In addition to industrial pollution, the almost constant dredging and deepening of navigation channels in New York harbor and the related disposal of dredge spoils caused siltation and corresponding problems for the shellfish industry. The illegal dumping of cellar dirt, street sweepings, garbage, and ballast also was a constant headache to oystermen (Blackford, 1887). As a result of the combined effect of human impacts, by 1905, the total commercial value of the annual oyster harvest in the lower estuary (perhaps one million bushels) was approximately $1 million, an amount far too small to warrant serious consideration in the total economy of New York City (Wood, 1904).

While chemical and industrial pollution, dredging, and illegal dumping received the most attention in this period, problems caused by pathogens associated with sewage pollution probably were of equal importance in the decline of oyster populations. City and State health officials were aware that sewage contamination was present in all New York harbor waters. Coliform counts of water and shellfish from the harbor were initiated during this period, but there was little recognition of the significance of coliform levels in oyster tissues. By the 1920's New York City discharged approximately 2.83×10^8 liters of raw, untreated sewage into harbor waters daily. The City's approach to dealing with human wastes was succinctly described as "...to conduct the sewage by the cheapest route to the nearest waterway, giving no thought whatever to its effect on the waterway and on adjacent waters" (New York Bay Pollution Commission, 1905: p. 16).

If industrial and sewage pollution were the major causes for the destruction of oyster beds in harbor waters, the final blow came with the linkage of typhoid to sewage-contaminated oysters from New York waters. As early as 1904, cases of typhoid were traced to oysters from Inwood (Jamaica Bay) (Anonymous, 1904). In 1915, 80% of the 150 typhoid cases in New York City were attributed to contaminated oysters from New York waters, and another major outbreak occurred between 1924 and 1925 (U.S. Works Progress Administration, 1939). Most New York City shellfish beds were closed for public health reasons by 1921, which marked the demise of New York City's oyster industry.

DISSOLVED OXYGEN

The role of sewage-derived organic matter in the decline of shellfish populations in the lower Hudson River was noted with alarm by officials in the late 1880's and led to the appointment of the New York Bay Pollution Commission in 1903 (Mann, 1940). This organization was absorbed by the Metropolitan Sewage Commission, which began sampling dissolved oxygen (DO) in harbor waters in 1909. Analyses have continued to the present and are performed under the auspices of various city, state, and interstate agencies. These continuous records of DO for a period of nearly seventy years provide the most useful, if indirect, picture of environmental change in the lower harbor. Average summer values of DO (expressed as percent saturation) for various sites between 1909 and 1935 are shown in Table 3. Gradual reductions in percent saturation occurred at all sites over time. For example, the DO in the Hudson River below Spuyten Duyvil declined from 72% in 1909 to 40% in 1935, with a minimum of 30% in 1921. Since these are average summer values, they mask the worst conditions, particularly in the lower East River and Hudson River, where DO frequently dropped to zero. As noted by Smith (1970), average summer values of percent saturation reached a nadir in the early 1920's and leveled off and remained at an average level of about 40% until the present. The construction of sewage treatment plants, which began in the early 1930's, appears to have stabilized DO levels in the face of continued increases in raw pollution load over the years (O'Connor, 1970).

Accounts of the destruction of oyster beds in New York harbor and the dramatic reduction in diversity of molluscan communities along the Staten Island shore leave little doubt that the marine environment throughout much of the harbor experienced critical deterioration in the years between 1880 and 1920 (O'Connor, 1970). While a constellation of factors is linked to this deterioration, including mismanagement of natural seed oyster beds, the impacts of industrial and sewage pollution are clearly of major importance. No comprehensive statistical data are available to evaluate total industrial inputs to the estuary during this period; however, historical accounts provide some idea of

Table 3. Average percent saturation of dissolved oxygen in the main branches of New York harbor from June 1 to October 1. (From Interstate Sanitation Commission, 1938: p. 17.)

Year	Number of samples	Hudson River below Spuyten Duyvil	Harlem River	Upper East River	Lower East River	Upper Bay	Kill Van Kull	The Narrows	Arthur Kill	Jamaica Bay
1909	404	72	55	86	65	67	79	83	–	–
1911	861	62	42	69	54	72	70	76	–	–
1912	150	58	–	–	49	64	65	71	–	–
1913	880	57	29	–	43	66	65	69	–	–
1914	473	50	30	50	40	71	–	68	–	–
1915	245	43	28	–	33	72	–	78	–	–
1916	176	46	24	–	26	64	–	63	–	–
1917	238	42	22	47	29	50	–	63	–	–
1918	54	54	23	50	21	56	–	61	–	–
1919	320	36	29	30	24	51	35	58	–	–
1920	264	44	23	50	27	43	42	52	–	–
1921	258	30	15	37	16	33	38	35	–	–
1922	280	44	26	51	26	51	51	60	–	–
1923	354	37	27	38	22	47	43	57	–	–
1924	643	44	26	45	26	48	48	73	–	–
1925	662	41	27	50	26	46	47	55	–	–
1926	396	23	14	37	13	26	29	40	26	66
1927	368	35	17	40	21	27	34	48	50	59
1928	433	37	28	41	23	47	41	44	56	69
1929	382	47	30	62	29	45	41	52	85	77
1930	332	37	25	40	20	38	39	51	49	73
1931	532	47	27	50	23	37	42	48	45	70
1932	582	41	20	38	16	43	47	51	71	75
1933	568	44	22	21	20	40	46	51	44	80
1934	751	47	22	39	16	39	50	52	55	69
1935	783	40	20	38	15	39	44	52	58	67

conditions in the harbor. For example, Allen (1921: p. 623) provided the following commentary:

> Large areas of the harbor covered with a film of oil may be mentioned as having caused much complaint recently. This condition has been prevalent along the Staten Island beaches and in the lower East River, where it frequently extends practically from shore to shore. The cause is due in part to extensive oil works on Newtown Creek and at Bayonne, New Jersey, but more especially by the discharge of bilge water from incoming oil-burning steamers.

While accounts of industrial pollution are dramatic, information provided by seventy years of DO records from the estuary suggests that the deterioration of water quality generally was correlated with sewage-related pollution.

COMPARISONS OF BENTHIC COMMUNITIES

Since bottom communities in lower New York harbor have been seriously perturbed for a period of 60-80 years, and no adequate data exist on conditions in the estuary prior to the onset of pollution and overfishing, biologists must utilize alternative approaches to assay the relative well-being of present and future benthic communities. The structure of benthic assemblages in the Hudson can be compared with that of less polluted estuaries nearby, although such comparisons introduce interpretive problems. The major Atlantic estuaries comparable in size to the Hudson also are perturbed to varying degrees, although smaller estuarine systems do exist that are much less affected by pollution. In the following sections, the species richness of the molluscan component of Raritan Bay's muddy/sand assemblage is compared with data from analogous molluscan associations in other Middle Atlantic coastal areas. While not definitive, this approach indicates the present impoverished condition of Raritan Bay and may prove useful for assaying future changes.

Status of the Hudson Estuary

The biological survey of the lower Hudson watershed, carried out by the State of New York Conservation Department (1937), constitutes the acknowledged "baseline" study of the mesohaline portion of the estuary (*i.e.*, that portion of the estuary characterized by salinities of 5-18 ‰. Recent analyses of invertebrate distributions and communities have been conducted by Hirschfield *et al.* (1966), Crandall (1977), and Ristich *et al.* (1977).

A sampling program carried out by Gross *et al.* (1971) in the lower estuary showed that 34 of 80 benthic samples from Newark Bay, the East, Harlem, and Hudson Rivers, and the Arthur Kill were totally devoid of benthic invertebrates common in similar but unperturbed environments. Of the remaining samples, nematodes and capitellid worms comprised most of the benthic fauna. With few exceptions, benthic communities were impoverished drastically or nonexistent. According to these authors, the distribution and abundance of pollution-tolerant capitellid worms indicated that 40% of the harbor area was affected severely by waste deposits.

Changes in benthic diversity and abundance in the saline waters of the Raritan River following limited pollution abatement (resulting from the construction of a major trunk sewer system) were analyzed by Dean and Haskin (1964). Species richness in the lower reaches of the River increased from six species prior to 1958, before the operation of the sewer, to 20 species in 1959. Densities of benthos increased during this period from approximately $100/m^2$ to $1,000/m^2$. Dean and Haskin were uncertain whether the observed succession of species was a result of pollution abatement or cyclic variations in populations.

More recently, McGrath (1974) and Pearce (1974) reported on the macrobenthos of the Raritan Bay area. As noted by Pearce, sediment composition in the Lower Bay is variable; salinity gradients are steep, and, hence, faunal lists must be interpreted carefully. Pearce reported a total of 78 benthic invertebrate species from intensive sampling of Raritan Bay, Lower Bay, and Sandy Hook Bay. In particular, the absence of gammarid amphipods was noted. The molluscan component comprised only 12 species, all but two of which (*Spisula solidissima* and *Lacuna vincta*) were widely distributed estuarine forms.

McGrath (1974) examined replicated 0.1-m^2 grab samples from 40 stations in Raritan Bay. His results indicate two benthic communities, a "sand" association dominated by the bivalve, *Tellina agilis,* and the polychaete, *Streblospio benedicti,* and a "mud" association dominated by the bivalve, *Mulinia lateralis.* Both assemblages appear to be extremely impoverished relative to those found in other estuaries and coastal embayments. Table 4 shows the number of molluscan species occurring in muddy/sand assemblages at several localities, including Buzzards Bay, Long Island Sound, the New York Bight apex, the Mullica River/Great Bay estuary, and Raritan Bay. Stable, high salinity embayments, such as Buzzards Bay and Fishers Island Sound, exhibit the highest diversity (23 and 24 molluscan species, respectively); polyhaline estuarine communities in central Long Island Sound and in the Mullica River contain 11-14 species. By comparison, the Raritan Bay association yielded only four species. Moreover, characteristic molluscan species of these communities, such as *Cylichnella (=Aceocina) canaliculata, Cylichnella oryza, Nucula proxima, Pitar morrhauna, Tellina agilis,* and *Yoldia limatula,* were absent from the Raritan Bay benthic community, although most of these occur in muddy/sand sediments in the Bight apex.

The data presented in Table 4 represent a qualitative approach to the assay of New York harbor benthic communities. Since these data refer only to a single sediment-controlled "community", they are of limited application to the lower estuary as a whole and certainly are not conclusive. Moreover, as only the molluscan component is treated, much significant information that could be provided by other major taxa is lacking. However, in the absence of reliable

Table 4. Molluscan species richness in muddy sand habitats.

	Total	Reference
Fishers Island Sound	24	Franz, 1976
Buzzards Bay (Station "R")	23	Sanders, 1960
Central Long Island Sound	12	Sanders, 1956
Central Long Island Sound	14	Franz, unpublished[a]
Central Long Island Sound	13	McCall, 1977
New York Bight apex	9	Franz, unpublished[b]
Raritan Bay	4	McGrath, 1974
Mullica River/Great Bay	11	Durand and Nadeau, 1972
New York Bight off southwestern Long Island	12	Steimle and Stone, 1973

[a]Data for 1972.
[b]Data for 1973.

historical data on benthic distributions and communities for the Raritan Bay area, this approach may provide a useful measure of the relative diversity of the Bay's benthic communities compared with those of nearby estuarine areas.

Mullica River/Great Bay Estuary

The nearest major estuary to New York harbor that can be considered relatively free of domestic or industrial pollution is the Mullica River/Great Bay estuary of New Jersey, which has been studied by Durand and Nadeau (1972). The macrobenthos was sampled by Petersen grab at 36 stations during the summer of 1968. Total macrobenthic invertebrate species richness was 143 species. Considering only the bay stations (*i.e.,* stations in which ebb tide salinity was usually greater than 20 ‰), 55 species made up 1% or more of the total number of individuals collected, 85 species comprised less than 1%, each, of the total, and five species, together, accounted for more than 20% of the total count. Most abundant was the gammarid amphipod, *Ampelisca abdita,* which occurred with maximum densities in excess of 5,000/m^2 at many stations.

The species richness of all invertebrates, based on five replicate Petersen grab hauls per station, exceeded 20 species over most of the Bay stations and reached a maximum of more than 40 species at some stations. Although biomass of macrobenthos was not measured, Durand and Nadeau (1972) noted that *A. abdita* was the dominant species, occurring at densities $\geq$ 1,000/m^2 over most of the Bay. Density of macrobenthos, excluding *A. abdita,* ranged from 100/m^2 to 1,000/m^2 at most stations in the western (upstream) two thirds of the Bay and from 1,000/m^2 to 5,000/m^2 at stations in a band across the eastern one third of the Bay, corresponding to ebb tide salinities of 23 ‰. As with all estuarine benthic communities, the Great Bay benthos was dominated by a few species, but the species richness was unusually high (85 species $\leq$ 1% of total count), reflecting perhaps the lack of environmental stress from organic and/or chemical pollution.

The total molluscan component of 35 species from the Mullica River/Great Bay estuary is shown in Table 5. Some relatively abundant inshore and estuarine species surprisingly are absent from this list, *e.g., Busycon carica, Nassarius trivittatus, Petricola pholadiformis, Polinices duplicatus,* and *Yoldia limatula.* However, as the macrobenthos was not the primary focus of this study, the absence of these species may reflect only random events in the sampling scheme.

In comparing the Bay fauna of Table 5 (Mullica River/Great Bay) with the shallow bay/inshore species recorded in 1887 from Staten Island (Table 1), it is evident that 30 of the 35 Great Bay species also occurred in Staten Island. This results in an overall Dice similarity index of approximately 65%, a level of similarity comparable to that expected between samples within a community (Sanders, 1960). Likewise, the Mullica River/Great Bay molluscan fauna overlaps the present molluscan fauna of Great South Bay (Long Island) at a level of 70% (Table 2). Thus, the molluscan assemblages of the Great Bay estuary generally are comparable to the present fauna of Great South Bay, and both are very

Table 5. Mollusca of the Mullica River/Great Bay, New Jersey estuary (from Durand and Nadeau, 1972) and Great South Bay, New York (from WAPORA, Inc., 1979).

Gastropoda	Mullica River/ Great Bay	Great South Bay
Anachis avara	+	+
Bittium alternatum	+	+
Busycon carica		+
Busycon canaliculatum	+	+
Cerithiopsis greeni		shell only
Crepidula convexa	+	+
Crepidula fornicata	+	+
Crepidula plana	+	+
Cylichna alba (probably *Cylichnella oryza*)	+	
Cylichnella (= *Acteocina*) *canaliculata*	+	+
Epitonium rupicolum	+	
Eupleura caudata	+	+
Haminoea solitaria		+
Hydrobia totteni	+	+
Ilynassa obsoleta	+	+
Lacuna vincta		+
Littorina littorea	+	+
Lunatia heros		+
Melampus bidentatus		+
Mitrella lunata	+	+
Nassarius trivittatus		+
Nassarius vibex	+	+
Odostomia producta		+
Odostomia impressa	+	
Odostomia trifida		+
Polinices duplicatus		+
Rictaxis punctostriatus	+	+
Seila adamsi		shell only
Triphora nigrocinta	+	+
Turbonilla interrupta		+
Turbonilla sp. (probably *interrupta* or *nivea*)	+	
Urosalpinx cinerea	+	+

similar to fauna of the Staten Island shallow bays and shore, as recorded by Smith in 1887. On the other hand, the depauperate molluscan fauna recently reported for Raritan Bay by McGrath (1974) and Pearce (1974) is related to the above estuarine assemblages at a level of only 40% (Table 2), reflecting primarily the reduced number of molluscan species reported by these workers (combined total of 14 species).

A striking difference between the present benthic assemblages of Raritan Bay and the Great Bay estuary is the absence of ampeliscid amphipods from the former (McGrath, 1974). McGrath suggested that the absence of amphipods may be caused by hydrocarbon residues in sediments. The susceptibility of benthic amphipods to toxicants contained in sludge as well as other materials disposed of in the Bight apex also was noted by Pearce (1970), Boesch (this volume), and Steimle *et al.* (this volume).

Table 5. (Continued)

Bivalvia	Mullica River/ Great Bay	Great South Bay
Aligena elevata		+
Anadara ovalis	+	+
Anadara transversa		
Anomia simplex		+
Argopecten irradians		+
Crassostrea virginica	+	+
Ensis directus	+	+
Gemma gemma	+	+
Laevicardium mortoni		+
Lyonsia hyalina	+	+
Macoma balthica		+
Macoma tenta	+	+
Mercenaria mercenaria	+	+
Mercenaria m. notata		+
Modiolus demissus	+	+
Mulinia lateralis	+	+
Mya arenaria	+	+
Mysella planulata		+
Mytilus edulis	+	+
Nucula proxima	+	+
Pandora gouldiana		+
Petricola pholadiformis		+
Pitar morrhuana		+
Solemya velum		+
Spisula solidissima	+	+
Tagelus plebeius		+
Tagelus divisus	+	
Tellina agilis	+	+
Yoldia limatula		+

McGrath (1974) indicated that, on the basis of densities, Raritan Bay assemblages are impoverished relative to those of other estuaries for which comparable data are available. Table 6 shows the average macrobenthos density in several Atlantic estuaries. McGrath's conclusion seems justified, although pollution is only one of a variety of factors affecting benthic density (Maurer *et al.*, 1978).

The interpretation of benthic faunal densities in relation to the relative "health" of an estuary without simultaneously considering the diversity is problematic. For example, Ristich *et al.* (1977) seemed to equate very high densities of benthic organisms in the lower Hudson estuary as evidence of the "health" of the estuary. However, since cultural eutrophication often is reflected by enhanced productivity of a relatively few tolerant species, high densities by themselves may not indicate a healthy (*i.e.*, normal) benthic community. This is not to deny the contention that "viable benthic communities" exist in the lower Hudson (Ristich *et al.*, 1977), but merely to suggest that this is not evidence that the Hudson is no longer a seriously perturbed estuary.

Table 6. Average macrobenthic density in selected Atlantic estuaries. Data based on 1-mm screen size. Mullica River/Great Bay density estimated from graphical data.

	Number/m²	Reference
Mystic River (Connecticut)	3,000	Rowe *et al.*, 1972
Moriches Bay (New York)	1,300	O'Connor, 1972
Raritan Bay (New Jersey)	109	McGrath, 1974
Mullica River/Great Bay (New Jersey)	4,000	Durand and Nadeau, 1972
Delaware Bay (New Jersey/Delaware)	722	Maurer *et al.*, 1978
Chesapeake Bay (eelgrass) (Virginia)	14,000	Orth, 1973
Tampa Bay (Florida)	510	Bloom *et al.*, 1972

In the absence of historical data, it would seem that the relative "health" of benthic communities in the Hudson estuary best could be determined by comparing the species structure and benthic productivity of Hudson estuary communities and those of analogous communities in less stressed estuaries. Specific comparisons between the lower Hudson estuary and the Mullica River/Great Bay estuary also may offer environmental managers an approach for examining the potential of a "rehabilitated" Hudson estuary.

ACKNOWLEDGEMENTS

Support for this project was provided by the MESA New York Bight Project (NOAA). I wish to thank Dr. Garry Mayer for editorial assistance and advice in the preparation of this report.

REFERENCES

Allen, K. 1921. The pollution of tidal harbors by sewage with especial reference to New York harbor. Am. Soc. Civ. Eng. Proc. Pap. Discuss. 47(10): 622-624.

Anonymous. 1904. Sewage-polluted oysters as a cause of typhoid and other gastrointestinal disturbances. U.S. Chem. Bur. Bull. 156: 28-44.

Blackford, E.G. 1885. Report of the commissioner of fisheries of the State of New York in charge of oyster investigations. Assembly of the State of New York, Vol. 6, Document No. 85. 137 pp.

Blackford, E.G. 1887. Second report of the oyster investigation and of the survey of oyster territory for the years 1885 and 1886. Assembly of the State of New York, Vol. 6, Document No. 28-36. 387 pp.

Bloom, S.A., J.L. Simon, and V.D. Hunter. 1972. Animal-sediment relation and community analysis of a Florida estuary. Mar. Biol. 13: 43-56.

Boesch, D.F. This volume. Ecosystem consequences of alterations of benthic community structure and function in the New York Bight region. pp. 543-568.

Botton, M.L. 1979. Effects of sewage sludge on the benthic invertebrate community of the inshore New York Bight. Estuarine Coastal Mar. Sci. 8: 169-180.

Crandall, M.E. 1977. Epibenthic invertebrates of Croton Bay in the Hudson River. N.Y. Fish Game J. 24(2): 178-186.

Dean, D. and H. Haskin. 1964. Benthic repopulation of the Raritan River estuary following pollution abatement. Limnol. Oceanogr. 9(4): 551-563.

Durand, J.B. and R.J. Nadeau. 1972. Water Resources Development in the Mullica River—Great Bay Estuary. New Jersey Water Resources Research Institute, Rutgers University, New Brunswick, NJ. 138 pp.

Esser, S.C. This volume. Long-term changes in some finfishes of the Hudson-Raritan estuary. pp. 299-314.

Franz, D.R. 1973. The ecology and reproduction of a marine bivalve, *Mysella planulata* (Erycinacea). Biol. Bull. 144(1): 93-106.

Franz, D.R. 1976. Benthic molluscan assemblages in relation to sediment gradients in northeastern Long Island Sound, Connecticut. Malacologia 15(2): 377-399.

Gross, M.G., J.A. Black, R.J. Kalin, J.R. Schramel, and R.N. Smith. 1971. Survey of marine waste deposits, New York metropolitan region. Marine Sciences Research Center, State University of New York, Stony Brook, Tech. Rep. 8. 72 pp.

Hall, A.V. 1969. Avoiding informational distortion in automatic grouping programs. Syst. Zool. 18: 318-329.

Hirschfield, H.J., J.W. Rachlin, and E. Leff. 1966. A survey of the invertebrates from selected sites of the lower Hudson River. *In:* Hudson River Ecology: Proceedings of a Symposium at Sterling Forest, Tuxedo, New York, Hudson River Valley Commission, Tarrytown, NY. pp. 220-257.

Hubbard, E.W. and S. Smith. 1865. Catalogue of the mollusca of Staten Island. Ann. Lyceum Nat. Hist. N.Y. 7: 151-154.

Ingersoll, E. 1881. The oyster industry. *In:* The History and Present Conditions of the Fisheries Industry, S. Baird and G. Goode (eds.), U.S. Government Printing Office, Washington, DC. 251 pp.

Interstate Sanitation Commission. 1938. Annual Report of the Interstate Sanitation Commission for the Year 1937 (Report to the Legislature of the State of New Jersey), MacCrellish and Quigley Company, Trenton, NJ. 80 pp.

Jacot, A. 1920. On the marine mollusca of Staten Island, New York. Nautilus 23(4): 111-115.

Kellogg, J.L. 1901. Clam and scallop industries of New York State. Bull. N.Y. State Mus. 43(8): 605-629.

Kochiss, J.M. 1974. Oystering from New York to Boston. *In:* The American Maritime Library, Vol. 7, Wesleyan University Press, Middletown, CT. 251 pp.

McCall, P.L. 1977. Community patterns and adaptive strategies of the infaunal benthos of Long Island Sound. Mar. Res. 35(2): 221-266.

McGrath, R.A. 1974. Benthic macrofaunal census of Raritan Bay—preliminary results, Pap. 24. *In:* Hudson River Ecology, Third Symposium on Hudson River Ecology, March 22-23, 1973, Hudson River Environmental Society, Inc. 40 pp.

Mann, R. 1940. Pollution abatement in New York harbor. The relation of marine borers to changing conditions of the harbor waters. *In:* Service Bureau American Wood-Preservers Association, Chicago, IL. 14 pp.

Maurer, D., L. Watling, P. Kinner, and C. Wethe. 1978. Benthic invertebrate assemblages of Delaware Bay. Mar. Biol. 45: 65-78.

Morison, S.E. 1965. The Oxford History of the American People. Oxford University Press, New York. 1150 pp.

National Marine Fisheries Service. 1972. The effects of waste disposal in the New York Bight. Final Report, Section 2: Benthic Studies. National Marine Fisheries Service, Middle Atlantic Coastal Fisheries Center, Sandy Hook Laboratory, Highlands, NJ. 63 pp.

New York Bay Pollution Commission. 1905. Report of the New York Bay Pollution Commission. New York State Legislature, Albany, NY. 135 pp.

O'Connor, D.J. 1970. Water quality analysis for the New York harbor complex. *In:* Water Pollution in the Greater New York Area, A.A. Johnson (ed.), Gordon and Breach, New York. pp. 121-144.

O'Connor, J.S. 1972. The macrofauna of Moriches Bay, New York. Biol. Bull. 142: 84-102.

Orth, R.J. 1973. Benthic infauna of eelgrass, *Zostera marina,* beds. Chesapeake Sci. 14: 258-269.

Pearce, J.B. 1970. The effects of solid waste disposal on benthic communities in the New York Bight. Paper FIR:mp/70/E-99. FAO Technical Conference on Marine Pollution, Rome, Italy. 12 pp.

Pearce, J.B. 1974. Invertebrates of the Hudson River estuary. Ann. N.Y. Acad. Sci. 250: 137-143.

Ristich, S.S., M. Crandall, and J. Fortier. 1977. Benthic and epibenthic macroinvertebrates of the Hudson River. I. Distribution, natural history and community structure. Estuarine Coastal Mar. Sci. 5: 255-266.

Rowe, G.T., P.T. Polloni, and J.I. Rowe, 1972. Benthic community parameters in the lower Mystic River. Int. Rev. Gesellshaft Hydrobiol. 57: 573-584.

Sanders, H.L. 1956. Oceanography of Long Island Sound, 1952-54. X. Biology of marine bottom communities. Bull. Bingham Oceanogr. Collect. 15: 345-414.

Sanders, H.L. 1960. Benthic studies in Buzzards Bay. III. The structure of the soft-bottom community. Limnol. Oceanogr. 5: 138-153.

Smith, E.T. 1970. Water quality of the New York harbor and Bight areas. *In:* Water Pollution in the Greater New York Area, A.A. Johnson (ed.), Gordon and Breach, New York, pp. 39-52.

Smith, S. 1887. Catalogue of the mollusca of Staten Island. Proc. Nat. Sci. Assoc. Staten Island 1: 50.

State of New York Conservation Department. 1937. Biological Survey of the Lower Hudson Watershed, 1936. Biological Survey No. 11, Supplement to Twenty-Sixth Annual Report. State of New York Conservation Department, Albany, NY. 373 pp.

Steimle, F.W. and R.B. Stone. 1973. Abundance and distribution of inshore benthic fauna off southwestern Long Island, New York. NOAA Tech. Rep. NMFS SSRF 673. 50 pp.

Steimle, F.W., J. Caracciolo, and J.B. Pearce. This volume. Impacts of dumping on New York Bight apex benthos. pp. 213-223.

U.S. Works Progress Administration, New York City. 1939. Report relative to shellfish industry in New York harbor. US-WPA Rep. No. 465-97-3-131. 264 pp.

WAPORA, Inc. 1979. Estuarine impact assessment for the Nassau-Suffolk Streamflow Augmentation Alternatives. U.S. Environmental Protection Agency Contract 68-01-4616.

Wood, B.F. 1904. The shellfisheries of New York State. Bull. For., Fish Game Comm., State of N.Y. 1904: 15-25.

DISTRIBUTION AND SEASONAL INCIDENCE OF "BLACK GILL" IN THE ROCK CRAB, *CANCER IRRORATUS*

Thomas K. Sawyer

U.S. Department of Commerce
National Oceanic and Atmospheric Administration
National Marine Fisheries Service
Northeast Fisheries Center
Oxford, Maryland 21654

Abstract. Rock crabs, *Cancer irroratus,* from coastal sewage dumpsites near New York, New Jersey, Delaware, and Maryland were examined for evidence of stress or disease attributable to the effects of ocean dumping. Excessive gill fouling by bacteria, debris, diatoms, copepods, amoebae, and stalked ciliates frequently was noted in microscopical studies but could not be used as evidence for environmental stress. Gills that were black over 50% or more of their tissue mass were observed in crabs collected from ocean disposal sites and appeared to be useful indicators of environmental degradation. The "black gill" condition is believed to be caused by the accumulation of black sludge sediment between the gill lamellae. Summarized data showed that blackened gills were present in 4.5% (134/2,920) of the crabs on a year-round basis, or in up to 10% when only intermolt animals were considered. Blackened gills were not observed within a second group of 268 crabs collected from control locations (Georges Bank; Montauk Point, New York; Sheepscott River, Maine). An additional 846 specimens were collected at stations from Cape Cod, Massachusetts to Cape Hatteras, North Carolina during surf clam assessment and scallop cruises. Eight percent (17/217) of the crabs from the surf clam stations and less than 1% (3/629) of the crabs from the scallop cruise had black gills. Station data from the two cruises showed that affected crabs usually were collected from locations in or near sewage disposal sites. Observations on the incidence of black gills in *C. irroratus* provided evidence that the condition occurs primarily in areas where bottom sediments are altered by the accumulation of sewage sludge deposits.

INTRODUCTION

Gross and microscopical observations on the effects of ocean dumping on crustacean health have shown that shell erosion and tissue necrosis are useful qualitative indicators of disease or stress in rock crabs, *Cancer irroratus,* and lobsters, *Homarus americanus.* Young and Pearce (1975) observed both conditions in crabs and lobsters collected near sewage and dredge spoil disposal sites in the New York Bight apex. Gopalan and Young (1975) also observed shell disease in up to 30% of the shrimp, *Crangon septemspinosa,* collected in the Bight apex. Shell erosion in shrimp sometimes was associated with necrosis of underlying epidermis and destruction of inner tissues, and with "burn spot" and rotted appendages. Cause and effect relationships between environmental pollution and crustacean diseases are circumstantial, since similar conditions have been described on earlier occasions. However, further study is likely to show that when factors such as molting cycles and seasonal differences are taken into account, the incidence of such pathology may be significantly higher in polluted than in nonpolluted environments.

Studies presented here summarize a five-year survey on the incidence and geographical distribution of "black gill" and associated histopathology in the rock crab. Background information suggests that severely diseased gills may interfere with respiratory function and may increase the susceptibility of stressed crabs to predation. The number of crabs examined in the current investigation for gross signs of gill blackening or discoloration and for microscopic evidence of gill disease was sufficiently large to support conclusions on the influence of geographical distribution, sex, size, and life history on the incidence of "black gill" and associated gill-fouling organisms.

METHODS

Rock crabs were captured by making 30-minute tows with an otter trawl, by removing crabs from clam or scallop dredges, or by collecting crabs from shallow water with lobster or crab pots. Specimens examined for survey purposes were taken by trawling or dredging, while others that were collected for incidental observations were caught in pots. Direct observations on gill condition were made by removing the dorsal carapace and recording gill condition as clean, disclored, or black. "Black gill" condition was noted on data sheets only when gills were black for 50% or more of their length. All crabs were measured across the dorsal carapace, and the width was recorded in centimeters. Records were made of sex, size, molting condition (intermolt, peeler, soft, papershell), and water depth. Gills to be saved for histological study were preserved in Davidson's fixing solution, embedded in paraffin, sectioned at 6 μm, and stained with Harris hematoxylin-eosin solution or by the Feulgen reaction. The number of crabs examined and the areas from which they were collected are summarized in Table 1. Collection sites were categorized as "polluted" on the basis of published accounts of ocean dumping activities, coliform bacterial contamination, and heavy metal or organic chemical burdens. Control sites were selected on the basis

Table 1. Geographical distribution and numbers of rock crabs, *Cancer irroratus*, examined for "black gill" condition.

Location	Number of stations	Total number examined	Number fixed for histology
New York Bight apex	3	1,447	472
Sandy Hook Bay-Lower Bay	3	888	317
Philadelphia-Camden sewage sites	26	585	381
Scallop survey, Mass.-N.C.	59	217	15
Surf clam survey, N.Y.-N.C.	139	629	39
Georges Bank, Atlantic Ocean	5	61	61
Sheepscott River, Me.	1	171	71
Montauk Point, N.Y.	1	36	36
Totals	237	4,034	1,392

of their location away from known sources of ocean dumping and their commercial use for the harvesting of fishes or shelfishes.

The incidence of "black gill" was studied by analyzing seasonal variations in crabs collected from Sandy Hook Bay-Lower Bay and the New York Bight apex, and comparing the data with results from other areas obtained at comparable seasons of the year. Periods of molting activity were noted to determine differences in male and female shedding cycles at near- and offshore stations, and the influence of such activity on the incidence of gill discoloration or blackening. Data from histological studies were categorized (microscopic debris, fouling bacteria, diatoms, stalked ciliates, amoebae, copepods), and the overall condition of each crab with black gills was compared with such findings to determine what association, if any, could be made between the microscopic fouling community and black discoloration. The incidence of each condition in crabs from the New York Bight apex and in crabs from Sandy Hook Bay-Lower Bay stations was compared by χ^2 test to analyze similarities or differences between the geographical distribution of the various fouling entities.

RESULTS

Distinctions between slightly discolored, discolored, and black gills were highly subjective early in the survey but became more reliable as larger numbers of crabs were evaluated. Clean gills were uniformly white, cream, or mahogany brown in color; disclored gills ranged from dark brown, to brown-black, to black at the tips or bases. The most conservative estimate on the incidence of "black gill" was accomplished by including in this category only those crabs with gills that were black for 50% or more of their length (Figure 1). Large numbers of animals had black discoloration at the tip or base of individual gills, but they were excluded from the "black gill" category because it was impossible to predict whether further blackening would occur. Therefore, the possible impact of sewage-related pollution may be significantly greater than is indicated by the data.

Five percent of the 2,335 rock crabs from the New York Bight apex and Sandy Hook Bay-Lower Bay stations were observed to have black gills, regardless of size, sex, molting activity, or season of the year (Table 2). The highest incidence in crabs from Sandy Hook Bay-Lower Bay stations was 9% (Table 3), which was observed during the April-June quarter following the adult male molting period. The highest incidence in New York Bight apex crabs was 10% (Table 3), which was observed during the October-December quarter and coincided with the annual migration of adult males into shallow waters prior to molting. Adult female crabs molted during late summer (August-September); such activity appeared to have only minimal effect on the incidence of "black gill."

Sampling regimes were adjusted to take advantage of the influence of molting and migratory behavior on crab abundance and gill condition. Collections of adult males in the estuary during the winter molt cycle revealed numerous

Figure 1. Black gills (arrow) in rock crab, *Cancer irroratus,* collected from Sandy Hook Bay, New Jersey, in June 1979; male specimen with carapace width of 6.5 cm.

specimens, mostly with clean gills. Collections from ocean waters during the late summer and fall, prior to shoreward migration and molting, provided numerous specimens and revealed the highest incidences of "black gill." Collections from the estuary during late spring sampled specimens that had not completed a successful molt. The number of organisms taken varied according to sampling intensity. Such specimens represented a portion of the estuarine population that either were slow to return to the ocean during seaward migration or failed to return at all (Table 3, Estuary: April-June). Such crabs were considered to be "stragglers" and often represented the "worst case" in terms of visual evidence of disease or stress. Thus, crabs with severely blackened gills and high incidences of fouling organisms were taken both in estuarine waters and in proximity to dumpsites. There was no qualitative difference between severely fouled gills from estuarine and apex waters.

Gill discoloration occasionally was observed in juvenile crabs; however, in most collections juveniles were excluded from consideration because of frequent molting activity during the year. Males were estimated to be subadults when carapace width measured less than 5 cm. Females could not be categorized similarly because some carried extruded egg masses but measured only 2-3 cm. Data obtained from studies in New York and New Jersey permitted estimations of the incidence and distribution of the "black gill" condition in *C. irroratus.*

Table 2. Incidence of "black gill" condition in rock crabs, *Cancer irroratus,* from near- and offshore collecting sites. Data not disaggregated to account for factors such as molting activity, size, and sex. Data presented as number of organisms with indicated condition followed, in parentheses, by percentage of organisms with that condition.

Location	Clean (%)	Discolored (%)	Black (%)	Total
New York Bight apex	842 (58)	521 (36)	84 (6)	1,447
Sandy Hook Bay-Lower Bay	636 (72)	214 (24)	38 (4)	888
Philadelphia-Camden sewage sites	341 (58)	232 (40)	12 (2)	585
Scallop survey, Mass.-N.C.	453 (72)	173 (24.5)	3 (0.5)	629
Surf clam survey, N.Y.-N.C.	115 (53)	85 (39)	17 (8)	217
Georges Bank, Atlantic Ocean	51 (84)	10 (16)	0	61
Sheepscott River, Me.	96 (56)	75 (44)	0	171
Montauk Point, N.Y.	29 (81)	7 (19)	0	36

Histological studies on approximately one third of the crabs collected from New York-New Jersey stations (789/2,335) showed that fouling by sediment particles, bacteria, diatoms, sessile ciliates, amoebae, and copepods often was associated with discolored (Table 4) or black (Table 5) gills. Gill filaments sometimes were "cemented" together by excessive fouling organisms, which caused small black foci to form between adjacent gill lamellae. Microscopic observations of the black foci showed they consisted of necrotic, blackened gill filaments that were melanized and brittle. Crabs with focal or mottled black gills were not included in the "black gill" category. Microscopic debris in the interlamellar spaces provided an indication of the type of sea bottom from which the crabs were collected. Animals from clean sandy bottom usually had large sand grains that often damaged the microtome knife blade used for

Table 3. Seasonal differences in incidence of "black gill" in rock crabs, *Cancer irroratus,* sampled from the New York Bight apex and Sandy Hook Bay-Lower Bay (estuary). Data presented as incidence followed, in parentheses, by percentage of organisms with "black gill." Low winter incidence reflects molting activity of adult males. Summer incidence influenced by molting activity of adult females and small number of crabs taken in estuary.

Location	Winter (%) Jan.-Mar.	Spring (%) Apr.-June	Summer (%) July-Sept.	Fall (%) Oct.-Dec.
Apex	1/200 (0.5)	55/780 (7)	11/290 (4)	17/177 (10)
Estuary	6/346 (2)	26/304 (9)	1/45 (2)	5/193 (3)
Totals	7/546 (1)	81/1,084 (7)	12/335 (4)	22/370 (6)

Table 4. Seasonal influence (%) of fouling conditions on gills of rock crabs, *Cancer irroratus,* collected in the New York Bight apex and in Sandy Hook Bay-Lower Bay (estuary). Data presented as percentage of organisms with indicated condition from the apex followed, in parentheses, by percentage with condition from the estuary.

Seasonal quarter	Discolored	Debris	Bacteria	Diatoms	Sessile ciliates	Amoeba	Copepod
Jan.-Mar.	22 (28)	28 (25)	37 (17)	15 (18)	4 (23)	11 (7)	13 (3)
Apr.-June	48 (50)	70 (57)	57 (56)	23 (21)	25 (47)	16 (9)	20 (9)
July-Sept.	63 (61)	65 (86)	50 (80)	21 (25)	13 (68)	18 (11)	18 (4)
Oct.-Dec.	80 (66)	79 (82)	60 (71)	30 (38)	10 (54)	35 (29)	13 (6)
Mean	53 (51)	61 (63)	51 (56)	22 (26)	13 (48)	20 (14)	16 (6)

sectioning the tissues; those from silty-clay sea bottoms were packed with fine-grained sediments and sectioned without difficulty. Gills that were judged "black" usually sectioned easily, contained fine-grained sediments, and harbored massive communities of fouling organisms. Specimens with gills judged to be 100% black often had external blackening and carapace erosion (Figure 2). Black gills often were melanized extensively with necrosis of gill cuticle or entire gill filaments.

Thin unidentified filamentous bacteria appeared to be the first organisms to attach to the gills of newly-molted specimens. Dense clumps of rod-shaped bacteria were observed to occlude completely the interlamellar spaces of seriously fouled gills, and the affected cuticle often was thick and melanized. The incidence of each category of fouling in crabs with blackened gills is summarized in Table 5.

Two percent of the rock crabs (12/585) taken from the Philadelphia-Camden sewage dumpsite stations had black gills (Table 2). All crabs with black gills were caught between 36 m and 54 m. Four of them were taken at the center of the sewage dumpsite; the remaining eight were caught within a radius of 24 km from the site. Station data further showed that approximately 25% of the 585 crabs were caught in the disposal site or within a radius of 8 km from its center. Although crabs were collected at all seasons of the year, the number of

Table 5. Incidence (%) of fouling conditions observed on rock crabs, *Cancer irroratus,* with "black gills" and collected from apex and estuary stations. Not included are data from sectioned or stained gills on which fouling was sparse or rare.

Number of crabs with black gills	Debris (%)	Bacteria (%)	Diatoms (%)	Sessile ciliates (%)	Amoebae (%)
59	(95)	(86)	(83)	(68)	(58)

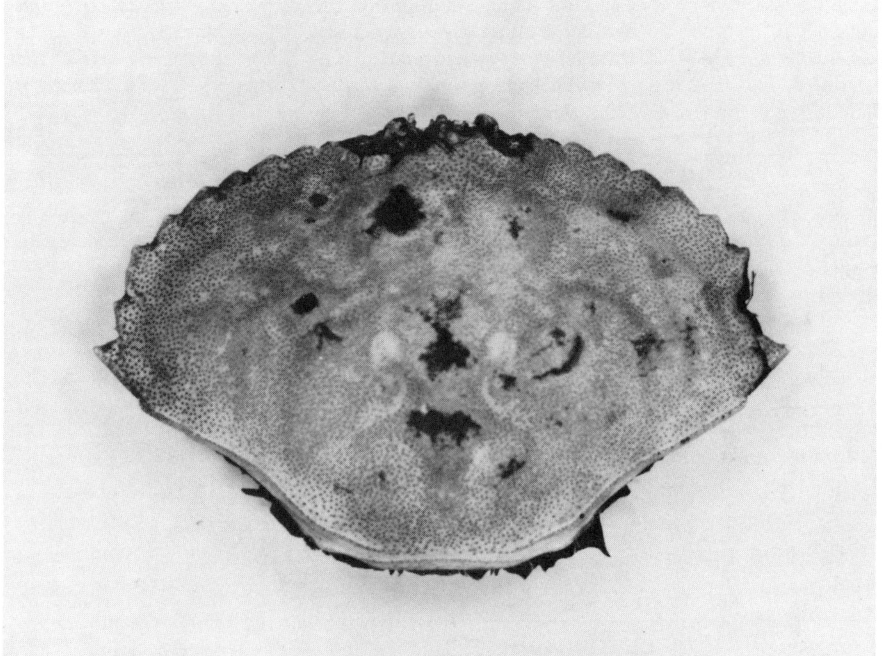

Figure 2. Blackened areas on carapace of rock crab, *Cancer irroratus,* collected from edge of a sewage dumpsite in New York Bight apex in Ocrober 1979 (note black margins around eyes); male specimen with carapace width of 8.5 cm.

collecting trips was too limited to support conclusions on seasonal differences (Table 6). Histological studies on 381 of the 585 animals showed that the microscopic fouling community was similar to that observed in New York-New Jersey crabs with two exceptions: (1) sessile ciliates rarely were observed, and, when present, they occurred in very small numbers, and (2) as many as 39 copepods were observed in a single section of gill tissue, while no more than nine (usually 1-5) occurred in the New York-New Jersey specimens.

Molting activity differed from that observed in crabs from the New York-New Jersey area in that both sexes appeared to molt simultaneously. Crabs collected in September 1978 made up a sample in which 208 adults were intermolt stages and 64 (28 males and 36 females) were peelers, soft crabs, or papershells. It was interesting to note that the incidence of "black gill" was the lowest (3/221) during the July-September quarter, which coincided with the observed molting period for adults. Summarized data on the Philadelphia-Camden dumpsite stations showed that within the total collection of 585 crabs, the gills of 58% were clean, 34% were discolored, 6% had slight blackening, and 2% had more than half of the gill area blackened and were classified under the "black gill" category.

Less than 1% (3/629) of the rock crabs collected during a scallop assessment cruise had black gills (Table 2). The stations sampled during the cruise ranged

Table 6. Seasonal incidence (%) of fouling conditions on gills of rock crabs, *Cancer irroratus*, collected in or near the Philadelphia-Camden dumpsites. Total sample=381 animals. Note reasonable agreement of seasonal data with overall or cumulative data. Data suggest that molting activity had little or not effect on seasonal observations (see text for explanation). Sample was composed of 220 males and 161 females.

Seasonal quarter	Discolored	Debris	Bacteria	Diatoms	Sessile ciliates	Amoeboe	Copepods
Jan.-Mar. 32 crabs	63	63	34	3	9	9	50
Apr.-June 31 crabs	42	55	40	0	3	3	50
July-Sept. 220 crabs	48	64	48	5	3	11	40
Oct.-Dec. 98 crabs	37	64	37	6	4	7	46
Mean	48	62	40	4	5	8	47

northward to Cape Cod, Massachusetts and southward to Cape Hatteras, North Carolina. The low incidence of "black gill" probably was because over half (59/89) of the stations were located more than 83 km from shore and were not in proximity to known sewage disposal sites. Seventy-three percent of the crabs from the collection had clean gills in contrast to 58%, as noted with the Philadelphia-Camden crabs and the New York Bight apex crabs. Stations sampled, with few exceptions, served as control sites for crabs that were not living in close proximity to sewage disposal sites. Microscopic observations on gills from 39/629 animals showed that the gill-fouling community and tissue pathology was similar to that observed in crabs from the other areas. Sessile ciliates and copepods were scarce, and, when present, they occurred in small numbers. Macroscopic observations showed that heavy infestations with unidentified gooseneck barnacles were frequent in crabs collected off southern Virginia and Cape Hatteras, North Carolina.

Visual observations on gills from 217 crabs collected during a surf clam assessment cruise showed that 8% (17/217) had black gills (Table 2). Stations ranged from Long Island, New York, to Cape Hatteras, North Carolina; most were considerably closer to shore than were those of the scallop cruise. Many of the stations were in or near the sewage disposal sites located near the New York metropolitan area, Delaware, and Maryland. Twelve of 59 stations yielded crabs with black gills. All but two stations were less than 74 km from shore; all but one were from depths between 19 m and 42 m. Clean gills were noted in only 53% of the crabs, discolored gills in 39%, and black gills in 8% (Table 2).

Table 7. Cumulative incidence (%) of gill-fouling conditions on rock crabs, *Cancer irroratus*, from New York Bight apex and Sandy Hook Bay-Lower Bay (estuary), (n=789) analyzed for χ^2 probability. Probability=likelihood that each set of numbers (% incidence) occurred by chance. High probability is significant when a given condition has a low overall incidence and/or when collections are made during the molting period.

Location	Discolored	Debris	Bacteria	Diatoms	Sessile ciliates	Amoeba	Copepods
Apex (estuary)	22 (28)	28 (25)	37 (17)	15 (18)	4 (23)	11 (7)	13 (3)
Probability	.59	.23	.003	.855	.013	.825	.62

Although the survey represented only one collection, the results were in agreement with those obtained from other collections in or near the sewage disposal sites. Microscopic observations on sectioned gills showed that similar categories of fouling microorganisms occurred in both "polluted" and "control" stations. Further studies are necessary to determine whether conspecific fouling organisms are present at all stations.

Collections from selected "control" sites were made at Montauk Point, Long Island, Sheepscot River, Maine, and the Georges Bank fishing grounds. Specimens from the first two locations were caught in lobster pots, and all were adults; those from Georges Bank were taken by trawling, and some were juveniles. "Black gill" was not observed in any of the 268 crabs in the control group (Table 2).

In summary, visual and microscopic examinations of rock crabs showed a positive relationship between the incidence of "black gill" and the presence of sewage-silt contamination of the sea bottom. The incidence of "black gill" also was influenced by molting activity of adult animals (Table 3). However, distributional analysis by the χ^2 square method (Table 7) showed that not all fouling conditions in the New York area were influenced by the source of the crabs (*i.e.*, apex vs. estuary). For example, differences between the incidence of diatom infestation in crabs from apex and crabs from the estuary largely appear to have been the result of chance (85.5% probability). On the other hand, differences between bacterial infestations probably were not the result of chance (0.3% probability) and thus reflect differences between the estuarine and apex environments and/or differences in the condition of crabs utilizing these areas. New information on the influence of specific pollutants on black gill condition and on microbial fouling is needed before cause and effect relationships can be evaluated further by statistical methods.

DISCUSSION

Intact gills from over 4,000 rock crabs were examined visually to estimate the incidence of the "black gill" condition on a geographical basis. Approximately

3,000 animals were collected from areas in or near established sewage sludge dumpsites, and approximately 1,000 were collected from areas that were not impacted similarly. The rock crab was selected for the survey because it walks rather than swims and, therefore, must live in direct contact with the sea bottom. During periods of inactivity, crabs rest on the sea bottom and circulate water and sand or silt over their gill surfaces; thus, the particulate matter on the seabed may accumulate between the gill lamellae and later may provide the scientist with direct visual evidence on the nature of bottom sediment deposits. The association between degraded ocean bottom and black gills in the rock crab is made here on the basis of a 4-5% incidence (151/3,137) in animals from impacted sites, and less than 1% incidence (3/897) among crabs examined from nonimpacted sites.

Stations located in the New York-New Jersey area were sampled repeatedly to obtain detailed information on the influence of adult molting cycles on the observed overall incidence of "black gill" condition. To a lesser extent, stations in or near the Philadelphia-Camden dumpsites were sampled for the same purpose. Seasonal cycles were important in scheduling collection trips, since up to 10% of the crabs were found to have black gills during intermolt seasons, while 2% or less had the condition during the molting period. Comparative data showed that offshore adult male and female crabs molted at approximately the same time (*i.e.,* in late summer). Therefore, the scallop and clam assessment collections, which were made during the month of April, fortuitously coincided with a season during which a high incidence of "black gill" would have been anticipated.

The frequency with which blackened gills were observed in rock crabs at the New York-New Jersey stations appeared related to the presence of seriously impacted bottom sediments such as those near the sewage disposal sites. The lower incidence of black gills in crabs from the Philadelphia-Camden stations (Table 2) reflects that sewage sludge disposal has occurred at this site for only five to six years and that sea bottom contamination may not be as serious as in the New York area. Young and Pearce (1975) studied crabs and lobsters collected near the New York sewage sludge and dredged materials dumping grounds and noted black deposits in the articulations of the appendages of affected animals. They also noted distinctive odors associated with animals obtained from sewage and dredge spoil sites. Specific chemical tests on the black deposits on crustacean gills and appendages have not been made, but it seems likely that they will prove related to settled waste deposits.

Black "spots" on the gills of lobsters, shrimp, and crabs sometimes cause focal or localized discoloration, which is distinct from "black gill" syndrome. Black spot does not necessarily indicate a specific response to stress or pollution since it may be caused by localized melanization of small groups of filaments or lamellae that are dead and necrotic (Lightner and Redman, 1977). Focal pigmentation or gill blackening has been induced experimentally by exposing crustaceans to cadmium (Couch, 1978) and by injecting turpentine into shrimp (Fontaine *et al.,* 1975). Focal melanization also has been observed in lobsters with fungal infections (Lightner and Fontaine, 1975; Fisher *et al.*, 1978) and in shrimp (*Pandalus borealis*) from the Gulf of Maine (Rinaldo and Yevich, 1974). Johnson and Bradbury (1976) surveyed 54 crustacean species and found that 24 had apostome ciliate infections of the gills. Infections were obvious by visual examination since encysted ciliates were enclosed by a ring of melanin. "Black spot" appears to be useful as a broad term to denote tissue death in crustacean species that respond to disease or injury by producing discrete foci of melanin. In contrast, "black gill" appears to be a useful descriptor only when all or most of the gill tissue is visibly affected.

A third category, for which the term "black mottled" is used in the present study, denotes the presence of narrow blackened bands or "stripes" horizontal to the gill axis. Black mottling was noted when all or most of the lamellar filaments were dead and melanized within narrow zones that were separated by otherwise clean and apparently healthy lamellae. Such gills were not included in the "black gill" category since the blackened areas did not affect 50% or more of the tissue. Since gill blackening may be caused by chemical or biological agents and has been noted in both natural and experimental environments (see above), the terms used to report such conditions often are misleading and confusing. However, observations from the present survey suggest that terms such as "black gill," "black spot," and "black mottled" are essential descriptors for recording visual evidence of gill disease during field surveys and when specific agents are not recognizable in advanced stages of tissue destruction and melanization.

Evidence from the present study suggests that the effects of sewage dumping on crustacean (*i.e.*, rock crab) health may include severe gill discoloration or gill blackening. Observations also suggest that studies with a single objective may be of limited value in assessing the overall effects of pollutants on benthic species. Recent efforts, for example, have shown that the distribution of coprostanol in bottom sediments may be used to map the spread of sewage-contaminated bottom sediments (Hatcher *et al.*, 1977). Similarly, fecal coliform counts (Babinchak *et al.*, 1977) may be useful indicators of the spread of sewage sludge laterally from known disposal sites. Watling *et al.* (1974) noted that the concentrations of zinc, lead, copper, chromium, nickel, and silver were 5 to 200 times greater in sewage sludge than in bottom sediments collected from dumpsites off of Delaware Bay. Greig *et al.* (1977) found elevated metal levels, especially for copper, in zooplankton collected from the New York Bight. Thus, a variety of chemical and biological data independently and collectively testify to the progressive deterioration of environments that continue to receive substantial amounts of pollutants such as may be contained in barge-delivered sewage sludge and dredged materials (O'Connor, 1975; Anderson, this volume; Franz, this volume; Lear *et al.*, this volume; Michael, this volume; Sindermann *et al.*, this volume; Wolfe *et al.*, this volume).

New information of a highly valid and reproducible nature might be obtained by conducting multidisciplinary research at a given moment in time and space. Correlations between the significance of coprostanol concentrations, bacterial numbers, heavy metal levels, and histopathology could be measured statistically if all samples for study were collected and examined at the same time. Specifically, individual animals could and should be processed for study by as many methodologies as practical. Similarly, sediments that accumulate between the gill lamellae of bottom-dwelling animals could be separated from tissue spaces by sonication and analyzed for direct comparisons with sediments present at collection sites. Measurements of total organic carbon in bottom sediments also provide reliable data on the extent of organic loading in clean or degraded benthic ecosystems. Thus, crustaceans collected from areas with unusually high sediment organic carbon might have discolored or blackened gills as a consequence of extremes in naturally occurring degradative processes, as well as of conditions following barged inputs.

Future studies designed around a multidisciplinary approach must be accomplished to obtain a more comprehensive data base. At the same time, further information is needed on the comparative biology and life history of *C. irroratus* populations in the New York Bight apex and in offshore communities; Gross and microscopic indications of disease or stress in *C. irroratus,* as recorded here, reflect only seasonal similarities or differences in terms of percent of affected specimens within each collection. New information on the prevalence of gross or microscopic pathological conditions in *C. irroratus* must await further studies on rock crab reproduction, growth, migratory behavior, and the effects of predation to assure that conclusions are representative of resident populations.

ACKNOWLEDGEMENTS

The author gratefully acknowledges support received from the NOAA MESA New York Bight Project, Stony Brook, New York, in the conduct of the "black gill" survey. Dr. Don Lear, Ms. Marria O'Malley, and Ms. Sue Smith, U.S. Environmental Protection Agency (EPA), Annapolis, Maryland, collected specimens and recorded gill condition data for all crabs caught in the Philadelphia-Camden sites. Mr. Earl J. Lewis, State of Maryland Office of Coastal Zone Management, Annapolis, Maryland, provided invaluable assistance during all phases of EPA cooperative research. Mr. Jeff Tinsman, University of Delaware, Lewes, Delaware, kindly collected and examined gill condition in all crabs sampled on Georges Bank. Ms. Sharon MacLean, Mr. Bruce Harke, and Mr. Austin Farley, National Marine Fisheries Service (NMFS), Oxford, Maryland, recorded all gill condition data obtained during the month-long resource assessment cruises (surf clams and scallops). The scope of the "black gill" survey and the large amount of numerical data that were available for analyses were made possible by the combined efforts of personnel from both state and federal agencies. Special thanks are expressed to Mr. John Ziskowski, NMFS, Sandy Hook Laboratory, Highlands, New Jersey, who coordinated and participated in all of the many field collections made at the New York and New Jersey stations. Further appreciation is expressed to the operators and crew of the Sandy Hook vessels RORQUAL, XIPHIAS, and KYMA who were responsible for the success of all trawling operations.

REFERENCES

Anderson, J.W. This volume. The transport of petroleum hydrocarbons from sediments to benthos and the potential effects. pp. 165-179.

Babinchak, J.A., J.T. Graikoski, S. Dudley, and M.F. Nitkowski. 1977. Distribution of faecal coliforms in bottom sediments from the New York Bight. Mar. Pollut. Bull. 8: 150-153.

Couch, J.A. 1978. Diseases, parasites, and toxic responses of commercial penaeid shrimps of the Gulf of Mexico and south Atlantic coasts of North America. Fish. Bull. 76: 1-44.

Fisher, W.S., E.H. Nilson, J.F. Steenbergen, and D.V. Lightner. 1978. Microbial diseases of cultured lobsters: a review. Aquaculture 14: 115-140.

Fontaine, C.T., R.G. Bruss, I.A. Sanderson, and D.V. Lightner. 1975. Histopathological response to turpentine in the white shrimp, *Penaeus setiferus.* J. Invertebr. Pathol. 25: 321-330.

Franz, D.R. This volume. An historical perspective on molluscs in lower New York harbor, with emphasis on oysters. pp. 181-197.

Gopalan, U.K. and J.S. Young. 1975. Incidence of shell disease in shrimp in the New York Bight. Mar. Pollut. Bull. 6: 149-153.

Greig, R.A., A. Adams, and D.R. Wenzloff. 1977. Trace metal content of plankton and zooplankton collected from the New York Bight and Long Island Sound. Bull. Environ. Contam. Toxicol. 18: 3-8.

Hatcher, P.H., L.E. Keister, and P.A. McGillivary. 1977. Steroids as sewage specific indicators in New York Bight sediments. Bull. Environ. Contam. Toxicol. 17: 491-497.

Johnson, C.A., III and P.C. Bradbury. 1976. Observations on the occurrence of the parasitic ciliate, *Synophrya,* in decapods in coastal waters off the southeastern United States. J. Protozool. 23: 252-256.

Lear, D.W., M.L. O'Malley, W.C. Muir, and G. Pence. This volume. Environmental effects of sewage sludge at the Philadelphia dumping site. pp. 481-493.

Lightner, D.V. and C.T. Fontaine. 1975. A mycosis of the American lobster, *Homarus americanus,* caused by *Fusarium* sp. J. Invertebr. Pathol. 25: 239-245.

Lightner, D.V. and R. Redman. 1977. Histochemical demonstration of melanin in cellular inflammatory processes of penaeid shrimp. J. Invertebr. Pathol. 30: 298-302.

Michael, A.D. This volume. The potential contribution of petroleum hydrocarbons to changes in benthic communities of the New York Bight. pp. 621-630.

O'Connor, J.S. 1975. Contaminant effects on biota of the New York Bight. Proc. Gulf Caribb. Fish. Inst. 28: 50-63.

Rinaldo, R.G. and P. Yevich. 1974. Black spot gill syndrome of the northern shrimp, *Pandalus borealis.* J. Invertebr. Pathol. 24: 224-233.

Sindermann, C.J., S.C. Esser, E. Gould, B.B. McCain, J.L. McHugh, R.P. Morgan, R.A. Murchelano, M.J. Sherwood, and P.R. Spitzer. This volume. Effects of pollutants on fishes and shellfishes. pp. 23-38.

Watling, L., W. Leathem, P. Kinner, C. Wethe, and D. Maurer. 1974. Evaluation of sludge dumping off Delaware Bay. Mar. Pollut. Bull. 5: 39-42.

Wolfe, D.A., D.F. Boesch, A. Calabrese, J.F. Lee, C.D. Litchfield, R.J. Livingston, A.D. Michael, J.M. O'Connor, M. Pilson, and L. Sick. This volume. Effects of toxic substances on communities and ecosystems. pp. 67-86.

Young, J.S. and J.B. Pearce. 1975. Shell disease in crabs and lobsters from New York Bight. Mar. Pollut. Bull. 6: 101-105.

IMPACTS OF DUMPING ON NEW YORK BIGHT APEX BENTHOS

Frank Steimle
Janice Caracciolo
John B. Pearce

National Marine Fisheries Service
Northeast Fisheries Center
Sandy Hook Laboratory
Highlands, New Jersey 07732

Abstract. The discharge of materials in the Hudson-Raritan estuary and New York Bight apex has altered some apex benthic habitats physically and chemically, introducing high concentrations of a variety of pollutants (*e.g.*, heavy metals, petroleum hydrocarbons, and organic carbon) that are known to be lethal or stressful to several groups of benthic invertebrates. Benthic invertebrate data from five survey cruises conducted during 1973-1974 at 66 stations in the apex are reviewed, with evidence of dumping impacts discussed. The benthic community, in the vicinity of two dumpsites receiving New York harbor dredge spoils and metropolitan area sewage sludge, exhibited a reduction in the number of species present, especially small amphipod crustaceans that are highly sensitive to pollution and are important food items of many fish species.

INTRODUCTION

The New York Bight apex, encompassing the coastal waters within 25 km of the mouth of New York harbor, is adjacent to one of the major human population and industrial centers of the world. Within the last one hundred years, the New York metropolitan area's use of the Hudson River, adjacent estuaries, and coastal waters has resulted in a decline in the quality of these waters and concurrent impacts to estuarine and marine organisms. Included was the elimination of a variety of valuable commercial fisheries from the area. Early documentation of these events is scarce, but a few reports do exist (*e.g.*, Goode, 1887). Few studies were conducted until the 1960's, when the Sandy Hook Laboratory and other research groups began investigations of man's impact on the apex (Steimle and Stone, 1973; Pearce *et al.*, 1976). The most extensive studies have occurred within the past ten years, many under the coordination of the National Oceanic and Atmospheric Administration's (NOAA) Marine EcoSystems Analysis (MESA) Program and National Marine Fisheries Service (NMFS) environmental programs.

The benthic invertebrate community is an important part of the apex ecosystem, contributing to food webs that support valuable fisheries and playing an important role in nutrient recycling. This group of organisms also is an indicator of environmental quality because its components are relatively immobile and often long-lived; they thus reflect the cumulative effects of exposure to environmental pollutants. In 1973 the NOAA MESA New York Bight Project sponsored a NMFS/Sandy Hook Laboratory survey of the benthic apex community. The survey was designed to investigate the impact of waste disposal practices, such as dumping, and other sources of contamination on the benthos of the inner Bight. This paper presents a summary of the results of the MESA-sponsored study and indicates that significant impacts to the benthic community have occurred.

STUDY AREA

The survey area consisted of an 805-km^2 rectangle defined by latitudes 40°10′N and 40°34′N and longitudes 73°36′W and 73°59′W (Figure 1). The area is bordered on the north by the barrier islands and beaches of southern Long Island, and on the west by the beaches of northern New Jersey. The entrance to New York harbor forms the northwest corner of the rectangle. The study area includes the Christiaensen Basin, which lies at the head of the Hudson Shelf Valley (a trough extending southeast across the continental shelf to the Hudson Canyon at the continental shelf-slope break).

Municipalities and industries have used the apex for waste disposal since the turn of the century. At the time of the study there were dumpsites within the apex accommodating dredge spoils, sewage sludge, acid wastes, cellar dirt, and derelict vessels. Only the first three sites are still active. The dredge spoil dumpsite, approximately 10 km east of Sea Bright, New Jersey, on the western slope of the Christiaensen Basin, accommodates material dredged from New York harbor, Raritan Bay, and the Newark Bay complex. This often consists of accumulations of highly contaminated material derived from outfalls and non-point sources discharging into the estuary, as well as material deposited from upstream sources. The sewage sludge dumpsite, approximately 18 km east of Sandy Hook, on the eastern edge of the Christiaensen Basin, accommodates the slurry produced by various metropolitan area treatment plants. For details concerning the constituents and volumes of materials deposited at these sites, see Gross (1976), Mueller *et al.* (1976), and U.S. Environmental Protection Agency (1978).

METHODS

This study is based on five sampling cruises conducted during August and October 1973, and January, March-April, and August-September 1974. During each cruise, five 0.1-m^2 Smith-McIntyre benthic grab samples were collected on a grid of 65-103 stations in the apex (see Pearce *et al.*, 1977 for location data). From each bottom sample, two 3.4-cm (inside diameter) cores were taken, one subsample for geological analyses and the other for heavy metals analyses. The remainder of the benthic sample was washed through a 1.0-mm screen. The retained material was preserved in 10% formalin and later transferred to a solution of 70% ethanol, with 5% glycerol added.

In the laboratory the preserved material was sorted; organisms were separated, identified to species, and counted. Geological sediment sub-samples were analyzed for grain size using a "Rapid Sediment Analyzer," and percentages of oxidizable organic material were calculated after hydrogen peroxide digestion (A. Cok, unpublished data). Heavy metals subsamples were analyzed

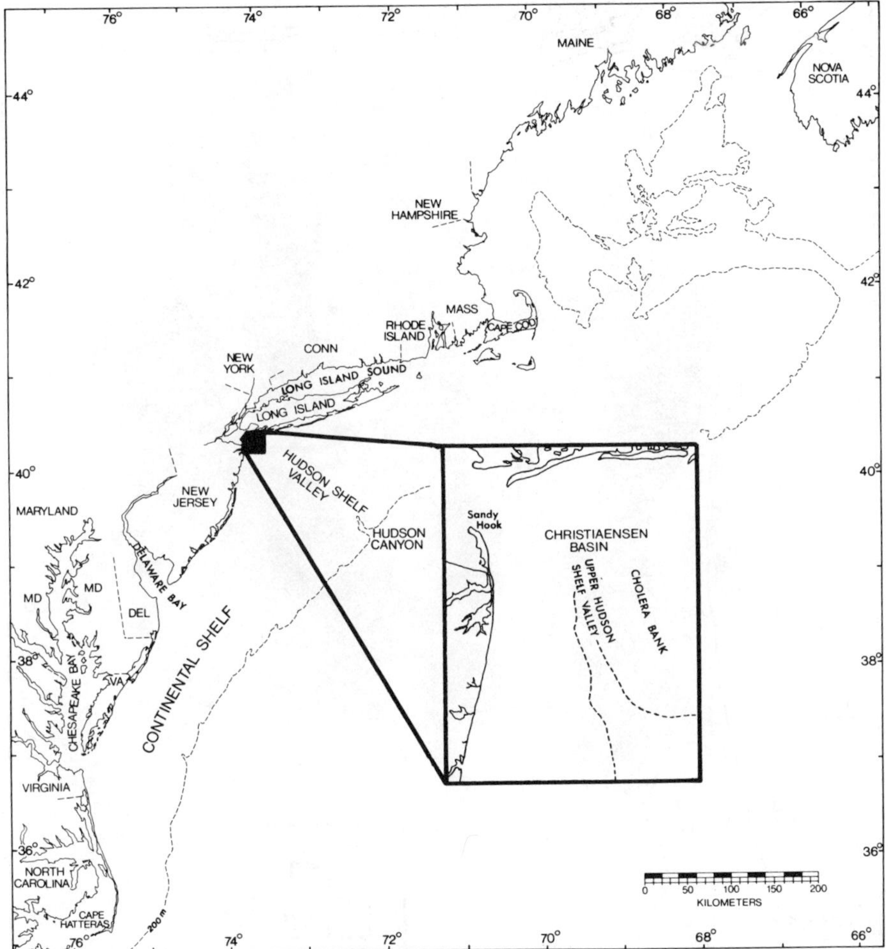

Figure 1. New York Bight apex (enlarged) and surrounding area.

by the Environmental Chemistry Investigation, NMFS Laboratory, Milford, Connecticut. For details of methods, see Pearce *et al.* (1977).

All data from the five MESA quarterly cruises have been previously reported by Pearce *et al.* (1977), where they were presented by sample as listings of species identifications and counts, total numbers of individuals, equitability and diversity, sediment grain size, total digestible carbon content, and heavy metals concentrations (Cr, Cu, Ni, Pb, and Zn). This report summarizes the data set and analyzes it for evidence of alterations to the benthic community.

RESULTS

The distribution of species diversities, H′ (Shannon and Weaver, 1963), averaged over the five quarterly cruises, is presented in Figure 2. Averaged values

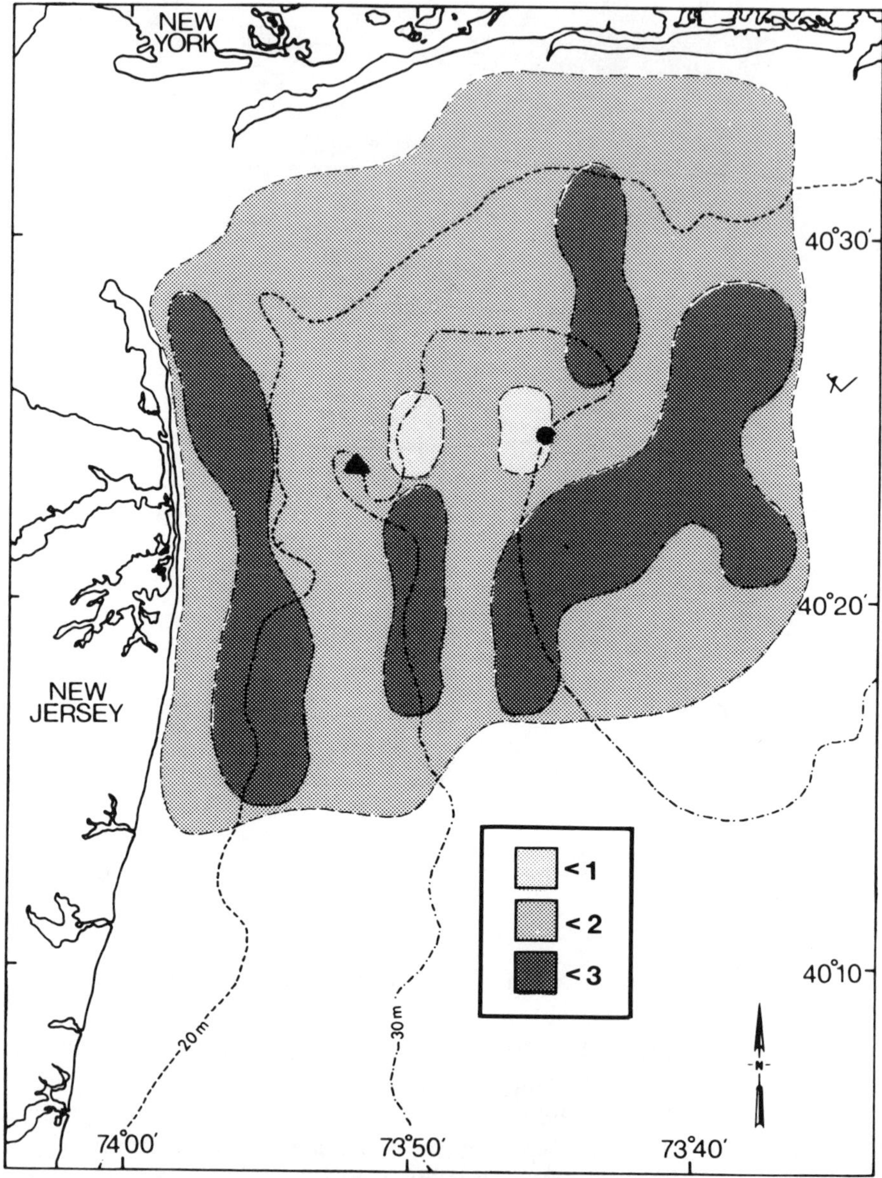

Figure 2. Average species diversity (H′) distribution. Locations of dredge spoil (▲) and sewage sludge (●) dumpsites are indicated.

for percentage of digestible organics and the distribution of lead in sediment are presented as Figures 3 and 4. Lead was chosen as an example of the type of distribution that was found for all five heavy metals (see Table 1).

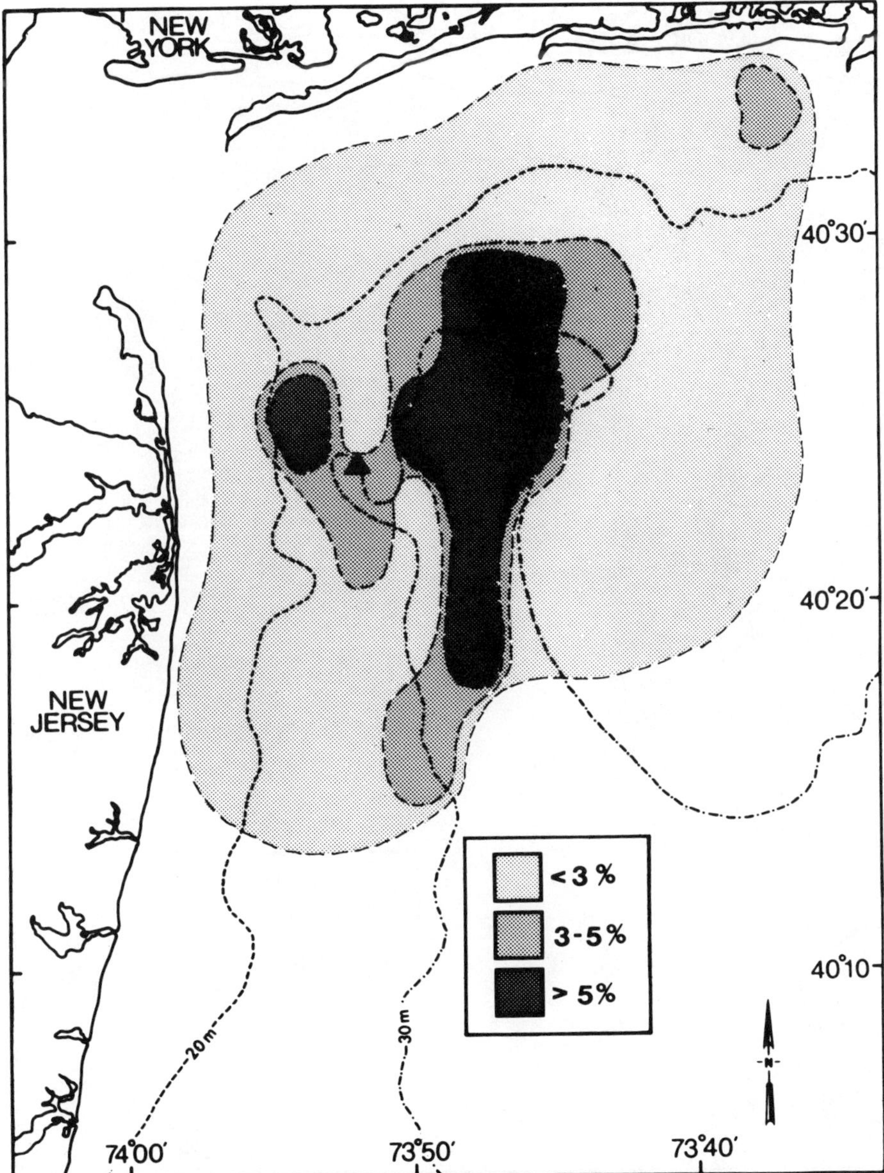

Figure 3. Distribution of digestible organic carbon (average %) in apex sediments. Locations of dredge spoil (▲) and sewage sludge (●) dumpsites are indicated.

DISCUSSION AND CONCLUSIONS

Examination of Figures 2-4 indicates that the benthic macroinvertebrate community and sediments have been altered significantly at the dredge spoil and

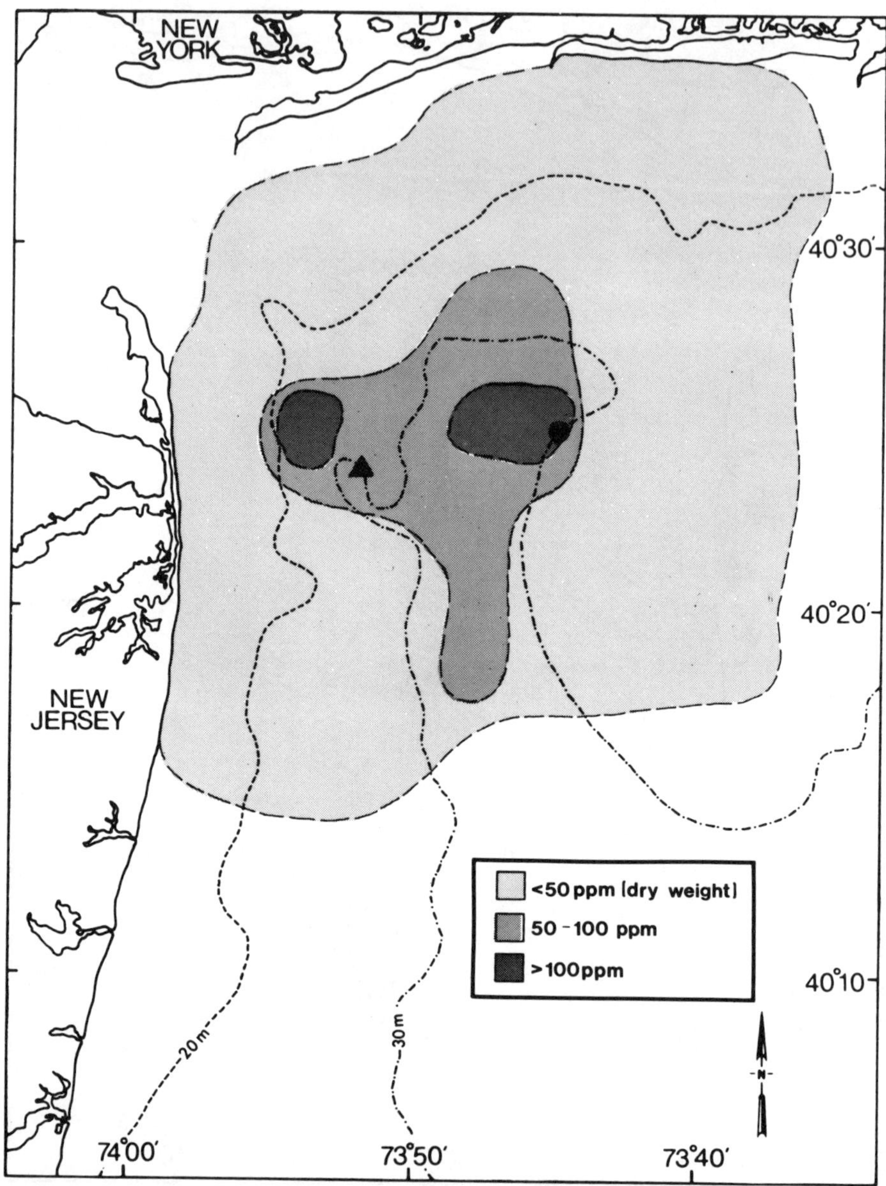

Figure 4. Average concentration levels of lead in the apex sediments. Location of dredge spoil (▲) and sewage sludge (●) dumpsites are indicated.

sewage sludge dumpsites in the apex. As indicated, in Figure 2, H′ diversity values are lowest within and in the immediate vicinity of both of these dumpsites. Low H′ values are most often associated with highly stressed environments, where a few opportunistic or tolerant species become dominant.

Table 1. Concentrations of five heavy metals in New York Bight sediments (ppm in dry sediments). The values outside of the apex are from Carmody *et al.* (1973).

	Cr	Cu	Pb	Ni	Zn
Within apex					
Sewage sludge dumpsite	183	220	157	29	365
Dredge spoil dumpsite	101	101	113	20	162
Inshore areas off Long Island and New Jersey	16	10	22	4	46
Outside apex					
Sandy sediments of New York Bight	6	3	12	3	18
Silty sediment of Hudson Shelf Valley	6	5	14	8	20

This results in a less stable, simpler community, usually consisting of a small number of species (Sanders, 1968). Tietjen (1980) also found low species diversities among nematode populations in these areas.

The benthic communities found in or near the dumpsites were dominated by polychaetes (57% by abundance), with a relative absence of crustaceans, especially amphipods (< 1% by abundance). This is in contrast to similar but unpolluted silty areas of the Middle Atlantic Bight, where amphipods generally comprise 30% of the benthic macrofauna, often being dominant, *i.e.*, ⩾ 70% (Pratt, 1973; Wigley and Theroux, 1976; Boesch *et al.*, 1977). Such alteration in community structure may be explained partially by the generally low tolerance of crustaceans to even low concentrations of pollutants (Blumer *et al.*, 1970; Sanders *et al.*, 1972).

Impacts to the benthos also are evident in the abundances of individual species. For example, *Capitella capitata,* a small deposit-feeding polychaete, was found in extremely high numbers at the sewage sludge dumpsite and was absent or in low abundance elsewhere in the apex. Although this taxon may consist of several species that have various ecological requirements (Grassle and Grassle, 1976), generally it is considered an opportunist, often associated with highly polluted environments (Reish, 1956; Schultz, 1969; Sanders *et al.*, 1972; Wolff, 1973; Warren, 1977).

Nucula proxima, a small (0.6 cm) bivalve, generally very abundant in the silty sediments of the Christiaensen Basin and upper Hudson Shelf Valley, and occurring widely in the rest of the apex, is absent in the vicinity of both dumpsites. *Nucula,* a surface deposit feeder, shows an apparent close correlation with high percentages of digestible organics in apex sediments (Figure 3), except at the sewage dumpsite. It is assumed that *Nucula* is intolerant of the high levels of toxic material found there (*e.g.*, Table 1).

Analyses of the sediments of the New York Bight apex and their contaminant burdens indicate an alteration by dumping (*e.g.*, elevated organic carbon, heavy metals, and hydrocarbon levels). The highest levels of organic carbon were found adjacent to both dumpsites (Figure 3). Hatcher and Keister (1976) suggested that material high in organic carbon, deposited in the Christiaensen Basin and in mud patches near Long Island, is predominantly of sewage origin, based on comparisons of total carbohydrates to total organic carbon ratios in the New York Bight.

In the 1973-1974 survey, high levels of five heavy metals were measured in the areas receiving both dredged materials and sewage sludge (Table 1). They are correlated, in general, with sediments of highest organic content (*e.g.*, see distribution of lead, Figure 4). Concentrations of these metals may be one to two orders of magnitude higher than values measured at apex stations away from the dumpsites and background levels in uncontaminated sands and silt.

Koons and Thomas (1979) report that total C_{15+} hydrocarbons are highest (3,600-6,500 ppm) in New York Bight areas where harbor dredge spoil and sewage sludge disposal occurs. In contrast, levels at the mouth of the Hudson-Raritan estuary are reported as low as 6-22 ppm, with concentrations of 82 ppm and 86 ppm reported at two locations approximately 80 km on the mid-continental shelf. Anderson's review (this volume) reports similar findings from other studies in the New York Bight.

Elevated levels of hydrocarbons and heavy metals are known to be toxic to marine life. They are lethal in high concentrations, but even in sublethal concentrations may cause pathological conditions, physiological disturbances, and behavioral changes. Exposure can affect the gill epithelia, liver, kidneys, and haemopoetic tissues of a variety of marine organisms, as well as cause spawning inhibition and other behavioral changes, thus making the organism less able to compete and/or more susceptible to predation. Larval stages are especially sensitive to these toxic materials and usually show increased abnormalities and slow growth rates (Shuster and Pringle, 1968; Portmann, 1970; Stirling, 1970; Connor, 1972; Vernberg *et al.*, 1973; Reish *et al.*, 1974, Calabrese *et al.*, 1977).

The Christiaensen Basin and upper Hudson Shelf Valley benthic environments also are subject to seasonal reductions in dissolved oxygen (DO), often to levels critical to many species of marine organisms common in the New York Bight (Pearce, 1972; Segar and Berberian, 1976; Thomas *et al.*, 1976; Steimle, 1977). Reduced DO levels during the summer months may result from the high sedimentary organic concentrations and consequent elevated biological oxygen demand from apparently flourishing micro- and meiofauna populations, and from chemical oxygen demand of sediments, as evident from high seabed oxygen consumption rates at the dredge spoil and sewage sludge dumpsites (Thomas *et al.*, 1976). However, summer oxygen depression may be related more to a larger scale phenomenon than to dumping, perhaps being more influenced by the discharge of the Hudson-Raritan estuary into the Bight (Steimle and Sindermann, 1978).

In summary, the benthic environment in the areas around the dredge spoil and sewage sludge dumpsites contains a variety of contaminants at elevated concentrations that have been reported to be stressful or lethal to many marine benthic organisms. These contaminants also may alter the habitat sufficiently to influence benthic community structure. The pattern of benthic community diversity and the severe reduction in expected abundance of sensitive amphipod crustaceans are good evidence that impacts have occurred.

With the planned termination of sewage sludge dumping in the area in 1981, it will be important to monitor the recovery of the apex ecosystem. The amount of time required for the dumpsites to recover is unknown. Dean and Haskin (1964) found that the benthic community, particularly the small amphipod crustaceans, showed marked recovery after pollution abatement at the nearby mouth of the Raritan River. Recolonization of contaminated dredge spoils has also been shown to be relatively rapid in areas adjacent to the New York Bight (Pratt *et al.,* 1973; Reid and Frame, 1977). However, little work has been done on the recovery of sewage sludge dumpsites.

ACKNOWLEDGEMENTS

We wish to thank Michele Cox for preparing illustrations and Virginia Boeckel and Catherine Noonan for typing the manuscript.

REFERENCES

Anderson, J.W. This volume. The transport of petroleum hydrocarbons from sediments to benthos and the potential effects. pp. 165-179.

Blumer, M., J. Sass, G. Souza, H. Sanders, F. Grassle, and G. Hampson. 1970. The West Falmouth oil spill; persistence of the pollution eight months after the accident. Woods Hole Oceanogr. Inst. Tech. Rep. WHOI-70-44. 32 pp.

Boesch, D., J. Kraeuter, and K. Serafy. 1977. Benthic ecological studies: megabenthos and macrobenthos, Chapter Six. *In:* Chemical and Biological Benchmark Studies on the Middle Atlantic Outer Continental Shelf. Final Report. Virginia Institute of Marine Science, Gloucester Point, VA. 194 pp.

Calabrese, A., J. MacInnes, D. Nelson, and J. Miller. 1977. Survival and growth of bivalve larvae under heavy metal stress. Mar. Biol. 41: 179-184.

Carmody, D., J. Pearce, and W. Yasso. 1973. Trace metals in the sediments of the New York Bight. Mar. Pollut. Bull. 4(9): 132-135.

Connor, P. 1972. Acute toxicity of heavy metals to some marine larvae. Mar. Pollut. Bull. 13: 190-192.

Dean, D. and H. Haskin. 1964. Benthic repopulation of the Raritan River estuary following pollutant abatement. Limnol. Oceanogr. 9: 551-563.

Goode, G. 1887. The Fisheries and Fishery Industries of the United States. Section 2. A Geographical Review of the Fisheries Industries and Fishery Committee for the year 1880. U.S. Government Printing Office, Washington, DC. 384 pp.

Grassle, J.F. and J.P. Grassle. 1976. Sibling species in the marine pollution indicator *Capitella*-Polychaeta. Science 192: 567-569.

Gross, M. 1976. Sources of urban wastes. Am. Soc. Limnol. Oceanogr. Spec. Symp. 2: 150-161.

Hatcher, P. and L. Keister. 1976. Carbohydrates and organic carbon in New York Bight sediments as possible indicators of sewage contamination. Am. Soc. Limnol. Oceanogr. Spec. Symp. 2: 240-248.

Koons, C. and J. Thomas. 1979. C_{15+} hydrocarbons in the sediments of the New York Bight. *In:* Proceedings of the 1979 Oil Spill Conference, American Petroleum Institute, Washington, DC. pp. 625-628.

Mueller, J., A. Anderson, and J. Jeris. 1976. Contaminants entering the New York Bight: sources, mass loads, significance. Am. Soc. Limnol. Oceanogr. Spec. Symp. 2: 162-170.

Pearce, J. 1972. The effects of solid waste disposal on benthic communities in the New York Bight. *In:* Marine Pollution and Sea Life, M. Ruivo (ed.), FAO Fishing News (books) Ltd., Surrey, England. pp. 404-411.

Pearce, J., J. Caracciolo, A. Frame, L. Rogers, M. Halsey, and J. Thomas. 1976. Distribution and abundance of benthic organisms in the New York Bight, August 1968-December 1971. NOAA Data Rep. ERL MESA-7, Boulder, CO. 114 pp.

Pearce, J., L. Rogers, J. Caracciolo, and M. Halsey. 1977. Distribution and abundance of benthic organisms in the New York Bight apex, five seasonal cruises, August 1973 through September 1974. NOAA Data Rep. ERL MESA-32, Boulder, CO. 803 pp.

Portmann, J. 1970. A discussion of the results of acute toxicity tests with marine organisms, using a standard method. Paper FIR:mp/70/E-31. FAO Technical Conference on Marine Pollution, Rome, Italy. 13 pp.

Pratt, S. 1973. Benthic fauna. *In:* Coastal and Offshore Environmental Inventory, Cape Hatteras to Nantucket Shoals, Marine Experiment Station, Graduate School of Oceanography, Univ. Rhode Island, Mar. Publ. Ser. No. 2. 70 pp.

Pratt, S., S. Saila, A. Gaines, Jr., and J. Krout. 1973. Biological effects of ocean disposal of solid wastes. Univ. Rhode Island, Mar. Tech. Rep. Ser. No. 9. 53 pp.

Reid, R. and A. Frame. 1977. Sediments and benthic macrofauna of river and disposal area. *In:* Physical, Chemical and Biological Effects of Dredging in the Thames River (CT) and Spoil Disposal at the New London (CT) Dumping Ground. NTIS Rep. No. AD 1044164. 44 pp.

Reish, D. 1956. An ecological study of Lower San Gabriel River, California, with special reference to pollution. Calif. Fish Game 42: 51-61.

Reish, D., F. Piltz, and J. Martin. 1974. Induction of abnormal polychaete larvae by heavy metals. Mar. Pollut. Bull. 5(8): 125-126.

Sanders, H. 1968. Marine benthic diversity: a comparative study. Am. Nat. 102: 243-282.

Sanders. H.L., J.F. Grassle, and G.R. Hampson. 1972. The West Falmouth oil spill. I. Biology. Woods Hole Oceanogr. Inst. Tech. Rep. WHOI-72-20. 48 pp.

Schultz, S. 1969. Benthos und Sediment in der Mecklenburger Bucht. Beitr. Meereskd. 24/25: 15-55.

Segar, D. and G. Berberian. 1976. Oxygen depletion in the New York Bight apex: causes and consequences. Am. Soc. Limnol. Oceanogr. Spec. Symp. 2: 220-239.

Shannon, C. and W. Weaver. 1963. The Mathematical Theory of Communication. University of Illinois Press, Urbana, IL. 117 pp.

Shuster, C. and B. Pringle. 1968. Effects of trace metals on estuarine molluscs. *In:* Proceedings of the First Mid-Atlantic Industrial Water Conference, 13-15 November 1967, University of Delaware, Newark, DE. pp. 285-304.

Steimle, F. 1977. The persistence and boundaries of the bottom water oxygen depletion problem of 1976, in the New York Bight. *In:* Oxygen Depletion and Associated Environmental Disturbances in the Middle Atlantic Bight in 1976. Tech. Series Rep. No. 3, NOAA/NMFS, Northeast Fisheries Center, Highlands, NJ. pp. 41-64.

Steimle, F. and C. Sindermann. 1978. Review of oxygen depletion and associated mass mortalities of shellfish in the Middle Atlantic Bight in 1976. Mar. Fish. Rev. 40(12): 17-26.

Steimle, F. and R. Stone. 1973. Abundance and distribution of inshore benthic fauna off southwestern Long Island, New York. NOAA Tech. Rep. NMFS SSRF-673. 50 pp.

Stirling, E. 1970. Some observations on the response of the benthic bivalve *Tellina tenuis* to pollutants. Proc. Int. Counc. Explor. Sea, C.M.1970/E:15, Fisheries Improvement Committee. 6 pp.

Thomas, J., W. Phoel, F. Steimle, J. O'Reilly, and C. Evans. 1976. Seabed oxygen consumption–New York Bight apex. Am. Soc. Limnol. Oceanogr. Spec. Symp. 2: 354-369.

Tietjen, J. 1980. Population structure and species composition of the free-living nematodes inhabiting sands of the New York Bight apex. Estuarine Coastal Mar. Sci. 10(1): 61-73.

U.S. Environmental Protection Agency. 1978. Environmental Impact Statement on the Ocean Dumping of Sewage Sludge in the New York Bight. Final, September 1978. U.S. Environmental Protection Agency, Region II, New York. 226 pp.

Vernberg, W., P. DeCoursey, and W. Padgett. 1973. Synergistic effects of environmental variables on larvae of *Uca pugilator*. Mar. Biol. 22: 307-312.

Warren, L. 1977. The ecology of *Capitella capitata* in British waters. J. Mar. Biol. Assoc. U.K. 57: 151-159.

Wigley, R. and R. Theroux. 1976. Macrobenthic Invertebrate Fauna of the Middle Atlantic Bight Region. Part II. Faunal Composition and Qualitative Distribution. NOAA/NMFS, Northeast Fisheries Center, Woods Hole, MA. 395 pp.

Wolff, W. 1973. The estuary as a habitat. An analysis of data on the soft-bottom macrofauna of the estuarine area of the Rivers Rhine, Meuse, and Scheldt. Zool. Verh. (Leiden) 126: 1-242.

POTENTIAL ROLES OF NEMATODES IN POLLUTED ECOSYSTEMS AND THE IMPACT OF POLLUTION ON MEIOFAUNA

John H. Tietjen

Department of Biology and
Institute of Marine and Atmospheric Sciences
City College of New York
Convent Avenue at 138th Street
New York, New York 10031

Abstract. Meiofauna are small benthic animals (usually less than 2 mm) whose densities on the shelf generally range from $10^5/m^2$ to $10^7/m^2$. Through their grazing of bacterial populations, they maintain the bacteria in a state of active growth and thus contribute significantly (along with the protozoa and macrofauna) to the organic decomposition process. Through direct excretion of phosphorus and nitrogen, meiofauna and other benthic animals also may contribute to the regeneration of nutrients, most of which are rapidly utilized by bacteria and microalgae.

In sediments with high organic enrichments, such as in the New York Bight apex and western Long Island Sound, meiofaunal densities and biomass are not depressed (compared to those of sediments with no organic enrichment). Where macrobenthic densities have been reduced by pollution, such as in the New York Bight apex, meiofauna and protozoa may become the dominant benthic animal influence in detritus decomposition and remineralization. Pollutants that reduce meiofaunal densities or alter the manner in which the animals function thus have an impact on the organic decomposition process.

In the New York Bight apex, increased levels of heavy metals in the dredge spoil and sewage sludge dumpsites are correlated with changes in the species composition and decreased species diversity of nematodes, the dominant taxon present. The replacement fauna contains species that have been shown in laboratory studies to have slower metabolic rates than the original fauna. Such metabolically less active animals could contribute to long-term accumulation of organic matter in the deeper basins of the New York Bight and Long Island Sound.

INTRODUCTION

The roles of the meiobenthos in benthic food webs and in the transfer of materials and energy both within the benthos and to the water column remain unclear (Coull and Bell, 1979; Tietjen, 1980a). Because the organisms are abundant (10^5-$10^7/m^2$ in most shelf sediments), taxonomically diverse, about five times more active metabolically per unit of biomass than the macrofauna (Gerlach, 1971), capable of more rapid generation and turnover times than the macrofauna (McIntyre, 1969), capable of competing with the macrofauna for common resources (McIntyre, 1969), and of potential significance in the regeneration of nutrients and decomposition of organic matter (Tenore *et al.*, 1977; Tietjen, 1980a), their inclusion in any study of benthic systems is recommended. Furthermore, their rapid generation time, general absence of

Contribution No. 113 from the Institute of Marine and Atmospheric Sciences, City University of New York.

pelagic larvae, and potential competitive interactions with the macrofauna have led to consideration of their use as indicators of environmental alteration (Pequegnat, 1975; Tietjen, 1977, 1980b). While macrofauna have been included in many studies of marine pollution, meiofauna have been included in but few. Taxonomic difficulties and the apparent absence of meiofauna from benthic food chains that culminate in commercially important finfish or shellfish are the main reasons for their absence from such studies.

The potential role of the meiofauna in benthic processes has been summarized by McIntyre (1969), Coull (1973), and others. Speculation on their activities focuses on the following main considerations:

Are the meiofauna important components in the diets of larger organisms?

Do the meiofauna compete with larger (macrobenthic) organisms for common resources, and with what results?

Do feeding and other activities of the meiofauna contribute to the decomposition of organic matter and regeneration of nutrients in the sea?

Can the meiofauna, in conjunction with other organisms, be used as indicators of environmental alterations, especially before such alterations become irreversible?

Although the first two items are important from the standpoint of trophic dynamics, they will not be discussed here. Attention will be focused on the latter two considerations because of their importance in evaluating the impact of pollution on benthic systems and processes.

ROLE OF MEIOFAUNA IN ORGANIC DECOMPOSITION

While there has been speculation on the importance of meiofauna in the decomposition of organic matter and nutrient cycling (Johannes, 1965; McIntyre, 1969; Gerlach, 1971; Coull, 1973; Fenchel and Jorgensen, 1977; Tenore *et al.*, 1977), few data on the quantitative impact of the meiofauna on organic decomposition are available. That there appear to be few predators of the meiofauna in offshore waters led McIntyre (1969) to the conclusion that "the main role of the smaller metazoans appears to be that of recycling nutrients some of which will eventually, with the winter turnover, reach the upper waters [p. 281]." That meiofauna function in some capacity in the breakdown of organic matter is evident from the study by Tenore *et al.* (1977), who found the oxidation and turnover rates of aged eelgrass (*Zostera marina*) in the presence of meiofauna to be double that of eelgrass that was meiofauna-free. They suggested that meiofaunal grazers mechanically break down detrital particles, thus rendering them more susceptible to bacterial colonization and microbial activity.

It is known that grazing activities by benthic animals ranging in size from protozoa (Fenchel and Harrison, 1976) to amphipods (Hargrave, 1970; Lopez *et al.*, 1977) may increase bacterial activity and the rate of organic decomposition. Also well established is that the nitrogen fraction of decomposing detritus increases with age, presumably as a result of increasing microbial biomass (Newell, 1965; Harrison and Mann, 1975; Fenchel and Harrison, 1976). It remains unclear how animal grazing increases microbial biomass, and what role benthic animals have in the nutrient regeneration process. Johannes (1965) suggested that bacterial excretion of nutrients (specifically phosphorus) relative to that by protozoans is insignificant. However, Barsdate *et al.* (1974) and Fenchel and Harrison (1976) present evidence to the contrary. They indicate that bacteria have high rates of dissolved phosphorus excretion, and that the role of protozoan grazers is modest compared to bacteria. Their evidence suggests that in the presence of heavy grazing pressure, bacterial biomass is reduced but bacterial uptake of phosphorus is increased, thus indicating an increase in bacterial turnover and metabolic activity. The purpose of this paper is not to discuss the relative importance of bacteria versus protozoa in nutrient dynamics, but merely to point out that animal grazing of detritus apparently does stimulate the metabolic activity of bacteria, and that the mechanism for the accompanying release of nutrients from the detritus is not well understood.

The influence of meiofaunal grazing on the release of nutrients from detritus is totally unknown. Whether meiofauna directly excrete significant quantities of dissolved nutrients into the water (as suggested by Johannes, 1965) or cause the bacteria (and fungi) to do so by stimulating their metabolic activities in response to meiofaunal grazing has not been established. However, some evidence suggests that nematodes may influence the concentration of dissolved inorganic phosphate in algal-nematode experimental systems.

In experimental flasks in which axenic cultures of the chlorophyte, *Chlorococcum* sp., and the diatom, *Cylindrotheca closterium*, were grown, Tietjen (1980a) observed that the addition of nematodes (*Chromadorina germanica*) significantly increased the levels of ortho-phosphate relative to that in nematode-free flasks (one-way ANOVA, $p < 0.05$) (Figure 1).

The phosphorus excretion rate (*sensu* Johannes, 1965) of *C. germanica* was calculated by measuring uptake and release of ^{32}P. The body equivalent excretion time, defined by Johannes as the time necessary to release an amount of dissolved phosphorus equal to the organism's own phosphorus content, was between one and two days. This is considerably shorter than that for the mussel, *Modiolus demissus* (1,540 hours), but longer than that of ciliates (12-202 minutes, Johannes, 1965). It is similar to that calculated by Johannes for mixed zooplankton and for the amphipod, *Lembos intermedium* (1.3-3 days).

Results of these preliminary experiments suggest that meiofauna may play a significant role in nutrient regeneration. Whether by direct excretion into the water or through stimulation of the metabolic activity of bacteria and other decomposers, meiofauna may play an important role in regulating the organic loads present in areas of high organic enrichment, such as the New York Bight apex and western Long Island Sound. Meiofaunal densities in these regions are apparently unaffected by the high organic loads present (Tietjen, 1977, 1980b).

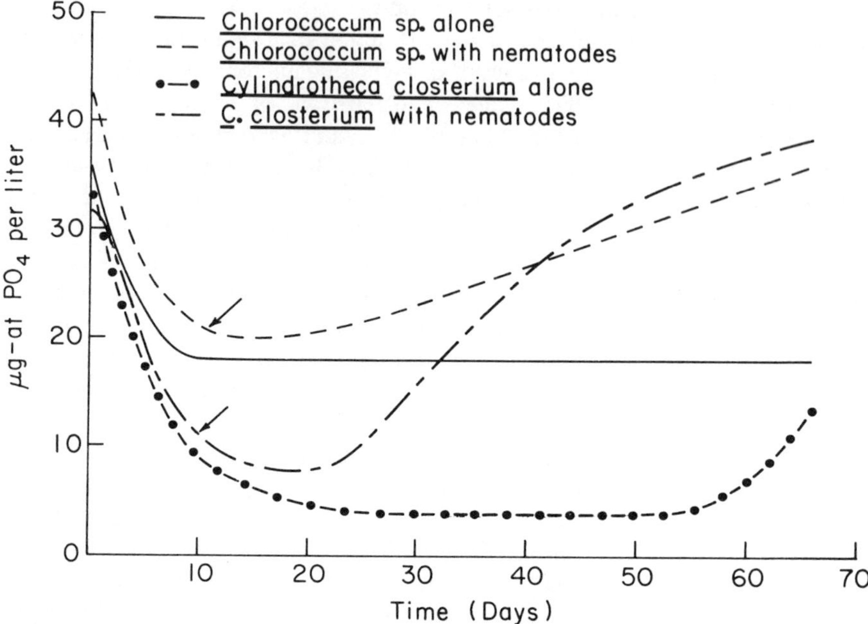

Figure 1. Concentration of dissolved inorganic phosphate in culture flasks containing algae alone (solid lines) and algae plus nematodes (dashed lines). Arrows indicate dates on which nematodes were added to algal cultures. Nematodes were added at initial densities of 25/10 cm^2; after 20 days, nematode densities in flasks were ~ 1,000/10 cm^2.

Therefore, in sediments where macrobenthic infauna have been impacted by organic pollution and the positive aspects of their grazing and excretory activities on the bacteria have been eliminated or severely inhibited, the meiofauna (and protozoa) may constitute the dominant animal influence on the detritus decomposition and remineralization processes occurring on the bottom.

IMPACT OF POLLUTION ON MEIOFAUNA

Systematic studies of meiofauna from polluted areas have been few. Marcotte and Coull (1974) observed a decrease in species diversity of harpacticoid copepods in response to increasing organic load in the Bay of Piran, Yugoslavia; Heip and Decraemer (1974) observed a similar phenomenon for nematodes in the southern North Sea. More recently, Tietjen (1977, 1980b) reported on the results of studies conducted on the nematodes of Long Island Sound and the New York Bight apex. Although organic enrichment also was studied, the principal pollutants examined were heavy metals. In Long Island Sound, a full range of sediments from muds to coarse sands was studied; in the Bight apex only sands and silty sands were considered. In both areas, total population densities, species composition, and species diversity were examined, both as a function of sediment type and also of heavy metal and organic load within sediment type. The results of the studies are summarized in Table 1.

In both Long Island Sound and the New York Bight apex, no discernible relationships between nematode densities and concentrations of heavy metals or organic carbon were found. Tietjen (1980b) speculated that heavy metal concentrations as high as 1,500 ppm and organic carbon concentrations of 7%, either singly or in concert, may not affect population densities of coastal marine nematodes. However, he also pointed to the paucity of studies on the effects of pollution on meiofauna.

As expected on the basis of studies of fine sediments from other areas (Ward, 1973; Lorenzen, 1974), the species composition of the finer sediments of Long Island Sound and the Bight apex was dominated by two species groups, the *Sabatieria pulchra* group and *Terschellingia longicaudata* group. Species diversity was low (relative to that of coarser sediments) as a result of the dominance of the fauna by a few species. No significant differences in species diversity or species composition were correlated with differences in sedimentary heavy metal loads.

In the non-impacted coarser sediments (medium-coarse sands) of both areas, the nematode fauna was more diverse, and the species composition differed from that of both impacted and non-impacted finer sediments. The *S. pulchra* and *T. longicaudata* species groups were absent or reduced; they were replaced by species of Chromadoridae, Monoposthiidae, and Desmodoridae. The distribution of species was significantly more equitable in sands than in finer sediments, and species richness was also higher. The reasons for this have been detailed previously (Wieser, 1960; Tietjen, 1977) and are related to the heterogeneous nature of sands, which provide a greater variety of microhabitats.

In both Long Island Sound and the New York Bight, it appears that the complex of species that normally inhabits the medium sands may be especially sensitive to elevated mass loads of heavy metals and/or organic carbon. Sufficiently high concentrations of these contaminants may result in the reduction (or perhaps the elimination) of species that belong to the families Chromadoridae, Monoposthiidae, and Desmodoridae, and the increased abundances of species normally associated with finer sediments (especially *S. pulchra*). The latter species may be able to thrive in stressed sands because it, as a normal inhabitant of mud, already is adapted for living under the natural "stresses" of low dissolved oxygen concentration and/or high organic content. For the same reason, the impact of elevated mass loads of heavy metals on species inhabiting finer sediments may be less, although some reduction in species diversity at high metal concentrations was seen in the Bight and also seems to have occurred in Long Island Sound (Tietjen, 1977).

In medium-coarse sediments, and perhaps also in fine sands that contain low amounts of silt-clay, a normal highly diverse nematode fauna under stress may undergo significant qualitative changes. A fauna more characteristic of muds (*i.e.*, less diverse and dominated by different species) may result despite the lack of change in sediment granulometry. Such change may significantly impact the

Table 1. Summary of population data on nematodes of Long Island Sound and the New York Bight apex. Data obtained from Tietjen (1977, 1980b). Density expressed as individuals/10cm². H′ is species diversity (Shannon and Weaver, 1963); SR is species richness (Margalef, 1958); J′ is equitability (Pielou, 1966).

	Fine sediments		Coarse sediments	
	Polluted	Nonpolluted	Polluted	Nonpolluted
Density	530-2710	560-1290	400-1319	110-5010
Mean H′	1.91	2.32	2.44	2.83
Mean SR	2.76	3.82	4.56	6.12
Mean J′	0.69	0.74	0.80	0.82
Faunal composition (dominant species and families)	*Sabatieria pulchra, Terschellingia* spp., *Tripyloides gracilis*	*Sabatieria pulchra, Terschellingia* spp., *Tripyloides gracilis*	*Sabatieria pulchra* and other Comesomatidae	*Microlaimus* spp. and other Desmodoridae, Chromadoridae, Cyatholaimidae, Monoposthiidae

macrofauna, because the nematodes that dominate muddy sediments are detritus feeders (Wieser, 1960; Tietjen, 1969, 1971, 1977; Warwick and Buchanan, 1970). While it is true that nonselective deposit-feeding macrobenthic species may ingest meiofauna along with sediment, the quantitative significance of meiofauna in the diets of macrofauna is still largely unknown (McIntyre, 1964; Marshall, 1970; Lee *et al.*, 1976; Bell and Coull, 1978). If meiofauna and macrofauna compete for organic matter, as suggested by McIntyre (1964), a larger meiobenthos population might lead to less efficient production of fish food (macrobenthos). Such a situation exists in the North Sea where, in the Fladen area, detritus-feeding meiofauna apparently "appropriate a greater share of the food than the macrobenthos [McIntyre, 1964: p. 673]."

IMPLICATIONS

The interactions among animals under pollution stress are poorly understood. Among the macrobenthos, organic enrichment normally results in lowered species diversity and dominance (*i.e.*, elevated abundance and biomass) by a few taxa. If benthic respiration resulting from macrobenthic metabolism and the decomposition of organic wastes is sufficiently high, dissolved oxygen levels may decrease to critical levels, possibly eliminating the local fauna (Pearce, 1972; Rowe *et al.*, 1972; Smith *et al.*, 1973; McIntyre, 1977).

Less is known about the effects of heavy metal enrichment. Pearce (1972) and Steimle *et al.* (this volume) report the presence of an impoverished benthic fauna at the New York Bight dredged materials dumpsite, while Bryan (1974), working with sediments collected in southwest England having metal concentrations as high as 10,000 ppm, suggested that animals such as *Nereis diverisicolor* may rapidly evolve a high tolerance to metal pollution. He found normal densities of *N. diversicolor* in metal-rich sediments and experimentally discovered that animals from such sediments had greater tolerances to the toxic effects of copper and zinc than animals from sediments not enriched with metals.

Therefore, among the macrobenthos there may be different effects of organic and metal enrichment on the fauna depending upon such factors as physiology, past environmental and evolutionary history, salinity (which affects ionic and osmotic regulation), and sediment granulometry. Totally unknown are how biological interactions among organisms (*e.g.*, competition and predation, etc.) are affected by pollution stress; the focus of pollution research has been mainly on the effects of one or, at most, a few pollutants on single organisms.

The significance of the presence of "pollution tolerant" species (such as the nematode, *Sabatieria pulchra*) in impacted New York Bight sediments has yet to be determined. Additional studies of the meiofauna from similarly polluted regions are needed. Whether the meiofauna is more or less tolerant of pollutants than the macrofauna is also unknown. This author is aware of but one published paper that reports the results of a systematic study of both macro- and meiofauna from a polluted area (*i.e.*, the Clyde Sea: McIntyre, 1977), and those results were preliminary. However, the results did conform to the general pattern: abundance and biomass increased in response to increased organic enrichment, up to a point where either low dissolved oxygen levels and/or toxic effects accompanying enrichment eliminated most of the fauna.

This paper has attempted to draw attention to the possibility that in heavily polluted systems nematodes and other meiofauna may be the only metazoans present. They, along with the protozoa, may play an important role in the decomposition of organic matter (Tenore *et al.*, 1977) and nutrient regeneration (Tietjen, 1980a). Information on the rates at which meiofauna regenerate or aid in the regeneration of nutrients is needed because none is available now. How certain pollutants may affect the rates at which different meiofaunal species contribute to the nutrient regeneration and organic matter decomposition processes is also unknown. If pollutants prove not only to affect the species composition of the meiofauna, but also to reduce the rates at which the meiofauna functions, one result could be a long-term accumulation of organic and other toxic materials in basins such as the New York Bight's Christiaensen Basin and several depressions in Long Island Sound.

REFERENCES

Barsdate, R.J., T. Fenchel, and R.T. Prentki. 1974. Phosphorus of model ecosystems: significance for decomposer food chains and effect of bacterial grazers. Oikos 25: 239-351.

Bell, S.S. and B.C. Coull. 1978. Field evidence that shrimp predation regulates meiofauna. Oecologia 35: 141-148.

Bryan, G.W. 1974. Adaptation of an estuarine polychaete to sediments containing high concentrations of heavy metals. *In*: Pollution and Physiology of Marine Organisms, F.J. Vernberg and W.B. Vernberg (eds.), Academic Press, New York. pp. 123-135.

Coull, B.C. 1973. Estuarine meiofauna: A review: Trophic relationships and microbial interactions. *In*: Estuarine Microbial Ecology, L.H. Stevenson and R.R. Colwell (eds.), University of South Carolina Press, Columbia, SC. pp. 499-512.

Coull, B.C. and S.S. Bell. 1979. Perspectives of marine meiofaunal ecology. *In*: Ecological Processes in Coastal and Marine Systems, R.J. Livingston (ed.), Plenum Press, New York. pp. 189-216.

Fenchel, T. and P. Harrison. 1976. The significance of bacterial grazing and mineral cycling for the decomposition of particulate detritus. *In*: The Role of Terrestrial and Aquatic Organisms in Decomposition Processes, J.M. Anderson and A. MacFadyen (eds.), Blackwell Scientific Publications, Oxford, England. pp. 285-299.

Fenchel, T. and B. Jorgensen. 1977. Detritus food chains of aquatic ecosystems: the role of bacteria. *In:* Advances in Microbial Ecology, Vol. I, M. Alexander (ed.), Plenum Press, New York. pp. 1-58.

Gerlach, S.A. 1971. On the importance of meiofauna for benthic communities. Oecologia 6: 176-190.

Hargrave, B.T. 1970. The effect of a deposit-feeding amphipod on the metabolism of benthic microflora. Limnol. Oceanogr. 15: 21-30.

Harrison, P.G. and K.H. Mann. 1975. Detritus formation from eelgrass (*Zostera marina*): the relative effects of fragmentation, leaching and decay. Limnol. Oceanogr. 20: 924-934.

Heip, C. and W. Decraemer. 1974. The diversity of nematode communities in the southern North Sea. J. Mar. Biol. Assoc. U.K. 54: 251-255.

Johannes, R.E. 1965. The influence of marine protozoa on nutrient regeneration. Limnol. Oceanogr. 12: 189-195.

Lee, J.J., K. Tenore, J. Tietjen, and C. Mastropaolo. 1976. An experimental approach toward understanding the role of meiofauna in a detritus based marine food web. Proc. 4th National Symp. Radioecology, Corvallis, OR, Hutchison, Dowden & Ross, Inc., Stroudsburg, PA. pp. 140-147.

Lopez, G.R., J.S. Levinton, and L.B. Slobodkin. 1977. The effect of grazing by the detritivore *Orchestia grillus* on *Spartina* litter and its associated microbial community. Oecologia 30: 111-127.

Lorenzen, S. 1974. Die Nematodenfauna der sublittoralen Region der Deutschen Bucht, insbesondere in Titan-Abwassergebiet bei Helgoland. Veroeff. Inst. Meeresforsch. Bremerhaven 14: 305-327.

McIntyre, A.D. 1964. Meiobenthos of sublittoral muds. J. Mar. Biol. Assoc. U.K. 44: 665-674.

McIntyre, A.D. 1969. Ecology of marine meiobenthos. Biol. Rev. 44: 245-290.

McIntyre, A.D. 1977. Effects of pollution on inshore benthos. *In*: Ecology of Marine Benthos, B.C. Coull (ed.), University of South Carolina Press, Columbia, SC. pp. 301-318.

Marcotte, B.M. and B.C. Coull. 1974. Pollution, diversity and meiobenthic communities in the north Adriatic (Bay of Piran, Yugoslavia). Vie Milieu 24: 281-300.

Margalef, R. 1958. Information theory in ecology. Gen. Systems 3: 36-71.

Marshall, N. 1970. Food transfer through the lower trophic levels of the benthic environment. *In*: Marine Food Chains, J.H. Steele (ed.), Oliver and Boyd, Edinburgh. pp. 52-66.

Newell, R. 1965. The role of detritus in the nutrition of two deposit feeders, the prosobranch *Hydrobia ulvae* and the bivalve *Macoma balthica*. Proc. Zool. Soc. London 14: 25-45.

Pearce, J.B. 1972. The effects of solid waste disposal on benthic communities in the New York Bight. *In*: Marine Pollution and Sea Life, M. Ruivo (ed.), Fishing News (books) Ltd., Surrey, England. pp. 404-411.

Pequegnat, W.E. 1975. Meiobenthos ecosystems as indicators of the effects of dredging. *In*: Estuarine Research, Vol. II, L.E. Cronin (ed.), Academic Press, New York. pp. 573-584.

Pielou, E.C. 1966. The measurement of diversity in different types of biological collections. J. Theor. Biol. 13: 131-144.

Rowe, G.T., P.T. Polloni, and J.I. Rowe. 1972. Benthic community parameters in the lower Mystic River. Int. Rev. Gasamten Hydrobiol. 57: 573-584.

Shannon, C. and W. Weaver. 1963. The Mathematical Theory of Communication. University of Illinois Press, Urbana, IL. 117 pp.

Smith, K.L., Jr., G.T. Rowe, and J.A. Nichols. 1973. Benthic community respiration near the Woods Hole sewage outfall. Estuarine Coastal Mar. Sci. 1: 65-70.

Steimle, F.J. Caracciolo, and J.B. Pearce. This volume. Impacts of dumping on New York Bight apex benthos. pp. 213-223.

Tenore, K.R., J.H. Tietjen and J.J. Lee. 1977. Effects of meiofauna on incorporation of aged eelgrass by polychaete, *Nephthys incisa.* J. Fish. Res. Board. Can. 34: 563-567.

Tietjen, J.H. 1969. The ecology of shallow water meiofauna in two New England estuaries. Oecologia 2: 251-291.

Tietjen, J.H. 1971. Ecology and distribution of deep-sea meiobenthos off North Carolina. Deep-Sea Res. 18: 941-957.

Tietjen, J.H. 1977. Population distribution and structure of the free-living nematodes of Long Island Sound. Mar. Biol. 43: 123-136.

Tietjen, J.H. 1980a. Microbial-meiofaunal interrelationships: a review. Microbiology 1980: 335-338.

Tietjen, J.H. 1980b. Population structure and species composition of the free-living nematodes inhabiting sands of the New York Bight apex. Estuarine Coastal Mar. Sci. 10: 61-73.

Ward, A.R. 1973. Studies on the sublittoral free-living nematodes of Liverpool Bay. I. The structure and distribution of the nematode populations. Mar. Biol. 22: 53-66.

Warwick, R.M. and J.B. Buchanan. 1970. The meiofauna off the coast of Northumberland. I. The structure of the nematode populations. J. Mar. Biol. Assoc. U.K. 50: 129-146.

Wieser, W. 1960. Benthic studies in Buzzards Bay. II. The meiofauna. Limnol. Oceanogr. 5: 121-137.

HEAVY METAL AND ANTIBIOTIC RESISTANCE IN *BACILLUS* AND *VIBRIO* FROM SEDIMENTS OF NEW YORK BIGHT

John F. Timoney
Jennifer Graves Port

Department of Microbiology
New York State College of Veterinary Medicine
Cornell University
Ithaca, New York 14853

Abstract. A large proportion of the sedimentary *Bacillus* from the waste disposal area of the New York Bight was resistant to 20 μg of mercury (Hg) per ml and to high concentrations of cadmium (Cd) and zinc (Zn). Concentrations of heavy metals were much higher in these than in control sediments further offshore where dumping of wastes has never been practiced and where heavy metal concentrations were low. Ampicillin resistance, mainly because of β-lactamase production, was significantly ($P < 0.001$) more prevalent in *Bacillus* from the dump area than from control sediments. Also, *Bacillus* with combined mercury and ampicillin resistance were six times more frequent in dump sediments than elsewhere.

Mercury resistance also was prevalent in *Vibrio* from dump sediments. The resistance pattern mercury-kanamycin-streptomycin was found in 10.2% (60 out of 586) of *Vibrio* or *Vibrio-Aeromonas* from dumpsite sediments, but was found in only 1.6% (8 out of 510) of similar strains from control sediments. This resistance was transferable by conjugation and was probably plasmid-mediated. An important mechanism of mercury resistance in both *Bacillus* and *Vibrio* was reduction to volatile metallic mercury (Hg°).

The results of this study suggest that heavy metal contamination of an ecosystem can exert a selection pressure for antibiotic resistance in bacteria in that environment.

INTRODUCTION

The toxicity and persistence of heavy metals in various environments have been the focus of considerable interest in recent years. Important studies have been centered on conversions of mercuric ions in soils (Simura and Miller, 1964) and in marine (Nelson and Colwell, 1975; Olson *et al.*, 1979) and freshwater sediments (Spangler *et al.*, 1973). These processes may lead to loss of the relatively nontoxic, volatile, inorganic mercury (Hg°) into the surrounding environment by conversion of more toxic compounds, such as methyl-mercury. Thus, bacterial interconversions can be of considerable ecologic benefit in depuration of ecosystems contaminated with mercury.

Heavy metal resistance has been shown to be genetically linked to antibiotic resistance in a number of bacterial genera (Richmond and John, 1964; McHugh *et al.*, 1975; Nakahara *et al.*, 1977). In most of these instances, the selection pressure for the plasmids encoding the resistances was antibiotic usage. Instances of the converse situation where heavy metals were the selection pressure for antibiotic resistance seldom have been described, yet conceivably may be much more common in nature.

It has been suggested that the high frequency of multiple-antibiotic and heavy metal-resistant *Escherichia coli* at one site in the sewage sludge dump area of the New York Bight was a result of the selection effect of heavy metals in the sediment (Koditschek and Guyre, 1974). Levels of metals in areas of the New York Bight where dumping of urban wastes and dredged materials has occurred for many years are elevated greatly (Carmody *et al.*, 1973). Such an environment would appear to serve as an excellent source of material to test the hypothesis that heavy metal contamination may influence the occurrence of antibiotic resistance in the bacterial flora of an ecosystem. In this paper, we present data on the distribution of mercury-resistant *Bacillus* and *Vibrio* sp. in sediments of the New York Bight, together with evidence of linkage of mercury and antibiotic resistance in this flora.

MATERIALS AND METHODS

Sampling

Sediment samples were collected during September 1975 and February 1976 at a series of stations in the inner and outer New York Bight (Figures 1, 2). Stations within four miles of the center (40°25'N, 73°45'W) of the sewage sludge dumpsite arbitrarily were classed as dumpsite stations. Samples from all other stations were classified as intermediate. A number of the intermediate stations lay in the upper Hudson Shelf Valley, a topographic depression that has accumulated silts and clays from the Hudson/Raritan estuary and the adjacent dumpsites and is known to have exceptionally high metal concentrations (Carmody *et al.*, 1973). Samples were collected in a Shipek or Smith-McIntyre grab. Only the top inch of sediment was sampled. Details of sample collection, processing, and storage are described elsewhere (Timoney *et al.*, 1978).

Enumeration, Isolation, and Identification of Bacteria

Tryptose-glucose-yeast agar containing 3.5% NaC1 and used alone or supplemented with either 2 μg or 20 μg Hg ($HgC1_2$)/ml was used for pour plate counts of total and mercury-resistant bacteria and for propagation of isolates. These cultures were incubated at 18°C and yielded mostly *Bacillus* sp. Percentages of these bacteria resistant to 20 μg Hg/ml on initial isolation were calculated for each sediment using the formula:

$$\frac{\text{Number of colonies on TGY + 20 } \mu\text{g Hg/ml}}{\text{Number of colonies on TGY}} \times 100$$

Sediments from the September 1975 cruise also were enriched in 10% NaC1-meat broth (10% w/v) and incubated at 18°C for 48 hours for the isolation of *Vibrio-Aeromonas* sp. The enrichment cultures were plated on tryptose-glucose-yeast (TGY) agar containing 3.5% NaC1. Between five and ten randomly selected colonies were subcultured for identification and antibiotic and heavy metal sensitivity testing.

Gram-negative organisms were identified to genus according to the scheme of Shewan *et al.* (1960). Gram-positive organisms were identified according to Cowan (1974).

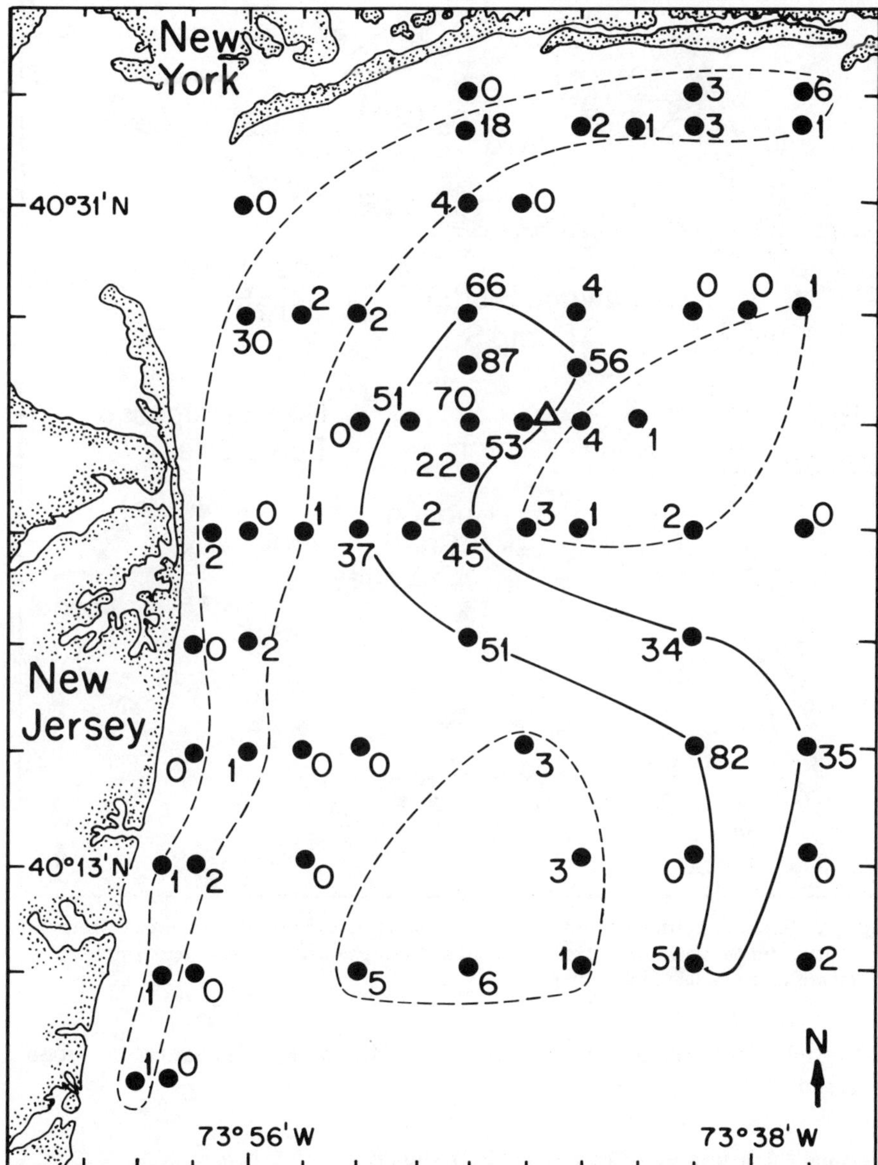

Figure 1. Sampling stations in inner New York Bight. Percentages of sediment bacteria resistant to 20 μg Hg/ml are shown as figures beside each station. Approximate isopleths are drawn at 1% resistance (– – –) and 30% resistance (——). The open triangle (△) indicates the location of the center of the sewage sludge site.

Antibiotic Sensitivity Testing

All isolates were replica plated on TGY agars containing ampicillin, kanamycin, streptomycin (50 μg/ml), chloramphenicol, and tetracycline

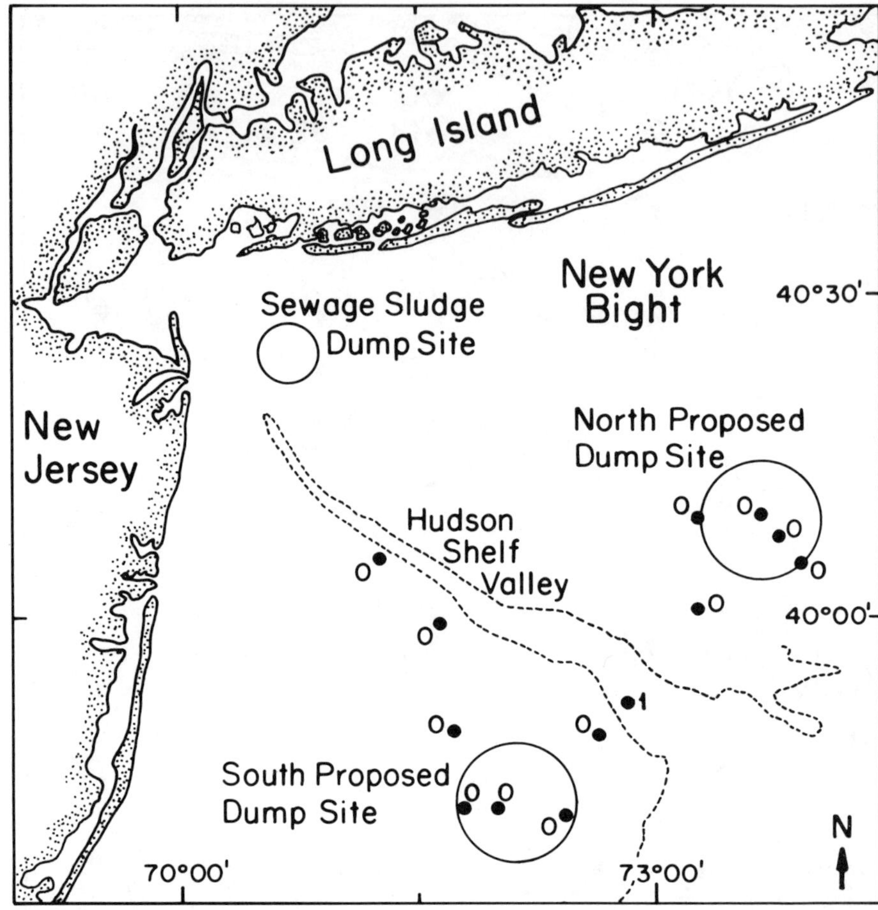

Figure 2. Sampling stations in outer New York Bight. Ocean dumping of urban wastes has never been practiced in these areas. Percentages of sediment *Bacillus* resistant to 20 μg Hg/ml are shown beside each station.

(30 μg/ml). Sensitivities were read after 48 hours incubation at room temperature.

Minimal Inhibition Concentration (MIC) Studies

The MIC of ampicillin, cadmium ($CdCl_2$), mercury ($HgCl_2$), and zinc ($ZnCl_2$) for *Bacillus* sp. were measured by replica plating of representative isolates on TGY agar containing dilutions of these substances as shown in Table 7.

Beta-lactamase

The Gots Test (Gots, 1945) was used for the detection of β-lactamase production by *Bacillus* strains.

Assay for Mercury Reduction

The dithizone assay of Summers *et al.* (1974) was used to demonstrate mercury reduction by representative mercury-resistant strains. X-ray film also was used to detect release of elemental mercury from culture plates. The film was placed over open culture plates in the dark for 24 hours.

Transfer of Mercury and Antibiotic Resistances from *Vibrio* sp. by Conjugation

A mutant of a *Vibrio* sp. from a non-dump sediment resistant to 100 μg nalidixic acid/ml was used as a recipient in conjugation experiments. Five ml each of 24-hour TGY broth cultures of donor and recipient were mixed thoroughly and drawn into 10-ml syringes; then they were forced through a 0.45-μ filter. The filters were transferred to TGY agar plates, incubated at room temperature for 24 hours, and transferred to tubes of sterile TGY broth that were shaken thoroughly to release all cells. Aliquots (0.1 ml) of the suspended cells were plated on TGY agars containing 100 μg nalidixic acid and either kanamycin, streptomycin, or mercury (20 μg/ml). After incubation at room temperature for 24 hours, the plates were checked for clones that had received any or all of the resistances of the donor strains.

Spontaneous Loss of Resistance to Mercury and Kanamycin in *Vibrio* sp.

Eight representative *Vibrio* strains that were resistant to mercury (20 μg/ml) and to kanamycin (50 μg/ml) were cultured in TGY broth at room temperature for 24 hours. Fifty-two clones of each strain were then replica-plated on TGY agars containing mercury (20 μg/ml) or kanamycin (50 μg/ml).

Metals Analysis

Representative sediments from each cruise were analyzed for mercury, cadmium, copper, and zinc by atomic absorption spectrophotometry at the NOAA National Marine Fisheries Service, Northeast Fisheries Center Laboratory, Milford, Connecticut.

RESULTS AND DISCUSSION

During the September 1975 and February 1976 cruises, sediments from 42 and 79 stations, respectively, were collected for the study. *Bacillus* was the most common genus in sediments examined by the pour-plate procedure (Table 1), whereas salt meat enrichment yielded mostly *Vibrio* or *Vibrio-Aeromonas* sp. The latter technique was not quantitative and may have allowed a selective advantage for some *Vibrio* strains. This technique was used only for samples from the September 1975 cruise. The pour-plate procedure involved a brief exposure of the sediments to 46°C, which would have been lethal to some *Vibrio* and *Pseudomonas* strains and so would have biased the counts in favor of the less heat-sensitive *Bacillus*.

Table 1. Bacterial genera isolated by a pour-plate technique from dumpsites and other sediments in the New York Bight.

Genus	Source of sediment					
	Dump stations		Intermediate stations		Control stations	
	September 1975	February 1976	September 1975	February 1976	September 1975	February 1976
Bacillus	94% [186/198]	99% [192/194]	76% [110/145]	93% [133/143]	99% [90/91]	89% [144/162]
Aeromonas, *Vibrio*, *Staphylococcus*, *Streptococcus*	6% [12/198]	1% [2/194]	24% [35/145]	7% [10/143]	1% [1/91]	11% [18/162]

It should be mentioned, in mitigation of the lack of sensitivity of our isolation procedures, that even the gentlest methods currently available for the culture and isolation of marine bacteria yield growths of no more than 1% of the total number known to be present by visual techniques. Thus, any method that could have been used would have been associated with considerable, but unavoidable, population biases.

The identification and enumeration data from the pour-plate procedure indicated that the *Bacillus* population did not vary to any significant degree from September to February for any of the three sediment categories. It therefore was decided to pool the enumeration data from both cruises. That *Bacillus* was the dominant organism isolated from dump, intermediate, and control sediments greatly enhanced the epidemiologic validity of comparisons of mercury-sensitive and resistant populations of this organism in the three sediment categories. Dump sediments had higher bacterial counts than intermediate or control sediments. This was probably a reflection of the greater organic content of dump sediments.

Sedimentary bacterial counts by the pour-plate method are shown in Table 2. The mean percentages of these populations resistant to 20 μg Hg/ml also are shown in this table and in Figures 1 and 2 for each station. Bacteria growing on medium containing 20 μg Hg/ml were regarded as mercury-resistant, although some of these bacteria appeared to lose this high level of resistance upon subculture.

As can be seen from the data in Tables 1 and 2, substantial proportions of the *Bacillus* flora of superficial sediments of the sewage sludge and dredge spoil dump areas of the New York Bight were resistant to mercury and other metals. Such resistant bacteria were rare in sediments at the proposed alternate dumpsites further offshore where ocean dumping of wastes never has been practiced. Furthermore, levels of cadmium, copper, mercury, and zinc (Table 3) were generally much higher in sediments from the dump area than elsewhere, although some sediments of, or close to, the upper Hudson Shelf Valley exhibited elevated levels of metals, as well as high counts of resistant bacteria. The overall distribution of resistant bacteria (Figures 1, 2) conforms well with the distribution of metals in the Bight, as reported by Carmody *et al.* (1973).

The high counts of resistant *Bacillus* in sediments with elevated concentrations of heavy metals from intermediate stations in, or near, the upper Hudson Shelf Valley may reflect inputs from dredge spoil and estuarine effluents as well as from sewage sludge disposal. The existence of resistant bacterial populations at these stations supports the possibility of *in situ* selection of resistant clones. In the case of dump sediments, it is possible that resistant *Bacillus* sp. were already present in the sludge at the time of dumping. However, since many resistant marine *Vibrio* and *Aeromonas* strains also were found in these sediments, it is certain that an *in situ* selection process also was taking place. Furthermore, the similarity of *Bacillus* populations that we observed in dump and control sediments is consistent with this conclusion.

Mercury was chosen for inclusion in primary selection media because prelimi-

Table 2. Mean counts of total and mercury (Hg)-resistant aerobic bacteria isolated by a pour-plate technique from sediments in the New York Bight. The numbers in parentheses are numbers of stations examined. Control stations were located at or near the projected alternate dumpsites. Mean percentages of mercury-resistant (20.0 μg Hg/ml) organisms were calculated from the separate counts on plain TGY agar for each station.

		Colony-forming units (C.F.U.)/g of wet sediment			
Cruise	Site	Plain TGY agar	TGY agar + 2.0 μg Hg/ml[a]	TGY agar + 20.0 μg Hg/ml[a]	% of total
September 1975	Dump (15)	6.2×10^3	6.4×10^3	1.9×10^3	23.2
	Intermediate (15)	1.9×10^3	1.8×10^3	3.5×10^1	0.9
	Control (12)	6.3×10^1	6.8×10^1	1×10^0	0.2
February 1976	Dump (13)	4.6×10^3	5.4×10^3	2.7×10^3	42.5
	Intermediate (47)	1.2×10^3	1.2×10^3	6.2×10^2	9.8
	Control	1.2×10^2	1.5×10^2	1×10^0	0.6

[a]Mercury was present as $HgCl_2$.

Table 3. Mean concentrations (μg/g, dry wt) of heavy metals in sediments of the New York Bight. Ranges of concentrations are shown beneath means. Numbers in parentheses are numbers of stations examined.

Metal	Control (7)	Intermediate (9)	Dump (15)
Cd	< 0.47	1.01 < 0.47 - 2.43	2.52 < 0.47 - 9.60
Cu	1.94	11.46 < 1.94 - 71.00	50.00 2.06 - 255.0
Hg	0.19 0.12 - 0.31	0.75 0.24 - 2.4	1.51 0.18 - 4.90
Zn	11.21 4.60 - 14.1	42.48 8.1 - 191.0	135.80 10.00 - 541.00

nary screening of different heavy metals indicated that it was the most toxic for sediment bacteria. This heavy metal was present in greatly elevated concentrations (1.61-4.9 μg/g dry sediment) in sediments of the dumpsite and also in some sediments of the upper Hudson Shelf Valley. Interestingly, another previous and more extensive survey of mercury levels made in 1974 (Pearce and Greig, unpublished data) also revealed elevated mercury levels in the same general area of the Bight as in our study. The source of mercury in contaminated sediments has not been defined, but presumably some of it is derived from sewage sludge since levels of mercury in sludge from metropolitan New York sewage plants have been shown to be as high as 15 μg/g dry wt (Furr *et al.*, 1976). Although mercury was probably an important selection pressure for mercury-resistant bacteria in sediments, it is likely that other metals, such as zinc and copper that are present in greatly elevated concentrations near waste disposal areas in the Bight and also to the southeast along the Hudson Shelf Valley (Carmody *et al.*, 1973), also may have exerted a selection pressure. This is because different metal resistances were frequently linked (Table 4), so selection for one tended to select for others that were present as well.

The antibiotic sensitivities of 1422 *Bacillus* clones are shown in Table 5. Ampicillin resistance in *Bacillus* sp. (mainly because of β-lactamase production) was much more frequent in dump (26%) than in intermediate (14%) or control (12%) sediments. This difference was highly significant ($p < 0.001$). Resistance to ampicillin in *Bacillus* from dump sediments was much more common (29%) in clones harvested from pour-plates containing 20 μg Hg/ml than in clones from pour-plates containing 2 μg Hg/ml, where only 8% were resistant to ampicillin. This difference was highly significant ($p < 0.001$). Thus, ampicillin and mercury resistance in *Bacillus* populations appeared to be linked.

Almost all (96%) *Vibrio* and *Vibrio-Aeromonas* sp. from all three sediment categories in the Bight were resistant to ampicillin. Since this was probably an

Table 4. Heavy metal resistance patterns of 366 strains of *Bacillus* spp. from sewage sludge dump sediments of the New York Bight. The criteria of resistance were: Hg ⩾ 5 μg/ml; Cd ⩾ 11 μg/ml; and Zn ⩾ 65 μg/ml.

Resistance pattern	% of total
Hg-Cd-Zn	24
Hg-Zn	24
Zn	20
Hg	14
Hg-Cd	3
Cd-Zn	2
Cd	–
Sensitive	12

intrinsic, generic type resistance, we omitted it from consideration in the study. Of the other antibiotics included in the survey, only resistance to kanamycin and streptomycin was detected in *Vibrio*. Interestingly, the multiple resistance pattern Kan^R-$Strep^R$-Hg^R occurred in 10.2% (60 out of 568) of *Vibrio* or *Vibrio-Aeromonas* from dumpsite sediments. Only 1.6% (8 out of 510) of these organisms isolated from control site sediments exhibited this resistance pattern. Transfer of this resistance pattern by conjugation was achieved only at a low frequency, and the resistances in the transconjugants proved to be unstable. Spontaneous loss of high level mercury resistance was common in *Vibrio* sp. (Table 6). Simultaneous loss of kanamycin resistance was much less common.

Metal resistance in many bacterial genera is known to be plasmid-mediated (Smith, 1967; Summers and Silver, 1972; Summers *et al.*, 1974; Olson *et al.*,

Table 5. Antibiotic resistances of 1422 *Bacillus* clones from sediments in the New York Bight.

	Source of sediment		
Resistance	Dump stations %	Intermediate stations %	Control stations %
Ampicillin	26.0	14.0	12.0
Streptomycin	2.7	12.0	8.0
Kanamycin	0.2	2.6	0.5
Tetracycline	0.2	0.4	0.5
Ampicillin and kanamycin	3.2	6.2	2.0
Other resistance combinations	1.1	2.6	0.5
Sensitive	65.0	66.0	77.0

Table 6. Spontaneous loss of resistance to mercury (Hg) (20 μg/ml) and kanamycin (50 μg/ml) in *Vibrio* sp. from New York Bight sediments.

Strain	Hg (20 μg/ml)	Kanamycin (50 μ/ml)	Hg and kanamycin
64d	12/52	0/52	0/52
92a	23/52	0/52	0/52
92c	40/52	0/52	0/52
96	29/52	0/52	0/52
22b	18/52	0/52	0/52
46b	17/52	0/52	2/52
61c	8/52	0/52	2/52
47a	39/52	0/52	0/52

1979). In our study, merucry, kanamycin, and streptomycin resistances were transferred from a resistant to a sensitive *Vibrio* by conjugation. This, together with the high rate of spontaneous loss of mercury resistance in this genus, is suggestive of plasmid-mediated resistance. Such plasmids already have been described in *Vibrio* sp. (Sizemore and Colwell, 1977). In the case of *Bacillus* sp., we recently have transformed a sensitive *Bacillus subtilis* with plasmid DNA from two mercury-resistant *Bacillus* from dump sediments. This is good evidence that metal resistance in these *Bacillus* is also plasmid-based.

The genetic basis of ampicillin resistance in the resistant *Bacillus* is as yet unknown. The gene for β-lactamase may be on the same plasmid as the gene for mercury resistance. It is possible, however, that certain populations of *Bacillus* that happen to have chromosomally-based genes for β-lactamase are better hosts for plasmids. Whatever the genetic basis of the linkage, it is clear that metal contamination increases the reservoir of antibiotic resistant genes in marine sediments.

The MIC studies (Table 7) suggest that the pool of ampicillin resistance genes is rather diverse and dispersed from a low through medium-to-high level of expression. In the case of genes for metal resistances, there was a pronounced tendency for MIC's to cluster around two levels of resistance, indicating less heterogenous gene populations.

Mercury reduction was observed in 45% (5 out of 11) of *Vibrio* strains resistant to 20 μg Hg/ml. Two strains sensitive to 2 μg Hg/ml were negative for mercury reduction. Almost half (4 out of 9) of a sample of strains that had spontaneously lost resistance to 20 μg Hg/ml still were able to reduce mercury, indicating that a decrease in level, rather than a complete loss of the resistance, had occurred. In the case of mercury-resistant *Bacillus* sp., 33% (6 out of 18) reduced mercury actively, 33% gave inconsistent results, and the remainder gave negative results. Thus, it would appear that more than one mechanism of mercury resistance was present in the *Bacillus* population. The effect of mercury release from agar medium on x-ray film during growth of a mercury-resistant *Bacillus* strain is shown in Figure 3.

Table 7. Minimum inhibitory concentrations (MIC) of mercury (Hg), cadmium (Cd), zinc (Zn), and ampicillin for *Bacillus* strains from sediments of the sewage sludge dumpsite in the New York Bight. Based on a total of 375 *Bacillus* strains.

Metal	MIC μg/ml	% of total
Hg ($HgCl_2$)	2	0.5
	5	36.2
	10	24.7
	15	17.3
	20	3.7
	50	7.4
	70	10.4
Cd ($CdCl_2$)	8	72.3
	11	2.4
	28	12.2
	56	10.6
	84	2.4
	112	0
Zn ($ZnCl_2$)	32	0
	49	22.0
	65	10.4
	162	25.3
	488	42.4
	650	0
Ampicillin	50	12 (69 strains)
	250	14
	1250	26
	6250	26
	6250	22

In a recent study of estuarine marine sediments of the Chesapeake Bay, Nelson and Colwell (1975) observed large numbers of mercury-reducing *Pseudomonas* in Hg-contaminated sediments. Mercury-resistant *Bacillus* also were observed, but were much less numerous than in our study. This difference may have been an artifact of isolation procedure since Nelson and Colwell used a surface plating procedure for primary isolation, whereas we used pour-plates that involved a brief exposure of the sediment to 46°C during preparation of pour-plates and that would have been deleterious to *Pseudomonas* and *Vibrio.* Also, in the latter we noted a marked tendency for colonies in mercury-supplemented media to be subsurface. A procedure that favored selection of surface colonies might, therefore, have discriminated against subsurface and generically different populations. It is clear, however, from both the Chesapeake Bay and New York Bight studies that mercury resistance can arise in generically different populations of marine sediment bacteria.

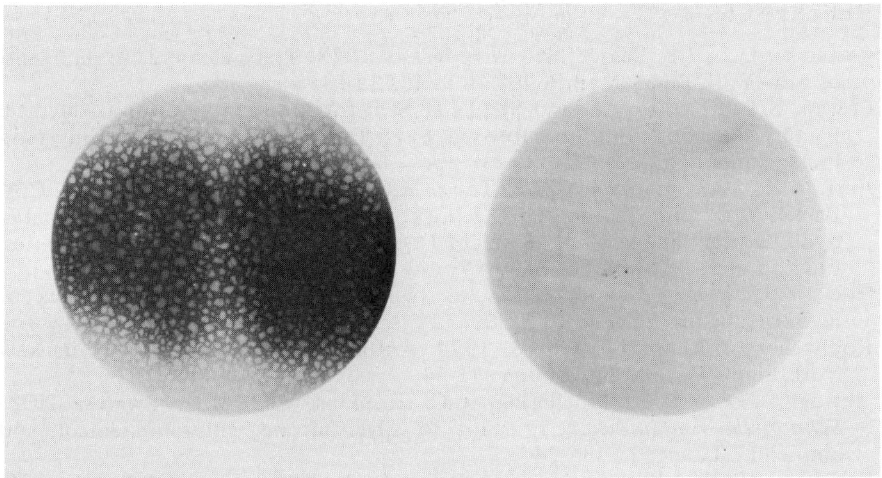

Figure 3. The culture on the left illustrates the effect on x-ray film of elemental mercury (Hg°) vapor released by a culture of mercury-reducing *Bacillus* sp. The medium was TGY agar containing 100 μg Hg/ml. The plate on the right contained the same concentration of mercury, but was not inoculated.

IMPLICATIONS

The selection of populations of mercury-reducing *Bacillus* and *Vibrio* sp. in contaminated marine sediments may have considerable ecological effect by promoting mobilization and loss of elemental mercury from sediments into the water column, where the element would be diluted and dispersed. Levels would be reduced in sediments, but the element would be spread into other phases of the marine ecosystem, albeit in low concentration. It also is possible, but less probable, that secondary methylation of inorganic mercury might occur (Olson *et al.,* 1979) and so result in the formation of a more toxic form of the element. Although ecologically advantageous, the capability of mercury reduction carries with it the inevitable consequence of an increase in the gene pool for antibiotic resistance, which, though not significant in the context of *Bacillus* sp., would have definite importance in the case of genera such as *Escherichia, Pseudomonas,* or *Vibrio.*

It is possible that metal-resistant bacteria can be used as an economic and rapid method of studying the distribution of metals in aquatic sediments and for monitoring the effects of abatement measures with regard to metals whose dumping can be controlled.

ACKNOWLEDGEMENTS

This work was supported by NOAA/ERL Contract No. 04-6-022-4409 from the Marine EcoSystems Analysis (MESA) New York Bight Project.

REFERENCES

Carmody, D.J., J.B. Pearce, and W.E. Yasso. 1973. Trace elements in sediments of New York Bight. Mar. Pollut. Bull. 4: 132-135.

Cowan, S.T. 1974. Cowan and Steel's Manual for the Identification of Medical Bacteria, Second Edition. Revised by S.T. Cowan, Cambridge University Press, London and New York. 238 pp.

Furr, A.K., A.W. Lawrence, S.S. Tong, M.C. Grandolfo, R.A. Hofstader, C.A. Bache, W.H. Gutermann, and D.J. Lisk. 1976. Multielement and chlorinated hydrocarbon analysis of municipal sewage sludges of American cities. Environ. Sci. Technol. 10: 683-687.

Gots, J.S. 1945. The detection of penicillinase producing properties of microorganisms. Science 102: 309.

Koditschek, L.K. and P. Guyre. 1974. Antibiotic resistant coliforms in New York Bight. Mar. Pollut. Bull. 5: 71-74.

McHugh, G.C., R.C. Moellering, C.C. Hopkins, and M.N. Swartz. 1975. *Salmonella typhimurium* resistant to silver nitrate, chloramphenicol, and ampicillin. Lancet 1: 235-240.

Nakahara, H., T. Ishikawa, Y. Sarai, I. Kondo, H. Kozukue, and S. Silver. 1977. Linkage of mercury, cadmium and arsenate and drug resistance in clinical isolates of *Pseudomonas aeruginosa.* Appl. Environ. Microbiol. 33: 975-976.

Nelson, J.D. and R.R. Colwell. 1975. The ecology of mercury-resistant bacteria in Chesapeake Bay. Microb. Ecol. 1: 191-218.

Olson, B.H., T. Barkay, and R.R. Colwell. 1979. Role of plasmids in mercury transformation by bacteria isolated from the aquatic environment. Appl. Environ. Microbiol. 38: 478-485.

Richmond, M.H. and M. John. 1964. Co-transduction by a staphylococcal phage of the genes responsible for penicillinase synthesis and resistance to mercury salts. Nature 202: 1360-1361.

Shewan, J.M., G. Hobbs, and W. Hodgkiss. 1960. A determinative scheme for the identification of certain genera of gram-negative bacteria with special reference to the Pseudomonadaceae. J. Appl. Bacteriol. 23: 379-390.

Simura, Y. and V.C. Miller. 1964. The degradation of organomercury fungicides in soil. J. Agric. Food Chem. 12: 253-257.

Sizemore, R.K. and R.R. Colwell. 1977. Plasmids carried by antibiotic resistant marine bacteria. Antimicrob. Agents Chemother. 12: 373-382.

Smith, D.H. 1967. R factors mediate resistance to mercury, nickel and cobalt. Science 156: 1114-1116.

Spangler, W.J., J.L. Spigarelli, J.M. Rose, and H.H. Miller. 1973. Methylmercury: bacterial degradation in lake sediments. Science 180: 192-193.

Summers, A.O., J. Schottel, D. Clark, and S. Silver. 1974. Plasmid-borne Hg (II) and organomercurial resistance. *In:* Microbiology–1974, D. Schlessinger (ed.), American Society of Microbiology, Washington, DC. pp. 219-393.

Summers, A.O. and S. Silver. 1972. Mercury resistance in a plasmid-bearing strain of *Escherichia coli.* J. Bacteriol. 112: 1228-1236.

Timoney, J.F., J. Port, J. Giles, and J. Spanier. 1978. Heavy-metal and antibiotic resistance in the bacterial flora of sediments of New York Bight. Appl. Environ. Microbiol. 36: 465-472.

A COMPARATIVE STUDY OF TRACE METAL CONTAMINATION IN THE SOUTHERN CALIFORNIA AND NEW YORK BIGHTS

David R. Young[1]

Southern California Coastal Water Research Project
646 West Pacific Coast Highway
Long Beach, California 90806

Abstract. Estimated annual inputs of trace metals to the New York Bight are from four to thirty-four times higher than those for the Southern California Bight. Typical seawater concentrations of dissolved and particulate metals in major harbors adjacent to these regions are remarkably similar and usually are at least several times higher than those for non-contaminated control sites. However, reported concentrations of dissolved metals above the New York Bight dumpsites are sometimes one or more orders of magnitude above values measured in the vicinity of submarine municipal wastewater discharges off southern California. In contrast, sediment concentrations of metals around the two major southern California outfalls are from two to five times higher than those at the New York dumpsites for dredged materials and sewage sludge. Metals levels in flocculent material inshore of one California outfall often are an order of magnitude above those at control sites; such particulates may be a significant vector for metals contamination of filter- and deposit-feeding invertebrates in this discharge zone. However, it is not known yet whether elevated tissue burdens are damaging to invertebrates from the vicinity of the outfalls. Fishes from the same area exhibit little or no metals contamination of their tissues; a similar finding is reported for the New York Bight. With the exception of mercury, concentrations of target trace metals do not appear to increase with trophic level in marine food webs.

INTRODUCTION

The New York Bight and the Southern California Bight are two of the most intensively studied ecosystems in the United States. The adjacent coastal areas are inhabitated by approximately 18 million and 12 million people, respectively, which together constitute almost 15% of the nation's population. Consequently, these marine ecosystems have received large inputs of waste materials, and there has been a growing interest in the biological effects that result from such waste loadings. This paper summarizes the major findings of the Southern California Coastal Water Research Project (SCCWRP) regarding trace metal inputs and their resultant distributions in seawater, sediment, and organisms feeding in or near the bottom off southern California. Wherever possible, these findings are compared to those obtained from similar studies within the New York Bight and Hudson-Raritan estuary, in the hope that conclusions may be reached regarding the impact of these trace contaminants on marine ecosystems.

[1] Present address: Dames and Moore, Marine Services Group, 1100 Glendon Avenue, Suite 1000, Los Angeles, California 90024.

Contribution number 146 of the Southern California Coastal Water Research Project.

INPUTS

The Southern California Bight is the open embayment of the Pacific Ocean extending along approximately 500 km of coastline between southern California (Figure 1) and northern Baja California (Mexico). The Bight's current system extends westward about 200 km, to a region where the water depth is approximately 3,000 m. Thus, the Bight has a surface area of about 100,000 km^2 (Southern California Coastal Water Research Project, 1973).

Two principal population centers are located along the shore of the Bight–in the Los Angeles/Orange County Basin and in San Diego County. About 95% of the municipal (domestic and industrial) wastewaters from these areas are discharged through four major submarine outfall systems whose locations are illustrated in Figure 1. Approximately 90% of this input occurs off Los Angeles and Orange Counties. A majority of the metals inputs received via other routes, summarized in Table 1, also occur in this area of the Bight (Young *et al.*, 1978a). For comparison, metals input data for the New York Bight (Mueller *et al.*, 1976) are listed in Table 2.

The estimated annual inputs listed in Table 1 indicate that submarine discharges of municipal wastewater (Schafer, 1979) are the dominant source of most target metals released to the Southern California Bight. Only in the case of

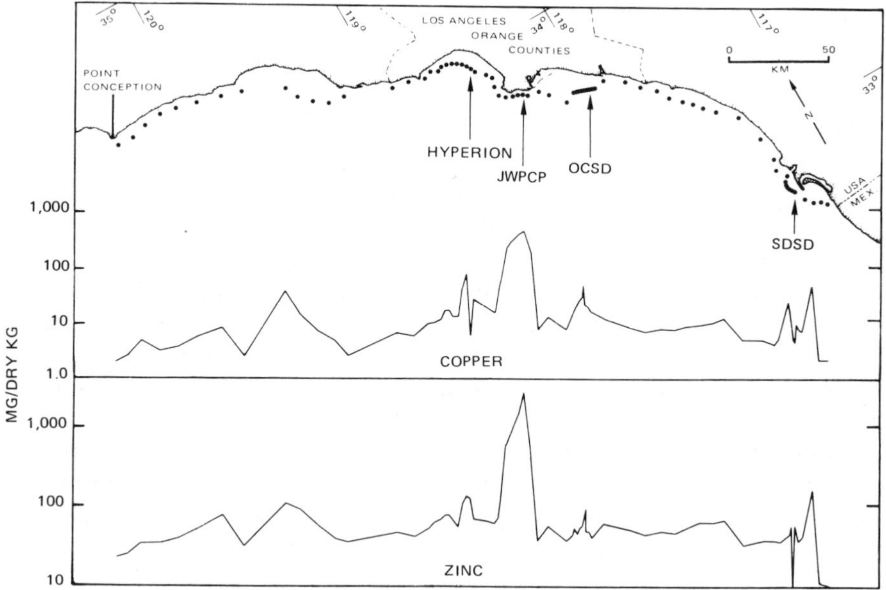

Figure 1. Concentrations of copper and zinc in surficial sediments along the 60-m isobath of the Southern California Bight. Major submarine outfall systems are Los Angeles City (Hyperion), Los Angeles County Joint Water Pollution Control Plant (JWPCP), Orange County Sanitation District (OCSD), and San Diego Sanitation District (SDSD). After Word and Mearns (1979).

Table 1. Estimated annual inputs (metric tons/yr) of trace metals to Southern California Bight. After Young *et al.* (1978a).

	Municipal wastewater 1977	Dry fallout[1] 1975	Storm runoff 1972-1973[2]	Thermal discharge[3] 1977	Total[4]
Ag	34	0.1	2.6	–	37
Cd	42	0.8	2.8	0.3	46
Cr	370	6.6	60	0.6	440
Cu	410	31	42	2.1	660
Hg	2.8	–	0.4	–	3.2
Ni	260	12	41	0.7	310
Pb	150	240	210	0.8	600
Zn	840	150	240	1.8	1,400

[1] 100 km x 100 km area off Los Angeles/Orange County Basin.

[2] Extrapolation of flow-weighted results from 1971-1972 storm survey (abnormally dry year). 1

[3] Southern California Edison cooling water (6.8 x 10^{12} liters/year).

[4] Includes approximately 180 metric tons/yr Cu used in vessel antifouling paints (probably an overestimate) and approximately 160 metric tons/yr Zn used in recreational vessel sacrificial anodes (probably an underestimate).

lead are inputs from atmospheric fallout (Young and Jan, 1977) and surface runoff (Young *et al.*, 1978a) comparable to those of submarine outfalls. Such inputs probably stem from the use of leaded gasoline. In addition, the use of copper in vessel antifouling paints (Young *et al.*, 1979) and zinc in sacrificial anodes (Southern California Coastal Water Research Project, 1973) appears to contribute significant quantities of these metals to harbor waters. The data summarized in Table 2 indicate that, for most metals surveyed, estimated annual inputs for the New York Bight are approximately an order of magnitude above those for the Southern California Bight. Although the majority of the estimates

Table 2. Estimated annual inputs (metric tons/yr) of trace metals to the New York Bight. After Mueller *et al.* (1976).

	Direct		Coastal zone				Total
			Wastewater		Runoff		
	Barged	Atmosph.	Muni.	Indust.	Gauged	Urban	
Cd	720	18	44	5	44	44	880
Cr	910	18	400	15	180	290	1,810
Cu	2,600	150	550	450	500	810	5,060
Hg	10	–	78	2	14	6	110
Pb	2,000	420	880	140	280	880	4,600
Zn	3,500	2,200	960	240	2,500	2,600	12,000

for inputs via municipal wastewater discharge are remarkably close for the two regions, input values for all of the other routes are generally much higher for the New York Bight. Ratios of estimated total annual inputs for the two regions (New York:southern California) are: cadmium–19; chromium–4.1; copper–7.7; mercury–34; lead–7.7; and zinc–8.6.

Comparisons of conditions in the New York and California Bights based upon data from Mueller *et al.* (1976) must be made with caution because of the nature of the estimates from the New York Bight. Mueller *et al.* (1976) considered contaminant inputs to the New York Bight to fall into two categories: direct Bight inputs (*i.e.*, sewage sludge dumping, dredged materials disposal, and atmospheric inputs) and coastal zone inputs (*i.e.*, runoff, municipal wastewater, and industrial wastewater). In estimating the magnitude of these inputs, the authors assumed that all pollutants entering the lower Hudson-Raritan estuary from outfalls, upstream sources, and runoff are washed into the Bight. This is an overestimate, since particle-associated contaminants settle out in the estuary.

The overestimate is compounded because contaminant-laden sediments are removed from the estuary and are disposed of at the dredged materials dumpsite in the Bight apex. Mueller *et al.* (1976) considered dredged materials to be a separate source of contaminants despite the fact that much of the contaminant burden of such spoils is of estuarine or riverine origin. The ratios of total annual metals inputs presented above for New York:southern California are probably higher than is actually the case, particularly for particle-associated metals.

DISTRIBUTIONS

Seawater

The metals loadings of the estuarine and coastal environments described above have resulted in measurable contamination of the water column in both regions. In fall 1974, Young and his colleagues at SCCWRP surveyed the concentrations of "dissolved" ($<0.4\ \mu$) and "particulate" ($>0.4\ \mu$) metals in triplicate sets of seawater samples collected at five intervals over a full tidal cycle from mid-channel and mid-depth at the entrances to three major southern California harbors (McDermott and Heesen, 1975; Young *et al.*, 1975; Young *et al.*, 1978a). Details of the sampling and analytical procedures (generally employing APDC-MIBK solvent extraction[1] and carbon-rod atomic absorption spectrometry) are described elsewhere (Young and Jan, 1975).

Alexander *et al.* (1978) conducted a similar survey in November 1973 over a five-station transect at the mouth of the Hudson-Raritan estuary. Again, the analytical procedures for dissolved metals included APDC-MIBK solvent extraction. The first three samplings of the transect, conducted between 0600 and 1300 on 8 November, encompassed the initial ebb tide of the survey and occurred during the portion of the survey judged by the authors to be the most

[1] Dissolved cadmium and zinc were concentrated on Chelex-100 ion exchange columns.

reliable. Therefore, these data have been selected for comparison with those obtained from the five sample sets collected at each of the three southern California harbor entrances.

For a given metal and phase (*i.e.*, dissolved vs. particulate), median values were determined for each station and discrete sampling period. Corresponding medians for each time series set at each station then were calculated. Finally, the overall medians for the entrances to the Hudson-Raritan estuary and the southern California harbors were computed. These results are presented in Table 3. The comparison shows a remarkable agreement between typical water column concentrations of target metals at the harbor mouths of the two regions.

To investigate further trace metal contamination of southern California harbors, two to three replicate samples of surface seawater were collected on 7 December 1976 from three sites in the vicinity of Newport harbor: at the entrance to the harbor; at a poorly-flushed inner harbor site situated 5 km from the entrance, near a large vessel repair facility; and at a site 3 km offshore of the harbor entrance and 12 km east of the Orange County Sanitation District outfall. Utilizing APDC-MIBK organic solvent extraction (except for cadmium and zinc, which again were concentrated on Chelex-100 ion exchange columns), the filtrates ($< 0.4\ \mu$) were analyzed for target metals. The results are presented in Table 4. These data indicate distinct contamination of seawater at the inner harbor station.

Waldhauer *et al.* (1978) reported dissolved copper levels of 12 μg/liter from surface collections at five inner (*i.e.*, western) stations from Raritan Bay sampled in September 1974. These data were obtained by anodic stripping voltammetry and are in remarkable agreement with the median value for dissolved copper observed in surface samples from inner stations of Newport harbor, California (8.6 μg/liter, n = 3; Table 4).

Comparisons of the dissolved and particulate metals occurring in the Los Angeles County Joint Water Pollution Control Plant (JWPCP) effluent, in the JWPCP outfall plume, and at a nearby control station are presented in Table 5. Data were obtained during a 1975 survey and are based on triplicate samples at

Table 3. Comparison of dissolved (diss.) and particulate (part.) metals in the harbors of southern California and New York. Data presented as μg/liter; for sources of data see text.

Metal	Phase	Southern California	New York
Cu	diss.	1.6	1.4
	part.	0.43	0.77
Ni	diss.	1.8	3.7
	part.	0.17	<0.2
Zn	diss.	4.8	6.5
	part.	1.3	2.0

Table 4. Sea surface metals concentrations (μg/liter) in the vicinity of Newport Harbor, California. See text for location of stations. Data previously unpublished.

Metal	Inner harbor	Harbor entrance	Offshore
Cd	0.35	0.06	0.02
Cu	8.6	2.2	0.6
Ni	1.1	0.3	0.2
Zn	22	1.5	<0.2

three plume stations and at a control site (Young and Jan, 1975). As is apparent from Table 5, trace metal contaminants from southern California municipal wastewater discharges are present primarily as particulates. Dissolved metals concentrations vary between metals. In two of the four cases (cadmium and chromium), the magnitudes of dissolved metals in the JWPCP outfall plume were essentially the same as those at the control site. However, the levels of dissolved copper and nickel in the plume were three to five times higher than those of the control station.

Segar and Cantillo (1976) reported a different situation in the New York Bight for total concentrations of dissolved metals, determined by a direct injection technique that they believe measured additional (perhaps largely colloidal) forms of dissolved metals not always recovered by conventional extraction procedures. Their values for total dissolved cadmium and copper averaged 0.82 μg/liter and 4.5 μg/liter, respectively, from ten to forty times above typical levels measured in the Southern California Bight (Young and Jan,

Table 5. Dissolved (diss.) and particulate (part.) metals in the vicinity of the Los Angeles County Joint Water Pollution Control Plant (JWPCP) effluent plume. Data represent median concentrations (μg/liter); data are from Young and Jan (1975).

Metal	Phase	JWPCP effluent	Outfall plume	Local control
Cd	diss.	2	0.07	0.05
	part.	36	0.17	0.02
Cr	diss.	35	0.24	0.23
	part.	640	1.9	0.03
Cu	diss.	25	0.49	0.10
	part.	480	1.2	0.02
Ni	diss.	180	1.0	0.29
	part.	150	0.23	0.02

1975; Martin *et al.*, 1976; Jan and Young, 1978; Vidal *et al.*, 1978; Young *et al.*, 1978a). Similarly, their average concentration of total dissolved zinc (32 μg/liter) was approximately one thousand times the concentration of dissolved zinc (0.01-0.03 μg/liter) measured off the southern California coast both by ion-exchange and organic extraction methods (Bruland *et al.*, 1978; Jan and Young, 1978). Three possible explanations for these differences come to mind: (1) levels of the dissolved metals in these regions are indeed very different; (2) the unique procedure of Segar and Cantillo (1976) systematically recovers much higher quantities of trace metals from the colloidal (or another) phase of the dissolved sample (filtrate) than do the organic solvent extraction or ion exchange procedures; or (3) the differences are mainly the result of contamination during collection and/or processing of the water samples. The relative importance of these (or additional) explanations cannot be ascertained from data presently available. However, in view of the potential toxicity of trace metals in the "dissolved" state (Eisler, 1973; Eisler and Wapner, 1975), it clearly is important that these differences be resolved if the ecological effect of such metals contamination is to be determined.

Sediments

The submarine discharge of municipal wastewater has caused distinct contamination of bottom sediments in the vicinity of the three largest outfall systems off southern California. This is illustrated by the distribution of two metals measured in surficial sediments (0-2 cm) collected in 1977 at 10-km intervals between Point Conception, California and the Mexican border (Figure 1) (Word and Mearns, 1979). These samples were obtained with a 0.1-m^2 modified Van Veen grab sampler at a depth of 60 m, the approximate depth of discharge for all the major outfalls except that of the Hyperion 11-km (7-mi) sludge line (which discharges at about 100 m).

It is clear from Figure 1 that the Palos Verdes Peninsula shelf around the JWPCP discharge is by far the most contaminated region along the 60-m contour. However, the bottom around the Hyperion sludge line is also highly contaminated with trace metals. In an effort to compare the levels of metals in sediments from the JWPCP and Hyperion monitoring zones (33 km^2 and 6 km^2, respectively) and those from the New York Bight dredged materials (n=9) and sewage sludge (n=12) dumpsites, maximum and mean metals concentrations for each of the four sites are presented in Table 6. Data from the California sites were taken by SCCWRP and are unpublished; data from the New York sites are taken from Carmody *et al.* (1973). Table 6 shows that, with the exception of lead (for which values are approximately equal), average concentrations of trace metals in the two southern California discharge zones are two to five times those measured at the two dumpsites in the New York Bight.

Another problem associated with the offshore discharge of waste materials is the resultant contamination of nearshore ecosystems. The greatest degree of trace metal contamination in southern California biota occurs in filter- and

Table 6. Concentrations of trace metals in surficial sediments (mg/kg, dry wt) from discharge or dumpsites in the Southern California and New York Bights. Hyperion = Los Angeles City 11-km (7-mi) outfall monitoring zone. JWPCP = Joint Water Pollution Control Plant monitoring zone. Data on the Southern California Bight from the Southern California Coastal Water Research Project (unpublished); data on the New York Bight from Carmody *et al.* (1973). New York Bight control values are the averages of non-dumpsite values reported for sandy and silty sediments by Carmody *et al.* (1973).

	Ag	Cd	Cr	Cu	Hg	Ni	Pb	Zn
So. Calif. Bight								
Hyperion								
Max.	42	91	1,300	1,000	6.1	170	600	–
Mean	8	24	280	300	1.7	58	130	–
JWPCP								
Max.	27	66	1,500	940	5.4	130	580	2,900
Mean	10	22	510	280	1.7	64	200	730
Control	0.4	0.4	23	9	0.04	12	7	42
New York Bight								
Sludge								
Max.	–	–	210	240	–	37	260	420
Mean	–	–	110	140	–	24	170	250
Dredge								
Max.	–	–	310	390	–	35	370	1,500
Mean	–	–	110	140	–	24	140	260
Control	–	–	6	4	–	6	13	19

deposit-feeding organisms (Young *et al.*, 1978b). Therefore, as part of an ecological survey directed by Grigg (1979), samples of flocculent material were collected by slurp gun from rocks located along five transects off the Palos Verdes Peninsula at depths of 7 m (20 ft) and 17 m (50 ft). Nearby surficial sediments (0-5 cm) were also sampled and analyzed in the SCCWRP laboratory. Median concentrations of seven metals measured in these samples are listed in Table 7. These data show that trace metal levels in the floc, which probably are typical of levels associated with particulates ingested by filter-feeding organisms, usually are several times higher than those on more compacted sediments routinely sampled in environmental surveys.

The poorly flushed portion of Newport harbor, discussed earlier in this paper, is one of the most contaminated areas of the Southern California Bight. This is reflected by metals levels in its sediments. Bottom sediments collected there by the City of Newport Beach (1972) contained the following metals concentra-

Table 7. Median concentrations (mg/kg, dry wt) of seven metals observed in floc and surface sediment (0-5 cm) samples collected from water depths of 7 m and 17 m during 1977 off Palos Verdes Peninsula. After Grigg (1979).

	Outfall				Control			
	7 m		17 m		7 m		17 m	
	Floc	Sediment	Floc	Sediment	Floc	Sediment	Floc	Sediment
Ag	<0.9	<0.3	0.8	<0.5	<0.3	<0.4	<0.6	<0.4
Cd	1.8	<0.5	2.2	0.8	<0.2	<0.3	0.6	<0.2
Cr	52	12	77	26	4.8	2.8	7.4	2.2
Cu	20	6.0	30	10	1.6	0.2	2.8	0.5
Ni	16	8.8	18	11	0.7	<1.2	<1.5	<1.1
Pb	30	5.2	38	14	4.0	<1.5	4.6	<1.5
Zn	65	33	89	44	14	8.5	32	12

tions (mg/kg, dry wt): silver–1.0; cadmium–2.3; chromium–29; copper–710; mercury–12; lead–44; and zinc–410. In comparison, maximum concentrations (mg/kg, dry wt) reported by Greig and McGrath (1977) for Raritan Bay were: cadmium–15; chromium–260; copper–1,230; lead–985; nickel–50; and zinc–815. In most cases the latter values are considerably above the estimated average baseline values for the two Bight areas (Table 6).

Tissues

In view of the relatively high levels of metals contamination found in Newport harbor and in the nearshore region of the Palos Verdes Peninsula, it is instructive to consider metals measured in tissues of the filter-feeding intertidal mussel, *Mytilus edulis,* collected in 1974 from inner Newport harbor and from the vicinity of the JWPCP outfalls (Young *et al.*, 1979). Selected results are summarized in Table 8. These data indicate the magnitude of biological contamination that can occur in tissues of the same organism in the vicinity of two different metals sources–vessel maintenance activities and municipal wastewater discharge.

For five of the eight metals reported in Table 8, concentrations were highest in specimens from Newport harbor. These organisms had been living closer to suspected contaminant sources and in water experiencing considerably less replenishment than mussels from the other areas. For example, zinc concentrations were as much as three to four times higher in harbor specimens than in coastal specimens. The suspected cause of elevated zinc levels was the dissolution of sacrificial zinc anodes on the hulls of pleasure boats moored in the harbor. In contrast, silver measurements in Table 8 reflected elevated concentrations in mussels growing inshore of the JWPCP outfalls, but no detectable elevations were observed in mussels from the recreational vessel anchorage. The outfalls represent a documented source of silver to coastal

Table 8. Mean concentrations (±1 SE) of trace metals (mg/kg, dry wt) in tissues of the intertidal mussel, *Mytilis edulis*, from inner Newport Harbor, a Newport Beach control site, the base of the Joint Water Pollution Control Plant (JWPCP) coastal outfalls, and five southern California coastal control sites. After Young *et al.* (1979).

Metal/tissue	Inner harbor	Newport Beach	JWPCP outfalls	Coastal controls
Ag				
Digestive gland	<0.3	<0.4	1.2 ± 0.3	<0.3
Gonad	<0.7	<0.4	2.2 ± 0.05	<0.6
Muscle	<0.4	<0.4	<0.6	<0.4
Cd				
Digestive gland	10 ± 1.9	<4.7	8.2 ± 1.1	<6.0
Gonad	9 ± 9	<2.8	<4.6	<3.5
Muscle	7.1 ± 3.4	<3.0	<3.7	<3.4
Cr				
Digestive gland	3.8 ± 0.6	3.7 ± 0.6	15 ± 0.9	4.8 ± 0.7
Gonad	2.0 ± 0.4	0.3 ± 0.1	1.3 ± 0.1	0.6 ± 0.1
Muscle	<0.6	<0.6	<1.0	<0.9
Cu				
Digestive gland	127 ± 18	16 ± 1.0	47 ± 2.8	20 ± 2.4
Gonad	93 ± 15	9.6 ± 0.2	42 ± 0.5	10 ± 0.8
Muscle	52 ± 10	5.7 ± 0.3	14 ± 1.0	11 ± 2.9
Ni				
Digestive gland	<2.7	3.6 ± 0.7	10 ± 1.4	4.1 ± 0.7
Gonad	<2.2	<0.9	<3.0	<1.3
Muscle	<1.3	<2.6	<1.2	<2.0
Pb				
Digestive gland	19 ± 2.7	5.5 ± 0.6	8.9 ± 1.0	7.5 ± 1.9
Gonad	13 ± 4.6	<0.9	3.2 ± 0.4	<1.1
Muscle	<1.3	<1.2	<1.2	<1.2
Sn				
Digestive gland	3.6 ± 0.8	1.4 ± 0.3	1.2 ± 0.3	1.5 ± 0.2
Gonad	5.4 ± 1.4	<0.3	3.4 ± 0.5	<0.4
Muscle	<0.5	<0.7	<0.4	<0.5
Zn				
Digestive gland	240 ± 26	80 ± 11	120 ± 14	110 ± 10
Gonad	360 ± 120	87 ± 3	120 ± 15	93 ± 12
Muscle	210 ± 66	79 ± 3	90 ± 42	130 ± 27

sediments (Table 6) and biota (Alexander and Young, 1976; Young and Jan, 1979), while the anchorage is not suspected as a source of silver.

It is not clear yet to what extent elevated levels of toxic metals in the dissolved and particulate phases of seawater are responsible for the elevated levels of metals in filter- and deposit-feeding invertebrates of the Southern California Bight (Alexander and Young, 1976; Eganhouse and Young, 1976, 1978; Young *et al.*, 1978a,b; Young *et al.*, 1979; Young and Jan, 1979). This is perhaps exemplified best by the zinc data shown in Tables 4 and 8, where at least a hundred-fold increase in zinc concentrations in the water column (<0.2 μg/liter vs. 22 μg/liter; Table 4) was accompanied by only a three- to four-fold increase in zinc concentrations of mussel tissues (Table 8). It also is not clear to what degree elevated tissue burdens of metals reflect damage to the organism or to its predators.

Relatively few corresponding studies of tissue contamination have been conducted for organisms living in contaminated regions of the Hudson-Raritan estuary of the New York Bight (see O'Connor and Rachlin, this volume, for a summary). Greig *et al.* (1977) observed a two- to three-fold elevation of silver in muscle tissue of the rock crab (*Cancer irroratus*) collected from the western Christiaensen Basin (including the dredged materials dumpsite). These authors also reported two- to six-fold elevations of silver, cadmium, and zinc in digestive glands of the channeled whelk (*Busycon canaliculatum*) from a Long Island Sound disposal site. These findings are similar to those reported for southern California by Young and his colleagues (cited above) and to the examples presented in Table 8.

In addition, studies conducted in both the Southern California and New York Bights have shown that, in contrast to the case for invertebrates, fishes collected from discharge zones exhibit relatively little or no detectable elevation of metals in their tissues (de Goeij *et al.*, 1974; Young *et al.*, 1975; McDermott *et al.*, 1976; Greig and Wenzloff, 1977; Greig *et al.*, 1977; Young *et al.*, 1978a,b). This finding is significant because it contradicts the commonly accepted theory that all (or most) pollutants are concentrated up the food web. As seen from the median muscle tissue concentrations of trace metals in seafood organisms living in or above highly contaminated sediments around the JWPCP outfalls (Table 9),

Table 9. Median concentrations (mg/kg wet wt) of trace metals in muscle tissue of seafood organisms collected from the food web of the Joint Water Pollution Control Plant discharge zone, 1975-1977. After Young and Mearns (1979).

Organisms	Trophic level	Ag	Cd	Cr	Cu	Hg	Ni	Zn
Scorpionfish/ bocaccio	IV-V	0.015	0.003	0.02	0.15	0.26	0.10	4.3
Crab/prawn/ sanddab	III-IV	0.005	0.004	0.03	2.0	0.08	0.06	9.8
Abalone/ scallop	II-III	0.016	0.42	0.60	1.8	0.03	0.36	13

six of the seven target metals exhibited no general increase of concentration with estimated trophic position . Mercury was the only exception, and even in this case the increase appears to be a natural one, as levels in fishes from the discharge zone are similar to those measured in control specimens (de Goeij *et al.*, 1974; Young *et al.*, 1978b).

ACKNOWLEDGEMENTS

I thank G. Patrick Hershelman of this Project for technical assistance provided in the preparation of this paper.

REFERENCES

Alexander, G.V. and D.R. Young. 1976. Trace metals in southern California mussels. Mar. Pollut. Bull. 7: 7-9.

Alexander, J.E., R. Hollman, and T. White. 1978. Heavy metal concentrations at the apex of the New York Bight. NOAA Tech. Memo. ERL MESA-34. 38 pp.

Bruland, K.W., G.A. Knauer, and J.H. Martin. 1978. Zinc in northeast Pacific water. Nature 271: 741-743.

Carmody, D.J., J.B. Pearce, and W.E. Yasso. 1973. Trace metals in sediments of New York Bight. Mar. Pollut. Bull. 4: 132-135.

City of Newport Beach. 1972. Trace metals in the sediments of Newport harbor, 1971. Report from the Harbor and Tidelands Administrator to the City Manager, 23 March 1973, City of Newport Beach, CA.

de Goeij, J.J.M., V.P. Guinn, D.R. Young, and A.J. Mearns. 1974. Neutron activation analysis trace element studies of Dover sole liver and marine sediments. *In:* Proceedings of the Symposium on Nuclear Techniques in Comparative Studies of Food and Environmental Contamination, International Atomic Energy Agency, Vienna. pp. 189-200.

Eganhouse, R.P. and D.R. Young. 1976. Mercury in tissues of mussels off southern California. Mar. Pollut. Bull. 7: 145-147.

Eganhouse, R.P. and D.R. Young. 1978. *In situ* uptake of mercury by the intertidal mussel, *Mytilus californianus.* Mar. Pollut. Bull. 9: 214-217.

Eisler, R. 1973. Annotated bibliography on biological effects of metals in aquatic environments. Ecological Research Series EPA-R3-73-007. National Environmental Research Center, Office of Research and Monitoring, U.S. Environmental Protection Agency, Corvallis, OR.

Eisler, R. and M. Wapner. 1975. Second annotated bibliography on biological effects of metals in aquatic environments. Ecological Research Series EPA-600/3-75-008. Office of Research and Development, Environmental Research Laboratory, U.S. Environmental Protection Agency, Narragansett, R.I.

Greig, R.A. and R.A. McGrath. 1977. Trace metals in sediments of Raritan Bay. Mar. Pollut. Bull. 8: 188-192.

Greig, R.A. and D.R. Wenzloff. 1977. Trace metals in finfish from the New York Bight and Long Island Sound. Mar. Pollut. Bull. 8: 198-200.

Greig, R.A., D.R. Wenzloff, A. Adams, B. Nelson, and C. Shelpuk. 1977. Trace metals in organisms from ocean disposal sites of the middle eastern United States. Arch. Environ. Contam. Toxicol. 6: 395-409.

Grigg, R.W. 1979. Long-term changes in rocky bottom communities off Palos Verdes. *In:* Coastal Water Research Project Annual Report for the Year 1978, Southern California Coastal Water Research Project, El Segundo, CA. pp. 157-184.

Jan, T.K., and D.R. Young. 1978. Determination of microgram amounts of some transition metals in seawater by methyl isobutyl ketone-nitric acid successive extraction and flameless atomic absorption spectrophotometry. Anal. Chem. 50: 1250-1253.

McDermott, D.J., G.V. Alexander, D.R. Young, and A.J. Mearns. 1976. Metal contamination of flatfish around a large submarine outfall. J. Water Pollut. Control Fed. 48: 1913-1918.

McDermott, D.J. and T.C. Heesen. 1975. Inputs of DDT, PCB, and trace metals from harbors. *In:* Coastal Water Research Project Annual Report for the Year 1975, Southern California Coastal Water Research Project, El Segundo, CA. pp. 133-138.

Martin, J.H., K.W. Bruland, and W.W. Broenkow. 1976. Cadmium transport in the California Current. *In:* Marine Pollutant Transfer, L. Windom and R.A. Duce (eds.), D.C. Heath and Co., Lexington, MA. pp. 159-184.

Mueller, J.A., J.S. Jeris, A.R. Anderson, and C.F. Hughes. 1976. Contaminant inputs to the New York Bight. NOAA Tech. Memo. ERL MESA-6. 347 pp.

O'Connor, J.M. and J. Rachlin. This volume. Perspectives on metals in New York Bight organisms: factors controlling accumulation and body burdens. pp. 655-673.

Schafer, H.A. 1979. Characteristics of municipal wastewater discharges, 1977. *In:* Coastal Water Research Project Annual Report for the Year 1978, Southern California Coastal Water Research Project, El Segundo, CA. pp. 97-101.

Segar, D.A. and A.Y. Cantillo. 1976. Trace metals in the New York Bight. *In:* Middle Atlantic Continental Shelf and the New York Bight, M.G. Gross (ed.), Am. Soc. Limnol. Oceanogr., Spec. Symp. 2: 171-198.

Southern California Coastal Water Research Project. 1973. The ecology of the Southern California Bight: implications for water quality management. Southern California Coastal Water Research Project Rep. TR 104. 531 pp.

Vidal, V.M.V., F.V. Vidal, J.D. Isaacs, and D.R. Young. 1978. Coastal submarine hydrothermal activity off northern Baja California. J. Geophys. Res. 83: 1757-1774.

Waldhauer, R., A. Matte, and R.E. Tucker. 1978. Lead and copper in the waters of Raritan and lower New York Bays. Mar. Pollut. Bull. 9: 38-42.

Word, J.Q. and A.J. Mearns. 1979. The 60-meter control survey. *In:* Coastal Water Research Project Annual Report for the Year 1978, Southern California Coastal Water Research Project, El Segundo, CA. pp. 41-56.

Young, D.R., G.V. Alexander, and D. McDermott-Ehrlich. 1979. Vessel-related contamination of southern California harbors by copper and other metals. Mar. Pollut. Bull. 10: 50-56.

Young, D.R. and T.K. Jan. 1975. Trace metals in nearshore seawater. *In:* Coastal Water Research Project Annual Report for the Year 1975, Southern California Coastal Water Research Project, El Segundo, CA. pp. 143-146.

Young, D.R. and T.K. Jan. 1977. Fire fallout of metals off California. Mar. Pollut. Bull. 8: 109-112.

Young, D.R. and T.K. Jan. 1979. Trace metal contamination of the rock scallop, *Hinnites giganteus,* near a large southern California municipal outfall. Fish. Bull. 76: 936-939.

Young, D.R., T.K. Jan, and T.C. Heesen. 1978a. Cycling of trace metal and chlorinated hydrocarbon wastes in the Southern California Bight. *In:* Estuarine Interactions, M. Wiley (ed.), Academic Press, New York. pp. 481-496.

Young, D.R., D.J. McDermott, T.C. Heesen, and T.K. Jan. 1975. Pollutant inputs and distributions off southern California. *In:* Marine Chemistry in the Coastal Environment, T.M. Church (ed.), American Chemical Society, Washington, DC. pp. 424-439.

Young, D.R. and A.J. Mearns. 1979. Pollutant flow through food webs. *In:* Coastal Water Research Project Annual Report for the Year 1978, Southern California Coastal Water Research Project, El Segundo, CA. pp. 185-202.

Young, D.R., M.D. Moore, G.V. Alexander, T.K. Jan, D. McDermott-Ehrlich, R.P. Eganhouse, and P. Hershelman. 1978b. Trace elements in seafood organisms around southern California municipal wastewater outfalls. Publication No. 60, California State Water Resources Control Board, Sacramento, CA. 104 pp.

CHLORINATED HYDROCARBON CONTAMINANTS IN THE SOUTHERN CALIFORNIA AND NEW YORK BIGHTS

David R. Young[1]

Southern California Coastal Water Research Project
646 West Pacific Coast Highway
Long Beach, California 90806

Abstract. Submarine discharge of municipal wastewater has been the dominant source of several important chlorinated hydrocarbons to the Southern California Bight. Seawater concentrations of PCB's and DDT's near the urban coast seldom exceed a few parts per trillion; in contrast, concentrations in surface sediments in the vicinity of Los Angeles outfalls range to 10 parts per million and 100 parts per million, respectively. Contaminated particulates apparently are the dominant mechanism of input and dispersion, and accumulations of these materials on the sea bottom are believed to control levels in the benthos. PCB concentrations in muscle tissue of seafood organisms around major outfalls range to approximately 1 ppm, and values for liver tissue are an order of magnitude higher. Mussels suspended at several depths near one diffuser system revealed gradients of PCB and DDT in the water column, with a ten-fold increase from the surface to the bottom. These chlorinated hydrocarbons are concentrated by organisms at higher tropic levels. Also, relatively high PCB concentrations have been reported in the New York Bight. The severe contamination of Hudson River sediments by PCB wastes may lead to extensive and persistent contamination of the New York Bight and the Hudson River estuary.

INTRODUCTION

Chlorinated hydrocarbons are of particular concern, because many of these compounds are manufactured in large quantities and are characterized by slow degradation in the environment (Young *et al.,* 1979a). Upon release into the environment, they may accumulate to toxic levels in organisms of an affected area. This paper summarizes observations made over the past decade by the Southern California Coastal Water Research Project (SCCWRP) on levels of chlorinated hydrocarbons in the Southern California Bight. Where possible, findings are compared to data from the New York Bight.

The marine ecosystem off Southern California, which is described elsewhere in this series of papers (Mearns and Word, this volume; Sherwood, this volume; Young, this volume), has been contaminated by at least three kinds of chlorinated hydrocarbons–DDT residues, polychlorinated biphenyls (PCB's), and certain chlorinated benzenes (Young and Heesen, 1978; Young *et al.,* 1980). The following description of coastal inputs, fates, and effects of these materials off Southern California provides information that is useful for evaluating similar inputs to the New York Bight and the Hudson River estuary.

[1] Present address: Dames and Moore Marine Services Group, 1100 Glendon Avenue, Suite 1000, Los Angeles, California 90024.

Contribution number 147 of the Southern California Coastal Water Research Project.

INPUTS

During the past decade, routes and rates of input of chlorinated hydrocarbons to the Southern California Bight have been systematically investigated (Southern California Coastal Water Research Project, 1973; Young and Heesen, 1974, 1978; Young *et al.*, 1975a, 1976b,c, 1980); estimated annual mass emissions are summarized in Table 1. These results show that during the early 1970's submarine discharge of municipal wastewater was the dominant known source of DDT's, dieldrin, and PCB's to the Bight. Subsequent studies suggest the same conclusion for di-, tri-, and hexachlorobenzene (Young and Heesen, 1976).

The predominant source of DDT residues was waste material released to the sewer system of Los Angeles County Sanitation Districts by a large pesticide manufacturer for subsequent discharge via the Joint Water Pollution Control Plant (JWPCP) outfalls, located off Palos Verdes Peninsula (Figure 1). The industrial release was terminated in the spring of 1970. Between 1971 and 1975 approximately 35 metric tons of total DDT, which had accumulated in sewer sediments, were flushed into the coastal ecosystem. Indirect evidence suggests that the total input from this source during the preceding one to two decades was at least an order of magnitude higher (Young *et al.*, 1978). By comparison, quantities of DDT residues released by the other major outfall systems in the Southern California Bight appear to have been insignificant. In contrast, significant amounts of PCB's were released by the Los Angeles County, Orange County, Los Angeles City, and San Diego City municipal outfalls (Figure 1). Estimated inputs to the Bight between 1972 and 1975 for the four systems were approximately 16, 10, 3.6, and 1.5 metric tons, respectively.

Table 1. Estimated annual inputs (kg/yr) of chlorinated hydrocarbons to the Southern California Bight.

Route	Year	Total DDT	Dieldrin	PCB 1242	PCB 1254
Municipal wastewater[1]	1972	6,490	100	⩾ 19,200	⩾ 260
Municipal wastewater	1973	3,920	⩽ 280	⩾ 1,900	1,510
Municipal wastewater	1974	1,580	95	4,270	1,020
Municipal wastewater	1975	1,270	–	2,400	680
Municipal wastewater	1976	970	–	2,210	610
Municipal wastewater	1977	780	–	1,250	300
Municipal wastewater	1978	1,050	–	330	260
Harbor industrial	1973-1974	40	10	70	⩽ 30
Antifouling paint	1973	< 1	–	1	⩽ 1
Surface runoff	1971-1972	100	20	100-170	90-110
Surface runoff	1972-1973	320	65	0-550	250-280
Aerial fallout[2]	1973-1974	1,400	–	–	1,100
Ocean currents	1973	⩽ 7,000	–	–	⩽ 4,000

[1] 1971 Joint Water Pollution Control Plant (JWPCP) discharge of total DDT: 21,600 kg.
[2] Inner coastal zone: 400 km x 50 km.

DISTRIBUTIONS

Seawater

Concentrations of DDT and PCB residues in surface seawater off southern and Baja California are extremely low. Although total (dissolved plus particulate) values for measurable DDT and PCB residues ranged to approximately 50 ppt and 30 ppt (ng/liter), respectively, near the JWPCP outfall diffusers in 1973, levels of these contaminants measured that year at the edge of the Bight (during a period of offshore winds) generally were below 1 ng/liter. A subsequent survey off the Mexican (Pacific) coast in 1975 indicated that DDT and PCB residues generally did not exceed 0.01 ng/liter and 0.1 ng/liter, respectively (Risebrough *et al.*, 1976b). In contrast, much higher levels were observed in a 1974 survey conducted at the mouths of Los Angeles and San Diego harbors. Average concentrations (±1 SE) of total DDT were 8.1 ± 0.7 ng/liter and 0.6 ± 0.2 ng/liter, respectively; corresponding values for PCB 1254 were 2.1 ± 0.5 ng/liter and 1.8 ± 0.3 ng/liter (Young *et al.*, in press). The values obtained for the mouth of Los Angeles harbor are in agreement with those reported by Risebrough *et al.* (1976a) for seawater collected during 1972 at the base of JWPCP outfalls, 8 km to the west of the harbor mouth (5.1 ng/liter p,p′-DDE and 1.2 ng/liter PCB).

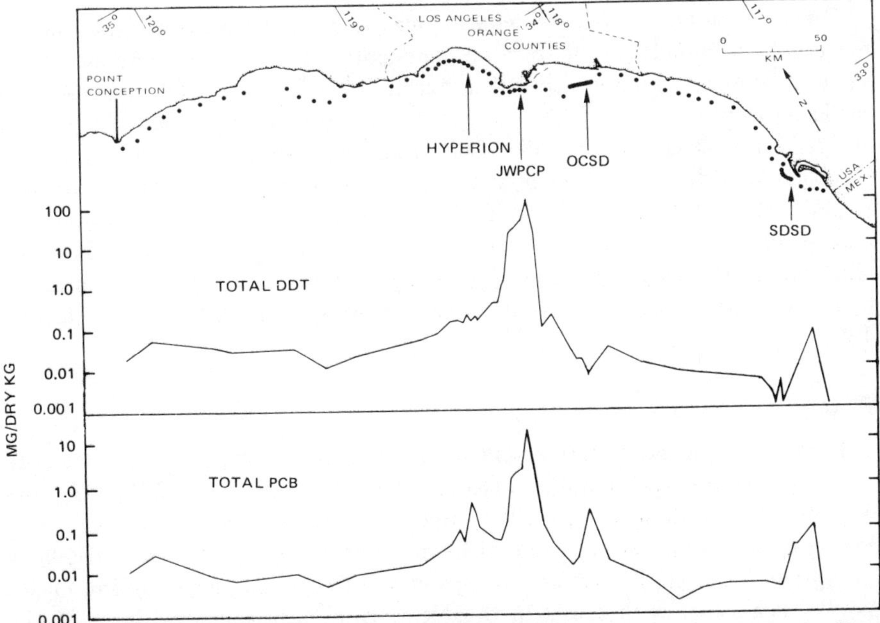

Figure 1. Concentrations of DDT and PCB residues in surficial sediments along the 60-m depth contour of the Southern California Bight. Major submarine outfall systems are Los Angeles City (Hyperion), Los Angeles County Joint Water Pollution Control Plant (JWPCP), Orange County Sanitation District (OCSD), and San Diego Sanitation District (SDSD). After Word and Mearns (1979).

Sediments

Distributions of total DDT and total PCB measured in surficial (0-2 cm) sediments collected during 1977 along the 60 m depth contour of the Southern California Bight (Figure 1) demonstrate the importance of the four major outfall systems as sources of one or both of these contaminants to the benthic environment. However, it should be noted that in the San Diego region the highest concentrations occur off the harbor mouth. The maximum concentrations of total DDT and total PCB in the JWPCP discharge zone in 1977 were 175 mg/kg, dry wt and 11 mg/kg, dry wt, respectively. Corresponding maxima obtained in a 1972 survey of this area were 206 mg/kg, dry wt and 13 mg/kg, dry wt (Young *et al.*, 1977). In addition, from the 1972 survey it was estimated that approximately 200 metric tons of total DDT were contained in the upper 30 cm of a 50-km² are of this discharge zone (Young *et al.*, 1976b).

The Hyperion seven-mile sludge effluent outfall discharges at a depth of about 100 m near the mouth of a major canyon in Santa Monica Bay. Consequently, the 60-m PCB values shown in Figure 1 are considerably lower than those observed in 1976 in a 6-km² area off the terminus of that outfall. The maximum concentration obtained during the 1976 survey was 5 mg/kg, dry wt. For comparison, Chen and Lu (1974) measured a maximum concentration of 3.5 mg/kg total PCB, dry wt, in surficial sediments from a highly industrialized section of Los Angeles harbor.

Particulate material appears to be an important factor for the distribution and possible bioaccumulation of trace contaminants. Therefore, floc and associated surficial sediment samples collected during 1977 by Grigg (1979) from the shallower depths of the JWPCP discharge zone (see Young, this volume) were analyzed for DDT's and PCB's. The results are presented in Table 2. For comparison, estimates of PCB and DDT concentrations on the particulate fraction of JWPCP final effluent during 1977 also are included. The estimates assume that 97% of the total PCB's and DDT's in the effluent were associated with the total suspended solids (averaging 220 mg/liter). This assumption is based on partitioning data from an earlier study of JWPCP effluent (Luthy, 1972).

Biota

Numerous studies of DDT and PCB contamination of marine biota have been conducted in Southern California (Risebrough, 1969; Burnett, 1971; Duke and Wilson, 1971; Delong *et al.*, 1973; McGregor, 1974; Anderson *et al.*, 1975; Young *et al.*, 1975c, 1977, 1980; McDermott *et al.*, 1976). These studies show clear gradients of DDT residues (principally p,p′-DDE) radiating away from Palos Verdes Peninsula, with levels in organisms from the general vicinity of the JWPCP outfalls exceeding those at control sites by two orders of magnitude (Young *et al.*, 1976b; Young and Heesen, 1978). Similarly, levels of total PCB's (principally PCB 1254) in organisms from around the three major outfall systems off the Los Angeles-Orange County Basin have exceeded control values

Table 2. Median concentrations (mg/kg, dry wt) of DDT and PCB residues observed in the 7-m and 17-m floc and surface sediment samples (0-5 cm) collected during 1977 off Palos Verdes Peninsula (after Grigg, 1979). Estimated DDT and PCB concentrations on the particulate fraction of Joint Water Pollution Control Plant (JWPCP) final effluent for 1977 are also listed. ND = not detected.

	JWPCP effluent particulates	Outfall				Control			
		7-m		17-m		7-m		17-m	
		Floc	Sediment	Floc	Sediment	Floc	Sediment	Floc	Sediment
Total DDT	7.0	0.88	0.08	1.1	0.66	0.04	ND	0.34	ND
Total PCB	3.0	0.07	ND	0.16	0.07	ND	ND	0.29	0.01

by one to two orders of magnitude (McDermott *et al.*, 1975, 1976). Muscle tissue concentrations of total DDT in fishes collected during this decade from the JWPCP discharge zone ranged from 1 mg/kg, wet wt to 90 mg/kg, wet wt (Young *et al.*, 1975b; Smokler *et al.*, 1979). The corresponding range for total PCB is from 0.1 mg/kg, wet wt to 2 mg/kg, wet wt, and levels of hexachlorobenzene measured in specimens collected near the outfall diffusers were approximately 0.002 mg/kg, wet wt (Young and Heesen, 1978). Liver tissue concentrations of these contaminants generally are an order of magnitude higher. As discussed by Sherwood (this volume), elevated levels of these chlorinated hydrocarbons have been associated with abnormal (enlarged, fatty) livers and with fin erosion disease in flatfishes from the Southern California Bight.

As shown by the values presented in Figure 1, as well as by those from the earlier SCCWRP studies cited above, bottom particulates in the JWPCP discharge zone have been highly contaminated by DDT's and PCB's. This is reflected by time series studies of these materials in the soft tissues of the intertidal mussel, *Mytilus californianus.* Studies were carried out during the summer of 1974 and involved the transfer of specimens of this filter-feeding invertebrate from a control site to the discharge zone. Organisms were maintained for 24 weeks in nylon bags suspended at five depths between the surface (0.5 m) and bottom (35 m), approximately 2 km northwest (down current) of the JWPCP outfall diffusers. Details of this study and preliminary findings have been described previously (Young *et al.*, 1976a).

The results of the full 24-week study (Figure 2) show a clear relationship between chlorinated hydrocarbon concentrations in the mussels and the proximity of the mussels to the bottom and to the wastewater plume trapped beneath the surface by the summer thermocline. A trend toward similar concentrations at the four subsurface depths, near the close of the survey (winter), may be associated with the disappearance of the thermocline. The absence of this feature may have enabled the wastewater plume and its associated contaminants to mix more equally throughout the water column. Subsequent studies (Risebrough *et al.*, 1976a; Young *et al.*, in press) suggest that *Mytilus* concentrates DDT and PCB residues on the order of 100,000 times above seawater concentrations.

As shown in Table 1, the JWPCP inputs of DDT and PCB residues both fell by an order of magnitude in the early 1970's. However, when concentrations of chlorinated hydrocarbons in surface sediments were compared for 19 JWPCP stations sampled in both 1972 and 1975, the median total DDT and PCB 1254 values were found to have decreased by factors of only 1.6 (from 19 mg/kg to 12 mg/kg total DDT, dry wt) and 1.2 (from 1.0 mg/kg to 0.81 mg/kg PCB 1254, dry wt), respectively. Corresponding reductions of median levels in the muscle of Dover sole (*Microstomus pacificus*) trawled from this sediment sampling area were 1.5 (from 17 mg/kg to 11 mg/kg total DDT, wet wt) and 1.3 (from 1.0 mg/kg to 0.76 mg/kg PCB 1254, wet wt). These findings indicate that when sediments are contaminated by chlorinated hydrocarbons such as DDT and PCB,

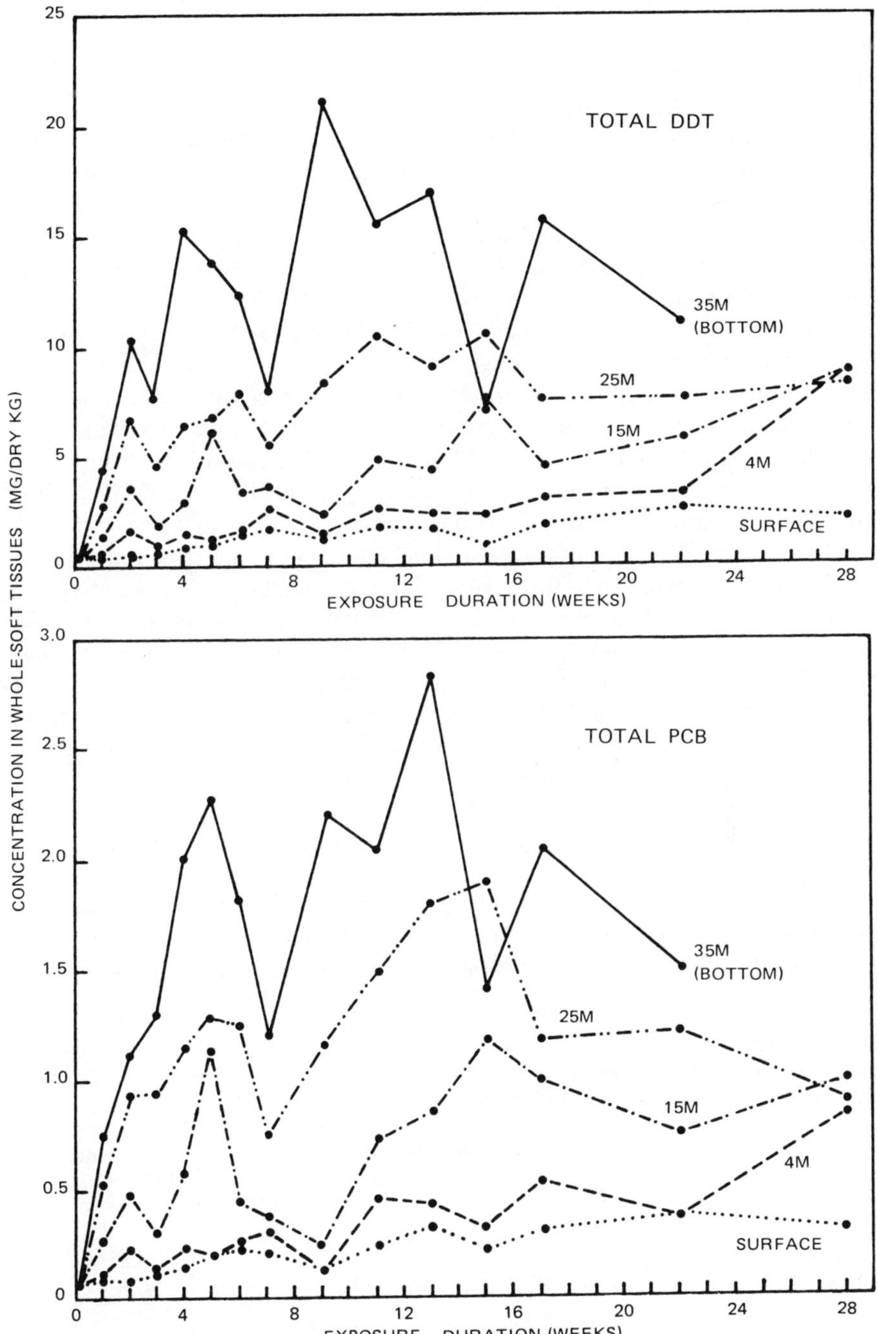

Figure 2. Uptake curves for DDT and PCB residues in *Mytilus californianus* translocated in June 1974 from a control site to the JWPCP outfall taut-line buoy system. Dry/wet weight ratio = 0.28 ± 0.02.

the compounds may persist in bottom-feeding fishes (and possibly other benthic organisms) long after major reductions have been made in the dominant contaminant inputs (Young *et al.*, 1977).

Similar results were reported by Sherwood (1976). Dover sole were collected from Southern California island and coastal control sites during 1974, and were maintained in the laboratory for up to ten months. Beginning in February 1975, several acclimated specimens were transferred to aquaria containing highly contaminated sediments collected within a few kilometers of the JWPCP outfall diffusers. Concentrations of total DDT and total PCB in the sediments ranged from 120 mg/kg, dry wt to 140 mg/kg, dry wt, and from 6 mg/kg, dry wt to 9 mg/kg, dry wt, respectively (Sherwood and Mearns, 1977). Other acclimated specimens were maintained in a similar fashion on beds of silica sand. (All fish were fed Tetramin Staple Tablet food; concentrations of total DDT and total PCB in the tablets were approximately 0.01 mg/kg, dry wt and 0.08 mg/kg, dry wt, respectively.) After 13-16 months, four test and five control specimens were sacrificed, and muscle and liver tissues were analyzed for DDT's and PCB's. The results are summarized in Table 3. Also listed are corresponding values obtained for 24 Dover sole collected during November 1976 within the same general area of the JWPCP discharge zone from which the test sediments had been obtained the previous year. The data indicate that initially uncontaminated flatfish, maintained on a relatively uncontaminated diet but exposed for a little over one year to highly contaminated outfall zone sediments, accumulated DDT and PCB residues in their livers to levels approaching those measured in specimens collected directly from the outfall zone. Although the effect in muscle tissue was not as large, distinct uptake above control values nevertheless was observed. These results illustrate the importance of DDT- and/or PCB-contaminated bottom sediments as indirect sources of chlorinated hydrocarbons to the benthos.

Recent findings from a study conducted in the JWPCP discharge zone support the concept of food web magnification of synthetic organics (Young and Mearns, 1979). As shown in Table 4, muscle tissue concentrations of total DDT

Table 3. Median concentrations (mg/kg, wet wt) of DDT and PCB residues in tissues of Dover sole exposed to contaminated (outfall zone) sediments or uncontaminated silica sand (1975-1976). After Sherwood (1976) and Sherwood and Mearns (1977).

	Outfall zone sediments		Silica sand
	Field	Laboratory	Laboratory
Total DDT			
Muscle	16	2.2	0.014
Liver	270	130	0.13
Total PCB			
Muscle	1.2	0.14	0.054
Liver	18	6.8	0.29

and PCB 1254 increased with assumed trophic level. Wet weight concentrations increased approximately 25-fold over a two-level step, with most of the increase occurring between the lower two levels. A relationship also was seen when results were expressed on the basis of lipid weight. Young *et al.* (1979b) reported on the deaths of captive gulls (*Laurus californicus*) and cormorants (*Phalacrocorax penicillatus* and *P. bougainvillii*) at the Los Angeles Zoo. The birds had been fed exclusively on a diet of queenfish (*Seriphus politus*) taken from the waters off Los Angeles. The total DDT concentration of whole fish was estimated as 4 mg/kg, wet wt. Median DDT concentrations in the breast muscles of four of the stricken birds was 220 mg/kg, wet wt. Corresponding levels measured in available brain tissue samples (220 mg/kg and 430 mg/kg, wet wt) indicated that DDT concentrations in this organ were sufficiently high to cause the death of the birds.

COMPARISON WITH THE NEW YORK REGION

Unfortunately, relatively few recent data on higher molecular weight chlorinated hydrocarbons in the New York Bight and Hudson River estuary have been published. Zooplankton collected in 1969 from the Bight about 100 km off the mouth of New York harbor contained approximately 2 mg/kg PCB, dry wt, more than 50 times the maximum value for total DDT; three nearby collections had almost twice this PCB concentration (Risebrough *et al.*, 1972).

In their summary of PCB levels in ocean fishes, Risebrough and de Lappe (1972) stated that the highest reported concentrations occurred in round herring, silverside, and mackerel from Long Island Sound (1.2-3.2 mg/kg, wet wt) and in sea bass from Tokyo Bay (1.7-3.4 mg/kg, wet wt). Zitko and Choi (1971) also reviewed reported levels of PCB's in whole specimens or muscle tissue of fishes from the coastal United States; some of the highest concentrations (approximately 1 mg/kg, wet wt) occurred off New Jersey. In addition, J. Foehrenbach (New York Department of Environmental Conservation, Stony Brook, New York) measured a median PCB concentration of 3 mg/kg, wet wt in muscle of six striped bass collected from the Bight (unpublished results).

Table 4. Median muscle tissue concentrations (mg/kg,wet wt) of total DDT and PCB 1254 in groups of organisms representing three levels of the marine food web in the Joint Water Pollution Control Plant (JWPCP) discharge zone off Palos Verdes Peninsula (1975-1977). After Young and Mearns (1979).

Organism	Trophic level	Total DDT	PCB 1254
Scorpionfish/Bocaccio	IV-V	2.1	0.23
Crab/prawn/sanddabs	III-IV	1.5	0.19
Abalone/scallop	II-III	0.08	0.01

However, Sherwood (this volume) found considerably lower median concentrations in muscle tissue of fin-eroded winter flounder (*Pseudopleuronectes americanus*) collected during 1977 from the Sandy Hook area of Raritan Bay (0.10 mg/kg, wet wt) and from the apex dump area (0.09 mg/kg, wet wt); corresponding median concentrations measured in liver tissue of the specimens were 4.2 mg/kg, wet wt and 2.8 mg/kg, wet wt. In comparison, median muscle and liver tissue concentrations for fin-eroded Dover sole collected in September 1977 from the sludge field around the Hyperion seven-mile outfall in the Southern California Bight were 0.22 mg/kg, wet wt and 5.4 mg/kg, wet wt, respectively (Sherwood, 1979). However, considerably higher levels of total PCB have been measured there in the past; median muscle tissue concentrations for specimens of this flatfish collected from the Hyperion discharge zone in 1971-1972 and 1974-1975 were 1.5 mg/kg, wet wt and 2.0 mg/kg, wet wt, respectively (McDermott-Ehrlich *et al.*, 1978).

Similar PCB levels in the diet of mammals have been shown to be chronically toxic. For example, when rhesus monkeys (*Macacus rhesus*) were fed a steady diet containing 2.5 mg/kg PCB 1248, wet wt, for six months, reproduction was seriously impaired. Only five of eight pregnancies were carried to full term; all infants were abnormally small, and within four months, three of six nursing infants died of PCB intoxication (Allen and Norback, 1976).

Relatively high total PCB concentrations were measured in bottom sediments collected in 1973 around the sewage sludge dumpsite in the New York Bight. The highest value (1.5 mg/kg, dry wt) occurred near the center of the dumpsite; concentrations along the New York and New Jersey coasts were much lower, generally by two or three orders of magnitude (West *et al.*, 1976). In comparison, the highest sediment concentration of total PCB measured in the sludge field around the Los Angeles County submarine outfalls in 1975 was 13 mg/kg, dry wt (Young and Heesen, 1978).

A study of sewage sludges from 15 United States cities indicates that PCB's are ubiquitous contaminants of such waste materials from major metropolitan areas; the median concentration measured was 3.8 mg/kg, dry wt (Furr *et al.*, 1976). Considerably higher values (unpublished) were measured by J. Foehrenbach (New York Department of Environmental Conservation) and I. Duedall (State University of New York at Stony Brook) in sludge from one New York City sewage treatment plant (Newtown Creek). The median PCB concentration in nine samples collected during July 1976 from a barge departing for the sludge dumpsite from this treatment plant was 49 mg/kg, dry wt.

Another example of chlorinated hydrocarbon contamination in this region is the serious pollution of the upper Hudson River by PCB wastes from General Electric plants at Hudson Falls and Fort Edwards, New York. Concentrations of the PCB 1016 in river sediments between these plants ranged up to 6,700 mg/kg, dry wt (Nadeau and Davis, 1976), and total PCB levels in local fishes ranged up to 560 mg/kg, wet wt (Spagnoli *et al.*, 1975). The average concentration in 68 fishes collected at or below Fort Edwards was 31 mg/kg, wet wt (U.S. Department of Health, Education, and Welfare, 1976). The corresponding value for shiner minnows was 78 mg/kg, wet wt. In addition, striped bass (*Morone saxatilis*) landed from the Hudson River near Albany had PCB concentrations ranging from 11 mg/kg, wet wt to 90 mg/kg, wet wt (New Jersey Department of Environmental Protection, 1977). Further, it has been estimated that more than 200 metric tons of PCB residues are contained in the bottom sediments of the upper Hudson River (Mount Pleasant, 1976). Thus, the gradual transport of these sediments to the New York Bight and lower Hudson River estuary may result in a persistent PCB contamination of these ecosystems.

ACKNOWLEDGEMENTS

I thank G.P. Hershelman for technical assistance provided in the preparation of this paper.

REFERENCES

Allen, J.E. and D.H. Norback. 1976. Pathobiological responses of primates to polychlorinated biphenyl exposure. *In:* Proceedings of the National Conference on Polychlorinated Biphenyls, 19-21 November 1975, Chicago, IL. EPA Report 560/6-75-004. U.S. Environmental Protection Agency, Office of Toxic Substances, Washington, DC. pp. 43-49.

Anderson, D.W., J.R. Jehl, Jr., R.W. Risebrough, L.A. Woods, Jr., L.R. Deweese, and W.G. Edgecomb. 1975. Brown pelicans: improved reproduction off the southern California coast. Science 190: 806-808.

Burnett, R. 1971. DDT residues: distribution of concentrations in *Emerita analoga* (Stimpson) along coastal California. Science 174: 606-608.

Chen, K.Y. and J.C.S. Lu. 1974. Sediment composition in Los Angeles-Long Beach harbors and San Pedro Basin. *In:* Marine Studies of San Pedro Bay, California, Part VII: Sediment Investigations, D.F. Soule and M. Oguri (eds.), The Allan Hancock Foundation, University of Southern California, Los Angeles. 117 pp.

Delong, R.L., W.G. Gilmartin, and J.G. Simpson. 1973. Premature births in California sea lions: association with high organochlorine pollutant residue levels. Science 181: 1168-1170.

Duke, T.W. and A.J. Wilson, Jr. 1971. Chlorinated hydrocarbons in livers of fishes from the northeastern Pacific Ocean. Pestic. Monit. J. 5: 228-232.

Furr, A.K., A.W. Lawrence, S.S.C. Tong, M.C. Grandollo, R.A. Hofstader, C.A. Bache, W.H. Gutenmann, and D.J. Lisk. 1976. Multi-element and chlorinated hydrocarbon analysis of municipal sewage sludges of American cities. Environ. Sci. Technol. 10: 683-687.

Grigg, R.W. 1979. Long-term changes in rocky bottom communities off Palos Verdes. *In:* Coastal Water Research Project Annual Report for the Year 1978, Southern California Coastal Water Research Project, El Segundo, CA. pp. 157-184.

Luthy, R.F., Jr. 1972. Trace constituents monitoring. *In:* Technical Services Department Monthly Report, February 1972, County Sanitation Districts of Los Angeles County, Whittier, CA. pp. 76-83.

McDermott, D.J., D.R. Young, and T.C. Heesen. 1975. Polychlorinated biphenyls in marine organisms off southern California. Southern California Coastal Water Research Project Rep. TM 223. NTIS No. PB 274464/AS, Springfield, VA. 45 pp.

McDermott, D.J., D.R. Young, and T.C. Heesen. 1976. PCB contamination of southern California marine organisms. *In:* Proceedings of the National Conference on Polychlorinated Biphenyls, 19-21 November 1975, Chicago, IL. EPA Report 560/6-75-004. U.S. Environmental Protection Agency, Office of Toxic Substances, Washington, DC. pp. 209-217.

McDermott-Ehrlich, D., D.R. Young, and T.C. Heesen. 1978. DDT and PCB in flatfish around southern California outfalls. Chemosphere 6: 453-461.

McGregor, J.S. 1974. Changes in the amount and proportions of DDT and its metabolites, DDE and DDD, in the marine environment off southern California, 1949-1972. Fish Bull. 72: 275-293.

Mearns, A.J. and J.Q. Word. This volume. Forecasting effects of sewage solids on marine benthic communities. pp. 495-512.

Mount Pleasant, R. 1976. Summary of PCBs in the upper Hudson River. *In:* New York State C.D.E. Exhibit 96, Phase II Hearings in the Matter of Alleged Violations of 17-0501, 17-0511, and 11-0503 of the Environmental Conservation Law of the State of New York by General Electric Company, Respondent. Interim Opinion and Order, File No. 2833, Albany, NY.

Nadeau, R.J. and R.A. Davis. 1976. Polychlorinated biphenyls in the Hudson River (Hudson Falls-Fort Edward, New York State). Bull. Environ. Contam. Toxicol. 16: 436-444.

New Jersey Department of Environmental Protection. 1977. PCBs and kepone in New Jersey: an interim report. Cancer and Toxic Substances Program, New Jersey Department of Environmental Protection, Trenton, NJ. 14 pp.

Risebrough, R.W. 1969. Chlorinated hydrocarbons in marine ecosystems. *In:* Chemical Fallout, M.W. Miller and G.G. Berg (eds.), Charles C. Thomas, Springfield, IL. pp. 5-23.

Risebrough, R.W. and B. de Lappe. 1972. Accumulation of polychlorinated biphenyls in ecosystems. Environ. Health Perspect. 1: 39-45.

Risebrough, R.W., B.W. deLappe, and T.T. Schmidt. 1976a. Bioaccumulation factors of chlorinated hydrocarbons between mussels and sea water. Mar. Pollut. Bull. 7: 225-228.

Risebrough, R.W., B.W. de Lappe, and W. Walker, II. 1976b. Transfer of higher molecular weight chlorinated hydrocarbons to the marine environment. *In:* Marine Pollutant Transfer, H.L. Windon and R.A. Duce (eds.), Lexington Books, D.C. Heath and Company, Lexington, MA. pp. 261-321.

Risebrough, R.W., V. Vreeland, G.R. Harvey, H.P. Miklas, and G.M. Carmignani. 1972. PCB residues in Atlantic zooplankton. Bull. Environ. Contam. Toxicol. 8: 345-355.

Sherwood, M.J. 1976. Fin erosion disease induced in the laboratory. *In:* Coastal Water Research Project Annual Report for the Year 1976, Southern California Coastal Water Research Project, El Segundo, CA. pp. 149-153.

Sherwood, M.J. 1979. The fin erosion syndrome. *In:* Coastal Water Research Project Annual Report for the Year 1978, Southern California Coastal Water Research Project, El Segundo, CA. pp. 203-221.

Sherwood, M.J. This volume. Fin erosion, liver condition, and trace contaminant exposure in fishes from three coastal regions. pp. 359-377.

Sherwood, M.J. and A.J. Mearns. 1977. Environmental significance of fin erosion in southern California demersal fishes. Ann. N.Y. Acad. Sci. 298: 177-189.

Smokler, P., D.R. Young, and K. Gard. 1979. DDTs in coastal marine fishes following termination of the dominant southern California industrial input. Mar. Pollut. Bull. 10: 331-334.

Southern California Coastal Water Research Project. 1973. The ecology of the southern California Bight: implications for water quality management.

Southern California Coastal Water Research Project Rep. TR 104. 531 pp.

Spagnoli, J.J., L.C. Skinner, and J. Dell. 1975. Monitoring of PCBs in fish taken from the Hudson River. New York State Department of Environmental Conservation Tech. Pap. No. 75-13.

U.S. Department of Health, Education, and Welfare. 1976. Final Report of the Subcommittee on the Health Effects of Polychlorinated Biphenyls and Polybrominated Biphenyls. U.S. Department of Health, Education, and Welfare, Washington, DC. 193 pp.

West, R.H., P.G. Hatcher, and D.K. Atwood. 1976. Polychlorinated Biphenyls and DDTs in Sediments and Sewage Sludge of the New York Bight. NOAA-ERL, Atlantic Oceanographic and Meteorological Laboratories, U.S. Department of Commerce, Miami, FL. 40 pp.

Word, J.W. and A.J. Mearns. 1979. 60-meter control survey off southern California. Southern California Coastal Water Research Project Rep. TM 229. 58 pp.

Young, D.R. This volume. A comparative study of trace metal contamination in the Southern California and New York Bights. pp. 249-262.

Young, D.R., G.W. Bowes, P. Moskowitz, and C. Wurster. 1979a. Halogenated hydrocarbons subpanel report. *In:* Chemical Pollutants of the New York Bight; Priorities for Research, J.S. O'Connor and H.M. Stanford (eds.), NOAA Special Report. pp. 33-43.

Young, D.R. and T.C. Heesen. 1974. Inputs and distributions of chlorinated hydrocarbons in three southern California harbors. Southern California Coastal Water Research Project Rep. TM 214. 27 pp.

Young, D.R. and T.C. Heesen. 1976. Inputs of chlorinated benzenes. *In:* Coastal Water Research Project Annual Report for the Year 1976, Southern California Coastal Water Research Project, El Segundo, CA. pp. 31-37.

Young, D.R. and T.C. Heesen. 1978. DDTs, PCBs, and chlorinated benzenes in the marine ecosystem off southern California. *In:* Water Chlorination: Environmental Impact and Health Effects, Vol. 2, R.L. Jolley, H. Gorchev, and D.H. Hamilton, Jr. (eds.), Ann Arbor Science, Ann Arbor, MI. pp. 267-290.

Young, D.R., T.C. Heesen, G.N. Esra, and E.B. Howard. 1979b. DDE-contaminated fish off Los Angeles are suspected cause in death of captive marine birds. Bull. Environ. Contam. Toxicol. 21: 534-590.

Young, D.R., T.C. Heesen, and R.W. Gossett. 1980. Chlorinated benzenes in southern California municipal wastewaters and submarine discharge zones. *In:* Water Chlorination: Environmental Impact and Health Effects, Vol. 3, R.L. Jolley, W.A. Brungs, and R.C. Cummings (eds.), Ann Arbor Science, Ann Arbor, MI. pp. 471-486.

Young, D.R., T.C. Heesen, and D.J. McDermott. 1976a. An offshore biomonitoring system for chlorinated hydrocarbons. Mar. Pollut. Bull. 7: 156-159.

Young, D.R., T.C. Heesen, and D. McDermott-Ehrlich. In press. Inputs and distributions of PCBs and DDTs in southern California harbors. Estuarine Coastal Mar. Sci.

Young, D.R., T.K. Jan, and T.C. Heesen. 1978. Cycling of trace metal and chlorinated hydrocarbon wastes in the Southern California Bight. *In:* Estuarine Interactions, M.L. Wiley (ed.), Academic Press, New York. pp. 481-496.

Young, D.R., D.J. McDermott, and T.C. Heesen. 1975a. Polychlorinated biphenyl inputs to the Southern California Bight. Southern California Coastal Water Research Project Rep. TM 224. 50 pp.

Young, D.R., D.J. McDermott, and T.C. Heesen. 1976b. DDT in sediments and organisms around southern California outfalls. J. Water Pollut. Control Fed. 49: 1919-1928.

Young, D.R., D.J. McDermott, and T.C. Heesen. 1976c. Marine inputs of polychlorinated biphenyls off southern California. *In:* Proceedings of the National Conference on Polychlorinated Biphenyls, 19-21 November 1975, Chicago, IL. EPA Report 560/6-75-004. U.S. Environmental Protection Agency, Office of Toxic Substances, Washington, DC. pp. 199-208.

Young, D.R., D.J. McDermott, T.C. Heesen, and D.A. Hotchkiss. 1975b. DDT residues in bottom sediments, crabs, and flatfish off southern California submarine outfalls. Calif. Water Pollut. Control Assoc. Bull. 12: 62-66.

Young, D.R., D.J. McDermott, T.C. Heesen, and T.K. Jan. 1975c. Pollutant inputs and distributions off southern California. *In:* Marine Chemistry in the Coastal Environment, T.M. Church (ed.), American Chemical Society, Washington, DC. pp. 424-439.

Young, D.R., D. McDermott-Ehrlich, and T.C. Heesen. 1977. Sediments as sources of DDT and PCB. Mar. Pollut. Bull. 8: 254-257.

Young, D.R. and A.J. Mearns. 1979. Pollutant flow through food webs. *In:* Coastal Water Research Project Annual Report for the Year 1978, Southern California Coastal Water Research Project, El Segundo, CA. pp. 185-202.

Zitko, V. and P.M.K. Choi. 1971. PCB and other industrial halogenated hydrocarbons in the environment. Fish. Res. Board Can. Tech. Rep. 272: 1-54.

CONTRIBUTED PAPERS ON THE EFFECTS OF POLLUTANTS ON FISHES

Opposite: Fish kill in the New York-New Jersey metropolitan area. (Courtesy New York Sea Grant Institute.)

EFFECTS OF TOXIC METALS IN MARINE ANIMALS OF THE NEW YORK BIGHT: SOME LABORATORY OBSERVATIONS

Anthony Calabrese
Edith Gould
Frederick P. Thurberg

National Marine Fisheries Service
Northeast Fisheries Center
Milford Laboratory
Milford, Connecticut 06460

Abstract. This report summarizes the results of eight years of experimental exposure studies on the effects of heavy metals in marine animals. The studies were conducted at the National Marine Fisheries Service Laboratory at Milford, Connecticut, and focused on organisms from the New York Bight. Of the metals tested, mercury and silver were most toxic to early-life stages (bivalve embryos and larvae, juvenile crustaceans), as determined by bioassay methods. Cadmium and mercury were the most toxic to adult teleosts and crustaceans, as determined by physiological and biochemical measurements. Tissue uptake was far greater for mercury than for cadmium. Lead was more toxic than cadmium for early-life forms but less toxic than either mercury or silver. Chronic exposure to 6 ppb cadmium or to 6 ppb mercury induced metabolic disturbances in the lobster (*Homarus americanus*), with somewhat more significant effects seen in cadmium-exposed animals, despite greater accumulation of mercury in body tissues. Analogous results were seen during chronic exposures of winter flounder (*Pseudopleuronectes americanus*) to 10 ppb of either cadmium or mercury; similar exposures of flounder to five times as much lead produced less metabolic disturbance. Low salinity enhanced or altered some effects and, in combination with suboptimal temperatures, increased metal toxicity to early-life forms. Chronic exposure of adult animals to sublethal concentrations of toxic metals drains energy reserves and may reduce capacity to adapt to and survive natural environmental stresses.

INTRODUCTION

Calabrese *et al.* (1977b) published a summary of work performed between 1972 and 1976 at the National Marine Fisheries Service, Northeast Fisheries Center, Milford (Connecticut) Laboratory on the effects of toxic metals in marine animals. Metal stress was examined in animals native to the New York Bight by means of bioassay methods and by physiological and biochemical measurements. The work was predominantly with cadmium, mercury, and silver—metals singled out for study because of their documented toxicity and occurrence in the New York Bight. The total daily mass loads of cadmium and mercury entering the Bight are 2.4 metric tons and 0.3 metric tons, respectively (Mueller *et al.*, 1976). Analogous data are not available for silver, but in a study of the Philadelphia dumpsite, Watling *et al.* (1974) reported a one-hundredfold greater concentration of silver in barge sludge than in surrounding sediments. New York Bight waters have been reported to contain from 5 ppb to 7 ppb cadmium in the approximate vicinity of the dredged materials dumpsite, as well as an average of 140 ppb lead and 80 ppb copper (Klein *et al.*, 1974, cited in Alexander and Alexander, 1977). Mercury and silver are found more frequently in sediments than in the water column. Since 1977, research designed to facilitate the interpretation of field observations has included other metals, as well as the effects of natural stresses (salinity, temperature) on metal-exposed animals. This report capsulizes work in these areas and updates the 1977 summary.

Experimental exposures under carefully controlled laboratory conditions are essential tools for understanding and interpreting both field observations and data on metal concentrations in the tissues of animals collected from "clean" and from polluted areas of the Bight. Information thus generated also will help in formulating baseline estimates of marine environmental quality, without which resource-oriented water-quality criteria cannot be established or enforced by regulatory agencies. To establish such criteria, the manner and degree of effect, as well as the influence of environmental conditions must be determined for various pollutants and for the life stages of marine animals of concern. The staggering quantities of polluted material that daily are poured, dumped, or rained into the coastal waters of the world make it imperative to discover how such activities affect the marine environment and its living resources. Although scientific inquiry has begun on an international scale, the scope of the problem is enormous.

Monitoring activities in the New York Bight and contiguous off-shore waters are helping to establish seasonal baseline information on metals levels in selected marine species and to evaluate laboratory exposure studies more accurately (Calabrese *et al.*, 1978; Gould, 1981). Conversely, laboratory studies aid the interpretation of abnormalities seen in field-collected animals.

Efforts at the Milford Laboratory have focused on the effects of toxic metals in fishes, molluscs, and crustaceans. Short-term tests were designed to measure mortality rates, to determine tolerance ranges, and to identify those metabolic systems most sensitive to metal stress. Long-term tests were designed primarily to study the effects of sublethal levels of metals on physiological and biochemical systems. Current work is examining how chronic exposure to sublethal concentrations of toxic metals can affect an animal's capacity to adapt to natural stress. No attempt is made in this paper to review the extensive scientific literature, as such material is covered in papers published previously and cited in this report.

METHODS

Adult test animals used in studies reported in this paper were collected from Long Island Sound and the adjacent New York Bight. Prior to toxic metal exposure, animals were acclimated in the laboratory for at least two weeks. Exposures were either short-term (hours to days) static tests or long-term (weeks to months) continuous-flow tests. At the end of each exposure period or at intervals during that time, animals were removed for evaluation of metal-induced stress. For low-salinity studies, seawater was diluted to the desired salinity with well water; pH and temperature were adjusted, and aeration was maintained throughout the holding period. Water was changed daily.

For studies of inhibited development in bivalve eggs and larvae, modified techniques of Davis and Calabrese (1964) were used. Physiological and biochemical changes in juvenile and adult organisms were measured and evaluated, as described by Thurberg *et al.* (1973), Calabrese *et al.* (1974, 1975), and Gould (1977, 1980). Except where otherwise noted, cadmium and mercuric chlorides and silver and lead nitrates were used. Studies with arsenic used sodium arsenite.

RESULTS AND DISCUSSION

Tables 1-3 summarize selected results from Milford Laboratory experiments on the effects of cadmium, mercury, and silver, respectively. Several inferences may be drawn from these data. Both the order and degree of metal toxicity vary, not only with such expected parameters as salinity and metal salt form, but also with life stage and with species. Early-life stages, for example, appear to be more sensitive to mercury and silver than to cadmium, as construed from short-term bioassay tests. This is true for the bivalve molluscs tested and also for juvenile lobsters (*Homarus americanus*). Other marine larval forms have been reported to display the same order of sensitivity (Waldichuk, 1974). Based on data from sublethal chronic exposure studies, the order of metals toxicity in adult animals is different from that observed in juvenile stages: cadmium produced more severe metabolic effects than either mercury or silver (Thurberg *et al.*, 1977; Dawson *et al.*, 1977; Gould, personal communication), with the exception of hematological parameters (Calabrese *et al.*, 1975), despite far lower rates of cadmium uptake in animal tissues (Calabrese *et al.*, 1975; Thurberg *et al.*, 1977). In assessing the potential hazards of metals to marine life, the National Academy of Sciences (1973) used toxicity and prevalence as criteria, and ranked mercury of greatest concern and cadmium and silver of next greatest concern. As determined by various enzyme measurements, the order of metals toxicity for adult lobsters from the New York Bight area is $CdCl_2 > HgCl_2 > AgNO_3 > Hg(NO_3)_2$; the order for adult winter flounder (*Pseudopleuronectes americanus*), also from this area, is $CdCl_2 > HgCl_2 > AgNO_3$ (Gould, personal communication). These adult animals apparently acquired a tolerance for mercury and silver (perhaps a sequestering mechanism) which is lacking in juvenile forms and which permits normal tissue metabolism in the presence of large body burdens of these metals.

More significant metabolic disturbances were observed from exposure to the chloride salt than to the nitrate, both for cadmium (Gould *et al.*, 1976) and for mercury (Gould, personal communication). As has been the case for all toxic metal experimentation at the Milford Laboratory, nominal metals concentrations in test tanks are reported as the amount of metal available in the salt solution and not of the total salt itself. Variations in toxicity among salts of the same metal, therefore, probably are attributable to differences in the bioavailability and properties of the metal compounds in seawater solution.

Table 1. Major findings from laboratory investigations of effects of cadmium on marine animals performed at NMFS Milford Laboratory. Concentration (third column) represents minimum level at which effects were noted.

Organism	Exposure period	Concentration (ppb)	Oxygen consumption	Osmoregulation	Enzyme activity	Other
Crassostrea virginica American oyster (embryos)	48 hr					LC_{50} = 3,800 ppb (Calabrese *et al.*, 1973)
Argopecten irradians Bay scallop (juveniles)	96 hr	940	Elevated (Nelson *et al.*, 1976)			LC_{50} = 1,480 ppb; significant Cd uptake but no mortality over 96-hr period (Nelson *et al.*, 1976)
Nassarius obsoletus Mud snail	72 hr	500	Elevated (MacInnes and Thurberg, 1973)			Distressed behavior (MacInnes and Thurberg, 1973)
Carcinus maenas Green crab	48 hr	500	Depressed (Thurberg *et al.*, 1973)	Disruption (Thurberg *et al.*, 1973)		
Cancer irroratus Rock crab	48 hr	120	Depressed (Thurberg *et al*, 1973)	No effect (Thurberg *et al.*, 1973)		
	96 hr	1,000			Chloride salt increased trans--aminase; nitrate salt no effect (heart) (Gould *et al.*, 1976)	Chloride salt increased serum Mg: nitrate salt no effect (Gould *et al.*, 1976)
	30 days	250	Depressed (Thurberg, pers. comm.)		Chloride salt increased transaminase; nitrate salt no effect (heart) (Gould, pers. comm.)	Serum Mg unchanged (Dawson, pers. comm.)

Eurypanopeus depressus Mud crab	72 hr	4,000	Depressed (Collier *et al.*, 1973)			LC_{50} = 4,900 ppb; LC_{100} = 11,000 ppb (Collier *et al.*, 1973)
Homarus americanus American lobster	30 days	3 - 6	Elevated (Thurberg *et al.*, 1977)	No effect (Thurberg *et al.*, 1977)	Enzyme induction (heart, tail muscle, antennal gland); lowered ligand sensitivity (heart, antennal gland) (Thurberg *et al.*, 1977; Gould, 1980)	
	60 days	3 - 6	Elevated (Thurberg *et al.*, 1977)	No effect (Thurberg *et al.*, 1977)		Significant Cd uptake in gill only; no significant uptake in digestive gland or muscle (Thurberg *et al.*, 1977)
Tautogolabrus adspersus Cunner	96 hr	3,000-48,000	Depressed (Thurberg and Dawson, 1974)	Disruption (Thurberg and Dawson, 1974)	Depressed transaminase, lowered ligand sensitivity (liver) (Gould and Karolus, 1974)	Some histopathological effects (Newman and MacLean, 1974); liver uptake 8.5x greater than gills (Greig *et al.*, 1974); in phagocytes of liver and spleen, increased bacterial uptake but decreased bactericidal rates; no effect on antibody production (Robohm and Nitkowski, 1974)
	30 days	50-100	Depressed (MacInnes *et al.*, 1977)	No effect (MacInnes *et al.*, 1977)	Depressed transaminase, higher shunt activity (liver) (MacInnes *et al.*, 1977)	

Table 1. Continued.

Organism	Exposure period	Concentration (ppb)	Oxygen consumption	Osmoregulation	Enzyme activity	Other
Tautogolabrus adspersus Cunner *(Continued)*	60 days	50	Depressed (MacInnes *et al.*, 1977)	No effect (MacInnes *et al.*, 1977)		
Morone saxatilis Striped bass	30 days	0.5-5.0	Depressed (Dawson *et al.*, 1977)		No effect (liver, skeletal muscle) (Dawson *et al.*, 1977)	
	90 days	5	No significant effect (Dawson *et al.*, 1977)		No effect (liver, skeletal muscle) (Dawson *et al.*, 1977)	90-day exposure and 30-day clearance; depressed transaminase and shunt activity (liver) (Dawson *et al.*, 1977)
Pseudopleuronectes americanus Winter flounder	60 days	5-10	Depressed (Calabrese *et al.*, 1975)		Enzyme induction, lowered ligand sensitivity (heart, kidney, gonad, skeletal muscle) (Gould, 1977; pers. comm.)	No detectable Cd uptake in blood or gills. No hematological or histopathological changes (Calabrese *et al.*, 1975)
	150 days	10	Elevated (Thurberg, pers. comm.)		Increased glycolysis and shunt activity, lowered ligand sensitivity (kidney, liver); enzyme induction (heart, gonad) (Gould, 1978)	No significant hematological changes; both control and exposed animals showed signs of anemia (Dawson, pers. comm.)

Table 2. Major findings from laboratory investigations of effects of mercury on marine animals performed at NMFS Milford Laboratory. Enzyme abbreviations: G6PdH, glucose-6-phosphate dehydrogenase, EC 1.1.1.49; GPI, glucosephosphate isomerase, EC 5.3.1.9; MDH, malate dehydrogenase, EC 1.1.1.37; LDH, lactate dehydrogenase, EC 1.1.1.27. Concentration (third column) represents minimum level at which effects were noted.

Organism	Exposure period	Concentration (ppb)	Oxygen consumption	Osmoregulation	Enzyme activity	Other
Crassostrea virginica American oyster						
(embryos)	48 hr					LC_{50} = 5.6 ppb (Calabrese *et al.*, 1973)
(larvae)	12 days					LC_{50} = 12.0 ppb (Calabrese *et al.*, 1977a)
Mercenaria mercenaria Hard clam						
(embryos)	48 hr					LC_{50} = 4.8 ppb (Calabrese and Nelson, 1974)
(larvae)	10 days					LC_{50} = 14.7 ppb (Calabrese *et al.*, 1977a)
Argopecten irradians Bay scallop (juvenile)	96 hr					LC_{50} = 89.0 ppb; significant Hg uptake and mortality in 24 hr (Nelson *et al.*, 1976)
Homarus americanus American lobster	30 days	6	No effect (Thurberg *et al.*, 1977)	No effect (Thurberg *et al.*, 1977)	Some enzyme induction, some loss of ligand sensitivity with chloride salt; no effect with nitrate salt (heart, antennal gland, gonad) (Thurberg *et al.*, 1977; Gould, pers. comm.)	Significant Hg uptake in digestive gland, gills, and tail muscle; gill uptake 30x that of Cd (Thurberg *et al.*, 1977)

Table 2. Continued.

Organism	Exposure period	Concentration (ppb)	Oxygen consumption	Osmoregulation	Enzyme activity	Other
Homarus americanus American lobster *(Continued)*	60 days	6	No effect (Thurberg *et al.*, 1977)	No effect (Thurberg *et al.*, 1977)		
Tautogolabrus adspersus Cunner	30 days	5	Elevated (Thurberg, pers. comm.)			
Morone saxatilis Striped bass	30 days	5	Depressed (Dawson *et al.*, 1977)		No effect (liver) (Dawson *et al.*, 1977)	
	60 days	10	Depressed (Dawson *et al.*, 1977)		No effect (liver) (Dawson *et al.*, 1977)	
	90 days	10	No significant effect (Dawson *et al.*, 1977)		No effect (liver) (Dawson *et al.*, 1977)	
Pseudopleuronectes americanus Winter flounder	60 days	5-10	Elevated (Calabrese *et al.*, 1975)		Some enzyme induction, some loss of ligand sensitivity, increased shunt activity (kidney, liver, heart, gonad) (Calabrese *et al.*, 1977b; Gould, pers. comm.)	Some blood changes; Hg uptake in blood and gills (Calabrese *et al.*, 1975)

	60 days + 60-day recovery	5-10	Elevated (Thurberg, pers. comm.)		G6PdH and GPI induction, and MDH repression after 60-day exposure; complete recovery of elevated enzyme activity after 60-day recovery, but no recovery from depressed MDH activity (liver); in kidney, no change in enzyme activity after 60-day exposure, but elevated activity after 60-day recovery (G6PdH, GPI, LDH) (Gould, pers. comm.)	Various hematological changes (Dawson, 1979)

Table 3. Major findings from laboratory investigation of effects of silver on marine animals performed at NMFS Milford Laboratory. Concentration (third column) represents minimum level at which effects were noted.

Organism	Exposure period	Concentration (ppb)	Oxygen consumption	Osmoregulation	Enzyme activity	Other
Crassostrea virginica American oyster						
(embryos)	48 hr					LC_{50} = 5.8 ppb (Calabrese *et al.*, 1973)
(larvae)	12 days					LC_{50} = 25.0 ppb (Calabrese *et al.*, 1977a)
(adults)	96 hr	100	Elevated (Thurberg *et al.*, 1974)			Ag uptake by gills (Thurberg *et al.*, 1974)
Mercenaria mercenaria Hard clam						
(embryos)	48 hr					LC_{50} = 21.0 ppb (Calabrese and Nelson, 1974)
(larvae)	10 days					LC_{50} = 32.4 ppb (Calabrese *et al.*, 1977a)
(adults)	96 hr	100	Elevated (Thurberg *et al.*, 1974)			Ag uptake by gills (Thurberg *et al.*, 1974)
Argopecten irradians Bay scallop						
(juveniles)	96 hr	22	Elevated (Nelson *et al.*, 1976)			LC_{50} = 33.0 ppb; significant Ag uptake and mortality in 24 hr (Nelson *et al.*, 1976)

Spisula solidissima Surf clam (larvae)	2-15 days	50	Elevated (Thurberg *et al.*, 1975)			100 ppb lethal; increased valve movement in adults (Thurberg *et al.*, 1975)
(juveniles)	96 hr	10	Elevated (Thurberg *et al.*, 1975)			
(adults)	96 hr	50	Elevated (Thurberg *et al.*, 1975)			
Mytilus edulis Blue mussel	96 hr	100	Elevated (Thurberg *et al.*, 1974)			Ag uptake by gills (Thurberg *et al.*, 1974)
Mya arenaria Soft-shell clam	96 hr	100	Elevated (Thurberg *et al.*, 1974)			Ag uptake by gills (Thurberg *et al.*, 1974)
C. virginica, M. mercenaria, M. edulis, Placopecten magellanicus	30, 60, 90 days	10	Elevated (Thurberg, 1977)			
Nassarius obsoletus Mud snail	72 hr	500	Depressed (MacInnes and Thurberg, 1973)			Distressed behavior (MacInnes and Thurberg, 1973)
Homarus americanus American lobster	30 days	6	No effect (Thurberg, pers. comm.)	No effect (Thurberg, pers. comm.)	Depressed transaminase (heart) some loss of ligand sensitivity (antennal gland), enzyme induction (gonad); all effects small (Gould, pers. comm.)	

Table 3. Continued.

Organism	Exposure period	Concentration (ppb)	Oxygen consumption	Osmoregulation	Enzyme activity	Other
Tautogolabrus adspersus Cunner	96 hr	120-150	Depressed (Thurberg and Collier, 1977)	No effect (Thurberg and Collier, 1977)	Depressed shunt activity (liver); changed ligand sensitivity (skeletal muscle) (Gould and MacInnes, 1977)	
Pseudopleuronectes americanus Winter flounder	60 days	10	No effect (Thurberg, pers. comm.)		Depressed transaminase (liver); very little effect (kidney, heart, gonad) (Gould, pers. comm.)	

Engel *et al.* (1981) recently observed that the toxicity and bioavailability of metals may be affected by chemical interactions in the test medium, and that caution must be used in interpreting test results. These authors suggest that biological response of test animals to toxic metals is a function of free metal-ion concentration, and not necessarily the total quantity of dissolved metal.

Studies of metal tolerance in striped bass (*Morone saxatilis*) and coho salmon (*Oncorhynchus kisutch*) confirmed the general observation that the metabolism of anadromous species adapts more readily to environmental change than that of strictly marine fishes. Striped bass, a facultative marine fish that moves to brackish or fresh water to spawn, adapted to long-term metal exposure with little sign of stress (Dawson *et al.*, 1977), in marked contrast to winter flounder and cunners (*Tautogolabrus adspersus*) (Calabrese *et al.*, 1975, 1977b; Dawson, 1979; Gould, 1977, 1978). A single study with coho salmon (Gould, personal communication) showed this fish also to be remarkably free of metabolic stress after long-term cadmium exposure (100 ppb for 60 days).

To examine how metal exposure affects an animal's capacity to adjust to normal environmental challenge, lobsters were exposed to 6 ppb cadmium for 30 days prior to being held for seven days in cadmium-free ambient- or low-salinity seawater (27 ‰ or 17 ‰). Low-salinity stress overshadowed the effects of sublethal cadmium stress, elevating respiration and altering glycolytic patterns in control animals, and modifying cadmium-induced enzymatic changes in metal-exposed animals (Gould, 1980).

Similar work with lead-exposed winter flounder (50 ppb for 60 days) that subsequently were held for two days at low salinity showed no change in gill-tissue respiration for either lead challenge or for low-salinity stress, and little or no detectable hematological change (Thurberg, personal communication). Biochemical testing revealed several lead-induced metabolic effects; most were variously affected by the flounders' exposure to low salinity, but two (some loss of ligand sensitivity in the kidney and lowered δ-aminolevulinate dehydratase activity in liver) remained unchanged by low-salinity stress (Gould, personal communication). In bioassay studies exposing oyster embryos (*Crassostrea virginica)* to lead and to other metals, lead proved to be twice as toxic as cadmium, but was two-hundredfold less toxic than mercury or silver (Calabrese *et al.*, 1973).

Exposure of adult rock crabs (*Cancer irroratus*) and lobsters to arsenic (100 ppb for 70 days) produced little or no effect on either species. Osmoregulation, gill-tissue respiration, heart rate, gill-bailer activity, and heart biochemistry were used as criteria in judging potential effects (Thurberg and Gould, personal communication).

Bioassay work on oyster embryos with use of combinations of metals (mercury, silver, copper, and zinc) demonstrated that the toxicities of all metals tested, whether singly or in combination, are enhanced by temperature stress (*i.e.*, temperatures either lower or higher than optimal) (MacInnes and Calabrese, 1978). Work with mercury-exposed juvenile bay scallops (*Argopecten irradians*) showed that metal concentration is the most important variable, followed by salinity and temperature. The toxicity of low concentrations of mercury is enhanced by a combination of high temperature and low salinity; neither one alone exerts a toxicity-enhancing effect (Nelson *et al.*, 1977).

Gould and Nitkowski (1979) reported that metal effects in the biochemistry of winter flounder ovarian tissue vary widely with the degree of gonad maturation. This was similarly true for female lobsters (Gould, personal communication). Whether this phenomenon results from some hormonally-stimulated protective mechanism (*e.g.*, sequestration or isoenzyme induction) or, more simply, from the greater amounts of inert protein in the ripe female gonad acting as a kind of blotter-buffer, remains to be discovered.

Two general sublethal metal effects are to be seen at the subcellular level: (1) the induction of enzymes involved in the mobilization of energy or in the production of metabolites for biosynthesis, and (2) the loss of ligand sensitivity, by which reaction rates of metabolic pathways are regulated. The former expends energy reserves and is evidenced most readily by the disruption of respiratory patterns. The latter results in a diminished ability of the organism to regulate its metabolism.

Establishing correlations between metal concentrations and physiological or biochemical abnormalities observed in tissues should be attempted only with great caution, as high tissue concentrations of a metal are not always related to metabolic disorders. Metals can be biochemically immobilized by a healthy animal's detoxifying mechanisms (*e.g.*, metallothionein or metallothionein-like compounds) and thus can be rendered metabolically inert (see O'Connor and Rachlin, this volume). This type of mechanism may have been responsible for observations such as those obtained by Thurberg *et al.* (1977), in which lobsters exposed for one month to 6 ppb mercury had significant uptake of that metal but only limited metabolic abnormalities. On the other hand, similar exposure to cadmium produced no detectable tissue uptake but produced metabolic abnormalities comparable to, or greater than those seen in the mercury-exposed animals. The choice of analytical procedures may further complicate the interpretation of metal data, for accumulated metals may be unevenly distributed within a single tissue or organ. For example, Sangalang and Freeman (1979) discovered that cadmium levels in the posterior kidney of cadmium-exposed brook trout (*Salvelinus fontinalis*) were consistently and significantly higher than those in the anterior kidney.

IMPLICATIONS

Bioassay work with early-life stages of marine animals has shown the sensitivity of these life forms to low levels of toxic metals, notably mercury and silver. Metal-induced sublethal stresses in adult marine animals, manifested by changes in respiration, osmoregulation, hematology, and enzyme chemistry, support the concept that toxic metals can place abnormal demands upon an animal's energy reserves. Perhaps more important is the attenuation of such physiological control mechanisms as ligand sensitivity, which impairs metabolic flexibility and reduces an animal's capacity for adapting to and surviving in a naturally changing environment. In addition to draining energy reserves, metal-induced stress may render the animal more susceptible to injury or death by environmental factors such as salinity, temperature, or disease.

The implications of such metal effects to fishery stock recruitment should not be dismissed lightly, as toxic-metal pollution is greatest in estuarine and inshore coastal waters. These highly polluted areas are the breeding grounds and nurseries for many marine species important to both commercial and sport fisheries, and early-life stages are the most vulnerable to toxic metals. The combination of such highly sensitive marine life stages with high concentrations of toxic pollutants places recruitment of fishery stocks in jeopardy.

As the oceans become increasingly polluted from rivers used as industrial and domestic sewers, by atmospheric fallout, and by ocean dumping, increasing stress will be put upon marine stocks. It is imperative that the steps required for maintaining a marine environment capable of supporting healthy, renewable fishery populations are rapidly and systematically determined, implemented, and monitored. The linking of specifically designed laboratory research to broader, more general field activities is one move in this direction.

REFERENCES

Alexander, J.E. and E.C. Alexander. 1977. Chemical Properties. MESA New York Bight Monograph 2, New York Sea Grant Institute, Albany, NY. 47 pp.

Calabrese, A., R.S. Collier, and J.E. Miller. 1974. Physiological response of the cunner, *Tautogolabrus adspersus*, to cadmium. I. Introduction and experimental design. NOAA Tech. Rep., NMFS SSRF 681. pp. 1-3.

Calabrese, A., R.S. Collier, D.A. Nelson, and J.R. MacInnes. 1973. The toxicity of heavy metals to embryos of the American oyster, *Crassostrea virginica*. Mar. Biol. 18: 162-166.

Calabrese, A., J.R. MacInnes, D.A. Nelson, and J.E. Miller. 1977a. Survival and growth of bivalve larvae under heavy-metal stress. Mar. Biol. 41: 179-184.

Calabrese, A., and D.A. Nelson. 1974. Inhibition of embryonic development of the hard clam, *Mercenaria mercenaria*, by heavy metals. Bull. Environ. Contam. Toxicol. 11: 92-97.

Calabrese, A., F.P. Thurberg, M.A. Dawson, and D.R. Wenzloff. 1975. Sublethal physiological stress induced by cadmium and mercury in the winter flounder, *Pseudopleuronectes americanus*. *In:* Sublethal Effects of Toxic Chemicals on Aquatic Animals, J.H. Koeman and J.J.T.W.A. Strik (eds.), Elsevier, Amsterdam. pp. 15-21.

Calabrese, A., F.P. Thurberg, and E. Gould. 1977b. Effects of cadmium, mercury, and silver on marine animals. Mar. Fish. Rev. 39: 5-11.

Calabrese, A., F.P. Thurberg, E. Gould, and J.T. Graikoski. 1978. Ocean pulse: some physiological, biochemical, and bacteriological activities. Int. Counc. Explor. Sea C.M.1978/L:36, Biological Oceanography Committee. 11 pp.

Collier, R.S., J.E. Miller, M.A. Dawson, and F.P. Thurberg. 1973. Physiological response of the mud crab, *Eurypanopeus depressus* to cadmium. Bull. Environ. Contam. Toxicol. 10: 378-382.

Davis, H.C. and A. Calabrese. 1964. Combined effects of temperature and salinity on development of eggs and growth of larvae of *M. mercenaria* and *C. virginica*. U.S. Fish Wildl. Serv. Fish. Bull. 63: 643-655.

Dawson, M.A. 1979. Hematological effects of long-term mercury exposure and subsequent periods of recovery on the winter flounder, *Pseudopleuronectes americanus*. *In:* Marine Pollution-Functional Responses, W.B. Vernberg, F.P. Thurberg, A. Calabrese, and F.J. Vernberg (eds.), Academic Press, New York. pp. 171-182.

Dawson, M.A., E. Gould, F.P. Thurberg, and A. Calabrese. 1977. Physiological response of juvenile striped bass, *Morone saxatilis*, to low levels of cadmium and mercury. Ches. Sci. 18:353-359.

Engel, D.W., W.G. Sunda, and B.A. Fowler. 1981. Factors affecting trace metal uptake and toxicity to marine organisms. I. Environmental parameters. *In:* Biological Monitoring of Marine Organisms, F.J. Vernberg, A. Calabrese, F.P. Thurberg, and W.B. Vernberg (eds.), Academic Press, New York. pp. 127-144.

Gould, E. 1977. Alteration of enzymes in winter flounder, *Pseudopleuronectes americanus*, exposed to sublethal levels of cadmium. *In:* Physiological Responses of Marine Biota to Pollutants, F.J. Vernberg, A. Calabrese, F.P. Thurberg, and W.B. Vernberg (eds.), Academic Press, New York. pp. 209-224.

Gould, E. 1978. Long-term response of the winter flounder, *Pseudopleuronectes americanus*, exposed to sublethal levels of cadmium. *In:* Abstracts of Papers of the 144th National Meeting, 12-17 February 1978, A. Herschman (ed.), AAAS, Washington, DC. p. 136.

Gould, E. 1980. Low-salinity stress in the American lobster, *Homarus americanus*, after chronic sublethal exposure to cadmium: biochemical effects. Helgol. Wiss. Meeresunters. 33: 174-184.

Gould, E. 1981. Monitoring sea scallops in the offshore waters of New England and the mid-Atlantic states: enzyme activity in adductor muscle. *In:* Biological Monitoring of Marine Pollutants, F.J. Vernberg, A. Calabrese, F.P. Thurberg, and W.B. Vernberg (eds.), Academic Press, New York. pp. 377-388.

Gould, E., R.S. Collier, J.J. Karolus, and S. Givens. 1976. Heart transaminase in rock crab, *Cancer irroratus*, exposed to cadmium salts. Bull. Environ. Contam. Toxicol. 15: 635-643.

Gould, E., and J. Karolus. 1974. Physiological response of the cunner, *Tautogolabrus adspersus*, to cadmium. V. Observations on the biochemistry. NOAA Tech. Rep., NMFS SSRF 681. pp. 21-15.

Gould, E. and J.R. MacInnes. 1977. Short-term effects of two silver salts on tissue respiration and enzyme activity in the cunner, *Tautogolabrus adspersus*. Bull. Environ. Contam. Toxicol. 18(4): 401-408.

Gould, E. and M. Nitkowski. 1979. Changing enzyme activities in maturing gonads of winter flounder, *Pseudopleuronectes americanus*. *In:* Cyclic Phenomena in Marine Plants and Animals, E. Naylor and R.G. Hartnoll (eds.), Pergamon Press, Oxford, England. pp. 231-238.

Greig, R.A., A.E. Adams, and B.A. Nelson. 1974. Physiological response of the cunner, *Tautogolabrus adspersus*, to cadmium. II. Uptake of cadmium by organs and tissues. NOAA Tech. Rep., NMFS SSRF 681. pp. 5-9.

MacInnes, J.R. and A. Calabrese. 1978. Response of embryos of the American oyster, *Crassostrea virginica*, to heavy metals at different temperatures. *In:* Physiology and Behaviour of Marine Organisms, D.S. McLusky and A.J. Berry (eds.), Pergamon Press, Oxford, England. pp. 195-202.

MacInnes, J.R., and F.P. Thurberg. 1973. Effects of metals on the behavior and oxygen consumption of the mud snail. Mar. Pollut. Bull. 4: 185-186.

MacInnes, J.R., F.P. Thurberg, R.A. Greig, and E. Gould. 1977. Long-term cadmium stress in the cunner, *Tautogolabrus adspersus*. Fish. Bull., U.S. 75: 199-203.

Mueller, J.A., J.S. Jeris, A.R. Anderson, and C.F. Hughes. 1976. Contaminant inputs to the New York Bight. NOAA Tech. Memo. ERL MESA-6. 347 pp.

National Academy of Sciences. 1973. Research Needs in Water Quality Criteria, 1972. National Academy of Sciences, National Academy of Engineering, Washington, DC. p. 38.

Nelson, D.A., A. Calabrese, and J.R. MacInnes. 1977. Mercury stress on juvenile bay scallops, *Argopecten irradians,* under various salinity-temperature regimes. Mar. Biol. 43: 293-297.

Nelson, D.A., A. Calabrese, B.A. Nelson, J.R. MacInnes, and D.R. Wenzloff. 1976. Biological effects of heavy metals on juvenile bay scallops, *Argopecten irradians,* in short-term exposures. Bull. Environ. Contam. Toxicol. 16(3): 275-282.

Newman, M.W. and S.A. MacLean. 1974. Physiological response of the cunner, *Tautogolabrus adspersus,* to cadmium. VI. Histopathology. NOAA Tech. Rep., NMFS SSRF 681. pp. 27-33.

O'Connor, J.M. and J.W. Rachlin. This volume. Perspectives on metals in New York Bight organisms: factors controlling accumulation and body burdens. pp. 655-673.

Robohm, R.A. and M.F. Nitkowski. 1974. Physiological response of the cunner, *Tautogolabrus adspersus,* to cadmium. IV. Effects on the immune system. NOAA Tech. Rep., NMFS SSRF 681. pp. 15-20.

Sangalang, G.B. and H.C. Freeman. 1979. Tissue uptake of cadmium in brook trout during chronic sublethal exposure. Arch. Environ. Contam. Toxicol. 8: 77-84.

Thurberg, F.P. 1977. Physiological response of marine bivalve molluscs to long-term silver exposure. Proc. Int. Union Physiol. Sciences 13: 753 (abstract).

Thurberg, F.P., W.D. Cable, M.A. Dawson, J.R. MacInnes, and D.R. Wenzloff. 1975. Respiratory response of larval, juvenile, and adult surf clams, *Spisula solidissima,* to silver. *In:* Respiration of Marine Organisms, J.J. Cech, Jr., D.W. Bridges, and D.B. Horton (eds.), TRIGOM Publications, South Portland, ME. pp. 41-52.

Thurberg, F.P., A. Calabrese, and M.A. Dawson. 1974. Effects of silver on oxygen consumption of bivalves at various salinities. *In:* Pollution and Physiology of Marine Organisms, F.J. Vernberg and W.B. Vernberg (eds.), Academic Press, New York. pp. 67-68.

Thurberg, F.P., A. Calabrese, E. Gould, R.A. Greig, M.A. Dawson, and R.K. Tucker. 1977. Response of the lobster, *Homarus americanus,* to sublethal levels of cadmium and mercury. *In:* Physiological Responses of Marine Biota to Pollutants, F.J. Vernberg, A. Calabrese, F.P. Thurberg, and W.B. Vernberg (eds.), Academic Press, New York. pp. 185-197.

Thurberg, F.P., and R.S. Collier. 1977. Respiratory response of the cunner to silver. Mar. Pollut. Bull. 8(2): 40-41.

Thurberg, F.P., and M.A. Dawson. 1974. Physiological response of the cunner, *Tautogolabrus adspersus*, to cadmium. III. Changes in osmoregulation and oxygen consumption. NOAA Tech. Rep., NMFS SSRF 681. pp. 11-13.

Thurberg, F.P., M.A. Dawson, and R.S. Collier. 1973. Effects of copper and cadmium on osmoregulation and oxygen consumption in two species of estuarine crabs. Mar. Biol. 23: 171-175.

Waldichuk, M. 1974. Some biological concerns in heavy metals pollution. *In:* Pollution and Physiology of Marine Organisms, F.J. Vernberg and W.B. Vernberg (eds.), Academic Press, New York. pp. 1-57.

Watling, L., W. Leathem, P. Kinner, C. Wethe, and D. Maurer. 1974. Evaluation of sludge dumping off Delaware Bay. Mar. Pollut. Bull. 5(3): 39-42.

LONG-TERM CHANGES IN SOME FINFISHES OF THE HUDSON-RARITAN ESTUARY

Stephen C. Esser
New Jersey Marine Sciences Consortium
Building 22
Fort Hancock, New Jersey 07732

Abstract. Long-term changes in abundance as expressed in catch statistics have occurred for most finfish species of the Hudson-Raritan estuary. The commercial catch of American shad within the Hudson-Raritan estuary has experienced significant fluctuations and decline since the end of the nineteenth century. Possible causes, including physical changes, pollution, and overfishing, are reviewed. An extensive sturgeon fishery flourished in the Hudson River during the 1800's but declined in the early 1900's. A similar decline occurred in the smelt fishery, which prospered in the Raritan River in the early 1800's. Only the striped bass population appears to have increased slightly in recent years. Other species, including bluefish, scup, and flounders, are caught both in the estuary and in the ocean, but the combined statistics make it impossible to determine the estuarine harvest alone. Most of these species use the estuary as a spawning or nursery area. Their populations are influenced by water quality of the Hudson and its tributaries.

INTRODUCTION

The Hudson-Raritan estuary receives its main freshwater input from the Hudson River Basin, a watershed with an area of over 37,000 km^2 (13,000 mi^2). This report focuses on fisheries supported by that portion of the River Basin below the Troy Dam, known as the lower Hudson River (Figure 1). Tidal influence extends throughout this 261-km (162-mi) stretch of River, and brackish water occasionally can be found as far north as Poughkeepsie (about 130 km). At its southern end, the Hudson receives several New Jersey rivers and drains into a series of bays leading to the ocean. The Passaic and Hackensack Rivers flow into Newark Bay, and the Raritan River drains into Raritan Bay. Raritan Bay forms the connection between the estuarine system and the ocean.

The fishery resources of the Hudson-Raritan estuary were important to the inhabitants of the area six thousand years before European explorers arrived, and early European arrivals were impressed by the fishes found in these waters. Robert Juet (1959), in his journal accounts of the voyage of Henry Hudson's HALF MOON, frequently described the abundant catches of fishes made during the voyage up the Hudson River.

Recent investigators have listed the fish species in various portions of the Hudson-Raritan estuary. A composite listing, in which the major species are grouped on the basis of migratory habits, is presented in Table 1. Because a number of migratory species use the Hudson-Raritan estuary for at least a portion of their lives, the estuary's finfish community varies with the seasonal movements of these species.

Most anadromous fishes in the Hudson-Raritan estuary spawn in the spring. Those species that spawn in the upper, less saline areas of the estuary include striped bass (*Morone saxatilis*), shad (*Alosa sapidissima*), alewife (*A. pseudo-*

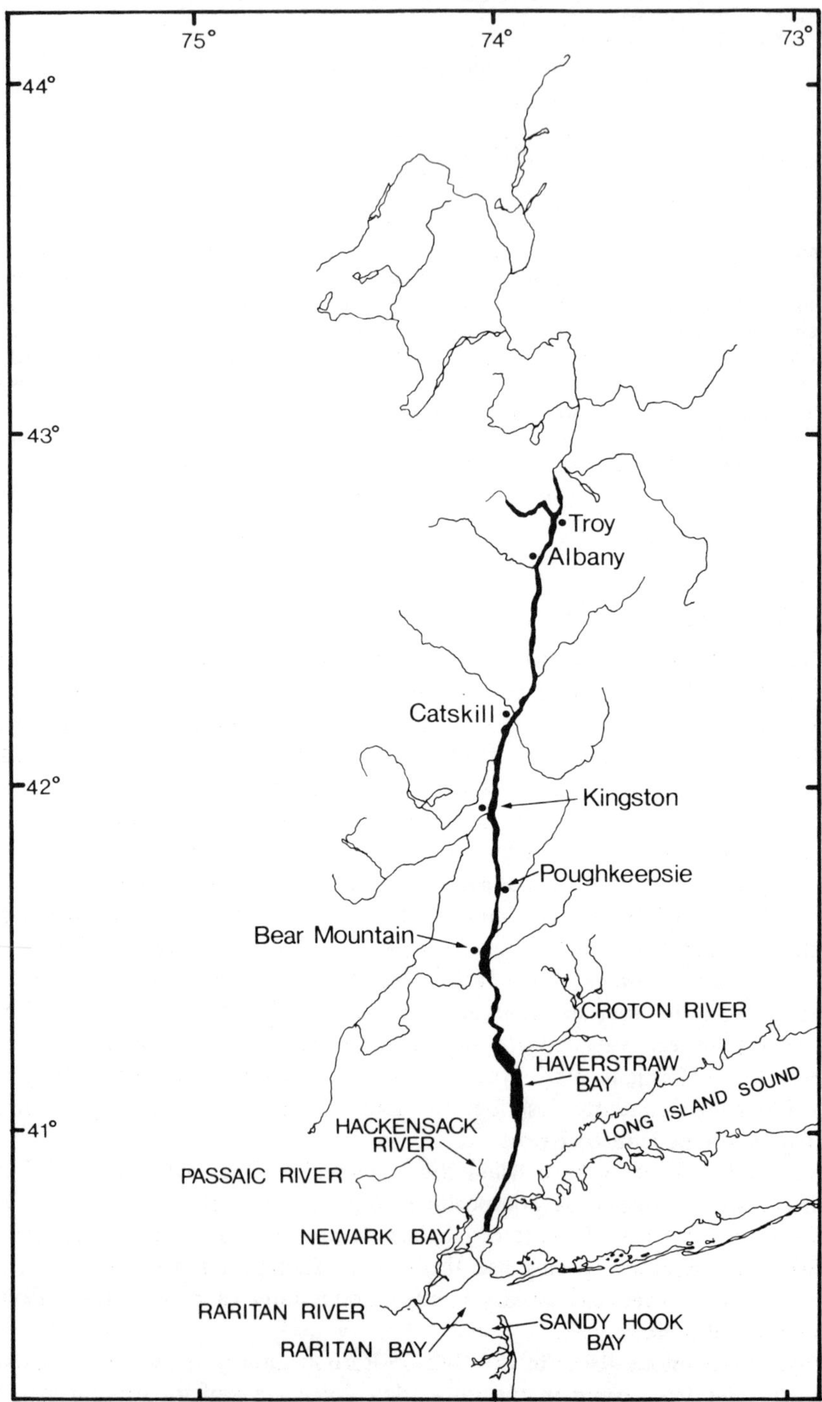

Figure 1. The lower Hudson River and New York harbor.

Table 1. Major fish species of the Hudson-Raritan estuary. From Greeley (1937), Perlmutter *et al.* (1972), and Dovel (1977). Resident brackish water species hatch and remain within the estuary. Resident freshwater species are found only in the upper reaches of the Hudson and its tributaries. Nursery species generally enter the estuary during the warm season to feed and mature. Diadromous species either migrate into fresh water from the sea (anadromous) or into the sea from fresh water (catadromous) to spawn.

Nonmigratory species			
Resident brackish water species		Resident fresh water species	
Fundulus heteroclitus	Mummichog	*Carassius auratus*	Goldfish
Fundulis majalis	Striped killifish	*Cyprinus carpio*	Carp
Fundulis diaphanus	Banded killifish	*Notemigonus crysoleucas*	Golden shiner
Apeltes quadracus	Fourspine stickleback	*Notropis atherinoides*	Emerald shiner
Menidia beryllina	Tidewater silverside	*Notropis hudsonius*	Spotted shiner
Menidia menidia	Atlantic silverside	*Catostomus commersoni*	White sucker
Hippocampus erectus	Spotted seahorse	*Ictalurus catus*	White catfish
Syngnathus fuscus	Northern pipefish	*Ictalurus nebulosus*	Brown bullhead
Roccus americanus	White perch	*Lepomis gibbosus*	Pumpkin seed
Perca flavescens	Yellow perch	*Lepomis macrochirus*	Bluegill
Scophthalmus aquosus	Windowpane	*Pomoxis nigromaculatus*	Black crappie
Acipenser brevirostrum	Shortnose sturgeon	*Etheostoma olmstedi*	Tessellated darter
Migratory			
Nursery species		Diadromous species	
Brevoortia tyrannus	Atlantic menhaden	*Acipenser oxyrhynchus*	Atlantic sturgeon
Anchoa mitchilli	Bay anchovy	*Alosa aestivalis*	Blueback herring
Pomatomus saltatrix	Bluefish	*Alosa pseudoharengus*	Alewife
Cynoscion regalis	Weakfish	*Alosa sapidissima*	Shad
Stenotomus chrysops	Scup	*Osmerus mordax*	Rainbow smelt
Paralichthys dentatus	Summer flounder	*Morone saxatilis*	Striped bass
Pseudopleuronectes americanus	Winter flounder	*Microgadus tomcod*	Atlantic tomcod
		Anguilla rostrata[a]	American eel

[a]Catadromous.

harengus), and sturgeon (*Acipenser oxyrhynchus*). After hatching, most of the young move downstream from the spawning areas and remain in the estuary until the fall. Some juveniles, such as those of striped bass and sturgeon, remain in the River for longer periods of time. Larval stages of these fishes are dependent for their development on the dense populations of forage species that occur in estuarine waters. Adults of most anadromous species return to the ocean in the summer or fall, although some striped bass may overwinter in the Hudson. Tomcod (*Microgadus tomcod*) and smelt (*Osmerus mordax*) represent two exceptions to the pattern for anadromous fishes. These species enter the estuary in the fall and remain until their spawning runs–early winter for tomcod and early spring for smelt. Adults return to more saline water after spawning.

The lower, more saline portions of the estuary are spawning areas for weakfish (*Cynoscion regalis*), scup (*Stenotomus chrysops*), and winter flounder (*Pseudopleuronectes americanus*). Spawning times for these fishes vary from late winter for winter flounder, to spring and early summer for weakfish, to early summer for scup. Adult and young scup remain in the coastal areas until fall, when they migrate offshore.

Other fish species spawn in the ocean but have larvae that enter the estuarine area. Species that follow this pattern include bluefish (*Pomatomus saltatrix*), menhaden (*Brevoortia tyrannus*), and summer flounder (*Paralicthys dentatus*). Young of the catadromous American eel (*Anguilla rostrata*) also migrate into the estuary. Juvenile eels (elvers) feed and grow in this environment for several years before reaching sexual maturity. The estuary provides ample food and a measure of protection from larger predators during the early life stages of these fishes.

Commercial fisheries have been well established in the Hudson-Raritan estuary since the mid-1800's. Total catch and fishing effort reached a peak in the 1940's and declined sharply in the 1960's. Commercial fishery statistical data are the basis for defining long-term trends in availability of fish species. If effort were constant, such records would reflect actual changes in abundance. However, catch records have other limitations aside from fluctuations in fishing effort. Most of the catch of species such as shad, sturgeon, and smelt occurs within the Hudson-Raritan estuary; data available on these fishes, therefore, can be related directly to this area. Other fish species, however, including weakfish, striped bass, bluefish, scup, and flounder, are sought by both commercial and recreational fishermen, and the catch is not restricted to estuarine areas. Ocean and estuarine catches were combined in early reports of landings, making an accurate determination of the estuarine catch alone impossible. The unreported recreational catch also precludes an estimate of the total annual estuarine harvest.

SHAD

Background

The American shad (*Alosa sapidissima*) is an anadromous fish. The spring passage to spawning grounds provides fishermen with an opportunity to intercept these fish, primarily with gill nets. The shad run in the Hudson begins in April and continues through June. Young shad spend their first summer in the Hudson and migrate to the ocean in the fall, after attaining a size of 76-150 mm. Shad remain in the ocean until maturity at four to five years, when they return to the Hudson to spawn. Shad in the Hudson do not die following spawning as they do in rivers south of North Carolina, but return to the ocean and may spawn annually as many as five times. Repeat spawners comprise half of the Hudson population (Talbot, 1954).

Fishery Data

Historical accounts of the New Jersey portion of the Hudson-Raritan system tell of the extensive shad fishery in Newark Bay and in the Passaic and Hackensack Rivers in the early and mid-1800's (Goode, 1887; Holmes, 1890; Brydon, 1974). Almost 500,000 nets, over half of them gill nets, were reported to be in use during the 1889 shad season within the whole estuarine system from Albany, New York to Sandy Hook, New Jersey (Commission of Fisheries of New York, 1890).

Commercial landings of shad have fluctuated widely during the period for which records are available. The period from 1880 to 1901 was the initial period of high production, with the highest production (1,714 metric tons) recorded in 1901. These high levels were established even though catches in some areas of New Jersey, such as the Passaic River, were declining because of pollution (Commission of Fisheries of New Jersey, 1885). This high production was followed by a period of low catches (ranging from approximately 18 metric tons per year to 414 metric tons per year) extending through the early 1930's (Table 2). Catches recorded between 1936 and 1946 show a second period of high production with a peak of 1,728 metric tons in 1944. Production levels have dropped since then to a level of about 106 metric tons per year during the early 1970's, at least an order of magnitude lower than the years of peak production (Medeiros, 1975).

Factors Involved in Shad Population Fluctuations

Several extensive studies have analyzed the Hudson River shad fishery. The most comprehensive study (Talbot, 1954) considered population fluctuations between 1915 and 1950 (with some references to earlier, scattered data) and discussed potential causes for the fluctuations. Burdick (1954) concentrated on the period between 1924 and 1951 and discussed pollution and net lift period as factors influencing shad abundance. Two more recent studies (Medeiros, 1975; Klauda *et al.*, 1977) analyzed the period between 1950 and the 1970's.

Physical Factors

A dam at Troy, New York (242 km upstream from New York City) restricts the shad run to that point, but no other physical barriers exist downstream. Talbot (1954) considered stream flow, water temperature, channel improve-

Table 2. Hudson River commercial shad landings and yield per effort indices of abundance. Landings are reported in metric tons. Yields per effort are reported in metric tons of fish/ m^2 net · hours fishing/week x 10^{-6}. Data after Medeiros (1975) and Klauda *et al.* (1977).

	Landings	Yield per effort		Landings	Yield per effort
1896	1,069	–	1944	1,728	16.1
1897	923	–	1945	1,577	16.0
1898	971	–	1946	1,348	11.1
1901	1,714	–	1947	899	6.1
1904	274	–	1948	1,068	8.6
1910	414	–	1949	783	6.6
1915	31	–	1950	458	4.3
1916	18	–	1951	347	7.1
1917	20	–	1952	488	10.4
1918	106	–	1953	426	11.4
1919	170	–	1954	567	12.9
1920	91	–	1955	685	18.8
1921	59	–	1956	762	22.2
1922	79	–	1957	679	20.1
1923	55	–	1958	474	16.3
1924	43	–	1959	531	19.0
1925	56	–	1960	328	12.1
1926	120	–	1961	267	10.6
1927	162	–	1962	239	9.2
1928	112	–	1963	158	8.3
1929	89	–	1964	82	4.7
1930	94	–	1965	108	6.8
1931	188	6.1	1966	53	4.1
1932	240	6.6	1967	80	6.3
1933	235	5.7	1968	115	7.7
1934	199	4.1	1969	110	8.3
1935	384	11.4	1970	105	7.2
1936	1,119	24.3	1971	77	6.6
1937	1,239	17.0	1972	131	10.5
1938	1,119	12.9	1973	114	8.5
1939	1,483	20.6	1974	105	5.7
1940	1,412	23.6	1975	102	3.8
1941	1,412	24.4	1976	96	–
1942	1,445	35.6	1977	83	–
1943	1,463	17.5			

ments, and ship traffic as factors that could have influenced shad population fluctuations. However, none of these factors were correlated with changes in shad abundance. All of these factors, especially dredging and ship traffic, may have produced localized effects in shad populations, but the shad stock was sufficiently large to allow sizeable catches in the 1930's and 1940's, despite these stresses.

Hatcheries

Shad fry hatched artificially at the town of Catskill, New York were stocked in the Hudson every year from 1869 through 1944. Hatchery production peaked at 18,000,000 fry in 1901, but this level did not maintain subsequent shad runs. Talbot (1954) compared hatchery production between 1911 and 1944 (production ranged from a high of 4,800,000 fry to a low of 2,000 fry) to the shad runs four and five years later, and found no correlation. He concluded that the number of eggs produced artificially was insignificant compared to natural production.

Chemical Factors

Chemical pollution frequently is mentioned as the factor that has influenced shad populations, but the absence of comparative water quality records earlier than 1936 makes a conclusion regarding the absolute effect of pollution difficult. Talbot (1954) investigated conditions in the River in 1951 and concluded that overall the River was not a good habitat for fishes. He found the most polluted area to extend from the Albany-Troy-Schenectady metropolitan area some 40 km downstream to the town of Hudson, New York. This area supported a commercial shad fishery through the 1800's and was a good spawning area, but it later became unsuitable for both fishing and spawning. In the 1940's and early 1950's, shad were not caught above the town of Hudson.

A report on dissolved oxygen (DO) levels in the Hudson River (Burdick, 1954) found the lowest readings to occur in the upper portions of the River. DO levels at stations where shad were collected suggested that oxygen concentrations as low as 4.0 ppm did not adversely affect the fishes. Since readings at very few of the surveyed stations fell below 4.0 ppm, Burdick concluded that low riverine DO levels did not impact shad populations. From a study of shad catches during 1924-1953, Burdick (1954) also found no significant relationships between chemical pollution and shad runs. His conclusion was reinforced by the absence of pollution-caused year-class failures, which would have led to sharp drops in abundance.

Overfishing

Most investigators of the Hudson shad fishery agree that the factor best correlated with shad abundance is the number of shad that escape the commercial fishery and are able to spawn. Talbot (1954) found a correlation between production in a particular year and escapement one year earlier (repeat spawners) and four or five years earlier (first-time spawners).

The importance of allowing a significant percentage of shad to complete their spawning run was known prior to the study cited above. New York has used net lift periods since 1915 to ensure the passage of sufficient fish to produce future generations (Medeiros, 1975). The regulations provide for specific times during the week when all nets are taken from the water to allow fish to complete their spawning run unmolested. Net lift periods of varying lengths have been in effect in most subsequent years.

Burdick's (1954) observations of shad abundance from 1924 to 1953 demonstrated a correlation with the length of time during which nets were removed from the River. Shad were more abundant following the removal of nets from the River for 60 hours per week; shad abundance declined when nets were removed for 36 hours or less.

Klauda *et al.* (1977) calculated annual abundance indices for the years from 1931 to 1975 from yield and fishing effort data (Table 2). These figures show that shad abundance increased during the mid-1930's to a high level in 1942, followed by a sharp decline lasting until the early 1950's. A recovery was experienced following this period, possibly because of increased lift times in 1951. Another peak occurred in 1956, and smaller peaks were experienced in 1965 and 1972. A change from linen and cotton nets to nylon nets in 1955 may account for a change in catching efficiency. In a study of the Chesapeake Bay shad fishery, Muncy (1960) found that nylon nets were twice as effective in capturing shad as linen or cotton nets. Assuming that this also is true for the Hudson, if the abundance index of Klauda *et al.* (1977) since 1955 is adjusted by a factor of one half, then the shad run of 1975 would be the smallest recorded since 1931.

A recent study of the Hudson River shad fishery by Medeiros (1975) agreed with earlier conclusions that low escapement was responsible for the early decline of the shad population. However, Medeiros felt that the more recent decline in the fishery involved other factors, including reduced fishing effort, as shown by decreases in the number of licensed nets and fishing intensity. Reduced landings and reduced fishing effort relate to the economics of the shad fishery. River runs are not simultaneous. The shad season in Florida begins in November, and shad not marketed in the south are sent to northern markets. River runs continue northward as the season progresses. The shad run in the Hudson occurs around the end of March when the market may be saturated with shad from southern fisheries. Prices for shad are down at this point, and the fishery, therefore, is less attractive. In addition, the popularity of shad has declined compared to other seafood such as shellfish and prepared fish products.

The claim by Medeiros (1975) that fishing effort has declined was challenged by Klauda *et al.* (1977), who said that fishing effort has increased since 1972. This latter claim was based on an index of effort determined by the total area of licensed gill nets and the hours of legal fishing per week. The change to more efficient nylon nets also may have increased the actual effective fishery effort.

Tainting of Shad by Oil Pollution

Fish in the Hudson-Raritan system, including shad, frequently have been reported to have a pollution-related gasoline or kerosene-like taste. Goode (1887) reported that shad from Newark Bay had an oily taste that rendered them unsaleable. This was presumably a result of coal oil pollution. Talbot (1954) reported observing large quantities of oil in several areas of the Hudson River, and traces of oil were sighted throughout the River. Shad collected during Talbot's study were tested for this off-flavor, and 18 out of 20 were reported to taste like gasoline or kerosene. In 1975 Medeiros indicated the shad were of high quality.

ATLANTIC STURGEON AND SHORTNOSE STURGEON

Background

The Atlantic sturgeon (*Acipenser oxyrhynchus*) is an anadromous species that was abundant in the Hudson-Raritan estuary during the 1800's. Adult Atlantic sturgeon migrate into the River to spawn, and the young remain in the River for up to five years after hatching (Murawski and Pacheco, 1977). The shortnose sturgeon (*A. brevirostrum*) is a smaller, nonmigratory species believed never to have been abundant in the Hudson system (Greeley, 1937).

Historical Data

In the 1800's large quantitites of sturgeon were caught in the Hudson and marketed under the name of "Albany beef" (Mearns, 1898). Sturgeon also were common in the Passaic River at that time (Holmes, 1890), and other accounts of this fishery tell of the capture of large sturgeon (1.5-2.0 m) weighing 115-225 kg (Brydon, 1974). Goode (1887) reported that the 1880 catch of sturgeon in the Hudson River was about 90 metric tons, but the fishery already had declined from earlier levels. The Report of the Commissioner of Fisheries for 1904 (U.S. Department of Commerce, 1905) commented on the severe decline of this fishery at the turn of the century. This decline was demonstrated by a drop in the value of landings from $46,573 in 1898 to $1,010 in 1904. The report forecast the imminent extinction of the fishery.

A survey of the lower Hudson watershed (Greeley, 1937) listed the Atlantic sturgeon as uncommon and seldom taken commercially. However, sturgeon continue to breed in moderate numbers and represent a potentially valuable resource (Dovel, 1977). An important factor limiting the establishment of larger populations is the destruction of small sturgeon in shad gear. Young sturgeon are more vulnerable to the Hudson River shad fishery than are other migratory fishes because the juveniles remain in these waters for four to five years.

The survey of the lower Hudson watershed by Greeley (1937) also suggested that the resident shortnose sturgeon, although not abundant at the time of the survey, may have some potential commercial importance. However, shortnose sturgeon now is classified as an endangered species and cannot be caught legally.

SMELT

The smelt (*Osmerus mordax*) is an anadromous species that spawns in streams of the Hudson-Raritan estuarine system in the spring. Early reports (Brown, 1857) indicated that a commercial fishery existed for these fish in the Passaic, Hackensack, and Raritan Rivers. Later accounts (Commission of Fisheries of New Jersey, 1872, 1878) stated that the major portion of the catch was obtained in the Raritan River. The declining catch of smelt in the 1870's resulted from overfishing and the destruction of spawning grounds by industrial pollution. Attempts were made in following years to replenish the stock through the placement of smelt eggs in the Hackensack River (8 million in 1896, 12 million in 1897) and the Passaic River (5 million in 1896, 11 million in 1897) (Fish and Game Commission of New Jersey, 1896, 1897). Later accounts of New Jersey county landings either do not include any landings of smelt or report landings of less than 0.2 metric tons, such as in 1921 (U.S. Department of Commerce, 1922). Greeley (1937) classified smelt as rare within the lower Hudson River.

WEAKFISH

Weakfish (*Cynoscion regalis*) commonly are found in Sandy Hook Bay, Raritan Bay, and Lower Bay during the warmer months and have been reported up the Hudson as far north as the Croton River (Boyle, 1969). The weakfish available in the Hudson-Raritan estuary are mostly products of nearshore spawning (Perlmutter *et al.*, 1956). Extensive spawning occurs southward and, to a lesser extent, in the New York Bight east into southern New England. Small weakfish (under three years old) commonly are available in Delaware and Chesapeake Bays and in other southern areas. A northern migration in the spring brings older weakfish from the south to the lower Hudson and other northern waters (Perlmutter *et al.*, 1956).

Middle Atlantic fisheries accounted for most of the weakfish landings from 1880 through 1920. At that time, the Virginia catch of smaller weakfish increased, reducing the number of larger fish reaching northern areas and decreasing the Middle Atlantic catch. Another period of increased southern catches in the 1940's and 1950's also was followed by decreased northern catches (Perlmutter *et al.*, 1956).

Some of the long-term fluctuations are apparent in landings of selected years from Monmouth County, New Jersey (Table 3). These figures represent fish principally caught in pound nets of the Sandy Hook-Raritan Bay area but may include some nearshore ocean catches. Comparison of the data is useful to document general trends in the fishery. The high level of production in the late 1800's contrasts dramatically with levels in 1967 and 1968. An increase in landings in 1969 and 1970 corresponds with McHugh's (1977) inclusion of weakfish among species that have experienced recoveries in abundance since 1969.

Table 3. Weakfish landings for Monmouth County, New Jersey from selected years. Data from Sandy Hook Marine Laboratory (1971).

Year	Landings (metric tons)	Year	Landings (metric tons)
1897	2,494	1967	2.0
1898	3,302	1968	1.2
1901	507	1969	4.1
		1970	7.4

STRIPED BASS

The Hudson River appears to be an important spawning area for striped bass (*Morone saxatilis*) (Clark, 1968). The portion of the River used for spawning is located between Bear Mountain and Kingston, New York; the principal spawning area is located at West Point (Rathjen and Miller, 1957).

The striped bass population within the Hudson-Raritan estuary is complex because of an admixture of contingents with different migratory histories. Clark's study (1968) of seasonal movements of striped bass indicates that one contingent summers in Long Island Sound and winters in the Hudson River; a second remains within the Hudson-Raritan estuary throughout the year; and a third winters and summers elsewhere but ascends the Hudson to spawn along with the other contingents.

Greeley (1937) classified striped bass as common within the lower Hudson River. He reported that young fish existed in great quantities during summer, and larger fish were available in commercial quantities during winter and spring.

Striped bass have been able to maintain and possibly increase their population levels within the Hudson system (Boyle, 1969; McHugh, 1977). Human disturbances of the estuarine environment thus far apparently have not affected growth and reproduction of the striped bass. It has been hypothesized that some modification of spawning areas by increased loading of silt, waste, and fertilizers may enhance egg and larval survival (Mansueti, 1961). Commercial landings of striped bass in the Middle Atlantic region show an irregular upward trend since 1951 (Table 4).

A hazard that faces all Hudson River fishes but one that has been studied most heavily regarding striped bass is the withdrawal of River water for cooling the numerous present and planned Hudson River power plants. The impacts of this use include the release of antifouling agents into the River and the impingement of adult fishes on intake screens. However, the most significant potential effect is the entrainment through cooling systems of fish eggs and larvae. Extensive modeling efforts have been made to determine the extent of this effect on fish populations (*e.g.*, Wallace, 1975; Hall, 1977). Controversy and litigation still remain over this point. Models predict the effect of egg and larval entrainment on the size of returning populations of spawning adults. There appears to be agreement that entrainment will reduce this population; disagreement continues over the magnitude of the reduction. Still unknown is whether a small reduction in the number of spawning adults will cause reductions in the number of young that survive to produce subsequent year-classes.

BLUEFISH

The bluefish (*Pomatomus saltatrix*) is an important summer visitor to the Hudson-Raritan estuary, particularly for recreational fishermen. Following the spawning of adults offshore in the spring, juvenile bluefish move into the estuary for the summer. Greeley (1937) commented on the presence of a large population of small bluefish in the lower Hudson River during late summer. This species varies considerably in annual abundance. The greatest commercial catch for the New York Bight was 7,398 metric tons reported in 1897 (McHugh and Ginter, 1978); however, this was followed by a downward trend. Some recovery was experienced around 1930 and has again been experienced since 1969 (McHugh, 1977). Commercial statistics are particularly poor indices of landings, since recreational catches of this popular fish are known to be far greater than commercial catches (McHugh and Ginter, 1978).

SCUP

Scup (*Stenotomus chrysops*), which spawns in the Sandy Hook-Raritan-Lower Bays of the Hudson-Raritan estuary, is another species that has experienced extreme fluctuations in abundance from historical times to the present (Jensen, 1974). Commercial landings of scup in New York and New Jersey reached a peak in the 1950's and 1960's (Table 4). This was followed by a sharp decline to a low around 1971; however, the catch has increased slowly since that time. The

Table 4. Scup and striped bass landings for the Middle Atlantic region, 1951-1974 (metric tons).

Year	Scup	Striped Bass
1951	8,353	445
1952	8,590	517
1953	11,807	464
1954	11,988	288
1955	9,381	285
1956	7,707	215
1957	8,907	318
1958	10,549	217
1959	11,856	338
1960	12,080	395
1961	11,674	568
1962	12,052	571
1963	9,994	668
1964	7,662	917
1965	7,544	695
1966	3,816	648
1967	3,314	917
1968	2,821	934
1969	2,383	856
1970	1,964	732
1971	1,537	686
1972	2,253	661
1973	2,663	1,403
1974	4,387	1,013

Source: Annual Statistical Digests, Bureau of Commercial Fisheries, Department of Interior and National Marine Fisheries Service, Department of Commerce.

sensitivity of scup to environmental conditions may have been involved in the decline of this species (McHugh and Ginter, 1978).

FLOUNDER

Early statistics on flounders did not differentiate among the various species common in the Hudson-Raritan estuary. These include summer flounder (*Paralichthys dentatus*), windowpane (*Scophthalmus aquosus*), and winter flounder (*Pseudopleuronectes americanus*). Since these flounders have significantly different migratory habits, this lack of specificity precludes historical comparison of fishery data. Flounders are caught extensively by commercial and sport fishermen. Both commercial and recreational catches of summer flounder declined between 1965 and 1970 but have increased sharply since that time (McHugh and Ginter, 1978). Commercial catches of winter flounder experienced a decline in the 1950's, followed by a recovery in the early 1960's and a second decline since 1966. The decline since 1966 may be related to increased foreign catches (McHugh and Ginter, 1978).

DISCUSSION

Assessment of estuarine conditions responsible for changes in abundance of the fisheries of the Hudson-Raritan estuary is complicated by the migratory nature of most of its important finfish species. However, estuarine use patterns by the various species provide evidence of their presence in the estuary at critical times in their life cycles, *i.e.*, spawning and subsequent egg, larval, and juvenile development. The abundance of some species depends on spawning success either in the Hudson River (shad, sturgeon, striped bass, and alewife) or in the lower bays of the estuary (weakfish, scup, and winter flounder).

For the first group, fishing pressure within the Hudson may seriously deplete spawning stock of some species on their way to spawning grounds. Pollution and physical disturbances have acted to reduce the spawning area available. However, much of the lost areas are being reclaimed by pollution control (Medeiros, 1975). Eggs and larvae are exposed to a water system that is intensively used for the disposal of human and industrial wastes. Some species (sturgeon and striped bass) appear to be more susceptible to the water quality problems of the estuary as a result of their extended residence times as juveniles. However, the effect on stock abundance is not uniform. Striped bass populations are doing well despite river pollution. Sturgeon populations have declined, but this seems to be mostly the result of overfishing. Shad populations also have declined as a result of overfishing. Some pollution may affect the fishery resources in other ways. Oil pollution, for example, may be the cause of the taste of oil reported for some fishes from the Hudson-Raritan estuary; however, factors other than oil pollution may cause these flavors (Stansby, 1978).

Complex ecological, health, and socioeconomic problems have arisen from the contamination of the Hudson-Raritan estuary by pollutants such as polychlorinated biphenyls (PCB's), pesticides, and heavy metals. It is difficult to estimate the relative impacts of these pollutants on the survival of fishes. Of particular concern is contamination by PCB's. These compounds are present at such levels in the tissues of fishes in the River that the New York Department of Environmental Conservation imposed a commercial fishing ban in 1975. Shad and sturgeon are excluded from the ban since adults of these species have a relatively short exposure time in the River and do not become heavily contaminated. PCB's, however, remain in sediments and are not readily broken down chemically. Unless some practical means of removing the contaminated sediments is devised, resident fishes may continue to be contaminated. Spawning and nursery areas affected by the contamination may result in decreased fish spawning and survival rates as well as increased occurrence of mutations and abnormalities.

Despite waste and misuse, the waters and fish stocks of the Hudson estuary remain potential providers of income and recreation to thousands within their range. It is up to the concerned interests–federal, state, local, and private–to protect this resource through imaginative, vigorous, and continuing management.

ACKNOWLEDGEMENTS

Funding for this study was provided to the New Jersey Marine Sciences Consortium by the MESA New York Bight Project of the National Oceanic and Atmospheric Administration, U.S. Department of Commerce. I wish to thank the New York Bight Project staff for their support and assistance during the course of the project. Funding originally was provided to the Consortium with Dr. Lionel Walford as the principal investigator. The present author took over the project following Dr. Walford's illness and death. The author has attempted to follow the expert guidance provided by Dr. Walford during the early stages of the project.

I would like to thank Mr. Anthony Pacheco and Dr. Robert Abel for their editorial assistance and suggestions. Assistance by Gene Hession, Sharon Gervolino, and Katherine Manger during the course of the literature search also is acknowledged gratefully.

REFERENCES

Boyle, R.H. 1969. The Hudson River–A Natural and Unnatural History. W.W. Norton and Company, New York. 304 pp.

Brown, J.J. 1857. The American Angler's Guide. D. Appleton and Company, New York. 250 pp.

Brydon, J.R. 1974. The Passaic River. Rutgers University Press, New Brunswick, NJ. 376 pp.

Burdick, G.E. 1954. An analysis of the factors, including pollution, having possible influence on the abundance of shad in the Hudson River. N.Y. Fish Game J. 1: 188-205.

Clark, J. 1968. Seasonal movements of striped bass contingent of Long Island Sound and the New York Bight. Trans. Am. Fish. Soc. 97: 320-343.

Commission of Fisheries of New Jersey. 1872. Second Annual Report of the

Commissioners of Fisheries of the State of New Jersey. State of New Jersey, Trenton, NJ. 22 pp.

Commission of Fisheries of New Jersey. 1878. Report of the Commissioners of Fisheries of the State of New Jersey for the Year 1878. State of New Jersey, Trenton, NJ. 33 pp.

Commission of Fisheries of New Jersey. 1885. Report of the Commissioners of Fisheries of the State of New Jersey for 1884 and 1885. State of New Jersey, Trenton, NJ. 45 pp.

Commission of Fisheries of New York. 1890. Eighteenth Report of the Commission of Fisheries. State of New York. Albany, NY. 151 pp.

Dovel, W.L. 1977. An Atlas of the Biologic Resources of the Hudson Estuary. Boyce Thompson Institute for Plant Research, Inc., Yonkers, NY. 104 pp.

Fish and Game Commission of New Jersey. 1896. Annual Report of the Board of Fish and Game Commissioners of the State of New Jersey for the Year Ending November 1, 1896. State of New Jersey, Trenton, NJ. 54 pp.

Fish and Game Commission of New Jersey. 1897. Annual Report of the Board of Fish and Game Commissioners of the State of New Jersey for the Year Ending October 31, 1897. State of New Jersey, Trenton, NJ. 57 pp.

Goode, G.B. 1887. The Fisheries and Fishery Industries of the United States. U.S. Government Printing Office, Washington, DC. 787 pp.

Greeley, J.R. 1937. A biological survey of the lower Hudson watershed. Fishes of the area with annotated list. Supplement to Twenty-sixth State of New York Conservation Department Annual Report, 1936. State of New York, Albany, NY. pp. 45-103.

Hall, C.A.S. 1977. Models and the decision making process: the Hudson River power plant case. *In:* Ecosystem Modeling in Theory and Practice, C.A.S. Hall and J.W. Day (eds.), John Wiley and Sons, New York. pp. 345-364.

Holmes, H. 1890. Reminiscenses of 75 Years of Belleville, Franklin and Newark. [n.p., no pub.] 96 pp. [Historical document available at Newark Public Library, New Jersey Section, Newark, NJ.]

Jensen, A.C. 1974. New York's fisheries for scup, summer flounder and black sea bass. N.Y. Fish Game J. 21: 126-134.

Juet, R. 1959. Juet's Journal: The Voyage of the HALF MOON from 4 April to 7 November 1609. New Jersey Historical Society, Newark. NJ. 37 pp.

Klauda, R.J., M. Nittel, and K.P. Campbell. 1977. The commercial fishery for American shad in the Hudson River: fishing effort and stock abundance trends. *In:* Proceedings of a Workshop on American Shad, U.S. Government Printing Office, Washington, DC. pp. 107-134.

McHugh, J.L. 1977. Limiting factors affecting commercial fisheries in the Middle Atlantic estuarine areas. *In:* Estuarine Pollution Control and Assessment. Proceedings of a Conference, Vol. 1, U.S. Environmental Protection Agency, Washington, DC. pp. 149-169.

McHugh, J.L. and J.J.C. Ginter. 1978. Fisheries. MESA New York Bight Atlas Monograph 16. New York Sea Grant Institute, Albany, NY. 129 pp.

Mansueti, R.J. 1961. Effects of civilization on striped bass and other estuarine biota in Chesapeake Bay and tributaries. Proc. Gulf Caribb. Fish. Inst. 14: 110-136.

Mearns, E.A. 1898. A study of the vertebrate fauna of the Hudson Highlands, with observations on the mollusca, crustacea, lepidoptera, and the flora of the region. Bull. Am. Mus. Nat. Hist. 10: 311-322.

Medeiros, W.H. 1975. The Hudson River shad fishery: background, management problems and recommendations. New York Sea Grant Institute, Albany, NY. 54 pp.

Muncy, R.J. 1960. A study of the comparative efficiency between nylon and linen gill nets. Chesapeake Sci. 1: 96-109.

Murawski, S.A. and A.L. Pacheco. 1977. Biological and fisheries data on Atlantic sturgeon, *Acipenser oxyrhynchus* (Mitchill). NOAA, NMFS, NEFC, Sandy Hook Laboratory Tech. Rep. 10. 69 pp.

Perlmutter, A., R.F. Heller, and H. Herma, Jr. 1972. Fluctuations of fish populations. *In*: Hudson River Ecology, G.P. Howells and G.L. Lauer (eds.), New York Department of Encironmental Conservation, Albany, NY. pp. 357-389.

Perlmutter, A., N.S. Miller, and J.C. Poole. 1956. The weakfish (*Cynoscion regalis*) in New York waters. N.Y. Fish Game J. 4: 1-43.

Rathjen, W.F. and L.C. Miller. 1957. Aspects of the early life history of the striped bass in the Hudson River. N.Y. Fish Game J. 4: 43-60.

Sandy Hook Marine Laboratory. 1971. Review of aquatic resources and hydrographic characteristics of Raritan, Lower New York, and Sandy Hook Bays. Report for the Battelle Institute. 86 pp.

Stansby, M.E. 1978. Flavors in fish from petroleum pickup. Mar. Fish. Rev. 40: 13-17.

Talbot, G.B. 1954. Factors associated with fluctuations in abundance of Hudson River shad. Fish Wildl. Ser. Fish. Bull. 56: 373-413.

U.S. Department of Commerce. 1905. Fishery Industries of the United States, 1904. U.S. Government Printing Office, Washington, DC. 129 pp.

U.S. Department of Commerce. 1922. Fishery Industries of the United States, 1921. U.S. Government Printing Office, Washington, DC. 187 pp.

Wallace, D.N. 1975. A critical analysis of the biological assumptions of Hudson River striped bass models and field survey data. Trans. Am. Fish. Soc. 75: 710-717.

EFFECTS OF PETROLEUM HYDROCARBONS ON SELECTED DEMERSAL FISHES AND CRUSTACEANS

Bruce B. McCain
Donald C. Malins

Environmental Conservation Division
National Marine Fisheries Service
Northwest and Alaska Fisheries Center
2725 Montlake Boulevard East
Seattle, Washington 98112

Abstract. Investigations dealing with the effects of petroleum hydrocarbons on marine fishes and crustaceans are described. In one study conducted in our laboratory, three species of flatfishes (of the family Pleuronectidae) were exposed to crude oil-contaminated sediments for periods ranging from one to four months. English sole (*Parophrys vetulus*) was the only species to undergo significant adverse effects, including weight loss, liver structure aberrations, and hematological changes. In another study conducted in our laboratory, the feeding behavior of spot shrimp (*Pandalus platyceros*) exposed to the seawater-soluble fraction of crude oil was monitored. Concentrations of as low as 100 ppb caused significant disruption of shrimp feeding behavior. Investigations by other researchers pertinent to possible adverse effects of hydrocarbons on species of fishes and crustaceans similar to those found in the New York Bight are also reported.

INTRODUCTION

The biological effects of discharging urban and industrial wastes into marine and estuarine environments are major concerns of our research unit. Using an interdisciplinary research approach, we have examined marine fishes and macroinvertebrates exposed to environmental contaminants under controlled laboratory conditions. In addition to evaluating biological effects (such as changes in behavior and size/weight, or the assessment of gross and/or microscopic pathological conditions), the exposure levels and tissue uptake and distribution of the pollutants were measured. The research described in this report provides two examples of experimental results dealing with the biological effects of aromatic hydrocarbons that are most applicable to problems in the New York Bight. In one study, flatfishes (of the family Pleuronectidae) were exposed to crude oil-contaminated sediment for periods of one to four months. The other study involved changes in the feeding behavior of shrimp exposed to seawater-soluble fractions (SWSF) of crude oil.

Although increased interest during the last few years in the effects of aromatic hydrocarbons on aquatic animals has resulted in a number of published reports, most studies have dealt with acute and subacute effects. Recent reviews of this literature have been presented by Craddock (1977), Hodgins *et al.* (1977), Johnson (1977), and Neff (1979). Pertinent examples of some of these studies will be presented below.

Exposure of pandalid shrimp (*Pandalus danae*) to SWSF of crude oil (containing 1-2 ppm hydrocarbons) for 96 hours has been reported to cause heightened random activity that sometimes was reversible (Swedmark *et al.*,

1973). Rice *et al.* (1979) have recently reported 96-hour median tolerance limits (TLm's) for several arctic and subarctic fish and crustacean species from the Pacific Ocean exposed to crude oil in static bioassay systems. Pacific herring (*Clupea pallasi*) and walleye pollock (*Theragra chalcogramma*) had 96-hour TLm's of 1.22 ppm and 1.73 ppm total aromatic hydrocarbons, respectively. The 96-hour TLm's of total aromatic hydrocarbons for two shrimp species were 0.87 ppm for *Crangon alaskensis* and 1.79 ppm for *P. goniurus.* Adults of another pandalid species (*P. danae*) exposed to concentrations of total hydrocarbons in 1.8 ppm of Prudhoe Bay crude oil (PBCO) had 50% mortality after an exposure time of 1.7 days (Anderson, 1979). Sanborn and Malins (1977) found that 100% of spot shrimp larvae were killed by exposure for 24-36 hours to 8-12 ppb of naphthalene or naphthalene bound to bovine serum albumin (BSA) in flow-through aquaria. Less than 1% of the control group exposed to the same concentration of BSA died during the same period.

The effects of long-term exposures of marine animals to aromatic hydrocarbons are not as well documented. Examples of recent research applicable to the New York Bight include two studies in which flatfishes were exposed to sediments contaminated with xenobiotics. In one study, Sherwood (1976) observed early signs of fin erosion in Dover sole (*Microstomus pacificus*) exposed for 13 months to marine sediment from a highly polluted area near Los Angeles. In the other study (McCain *et al.*, 1978), a portion of which will be reported here, English sole (*Parophrys vetulus*) were exposed to sediment-associated PBCO for four months. Compared to control sole residing on clean sediment, the oil-exposed sole took up substantial amounts of petroleum hydrocarbons, developed liver abnormalities, lost weight, and had a higher mortality rate. Although the species used in these studies are not present in the New York Bight, representatives of the same families are found in or near the Bight area (Smith, 1974).

The concentrations of selected polynuclear aromatic hydrocarbons (PAH) in sediment and tissues of animals from Raritan Bay also are presented. These data are the results of chemical analyses recently performed by the National Oceanic and Atmospheric Administration National Analytical Facility (NAF) for the MESA New York Bight Project (MacLeod and Ramos, personal communication).

METHODS

Flatfishes

Flatfishes in one tank supplied with flowing seawater were exposed to sediment contaminated with PBCO; controls were held in a second tank containing the same type of sediment without added petroleum. These experiments involved two basic types of sediment (sandy and silty), three initial concentrations of PBCO (0.2, 0.5, and 1.0%, v/v), three species of 1- to 2-year old flatfishes (English sole [*Parophrys vetulus*], starry flounder [*Platichthys stellatus*], and rock sole [*Lepidopsetta bilineata*]), and exposure times ranging from three weeks to four months. The silt content of the sandy and silty sediment types was 1.0% and 47.4%, respectively (Table 1). The sediment-seawater mixtures with and without PBCO, after blending in a cement mixer, were placed in each aquarium to a depth of 5 cm (see McCain *et al.*, 1978 for a diagram and further details). The sediments were rinsed overnight by allowing seawater to flow through the tanks at a rate of 10-12 liters/minute; then the experimental and control fishes were added. The water temperature averaged 8.5°C, and the salinity average was 27 ⁰/oo. The fishes were fed to satiation with a combination of frozen euphausids and live earthworms, and the aquaria were examined daily for dead or moribund fishes.

At intervals of two weeks to one month, all of the fishes were examined for externally visible abnormalities and measured for weight and length. At the same time, three to six fishes from each group were sacrificed, and tissue samples were removed for histological, hematological, and chemical analyses. For histological examination of each fish, pieces of gill, gonad, fin, skin, gastrointestinal tract, kidney, liver, and eye lens were removed and preserved in appropriate fixatives. Specimens to be examined by light microscopy were embedded in paraffin, sectioned, and stained by a variety of histochemical stains and methods. Hematological tests were performed for hemoglobin, hematocrit, total red blood cells, leukocytes, and differential leukocytes. Concentrations of serum albumin and total proteins were measured by standard spectrophotometric methods. Samples of muscle, skin, and liver of the fishes were collected for analyses of aromatic hydrocarbons. Muscle samples from each fish were analyzed individually, and skin and liver samples from the three fish in each group were pooled. Tissue samples, frozen at -20°C immediately after they were taken, were thawed immediately before chemical analysis.

Samples of sediment, interstitial water, and water immediately above the sediment also were collected at each sampling interval for chemical analyses. Sediment samples were taken from the top 2 cm of sediment with a coring device. Interstitial water samples were obtained by inserting a 2-cm diameter glass cylinder, with a mesh screen attached to the bottom, into the top 3 cm of the sediment and collecting the water that accumulated in the cylinder with a syringe. The above-sediment water sample was collected by placing a screw cap bottle on its side on top of the sediment, releasing the cap, and allowing the bottle to fill.

The main test conditions for the five experiments are given in Table 1.

Shrimp

Observations on the feeding reponse of adult spot shrimp (*Pandalus platyceros*) were made during six-day exposures to the SWSF of PBCO (Miller, 1979). Seawater was serially filtered through 5- and 1-μm polypropylene filter bags into a 1,000-liter fiberglass head box that supplied both the test chambers and an oil solubilizer system (Roubal *et al.*, 1977), the effluent of which could be routed to the chambers as desired under continuous flow conditions. The composition of this mixture is described elsewhere (Roubal *et al.*, 1977). Samples

Table 1. Conditions of experiments concerned with pathological changes in flatfishes from exposure to oil-contaminated sediment.

Experiment number	Initial amount of PBCO (% [v/v])	Fish species	Initial number of animals per group	Sediment type	Duration experiment (days)	Sampling times (days)
1	0.2	English sole	35	High sand	148	0, 2, 7, 16, 30, 51, 148
2	0.5	Rock and English sole	41 41	High silt	29	0, 14, 29
3	0.5	Rock and English sole	20 30	High sand	21	0, 14, 21
4	0.5	Starry flounder	50	High sand	62	0, 13, 28, 62
5	1.0	English sole	44	High silt	90	0, 28, 61, 90

of the water supplying the chambers were collected every two days and subjected to chemical analyses. To obtain different levels of exposure, the SWSF from the oil solubilizer (250 ppb of total extractable petroleum hydrocarbons [TEPH]) were diluted in the mixing box of each shrimp chamber, while a total flow of 300 ml/minute was maintained.

Each test consisted of three, 3-minute observations of feeding behavior. A positive feeding response was scored when shrimp contacted the stimulus source outlet with their chelipeds. The shrimp chambers were shielded with black plastic, and observations on feeding behavior were made through one-way mirrors. The stimulus source outlet was located at the upstream end of the chamber, and the stimulus, consisting of squid extract (Mackie, 1973) or seawater control, was released at a flow rate of 10 ml/minute. Each test included observations of background feeding behavior under control conditions (seawater was released from the stimulus source outlet). A total of 33 shrimp were used for the experiment, with an average of four used per each concentration of the SWSF.

Chemical Analyses

Extraction of water samples for analysis followed standard procedures (National Analytical Facility, 1978). Hydrocarbon analyses of sediment, water, and tissue were performed by the NAF using a modification of procedures described by MacLeod *et al.* (1977) and Brown *et al.* (1979) involving gravimetric determination of total extractable hydrocarbons and gas-liquid chromatography (GLC) for alkanes and arenes. By subtracting the concentrations of total extractable hydrocarbons in the non-oil-contaminated sediment, the value of TEPH in the oil-contaminated sediment was determined. The non-oil-contaminated sediment contained no detectable aromatic hydrocarbons.

RESULTS AND DISCUSSION

Flatfishes

Of the three species of flatfishes exposed to PBCO-contaminated sediments (Table 1), adult rock sole (experiments 2 and 3) and starry flounder (experiment 4) exhibited no detectable petroleum-associated adverse effects. However, in adult English sole (experiment 1) histological, hematological, and physiological changes were observed. The effects of exposure to high-sand sediment contaminated with 0.2% (v/v) PBCO for four months included the following apparently oil-related changes: (1) During the first month, when sediment and tissue levels of aromatic hydrocarbons were highest, 50% of the exposed fishes and none of the controls had severe hepatocellular lipid vacuolization (HLV), with greater than approximately two thirds of the cytoplasm of hepatocytes occupied by lipid vacuoles. (2) The number of fishes weighing less than their initial weight was higher in the experimental (12 of 17 fishes) than in the control (5 of 16 fishes) groups ($P = 0.05$) after four months of exposure. (3) Although no control fishes died (except for those purposely sacrificed) and all control fishes looked normal during the four-month experiment, 18% (3 of 17 fishes) of the oil-exposed fishes died or were moribund and appeared extremely emaciated (McCain *et al.*, 1978).

Two other experiments were performed in which English sole and rock sole were exposed to silty sediment containing 0.5% and 1.0% PBCO, corresponding to experiments 2 and 5, respectively (see Table 1). The levels of TEPH during experiment 2 are given in Figure 1; the aromatic hydrocarbon composition of the sediment and the levels of these compounds taken up by the fishes are shown in Figure 2. After 29 days of exposure, severe HLV was observed in 83% of the oil-exposed English sole, but, in contrast to the previous studies (McCain *et al.*, 1978), it was also found in 40% of the control English sole. Some statistically significant hematological changes also were observed in the oil-exposed fishes. At 14-days exposure, hematocrit and hemoglobin values and numbers of white blood cells (15.7%, 3.4 g/dl, and 8.3 WBC/mm^3 x 10^4, respectively) were all significantly lower ($P < 0.025$) in oil-exposed English sole than in controls (21.0%, 4.8 g/dl, and 22.1 WBC/mm^3 x 10^4, respectively). However, by one month, the hematological values for the control group also had become lower and were not significantly different from the oil-exposed group. These data suggest that exposure of English sole to PBCO-contaminated sediments accelerated the development of hematological changes that also were observed in control fishes approximately two weeks later.

In experiment 5 (see Table 1) the initial sediment TEPH concentration was 4,088 μg/g (dry wt), and at one month the concentration was 3,806 μg/g, a decrease of only 7%. The most abundant aromatic hydrocarbons were similar to those shown in Figure 2, with 1- and 2-methylnaphthalenes present at 7.2 μg and 14.8 μg, respectively. The oil-exposed group had an 83% frequency of severe

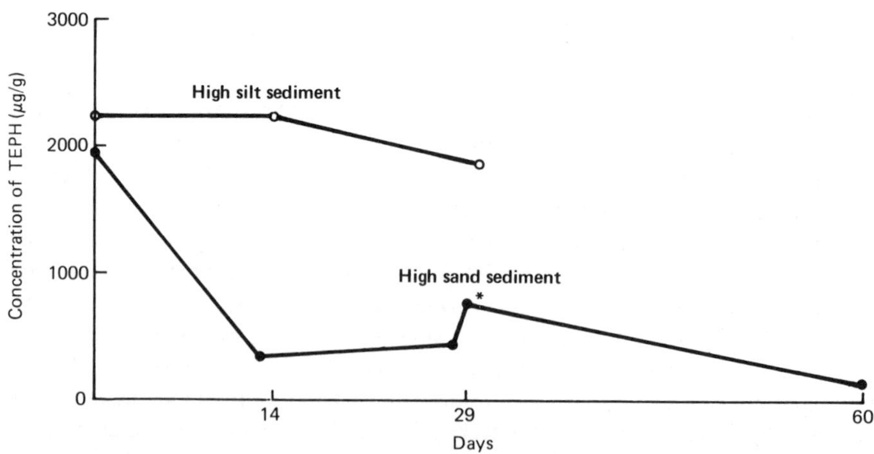

Figure 1. Concentrations of TEPH in PBCO-contaminated sediments from experiment 2 (silty sediment) and experiment 4 (sandy sediment). Both experiments had initial PBCO concentrations of 0.5% (v/v). *Indicates remixing of sediment.

HLV; however, 50% of the controls also had severe HLV. In contrast to experiment 2, English sole after one month exposure to oiled sediments in experiment 5 were found to have the following hematological differences that

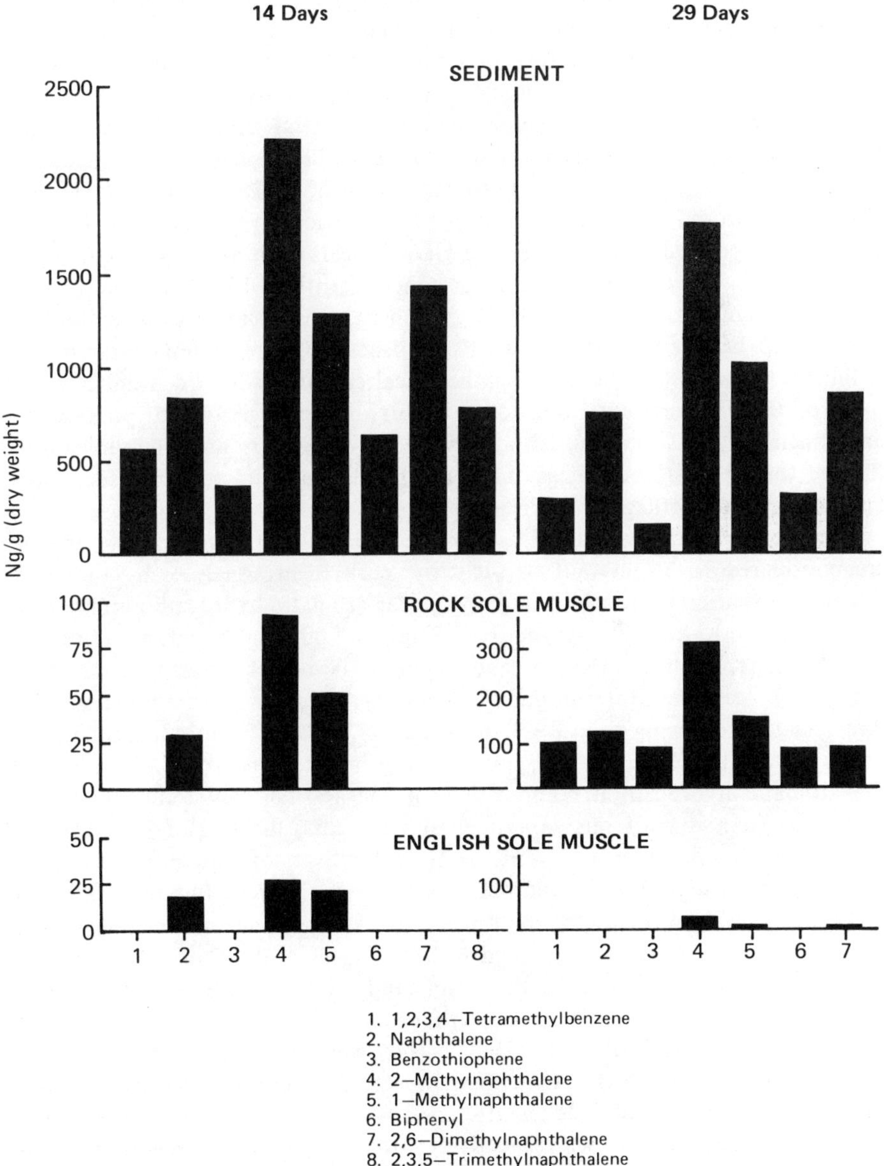

Figure 2. Concentrations of major aromatic hydrocarbons in PBCO-contaminated sediments and in the English sole and rock sole associated with the sediments at 14 and 29 days. Tissue levels are an average of two replicate analyses of 10 g muscle samples from two fish. Control fish tissue did not have detectable levels of PAH. (Experiment 2).

varied significantly from those of controls: (1) hematocrit, 18.7% for oil-exposed and 21.2% for control ($P < 0.001$); (2) percent lymphocytes in blood smears, 71.1% for oil-exposed and 78.5% for controls ($P < 0.025$); and (3) hemoglobin, 4.08 g/dl for oil-exposed and 4.4 g/dl for controls ($P < 0.05$). No significant differences were found between the lengths or body weights of oil-exposed and control fishes in experiments 2 and 5.

It is clear from these studies that severe HLV in the species of flatfishes examined is not a specific indicator of long-term exposure to petroleum hydrocarbons. In fishes, HLV appears to be a general stress response and has been attributed to a variety of causes, including nutritional deficiencies (Snieszko, 1972), exposure to pesticides (Couch, 1975), and contact with polluted environments (Pierce *et al.*, 1978). Also, under the experimental conditions described above, the severe HLV appears to be a reversible condition since the degree of HLV decreased concomitantly with the reduction of petroleum hydrocarbon levels in the sediment. Therefore, a cause-and-effect relationship between contact of these flatfishes with petroleum-contaminated sediments and the occurrence of pathological conditions has been indicated in some of these experiments, but not in others. Interestingly, recent preliminary experimental results from our laboratory (D.D. Weber, personal communication) showed that at least one species of flatfish, English sole, did not avoid sediments containing up to 8,500 ppm TEPH.

These experiments with flatfishes exposed to PBCO-contaminated sediments are considered to be relevant to the New York Bight. Analyses by the NAF established that concentrations of polynuclear aromatic hydrocarbons (PAH) in sediment from Raritan Bay ranged from 5 ng/g to 500 ng/g dry wt, with fluoranthene, pyrene, and benz(a)anthracene found at the higher concentrations. Also, liver tissue from five winter flounder (*Pseudopleuronectes americanus*) from this Bay had the following four PAH: naphthalene, 1- and 2-methylnaphthalenes, and biphenyl at 500, 1,000, 700, and 200 ng/g dry wt, respectively. Even though the aromatic hydrocarbon composition of the Raritan Bay sediment differs somewhat from that of our experimental sediments, the PAH detected in the livers of winter flounder also were found in the livers of experimental English sole (McCain *et al.*, 1978) at approximately the same levels. In winter flounder livers, these levels also were of the same order of magnitude as was reported in the livers of starry flounder exposed to 0.9 ± 0.1 ppm SWSF of PBCO for one week (Roubal *et al.*, 1978). In the latter study, juvenile fish were exposed to SWSF using the above-mentioned crude oil solubilizer and flow-through aquarium system (Roubal *et al.*, 1977). They were then placed in clean water and allowed to depurate for up to two weeks. Concentrations of naphthalene and 1- and 2-methylnaphthalenes after one week of oil exposure were 1,500, 2,400, and 3,000 ng/g dry wt, respectively; after two weeks of depuration, these compounds were not detected in livers. In these experiments only parent PAH were analyzed.

It is important to note that flatfishes extensively metabolize PAH (Varanasi *et al.*, 1979). The analyses performed in the present study on livers of oil-exposed flatfishes are limited to the parent PAH and do not reveal the degree or type of metabolites accumulated in this organ.

Shrimp

The results of experiments on the feeding response of spot shrimp indicated that exposure to the SWSF of PBCO causes a decrease in feeding activities, particularly those involving searching. Figure 3 shows the percentage of shrimp that contacted and began feeding on the stimulus source outlet in the upstream end of the test chamber as a function of SWSF concentration. These results represent a composite of daily observations taken over a six-day exposure period at each concentration of SWSF. Upon addition of squid extract stimulus, greater than 70% of the shrimp exhibited feeding behavior when the SWSF were not present; there was no feeding activity at a continuous exposure to a concentration of 287 ppb SWSF. The concentration at which the feeding response of adult spot shrimp was reduced by 50% (EC_{50}) was between approximately 20 ppb and 75 ppb SWSF of PBCO.

Tissue levels of aromatic hydrocarbons in pandalid shrimp in the New York Bight area have not been reported. However, another crustacean, the American lobster (*Homarus americanus*), was analyzed by the NAF; several high molecular weight PAH were detected in digestive glands. Examples of some of the PAH at highest concentrations (ng/g dry wt) were pyrene (1,500), benz(a)anthracene

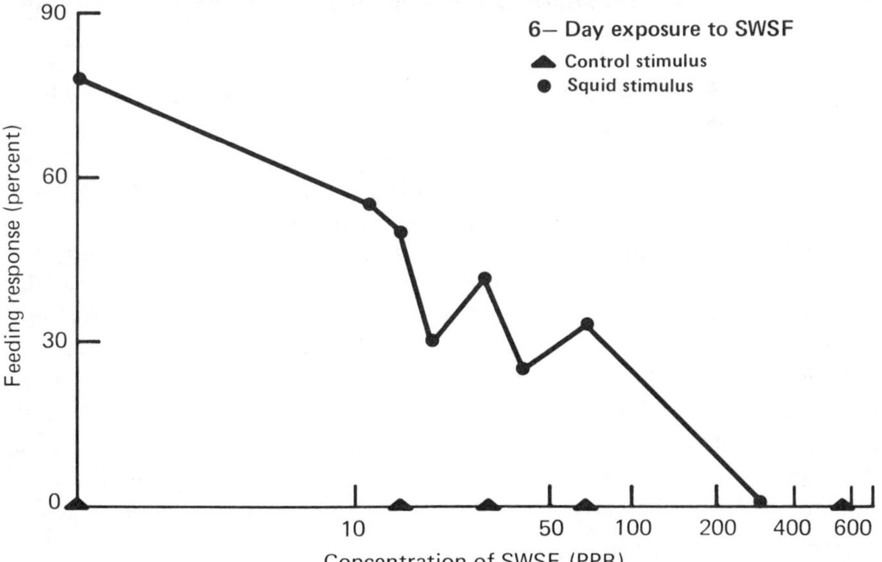

Figure 3. The percentage of spot shrimp that contacted and began feeding on the stimulus source outlet. Each data point is the average of three, 3-minute observations of background activity, response to seawater control, and/or response to a 1:10 dilution of squid extract. A total of six control and 27 oil-exposed shrimp were tested. SWSF refers to the seawater-soluble fraction of Prudhoe Bay crude oil.

(600), chrysene (800), and benzo(a)pyrene (300). These data suggest that crustaceans are being exposed to high levels of polynuclear aromatic hydrocarbons, either from contact with contaminated water or sediment, or through their diet.

Although not directly applicable to the crustaceans and flatfishes in New York Bight, the results presented here demonstrate examples of laboratory-induced biological effects that could occur in marine environments heavily contaminated with aromatic hydrocarbons. Disruption of chemically-mediated feeding behavior in crustaceans by seawater-soluble aromatic hydrocarbons, which was reported here in spot shrimp at levels of 100 ppb or less, would have a serious impact on the health and survival of affected animals. Although concentrations of aromatic hydrocarbons in seawater as high as 100 ppb are not commonly measured, concentrations of 1,000-3,000 ppb of aromatic hydrocarbons have been reported for the surface waters of the New York Bight (Brown *et al.,* 1973). In addition, our observation that adult English sole developed pathological conditions (*i.e.,* weight loss, liver structure aberrations, and hematological changes) in some experiments, when held for several weeks with sediments contaminated with about 500 ppm dry wt of TEPH, offers more examples of abnormalities that might occur in other species of feral flatfishes associated with sediments contaminated with similar or higher levels of aromatic hydrocarbons. In fact, since at least one species of flatfish (English sole) will not avoid sediment contaminated with almost twenty times higher amounts of TEPH (8,500 ppm), even more severe abnormalities might be expected in highly impacted areas.

In view of these data and the high levels of PAH detected by the NAF in the sediment, winter flounder, and lobsters from the New York Bight, there is little doubt that many species of marine animals in this area are residing in environments contaminated to the extent that their survival may be threatened. The high frequency (7.6% in 1973 and 1974) of fin erosion in winter flounder (Ziskowski and Murchelano, 1975) from Raritan Bay also supports the supposition that this species is being exposed to toxic chemicals. Although the cause of this disease is not known, microbial agents have been essentially eliminated as the primary cause (Murchelano and Ziskowski, this volume), leaving one or more toxic chemicals as probable etiological agents (Mearns and Sherwood, 1974; Wellings *et al.,* 1976). Other species of fishes and invertebrates, particularly the early life stages, also may be at risk from exposure to toxic levels of contaminants in the New York Bight. Since "as much as 80% of the traffic on the Bight is petroleum-related" (Purrett-Carroll, 1979), petroleum-associated aromatic hydrocarbons are, and will continue to be, likely causes of degraded ecosystems in this area.

ACKNOWLEDGEMENTS

Special thanks go to D.D. Weber and S.E. Miller for their research data used in this report, and to H.O. Hodgins, U. Varanasi, and F.G. Johnson for their reviews of this manuscript.

This study was supported by the Bureau of Land Management through interagency agreement with the National Oceanic and Atmospheric Administration, under which a multi-year program responding to needs of petroleum development of the Alaskan Continental Shelf is managed by the Outer Continental Shelf Environmental Assessment Program Office.

REFERENCES

Anderson, J.W. 1979. Long-term effects of hydrocarbons on selected ecosystems and associated organisms. *In:* Pacific Northwest Laboratory Annual Report for 1978 to the DOE Assistant Secretary for Environment, Part 2, Ecological Sciences, under Contract EY-76-C-06-1830, Richland, WA. pp. 4.5-4.12.

Brown, D.W., L.S. Ramos, A.J. Friedman, and W.D. MacLeod, Jr. 1979. Analysis of trace levels of petroleum hydrocarbons in marine sediments using a solvent/slurry extraction procedure. *In:* Trace Organic Analysis: a New Frontier in Analytical Chemistry, S.N. Chesler and H.S. Hertz (eds.), Natl. Bur. Stand. Spec. Publ. 519. pp. 161-168.

Brown, R.A., T.D. Searl, J.J. Elliott, B.G. Phillips, D.E. Brandon, and P.H. Monaghan. 1973. Distribution of heavy hydrocarbons in some Atlantic Ocean waters. *In:* Proceedings of 1973 Joint Conference on Prevention and Control of Oil Spills, American Petroleum Institute, Washington, DC. pp. 505-519.

Couch, J.A. 1975. Histopathological effects of pesticides and related chemicals on the livers of fishes. *In:* The Pathology of Fishes, W.E. Ribelin and G. Magaki (eds.), University of Wisconsin Press, Madison, WI pp. 559-584.

Craddock, D.R. 1977. Acute toxic effects of petroleum on arctic and subarctic marine organisms. *In:* Effects of Petroleum on Arctic and Subarctic Marine Environments and Organisms, Vol. II, D.C. Malins (ed.), Academic Press, New York, pp. 1-93.

Hodgins, H.O., B.B. McCain, and J.W. Hawkes. 1977. Marine fish and invertebrate diseases, host disease resistance, and pathological effects of petroleum. *In:* Effects of Petroleum on Arctic and Subarctic Marine Environments and Organisms, Vol. II, D.C. Malins (ed.), Academic Press, New York. pp. 95-173.

Johnson, F.G. 1977. Sublethal biological effects of petroleum hydrocarbon exposures: bacteria, algae, and invertebrates. *In:* Effects of Petroleum on Arctic and Subarctic Marine Environments and Organisms, Vol. II, D.C. Malins (ed.), Academic Press, New York. pp. 271-318.

McCain, B.B., H.O. Hodgins, W.D. Gronlund, J.W. Hawkes, D.W. Brown, M.S. Myers, and J.H. Vandermeulen. 1978. Bioavailability of crude oil from experimentally oiled sediments to English sole (*Parophrys vetulus*), and pathological consequences. J. Fish. Res. Board Can. 35:657-664.

Mackie, A.M. 1973. The chemical basis of food detection in the lobster *Homarus gammarus*. Mar. Biol. 21: 103-108.

MacLeod, W.D., Jr., D.W. Brown, R.G. Jenkins, L.S. Ramos, and V.D. Henry. 1977. Petroleum hydrocarbons in the northern Puget Sound area—a pilot design study. NOAA Tech. Memo. ERL MESA-8. 53 pp.

Mearns, A.J. and M. Sherwood. 1974. Environmental aspects of fin erosion and tumors in Southern California Dover sole. Trans. Am. Fish. Soc. 103: 799-810.

Miller, S.C. 1979. Feeding behavior of marine decapod crustaceans, and observations of behavioral responses of spot shrimp, *Pandalus platyceros* (Brandt), to food stimulants and petroleum hydrocarbons. M.S. Thesis, University of Washington. 157 pp.

Murchelano, R.A. and J. Ziskowski. This volume. Fin rot disease in the New York Bight (1973-1977). pp. 347-358.

National Analytical Facility. 1978. Laboratory Manual. Processed Report. NOAA, NMFS, Northwest and Alaska Fisheries Center, Seattle, WA. 21 pp.

Neff, J.M. 1979. Polycyclic Aromatic Hydrocarbons in the Aquatic Environment: Sources, Fates and Biological Effects. Applied Science Publishers Ltd., London. 262 pp.

Pierce, K.V., B.B. McCain, and S.R. Wellings. 1978. Pathology of hepatomas and other liver abnormalities in English sole (*Parophrys vetulus*) from the estuary of the Duwamish River, Seattle, Washington. J. Natl. Cancer Inst. 60(6): 1443-1453.

Purrett-Carroll, L. 1979. New light on the New York Bight. NOAA 9: 44-47.

Rice, S.D., A. Moles, T.L. Taylor, and J.F. Karinen. 1979. Sensitivity of 39 Alaskan marine species to Cook Inlet crude oil and No. 2 fuel oil. *In:* Proceedings of 1979 Oil Spill Conference (Prevention, Behavior, Control, Cleanup), American Petroleum Institute, U.S. Environmental Protection Agency, and U.S. Coast Guard, Washington, DC. pp. 549-554.

Roubal, W.T., D.H. Dovee, T.K. Collier, and S.I. Stranahan. 1977. Flow-through system for chronic exposure of aquatic organisms to seawater-soluble hydrocarbons from crude oil: construction and applications. *In:* Proceedings of 1977 Oil Spill Conference (Prevention, Behavior, Control, Cleanup), American Petroleum Institute, Washington, DC. pp. 551-555.

Roubal, W.T., S.I. Stranahan, and D.C. Malins. 1978. The accumulation of low molecular weight aromatic hydrocarbons of crude oil by coho salmon (*Oncorhynchus kisutch*) and starry flounder (*Platichthys stellatus*). Arch. Environ. Contam. Toxicol. 7: 237-249.

Sanborn, H.R. and D.C. Malins. 1977. Toxicity and metabolism of naphthalene: a study with marine larval invertebrates. Proc. Soc. Exp. Biol. Med. 154: 151-155.

Sherwood, M. 1976. Fin erosion disease induced in the laboratory. *In:* Annual Report of the Southern California Coastal Water Research Project, Southern California Coastal Water Research Project, El Segundo, CA. pp. 143-148.

Smith, R.I. 1974. Keys to Marine Invertebrates of the Woods Hole Region. Marine Biological Laboratory, Woods Hole, MA. 208 pp.

Snieszko, S.F. 1972. Lipoid liver degeneration and viral hemorrhagic septicemia. *In:* Fish Nutrition, J.E. Halver (ed.), Academic Press, New York. pp. 422-439.

Swedmark, M., A. Granmo, and S. Kollberg. 1973. Effects of oil dispersants and oil emulsions on marine animals. Water Res. 7: 1649-1672.

Varanasi, U., D.J. Gmur, and P.A. Treseler. 1979. Influence of time and mode of exposure on biotransformation of naphthalene by juvenile starry flounder (*Platichthys stellatus*) and rock sole (*Lepidopsetta bilineata*). Arch. Environ. Contam. Toxicol. 8: 673-692.

Wellings, S.R., C.E. Alpers, B.B. McCain, and B.S. Miller. 1976. Fin erosion disease of starry flounder (*Platichthys stellatus*) and English sole (*Parophrys vetulus*) in the estuary of the Duwamish River, Seattle, Washington. J. Fish. Res. Board Can. 33: 2577-2586.

Ziskowski, J. and R. Murchelano. 1975. Fin erosion in winter flounder (*Pseudopleuronectes americanus*) from the New York Bight. Mar. Pollut. Bull. 6: 26-28.

SOME POLLUTION-ASSOCIATED DISEASES AND ABNORMALITIES OF MARINE FISHES AND SHELLFISHES: A PERSPECTIVE FOR THE NEW YORK BIGHT

Robert A. Murchelano

National Marine Fisheries Service
Northeast Fisheries Center
Oxford Laboratory
Oxford, Maryland 21654

Abstract. In recent years, there has been a tendency to associate environmental degradation with diseases of marine fishes and shellfishes. A brief review is provided of pollution-associated diseases caused by viruses and bacteria and of a number of pollution-associated diseases whose etiology is uncertain. Microbial diseases discussed include lymphocystis, herpeslike virus, *Baculovirus,* vibriosis, and shell disease; diseases of uncertain etiology include fin rot, skeletal anomalies, and "black gill"; neoplastic diseases include epidermal papilloma, stomatopapilloma, hepatoma, and several neoplastic diseases of shellfishes. The prevelance of these diseases varies geographically, and in any one area usually has not been statistically compared to an acceptable reference area. The prognosis for the animal also varies; some diseases are fatal and others have little effect on their hosts. The only pollution-associated diseases that are numerically prevalent in the New York Bight are shell disease, fin rot, and "black gill" disease.

INTRODUCTION

Recently, infectious and noninfectious diseases of marine fishes, crustaceans, and molluscs have received increased interest from the scientific community. This partly is because of the impact of disease on aquaculture, but also because of the awareness that environmental degradation may affect aquatic animal health. Disease is a likely consequence when an unfavorable environment impairs fundamental behavioral and metabolic processes.

Assessment of the prevalence of fish and shellfish diseases in animals from pristine and degraded environments is difficult. The requisite spatial and temporal sampling can be accomplished only with ample personnel and financial resources. Although some diseases can be diagnosed on the basis of gross lesions alone, others require intensive laboratory study. Assessments should be limited initially to nonmigratory fishes or sessile invertebrates; otherwise, it is difficult to establish the specific loci where disease induction occurs.

Many diseases of marine fishes and shellfishes are caused by microorganisms. The interactions between host, pathogen, and environment are fundamental determinants of the diseases caused by viruses, bacteria, fungi, and protozoa (Snieszko, 1974). In general, the probability of microbial disease is enhanced by pollution in two ways. Pollutants that increase amounts of metabolizable substrates or increase ambient temperature may favor the growth of microbial pathogens, thereby increasing the probability of infection in susceptible hosts. Pollutants that physiologically stress fishes and shellfishes may decrease host resistance by impairing mechanisms of cellular and humoral immunity. For most of the pollution-associated microbial diseases of marine animals, the specific operative mechanisms largely are unknown.

The prevalence of several infectious diseases, several diseases of uncertain etiology, and several neoplastic diseases appears to be higher in environments that have been anthropogenically altered. Presently, however, evidence of a relationship between increased disease prevalence and environmental degradation is only circumstantial. Additional studies are necessary to evaluate statistically differences in prevalence, to evaluate the role of stress in disease induction, and to establish the etiology of those diseases for which it presently is unknown.

This paper will review existing literature on pollution-associated diseases and abnormalities of marine fishes, crustaceans, and molluscs. Since the New York Bight is a severely polluted aquatic environment, the perspective presented in this brief review is valuable to environmental and resource managers in the Bight area.

MICROBIAL DISEASES–VIRUSES

Lymphocystis

Lymphocystis is a viral disease that affects many freshwater and marine fishes (Nigrelli and Ruggieri, 1965; Lawler *et al.*, 1977). The disease primarily affects the integument of host species, and gross lesions are recognized easily (Figure 1). Diagnosis usually can be made on gross observations alone. Several studies have associated increased prevalence of lymphocystis disease with environmental degradation (Christmas and Howse, 1970; Perkins *et al.*, 1972; Dethlefsen, 1978), but one study found no association (Möller, 1978).

Lymphocystis has been noted in flatfishes from the northeast Irish Sea, an area used for dumping industrial wastes (Perkins *et al.*, 1972). Depending on the location of trawl stations, the prevalence of lymphocystis varied from 0.8% to 5.9% in plaice, *Pleuronectes platessa,* and from 0% to 7.8% in dab, *Limanda limanda.* Lesions were found more frequently in fishes larger than 25 cm.

Figure 1. Lymphocystis disease in winter flounder (*Pseudopleuronectes americanus*). Small lymphocystis nodules are present on body and fins. Fin nodules are more visible (arrows).

Lymphocystis was noted in dab from sewage sludge and titanium dioxide dumping sites on the German coast (Dethlefsen, 1978). The prevalence of the disease in dab from the central area of the sewage sludge site was 1.0% and 1.5% in the control area. Because of the high variability of the data from the sludge site, differences in prevalence were not statistically significant. Disease prevalence assessed at five locations at the titanium dioxide site averaged 1.9% and was significantly greater than at the control site. In another study in the German Bight, the prevalence of lymphocystis disease in dab was not significantly higher in polluted areas than nonpolluted areas (Möller, 1978). Disease prevalence was highest in dab from the Dogger, an area outside the influence of coastal pollution.

Lymphocystis has been described in Atlantic croaker, *Micropogon undulatus,* and sand sea trout, *Cynoscion arenarius,* from the Mississippi Gulf coast (Christmas and Howse, 1970). Although only 9 out of 20,465 (0.04%) croaker and 3 out of 12,153 (0.02%) sea trout had lymphocystis, the amount of industrial pollution was much greater in areas where the disease was noted. The prevalence of the disease in silver perch, *Bairdiella chrysura,* from Mississippi Sound in the Gulf of Mexico is increasing and may coincide with the introduction of pollutants from excessive freshwater runoff (Edwards and Overstreet, 1976).

Lymphocystis disease was noted in striped bass, *Morone saxatilis,* caught in Long Island Sound in an area adjacent to the discharge plume of a fossil-fuel electric power generating station (Zawacki, personal communication). Discharge water heats a topographic basin of limited size containing fishes that have aborted their seasonal migrations and overwinter in the area. Crowding probably favors disease transmission.

Winter flounder, *Pseudopleuronectes americanus,* with lymphocystis have been noted in the New York Bight (Ziskowski, personal communication). It presently is unknown whether the prevalence of the disease is higher in the Bight than in adjacent unpolluted areas.

Although the studies cited tend to emphasize a relationship between pollution and lymphocystis, recent disease surveys conducted in the Bering Sea (McCain *et al.,* 1978, 1979) revealed that 2.1% and 2.8% of yellowfin sole, *Limanda aspera,* had lymphocystis in 1975 and 1976, respectively. The Bering Sea cannot be considered polluted. Disease surveys in progress in the western North Atlantic indicate that lymphocystis affects both American plaice, *Hippoglossoides platessoides,* and winter flounder, *P. americanus* (Murchelano, unpublished data). The area surveyed is not polluted.

Herpeslike Virus and *Baculovirus*

Only two viral diseases of shellfishes have been associated with pollution, one with a bivalve mollusc and one with a decapod crustacean. Farley *et al.* (1972) found the hemocytes of moribund American oysters, *Crassostrea virginica,* from the Piscataqua River, Maine to contain intranuclear inclusions. Electron microscopic examination revealed that the inclusions were composed of a herpeslike virus (Figure 2). Mortalities occurred when the animals were placed in a grow-out area that received heated water from a fossil-fuel electric power generating station. Although a viral cause for the observed mortalities has not yet been demonstrated experimentally, in the heated water area there was a good correlation between the presence of viral inclusions and the number of moribund oysters.

A viral disease has been described in pink shrimp, *Penaeus duorarum,* from the Gulf of Mexico (Couch, 1974). Patent virus infections were noted in shrimp chronically exposed to polychlorinated biphenyls (PCB's) in the laboratory. In a subsequent study, naturally infected wild shrimp were exposed to Aroclor 1254 at 3 ppb for 30 days, but no consistent increase in the number of infected shrimp was noted (Couch, 1976). In yet another study with naturally infected wild shrimp, there were more infected animals after exposure to Aroclor 1254 at 0.7 ppb for 35 days (Couch and Courtney, 1977). At the end of the exposure period, patent infections occurred in 45.5% of the control shrimp and 75.0% of the exposed shrimp. At present, the results of these three laboratory studies are equivocal; there also is no evidence linking mortalities of wild, virus-infected pink shrimp to PCB's.

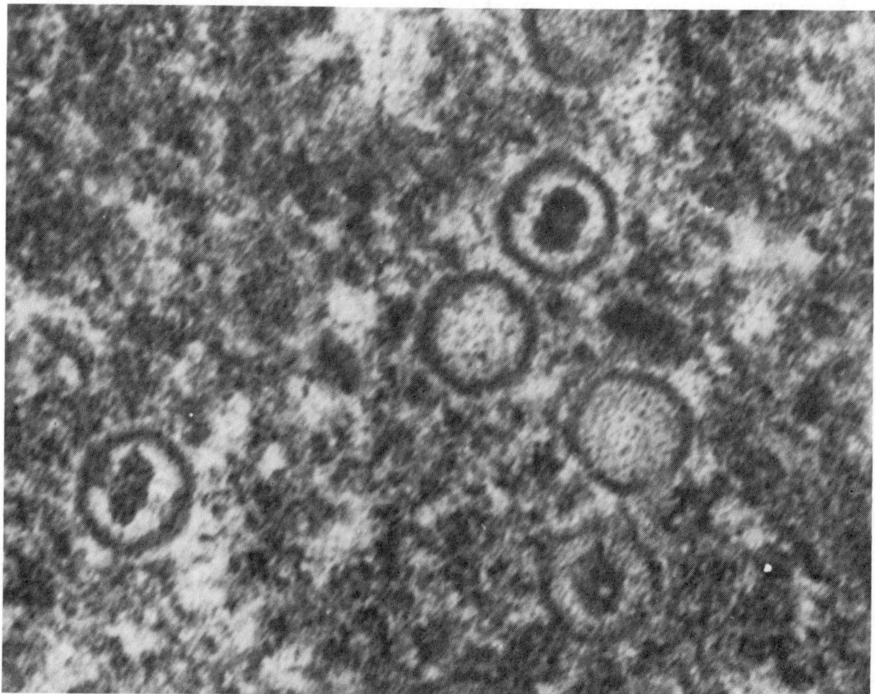

Figure 2. Electron micrograph of herpeslike virus of the American oyster (*Crassostrea virginica*). Virus particles are hexagonal in appearance and are found in the nuclei of hemocytes. (Photograph courtesy of C.A. Farley.)

MICROBIAL DISEASES–BACTERIA

Vibriosis

Bacteria of the genus *Vibrio* are ubiquitous in seawater and frequently have been implicated in diseases of fishes and shellfishes. *Vibrio anguillarum* is a significant pathogen of both cultured and wild marine fishes, and outbreaks of vibriosis have been associated with increases in water temperature (Anderson and Conroy, 1970).

Two studies have associated ulcerative lesions caused by *V. anguillarum* with pollution; one has shown no association. Ulcerative lesions have been described from cod, *Gadus morhua,* and dab, *Limanda limanda,* from the North Sea (Möller, 1978), and cod from Danish coastal waters (Jansen and Larsen, 1978; Larsen and Jensen, 1978). In an extensive study of cod from Danish coastal areas, ulcerative lesions were more common in cod from areas polluted with sewage and industrial wastes than in those from nonpolluted areas. Of 9,103 cod examined, 22% (1,080/4,906) from polluted areas had lesions, while less than 1% (2/4,197) from control areas had lesions. Over 90% of the ulcers were noted in 1.5 to 2.5 year-old cod measuring 25-35 cm. Only 2.3% of dab from industrially polluted areas of the German Bight had *Vibrio* infections, whereas 8.6% of dab from the unpolluted Dogger had ulcerative lesions. The higher prevalence of vibriosis in dab from the Dogger correlated with the generally poor condition of the fishes examined.

Recent trawl surveys conducted at the sewage sludge dumpsite in the New York Bight have revealed ulcers in red hake, *Urophycis chuss* (Ziskowski, unpublished date). Red hake migrate inshore in winter and are found in abundance in the soft bottoms of the dumpsite area. In November 1978 and January 1979, 396 red hake from the dumpsite were examined for the presence of integumental lesions; 25 fish (6.3%) had ulcers. None of 98 red hake examined from offshore waters in the western North Atlantic had ulcers. Positive blood cultures were obtained from diseased fish; however, the identity of the organisms isolated has not been determined. Additional surveys of inshore and offshore red hake are needed to evaluate further the possible correlation between increased prevalence of ulcers in hake from the dumpsite and pollution.

Shell Disease

Only one bacterial disease of shellfishes has been associated with pollution. Various chitinoclastic bacteria and fungi are the cause of shell disease (Rosen, 1970). The disease affects aquatic crustaceans and causes progressive necrosis and lysis of the chitinous exoskeleton. Shrimp (Gopalan and Young, 1975) and crabs and lobsters (Young and Pearce, 1975) with shell disease have been found in the New York Bight. The Bight is polluted with harbor dredge spoils, industrial wastes, and sewage sludge. In one study, 30% of 828 caridean shrimp, *Crangon septemspinosa,* from certain areas of the Bight had shell lesions. Few shrimp (number not given) with exoskeletal lesions were observed in the 48 shrimp collected from unpolluted areas in Massachusetts and North Carolina. In the other New York Bight study, shell disease was noted in crabs (*Cancer irroratus*) and lobsters (*Homarus americanus*). Animals were collected within and near the sewage sludge/dredged materials dumpsite areas and from areas remote from the influence of dumping. Although no numeric information was given, crabs and lobsters from dumpsites exhibited exoskeletal lesions on pereiopods, chelipeds, and other areas of the exoskeleton. Similar lesions developed on crabs and lobsters held for six weeks in laboratory aquaria containing either sewage sludge or dredge spoils. Lesions did not appear on lobsters held for six weeks in aquaria with uncontaminated sediments.

DISEASES OF UNCERTAIN ETIOLOGY

Fin Rot

Fin rot is a disease of demersal and pelagic marine fishes that has been studied extensively in various parts of the United States and that also has been described in fishes from Great Britain, Europe, and Japan. The disease has been associated with pollution in the New York Bight (Mahoney *et al.*, 1973; Murchelano, 1975; Ziskowski and Murchelano, 1975; Murchelano and Ziskowski, this volume; Sherwood, this volume), sewage outfall areas in southern California (Mearns and Sherwood, 1974), and an industrialized tributary of Puget Sound, Washington (Wellings *et al.*, 1976). It also has been noted in polluted areas (industrial wastes, sewage) of the Irish Sea (Perkins *et al.*, 1972), North Sea (Dethlefsen, 1978), and coastal Japan (Nakai *et al.*, 1973). Fin rot is characterized by progressive necrosis of fin tissues; in flatfishes, lesions of the anal and dorsal fins (those that contact sediments) predominate. From 1967 to 1971, 22 demersal and pelagic fish species from the New York Bight were noted with the disease (Mahoney *et al.*, 1973). The principal species affected were bluefish, *Pomatomus saltatrix;* summer flounder, *Paralichthys dentatus;* weakfish, *Cynoscion regalis;* and winter flounder, *Pseudopleuronectes americanus.* Of 3,584 bluefish examined, 866 (24.1%) had fin rot. The disease also was noted in 135 out of 848 (15.9%) summer flounder, 132 out of 1,229 (10.7%) weakfish, and 100 out of 1,318 (7.5%) winter flounder.

Intensive trawl surveys conducted within the New York Bight, and adjacent coastal areas to the northeast and south during 1973 and 1974 revealed that 371 out of 2,632 (14.1%) winter flounder from the inner Bight and 36 out of 1,857 (1.9%) flounder from adjacent, nonpolluted areas had fin rot (Ziskowski and Murchelano, 1975). Only 19 out of 964 (1.9%) winter flounder from Great Bay, New Jersey, a pristine environment, had the disease.

Trawl surveys conducted in southern California from 1969 to 1972 indicate that fin rot disease is more prevalent in fishes collected adjacent to major municipal wastewater discharge sites (Mearns and Sherwood, 1974). Disease prevalence was 41% in Dover sole, *Microstomus pacificus,* collected from 1970 to 1972 near Palos Verdes where the wastewater outfall of Los Angeles County is located. In adjacent areas to the north and south without wastewater outfalls, Santa Monica and San Pedro Bays, disease prevalence was 7.1% and 1.6% respectively. Both the spatial distribution of diseased fish and the location of fin lesions (anal and dorsal fins) suggest that the disease is initiated by contact with sediments adjacent to outfalls.

Two species of flatfishes from the Duwamish River, an industrially polluted tributary of Puget Sound, exhibit fin rot disease (Wellings *et al.*, 1976). In 1974, surveys revealed that 168 out of 2,099 (8.0%) starry flounder, *Platichthys stellatus,* examined had fin rot; the disease was substantially less prevalent in English sole, *Parophrys vetulus.* Bottom sediments from the Duwamish River are a continuous source of PCB's, and livers of starry flounder contain abnormally high PCB levels.

Fin rot disease accompanied by high mortality was induced in spot, *Leiostomus xanthurus,* exposed to 3-5 ppm of Aroclor 1254 (Couch and Nimmo, 1974; Couch, 1975). Dover sole developed fin rot when held in laboratory aquaria containing sediments from areas adjacent to wastewater outfalls (Sherwood, 1976; Sherwood and Mearns, 1977). The lesions observed in four out of five exposed fish resembled those in fish from outfall areas. DDT and PCB levels in the livers of the exposed fish were substantially higher than in controls; no major differences in heavy metal concentrations were noted. Attempts to induce fin rot in winter flounder caged in the sewage sludge area of the New York Bight were only partially successful (Murchelano and Ziskowski, 1976). Acute anal, dorsal, and caudal fin lesions were present on fish caged both at the sludge and control sites; however, the severity of the lesions and condition of the fish invariably were worse at the sludge site. Survival of fish also was poor at the sludge site.

Fin rot was observed in mullet, *Mugil cephalus,* held in estuarine ponds that were deliberately contaminated with crude oil (Minchew and Yarbrough, 1977). After the fish were permitted to acclimate to pond conditions for six months, the ponds were dosed with crude oil to achieve a concentration of 4.0-5.0 ppm. Mullet from control and experimental ponds were sampled 13, 21, 27, 34, and 56 days after the oil was introduced. After 56 days, 4 out of 70 (5.7%) mullet from the control ponds and 59 out of 69 (85.5%) from the oil-contaminated ponds had fin rot. Acute fin lesions were prominent on caudal fins, but also were noted on pectoral and pelvic fins. Bacteriologic culture yielded an organism tentatively identified as a *Vibrio* sp. The authors suggested that oil exerted a nonspecific stress on the mullet, reduced the numbers of zooplankton foraged, and provided a substrate for bacterial growth, all of which contributed to produce fin rot.

A bacterial etiology for fin rot disease of wild marine fishes, although suspected, has not been demonstrated. Histopathologic examination of fin and other tissues from diseased fishes has not implicated bacteria in disease pathogenesis (Klontz and Bendele, 1973; Murchelano, 1975; Wellings *et al.*, 1976). The types of bacteria isolated from Dover sole with and without fin rot did not differ significantly (Manfredi, 1976). Bacteria belonging to the genera *Achromobacter, Flavobacterium, Pseudomonas,* and *Vibrio* were found in tissues from both diseased and normal fishes. No dominant type, as would be expected if the lesions were caused by bacterial activity, was found in diseased fishes. Serum agglutinin titers were assessed against bacteria isolated from fin rot lesions of winter flounder from the New York Bight (Robohm, personal communication). No consistent, elevated agglutinin titers were detectable in fishes with fin rot. Heavy metals and PCB's both have been implicated in the etiology of fin rot disease. Concentrations of heavy metals and PCB's in flatfishes with the disease were higher than in healthy fishes in California, New York, and Washington (Sherwood *et al.*, 1978). In general, values were highest in California and lowest in New York. Presently, although several studies have demonstrated higher fin rot disease prevalence in fishes from polluted environments, none have established a specific etiology for the disease.

Skeletal Anomalies

There are many reports on anatomic anomalies of fishes (Dawson, 1964, 1966, 1971). With rare exception, however, they are unique to individual fishes and are not numerically consequential.

A high prevalence of skeletal deformities was noted in sand bass, *Paralabrax nebulifer,* from southern California (Valentine and Bridges, 1969). Of 104 bass examined for external deformities, 56% had deformities of gill rakers and ventral fins. Examination of 40 bass from Mexico revealed no deformities. Asymmetry analyses were made of sand bass, grunion (*Leuresthes tenuis*), and surf perch (*Amphistichus argentus*) from southern California (Valentine *et al.*, 1973). Seven morphological characteristics were examined to assess the extent of the deviations from perfect symmetry. Counts of pectoral fin rays and gill rakers of grunion and surf perch from one locality near Los Angeles showed increasing asymmetry over two decades. Asymmetry values of all three fish species are highest in populated areas of southern California and decrease to the north and south. Increases in asymmetry correlate well with the known distribution of toxicants. Both chlorinated hydrocarbons and heavy metals interfere with calcium metabolism and may cause skeletal deformation; both are abundant in coastal waters of southern California. In another study of the sand bass, skeletal anomalies involving gill rakers, cranial, opercular and vertebral bones, and fin rays were more prevalent in southern California than Baja, California (Valentine, 1975). An experimental study was conducted to assess the deviation in symmetry subsequent to exposure of developing fish to a chlorinated hydrocarbon (Valentine and Soulé, 1973). Eggs of grunion were hatched in flowing seawater containing from 0.001 ppb to 500 ppb DDT; fry were examined 38-61 days after egg collection. Statistically significant differences in pectoral fin ray asymmetry were noted with increasing concentrations of DDT.

Skeletal deformities consisting of dwarfism, pugheadedness, and spinal curvatures were noted in a variety of fishes from sewage sludge and titanium dioxide dumpsites in the German Bight, but were most abundant in cod, *Gadus morhua* (Dethlefsen, 1978). Deformities were noted in 0.7% of 2,191 cod from the central area of the titanium dioxide site and 1.9% of 419 cod from the central area of the sewage sludge dumpsite. No skeletal deformities were noted in 138 cod from a comparison area approximately 120 miles northwest from the two sites.

Vertebral deformities in herring, *Clupea harengus,* have been used as indicators of the quality of the environment in which the fish spawn (van de Kamp, 1977). Presumably, deformities develop during the egg or early larval stage and consist of groups of two or three incomplete vertebrae. The numbers of deformities were noted in herring from different areas (Atlantic, north and south of Ireland, and north, central, and south North Sea) and in different year classes in the same area. The highest percentages of deformities occurred in herring from the areas with the most pollution; there also was a slight yearly increase in the numbers of deformities in these areas. In the Atlantic, north and south of Ireland, the frequencies of deformities were 2.1% and 1.3%, respectively; in the north, central, and south North Sea, they were 2.4%, 2.7%, and 3.7%, respectively. It was noted that earlier, when pollution was substantially less, for at least one area (Atlantic, south of Ireland), the numbers of deformities were not significantly lower than at present.

Coastal industrial and organic pollution in Japan also may be responsible for skeletal anomalies in fishes. Broken, compressed, and fused vertebrae, deformed cranium and operculum, and loss of caudal fin have been noted in nine species of fishes from the Inland Sea near Hiroshima (Matsusato, 1973). Most of the lesions were found in mullet, *Mugil cephalus.*

A bent-fin disease has been noted in winter flounder from the New York Bight (Ziskowski *et al.,* 1980). Affected fish exhibit a line of dextral flexure typically at the middle posterior aspect of the anal and dorsal fins (Figure 3). Warping of the distal aspect of the caudal fin is also conspicuous. Fish with bent fins have significantly more axial skeletal abnormalities, such as vertebral fusions and accessory processes. Winter flounder with bent fin disease have been found only in the inner (*i.e.,* more polluted) portion of the New York Bight.

"Black Gill"

In addition to shell lesions, crabs (*Cancer irroratus*) and lobsters (*Homarus americanus*) from the New York Bight also have discolored or blackened gills (Young and Pearce, 1975; Sawyer, this volume). Gross and microscopic observations were made on gills of 2,091 rock crabs and 202 lobsters from the Bight and coastal waters to the northeast and south to determine if the "black gill" condition is associated with ocean dumping of sewage sludge (Sawyer *et al.,* 1979). Although histologic sections of gills revealed debris, bacteria, and a variety of protozoan and metazoan epibionts, it was not possible to determine if the black discoloration was related principally to ocean dumping. Whether gills were clean, discolored, or blackened depended primarily on the particular stage of molt. Specific effects of ocean dumping on gill condition were difficult to assess because gill fouling was noted in crabs from all areas sampled.

NEOPLASTIC DISEASES–FISHES

Many kinds of neoplastic diseases have been described in marine fishes (Lucke and Schlumberger, 1949; Wellings, 1969; Madesley-Thomas, 1975). Most of the pollution-associated neoplasias are integumental.

Epidermal Papilloma

Epidermal papillomas are benign epidermal tumors that have been found in many species of flatfishes (Figure 4). There is extensive literature pertaining to their epizootiology and etiology, particularly in flatfishes from the Pacific coast of the United States. The geographic distribution of flatfishes with epidermal papillomas has been summarized and the feasibility of their use as indicators of the presence of environmental carcinogens discussed (Stich *et al.*, 1977).

Epidermal papillomas were found in as many as 58.6% of English sole, *Parophrys vetulus,* from collecting sites near Vancouver, British Columbia (Stich *et al.*, 1976); 54% of starry flounder, *Platichthys stellatus,* from Bellingham Bay, Washington (McArn and Wellings, 1971); 43.4% of sand sole, *Psettichthys melanostictus,* from northern Hecate Strait, British Columbia (Nigrelli *et al.*, 1965); 15% of flathead sole, *Hippoglossoides elassodon,* from East Sound, Washington (Miller and Wellings, 1971); and 9.1% of Dover sole, *Microstomus pacificus,* from areas of the Pacific coast adjacent to Los Angeles, California (Mearns and Sherwood, 1974). They also have been found in starry flounder, *Limanda schrenki,* and *Verasper moseri* from areas on the eastern coast of

Figure 3. Bent-fin disease of winter flounder (*Pseudopleuronectes americanus*). Anal and dorsal fins have been fin rays. Caudal fin is extensively damaged and has both bent and missing fin rays.

Hokkaido, Japan (Oishi *et al.*, 1976). Tumor prevalence ranged from 0.34% to 8.29% in flatfishes obtained from essentially unpolluted waters. Dab from the German Bight have been noted with epidermal papillomas (Dethlefsen, 1978). Tumor prevalence varied from 2.0% to 6.6% in dab from a titanium dioxide dumpsite, from 0% to 1.0% at a sewage sludge dumpsite, and was 0.6% at an unpolluted site. Monthly collections of 0-year-class winter flounder from the inner New York Bight in 1978 provided 3,155 fish for examination (Murchelano and Ziskowski, unpublished data). None of the fish examined had gross lesions characteristic of developing or mature epidermal papillomas. As yet, flatfishes with epidermal papillomas have not been found on the Atlantic coast of the United States.

Epidermal papillomas were found in rock sole, *Lepidopsetta bilineata,* from the Bering Sea (McCain *et al.*, 1978, 1979). Tumor prevalence was 1.0% in 1975 and 1.4% in 1976.

Stomatopapilloma

Another papilloma that has been associated with pollution is the stomatopapilloma of the European eel, *Anguilla anguilla.* Commonly known as "cauliflower" disease because of the gross appearance of the lesion, eel stomatopapillomas first were observed in the Baltic, but now are found westward in eels from coastal waters of Belgium, Denmark, Germany, and Holland (Deys, 1969). In the industrially polluted Elbe estuary in Germany, prevalences as high as 28% have been noted (Peters, 1975). No specific etiology for this lesion has been established; however, viruses and chemical carcinogens may be involved (McAllister *et al.*, 1977). Seasonal increase in water temperature favors tumor prevalence and growth (Peters and Peters, 1977).

Hepatoma

The only non-integumental pollution-associated tumors of fishes are the hepatoma of the Atlantic hagfish, *Myxine glutinosa* (Falkmer *et al.*, 1977), and

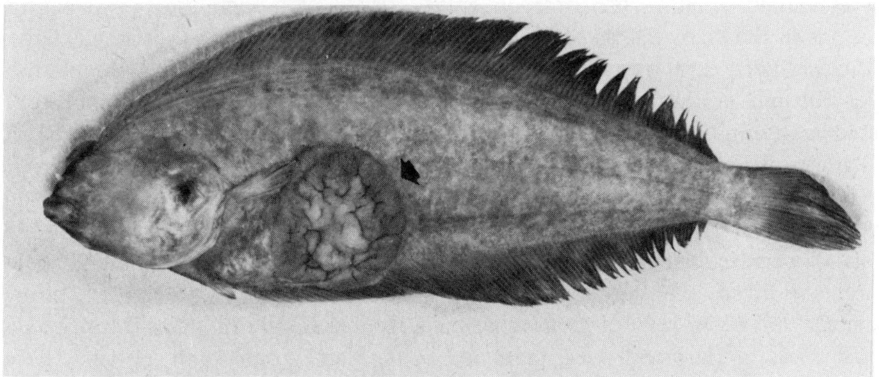

Figure 4. Large epidermal papilloma (arrow) on the non-pigmented side of a Dover sole. (Photograph courtesy of M. Sherwood.)

the hepatoma of the English sole (McCain *et al.*, 1977). Approximately 28,000 hagfish from the Gullmar Fjord on the Swedish west coast were examined for liver tumors over a five-year period. The incidence of hepatomas in the Fjord decreased from 5.8% in 1972, to 2.9% in 1973, and to 0.6% from 1974 to 1976. Their incidence in hagfish caught outside the Fjord decreased from 2.8% in 1972 to 0.9% in 1974. Livers of hagfish from the Gullmar Fjord contained approximately 5 mg/kg wet wt of PCB's; livers of hagfish from outside the Fjord contained approximately 0.2 mg/kg wet wt of PCB's. Although muds from the Fjord did not contain PCB's, livers of cod from the Fjord contained approximately 3 mg/kg wet wt of PCB's. Dead cod presumably are a food source for bottom-feeding hagfish.

Liver tumors were found in English sole from the Duwamish River, Washington. The tumors were noted in 20 out of 62 (32%) sole collected from July 1975 to January 1976. Most of the hepatomas were basophilic nodules consisting of minimally deviated hepatocytes. Many nodules appeared to be invasive. The cause of these lesions is not known; PCB's are suspected, however, since Duwamish River sole contain high levels of PCB's.

NEOPLASTIC DISEASES-SHELLFISHES

Epizootic neoplastic diseases have been noted in mussels, *Mytilus edulis* (Farley, 1969), and oysters, *Ostrea lurida* (Farley and Sparks, 1970), from Yaquina Bay, Oregon. In specific areas of the Bay, 14% of the mussels and 12% of the oysters examined had hematopoietic neoplastic lesions (Farley, 1976). The only significant pollutants in the Bay are pulp mill wastes and petroleum hydrocarbons introduced from small boat exhausts. Although both substances are suspected to be carcinogens, neither has been confirmed as causative. From 1972 to 1974, only 1.9% of 262 native oysters and none of 1,426 hatchery-reared, introduced oysters from the Bay exhibited the disease (Mix *et al.*, 1977).

A gill neoplasm (Figure 5) was noted in the duck clam, *Macoma balthica,* from the Tred Avon River, Maryland, an estuary that, like Yaquina Bay, Oregon, is essentially pristine (Christensen *et al.*, 1974). Although the prevalence of lesions in the clam was as high as 12%, adjacent populations of soft-shell clams, *Mya arenaria,* and oysters, *Crassostrea virginica,* did not contain neoplasms. Agricultural pesticides are used extensively in the area; however, laboratory studies attempting tumor induction with specific pesticides were unsuccessful (Farley, 1977).

Over the period 1967-1977, more than 14,000 soft-shell clams from coastal waters in the states of California, Maine, Maryland, Massachusetts, and Rhode Island were examined histologically (Yevich and Barszcz, 1977). Clams were obtained from unpolluted areas as well as areas contaminated by oil, heavy metals, and other pollutants. Neoplasms were noted only in clams from two oil spill areas in Harpswell Neck and Searsport, Maine (Barry and Yevich, 1975; Yevich and Barszcz, 1976, 1977). At Harpswell Neck, hematopoietic neoplasms were noted in 48 out of 471 (10.19%) of the clams examined; at Searsport, gonadal neoplasms were noted in 68 out of 1,185 (5.74%) of the clams. Harpswell Neck was the site of a JP-4 oil spill and Searsport of a No. 2 and a JP-5 oil

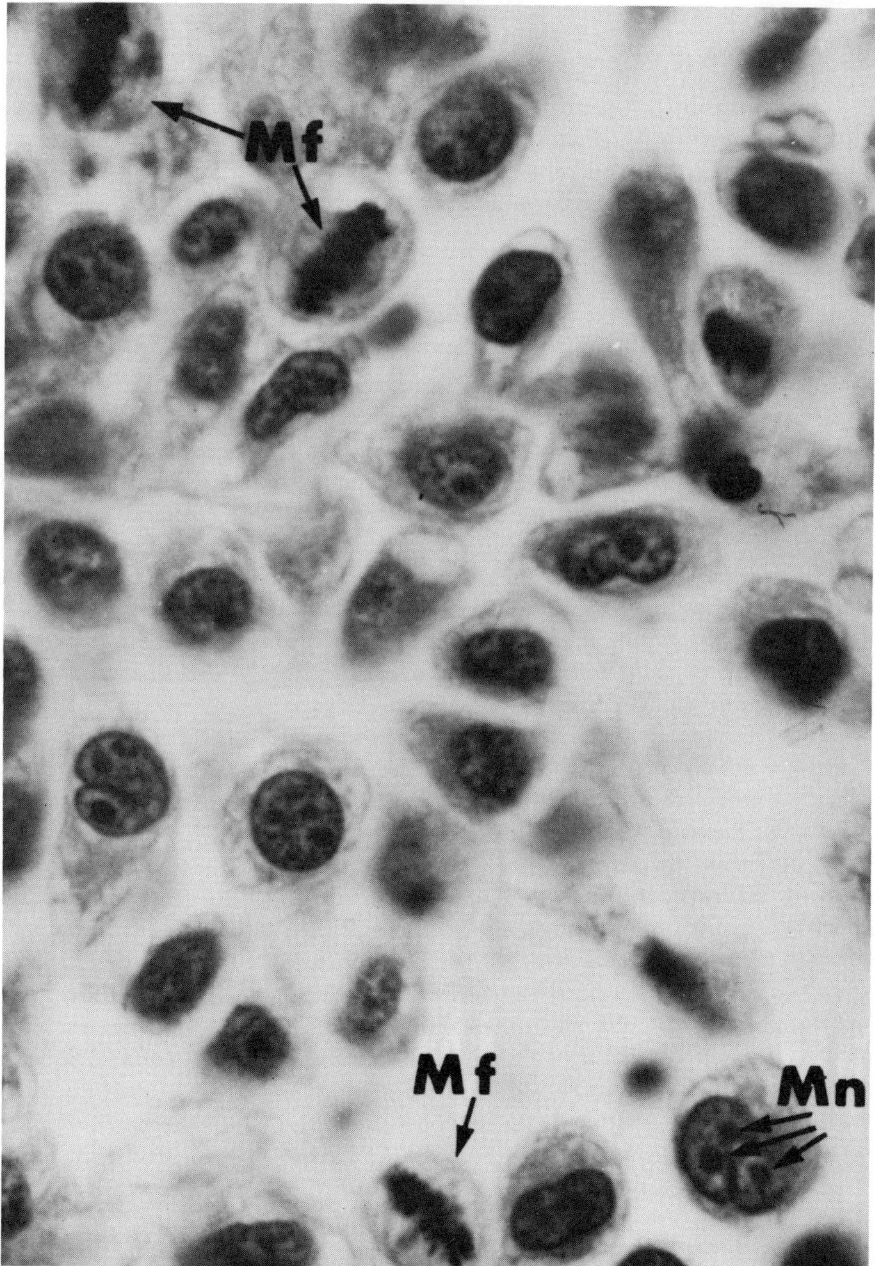

Figure 5. Neoplastic cells in the gill of the duck clam (*Macoma balthica*). Mitotic figures (Mf) and multiple nucleoli (Mn) are frequently seen. (Photograph courtesy of D. Christensen.)

spill. In another study of neoplasia in the soft-shell clam, hematopoietic and gonadal lesions were noted in 159 out of 1,325 (12.0%) of the clams sampled from ecologically diverse environments in Maine and Rhode Island (Brown *et al.*, 1977). The prevalence of neoplasms at various sites varied from 0% to 64%. Clams from both oil-contaminated and oil-free sites had neoplastic disease.

A high prevalence of hematopoietic neoplasms, 31 out of 369 or 8.40%, has been noted in inbred American oysters, *C. virginica,* from several estuaries of the Chesapeake Bay in Virginia (Frierman, 1976). The prevalence of neoplastic lesions in indigenous and other inbred oysters was low, indicating a possible genetic rather than an environmental basis for the disease.

DISCUSSION

In the literature reviewed, increased prevalence of a particular disease has been associated only with poor environmental quality. Few studies were undertaken specifically to assess disease prevalence in environments that differed substantially in their pollutant content. Usually, the association with environmental change was made subsequent to finding diseased animals in the population sampled. Evaluation of the prevalence of a specific disease in dissimilar environments at a designated confidence level has been done only with fin rot disease in the New York Bight and skeletal anomalies in southern California. Although surveys that establish significant differences in disease prevalence between polluted and reference environments make the influence of pollution suspect, they neither establish disease etiology nor define the specific role of pollution in disease pathogenesis.

Although the etiology of some pollution-associated diseases of fishes and shellfishes is known (*e.g.,* those diseases caused by bacteria and viruses), it is uncertain how pollution augments the prevalence of these diseases. Eutrophication and increased temperature may favor the growth of pathogenic microorganisms, thereby increasing infection pressure. Concomitant stresses imposed by other types of pollutants may decrease disease resistance by impairing essential immunologic and physiologic activities. The role of stress in altering disease resistance of fishes has been documented (Wedemeyer, 1970). Stresses that affect steroid hormone metabolism, inflammation, and interferon and antibody production all increase susceptibility to infectious diseases. Most of the research evaluating stress in fish diseases has been conducted with cultured, freshwater fishes; however, the findings may be relevant to wild fishes in the marine environment. In either situation, the interactions of host, pathogen, and environment (Snieszko, 1974) are fundamental determinants of disease and its consequences. Unfortunately, except in a general way, the role of stress in shellfish diseases is poorly understood.

As has been frequently demonstrated in human and veterinary animal medicine, determining etiology is an iterative process that is time-consuming, demanding in skills, and costly. For over ten years, research to determine the cause of fin rot disease and epidermal papillomas in flatfishes, stomatopapillomas in eels, and several neoplastic diseases of bivalve molluscs has been largely inconclusive. Even when the definitive etiology of these diseases is established, it still will be necessary to substantiate the role of pollution in pathogenesis. Despite extensive research on fin rot, the etiology of the disease remains unknown; but,

> . . . as in other disease processes in any species, the observations best fit a multifactorial hypothesis of etiology, *i.e.* it is logical to assume that multiple environmental variables (such as chemical pollutants, physical factors, mechanical injury, etc.) act in some combination on a susceptible genetic background to produce a particular incidence of disease in the population at risk [Wellings *et al.*, 1976: p. 2584].

Epizootic epidermal papillomas of flatfishes, because of their high prevalence and the presence of "X" cells within the lesion, have been studied extensively. As with the eel stomatopapilloma, a virus etiology was suspected. Recent studies reveal that "X" cells are not transformed fish cells, but may be unicellular parasites, possible amoebae (Dawe, personal communication). The neoplasms of bivalve molluscs have been subjected to numerous examinations involving a variety of scientific disciplines. Again, an infectious etiology is suspected, but as yet an oncogenic virus has not been demonstrated. Preliminary experiments recently conducted with soft-shell clams suggest an infectious etiology (Brown, personal communication); however, the results of these studies must be interpreted as preliminary.

One of the more important aspects of disease is how it affects resource animals. The consequences of the pollution-associated diseases discussed vary considerably, and it is impossible to state categorically that any one disease is either insignificant or invariably fatal for its host. Lymphocystis generally is benign, but in some instances may be fatal. Vibriosis, on the other hand, is a systemic disease that usually is fatal. Fin rot and skeletal anomalies probably cause behavioral modifications that affect survival by affecting the ability to capture food or avoid predation. Epidermal papillomas, although histologically benign, also may affect survival through behavioral modification. Whether the herpeslike virus disease of oysters and the *Baculovirus* disease of pink shrimp cause extensive mortalities in wild populations is unknown. Limited field and laboratory studies suggest, however, that both diseases may be fatal for their respective hosts. Shell and black gill diseases of crustaceans, except in unusual circumstances, probably are not fatal since both the exoskeleton and gill cuticle are replaced when the animal molts. Since the neoplastic diseases of bivalve molluscs exhibit histologic characteristics similar to those in malignant lesions of homeotherms, these diseases may be fatal to their hosts.

A major deficiency of most of the disease surveys conducted to date is that they usually focus on adult animals only. Admittedly, adults are easier to sample and work with in field and laboratory studies. It is the pre-adult stages, however, that may be more sensitive to environmental perturbation and that, therefore, merit attention. Although disease may affect the longevity and reproductive potential of adults, in pre-adults it may threaten survival itself. The consequences for the resource are obvious.

It commonly is alleged that the New York Bight, specifically the coastal waters bordering New York and New Jersey, is one of the most severely polluted marine environments in the United States. Why then are not all the pollution-associated fish and shellfish diseases discussed herein numerically prevalent in the Bight? It appears logical to hypothesize that disease prevalence will be greater where the environment has been altered most. The spectrum of pollutants in the Bight certainly is extensive; furthermore, their concentrations at any time and place are likely to exceed those found in many other areas. If the presence of a particular disease depends on the presence of a particular species, then the *Baculovirus* disease of pink shrimp, epidermal papilloma of pleuronectids, and hepatoma of hagfish and English sole should not be expected in the Bight since the target species are not there. For infectious diseases, it also is possible that the infective agent is not in the environment, or that its transmissibility is limited by inappropriate vectors, environmental conditions, or a variety of other factors.

The apparent absence of many pollution-associated fish and shellfish diseases in the New York Bight may also reflect the minimum sampling effort expended to date. There have been few continuous and comprehensive surveys of fish and shellfish diseases in the Bight. Although numerous, trawl surveys have not adequately sampled the Bight either spatially or temporally; even less effort has been made to sample shellfish resources. Increased awareness of the variety of forms of pollution-associated diseases and more systematic sampling may profoundly change the presently held impressions of disease prevalence. The availability of accurate and synoptic data should be useful to those engaged in environmental and resource management in the New York Bight.

ACKNOWLEDGEMENTS

I wish to acknowledge the financial support of the MESA New York Bight Project during the preparation of this paper. I am especially grateful to Drs. Garry F. Mayer and Joel O'Connor for their many helpful comments recently and over the years.

REFERENCES

Anderson, J.I.W. and D.A. Conroy. 1970. *Vibrio* disease in marine fishes. *In:* A Symposium on Diseases of Fishes and Shellfishes, S.F. Snieszko (ed.), Am. Fish. Soc. Spec. Publ. No. 5. pp. 266-272.

Barry, M. and P.P. Yevich. 1975. The ecological, chemical and histopathological evaluation of an oil spill site. Part III. Histopathological studies. Mar. Pollut. Bull. 6: 171-173.

Brown, R.S., R.E. Wolke, S.B. Saila, and C.W. Brown. 1977. Prevalence of neoplasia in 10 New England populations of the soft-shell clam (*Mya arenaria*). Ann. N.Y. Acad. Sci. 298: 522-534.

Christensen, D.J., C.A. Farley, and F.G. Kern. 1974. Epizootic neoplasms in the clam *Macoma balthica* (L.) from Chesapeake Bay. J. Natl. Cancer Inst. 52: 1739-1749.

Christmas, J.Y. and H.D. Howse. 1970. The occurrence of lymphocystis in *Micropogon undulatus* and *Cynoscion arenarius* from Mississippi estuaries. Gulf Res. Rep. 3: 131-154.

Couch, J.A. 1974. An enzootic nuclear polyhedrosis virus of pink shrimp: ultrastructure, prevalence, and enhancement. J. Invertebr. Pathol. 24: 311-331.

Couch, J.A. 1975. Histopathological effects of pesticides and related chemicals on the livers of fishes. *In:* The Pathology of Fishes, W.E. Ribelin and G. Migaki (eds.), University of Wisconsin Press, Madison, WI. pp. 559-584.

Couch, J.A. 1976. Attempts to increase *Baculovirus* prevalence in shrimp by chemical exposure. Prog. Exp. Tumor Res. 20: 304-314.

Couch, J.A. and L. Courtney. 1977. Interaction of chemical pollutants and virus in a crustacean: a novel bioassay system. Ann. N.Y. Acad. Sci. 298: 497-504.

Couch, J.A. and D.R. Nimmo. 1974. Detection of interactions between natural pathogens and pollutants in aquatic animals. *In:* Proceedings of Gulf Coast Regional Symposium on Diseases of Aquatic Animals, R.L. Amborski, M.A. Hood, and R.R. Miller (eds.), Center for Wetland Resources, Louisiana State University, Baton Rouge, LA. pp. 261-265.

Dawson, C.E. 1964. A bibliography of anomalies of fishes. Gulf Res. Rep. 1: 308-399.

Dawson, C.E. 1966. A bibliography of anomalies of fishes. Suppl. 1. Gulf Res. Rep. 2: 169-176.

Dawson, C.E. 1971. A bibliography of anomalies of fishes. Suppl. 2. Gulf Res. Rep. 3: 215-239.

Dethlefsen, V. 1978. Occurrence and abundance of some skeletal deformities, diseases and parasites of major fish species in dumping areas off the German coast. Int. Counc. Explor. Sea C.M. 1978/E:8. 17 pp.

Deys, B.F. 1969. Papillomas in the Atlantic eel, *Anguilla vulgaris. In:* A Symposium on Neoplasms and Related Disorders of Invertebrate and Lower Vertebrate Animals, C.J. Dawe and J.C. Harshbarger (eds.), Natl. Cancer Inst. Monogr. 31. pp. 187-193.

Edwards, R.H. and R.M. Overstreet. 1976. Mesenchymal tumors of some estuarine fishes of the northern Gulf of Mexico. I. Subcutaneous tumors, probably fibrosarcomas, in the striped mullet, *Mugil cephalus*. Bull. Mar. Sci. 26: 33-40.

Falkmer, S., S. Marklund, P.E. Mattsson, and C. Rappe. 1977. Hepatomas and other neoplasms in the Atlantic hagfish (*Myxine glutinosa*): a histopathologic and chemical study. Ann. N.Y. Acad. Sci. 298: 342-355.

Farley, C.A. 1969. Sarcomatoid proliferative disease in a wild population of blue mussels (*Mytilus edulis*). J. Natl. Cancer Inst. 43: 509-516.

Farley, C.A. 1976. Ultrastructural observations on epizootic neoplasia and lytic virus infection in bivalve mollusks. Prog. Exp. Tumor Res. 20: 283-294.

Farley, C.A. 1977. Neoplasms in estuarine mollusks and approaches to ascertain causes. Ann. N.Y. Acad. Sci. 298: 225-232.

Farley, C.A., W.G. Banfield, G. Kasnic, Jr., and W.S. Foster. 1972. Oyster herpes-type virus. Science 178: 759-760.

Farley, C.A. and A.K. Sparks. 1970. Proliferative diseases of hemocytes, endothelial cells, and connective tissue cells in mollusks. Bibl. Haematol. 36: 610-617.

Frierman, E.M. 1976. Occurrence of hematopoietic neoplasms in Virginia oysters (*Crassostrea virginica*). Mar. Fish. Rev. 38: 34-36.

Gopalan, U.K. and J.S. Young. 1975. Incidence of shell disease in shrimp in the New York Bight. Mar. Pollut. Bull. 6: 149-153.

Jensen, N.J. and J.L. Larsen. 1978. The ulcus-syndrome in cod in Danish coastal waters. Int. Counc. Explor. Sea C.M. 1978/E:28. 15 pp.

Klontz, G.W. and R.A. Bendele. 1973. Histopathological analysis of fin erosion in southern California marine fishes. Southern California Coastal Water Research Project Rep. TM 203. 8 pp.

Larsen, J.L. and N.J. Jensen. 1978. Bacterial species implicated in the ulcus-syndrome of cod. Int. Counc. Explor. Sea C.M. 1978/E:29. 21 pp.

Lawler, A.R., J.T. Ogle, and C. Donnes. 1977. *Dascyllus* spp.: New hosts for lymphocystis, and a list of recent hosts. J. Wildl. Dis. 13: 307-312.

Lucke, B. and H.G. Schlumberger. 1949. Neoplasia in cold-blooded vertebrates. Physiol. Rev. 29: 91-126.

McAllister, P.E., T. Nagabayashi, and K. Wolf. 1977. Viruses of eels with and without stomatopapillomas. Ann. N.Y. Acad. Sci. 298: 233-244.

McArn, G.E. and S.R. Wellings. 1971. A comparison of skin tumors in three species of flounders. J. Fish. Res. Board Can. 28: 1241-1251.

McCain, B.B., W.D. Gronlund, M.S. Myers, and S.R. Wellings. 1979. Tumours and microbial diseases of marine fishes in Alaskan waters. J. Fish Dis. 2: 111-130.

McCain, B.B., M.S. Myers, W.D. Gronlund, S.R. Wellings, and C.E. Alpers. 1978. The frequency, distribution, and pathology of three diseases of demersal fishes in the Bering Sea. J. Fish Biol. 12: 267-276.

McCain, B.B., K.V. Pierce, S.R. Wellings, and B.S. Miller. 1977. Hepatomas in marine fish from an urban estuary. Bull. Environ. Contam. Toxicol. 18: 1-2.

Mahoney, J.B., F.H. Midlige, and D.G. Deuel. 1973. A fin rot disease of marine and euryhaline fishes in the New York Bight. Trans. Am. Fish. Soc. 102: 596-605.

Manfredi, E.T. 1976. A bacteriological study of fin erosion disease in Dover sole (*Microstomus pacificus*). M.S. Thesis, California State University, Long Beach. 94 pp.

Matsusato, T. 1973. On the skeletal abnormalities in marine fishes. I. The abnormal marine fishes collected along the coast of Hiroshima pref. Bull. Nansei Reg. Fish. Res. Lab. 6: 17-58.

Mawdesley-Thomas, L.E. 1975. Neoplasia in fish. *In:* The Pathology of Fishes, W.E. Ribelin and G. Migaki (eds.), University of Wisconsin Press, Madison, WI. pp. 805-870.

Mearns, A.J. and M. Sherwood. 1974. Environmental aspects of fin erosion and tumors in southern California Dover sole. Trans. Am. Fish. Soc. 103: 799-810.

Miller, B.S. and S.R. Wellings. 1971. Epizootiology of tumors on flathead sole (*Hippoglossoides elassodon*) in East Sound, Orcas Island, Washington. Trans. Am. Fish. Soc. 100: 247-266.

Minchew, C.D. and J.D. Yarbrough. 1977. The occurrence of fin rot in mullet (*Mugil cephalus*) associated with crude oil contamination of an estuarine pond-ecosystem. J. Fish Biol. 10: 319-323.

Mix, M.C., H.J. Pribble, R.T. Riley, and S.P. Tomasovic. 1977. Neoplastic disease in bivalve mollusks from Oregon estuaries with emphasis on research on proliferative disorders in Yaquina Bay oysters. Ann. N.Y. Acad. Sci. 298: 356-373.

Möller, H. 1978. Geographical and seasonal variation in fish pests in North Sea and Baltic. Int. Counc. Explor. Sea C.M. 1978/E:9. 14 pp.

Murchelano, R.A. 1975. The histopathology of fin rot disease in winter flounder from the New York Bight. J. Wildl. Dis. 11: 263-268.

Murchelano, R.A. and J. Ziskowski. 1976. Fin rot disease studies in the New York Bight. *In:* Proceedings of the Symposium on the Middle Atlantic Continental Shelf and the New York Bight, M.G. Gross (ed.), Am. Soc. Limnol. Oceanogr. Spec. Symp. 2: 329-336.

Murchelano, R.A. and J. Ziskowski. This volume. Fin rot disease in the New York Bight (1973-1977). pp. 347-358.

Nakai, Z., M. Kosaka, S. Kudoh, A. Nagai, F. Hayashida, T. Kubota, M. Ogura, T. Mizutani, and I. Votani. 1973. Summary report on marine biological studies of Suruga Bay accomplished by Tokai University 1964-1972. J. Fac. Mar. Sci. Technol. Tokai Univ., 10th Ann. Commem. Vol. pp. 63-118.

Nigrelli, R.F., K.S. Ketchen, and G.D. Ruggieri. 1965. Studies on virus diseases of fishes. Epizootiology of epithelial tumors in the skin of flatfishes of the Pacific coast, with special reference to the sand sole (*Psettichthys melanosticus*) from northern Hecate Strait, British Columbia, Canada. Zoologica 50: 115-122.

Nigrelli, R.F. and G.D. Ruggieri. 1965. Studies on virus diseases of fishes. Spontaneous and experimentally induced cellular hypertrophy (lymphocystis disease) in fishes of the New York Aquarium, with a report of new cases and an annotated bibliography (1874-1965). Zoologica 50: 83-96.

Oishi, K., F. Yamazaki, and T. Harada. 1976. Epidermal papillomas of flatfish in the coastal waters of Hokkaido, Japan. J. Fish. Res. Board Can. 33: 2011-2017.

Perkins, E.J., J.R.S. Gilchrist, and O.J. Abbott. 1972. Incidence of epidermal lesions in fish of the North-East Irish Sea area, 1971. Nature 238: 101-103.

Peters, G. 1975. Seasonal fluctuations in the incidence of epidermal papillomas of the European eel, *Anguilla anguilla* L. J. Fish Biol. 7: 415-422.

Peters, G. and N. Peters. 1977. Temperature-dependent growth and regression of epidermal tumors in the European eel (*Anguilla anguilla* L.). Ann. N.Y. Acad. Sci. 298: 245-260.

Rosen, B. 1970. Shell disease of aquatic crustaceans. *In:* A Symposium on Diseases of Fishes and Shellfishes, S.F. Snieszko (ed.), Am. Fish. Soc. Spec. Publ. No. 5. pp. 409-415.

Sawyer, T.K. This volume. Distribution and seasonal incidence of "black gill" in the rock crab, *Cancer irroratus.* pp. 199-211.

Sawyer, T.K., S.A. MacLean, J.E. Bodammer, and B.A. Harke. 1979. Gross and microscopical observations on gills of rock crabs (*Cancer irroratus*) and lobsters (*Homarus americanus*) from nearshore waters of the eastern United States. *In:* Proceedings of the Second Biennial Crustacean Health Workshop, Texas A&M Univ. Sea Grant Publ. TAMSG 79-114. pp. 68-91.

Sherwood, M.J. 1976. Fin erosion disease induced in the laboratory. *In:* 1976 Annual Report of the Southern California Coastal Water Research Project, Southern California Coastal Water Research Project, El Segundo, CA. pp. 149-153.

Sherwood, M.J. This volume. Fin erosion, liver condition, and trace contaminant exposure in fishes from three coastal regions. pp. 359-377.

Sherwood, M.J. and A.J. Mearns. 1977. Environmental significance of fin erosion in southern California demersal fishes. Ann. N.Y. Acad. Sci. 298: 177-189.

Sherwood, M.J., A.J. Mearns, D.R. Young, B.B. McCain, and R.A. Murchelano. 1978. A comparison of trace contaminants in diseased fishes from three areas. *In:* Report from Southern California Coastal Water Research Project, El Segundo, California to MESA New York Bight Project Office, State University of New York, Stony Brook, NY. 111 pp.

Snieszko, S.F. 1974. The effects of environmental stress on outbreaks of infectious diseases of fishes. J. Fish Biol. 6: 197-208.

Stich, H.F., A.B. Acton, and C.R. Forrester. 1976. Fish tumors and sublethal effects of pollutants. J. Fish. Res. Board Can. 33: 1993-2001.

Stich, H.F., A.B. Acton, K. Oishi, F. Yamazaki, T. Harada, T. Hibino, and H.G. Moser. 1977. Systematic collaborative studies on neoplasms in marine animals as related to their environment. Ann. N.Y. Acad. Sci. 298: 374-388.

Valentine, D.W. 1975. Skeletal anomalies in marine teleosts. *In:* The Pathology of Fishes, W.E. Ribelin and G. Migaki (eds.), University of Wisconsin Press, Madison, WI. pp. 695-718.

Valentine, D.W. and K.W. Bridges. 1969. High incidence of deformities in the serranid fish, *Paralabrax nebulifer,* from southern California. Copeia 1969: 637-638.

Valentine, D.W. and M.E. Soulé. 1973. Effect of p,p'-DDT on developmental stability of pectoral fin rays in the grunion, *Leuresthes tenuis.* Fish. Bull. 71: 921-926.

Valentine, D.W., M.E. Soulé, and P. Samollow. 1973. Asymmetry analysis in fishes: a possible statistical indicator of environmental stress. Fish. Bull. 71: 357-370.

van de Kamp, G. 1977. Vertebral deformities in herring around the British Isles and their usefulness for a pollution monitoring programme. Int. Counc. Explor. Sea C.M. 1977/E:5. 9 pp.

Wellings, S.R. 1969. Neoplasia and primitive vertebrate phylogeny: echinoderms, prevertebrates, and fishes—a review. Natl. Cancer Inst. Monogr. 31: 59-128.

Wellings, S.R., C.E. Alpers, and B.B. McCain. 1976. Fin erosion disease of starry flounder (*Platichthys stellatus*) and English sole (*Parophrys vetulus*) in the estuary of the Duwamish River, Seattle, Washington. J. Fish. Res. Board Can. 33: 3577-3586.

Yevich, P.P. and C.A. Barszcz. 1976. Gonadal and hematopoietic neoplasms in *Mya arenaria.* Mar. Fish. Rev. 38(10): 42-43.

Yevich, P.P. and C.Z. Barszcz. 1977. Neoplasia in soft-shell clams (*Mya arenaria*) collected from oil-impacted sites. Ann. N.Y. Acad. Sci. 298: 409-426.

Young, J.S. and J.B. Pearce. 1975. Shell disease in crabs and lobsters from New York Bight. Mar. Pollut. Bull. 6: 101-105.

Ziskowski, J. and R. Murchelano. 1975. Fin erosion in winter flounder. Mar. Pollut. Bull. 6: 26-29.

Ziskowski, J.J., V.T. Anderson, Jr., and R.A. Murchelano. 1980. A bent fin ray condition in winter flounder, *Pseudopleuronectes americanus*, from Sandy Hook and Raritan Bays, New Jersey, and Lower Bay, New York. Copeia 1980(4): 895-899.

FIN ROT DISEASE IN THE NEW YORK BIGHT (1973-1977)

Robert A. Murchelano

National Marine Fisheries Service
Northeast Fisheries Center
Oxford Laboratory
Oxford, Maryland 21654

John Ziskowski

National Marine Fisheries Service
Northeast Fisheries Center
Sandy Hook Laboratory
Highlands, New Jersey 07732

Abstract. The results of research on the prevalence and etiology of fin rot disease of New York Bight winter flounder, *Pseudopleuronectes americanus,* are presented. Except for 1975 and 1976, significant differences ($P < 0.05$) in disease prevalence were found between the apex and Sandy Hook/Raritan Bays, the apex and control area, and Sandy Hook/Raritan Bays and control area. The apex had the highest prevalence of the disease (7.4%) for the five-year period from 1973 to 1977. It was not possible to determine whether the etiology of the fin rot is infectious or noninfectious. Bacteria do not appear to be involved in the pathogenesis of the disease, and the concentrations of trace metals and polychlorinated biphenyls (PCB's) in tissues were not elevated consistently. Based on bacteriologic, histopathologic, and immunologic studies, the disease probably is not fatal; however, behavioral modification may result in some mortalities. Negligible mortality and the absence of an extensive winter flounder fishery in the New York Bight both make it unlikely that there is a major effect on the population dynamics of the species. The disease is unlikely to reduce the marketability of fish since muscle tissue is not affected and since the fish are not sold in the round. The most important aspect of the presence of fin rot disease in Bight winter flounder rests in its value as an indicator of environmental stress.

INTRODUCTION

Fin rot (fin erosion) aptly describes a disease of fishes characterized by the destruction of fin tissues. The disease is well-known by freshwater fish culturists and tropical fish hobbyists (Davis, 1953; Snieszko, 1953, 1958; Bullock, 1968; Amlacher, 1970; Bullock and Snieszko, 1970; Bullock *et al.* 1971); it also occurs in wild marine fishes (Sindermann and Rosenfield, 1954; Oppenheimer, 1958; Sindermann, 1966). Epizootics of fin rot disease in marine fishes have been associated with pollution and have been reported from the east coast (Mahoney *et al.,* 1973; Ziskowski and Murchelano, 1975; Murchelano and Ziskowski, 1976) and west coast (Mearns and Sherwood, 1974; Wellings *et al.,* 1976) of the United States. The disease is more prevalent in flatfishes, and attempts have been made to determine its specific etiology (Klontz and Bendele, 1973; Murchelano, 1975; Manfredi, 1976; Wellings *et al.,* 1976).

Studies of fin rot disease in winter flounder, *Pseudopleuronectes americanus,* have been conducted in the New York Bight since 1973. The emphasis of these studies was to determine the spatial and temporal distribution of diseased winter flounder and, if possible, the cause of the disease. The results of these studies and their implication for environmental and resource managers are presented here.

MATERIALS AND METHODS

Surveys

The prevalence of fin rot disease in winter flounder was determined principally by otter trawl surveys in Sandy Hook/Raritan Bays, the apex, and control areas consisting of Great Bay and offshore waters of the Bight (Figure 1). A Yankee otter trawl with 18.7-m headrope and 25-m sweep was used on larger vessels offshore, and another trawl with 7.8-m headrope and 9.3-m sweep was used on smaller vessels nearshore. The large net usually was towed for 30 minutes and the small net for 15-30 minutes; both nets were fitted with a 1.2-cm mesh nylon liner in the cod end.

Trap Studies

Caged winter flounder were placed at the apex sewage sludge dumpsite and at a control site southeast of Jones Inlet, Long Island (Murchelano and Ziskowski, 1976). The cages measured 1.5 x 1.2 x 0.6 m and were constructed of a welded frame of 2.5-cm concrete reinforcing bar stock enclosed with 5.0-cm stretch mesh nylon netting. The cages were deployed in chains of two or three with anchors and buoys. When on the bottom, the cages rested on a 1.5 x 1.2-m side. From 10 to 20 winter flounder measuring 22-27 cm (SL) were placed in each cage; the cages were placed at the designated sites for 12-29 days.

Histopathology

Tissues from winter flounder with fin rot were fixed in 10% seawater-formalin. Fin tissues were decalcified in RDO (DuPage Kinetic Laboratories[1]) before sectioning. All tissues were processed routinely for examination by light microscopy (Murchelano, 1975).

Bacteriology and Immunology

Fin tissues from winter flounder with fin rot were cultured bacteriologically. Blood was obtained from fish with advanced lesions and serum titrated against bacterial strains isolated from fin lesions and a variety of other bacterial strains. Details and results of the procedures employed are given elsewhere (Robohm *et al.*, 1979).

Chemistry

Levels of chlorinated hydrocarbons and trace metals were determined in winter flounder from three locations in the New York Bight and adjacent

[1] Reference to trade names does not imply endorsement of commercial products by the National Marine Fisheries Service.

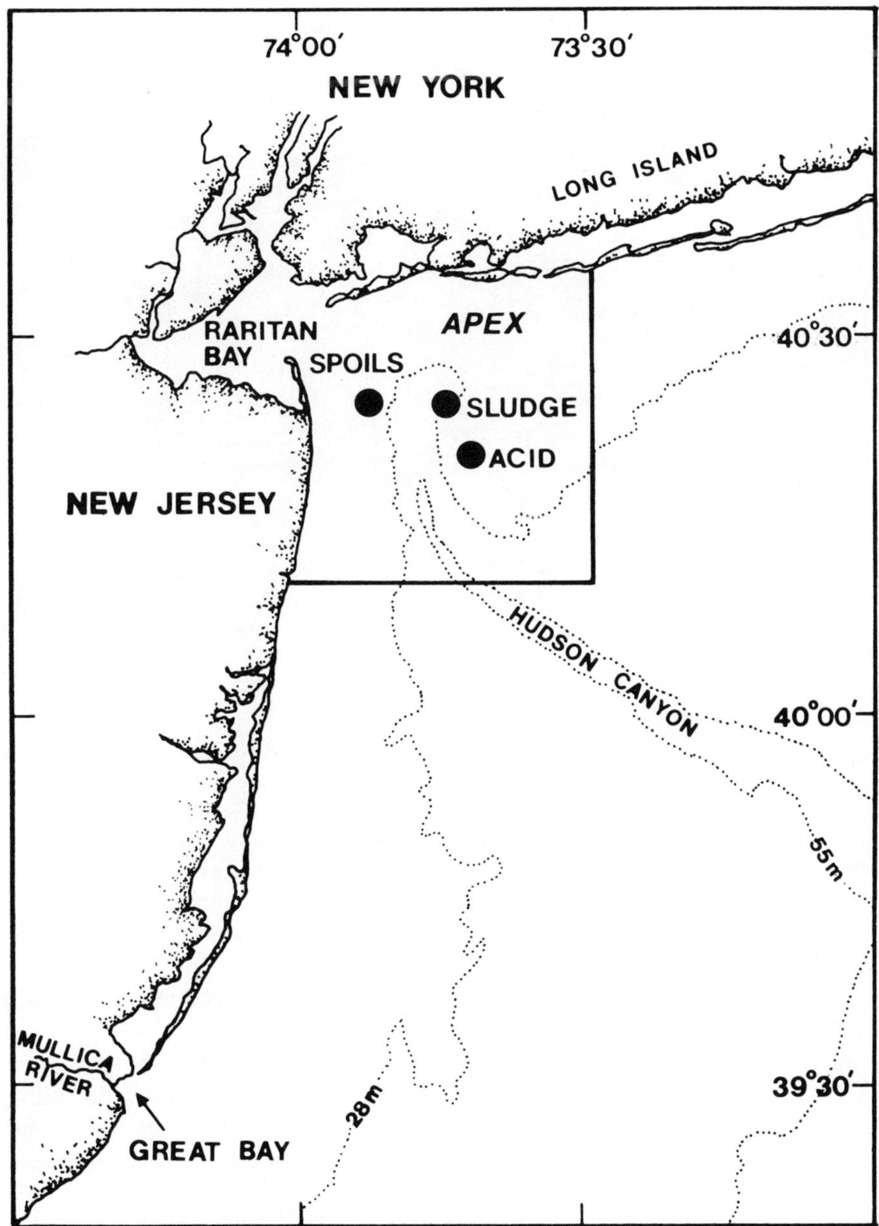

Figure 1. Fin rot prevalence study areas: Raritan Bay, apex, Great Bay, and portion of offshore area. Also indicated are the relative locations of the sewage sludge, dredged materials, and acid waste dumpsites in the New York Bight apex.

waters—Great Bay, the apex, and Sandy Hook/Raritan Bays. Details and results of the procedures employed are given elsewhere (Sherwood *et al.*, 1978).

Figure 2. Winter flounder with extensive fin rot of dorsal and anal fins.

RESULTS AND DISCUSSION

The Disease

Fins of fish with fin rot disease appear to be rotted or eroded (Figure 2). In its early stages, the disease affects the distal margin of the fin; however, as the disease progresses, more fin tissue is lost medially. Fin rot affects all fins—anal, caudal, dorsal, pectoral, and pelvic—but in flatfishes it is found more frequently on the anal and dorsal fins. In some fishes, destruction of fin epithelium proceeds more rapidly than destruction of fin rays, resulting in a fin stump with bare rays projecting beyond its distal margin. Hemorrhage and reparative tissue (granulation tissue) are present in varying degrees.

The most consistent histologic findings are epidermal hyperplasia and dermal fibrosis, accompanied by hemorrhage, congestion, and hyperplasia of eosinophilic granule cells and mucus cells (Murchelano, 1975). These observations are compatible with a chronic disease. Occasional inflammatory foci were observed, but no consistent leukocytic response was noted. It was not possible to demonstrate the presence *in situ* of bacteria, fungi, or other parasites in any of the histologic preparations of fin tissues examined. Examination of gills and viscera revealed no significant histopathology in the fish examined.

The Cause of the Disease

Histologic examination of fin tissues of winter flounder consistently has failed to reveal the presence of microorganisms. This, however, does not rule out the possibility of a microbial involvement. Unless present in substantial numbers, bacteria are difficult to see in histologic sections and could be missed. Winter flounder exhibit chronic, proliferative lesions that appear to have healed, and bacteria are not likely to be present in such lesions. Bacteria were observed in fin lesions of several summer flounder, *Paralichthys dentatus,* collected from the Bight (Murchelano, 1977). The fin lesions were characterized by loss of epidermal, dermal, and ray tissues, and exhibited a ragged or frayed appearance. When examined microscopically, these acute, nonproliferative lesions contained numerous bacteria and, unlike the chronic, proliferative lesions of winter flounder, also contained inflammatory cells. In fishes, leukocytic proliferation is affected by temperature and type of infectious agent; however, the absence of inflammatory cells generally signifies a nonmicrobial disease.

The bacteria isolated from fin lesions of winter flounder belong to the genera *Achromobacter, Flavobacterium, Pseudomonas,* and *Vibrio*—part of the normal flora of seawater. None of the lesions cultured contained known bacterial fish pathogens, nor were the lesions dominated by a single organism. There are no major differences in the types of bacteria present on the surfaces of healthy or eroded fins. Bacterial agglutinin titers in sera from winter flounder with fin rot disease were not elevated consistently against bacteria isolated from eroded fin tissues (Robohm *et al.,* 1979). Titers against a collection of known fish pathogens also were not elevated. No bacteria could be isolated from viscera of winter flounder with fin rot.

Trace contaminants, heavy metals, and chlorinated hydrocarbons were measured in winter flounder from the apex, Sandy Hook/Raritan Bays, and Great Bay (Sherwood *et al.,* 1978; Sherwood, this volume). Concentrations of cadmium, chromium, copper, lead, nickel, silver, and zinc were measured in liver and muscle by atomic absorption spectroscopy and expressed as mg/kg wet wt of tissue. In liver, concentrations of cadmium were significantly lower ($P < 0.05$) in nondiseased winter flounder from the apex (0.028) than nondiseased flounder from the Great Bay control site (0.082). In muscle, concentrations of copper were significantly higher ($P < 0.05$) in winter flounder with fin erosion from Sandy Hook/Raritan Bays (0.12) than in nondiseased flounder from Great Bay (0.10). Overall, concentrations of heavy metals from normal and diseased winter flounder from the three sites were within the range reported for this species (Greig and Wenzloff, 1977).

Concentrations of DDT and PCB's (mg/kg wet wt) were measured in liver and muscle by electron-capture gas chromatography. Values were considered significantly different if $P < 0.05$. In liver, concentrations of total PCB's were significantly higher in Sandy Hook/Raritan Bay winter flounder with fin rot (4.2) than in normal-appearing flounder from Great Bay (0.89). Concentrations of total PCB's, however, were significantly lower in livers of diseased winter flounder from Sandy Hook/Raritan Bays (4.2) than in livers of normal fish from the Bays (5.8). Concentrations of total PCB's were significantly higher in livers of normal apex winter flounder (4.2) than in livers of normal Great Bay flounder (0.89). In muscle, concentrations of total PCB's were significantly higher in

Sandy Hook/Raritan Bay winter flounder with fin rot (0.10) than in normal flounder from Great Bay (0.041). Concentrations of total PCB's also were significantly higher in muscle of normal apex winter flounder (0.11) than in muscle of normal Great Bay flounder (0.041). No differences were detected in concentrations of DDT from normal and diseased winter flounder from the three sites.

Attempts to induce fin rot disease in winter flounder by placing caged fish at the sewage dumpsite were largely unsuccessful (Murchelano and Ziskowski, 1976). Although the number of mortalities was less, and the condition of caged fish was better at the control site than the sewage dumpsite, the caudal fin lesions that developed on fish from both sites probably were due to the cages rather than the environment. Caudal fin lesions rarely are found in winter flounder from the Bight; most fin lesions involve the larger anal and dorsal fins. It is difficult to confine wild fishes in cages and expect them to respond to the environment rather than the cage.

The etiology of fin rot disease in winter flounder is not apparent from any of the research conducted to date. The location of fin lesions (approximate midpoints of anal and dorsal fins) suggests that the disease is initiated by contact with bottom sediments. A variety of microorganisms and noxious chemicals are present in bottom sediments of the New York Bight. Available data suggest that the etiology of fin rot disease is not infectious. It yet must be determined if noxious chemicals present in sediments are responsible for initiating the disease.

The Prevalence of the Disease

In addition to winter flounder, fin rot disease has been found in many benthic and pelagic fishes inhabiting the New York Bight (Mahoney *et al.*, 1973). As a consequence of survey cruises conducted from 1973 to 1977, the disease also has been noted in cunner, *Tautogolabrus adspersus;* four-spotted flounder, *Paralichthys oblongus;* tautog, *Tautoga onitis;* and yellowtail flounder, *Limanda ferruginea.* Survey activities within the Bight have focused on the winter flounder since the disease is most prevalent in this species.

Figure 3 presents data on the prevalence of fin rot disease in winter flounder from Sandy Hook/Raritan Bays, the apex, Great Bay, and outside the apex (control area). Examination of the data reveals that the highest disease prevalence was noted in 1973. At that time, 15.0% of apex winter flounder examined had fin rot, 8.0% of Sandy Hook/Raritan Bay flounder had the disease, and only 2.1% of flounder from the control area were diseased. In all years except 1975 and 1976, the highest prevalence was in the apex. Since 1973, there has been a decline in disease prevalence in all areas. There is no apparent reason for this decrease in prevalence. It is noteworthy that the volume of sewage sludge and dredged materials dumped in the Bight has not changed substantially during the five-year period (U.S. Environmentl Protection Agency, 1978).

Yearly differences in apex disease prevalence from 1973 to 1977 were

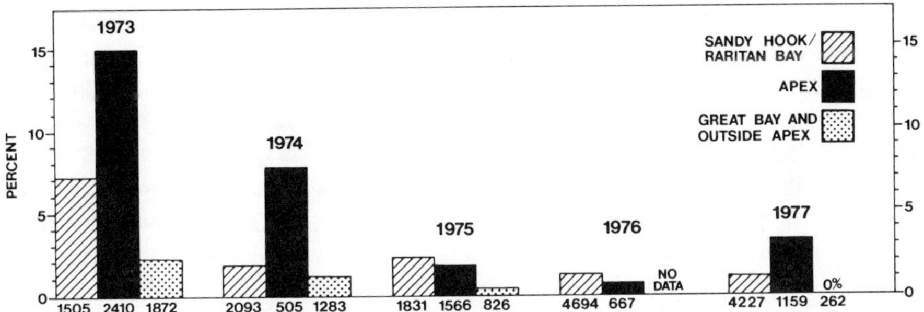

Figure 3. Fin rot prevalence in winter flounder from Raritan Bay, the apex, Great Bay, and offshore outside the apex, 1973-1977.

significant ($P < 0.05$). In Sandy Hook/Raritan Bays, yearly differences in disease prevalence were significant only between 1973 and 1974 and between 1976 and 1977. In 1973, 1974, and 1977, significant differences in disease prevalence were noted between the apex and Sandy Hook/Raritan Bays, the apex and control area, and Sandy Hook/Raritan Bays and control area. Averaging the prevalence of the disease from 1973 to 1977 by area reveals that the apex has the highest overall prevalence at 7.4%; summing the prevalence for all areas for the same time period reveals that of 21,760 winter flounder examined, 787 or 3.6% in the areas sampled had fin rot.

During the conduct of the field survey, 405 of the 787 winter flounder with fin rot disease were measured. Diseased fish were assigned to one of four size categories ranging from 1 cm to greater than 30 cm (SL). Ten flounder (2.5%) measured from 1 cm to 10 cm, 144 (35.5%) from 11 cm to 20 cm, 228 (56.3%) from 21 cm to 30 cm, and 23 (5.7%) greater than 30 cm. (Although 362 diseased winter flounder were not measured, there is no reason to believe they would have differed substantially in size from the sample measured.) From the resulting size-frequency distribution, it would appear that most of the winter flounder with fin rot were two- to three-year-old fish (56.3%). It is possible, however, that the fish were older than their size indicated, since pollution may have stunted growth.

IMPLICATIONS

For the Fish

Fin rot *per se* does not appear to be a fatal disease for winter flounder; however, it is not possible to state categorically that mortalities do not occur. Since bacteria are not the primary agents involved in pathogenesis, the sequence of events possible from acute bacterial infection is unlikely. Although some winter flounder with fin rot die when placed in laboratory aquaria, the aquarium environment is stressful to wild fish and does not optimize chances for survival. Even some apparently healthy winter flounder die when placed in laboratory aquaria.

The histologic appearance of winter flounder fin lesions does not suggest a progressive disease. The most constant characteristic of fin rot in Bight winter flounder is the presence of fin lesions that are undergoing resolution. If chemical irritants in sediments initiate such lesions, the necrosis produced is resolved through a healing process that restores the integrity of the damaged fin tissue.

Mortality could occur indirectly as a result of behavioral modifications. Flounder, in general, utilize their large anal and dorsal fins in swimming, moving across the bottom, burying, and assuming various postural stances involved with rest and capture of prey. If large areas of the anal and dorsal fins are damaged, activities dependent on their use may be impaired, particularly the ability to capture prey and avoid predators. Presently, any predictions of the outcome of fin rot disease in winter flounder are speculative. It is noteworthy that inferences on mortality have not been made for two other pleuronectids in which the disease is prevalent–Dover sole, *Microstomus pacificus,* from southern California (Mearns and Sherwood, 1974) and starry flounder, *Platichthys stellatus,* from the Duwamish River, Washington (Wellings *et al.,* 1976). Histologic observations of fin and other tissues were not consistent with an infectious disease, and experimental studies were not conducted to estimate mortality of diseased fishes.

For the Fishery

It is unlikely that fin rot disease in winter flounder presents a significant problem for the fishery in the New York Bight. The commercial fishery in the Bight is not extensive (McHugh, 1977); winter flounder is a minor species south of New York. The Maine to New York commercial catches of winter flounder from 1973 to 1975 were estimated at 8,716, 7,185, and 7,800 metric tons, respectively. For the same period, New Jersey to North Carolina landings were appreciably less at 75, 66, and 48 metric tons, respectively. Further evidence of the size of the winter flounder fishery in the Bight is provided by landings in New York and New Jersey. From 1973 to 1975, commercial landings in New York were 529, 253, and 266 metric tons, whereas in New Jersey they were 72, 64, and 48 metric tons. Contrasting the New Jersey to North Carolina landings with those from New Jersey alone, it is obvious that a commercial fishery of any size does not exist south of New Jersey.

In spite of increasing pollution, commercial landings of winter flounder have remained fairly constant for more than 20 years (Wise, 1974) and do not suggest that mortality from fin rot disease has caused significant population decline. From 1950 to 1974, the total winter flounder annual commercial catch has fluctuated between approximately 5,430 and 13,600 metric tons. It has been suggested that the catch size is maintained by increased fishing; however, this appears inconsistent with the general decline in United States coastal fisheries. Although commercial landings do not attest to significant recent mortality of winter flounder, catch data that do not consider catch per unit effort may be misleading.

Fishermen probably do not discard winter flounder with fin rot disease.

Although there is no accurate information on how winter flounder are marketed in the New York/New Jersey area, one can assume that they are marketed both in the round and filleted. Whether the filleting is done by fishermen or processors on shore, winter flounder with fin rot probably are not discarded since they yield fillets as marketable as nondiseased fish. If fin rot disease progressed to involve adjacent muscle tissue, then marketability might be affected. None of the Bight flounder examined had lesions that involved other than fin tissue; however, in a study conducted in Narragansett Bay, Rhode Island, winter flounder with fin and muscle lesions were noted (Levin *et al.*, 1972).

In addition to the commercial fishery, the shallow coastal waters of New York and New Jersey form the basis of an extensive winter flounder recreational fishery. In 1960 the recreational catch from New Jersey to North Carolina was 5,616 metric tons, in contrast to a commercial catch in the same area of only 56 metric tons. In 1965 and 1970, the recreational vs. commercial catches from New Jersey to North Carolina were 3,145 vs. 227 metric tons and 5,843 vs. 136 metric tons, respectively. Unfortunately, catch data by individual states is not available.

Recreational catches of winter flounder with fin rot disease may affect the fishery by altering its image. While it is unlikely that commercial fishermen will jeopardize the sale of their product by calling attention to a disease that does not limit its marketability, recreational fishermen may take a different perspective. Since income is not at stake, they may attempt to awaken environmental consciousness by calling attention to the disease and linking it to environmental degradation. If the public is aroused, and knowledgeable individuals fail to place the disease in its proper perspective, then the fishery may be adversely affected economically.

For the Consumer

It is unlikely that fin rot disease of winter flounder has any significant effect on the consumer. There would be obvious reluctance for consumers to purchase winter flounder with fin rot in the round. Although wholesome, such fish are not aesthetically attractive. Since the appearance of fishes sold in the round is a major determinant of marketability, fishermen selling fin rot-diseased fishes eventually would find their products unsaleable. The simple expedient of filleting the fish provides an acceptable solution to the aesthetic problem. Filleting diseased fishes eliminates a potential adverse image for the fishery and enables fishermen to effect the sale of diseased fishes. There is no known health hazard from the consumption of muscle tissue from winter flounder with fin rot disease. If future analyses for substances yet unspecified but hypothesized to be involved in the etiology of the disease prove to be beyond acceptable limits for human consumption, then there is a potential threat to the consumer and fishery. On the basis of present Food and Drug Administration standards, the trace contaminants that have been found in diseased winter flounder pose no threat to the consumer (see Sherwood, this volume).

For the Environmental Manager

The presence of fin rot disease in winter flounder and other species of demersal and pelagic fishes from the New York Bight is of importance to individuals engaged in environmental and resource management. The real value of these observations is that they indicate that the environment is affecting fish health. The most significant aspect of the research on fin rot disease, whether in the New York Bight, southern California, or Washington, is documentation that the disease is numerically prevalent in degraded environments. The significance of this is somewhat mitigated by the lack of information on prevalence before dumping activities commenced in the Bight, before ocean sewage outfalls were employed in southern California, and before the Duwamish River in Washington became a sink for the discharge of pollutants from shore-based industry. Despite this lack of historic perspective, however, when the prevalence of fin rot disease in degraded environments is contrasted with its prevalence in ecologically and hydrographically similar pristine environments, an environmental influence becomes suspect.

Fin rot disease appears to be a symptom of environmental stress. Although the disease does not appear to be consequential for the fish, fishery, or consumer, it signals that other events with more far-reaching effects may be occurring. Environments that are conducive to fin rot disease also may cause other diseases, some lethal and others sublethal. One can only speculate on the effects of sublethal disease; however, behavior, growth rate, and reproductive potential all may be affected. Unfortunately, the criteria for assessing fish health are rather limited, and mass mortality frequently provides the only clue of an unfavorable interaction between fishes and their environment. Although attrition from disease is occurring constantly, moribund fishes are not likely to be noted in most resource surveys.

There is little doubt that the ecology of the New York Bight has been altered by the diverse spectrum of pollutants it receives daily. Fin rot disease probably is a consequence of the collective presence of these pollutants. Although presently it is not possible to identify the specific pollutants responsible for fin rot, a high prevalence of the disease in a particular environment should alert environmental and resource managers to the need for surveillance. In the Bight both the spectrum and quantity of pollutants potentially significant to the etiology of fin rot and other diseases are so substantial that it virtually is impossible to implicate one over another. Therefore, it is unlikely that managers can advocate remedial measures based only on a demonstrated high prevalence of fin rot disease. In its proper perspective, and in environments that have not been so extensively polluted, fin rot disease provides a warning that events consequential to fish health are occurring. Together with other indicators of environmental quality, the disease provides further evidence of the effects of pollutants on living marine resources.

ACKNOWLEDGEMENTS

The authors wish to acknowledge the continued financial support of the MESA New York Bight Project Office during the conduct of this study. We also wish to thank Mr. Henry Fisler, Mr. John LeBaron, and Mr. Anthony Pacheco of the NMFS Sandy Hook Laboratory for their assistance in the conduct of field studies and in data analysis. We are indebted to Ms. Marjorie Sherwood of the Southern California Coastal Water Research Project for use of her data on chemical analyses for chlorinated hydrocarbons and heavy metals.

REFERENCES

Amlacher, E. 1970. Textbook of Fish Diseases. T.F.H. Publications, Inc., Jersey City, NJ. 302 pp.

Bullock, G.L. 1968. The bacteriology of brook trout with tail rot. Prog. Fish Cult. 30: 19-22.

Bullock, G.L., D.A. Conroy, and S.F. Snieszko. 1971. Diseases of Fishes. T.F.H. Publications, Inc., Jersey City, NJ. 151 pp.

Bullock, G.L. and S.F. Snieszko. 1970. Fin rot, cold water disease, and peduncle disease of salmonid fishes. U.S. Bur. Sport Fish. Wildl., Fish. Dis. Leafl. 25. 2 pp.

Davis, H.S. 1953. Culture and Disease of Game Fishes. University of California Press, Berkeley and Los Angeles. 332 pp.

Greig, R.A. and D.R. Wenzloff. 1977. Trace metals in finfish from the New York Bight and Long Island Sound. Mar. Pollut. Bull. 8: 198-204.

Klontz, G.W. and R.A. Bendele. 1973. Histopathological analysis of fin erosion in southern California marine fishes. Southern California Coastal Water Research Project. Rep. TM 203. 8 pp.

Levin, M.A., R.E. Wolke, and V.J. Cabelli. 1972. *Vibrio anguillarum* as a cause of disease in winter flounder (*Pseudopleuronectes americanus*). Can. J. Microbiol. 18: 1585-1592.

McHugh, J.A. 1977. Fisheries and fishery resources of New York Bight. NOAA Tech. Rep. NMFS Circ. 401. 50 pp.

Mahoney, J.B., F.H. Midlige, and D.G. Deuel. 1973. A fin rot disease of marine and euryhaline fishes in the New York Bight. Trans. Am. Fish. Soc. 102: 596-605.

Manfredi, E.T. 1976. A bacteriological study of fin erosion disease in Dover sole (*Microstomus pacificus*). M.S. Thesis, California State University, Long Beach. 94 pp.

Mearns, A.J. and M. Sherwood. 1974. Environmental aspects of fin erosion and tumors in southern California Dover sole. Trans. Am. Fish. Soc. 103: 799-810.

Murchelano, R.A. 1975. The histopathology of fin rot disease in winter flounder from the New York Bight. J. Wildl. Dis. 11: 263-268.

Murchelano, R.A. 1977. Histopathology of an acute fin lesion in the summer flounder, *Paralichthys dentatus,* and some speculations on the etiology of fin rot disease in the New York Bight. J. Wildl. Dis. 13: 103-106.

Murchelano, R.A. and J. Ziskowski. 1976. Fin rot disease studies in the New York Bight. *In:* Proceedings of the Symposium on the Middle Atlantic Continental Shelf and the New York Bight, M.G. Gross (ed.), Am. Soc. Limnol. Oceanogr. Spec. Symp. 2: 329-336.

Oppenheimer, C.H. 1958. A bacterium causing tail rot in the Norwegian codfish. Publ. Inst. Mar. Sci. Univ. Tex. 5: 160-162.

Robohm, R.A., C. Brown, and R.A. Murchelano. 1979. Comparison of antibodies in marine fish from clean and polluted waters of the New York Bight: relative levels against 36 bacteria. Appl. Environ. Microbiol. 38: 248-257.

Sherwood, M.J. This volume. Fin erosion, liver condition, and trace contaminant exposure in fishes from three coastal regions. pp. 359-377.

Sherwood, M.J., A.J. Mearns, D.R. Young, B.B. McCain, and R.A. Murchelano. 1978. A comparison of trace contaminants in diseased fishes from three areas. *In:* Report from Southern California Coastal Water Research Project, El Segundo, California to MESA New York Bight Project Office, State University of New York, Stony Brook, NY. 161 pp.

Sindermann, C.J. 1966. Diseases of marine fishes. Mar. Biol. 4: 1-89.

Sindermann, C.J. and A. Rosenfield. 1954. Diseases of fishes of the western North Atlantic. I. Diseases of the sea herring (*Clupea harengus*). Res. Bull. Maine Dep. Sea Shore Fish. 18. 23 pp.

Snieszko, S.F. 1953. Therapy of bacterial fish diseases. Trans. Am. Fish. Soc. 83: 313-330.

Snieszko, S.F. 1958. Fin rot and peduncle disease of salmonid fishes. U.S. Fish Wildl. Serv. Fish. Leafl. 462. 2 pp.

U.S. Environmental Protection Agency. 1978. Environmental impact statement on the ocean dumping of sewage sludge in the New York Bight. U.S. Environmental Protection Agency, Region II, New York. 226 pp.

Wellings, S.R., C.E. Alpers, and B.B. McCain. 1976. Fin erosion disease of starry flounder (*Platichthys stellatus*) and English sole (*Parophrys vetulus*) in the estuary of the Duwamish River, Seattle, Washington. J. Fish. Res. Board Can. 33: 2577-2586.

Wise, F.P. (ed.). 1974. The United States Marine Fishery Resource. NOAA/ NMFS MARMAP Contrib. 1. 379 pp.

Ziskowski, J. and R. Murchelano. 1975. Fin erosion in winter flounder. Mar. Pollut. Bull. 6: 26-29.

FIN EROSION, LIVER CONDITION, AND TRACE CONTAMINANT EXPOSURE IN FISHES FROM THREE COASTAL REGIONS

Marjorie J. Sherwood

Southern California Coastal Water
Research Project
646 West Pacific Coast Highway
Long Beach, California 90806

Abstract. Fin erosion in demersal fishes is one example of a response common to several regions receiving waste discharges. The regions include southern California, Puget Sound, and New York. Although the precise causes of the disease remain undetermined, fin erosion appears to result from the exposure of susceptible species to contaminated sediments. The present study demonstrates that in these three regions, the levels of total PCB's in muscle, liver, and brain tissues were higher in fishes with fin erosion from contaminated sites than in apparently unaffected specimens from control sites. PCB exposure may contribute to the development of the disease.

A second response common to the three regions involves changes in the size and/or lipid content of the liver. This response also may be related to chlorinated hydrocarbon exposure.

INTRODUCTION

Populations and communities of marine organisms are being studied to determine their responses to discharged wastes and to establish criteria for evaluating environmental health. The identification of similar responses in regions with different waste discharge histories suggests that a comparison of conditions in the regions might help to determine causative agents.

Fin erosion in demersal fishes is one example of a response common to several regions receiving waste discharges. The regions include southern California, Puget Sound, and New York. Studies in these three regions have revealed similarities in the distribution and nature of the disease. The prevalence of affected individuals was higher at contaminated sites than at control sites, and the fin lesions in the most frequently affected species did not appear to be the result of a microbial process.

The objectives of this paper are to compare liver condition and trace contaminant exposure of fishes with and without fin erosion from southern California, Puget Sound, and New York. The discussion focuses on studies conducted in southern California.

BACKGROUND

The coastal waters off southern California receive much of the liquid waste generated by the area's human population. There are five major sources of municipal wastewaters. These are the City of Los Angeles, which discharges wastewater effluent at a rate of 1,210 x 10^6 liters/day (319 mgd) through a 60-m deep, 8-km (5-mile) outfall, and 17 x 10^6 liters/day (4.6 mgd) of digested

Contribution 148 of the Southern California Coastal Water Research Project.

sludge through a 100-m deep, 11-km (7-mile) outfall into Santa Monica Bay; the County Sanitation Districts of Los Angeles County, which discharge approximately 1,270 x 10^6 liters/day (335 mgd) on the Palos Verde shelf; the County Sanitation Districts of Orange County, which discharge 680 x 10^6 liters/day (180 mgd) in San Pedro Bay; the City of San Diego, which discharges 438 x 10^6 liters/day (116 mgd) off Point Loma; and Ventura County, which discharges 42.3 x 10^6 liters/day (11.2 mgd) off Oxnard (approximately 40 km north of Santa Monica Bay) (Schafer, 1979) (Figure 1). Three major harbors also are located in southern California–the Los Angeles/Long Beach complex, which empties into northern San Pedro Bay, Newport Harbor, which empties into southern San Pedro Bay, and San Diego harbor.

In southern California, fin erosion was evident in 33 of 151 fish species collected from 1972 to 1976 (Sherwood and Mearns, 1977). The Dover sole (*Microstomus pacificus*) was the most frequently affected species. Other species frequently affected were the rex sole (*Glyptocephalus zachirus*), slender sole (*Lyopsetta exilis*), greenstriped rockfish (*Sebastes elongatus*), and vermilion rockfish (*S. miniatus*) (Sherwood and Mearns, 1977).

For most species with fin erosion, frequencies of the disease were highest in the vicinity of the wastewater discharge on the Palos Verdes shelf. Thirty-nine percent of the Dover sole collected there between 1972 and 1976 had eroded fins. In more recent surveys (1977), 28% of the calico rockfish (*S. dalli*) taken on the Palos Verdes shelf had fin erosion (Sherwood, 1979).

The disease appeared to develop on the Paos Verdes shelf. At a 137-m station northwest of the outfall pipes, less than 1% of the recently-settled Dover sole (on the bottom for about two months) had eroded fins. After almost eight months on the bottom, over 70% of the recently-settled individuals sampled exhibited the disease (Mearns and Sherwood, 1974; Sherwood and Mearns, in press).

Although the Palos Verdes shelf appeared to be the primary site at which fin erosion was being initiated, the disease also appeared to develop in Santa Monica Bay at the end of the 11-km (7-mile) outfall (Sherwood, 1979). The disease was not seen in fishes taken near the discharges of the City of San Diego and the County of Ventura, nor in the area of multiple, natural oil seeps around Coal Oil Point, just south of Point Conception.

Erosion in the Dover sole generally was restricted to the fins. However, some specimens also had slight gill hyperplasia (Klontz and Bendele, 1973). Fin lesions were characterized by epidermal hyperplasia, papillary folding of the epidermis and underlying dermis, few mucocytes and eosinophilic granular cells in advanced stages, a proliferation of dermal fibroblasts and fibrosis, an abnormal distribution of melanophores, focal hyperemia and hemorrhages, few inflammatory cells, and large basophilic cells associated with fin ray resorption (Pierce, personal communication).

In the Puget Sound area, fin erosion was confined to the Duwamish River estuary and was absent at nearby municipal outfall sites (Miller *et al.*, 1977) (Figure 2). The Duwamish estuary receives agricultural, industrial, and municipal wastes. Fin erosion was identified in two species of flatfishes–the starry

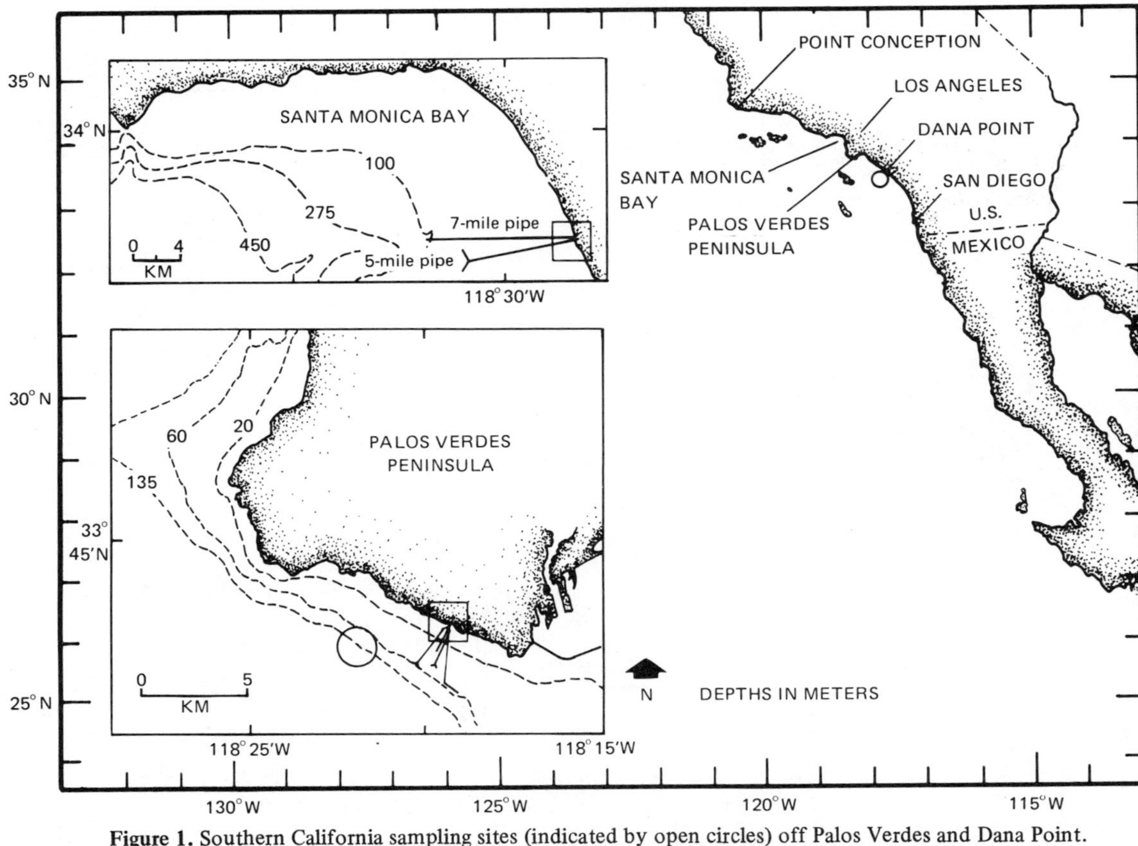

Figure 1. Southern California sampling sites (indicated by open circles) off Palos Verdes and Dana Point. Submarine outfall sites off Palos Verdes and in Santa Monica Bay indicated by open squares.

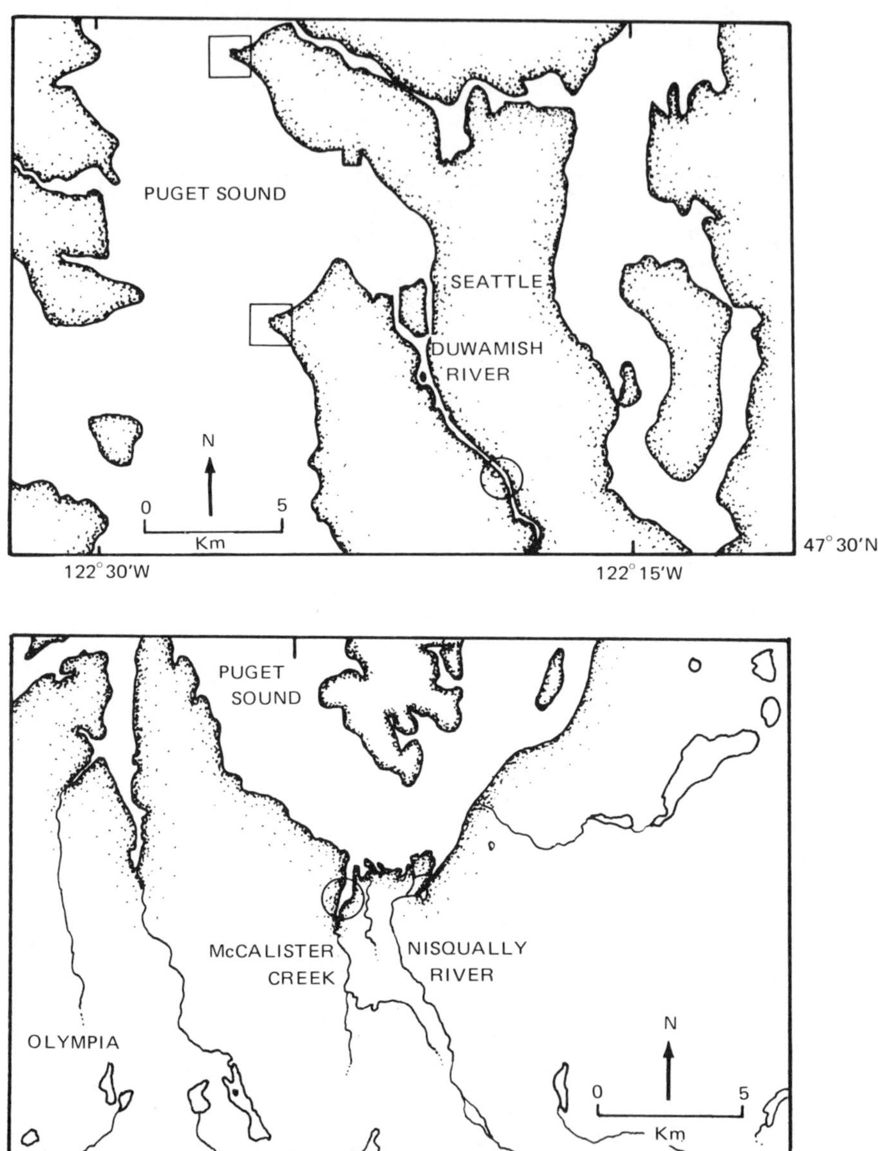

Figure 2. State of Washington sampling sites (indicated by open circles) in the Duwamish River, near Seattle, and in McCalister Creek, near Olympia. Submarine municipal wastewater discharge sites near Seattle indicated by open squares.

flounder (*Platichthys stellatus*) and the English sole (*Parophrys vetulus*) (Wellings *et al.*, 1976). In 1974 prevalence of the disease in starry flounder ranged from 1.6% in July to 14.6% in March, with an average monthly prevalence of 8%. Average monthly prevalence of the disease in English sole was 0.5%. In the starry flounder, the fin lesions were largely those of a

> chronic fibrosing disease with epidermal hyperplasia, proliferation of dermal fibroblasts, fibrosis, focal infiltrations of cells interpreted as lymphocytes, mucous cell changes, aggregation of melanophores in the vicinity of resorbing fin rays and around sclerotic blood vessels, patchy loss of melanophores from the dermal-epidermal junction, and an increase and spreading of eosinophilic granular cells in the epidermis [Wellings *et al.*, 1976, p. 2578].

Studies on the occurrence of fin rot in bottom fishes in the waters of the New York Bight were initiated in 1967 by Mahoney *et al.* (1973). Studies were continued in 1973 by Ziskowski and Murchelano (1975). Of five flatfish species in which fin erosion was observed, winter founder (*Pseudopleuronectes americanus*) was the most frequently affected. Prevalence of the disease was greater within the apex waste dumping area than outside the area (Figure 3). Within the apex, the prevalence of fin erosion was higher at stations with high sediment organic content (Murchelano and Ziskowski, 1976). At the enriched stations, 19.5% of the winter flounder had eroded fins in 1973, and 5.1% were affected in 1974-1975. At the stations with less organic carbon, 6.1% were affected in 1973 and 3.0% in 1974-1975. Prevalence of the disease also was higher in Sandy Hook/Raritan Bay than in the Great Bay, New Jersey control area.

In the winter flounder, evidence of the disease was restricted to the fins; anal, caudal, and dorsal fins were the most frequently affected.

> All lesions were characterized by epidermal hyperplasia and dermal fibrosis. The thickness of the hyperplastic epidermis and fibrotic dermis varied with the severity of the disease and the time of the year. Hyperemia and focal hemorrhage were common in the dermis of the fin stump. Occassionally a focal leukocytic inflammatory response was observed.
>
> Melanophores were very prominent around blood vessels in the fibrotic dermis, especially deep within the tissue. Eosinophilic granule cells (EGC), normally seen in the epidermis adjacent to the basement membrane, appeared to be more abundant in diseased tissues . . . [Murchelano, 1975, pp. 264-266].

METHODS

A comparison of trace contaminants in diseased and apparently healthy flatfishes from southern California, Puget Sound, and New York was conducted in 1976-1977. Analyses were performed on the species from each region most severely affected with fin erosion. For each species, specimens in three categories

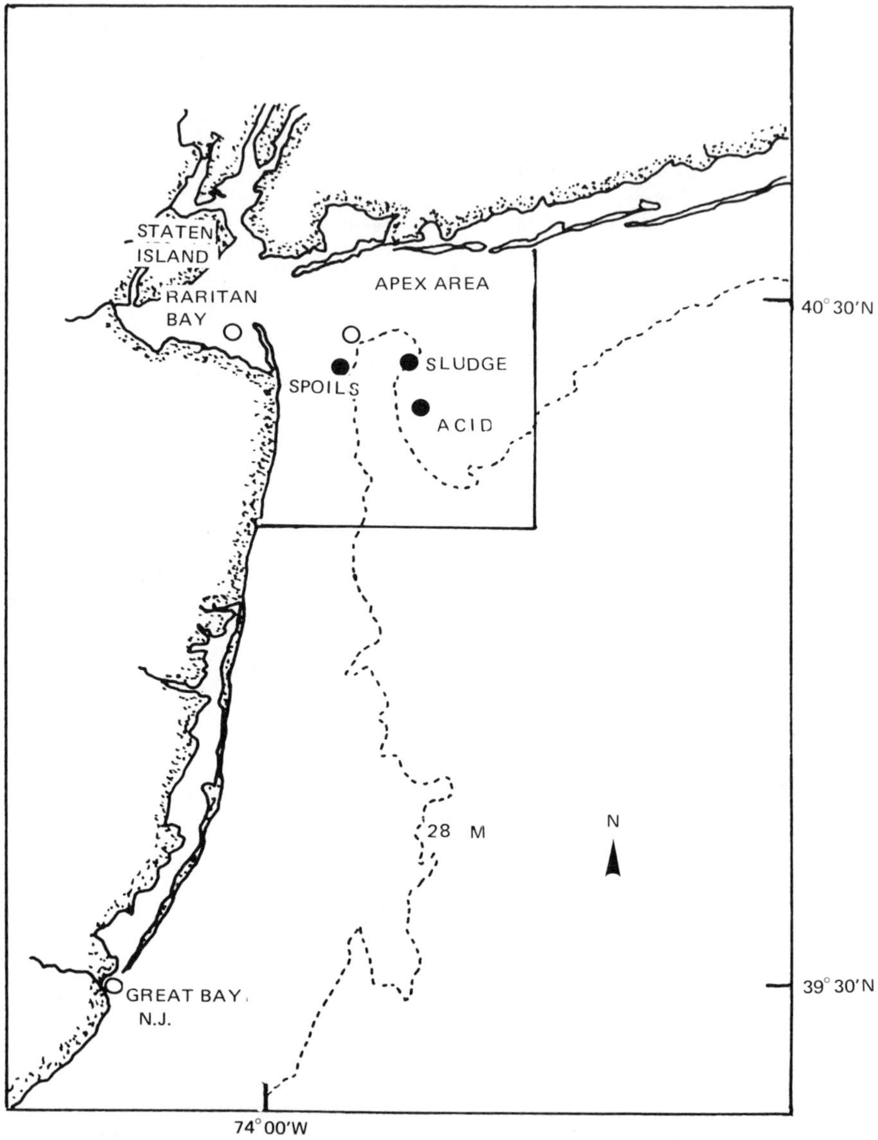

Figure 3. Sampling sites (indicated by open circles) in Raritan Bay, the New York Bight apex waste dumping area, and Great Bay, New Jersey. Locations of the dredged materials (spoils), sewage sludge, and acid wastes dumpsites indicated as closed circles.

were collected by small otter trawl–diseased individuals and apparently unaffected individuals from a contaminated site, and apparently unaffected individuals from a control site; the species and collection sites are listed in Table 1. Liver size was measured in each individual, and muscle, liver, and brain tissues were analyzed for chlorinated hydrocarbons. Total lipid content (expressed as a

Table 1. Species and collection sites examined in the comparison of trace contaminants.

Region	Species	Collection Sites	
		Contaminated	Control
Southern California	Dover sole *(Microstomus pacificus)*	Palos Verdes shelf	Dana Point
Puget Sound	Starry flounder *(Platichthys stellatus)*	Duwamish River	McCalister Creek
New York	Winter flounder *(Pseudopleuronectes americanus)*	Apex area and Sandy Hook/ Raritan Bay	Great Bay, N.J.

percentage by weight) was measured in liver tissue. Trace elements were measured in muscle and liver using atomic absorption spectroscopy. Tissues from 12 individuals in each category were analyzed whenever possible. Brain samples consisted of composites from three individuals. In addition, three sediment samples were taken at each fish collection site, and subsamples of the top 5 cm were analyzed for chlorinated hydrocarbons, metals, total acid-volatile sulfides, and total organic carbon.

The methods used were generally those employed in studies on the occurrence of fin erosion in southern California; these are described by Bligh and Dyer (1959), Young *et al.* (1976), Young and Heesen (1978), and Sherwood *et al.* (1980). Raw data for the tables presented in this paper are available in Sherwood *et al.* (1980).

RESULTS

Liver Condition

In winter flounder there were no statistically significant differences among liver-somatic index values (ratios of liver weight to body weight multiplied by 100) of fishes collected in Great Bay (New Jersey), Sandy Hook/Raritan Bay, or the apex dump area; however, there were differences in percent lipid ($p \leqslant 0.05$, two-tailed Mann-Whitney U-test) (Table 2). Specimens without eroded fins from Sandy Hook/Raritan Bay and the apex dump area contained greater amounts of lipid in the liver than specimens from Great Bay. In specimens with eroded fins, the percent lipid was comparable to that of Great Bay specimens.

In starry flounder, both liver-somatic index and percent lipid were higher in specimens from the contaminated Duwamish River than in specimens from the control site (McCalister Creek)—the index by a factor of about two and the percent lipid by a factor of about 5.5. Liver responses were present in specimens with

Table 2. Liver-somatic index and percent lipid in the livers of winter flounder, starry flounder, and Dover sole, 1976-1977.

Collection site and species	Number of specimens	Liver-somatic index, median and range	% lipid in liver, median and range	Specimen total weights, range (gm)
Winter flounder				
Great Bay, N.J.	12	2.2 (1.2-3.6)	12 (6.0-19)	213.8-424.6
Apex area				
No apparent fin erosion	12	2.2 (1.3-2.7)	18 (7.5-30)	116.8-149.4
Fin erosion	3	1.6 (1.5-2.6)	9.1 (7.2-20)	123.2-141.7
Sandy Hook/ Raritan Bays				
No apparent fin erosion	12	2.6 (1.7-3.3)	22 7.1-27)	112.8-293.4
Fin erosion	12	2.1 (1.7-3.1)	9.2 (6.0-29)	92.5-274.4
Starry Flounder				
McCalister Creek	12	1.5 (1.3-1.9)	4.5 (3.2-7.6)	139.6-284.2
Duwamish River				
No apparent fin erosion	12	3.0 (2.3-4.1)	26 (16-47)	82.5-168.5
Fin erosion	12	2.8 (2.1-3.9)	24 (12-34)	61.8-235.2
Dover Sole				
Dana Point	12	0.8 (0.6-1.2)	9.0 (0.6-1.2)	63.3-106.8
Palos Verdes shelf				
No apparent fin erosion	12	2.2 (0.9-3.0)	29 (8.8-35)	64.8-100.0
Fin erosion	12	2.6 (1.6-3.8)	28 (19-40)	75.5-101.4

and without eroded fins; however, there were no statistically significant differences in percent lipid in the livers of starry flounder from the Duwamish River with and without eroded fins.

In Dover sole, both liver-somatic index and percent lipid were higher in specimens from the Palos Verdes shelf than in specimens from Dana Point—both by a factor of about three. Liver responses were present in specimens with and without eroded fins from the contaminated area, as in starry flounder. Liver lipid contents of Dover sole from one station on the Palos Verdes shelf and one station off the Dana Point control area were compared with total PCB's in the muscle tissue of the specimens; the results suggested that increases in tissue burdens of chlorinated hydrocarbons were associated with changes in the magnitude of lipid infiltration of the liver (Table 3).

Trace Contaminants

There was no consistent pattern of metal concentrations in tissues of fishes from the three regions (Tables 4, 5, 6). In Dover sole, chromium and copper were significantly higher in liver samples of specimens with eroded fins from the contaminated site. In starry flounder, cadmium, copper, and zinc concentrations were lower, and silver was higher in liver samples of diseased specimens from the contaminated site. In winter flounder, there were too few specimens with fin erosion from the apex waste dumping area for a comparison. In specimens without fin erosion from that site, the cadmium concentration in liver tissue was significantly lower; however, there were no significant differences between the levels of metals in the liver samples of Raritan Bay specimens with eroded fins and those of apparently healthy specimens from Great Bay, New Jersey. Although patterns of trace metals in fish tissues varied among southern California, Washington, and New York, highly contaminated sediments were present in each of the geographical regions from which diseased specimens were taken (Table 7) (Carmody *et al.*, 1973; Wellings *et al.*, 1976; Greig and McGrath, 1977). The values presented in Table 7 may not represent the highest levels present in the contaminated sediments of each region.

Table 3. Association of liver lipid content with total PCB's in the muscle tissue of Dover sole from the Palos Verdes shelf and Dana Point, 1976.

Total PCB's (mg/kg, wet wt)	Number of specimens	Percent lipid (median and range)
0-0.1	14	9 (6.2-16)
> 0.1-0.5	3	23 (19-23)
> 0.5-1.0	6	28 (24-35)
> 1.0-1.5	7	32 (27-38)
> 1.5-2.0	4	30 (24-40)
> 2.0-2.5	2	28 (22-33)

Table 4. Median concentrations (mg/kg, wet wt) and ratios by area of trace metals analyzed by atomic absorption spectroscopy in the muscle and liver tissue of Dover sole from southern California, November 1976.

	Muscle			Liver		
	Dana Point (A) N=6	Palos Verdes fin erosion (C) N=6	C/A*	Dana Point (A) N=6	Palos Verdes fin erosion (C) N=6	C/A*
Silver**	< 0.005 to < 0.007	< 0.005 to < 0.006	–	0.006	0.11	–
Cadmium	0.002	0.004	–	0.86	0.52	–
Chromium	0.010	0.010	–	0.056	0.20	3.6
Copper	0.070	0.078	–	2.3	3.8	1.7
Nickel**	< 0.037 to < 0.047	< 0.026 to < 0.043	–	< 0.13 to < 0.32	< 0.05 to < 0.16	
Lead**	< 0.070 to < 0.087	< 0.069 to < 0.080	–	0.38	0.14	–
Zinc	1.9	2.1	–	23	26	–

*Ratios shown in cases where differences between collection sites were significant ($p \leqslant 0.05$) (two-tailed Mann-Whitney U-test).

**Upper-limit values influenced by sample weight. Ranges shown instead of median.

Total DDT levels in the tissues of diseased fishes from the contaminated area off Palos Verdes were two to three orders of magnitude higher than levels in the tissues of diseased specimens from contaminated sites in the other two regions (Table 8). However, total PCB levels in the diseased fishes from the three areas were similar. There were no statistically significant differences in brain tissue concentrations; median liver concentrations differed by factors of less than seven; and median muscle concentrations differed by factors of less than ten (Table 9). Levels were substantially above those in fishes from control areas (Table 10).

DISCUSSION

Fin Erosion

A number of possible causes of fin erosion disease in southern California demersal fishes were investigated. Among the first considered were infectious agents such as bacteria, fungi, and protozoa. By histological examination, fin

Table 5. Median concentrations (mg/kg, wet wt) and ratios by area of trace metals analyzed by atomic absorption spectroscopy in the muscle and liver tissue of starry flounder from Washington, March-April 1977.

	Muscle			Liver		
	McCalister Creek (D) N=6	Duwamish R. estuary fin erosion (F) N=6	F/D*	McCalister Creek (D) N=6	Duwamish R. estuary fin erosion (F) N=6	F/D*
Silver**	< 0.005 to < 0.006	< 0.005	–	0.068	0.14	2.1
Cadmium	0.007	0.002	0.29	0.53	0.16	0.30
Chromium**	< 0.008 to 0.014	< 0.008 to 0.015	–	0.045	0.040	–
Copper	0.16	0.16	–	6.4	4.4	0.69
Nickel**	< 0.030 to < 0.047	< 0.033 to < 0.044	–	< 0.05 to < 0.19	< 0.07 to < 0.14	–
Lead**	< 0.068 to < 0.083	< 0.073 to < 0.088	–	0.14	0.22	–
Zinc	3.4	3.8	–	36	28	0.78

*Ratios shown in cases where differences between collection sites were significant ($p \leqslant 0.05$) (two-tailed Mann-Whitney U-test).

**Upper limit values influenced by sample weight. Ranges shown instead of median.

lesions were characterized as having a minimal inflammatory response; bacteria, fungi, and protozoa were not seen in histological sections or in a small number of photographs taken by scanning electron microscopy (Klontz and Bendele, 1973; Cundell, 1976; Pierce, personal communication). Bacterial culture studies of specimens with and without eroded fins, from contaminated and control areas, revealed no systemic infectious agent (Manfredi, 1976). Although pathogens were isolated from all specimens, there were no specific pathogens in abundance at the site of the lesions.

A number of noninfectious processes also were considered as possible causes of fin erosion disease. In southern California specimens, these included irritation by macroparasites, fin-nipping, and exposure to sediments. No macroparasites were reported in association with fin lesions in Dover sole collected in trawl surveys. In the laboratory, fin-nipping did occur, but observations indicated that individuals most often were nipped in the head region or in the area of the caudal fin and not on the fins most frequently eroded in field specimens. The

Table 6. Median concentrations (mg/kg, wet wt) and ratios by area of trace metals analyzed by atomic absorption spectroscopy in the muscle and liver tissue of winter flounder from the waters of the New York area, 1977. Only three specimens with fin erosion were collected from apex dump area.

	Great Bay (G) N=6	Raritan Bay fin erosion (I) N=6	Apex dump area (J) N=6	I/G*	J/G*
Muscle					
Silver	0.002	0.004	0.002	–	–
Cadmium	0.001	< 0.001	< 0.001	–	–
Chromium**	< 0.005	0.010	0.008	–	–
Copper	0.10	0.12	0.11	1.2	–
Nickel**	< 0.019 to < 0.023	< 0.019 to < 0.023	< 0.020 to < 0.021	–	–
Lead**	< 0.022 to < 0.027	< 0.022 to < 0.030	< 0.018 to < 0.024	–	–
Zinc	4.3	4.2	3.7	–	–
Liver					
Silver	0.10	0.14	0.096	–	–
Cadmium	0.082	0.14	0.028	–	0.34
Chromium**	< 0.018 to 0.047	< 0.025 to 0.033	< 0.032 to 0.066	–	–
Copper	3.5	4.1	4.0	–	–
Nickel**	< 0.030 to < 0.11	< 0.094 to < 0.26	< 0.12 to < 0.18	–	–
Lead**	< 0.061 to 0.39	< 0.076 to 0.21	< 0.099 to 0.21	–	–
Zinc	28	27	30	–	–

*Ratios shown in cases where differences between collection sites were significant ($p \leqslant 0.05$) (two-tailed Mann-Whitney U-test).

**Upper limit values influenced by sample weight. Ranges shown instead of median.

Table 7. Metals (mg/kg, dry wt) in sediments collected off California, Washington, and New York, 1976-1977.

	Palos Verdes shelf[a]	Duwamish R. estuary[b]	Sandy-Hook/ Raritan Bay[c]	Apex dump area[d]	Dana Point[e]	McCalister Creek[f]	Great Bay, N.J.[g]
Silver							
Median	15			0.9			
Range	14-15	< 1.5-< 2.8	< 1.5-1.4	0.9-1.1	< 1.2-< 1.4	< 0.5-< 1.0	< 0.7-< 1.0
Cadmium							
Median	52	1.1	0.9	1.3			
Range	47-57	1.0-1.5	0.8-1.0	1.1-1.3	< 0.8-< 0.9	< 0.3-< 0.6	< 0.5-< 0.6
Chromium							
Median	1,100	43	53	52	18	8.5	20
Range	1,000-1,300	29-57	38-55	44-55	13-19	7.7-8.8	18-44
Copper							
Median	620	50	38	36	8.3	10	9.2
Range	570-650	49-52	37-43	31-42	8.1-9.4	8.4-12	9.0-21
Nickel							
Median	81	21	13	11	12	5.9	11
Range	76-91	20-25	12-14	8.5-12	10-12	4.8-6.4	10-25
Lead							
Median	410	80	45	57	3.5	2.8	13
Range	360-440	73-81	41-51	54-59	3.0-5.0	0.6-3.6	12-31
Zinc							
Median	1,700	130	100	100	44	22	49
Range	1,600-1,900	110-140	100-110	91-110	44-46	21-23	48-115

a, e Collected 5 and 18 November 1976, respectively, with a modified Van Veen II grab.
b, f Collected 25 and 31 March 1977, respectively, with an Eckmann grab.
c Collected 13 July 1977 with a Smith-McIntyre grab.
d Collected 5 October 1977 with a Smith-McIntyre grab.
g Collected 11 November 1976 with a Smith-McIntyre grab.

Table 8. A comparison of total DDT concentrations in the tissues of Palos Verdes Dover sole with fin erosion (PV), Duwamish River starry flounder with fin erosion (DR), and Raritan Bay winter flounder with fin erosion (RB).

	PV/DR*	PV/RB*
Muscle (N = 12)	500	890
Liver (N = 12)	190	500
Brain (N = 4)	340	200

*Ratios shown only if differences between sample sets were statistically significant ($p \leqslant 0.05$) (two-tailed Mann-Whitney U test).

Table 9. Levels of total PCB's (mg/kg, wet wt) in the tissues of Dover sole, starry flounder, and winter flounder with fin erosion, 1976-1977.

	Dover sole (Palos Verdes shelf)	Starry flounder (Duwamish R. estuary)	Winter flounder (Sandy Hook/ Raritan Bays)	Winter flounder (Apex area)
Muscle				
Number of specimens	12	12	12	3
Median value	0.98	0.41	0.10	0.093
Range	0.16-2.4	0.16-2.1	0.073-0.14	0.079-0.11
Liver				
Number of specimens	12	12	11	3
Median value	20	16	4.2	2.8
Range	7.4-56	9.0-160	1.4-7.4	2.0-10
Brain				
Number of composite samples	4	4	4	1
Median value	1.0	1.1	1.8	0.47
Range	0.83-1.3	0.67-3.1	0.46-2.0	–

data did suggest that exposure to contaminated sediments was a factor in the development of the disease for the following reasons:

- The fins contacting the bottom most frequently–the dorsal and anal fins (especially the midportion, upon which the Dover sole supports itself for feeding)–were the most frequently and severely affected (Mearns and Sherwood, 1974). The pectoral fin on the blind side, which extends into

Table 10. Levels of total PCB's (mg/kg, wet wt) in the tissues of Dover sole, starry flounder, and winter flounder without fin erosion from control areas, 1976-1977. (N.D. = not detected).

	Dover sole (Dana Point)	Starry flounder (McCalister Creek)	Winter flounder (Great Bay, N.J.)
Muscle			
Number of specimens	12	12	12
Median value	0.008	0.026	0.041
Range	N.D.-0.021	0.007-0.076	0.005-0.14
Liver			
Number of specimens	12	12	12
Median value	0.048	0.15	0.89
Range	N.D.-0.18	0.025-1.1	0.25-2.4
Brain			
Number of composite samples	4	4	4
Median value	0.12	0.23	0.16
Range	0.025-0.14	0.056-0.29	0.035-0.28

the sediment, was eroded more often than the one that projects into the water column.

- Apparently healthy Dover sole from the Dana Point and Santa Catalina Island control areas were exposed in the laboratory to sediments from the most contaminated region of the Palos Verdes shelf. At the end of 1.5 years, three of the four test animals had a fin disease that resembled early stages of fin erosion seen in field specimens; none of the five control individuals were affected (Sherwood, 1976; Sherwood and Mearns, 1977).
- In recently-settled Dover sole, prevalence of fin erosion was highest where bottom sediment contamination was highest on the Palos Verdes shelf (Sherwood, 1979).

For the contaminants considered, the common factors among the four disease sites were exposure to PCB's, as reflected in tissue concentrations, and exposure to metals in sediment that were not readily available for uptake. The comparisons suggest that PCB exposure may contribute to the development of fin erosion in bottom fishes. However, degree of fin erosion did not correlate with the concentration of PCB's in the tissues (Sherwood *et al.,* 1980).

Liver Condition

In winter flounder, it appeared that fin erosion was associated with a reduction in liver lipid content, compared to apparently unaffected specimens from the disease areas. This difference may be a function of reduced feeding in the fish with eroded fins. Tyler and Dunn (1976) measured liver weight and fat content (determined by the ether extraction method of Leim, 1957) in winter flounder fed different rations in the laboratory and in specimens taken directly from Passamaquoddy Bay, New Brunswick. Liver weight was associated with food consumption in the laboratory fish. At most, fat was 12% of the total liver weight, and changes in fat content alone could not account for most of the ten-fold change in liver size. In fish taken directly from the Bay, percent liver fat and liver index (in eviscerated specimens) increased in the summer and decreased in early September with ovarian development.

Abnormal livers had been reported previously by Miller *et al.* (1976) in 90% of the starry flounder collected in 1975 from the Duwamish River. Livers were yellow in color and enlarged. Mean index values were 2.42 and 1.07 for specimens from the Duwamish and Nisqually Rivers (Figure 2), respectively. Histologically, the livers had vacuolated hepatocytes and loss of normal structure. Fish from the Duwamish River also had high levels of serum cholesterol, which suggest liver damage (Wellings *et al.*, 1976). In specimens collected between May 1975 and January 1976, Pierce (1976) reported liver color abnormalities in 92% of the starry flounder from the Duwamish River. All individuals with fin erosion that were examined had liver color abnormalities; 90% of those without eroded fins had abnormal liver color. In McCalister Creek, 13% of the apparently unaffected organisms had liver color abnormalities. Histologically, the livers of specimens with eroded fins were among thos most severely damaged.

Liver abnormalities had been reported previously in Palos Verdes Dover sole. In addition to increases in liver size and lipid content, histological changes also were observed (Pierce *et al.*, 1977). There was an increase in the size and irregularity of fatty vacuoles in liver cells. The structure of the tissue showed evidence of disarray, and, in some cases, the disappearance of sinusoidal spaces. Melano-macrophage centers (collections of cells containing yellow-brown to black pigment) were larger and more numerous; there appeared to be an increase in the amount of interstitial fibrotic tissue present, and there were areas of necrosis.

Although liver size and fat content are affected by factors such as diet, reproductive condition, and disease, they also appear to change in relation to exposure to chlorinated hydrocarbons (Kimbrough, 1974; Couch, 1975).

CONCLUSIONS AND IMPLICATIONS

The precise causes of the fin erosion disease remain undetermined. However, the erosion pattern of the fins and the association of higher prevalences of fin erosion with greater degrees of sediment contamination suggest that fin-sediment contact in an area where toxic wastes have accumulated on the bottom is an important factor in disease development. The present study demonstrated that, in all three regions, the levels of total PCB's in muscle, liver, and brain tissues were higher in fishes with fin erosion from the contaminated areas than in apparently unaffected specimens from the control areas. It may be that PCB exposure of susceptible species is a factor in the occurrence of the disease in a given area.

Changes in liver size and/or lipid content were recorded in specimens with and without eroded fins from the contaminated areas. Although these types of changes can be affected by diet, reproductive condition, and disease, they also appear to take place as a result of exposure to chlorinated hydrocarbons. Increases in liver size and/or lipid content documented in this study may be a response to DDT and/or PCB exposure.

The significance of fin erosion is as an indicator of diseased fishes and degraded environments. For the most frequently affected species in each region, the disease does not appear to alter the success of the population as a whole.

In all three regions, but particularly at sites in southern California and Puget Sound, the occurrence of eroded fins is associated with liver abnormalities. The lesions are extensive in certain individuals, and, as a result, these individuals may be incapable of normal growth and reproduction.

The occurrence of fin erosion appears to be associated with the exposure of susceptible species to sediments contaminated with chemical wastes. Although the etiology is unknown, the disease does signal the existence of a degraded environment.

Although the fin condition itself does not pose a threat to public health, consumption of fishes with eroded fins may be a problem. In all three regions, fishes with eroded fins had elevated levels of contaminants, and some southern California and Puget Sound specimens exceeded the Food and Drug Administration guidelines for total DDT and/or total PCB's (Table 9) (Sherwood *et al.*, 1980).

ACKNOWLEDGEMENTS

This program depended on the cooperation and efforts of people from several laboratories. I thank Project staff members Dr. Alan Mearns, Dr. David Young, Tsu-Kai Jan, Theadore Heesen, Richard Gossett, Jean Wright, Harold Stubbs, and Michael Moore. I also thank staff members of the Los Angeles and Orange County Sanitation Districts and the City of Los Angeles. Specimens were collected from Puget Sound with the cooperation of Dr. Bruce McCain and Ken Pierce, National Marine Fisheries Service, and from the New York Bight, with the cooperation of Dr. Robert Murchelano and John Ziskowski, National Marine Fisheries Service.

This work was supported in part by Grant No. R801152 from the U.S. Environmental Protection Agency and Grant No. 04-7-022-44002 from the National Marine Fisheries Service, MESA/NOAA New York Bight Project.

REFERENCES

Bligh, E.G. and W.J. Dyer. 1959. A rapid method of total lipid extraction and purification. Can. J. Biochem. Physiol. 37: 911-917.

Carmody, D.J., J.B. Pearce, and W.E. Yasso. 1973. Trace metals in sediments of the New York Bight. Mar. Pollut. Bull. 4: 132-135.

Couch, J.A. 1975. Histopathological effects of pesticides and related chemicals on the livers of fishes. *In:* The Pathology of Fishes, W.E. Ribelin and G. Migaki (eds.), University of Wisconsin Press, Madison, WI. pp. 559-584.

Cundell, A. 1976. [Photograph] *In:* Annual Report, 1976 of the Coastal Water Research Project, Southern California Coastal Water Research Project, El Segundo, CA. 142 pp.

Greig, R.A. and R.A. McGrath. 1977. Trace metals in sediments of Raritan Bay. Mar. Pollut. Bull. 8: 188-192.

Kimbrough, R.D. 1974. The toxicity of polychlorinated polycyclic compounds and related chemicals. *In:* Critical Reviews Toxicology, Vol. 2, CRC Press, Inc., West Palm Beach, FL. pp. 445-498.

Klontz, G.W. and R.A. Bendele. 1973. Histopathological analysis of fin erosion in southern California marine fishes. Southern California Coastal Water Research Project Rep. TM 203. 8 pp.

Leim, A.H. 1957. Fatness of herring in Canadian Atlantic waters. Bull. Fish. Res. Board Can. 111: 117-184.

Mahoney, J.B., F.H. Midlige, and D.G. Deuel. 1973. A fin rot disease of marine and euryhaline fishes in the New York Bight. Trans. Am. Fish. Soc. 102: 596-605.

Manfredi, E. 1976. A bacteriological study of fin erosion disease in Dover sole (*Microstomus pacificus*). M.S. Thesis, California State University, Long Beach. 94 pp.

Mearns, A.J. and M.J. Sherwood. 1974. Environmental aspects of fin erosion and tumors in southern California Dover sole. Trans. Am. Fish. Soc. 103: 799-810.

Miller, B.S., B.B. McCain, R.C. Wingert, S.F. Borton, and K.V. Pierce. 1976. Ecological and disease studies of demersal fishes near METRO-operated sewage treatment plants in Puget Sound and the Duwamish River. One-year Progress Rep. to METRO, Contract No. CR 2231. 135 pp.

Miller, B.S., B.B. McCain, R.C. Wingert, S.F. Borton, K.V. Pierce, and D.T. Griggs. 1977. Ecological and disease studies of demersal fishes in Puget Sound near METRO-operated sewage treatment plants and in the Duwamish River. Final Report to METRO, Contract No. CR 2231. 164 pp.

Murchelano, R.A. 1975. The histopathology of fin rot disease in winter flounder from the New York Bight. J. Wildl. Dis. 11: 263-267.

Murchelano, R.A. and J. Ziskowski. 1976. Fin rot disease studies in the New York Bight. Am. Soc. Limnol. Oceanogr. Spec. Symp. 2: 329-336.

Pierce, K.V. 1976. Description and incidence of abnormal livers and other organs in starry flounder and English sole from the Duwamish River, Seattle, Washington, and other nearby Puget Sound sites. M.S. Thesis, University of Washington. 108 pp.

Pierce, K.V., B.B. McCain, and M.J. Sherwood. 1977. Histology of liver tissue from Dover sole. *In:* Annual Report, 1977 of the Coastal Water Research Project, Southern California Coastal Water Research Project, El Segundo, CA. pp. 207-212.

Schafer, H. 1979. Characteristics of municipal wastewater discharges, 1977. *In:* Annual Report, 1977 of the Coastal Water Research Project, Southern California Coastal Water Research Project, El Segundo, CA. pp. 97-102.

Sherwood, M. 1976. Fin erosion induced in the laboratory. *In:* Annual Report, 1976 of the Coastal Water Research Project, Southern California Coastal Water Research Project, El Segundo, CA. pp. 149-154.

Sherwood, M. 1979. The fin erosion syndrome. *In:* Annual Report, 1978, Coastal Water Research Project, Southern California Coastal Water Research Project, El Segundo, CA. pp. 203-221.

Sherwood, M.J. and A.J. Mearns. 1977. Environmental significance of fin erosion in southern California demersal fishes. Ann. N.Y. Acad. Sci. 298: 177-189.

Sherwood, M.J. and A.J. Mearns. In press. Fate of post-larval bottom fishes in a highly urbanized coastal zone. ICES Symposium on the Early Life History of Fish, 2-5 April 1979.

Sherwood, M.J., A.J. Mearns, D.R. Young, B.B. McCain, and R.A. Murchelano. 1980. A comparison of trace contaminants in diseased fishes from three areas. *In:* Report from Southern California Coastal Water Research Project, El Segundo, California to MESA New York Bight Project Office, State University of New York, Stony Brook, NY. 131 pp.

Tyler, A.V. and R.S. Dunn. 1976. Ration, growth, and measures of somatic and organ condition in relation to meal frequency in winter flounder, *Pseudopleuronectes americanus,* with hypotheses regarding population homeostasis. J. Fish. Res. Board Can. 33: 63-75.

Wellings, S.R., C.E. Alpers, B.B. McCain, and B.S. Miller. 1976. Fin erosion disease of starry flounder (*Platichthys stellatus*) and English sole (*Parophrys vetulus*) in the estuary of the Duwamish River, Seattle, Washington. J. Fish. Res. Board Can. 33: 2577-2586.

Young, D.R. and T.C. Heesen. 1978. DDT's, PCB's and chlorinated benzenes in the marine ecosystem off southern California. *In:* Water Chlorination: Environmental Impacts and Health Effects, Vol. 2, R.L. Jolley, H. Gorshev, and D.H. Hamilton, Jr. (eds.), Ann Arbor Science Publishers, Inc., Ann Arbor, MI. pp. 267-290.

Young, D.R., D.J. McDermott, and T.C. Heesen. 1976. DDT in sediments and organisms around southern California outfalls. J. Water Pollut. Control Fed. 48: 1919-1928.

Ziskowski, J. and R. Murchelano. 1975. Fin erosion in winter flounder. Mar. Pollut. Bull. 6: 26-29.

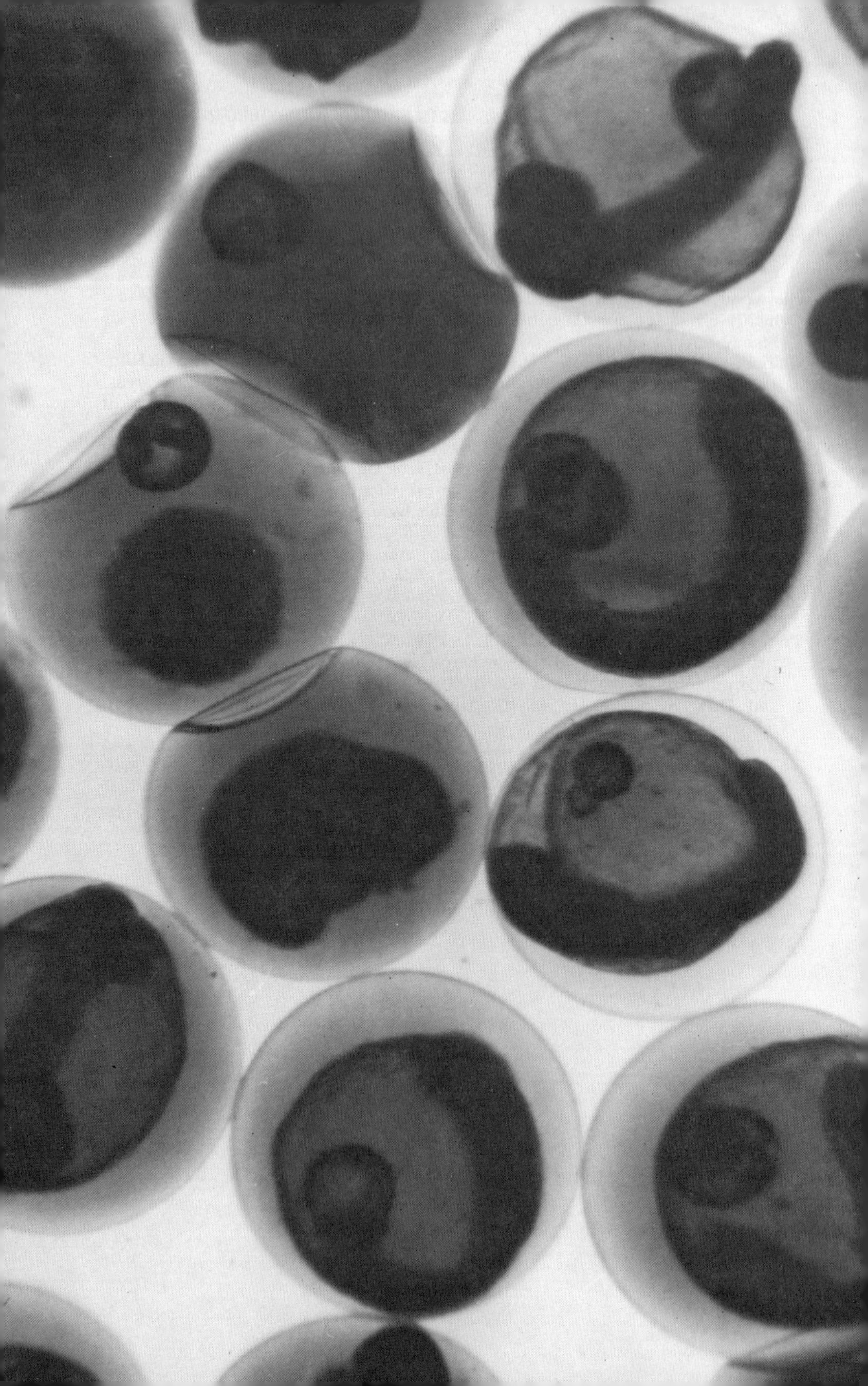

CONTRIBUTED PAPERS ON THE EFFECTS OF POLLUTANTS ON PLANKTON AND NEUSTON

Opposite: A collection of planktonic Atlantic mackerel eggs from surface waters of the New York Bight. Investigations suggest that the development of such eggs may be affected by toxicants in the Bight water column. (Courtesy A.C. Longwell, NOAA National Marine Fisheries Service.)

CYTOLOGIC, CYTOGENETIC, AND EMBRYOLOGIC STATE OF ATLANTIC MACKEREL EGGS FROM SURFACE WATERS OF THE NEW YORK BIGHT IN RELATION TO POLLUTION

A. Crosby Longwell
J.B. Hughes

National Marine Fisheries Service
Northeast Fisheries Center
Milford Laboratory
Milford, Connecticut 06460

Abstract. Cytologic, cytogenetic, and embryologic investigations were conducted on developing Atlantic mackerel (*Scomber scombrus*) eggs collected from the surface waters of the New York Bight. Samples were taken in May 1974 and May 1977. Cleavage through tail-free embryo stages was examined.

Development rates (as calculated by mitotic index), viability (as calculated from early indicators of cell death), and chromosome-mitotic abnormalities in mackerel embryos varied widely over the sampling area in both 1974 and 1977. In 1974, sufficient sites were sampled to enable their grouping for statistical tests of area differences in mackerel egg health. The 1974 stations showed significantly lower egg viability in areas of the Bight that generally are impacted by contaminants from the Hudson-Raritan estuary and apex dumpsites.

In 1977, gross embryo malformations were frequent but geographically variable. The 1977 data exhibited statistically significant associations between sytologic, cytogenetic, and embryologic measures of mackerel egg health, toxic metal levels in surface water samples, and toxic metal and toxic hydrocarbon levels in zooplankton samples from the Bight.

INTRODUCTION

Cytological and cytogenetic criteria are sensitive indicators of embryo health because normal chromosome and cell division and cell differentiation must occur at prescribed rates for development to be successful. In 1974, research was begun to examine the relationship between the embryological health of Atlantic mackerel (*Scomber scombrus*) and pollution in the New York Bight. This paper summarizes the results of this investigation and reviews the recent literature in relation to field findings on the Atlantic mackerel. A more comprehensive discussion of this research is provided by Longwell and Hughes (1980).

RESULTS OF MACKEREL EGG STUDIES

Collections of planktonic mackerel eggs were made in the New York Bight during the 1974, 1977, and 1978 spawning seasons. To date, only the data from the 1974 and 1977 studies have been completely analyzed.

Initial cytologic-cytogenetic analyses involved 4,030 embryos collected in May 1974 from surface plankton samples at 50 sites in the New York Bight. Seven developmental stages were represented in the collections: cleavage, morula, blastula, gastrula, early embryo, tail-bud embryo, and tail-free embryo. These were examined for general cell condition, the state of chromosomes and chromosome divisions in dividing cells, and the rate of chromosome divisions (mitotic index). Moribundity estimates were calculated on the basis of the seriousness of mitotic disarray of chromosomes, cessation of mitosis, and obvious cell lysis (Longwell and Hughes, 1980).

Cytological and cytogenetic abnormalities frequently were observed in the 1974 samples from some Bight sample sites. Many samples of normal-appearing embryos exhibited high incidences of aberrant chromosome divisions. In some embryos all mitoses were highly irregular. Fewer abnormalities were present at and after gastrulation than before. The mitotic index of embryos varied widely across developmental stages and between sample sites, with no relationship to temperature. However, highly significant correlation coefficients for moribundity and mitotic index were observed between all development stages examined except cleavage (Longwell and Hughes, 1980).

Mortality-moribundity data for gastrula and early-embryo stages were available from more stations (35) than for any other stage studied. Data from these stages therefore were used to ascertain whether mackerel egg mortality (calculated on the basis of cytological criteria) varied in 1974 across the New York Bight. Graphic representations of Stein estimates of mortality-moribundity for the stations are presented in Figure 1 (see Longwell and Hughes, 1980, for a discussion of statistical techniques).

Subsequent analyses grouped the sample stations according to geography, proximity to major dumping areas, surface and 10-m temperature contours, and general water circulation patterns in the Bight. Eight groups of stations were identified (I-VIII, Figure 1). The Kruskal-Wallace test was applied to determine whether overall differences in the Stein estimates of mortality-moribundity existed among the eight areas. Station groups IV, V, VII, and VIII exhibited the least mortality (most normal mitoses and highest mitotic indices). With the exception of station groups IV and V in the Long Island coastal area, along which flows a current of clean water, groups with the lowest mortalities were those most distant from coastal zones and from major dumping sites. This is despite the fact that temperature and salinity optima for mackerel development generally occur in near-shore rather than off-shore areas (Longwell and Hughes, 1980).

In May 1977, developing mackerel embryos were collected from the New York Bight along with samples of sea surface microlayer, subsurface water, and plankton. Water samples were analyzed for toxic metals; plankton samples were analyzed for both toxic metals and a suite of aromatic and chlorinated hydrocarbons (see Longwell and Hughes, 1980, for details).

Canonical correlations between principal components of environmental variables (temperature, salinity, toxic hydrocarbons, and toxic metals) and cytological variables showed associations between the toxic metal load of surface water, salinity, level of aromatic and chlorinated hydrocarbons in plankton, abnormal states of cell differentiation in early-stage embryos, and mitotic abnormalities (Table 1). Other associations also were observed between cytological measures of egg health at various development stages, contaminant levels, temperature, and salinity. The association between toxic hydrocarbons, abnormal cell differentiation, and chromosome-mitotic aberrations was at the 0.05 level of significance (Longwell and Hughes, 1980).

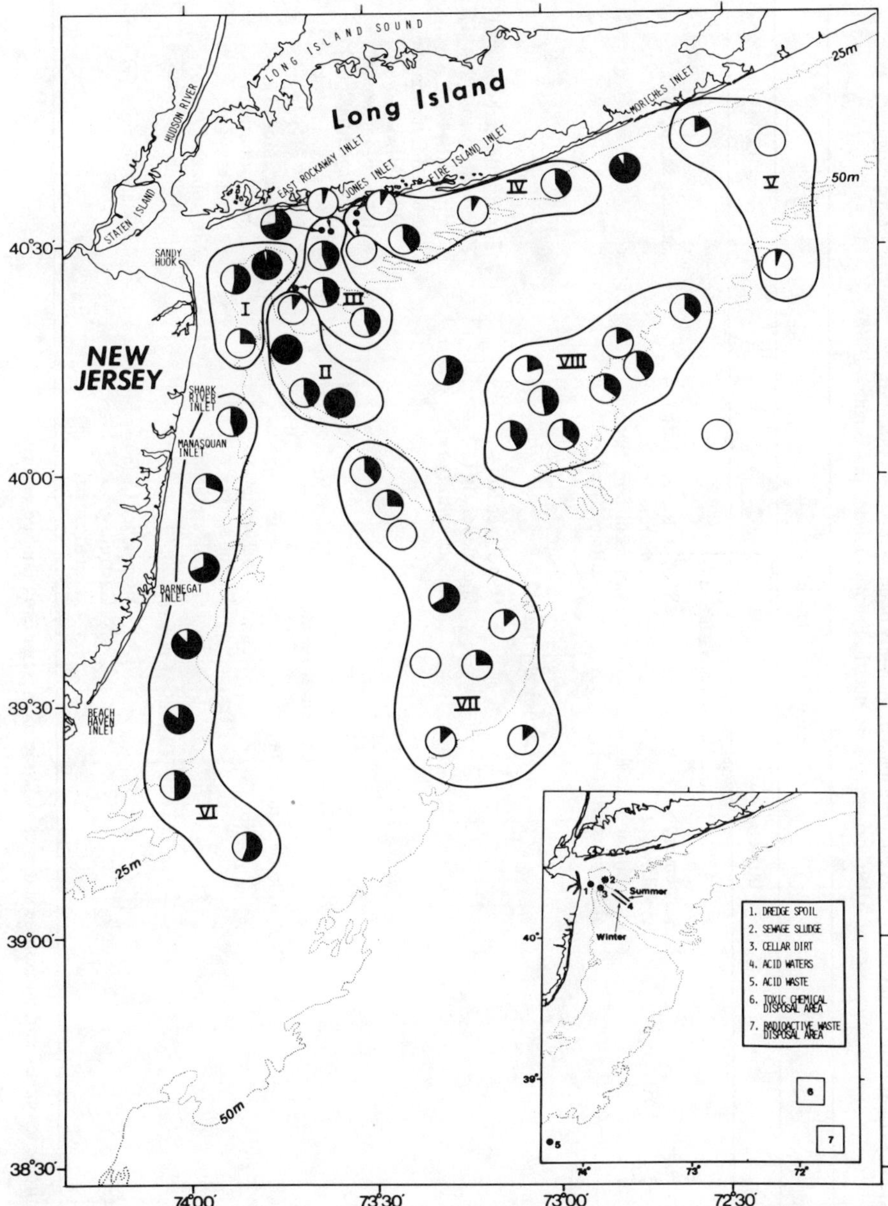

Figure 1. Stein estimates of mortality-moribundity in gastrula-early embryo stages of Atlantic mackerel eggs collected from the New York Bight in May 1974. Circles indicate station positions; shaded portions of circles show percent embryos dead or moribund, as determined by cytologic-cytogenetic appraisal. Stations are grouped into eight areas (I-VIII) for statistical treatment. From Longwell and Hughes (1980).

Table 1. Canonical correlations between principal components of environmental variables and cytologic-embryologic measures of Atlantic mackerel egg health. From Longwell and Hughes (1980).

Significance level			
.002	.005	.02	.05
Canonical variate 1			
No. cleavage embryos (+) Cleavage viability (+) Tail-bud embryo viability (+) Tail-bud mitotic index (+)	Morula embryo viability (+) Mitotic-chromosomal abnormalities[1] (–)	Surface temperature (–) No. morula (+) Morula differentiation difficulties (–)	Surface salinity (+)
Canonical variate 2			
Microlayer Cd (–)		Microlayer Ni (–) Microlayer Hg (–) Surface salinity (+)	Toxic hydrocarbons level (–) Morula differentiation difficulties (–) Tail-bud differentiation difficulties (–) Mitotic-chromosomal abnormalities[1] (–)
Canonical variate 3			
Yolk sac viability (+) Yolk sac mitotic index (+)	No. yolk sac mitoses (+)	Early embryo gross malformation (–) Tail-bud gross malformation (–)	Gastrula embryo gross malformation (–) No. tail-bud (+) No. yolk sacs (+)
Canonical variate 4			
No. gastrula (+) No. early embryo (+)		Microlayer Cr (+) Gastrula gross malformation (+)	Yolk sac viability (+)

[1] As measured in mitotic telophase and metaphase stages of yolk sac membrane from tail-bud embryo.

A weakness of the 1977 study was the limited number of stations for which hydrocarbon and toxic metal analyses were undertaken. Additional samples were obtained from a sampling cruise conducted in May 1978, and their biological and chemical analyses have been completed. Biological, physical, and analytical chemical data from 1977 and 1978 will be pooled for a re-analysis of statistical associations.

DISCUSSION

The literature on non-aquatic species contains many reports of cytogenetic effects induced by toxic chemicals. There is ample evidence that toxic metals may cause chromosome-breakage, may affect the mitotic spindle apparatus, thus causing abnormal distributions of chromosomes to daughter cells (*e.g.*, Leonard and Lauwerys, 1980), and may adversely affect cells and embryos (*e.g.*, Westernhagen *et al.*,1975). Pesticides also have adverse cytologic effects (Epstein and Legator, 1971; Ahmed and Grant, 1972).

Fishes, crustaceans, and shellfishes are similar to mammals in that they appear to possess systems for the metabolism of polynuclear aromatic hydrocarbons to direct mutagens (Ahokas, 1979). Hooftman (1981) reported that the freshwater fish, *Notobranchius rachowi* (a tooth carp), was able to convert benzo(a)pyrene to its genetically active metabolite. A six-day exposure to 0.1 μg/l benzo(a)pyrene was sufficient to induce chromosome and chromatid breaks, chromosome gaps, and chromosome interchanges in gill tissue at frequencies that were highly significant.

Mudminnows (*Umbra pygmaea*) held in water from the Rhine River showed higher incidences of chromosome mutations in gill cells than did fish held in untreated ground water (Prein *et al.*, 1978). Fish exposed to Rhine water for three and 11 days showed sister-chromatid exchange rates two and three times that of fish exposed to ground water of drinking quality (Alink *et al.*, 1980). Analyses indicated that many classes of chemical compounds may have been involved in inducing the cytogenetic effects. Of the 92 compounds identified in the river water, 28 were aromatic hydrocarbons, 5 were acids, 12 were esters, 5 were aldehydes, 16 were phenolic compounds, and 16 were aromatic bases.

A sister-chromatid exchange rate of 2.2 exchanges per cell was determined for kidney cells of seven English sole (*Parophrys vetulus*) collected from a presumed relatively unpolluted site in Puget Sound (Stromberg *et al.*, 1981). Increased rates occurred in fish from the same site intraperitoneally injected with 0.05, 1.5, and 5 μg/g body weight benzo(a)pyrene and held at 14°C. These rates were 3.0, 4.1, and 5.9 sister-chromatid exchanges/cell, respectively, all significantly higher than the control. Ten English sole from a polluted river had a rate of 3.3, also significantly higher than the assumed baseline, and not significantly different from the rate in fish given the experimental intraperitoneal injection of 0.05 μg/g benzo(a)pyrene. Chemical analyses of sediment samples at this site revealed high concentrations of several mutagenic chemicals with benz(a)-

anthracene (2,700 μg/g, dry wt.) and benzo(a)pyrene (1,500 μg/g, dry wt.) being particularly high (Malins *et al.*, 1980).

Observations of cytogenetic irregularities in developing embryos of Atlantic mackerel from the polluted New York Bight thus should not be entirely unexpected. However, there seems to be only one other aquatic field study that suggests an association between environmental contamination and chromosome effects in a reproductive stage. A population of royal fern (*Osmunda regalis*) growing in a Massachusetts river that was heavily polluted with industrial waste was found to have a far higher frequency of chromosome mutations in developing germ cells than did populations from nearby non-polluted areas (Klekowski and Berger, 1976; Klekowski, 1978).

Eggs of three species of flatfish were experimentally exposed to concentrations of benzo(a)pyrene comparable to levels found in polluted harbors, 0.10 - 4.2 ppb (Hose *et al.*, in press). Benzo(a)pyrene-treated eggs of the sand sole (*Psettichthys melanostictus*) incubated at 10°C with the hydrocarbon had lower incidences of hatching and more developmental anomalies than control eggs. Nuclear pycnosis and general disruption of the neural and ocular tissues occurred in the larvae from exposed eggs of the flathead sole (*Hippoglossoides elassodon*), but there was no increase in gross morphological abnormalities or in hatching success. Eggs and larvae of the English sole also were unaffected by these measures. However, preliminary cytogenetic analyses of the English sole eggs just after gastrulation, using the methods of Longwell and Hughes (1980), revealed highly significant increases in chromosome abnormalities over the control eggs.

Accumulation of hydrophobic PAH as benzo(a)pyrene in lipid-rich gonads of fish could further increase chromosome mutations as observed in spawned, developing eggs, with subsequent reduced rates of successful fertilization, development, and hatching. Twenty-four hours after intraperitoneal injection of benzo(a)pyrene (80 mg BaP plus 200 μCi[^{3}H] BaP per kg body weight), its unmetabolized form and its polar metabolites were detected in the ovary, oocytes, and milt of treated fish. Oocytes and milt concentration of benzo(a)pyrene and its metabolites were three to eleven times greater than those in gonadal tissue as a whole (Hose *et al.*, 1981).

As is the case for many fish stocks, mackerel numbers are estimated to have dropped dramatically over recent decades (ICES, 1978). Aside from the question of the present importance of the New York Bight as a mackerel nursery (Sette, 1950, 1952; ICES, 1978), the ultimate impact to the fishery stock of general zygote loss in the Bight is fully interpretable only in reference to the ecosystem as a whole. This touches on basic fishery biology problems, particularly as they relate stock to recruitment (Hunter, 1976; Dahlberg, 1979). Specifically, the significance of pollution-related zygote losses in the Bight to the eventual size of fish populations revolves about the relevance of zygote loss to fishery recruitment over a period of years. Recruitment, in turn, is related to fluctuations in zygote, larval, and juvenile losses as a result of natural environmental variations, to predation at later stages, to fishing intensity, to variations in fecundity, and, perhaps, to other factors.

The portion of zygote loss attributable to mutation should not be regarded as altogether similar to losses caused by such factors as increased predation. In many instances, the same substances effecting genetic death in zygotes may have mutagenic effects in the germ line of adults. Mutagens may limit the ultimate reproductive capacity of surviving larvae by limiting number and normalcy of germ-line primordia. These differentiate as early as the gastrulation stage of embryo development (Egami and Ijiri, 1979).

The New York Bight mackerel egg study estimated both genetic and other zygotic death. Nevertheless, it indicates that chromosome abnormalities were a factor and suggests an association between pollution and the occurrence of such abnormalities.

ACKNOWLEDGEMENTS

Mackerel studies and methods development have been supported by the Marine EcoSystems Analysis Program and Northeast Fisheries Center, National Oceanic and Atmospheric Administration, U.S. Department of Commerce.

The technical assistance of the following individuals is acknowledged: J. Kubinski, E. Flynn, and A. Hebert, with special credit to D. Perry. R. Riccio provided considerable editorial and secretarial assistance in preparing this manuscript.

REFERENCES

Ahmed, M. and W.F. Grant. 1972. Cytological effects of the pesticides phosdrin and bladex on *Tradescantia* and *Vicia faba*. Can. J. Genet. Cytol. 14: 157-165.

Ahokas, J.T. 1979. Cytochrome P-450 in fish liver microsomes and carcinogen activation. *In:* Pesticide and Xenobiotic Metabolism in Aquatic Organisms, M.A.Q. Khan, J.J. Lech, and J.J. Menn (eds.), American Chemical Society, Washington, DC. pp. 279-296.

Alink, G.M., E.M.H. Frederix-Wolters, M.A. van der Gaag, J.F.J. van de Kerkhoff, and C.L.M. Poels. 1980. Induction of sister-chromatid exchanges in fish exposed to Rhine water. Mutat. Res. 78: 369-374.

Dahlberg, M.D. 1979. A review of survival rates of fish eggs and larvae in relation to impact assessments. Mar. Fish. Rev. 41: 1-12.

Egami, N. and K. Ijiri. 1979. Effects of irradiation on germ cells and embryonic development in teleosts. Int. Rev. Cytol. 59: 195-248.

Epstein, S.S. and M.S. Legator. 1971. The mutagenicity of pesticides, concepts and evaluation. M.I.T. Press, Cambridge, MA. 220 pp.

Hooftman, R.N. 1981. The induction of chromosome aberrations in *Notobranchius rachowi* (Pisces: Cyprinodontidae) after treatment with ethyl methanesulphonate or benzo(a)pyrene. Mutat. Res. 91: 347-352.

Hose, J.E., J.B. Hannah, D. DiJulio, M.L. Landolt, B.S. Miller, W.T. Iwaoka, and S.P. Felton. In press. Effects of benzo(a)pyrene on early development of flatfish. Arch. Environ. Contam. Toxicol. 11.

Hose, J.E., J.B. Hannah, M.L. Landolt, B.S. Miller, S.P. Felton, and W.T. Iwaoka. 1981. Uptake of benzo(a)pyrene by gonadal tissue of flatfish (family Pleuronectidae) and its effects on subsequent egg development. J. Toxicol. Environ. Hlth. 7: 991-1000.

Hunter, J.R. (ed.). 1976. Report of a colloquium on larval fish mortality studies and their relation to fishery research. NOAA Tech. Rep. NMFS Circ. 395. 5 pp.

ICES. 1978. Report to the International Council for the Exploration of the Sea on the mackerel stocks. International Council for the Exploration of the Sea, Copenhagen, Denmark.

Klekowski, E.J. Jr. 1978. Detection of mutation damage in fern populations: an *in situ* bioassay for mutagens in aquatic ecosystems. *In:* Chemical Mutagens, Principles and Methods for Their Detection, Vol. 5, A. Hollaender and F.J. de Serres (eds.), Plenum Press, New York. pp. 79-99.

Klekowski, E.J. Jr. and B.B. Berger. 1976. Chromosome mutations in a fern population growing in a polluted environment: a bioassay for mutagens in aquatic environments. Am. J. Bot. 63: 239-246.

Leonard, A. and R.R. Lauwerys. 1980. Carcinogenicity, teratogenicity and mutagenicity of arsenic. Mutat. Res. 75: 49-62.

Longwell, A. Crosby and J.B. Hughes. 1980. Cytologic, cytogenetic, and developmental state of Atlantic mackerel eggs from sea surface waters of the New York Bight, and prospects for biological effects monitoring with ichthyoplankton. Rapp. R.-V. Reun. Cons. Int. Explor. Mer. 179: 275-291.

Malins, D.C., B.B. McCain, D.W. Brown, A.K. Sparks, and H.O. Hodgins. 1980. Chemical contaminants and biological abnormalities in central and southern Puget Sound. NOAA Tech. Memo. OMPA-2. 295 pp.

Prein, A.E., G.M. Thie, G.M. Alink, C.L.M. Poels, and J.H. Koeman. 1978. Cytogenetic changes in fish exposed to water of the river Rhine. Sci. Tot. Environ. 9: 287-292.

Sette, O.E. 1950. Biology of the Atlantic mackerel (*Scomber scombrus*) of North America. Part I–Early life history, including growth, drift, and mortality of the egg and larval populations. U.S. Fish Wildl. Serv. Fish. Bull. 50: 149-237.

Sette, O.E. 1952. Biology of the Atlantic mackerel (*Scomber scombrus*) of North America. Part II–Migrations and habits. U.S. Fish Wildl. Serv. Fish. Bull. 51: 251-358.

Stromberg, P.T., M.L. Landolt, and R.M. Kocan. 1981. Alterations in the frequency of sister chromatid exchanges in flatfish from Puget Sound, Washington, following experimental and natural exposure to mutagenic chemicals. NOAA Tech. Memo. OMPA-10. 43 pp.

Westernhagen, H., M. Dethlefsen, and H. Rosenthal. 1975. Combined effects of cadmium and salinity on development and survival of garpike eggs. Helgol. Wiss. Meeresunters. 27: 268-282.

FACTORS INFLUENCING THE FATE OF SEWAGE-DERIVED NUTRIENTS IN THE LOWER HUDSON ESTUARY AND NEW YORK BIGHT

Thomas C. Malone[1]

Lamont-Doherty Geological Observatory
Palisades, New York 10964

Abstract. The lower Hudson River estuary receives large quantities of sewage wastes and associated plant nutrients. Most of the nutrient load is transported into adjacent coastal waters where it is taken up by phytoplankton within a relatively small area (< 1,500 km²). A better understanding of the effects of freshwater flow of the Hudson River on this nutrient flux is required before urban use of fresh water is increased above current levels. Arguments based on the dynamics of phytoplankton productivity and nutrient uptake lead to the conclusion that secondary sewage treatment is having a beneficial effect on water quality in the lower estuary and apex of the New York Bight, but that tertiary treatment could have a negative impact on commercial and sport fisheries of the region.

INTRODUCTION

Cultural eutrophication and related water quality problems occur as a consequence of inputs of organic and inorganic nutrients of sewage origin. These inputs often give rise to accumulations of organic matter in the form of sewage wastes or phytoplankton biomass and phytodetritus. Such accumulations influence water quality, depending on where, when, and how rapidly organic matter is oxidized, usually by living organisms.

This review describes the effects of municipal wastes discharged into the lower Hudson River estuary (the salt intruded reach) on phytoplankton productivity and water quality in the New York Bight. I will focus on the influence of estuarine processes on nutrient utilization and resultant production of particulate organic matter by phytoplankton. Because of the limited amount of information currently available, conclusions will be of a qualitative nature, but will provide guidelines for policy decisions concerning the use of fresh water in the drainage basin of the Hudson Valley and the treatment and disposal of domestic wastes.

THE LOWER HUDSON ESTUARY

The salt intruded reach of the Hudson estuary receives about 8.6×10^6 m³/day of sewage wastes, more than 95% of which is discharged into the Upper Bay. This represents 1.6×10^5 kg N/day and 1.3×10^4 kg P/day and produces high concentrations of dissolved inorganic nitrogen and phosphorus throughout the year (Howells *et al.*, 1970; Garside *et al.*, 1976; Simpson *et al.*, 1977) (Table 1).

[1] Present address: Oceanographic Sciences Division, Brookhaven National Laboratory, Upton, New York 11973.

Table 1. Mean concentration of dissolved inorganic nitrogen (μg-at/liter, nitrate + nitrite + ammonia) at the surface (S) and near-bottom (B), mean suspended solids content of the water column (g/m^2; water column depth about 13 m in Upper Bay and 25 m in apex), and the mean organic content of suspended solids as percent of total suspended solids (C = coefficient of variation). JFM = January, February, March; AMJ = April, May, June; JAS = July, August, September; OND = October, November, December.

		JFM	C	AMJ	C	JAS	C	OND	C
Nitrogen									
Upper Bay	S	40.8	(11)	40.5	(49)	53.3	(19)	50.2	(16)
	B	29.5	(37)	29.7	(63)	44.3	(78)	36.8	(15)
Apex	S	6.8	(86)	4.0	(95)	2.3	(97)	9.0	(93)
	B	5.7	(31)	4.8	(92)	5.2	(48)	6.4	(60)
Suspended Solids									
Upper Bay		160	(28)	180	(29)	230	(59)	320	(25)
Apex		95	(20)	95	(31)	130	(32)	185	(67)
Percent organic									
Upper Bay		20		22		22		9	
Apex		25		38		30		19	

Despite the high rates of nutrient supply, phytoplankton biomass and productivity in the lower estuary are low. The chlorophyll *a* content of the euphotic zone is generally less than 30 mg/m^2, and productivity has not been observed to exceed 2.5 g C/m^2/day (Malone, 1977a, unpublished data).

Phytoplankton productivity accounts for 5-50% of total organic carbon inputs to the lower estuary, with an annual mean of 15%. This is primarily a consequence of two factors: (1) high concentrations of suspended sediments that limit the euphotic zone to the surface layer of net seaward flow, and (2) high dilution rates of the surface layer due to freshwater flow through the estuary (Malone, 1977a). Phytoplankton growth rates appear to be light- and temperature-dependent. Thus, the suspended sediment load of the surface layer is a critical factor regulating phytoplankton productivity, and reduction of this load will result in higher phytoplankton biomass and productivity within the estuary. A second critical factor is the flushing rate of the surface layer, which is related in a complex way to freshwater flow. During the winter when phytoplankton growth rates are much less than flushing rates, increases in phytoplankton biomass are usually related to the advection of phytoplankton into the estuary with intermediate and bottom water from Raritan Bay and the apex of the New York Bight, respectively. During the summer, when growth

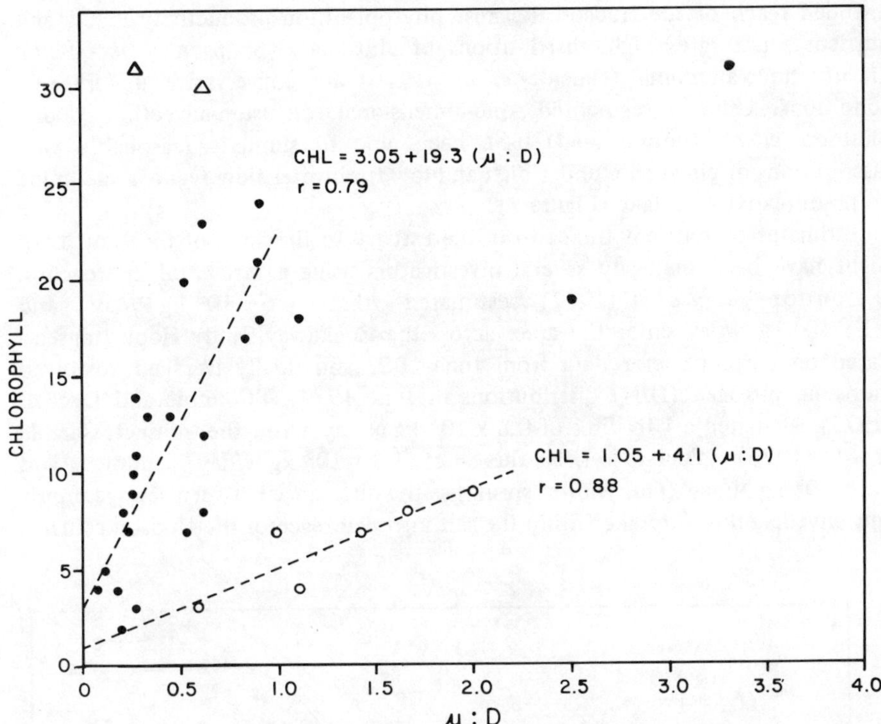

Figure 1. Relationship between the chlorophyll *a* content of the euphotic zone (mg/m^2) and the ratio of phytoplankton growth rate (μ = doublings/day) to the dilution rate (D = dilutions/day) of the euphotic zone in the Upper Bay (● – reflects steady state conditions when chlorophyll *a* is primarily due to variations in μ:D; △ – reflects exceptionally high chlorophyll *a* concentrations in the apex; ○ – reflects high grazing rates by zooplankton; r = correlation coefficient).

rates are high and flushing rates are low, increases in phytoplankton biomass frequently occur because of growth within the lower estuary (Malone, 1977a). Thus, phytoplankton biomass (chlorophyll *a*) in the surface layer increases as the ratio of phytoplankton growth rate (μ) to the dilution rate of the surface layer (D) increases.

A significant ($P < 0.01$) linear correlation was found for the regression of euphotic zone chlorophyll *a* on μ:D when μ was less than D (Figure 1). Large deviations are associated with high chlorophyll *a* concentrations in offshore source waters (Malone, 1976, 1977b) or with high grazing rates within the estuary (Malone, 1977a; Chervin, 1978). Concentrations seem to plateau near 30 mg/m^2, suggesting that this is near the maximum amount of phytoplankton biomass that can be sustained in the euphotic zone under the conditions of flow and water transparency that prevailed during 1973-1976.

Phytoplankton growth is not only nutrient-saturated, but also nutrient uptake by phytoplankton has little impact on nutrient concentrations in the salt intruded reach of the Hudson. Because phytoplankton productivity is low and nutrient input rates high, distributions of phosphate (Simpson *et al.*, 1977), nitrate, and ammonia (Garside *et al.*, 1976) are conservative under most conditions. Using a segmented, one-dimensional, diffusion-advection model, Simpson *et al.* (unpublished) have been able to simulate reasonably well distributions of phosphate under high and low freshwater flow regimes, assuming no phytoplankton uptake (Figure 2).

Estimates of nutrient fluxes from the estuary to the apex of the New York Bight have been made by several investigators using a variety of approaches. O'Connor *et al.* (1977) estimated that 4.6×10^4 kg P/day and 2.1×10^5 kg N/day enter the apex across the Rockaway-Sandy Hook transect. Based on current meter data from June 1952 and tidally-averaged, dissolved inorganic nitrogen (DIN) distributions in June 1974, O'Connors and Duedall (1975) calculated a DIN flux of 0.8×10^5 kg N/day across the transect. Garside *et al.* (1976) calculated DIN fluxes of 1.2×10^5 kg N/day (summer) and 1.6×10^5 kg N/day (fall, winter, spring) as the difference between sewage inputs and phytoplankton uptake within the salt intruded reach of the Hudson estuary.

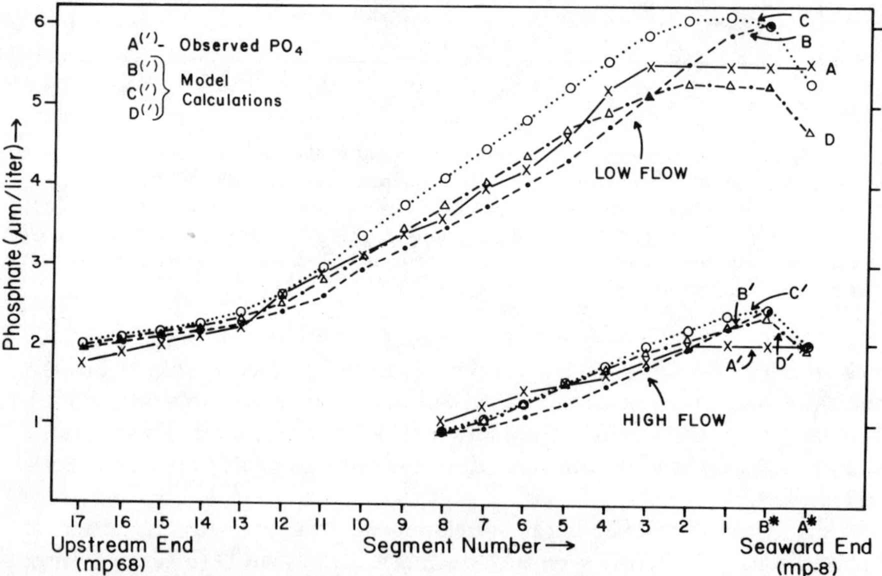

Figure 2. Comparison of observed phosphate concentration in the lower Hudson estuary with several model calculations. Observed values (A′ and A) for high (1,130 m^3/sec) and low (220 m^3/sec) flows are means of four-mile segments averaged vertically over one tidal cycle. Curves B′ and B reflect the assumption that all sewage phosphate is discharged into segment B* (Upper Bay); curves C′ and C reflect a distribution of sewage loading that approximates the actual pattern of discharge; curves B, B′ and C, C′ reflect small phosphate inputs from the sediments and the oxidation of suspended organic matter. Curves D′ and D reflect the effects of distributed sewage loading without any additional inputs. A* = Lower Bay; mp = mile point above (+) and below (–) the Battery.

Simpson *et al.* (1977) used a one-box model for phosphate in the Upper Bay to estimate the phosphate fluxes into and out of the Upper Bay. They calculated a phosphate flux of 1.8×10^4 kg P/day from the Upper Bay across the Verrazano Narrows. The mean molar N:P ratio in the Upper Bay is 21 (Malone, 1976), giving an equivalent DIN flux of 1.6×10^5 kg N/day. On the basis of these estimates, the mean flux of DIN to the apex will be assumed to be about 1.5×10^5 kg N/day or 94% of the input of sewage-nitrogen. Since sewage wastes are the major source of nitrogen and phosphorus in the salt intruded reach of the estuary, fluxes out of the estuary remain relatively constant, even though concentrations vary, *i.e.*, changes in concentration are compensated by changes in flow.

NEW YORK BIGHT

In contrast to the estuary, phytoplankton productivity in the apex accounts for 70-80% of the input of particulate organic matter (Table 2) (Segar and Berberian, 1976; Garside and Malone, 1978). Geographically, productivity tends to be highest within 20 km of the mouth of the estuary and decreases with distance as nutrient and chlorophyll *a* concentrations decrease and salinity increases (Malone, 1976; unpublished data). Seasonal variations are characterized by two bloom periods: (1) February-April, when phytoplankton biomass is highest and chain-forming diatoms usually dominate, and (2) June-August, when phytoplankton productivity is highest and small, solitary algae dominate (Malone, 1976, 1977b; Malone and Chervin, 1979). Like the estuary, phytoplankton growth appears to be light- and temperature-dependent under most conditions. Since these factors vary seasonally and exhibit little geographic variability, spatial variations in productivity are mainly a function of phytoplankton biomass in the surface layer, as determined by mixing, sinking, and grazing rates. During winter and early spring, mixing and sinking are the most important processes limiting accumulations of biomass (Malone and Chervin, 1979). Most of the biomass produced prior to thermal stratification sinks from the water column, and a small proportion is transported back into the estuary with bottom water. This flux into the benthos could be reflected in subpycnocline biological oxygen demand (BOD) during late spring and summer when the water column is well-stratified and bottom temperatures are 2°-4°C warmer.

As the surface layer warms and the water column becomes thermally stratified, most primary productivity occurs in and above the pycnocline, and an increasing proportion is grazed by zooplankton (Malone, 1976; Chervin, 1978; Malone and Chervin, 1979). Copepods alone assimilate an average of 30% of phytoplankton productivity during the summer. In contrast to the estuary, organic detritus forms a large portion (30-80%) of the particulate matter assimilated by copepods, and the production and sinking of fecal material by zooplankton may be the major mechanism by which organic matter produced by phytoplankton is transported below the pycnocline.

Table 2. Particulate organic carbon inputs to the apex and respiratory demand based on data presented by (1) Garside and Malone (1978) and (2) Thomas *et al.* (1976) for an apex area of 1,250 km².

		10^8 kg C/yr	
		(1)	(2)
A. Inputs			
Primary production[a]		5.43	5.43
Estuarine runoff		1.24[b]	0.55[c]
Dredge spoils		--	1.39
Sewage sludge		0.36	0.86
	Total	7.33	8.23
B. Respiration			
Water column		3.70[d]	11.42
Benthos		3.64	2.30
	Total	7.34	13.72
C. (A) - (B)		-0.01	-5.49

[a] Based on a three-year mean of 430 g C/m²/yr (Malone, 1977).

[b] Includes dredge spoils.

[c] The original estimate of Thomas *et al.* (1976) included dissolved organic carbon. The value given here is for particulate organic carbon, assuming that DOC = 10 x POC.

[d] Recalculated without the negative rates reported by Garside and Malone (1978).

Monthly oxygen and carbon budgets appear to be strongly influenced by the seasonal cycles of phytoplankton productivity and zooplankton respiration (Garside and Malone, 1978). Benthic oxygen demand ranges from 0.3 g C/m²/day to 1.0 g C/m²/day, with low rates during February-April and high rates during August-September (Thomas *et al.*, 1976; Garside and Malone, 1978). High rates also are associated with the sewage sludge and dredge spoil dumpsites and with periods of high estuarine runoff. High phytoplankton productivity during the summer is accompanied by a large increase in water column respiration relative to benthic respiration (Garside and Malone, 1978). Chervin's (1977) estimates of copepod respiration suggest that copepods alone could account for as much as 40% of the water column respiration calculated by Garside and Malone. Ocean dumping and estuarine runoff appear to have a greater direct impact on benthic respiration, while phytoplankton productivity seems to have a greater impact on water column respiration, especially in and above the pycnocline during the summer. Since bacterial biomass is usually several orders of magnitude higher in the benthos than in the water column (Litchfield, personal communications), organic matter reaching the benthos has a higher probability of being metabolized by bacteria, while that remaining in the water column has a higher probability of being metabolized by metazoan particle grazers, such as copepods, cladocerans, and tunicates.

The carbon budget of the apex is balanced, depending on which estimates of respiration are most accurate (Table 2). The estimates of Thomas *et al.* (1976), based on a limited number of direct measurements, indicate that water column respiration accounts for 80% of total annual respiration, while the calculations of Garside and Malone (1978) indicate that annual respiration is equally partitioned between the water column and benthos. Consequently, the carbon budget of Garside and Malone is balanced, while the Thomas *et al.* budget has a deficit of 5.5×10^8 kg C/yr (1.2 g C/m^2/day). Such a deficit could reflect a sampling bias in the data of Thomas *et al.* Most estimates of water column respiration were made off the coast of New Jersey in the plume of the Hudson River where respiration is likely to be much higher than elsewhere in the apex. Thus, the estimate of Garside and Malone is probably representative of the apex as a whole, while that of Thomas *et al.* is more representative of the plume *per se.*

These patterns of water column stratification and respiration give rise to a seasonal cycle of dissolved oxygen in bottom water. During the winter months, the water column is well-mixed, and dissolved oxygen levels are near saturation (6-8 ml/liter). As the water column begins to stratify during April-May, dissolved oxygen concentrations in the subpycnocline layer begin to decline so that by July-August, concentrations are usually 10-40% of saturation (2-4 ml/liter). Local anoxic conditions often develop below the pycnocline during the summer near the sludge dumpsite. The more widespread development of an oxygen minimum layer appears to reflect BOD based on the following inputs: (1) accumulations of organic matter in sediments due to sinking of phytoplankton biomass produced during the winter and early spring, (2) sinking of fecal material produced by zooplankton during the summer, (3) organic detritus of sewage origin transported into the apex from the Hudson-Raritan estuaries and through outfalls along the New Jersey coast, and (4) sludge and dredge spoil dumping. The relative importance of these inputs cannot be evaluated at this time. However, the data suggest that organic matter packaged as phytoplankton or derived from phytoplankton has a greater tendency to enter metazoan food chains, while organic matter of sewage origin tends to be metabolized by bacteria, especially during the summer. Thus, reductions in the organic content of sewage effluents and sludge will have a greater impact per unit of nitrogen than will equivalent reductions in the inorganic nitrogen content of sewage.

A dissolved inorganic nitrogen budget can be calculated for the apex based on these estimates of respiration (assuming excretion of N:C = 0.05 by weight), the DIN flux from the estuary, and phytoplankton productivity (Table 3). Inputs of

Table 3. Comparison of dissolved inorganic nitrogen inputs to the apex and phytoplankton uptake. Inputs include estuarine runoff and remineralization; the latter was calculated from respiration rates given in Table 2, assuming a N:C excretion ratio of 0.05. Phytoplankton uptake was calculated from primary productivity assuming a N:C ratio of 0.14. Data are from (1) Garside and Malone (1978) and (2) Thomas *et al.* (1976), as indicated in Table 2.

		10^7 kg N/yr	
		(1)	(2)
A. Inputs			
Estuarine runoff		5.5	5.5
Remineralization		4.2	8.0
	Total	9.7	13.5
B. Uptake		7.9	7.9
C. (A) - (B)		1.8	5.6

DIN exceed uptake by phytoplankton, and remineralization exceeds estuarine runoff, regardless of the value of respiration used to calculate the rate of nitrogen regeneration within the apex. Since DIN concentrations generally exceed 5 mg/m^3 and average 77 mg/m^3 in the apex (Malone, 1976), an excess of input over uptake is consistent with observed concentrations.

These budget calculations indicate that the influence of sewage-derived nitrogen on phytoplankton productivity is limited to the area of the apex. Two approaches were used to test this hypothesis. Based on the DIN fluxes calculated by Garside *et al.* (1976), measurements of phytoplankton productivity and the mean C:N ratio of particulate organic matter in the apex (7 by weight), the area required by phytoplankton to assimilate DIN inputs from estuarine runoff was calculated by season (Table 4). The required area varied from 670 km^2 during the summer to 1,350 km^2 during the winter, or 54% and 108% of the 1,250 km^2 area of the apex. Second, regressions of DIN concentration on the proportion of fresh water in the surface layer were used to calculate the salinity at which DIN would reach zero (Table 5). Similar regressions were run for silicate. Note that the salinity of zero nutrient concentration is less than or equal to the mean salinity of the outer stations of the apex and increases as the area required to assimilate DIN increases. These calculations are based on independent data sets and support the conclusion that the influence of nutrients of estuarine (sewage) origin is limited to the apex.

FUTURE IMPACTS OF MUNICIPAL WASTES

The major change in contaminant inputs to the New York Bight during the next 20 years will occur as a consequence of increased secondary treatment of

Table 4. Seasonal comparisons of dissolved inorganic inputs from estuarine runoff and uptake by phytoplankton (A = area of apex required to assimilate all DIN given mean primary productivity for each three-month period). JFM = January, February, March; AMJ = April, May, June; JAS = July, August, September; OND = October, November, December.

	10^5 kg N/day		km²
Months	Input	Uptake	A
JFM	1.6	2.2	900
AMJ	1.6	2.7	700
JAS	1.2	2.1	670
OND	1.6	1.5	1,350

municipal wastes. As population size increases and more secondary treatment plants become operational, the volumes of waste effluent and sludge also will increase. These increases for nitrogen will be on the order of 12 x 10^3 kg N/day for effluent inputs to the estuary and 55 x 10^3 kg N/day for sludge inputs to the apex, if dumping were to continue beyond 1981 (Anderson and Mueller, 1976).

Since this will reduce the amount of organic matter entering the estuary with sewage effluents, it could result in increased light penetration and higher primary productivity in the estuary. The rate of attenuation of downwelling radiation is significantly ($P < 0.01$) correlated with the concentration of suspended solids in both the lower Hudson estuary (Figure 3) and the apex of the New York Bight (Malone, 1976). Chlorophyll *a* accounts for less than 10% of light attenuation in the lower estuary, compared to less than 50% in the apex. Likewise, suspended

Table 5. Results of linear correlation and regression analysis of nutrient concentration (μg-at/liter) on the proportion of fresh water in the sample ($1 - S_X/S_O$ where S_X = salinity of sample and S_O = salinity of source water, assumed to be 33°/₀₀ for these calculations); n = number of measurements; r = correlation coefficient; a = y intercept; b = slope; S_O' = salinity at which nutrient concentration is zero; and $\bar{S}$ = mean salinity of outer apex).

Nutrient	Months	n	r	a	b	S_O'	$\bar{S}$
SiO_4	JFM	21	0.87	- 3.2	76	30.6	31.0
	AMJ	38	0.83	- 3.0	53	30.4	30.7
	JAS	26	0.85	- 2.5	48	30.4	30.7
	OND	30	0.78	+0.6	62	32.4	31.7
DIN	JFM	21	0.79	- 5.0	161	30.9	31.0
	AMJ	38	0.89	- 9.8	157	30.2	30.7
	JAS	26	0.94	- 14.2	195	29.7	30.7
	OND	30	0.93	- 3.7	265	31.7	31.7

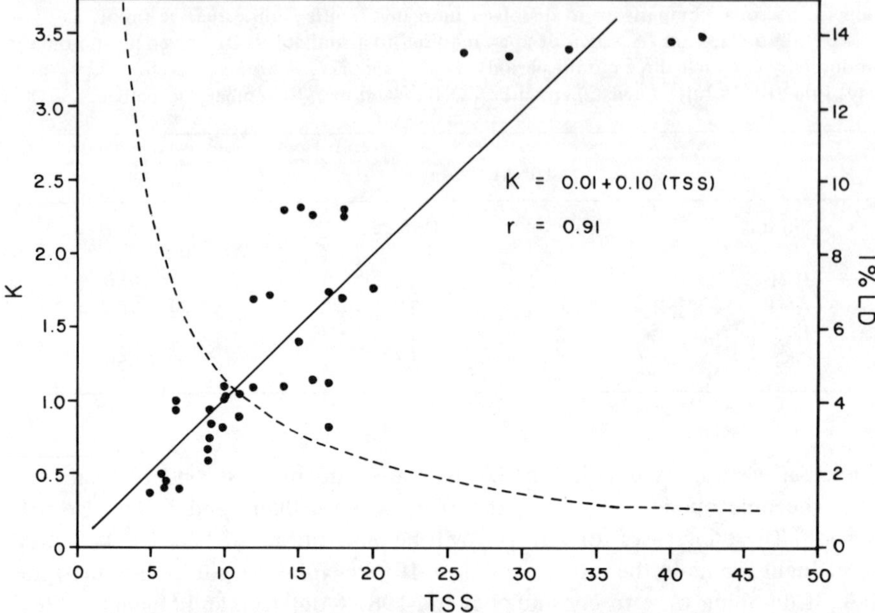

Figure 3. Relationship between the rate of attenuation of downwelling light (k = extinction coefficient, per m) and the concentration of total suspended solids (TSS, mg/liter) at the surface in the lower Hudson estuary. The dashed line shows the relationship of the 1% light depth (1% LD) to TSS derived from the least squares regression line (r = correlation coefficient).

inorganic solids account for 78-91% (seasonal means) of total solids in the Upper Bay, compared to 62-81% in the apex (Table 1). River-borne suspended sediments appear to be the dominant source of light extinction in the lower estuary. Consequently, upgrading sewage treatment to remove increasing amounts of organic matter will have little effect on transparency in the estuary and, therefore, on productivity. Thus, most of the effluent inputs to the estuary will be transported to the apex, resulting in an increase of about 67×10^3 kg N/day. Since current nitrogen inputs are in the range of $3 - 6 \times 10^5$ kg N/day, the maximum increase in productivity in the apex, assuming no grazing response, will be about 15%.

Since phytoplankton growth in the apex appears to be mainly a function of light and temperature, any increase in nitrogen inputs will result in higher phytoplankton productivity within the apex only if biomass increases. This will depend on the flushing rate of the apex and the response of grazers. The most critical period will be summer because growth rates exceed flushing rates, and grazing is the major factor limiting phytoplankton biomass in the apex. Grazing experiments with copepods indicate that increases in phytoplankton productivity will result in proportionate increases in grazing rates (Chervin, 1978, unpublished), so that any increase in phytoplankton biomass should be small relative to the increase in nitrogen supply. This would minimize the impact of increased enrichment on water quality.

The degree or type of sewage treatment may be more important than the absolute quantities discharged. Organic material processed by metazoan food chains will have a smaller impact on dissolved oxygen than if organic material is processed by bacteria, especially if bacterial decomposition tends to occur below the pycnocline and grazing by metazoans above the pycnocline. If this is the case, once all sewage treatment is upgraded to secondary treatment in the New York/New Jersey metropolitan area, water quality in the apex should improve. Tertiary treatment will reduce nutrient inputs and, therefore, phytoplankton productivity. However, such treatment is probably unnecessary and may even have a detrimental impact on local fisheries.

In this regard, increased urban use of fresh water from the Hudson River drainage basin should be coupled with an increase in the capacity of secondary treatment plants. A substantial reduction in freshwater flow of the Hudson will be accompanied by increases in phytoplankton biomass in the lower Hudson, depending on where and with that efficiency the fresh water is returned to the River. Such increases will result in lower nutrient and higher biomass fluxes to the apex, but should have little impact on water quality in the apex.

While high phytoplankton productivity in the apex is a consequence of nutrient fluxes from the estuary, the 1976 bloom of *Ceratium tripos* is an important exception (Malone, 1978). The bloom began throughout the New York Bight in January, with maximum cell densities developing in the mid-shelf to shelf-break region in late March, prior to the onset of thermal stratification. Within the apex, cell densities increased with distance from the mouth of the estuary and were inversely related to the proportion of fresh water present. As the water column stratified, the distribution of *C. tripos* in the Bight shifted so that by mid-May an inshore-offshore decrease in abundance was observed with maximum cell densities at the head of the Hudson Shelf Valley in high salinity water. As thermal stratification continued to develop, the center of maximum abundances shifted to the southeast. This temporal pattern in the spatial distribution of *C. tripos* indicates that the bloom did not develop in response to local nutrient enrichment within the apex and that as the water column stratified, an onshore transport of cells occurred, shifting the center of maximum abundances to New Jersey coastal waters. The *C. tripos* bloom resulted in a gradual accumulation of particulate organic matter (as biomass) that did not enter pelagic food chains. Respiration of this biomass and its decay below the pycnocline were probably major factors in the development of oxygen-poor bottom waters in June, July, and August. Localization of the oxygen minimum off the New Jersey coast probably reflects the two-layered estuarine circulation of the shelf and perhaps the assimilation of particulate organic matter of estuarine origin by *C. tripos* populations.

ACKNOWLEDGEMENTS

This work was supported by the National Science Foundation (Grant OCE 76-80883) and the New York Bight Marine EcoSystems Analysis Program at the National Oceanic and Atmospheric Administration (Grant No. 04-022-44003).

REFERENCES

Anderson, A.R. and J.A. Mueller. 1976. Estimates of New York Bight Future (2000) Contaminant Inputs. Final Report Submitted to MESA New York Bight Project for NOAA Grant No. 04-6-022-44027. 31 pp.

Chervin, M. 1977. The assimilation of particulate organic carbon by estuarine and coastal copepods. Ph.D. Dissertation, City University of New York. 197 pp.

Chervin, M. 1978. Assimilation of particulate organic carbon by estuarine and coastal copepods. Mar. Biol. 49: 265-275.

Garside, C. and T.C. Malone. 1978. Monthly oxygen and carbon budgets of the New York Bight apex. Estuarine Coastal Mar. Sci. 6: 93-104.

Garside, C., T.C. Malone, O.A. Roels, and B.A. Sharfstein. 1976. An evaluation of sewage-derived nutrients and their influence on the Hudson estuary and New York Bight. Estuarine Coastal Mar. Sci. 4: 281-289.

Howells, G.P., T.J. Kneip, and M. Eisenbud. 1970. Water quality in industrial areas: profile of a river. Environ. Sci. Technol. 4: 26-35.

Malone, T.C. 1976. Phytoplankton productivity in the apex of the New York Bight: environmental regulation of productivity/chlorophyll *a*. *In:* The Middle Atlantic Shelf and New York Bight, M.G. Gross (ed.), Am. Soc. Limnol. Oceanogr. Spec. Symp. 2. pp 260-272.

Malone, T.C. 1977a. Environmental regulation of phytoplankton productivity in the lower Hudson estuary. Estuarine Coastal Mar. Sci. 5: 157-171.

Malone, T.C. 1977b. Light-saturated photosynthesis by phytoplankton size fractions in the New York Bight, USA. Mar. Biol. 42: 281-292.

Malone, T.C. (ed.). 1978. The 1976 *Ceratium tripos* bloom in the New York Bight: causes and consequences. NOAA Tech. Rep. NMFS Circ. 410. 14 pp.

Malone, T.C. and M. Chervin. 1979. The production and fate of phytoplankton size fractions in the plume of the Hudson River, New York Bight. Limnol. Oceanogr. 24(4): 683-696.

O'Connor, D.J., R.V. Thomann, and H.J. Salas. 1977. Water Quality. MESA New York Bight Atlas Monograph 27, New York Sea Grant Institute, Albany, NY. 104 pp.

O'Connors, H.B. and I.W. Duedall. 1975. The seasonal variation in sources, concentrations and impacts of ammonium in the New York Bight apex. *In:* Marine Chemistry in the Coastal Environment, T.M. Church (ed.), American Chemical Society Symposium Series No. 18. pp. 636-663.

Segar, D.A. and G.A. Berberian. 1976. Oxygen depletion in the New York Bight apex: causes and consequences. *In:* The Middle Atlantic Shelf and New York Bight, M.G. Gross (ed.), Am. Soc. Limnol. Oceanogr. Spec. Symp. 2. pp. 220-239.

Simpson, H.J., S.C. Williams, C.R. Olsen, and D.E. Hammond. 1977. Nutrient and particulate matter budgets in urban estuaries. *In:* Estuaries, Geophysics, and the Environment, National Academy of Sciences, Washington, DC. pp. 94-103.

Thomas, J.P., W.C. Phoel, F.W. Steimle, J.E. O'Reilly, and C.A. Evans. 1976. Seabed oxygen consumption-New York Bight apex. *In:* The Middle Atlantic Shelf and New York Bight, M.G. Gross (ed.), Am. Soc. Limnol. Oceanogr. Spec. Symp. 2. pp. 354-369.

EVOLUTION OF RESISTANT PHYTOPLANKTON STRAINS THROUGH EXPOSURE TO MARINE POLLUTANTS

Lynda S. Murphy
Bigelow Laboratory for Ocean Sciences
West Boothbay Harbor, Maine 04575

Robert R.L. Guillard
Woods Hole Oceanographic Institution
Woods Hole, Massachusetts 02543

Jerome Gavis
Department of Geography and Environmental Engineering
The Johns Hopkins University
Baltimore, Maryland 21218

Abstract. We evaluated the responses of 17 clones of marine diatoms, isolated from a range of environments, to two industrial wastes now disposed of at deep-ocean dumpsites. The clones showed a wide range of sensitivity to the stress, and there were no correlations with latitude of isolation, distance from shore, temperature regime, salinity, or general environmental stability. Three neritic clones from heavily polluted waters were much less sensitive than were other neritic and oceanic clones. Further, we surveyed cupric ion sensitivity in 15 clones of phytoplankton isolated from different neritic environments. All extremely resistant clones came from regions known or presumed to be heavily polluted. These data suggest that some phytoplankters can develop strains resistant to a marine pollutant.

INTRODUCTION

When pollutants are introduced into a phytoplankton community, a change in community structure occurs that may or may not be accompanied by a change in biomass. Thus, Parsons *et al.* (1976) found an increase in microflagellates following low level addition of mineral hydrocarbons to Controlled Ecosystem Pollution Experiment (CEPEX) enclosures, and Thomas and Seibert (1977) observed that copper addition led to dominance of microflagellates and certain pennate diatoms and the disappearance of centric diatoms and dinoflagellates. O'Connors *et al.* (1978) noted a decrease in average cell size in natural communities contained in dialysis bags and subjected to polychlorinated biphenyls (PCB's).

Changes in the structure of phytoplankton assemblages could result from the direct action of the toxic substance(s) on the algae or from a secondary effect caused by alteration in the predator community. Controlled laboratory experiments with clonal cultures have indicated that the direct effect can occur. Differential sensitivities have been demonstrated, with diatoms and dinoflagellates showing greater sensitivities than microalgae. Dunstan *et al.* (1975) found the coccolithophore, *Cricosphaera carterae* (clone 156), and the green microflagellate, *Dunaliella tertiolecta* (clone Dun), less sensitive to aromatic hydrocarbons than the diatom, *Skeletonema costatum* (clone Skel), or the dinoflagellate,

Amphidinium carterae (clone Amphi 137). Menzel *et al.* (1970) found the diatom, *Thalassiosira pseudonana* (formerly *Cyclotella nana*) (clone 13-1), far more sensitive to chlorinated hydrocarbons than *D. tertiolecta* (clone Dun). *S. costatum* (clone Skel) and the coccolithophore, *Coccolithus huxleyi* (clone BT-6), showed intermediate responses. Mosser *et al.* (1972) tested the effects of chlorinated hydrocarbons on mixed cultures of *D. tertiolecta* (clone Dun) and *T. pseudonana* (clone 3H). Although the diatom outgrew the green microflagellate in the control cultures, the latter predominated in treated cultures, even at concentrations that did not inhibit 3H in pure culture.

In response to pollution, species that are capable of developing resistance to various pollutants appear to emerge as dominants. The development of such resistance may involve physiological adaptation in a species whose members are capable of such flexibility, or genetic adaptation by selection of resistant individuals within a polymorphic species. The two are not mutually exclusive, but they are distinct; therefore, it is important that they be considered as separate phenomena. Harrison *et al.* (1977) observed assemblages that dominated in copper-treated CEPEX enclosures were more resistant to copper than were the assemblages that developed in control enclosures. Some change had occurred, but it is not clear whether the increased resistance was physiological or genetic. Jensen *et al.* (1974) determined that a clone of *S. costatum* isolated from a zinc-polluted fjord was two orders of magnitude more resistant to zinc than a conspecific clone isolated from an unpolluted fjord. This intraspecific difference in sensitivity may be an example of short-term evolution of a resistant genotype.

We report here the results of a series of experiments designed to test the potential for genetic selection by comparing the responses of a series of clones in several species of marine phytoplankton to two classes of industrial wastes and to copper.

Industrial Wastes

In this portion of our study, we compared estuarine clones from polluted and relatively unpolluted environments with oceanic clones of the same species. Two chemical stresses were used. One was a complex industrial waste product from the DuPont-Grasselli Plant (Linden, New Jersey), normally disposed of at Deep Water Dumpsite (DWD) 106 (*ca* 196 km [106 nmi] southeast of the apex of the New York Bight and lying over the continental slope). The second was a pharmaceutical waste disposed of at the Puerto Rican dumpsite (*ca* 81 km [44 nmi] north of Arecibo, Puerto Rico).

The sites lie in the transition zone between neritic and oceanic environments, and the environmental boundary separates distinct phytoplankton assemblages. Even those phytoplankton species that appear to cross the boundary consist of distinct ecological races (or semi-species), each with its own genetically-based physiological characteristics (Murphy and Guillard, 1976). An evaluation of toxic effects in such regions should take into account the environmental origins of the organisms being tested.

The Grasselli waste, generated in the synthesis of N,O dimethyl hydroxylamine (DMHA) and anisole, is a clear liquid with a density of 1.15 g/ml and a pH of *ca* 12.5. Of the organic carbon fraction (0.5-1.0%), up to 70% is in the form of sodium methyl sulfate, 20% is in the form of methanol, 1% is in the form of DMHA and other amines, and the remainder is mainly phenol and anisole. Vaccaro and Dennett (1977) demonstrated that the toxic moiety of the waste is organic.

Sargasso Sea medium (SS) with a pH of 8.4 was used in these experiments. This pH is realistic for surface waters in the open ocean. Addition of Grasselli waste to a concentration of 100 ppm did not alter the pH. A 500 ppm concentration had a pH of 8.6, and the 1,000 ppm concentration had a pH of 8.8 in the SS medium. The waste persisted in the wake of the barge at a concentration up to 100 ppm for at least 17 hours following a dump (Kohn and Rowe, 1978). We thus used concentrations up to one order of magnitude above this ambient level.

The pharmaceutical waste consists of aqueous wastes from seven pharmaceutical firms. They contain a diverse array of organic chemicals used for drug synthesis and purification. The metal concentrations are low (less than 1 ppm), and organic carbon content is high (1-4%). Recent studies indicate dilution rates are similar to those at DWD 106 (10-100 ppm ten hours after dumping; T.P. O'Connor, personal communication).

Copper

The purpose of this portion of our study was to survey cupric ion sensitivity in 15 clones of phytoplankton isolated from different neritic environments, some of which are known or presumed to be heavily polluted. Four of the clones are in the *"Thalassiosira pseudonana"* group, and three are *Skeletonema costatum.* The others are species isolated from Great South Bay, New York, the source of 3H, one of the most widely used clones in phytoplankton research. (The copper response of oceanic phytoplankton is influenced strongly by other metals, notably iron. Studies on oceanic clones will be reported elsewhere.)

Recent studies (Sunda and Guillard, 1976; Anderson and Morel, 1978; Jackson and Morgan, 1978) have shown conclusively that phytoplankton respond only to the activity of cupric ion in solution. The presence of natural organic chelators in seawater is likely to be higher in areas heavily polluted with copper (and other trace metals). Thus, the very high concentrations of copper found in some polluted bays and estuaries may not be associated with high levels of cupric ion activity.

METHODS

Industrial Wastes

Description of Clones

Experiments were conducted using clones of two species of marine diatoms from the Bigelow Laboratory for Ocean Sciences culture collection (Table 1).

Table 1. Origin of diatom clones used in industrial waste studies.

Species	Clone	Origin	Date	Isolator
Thalassiosira pseudonana	3H	Great South Bay Long Island, New York	1958	R. Guillard
	W	Wümme River estuary Bremen, Germany	1973	E. Paasche
	C5	Chincoteague Island lagoon Virginia	1964	D. Wilson
	STX-97	St. Croix reef Virgin Islands	1969	K. Haines
	Swan-1	Swan River estuary S.W. Australia	1965	R. Davis
	13-1	33°11′N, 65°15′W Sargasso Sea	1958	R. Guillard
	58-102	39°05′N, 71°56′W Continental slope	1976	L. Murphy
	35-10b	39°13′N, 69°20′W Continental slope	1977	P. Hoar/ L. Murphy
	35-11a	39°13′N, 69°20′W Continental slope	1977	P. Hoar/ L. Murphy
	35-81	36°11′N, 69°35′W Sargasso Sea	1977	P. Hoar/ L. Murphy
	35-128b	39°20′N, 71°00′W Continental slope	1977	P. Hoar/ L. Murphy
Skeletonema costatum	Skel	Long Island Sound New York	1956	R. Guillard
	FHS21	Friday Harbor Washington	before 1964	C. McAllister/ R. Guillard
	Fry-2	Freeport, Texas	1972	G. Fryxell
	FH-1	Falmouth harbor Massachusetts	1974	L. Murphy
	G44S	Sargasso Sea	1968	P. Hargraves
	35-24a	39°12′N, 69°20′W Continental slope	1977	P. Hoar/ L. Murphy

Clone Maintenance

The growth medium was prepared from 0.8 μm filtered surface Sargasso Sea (SS) water, autoclaved in teflon, and allowed to equilibrate overnight. The water was then enriched by sterile addition of nutrients from sterile stocks as "F/5" (Guillard and Ryther, 1962), but enriched with silicate to the level of "F/2." In this medium, phosphate was the limiting nutrient. The medium was dispensed aseptically into sterile, polypropylene-capped 25 x 120-mm pyrex test tubes.

Stock clones were monitored daily and transferred to fresh medium before the growth rate (μ) slowed. Thus, the cells were never allowed to reach nutrient limitation. The clones were maintained in this semi-continuous culture in SS medium and grown in a constant temperature room at 21° ± 0.5°C and 12,000 lux provided by (Sylvania) cool-white fluorescent tubes on a 14:10 L:D cycle.

Growth was monitored in a Turner Designs model 10 fluorometer adapted for 25 x 120-mm test tubes and equipped for the measurement of *in vivo* chlorophyll fluorescence. Fluorescence was proportional to the cell count, both under normal (control) conditions and in the presence of waste.

Bioassay

For each experiment, triplicate, clean, sterile tubes were filled with SS medium, and 1% v/v of the appropriate dilution of DuPont-Grasselli waste or pharmaceutical waste was added to the "treated" tubes. An equal volume of sterile distilled water was added to "control" tubes. Tubes were agitated and then inoculated with cells from log phase cultures that had been acclimatized to the appropriate experimental conditions. The volume of the inoculum was calculated to give an initial reading of 1.0 ± 0.3 fluorometric units. After reading in the fluorometer, the tubes were returned to the culture room. At 24-hour intervals, they were removed, allowed to equilibrate five minutes, read in the fluorometer, and returned to the culture room. This procedure was repeated for four days or until the end of log phase growth.

For each tube, the daily fluorescence readings were normalized to an initial (t_o) value of 1.0 unit. The ratio of the normalized fluorescence at 72 hours of the treated to the control experiment was expressed as Fl_T/Fl_C. This ratio approximates the reduction in chlorophyll due to the waste.

Copper

Description of Clones

The experiments were conducted using 15 clones of marine phytoplankton maintained in the culture collections of the Bigelow Laboratory for Ocean Sciences (diatom clones) and the Woods Hole Oceanographic Institution (other clones) (Table 2). Eight of these clones were isolated from Great South Bay, New York, a Long Island embayment known to be high both in metals (from marine paints) and in organic material (from duck farm wastes and, presumably, pesticides).

Table 2. Origin of clones used in copper study.

Species	Clone	Origin	Date	Isolator
Bacillariophyceae				
Thalassiosira pseudonana	3H	Great South Bay Long Island, New York	1958	R. Guillard
	W	Wümme River estuary Bremen, Germany	1973	E. Paasche
	C5	Chincoteague Island lagoon Virginia	1964	D. Wilson
	Swan-1	Swan River estuary S.W. Australia	1965	R. Davis
Skeletonema costatum	Skel	Long Island Sound Milford, Connecticut	1956	R. Guillard
	FH-1	Falmouth harbor Massachusetts	1974	L. Murphy
	EP-1	Eel Pond Falmouth, Massachusetts	1974	L. Murphy
Bellerochea polymorpha	Say 7	Great South Bay Long Island, New York	1965	R. Guillard
Synedra fragilarioides	Say 10	Great South Bay Long Island, New York	1965	R. Guillard
Phaeodactylum tricornutum	Phaeo	Great South Bay Long Island, New York	1952	J. Ryther
Chlorophyceae				
Nannochloris atomus	GSB Nanno	Great South Bay Long Island, New York	1952	J. Ryther
Eustigmatophyceae				
Monallantus salina (?)	GSB Sticho	Great South Bay Long Island, New York	1952	J. Ryther
	Say 2	Great South Bay Long Island, New York	1965	R. Guillard
	Say 3	Great South Bay Long Island, New York	1965	R. Guillard
Prymnesiophyceae				
Pavlova (Monochrysis) lutheri	Mono	Finnish Tide pool	1956	M. Droop

Clone Maintenance

The growth medium was a modified "F/2" made in aged, filtered Sargasso Sea water, with iron and Na-EDTA levels reduced to 1 μM each and no copper added. Tris buffer was added to a concentration of 10 mM. Medium was dispensed into polypropylene-capped 25 x 120-mm pyrex test tubes and autoclaved. Autoclaving did not alter the pH, which remained at 8.1. Clones were maintained and growth monitored as for the industrial waste procedures above.

Bioassay

Growth rates were determined at 16 different cupric ion activities between the blank ($10^{-13.95}$M = pCu 13.95) and $10^{-8.5}$M = pCu 8.5. In an experiment, 0.1 ml of each stock copper solution was added aseptically to each of 15 tubes of previously autoclaved medium; the sixteenth tube was a blank. The pH was adjusted where necessary (the four highest concentrations) by sterile addition of NaOH solution. Monitoring procedures were as above (industrial waste procedures).

Data Analysis

The growth rate, μ, was determined by means of linear regression of the natural logarithm of the fluorescence against time by digital computation.

RESULTS

Industrial Wastes

The Grasselli Waste

We tested 11 clones of *Thalassiosira pseudonana* and six clones of *Skeletonema costatum* with the Grasselli waste at 100 ppm and 1,000 ppm. There is considerable variability among the oceanic and neritic clones in the ranges of fluorescence reduction (Table 3). No significant reduction in fluorescence occurred in any clones exposed to the lowest concentration, 100 ppm. However, at 1,000 ppm all of the clones except 3H, W, and Skel showed fluorescence less than 60% of the controls. Except for these extremely resistant clones, there was no correlation between resistance and place of isolation.

The Pharmaceutical Waste

We tested 11 clones of *T. pseudonana* with pharmaceutical waste at 100 ppm and 1,000 ppm. This waste was more toxic, and all clones died at 1,000 ppm. As with the Grasselli waste, there was considerable variability among the oceanic and neritic clones in the ranges of fluorescence reduction at 100 ppm (Table 4), and no obvious correlation with general environmental variability. Only clone 3H (from Great South Bay) was highly resistant at this concentration.

Table 3. Relative fluorescence (Fl_T/Fl_C) at 72 hours as affected by Grasselli waste.

Clone	Concentration of waste 100 ppm	1,000 ppm
Polluted neritic		
3H	1.05 ± 0.09	0.87 ± 0.07
W	0.98 ± 0.07	0.98 ± 0.10
Skel	1.00 ± 0.04	0.91 ± 0.09
Other neritic		
C5	0.94 ± 0.21	0.41 ± 0.17
STX-97	0.99 ± 0.03	0.28 ± 0.05
Swan-1	0.88 ± 0.08	0.22 ± 0.02
FHS 21	0.93 ± 0.10	0.22 ± 0.04
Fry-2	1.05 ± 0.13	0.44 ± 0.17
FH-1	1.01 ± 0.16	0.56 ± 0.17
Oceanic		
13-1	0.93 ± 0.13	0.25 ± 0.03
58-102	0.73 ± 0.11	0.09 ± 0.01
35-10b	0.84 ± 0.14	0.25 ± 0.04
35-11a	0.79 ± 0.09	0.35 ± 0.06
35-81	0.96 ± 0.13	0.19 ± 0.03
35-128b	0.84 ± 0.11	0.01 ± 0.003
G44S	0.98 ± 0.06	0.53 ± 0.02
35-24a	0.94 ± 0.12	0.01 ± 0.001

Table 4. Relative fluorescence (Fl_T/Fl_C) at 72 hours as affected by the pharmaceutical waste (100 ppm).

Clone	
Polluted neritic	
3H	0.98 ± 0.12
W	0.71 ± 0.09
Other neritic	
C5	0.72 ± 0.12
STX-97	0.53 ± 0.07
Swan-1	0.40 ± 0.08
Oceanic	
13-1	0.32 ± 0.03
58-102	0.16 ± 0.02
35-10b	0.30 ± 0.05
35-11a	0.21 ± 0.04
35-81	0.74 ± 0.07
35-128b	0.49 ± 0.14

Table 5. Levels of pCu affecting growth rate.

Clone	pCu Threshold response	Zero growth
Polluted neritic		
3H	11.5	8.5
W	11.0	9.8
Skel	9.6	8.5
Say 7	10.4	9.0
Say 10	9.6	8.5
Phaeo	9.0	8.5
GSB Nanno	9.4	8.5
GSB Sticho	8.7	8.5
Say 2	none	8.5
Say 3	9.1	8.7
Other neritic		
C5	10.4	9.0
Swan-1	10.4	9.6
FH-1	9.8 +	9.6
EP-2	10.4	9.0
Mono	10.3	9.6

Copper

The threshold of response for other than exceptionally resistant clones lies between pCu 9 and pCu 11 and does not correlate in any obvious way with systematic position. This is shown in Table 5. Certain clones are exceptionally resistant to copper, growing slowly or even at full growth rate up to the limiting equilibrium concentration of free cupric ion in seawater (pCu 8.5, at least). The extremely resistant clones all came from regions known or presumed to be heavily polluted (*e.g.*, Milford Harbor, Connecticut or Great South Bay, New York). Clone W, from Bremen, Germany, is a copper-sensitive clone.

DISCUSSION

Fifty years ago, Baas-Becking (1928) noted that phytoplankton from hypersaline environments were highly resistant to metals and temperature flux. He suggested a possible interrelationship, but noted that the hypersaline environments were also high in arsenates and borates. Slobodkin and Sanders (1969) hypothesized that stress reactions are generalized and that organisms adapted to stable, predictable environments should be more vulnerable to environmental changes than organisms adapted to unstable, unpredictable environments.

Fisher *et al.* (1973) and Fisher (1977) observed that diatom clones established from oceanic environments were more sensitive to pollution stress than were conspecific clones isolated from polluted estuarine environments. This increased resistance of estuarine clones may indicate a generalized genetic adaptation to change *per se,* or it may be the result of a more specific genetic adaptation to pollution stress. The rapid response time of microalgae to pollution may be an intrinsic characteristic of these organisms, it may be characteristic of individuals from unstable environments, or it may reflect the more rapid selection of pollution-resistant genotypes that can occur in haploid organisms. (Diatoms are diploid, and dinoflagellates, though haploid in the vegetative stage, have a long-lived diploid cyst. Sexual reproduction appears to occur regularly in nature. The minute flagellates and coccoid algae are haploid, except for a fleeting pre-meiotic zygote stage, and probably can reproduce asexually indefinitely.)

In this study we have compared the pollution responses of conspecific clones of marine diatoms isolated from polluted neritic, unpolluted neritic, and oceanic waters. The responses of clones from unpolluted neritic and oceanic waters were broadly overlapping and did not correlate with general environmental stability. All extremely resistant clones came from polluted waters. Not all clones from polluted waters were resistant, and not all resistant clones were resistant to all chemical stresses tested; however, resistant clones always came from areas of high pollution. Clone W, for example, is highly resistant to the Grasselli waste, fairly resistant to the pharmaceutical waste, and copper-sensitive. This suggests a certain degree of adaptive specificity, and is most probably related to previous exposure to specific or related chemicals. Not all areas high in halogenated hydrocarbons (for example) are also high in copper.

The microalgae may show a similar pattern. Clone Mono, isolated from an unpolluted tide pool, was as sensitive to cupric ion as diatoms isolated from unpolluted environments. Although more work needs to be done with the microalgae, it appears that pollution resistance is not an innate physiological property of this group, and that the apparent takeover capacity following introduction of pollutants may be because of their haploid genome, which is more directly exposed to selection.

It is clear that the short-term pollution history of an environment is at least as important in determining resistance to pollution as is the long-term stability and predictability of that environment. Further, some phytoplankters can and do evolve resistance to moderate levels of pollution, both organic and inorganic. The smaller, haploid forms respond first, but the diploid diatoms also have evolved resistant strains.

CONCLUSIONS AND RECOMMENDATIONS

Resistance may be because of an ability to avoid uptake or an ability to metabolize or partition the material after uptake. We do not know which mechanism is operative. We also do not know what side effects (pleiotrophic or linkage effects) such resistance may confer. Fisher (1977) suggested that resistance to chemical stress may involve changes in membrane structure and permeability. Such an alteration may limit uptake of other, required chemicals. Mutations conferring resistance may be linked closely to other mutations, and selection for resistance could alter many characteristics of a population. Resistant strains may concentrate toxicants, while remaining relatively unaffected by them. The toxicants could then be passed up the food chain.

Short-term evolution of resistant genotypes apparently has occurred in local waters and obviously will continue to do so as long as selection forces (*i.e.*, pollution) remain strong. The evolution of resistant genotypes is not, in itself, a cause for concern. It is necessary, however, to study the nature of these changes and to evaluate the possibility that resistance is accompanied by other, less desirable, changes. We strongly recommend a study of the nature of resistance and potential side effects of such selection.

ACKNOWLEDGEMENTS

We thank R.A. Belastock and B. Woodward for their excellent technical assistance. This work was supported by NOAA grant 04-8-MO1-41 and NSF grant OCE78-08858.

REFERENCES

Anderson, D.M. and F.M.M. Morel. 1978. Copper sensitivity of *Gonyaulax tamarensis.* Limnol. Oceanogr. 23: 283-295.

Baas-Becking, L.G.M. 1928. On organisms living in concentrated brine. Tijdschr. Ned. Dierkd. Ver. 3: 6-9.

Dunstan, W.M., L.P. Adkinson, and J. Natoli. 1975. Stimulation and inhibition of phytoplankton growth by low molecular weight hydrocarbons. Mar. Biol. 31: 305-310.

Fisher, N.S. 1977. On the differential sensitivity of estuarine and open-ocean diatoms to exotic chemical stress. Am. Nat. 111: 871-895.

Fisher, N.S., L.B. Graham, E.J. Carpenter, and C.F. Wurster. 1973. Geographic differences in phytoplankton sensitivity to PCBs. Nature 241: 548-549.

Guillard, R.R.L. and J.H. Ryther. 1962. Studies of marine planktonic diatoms. I. *Cyclotella nana* Hustedt and *Detonula confervacea* (Cleve) Gran. Can. J. Microbiol. 8: 229-239.

Harrison, W.G., R.W. Eppley, and E.H. Renger. 1977. Phytoplankton nitrogen metabolism, nitrogen budgets, and observations on copper toxicity: Controlled Ecosystem Pollution Experiment. Bull. Mar. Sci. 27: 44-57.

Jackson, G.A. and J.J. Morgan. 1978. Trace metal-chelator interactions and phytoplankton growth in seawater media: theoretical analysis and comparison with reported observations. Limnol. Oceanogr. 23: 268-282.

Jensen, A., B. Rystad, and S. Melsom. 1974. Heavy metal tolerance of marine phytoplankton. I. The tolerance of three algal species to zinc in coastal sea water. J. Exp. Mar. Biol. Ecol. 15: 145-157.

Kohn, B. and G.T. Rowe. 1978. Dispersion of Two Liquid Industrial Wastes at Deep Water Dumpsite 106, off the Coast of New Jersey, U.S.A. Final Report Submitted to the Ocean Dumping Program, NOAA, Grant No. 04-6-158-44072. 35 pp.

Menzel, D.W., J. Anderson, and A. Randke. 1970. Marine phytoplankton vary in their response to chlorinated hydrocarbons. Science 167: 1724-1726.

Mosser, J.L., N.S. Fisher, and C.F. Wurster. 1972. Polychlorinated biphenyls and DDT alter species composition in mixed cultures of algae. Science 176: 533-535.

Murphy, L.S. and R.R.L. Guillard. 1976. Biochemical taxonomy of marine phytoplankton by electrophoresis of enzymes. I. The centric diatoms *Thalassiosira pseudonana* and *T. fluviatilis.* J. Phycol. 12: 9-13.

O'Connors, H.B., Jr., C.F. Wurster, C.D. Powers, D.C. Biggs, and R.G. Rowland. 1978. Polychlorinated biphenyls may alter marine trophic pathways by reducing phytoplankton size and production. Science 201: 737-739.

Parsons, T.R., W.K.W. Li, and R. Walters. 1976. Some preliminary observations on the enhancement of phytoplankton growth by low levels of mineral hydrocarbons. Hydrobiologia 51: 85-89.

Slobodkin, L.B. and H.L. Sanders. 1969. On the contribution of environmental predictability to species diversity. Brookhaven Symp. Biol. 22: 82-95.

Sunda, W. and R.R.L. Guillard. 1976. The relationship between cupric ion activity and the toxicity of copper to phytoplankton. J. Mar. Res. 34: 511-529.

Thomas, W.H. and D.L.R. Seibert. 1977. Effects of copper on the dominance and the diversity of algae: Controlled Ecosystem Pollution Experiment. Bull. Mar. Sci. 27: 23-33.

Vaccaro, R.F. and M.R. Dennett. 1977. The Environmental Responses of Marine Bacteria to Waste Disposal Activities at Deep Water Dumpsite 106. Final Report Submitted to the Ocean Dumping Program, NOAA, Grant No. 04-7-158-4405. 15 pp.

THE LONG-TERM EFFECTS OF SUSPENDED PARTICULATES ON SURVIVAL AND REPRODUCTION OF THE MYSID SHRIMP, *MYSIDOPSIS BAHIA*, IN THE LABORATORY

D.R. Nimmo

Environmental Research and Technology, Inc.
1716 Heath Parkway
Fort Collins, Colorado 80522

T.L. Hamaker
E. Matthews

Environmental Research Laboratory
U.S. Environmental Protection Agency
Sabine Island
Gulf Breeze, Florida 32561

W.T. Young

State of Florida
Department of Environmental Regulation
160 Government Center
Pensacola, Florida 32501

Abstract. Methods are described for conducting short- and long-term laboratory tests to determine effects of suspended particulates on the life cycle of the mysid, *Mysidopsis bahia.* No apparent effects were observed on adults within 96 hours; however, total populations were reduced to about 75% of controls within 28 days, if tests were begun with a mixture of juveniles and gravid adults. At least two effects were noted: reduction in the number of juveniles released and mortality of the initial animals with time. Factors responsible for reduced populations included interference with feeding or mating, clogging of respiratory apparatus, and disorientation in water currents. These findings are discussed with respect to turbidity produced by mining of fossilized shells or dredging in estuaries.

INTRODUCTION

Dredging operations in estuaries, such as mining fossilized shell beds or maintaining channels and harbors, have been significant activities of man for many years. Buried shell is one of the natural resources found in estuaries, and demand for this material is increasing. Similarly, hydraulic dredging is important to the commerce and economy of all coastal regions in the United States. According to May (1973), hydraulic channel dredging and shell dredging use similar equipment although the operations differ. Operators of channel dredges construct and maintain channels and harbors. Substratum and water are pumped from the channel bottom, and the dredged material is discharged some distance away from the channel. In contrast, shell dredging operations take place outside navigation channels in the open estuary. The material is screened and washed, and waste is returned overboard, usually in the immediate vicinity of the dredge operation.

From the viewpoint of the biologist, the question of short- or long-term effects on aquatic biota in the vicinity of the dredge is paramount, but few studies actually have documented deleterious effects on biota. Studies usually suggest that dredged areas become repopulated within several months (Harrison, 1967; Flemer *et al.*, 1968; Cronin *et al.*, 1970; May, 1973; Diaz and Boesch, 1977). May (1973) pointed out that short-term effects of dredging in Mobile Bay, Alabama resulted in a temporary increase in suspended solids over a relatively small area; the sediment discharged by dredges quickly settled to the bottom and formed a turbidity current out to a distance of about 500 m or less. He also concluded that the mud flow outside dredge cuts consolidated to form a layer from about 2-3 cm to several centimeters thick. The newly-formed layer could suffocate some benthic organisms such as worms and small molluscs, but the loss probably was transitory because of recolonization.

The effects of suspended material on marine plankton or nekton have been regarded as limited based on laboratory studies. Most studies have been relatively short-term, particularly when the length of the test is compared to the longevity of the species tested. In addition, most studies have tested only adult stages. Wilber (1971) estimated that there are approximately 9,000 species of marine invertebrates with free swimming larvae, and that most of these species spend a portion of their life cycle in estuaries, *i.e.*, areas of relatively high concentrations of suspended solids. Other limitations exist because tests were conducted under static conditions where changes in water quality during the tests could have affected the results. Many investigators used mixtures of kaolin, Fuller's earth, and chalk, and thus may not have fully simulated naturally occurring sediments.

Some studies using experimentally produced turbidities have been conducted on the developing eggs of oysters and clams, and on adult copepods, other invertebrates, and several fishes. Deleterious effects of turbidity on eggs of the oyster, *Crassostrea virginica,* and the clam, *Mercenaria (=Venus) mercenaria,* were documented by Davis (1960) and Loosanoff (1961). Paffenhofer (1972) found that the copepod, *Calanus helgolandicus,* ingested particles of suspended "red mud" along with food and probably obtained insufficient nutritive material. A flow-through bioassay using suspensions of naturally occurring sediments showed that several fishes and invertebrates are able to survive continuous exposures to particulates at concentrations of grams or tens of grams per liter (Peddicord and McFarland, 1978).

In this study, a flow-through test was used to evaluate the effects of suspended solids throughout the life cycle of a small estuarine mysid, *Mysidopsis bahia.* This mysid has been used successfully to determine the long-term effects of a variety of chemicals (Nimmo *et al.*, 1977; Nimmo *et al.*, 1978c; Nimmo *et al.*, 1980). It is used currently in ecological evaluations of dredged material proposed for discharge into ocean waters (Peddicord, 1977) and to determine the adverse effects of chemical substances that might enter marine or estuarine environments (Anonymous, 1979).

Research was initiated in response to proposed shell dredging in East Bay (near Pensacola), Florida. East Bay is a relatively uncontaminated shallow estuary; its waters are categorized as Class II waters (*i.e.*, productive fish and shellfish areas). The quantities of shell to be dredged, duration of the project, and proximity to the U.S. Environmental Protection Agency's Environmental

Research Laboratory provided impetus for the study. This report includes a description of the test apparatus, procedures for using the mysid in testing the effects of suspended solids, and data on apparent effects of suspended solids from a natural sediment taken from East Bay.

MATERIALS AND METHODS

Mysids were obtained from cultures maintained at the Environmental Research Laboratory, Gulf Breeze, Florida. Culture containers were 38-liter glass aquaria supplied with filtered (5μ) flowing water (15-30°/oo and 22°-26°C). Mysids were fed 48-hour old *Artemia salina* larvae daily (Nimmo *et al.*, 1978a).

Procedures for conducting tests with suspended sediment were similar to those described for toxicity tests with materials such as pesticides (Nimmo *et al.*, 1978b). The test apparatus (Figure 1) maintained a continuous flow of

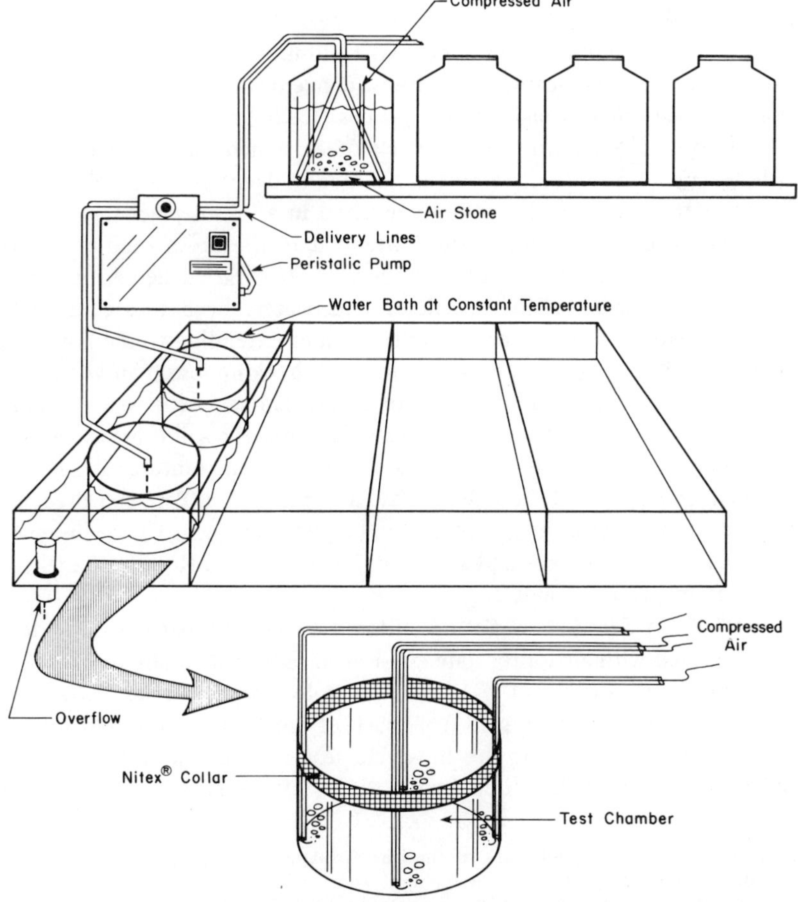

Figure 1. Schematic of apparatus design used to test the effects of suspended solids on *Mysidopsis bahia.*

Table 1. Analysis of East Bay, Florida sediment.
Percent water = 76.83; percent volatile solids = 13.34.

Particle size	Percentage
< 0.0625 mm	97.87
0.0625 - 0.125 mm	1.47
0.125 - 0.500 mm	0.37
0.500 - 2.0 mm	0.06
> 2.0 mm	0.23

suspended sediment through the chambers where the mysids were contained. Four 19-liter glass carboys were used to contain the sediment-water suspensions and control water without suspended sediment. Compressed air, delivered through airstones (20 cm in length), was used to stir the sediment and water vigorously. From the four carboys, glass tubes delivered the suspensions and control water to a Harvard[1] multichannel peristalic pump, which, in turn, fed each of eight test chambers. Duplicates for each test concentration (three concentrations and control) necessitated eight delivery lines. Each line delivered either a test concentration or control water to a test chamber, which was a large crystallizing dish (dimensions 190 mm in diameter, 100 mm deep, for a total volume of 2,600 ml). A flow rate of 300 ml/hour resulted in a 95% replacement every 24 hours. A No. 210 Nitex[2] mesh collar attached to the lip of each crystallizing dish with silicone cement prevented the loss of *Artemia* larvae and mysids but permitted the passage of fine particles and water. Each test chamber was equipped with four equally spaced capillary tubes extending to the bottom and cemented in place around the perimeter of the dish. Compressed air was supplied via these tubes. The continuous supply of suspended sediment to the chambers and the vigorous agitation by the compressed air kept the solids in suspension. A water bath surrounding the test chambers maintained temperatures at 24° ± 2°C.

Sediment was obtained from East Bay, Florida with a Peterson dredge and transported to the laboratory in 4.5-liter (1-gal) glass jars. The sediment was mixed, replaced into the jars, and stored at 7°C until used. Results of analysis of particle sizes are given in Table 1.

Seawater from Santa Rosa Sound, filtered to remove particulates 5 μm or larger, was mixed with an appropriate amount of sediment to obtain 2.0%, 0.5%, and 0.1% suspensions (v/v wet wt) and a control without sediment. Gravimetric analysis of these mixtures sampled from each of the duplicate concentrations on five separate occasions during the life-cycle test showed the average sediment concentrations to be 1.02 g/liter (range 0.53-1.59 g/liter), 0.23 g/liter (range

[1] Harvard is a registered trademark of the Harvard Apparatus Company, Inc., 150 Dover Road, Millis, Massachusetts 02054. Model used was 1203 Series 1200. Reference to commercial products does not constitute endorsement by ERT, EPA, or Florida DER.

[2] Nitex is a registered trademark of Tobler, Ernst, and Trabor, Inc., New York, New York.

0.17-0.31 g/liter), and 0.045 g/liter (range 0.032-0.064 g/liter), respectively.

The sediment and water used in the study were analyzed by gas chromatography for pollutants such as polychlorinated biphenyls (PCB's) and pesticides at the Environmental Research Laboratory, U.S. Environmental Protection Agency, Gulf Breeze, Florida. Concentrations were below the limits of detection, *i.e.,* 0.004 mg/liter for DDT and metabolites, 0.08 mg/liter for PCB's, and 0.2 mg/liter toxaphene in sediment (dry wt); 0.01 μg/liter for DDT and dieldrin, 0.04 μg/liter for chlordane, 0.08 μg/liter for PCB's, and 0.2 μ/liter for toxaphene in water (Dr. Richard Garnas, personal communication).

Two flow-through tests were conducted to determine effects of suspended solids: a 96-hour acute test and a 28-day life-cycle test. For the 96-hour test, 10 adult mysids were added to each test chamber after the peristalic pump had delivered suspended sediment for 24 hours. The life-cycle test was initiated by isolating 250-300 gravid females in a 5-liter glass battery jar. Salt water entered the jar at approximately 10 ml/min. The overflow exited over the top of the jar through a Nitex (No. 210) collar, which prevented the loss of mysids and *Artemia* larvae. Twenty-five juveniles (24 hours old or less) were added to each test chamber, and the peristalic pump was started. During both experiments, the mysids were fed 48-hour old *Artemia* larvae *ad libitum.* For the 96-hour acute test, water temperature was 22°C and salinity 29-31 °/oo. For the 28-day life cycle test, the temperature range was 23°-26°C; salinity ranged from 23.5 °/oo to 31 °/oo.

Before recoving and counting mysids from the test containers, a gentle stream of water at the same salinity and temperature as the test water was used to wash the finer suspended solids through the Nitex collar and to separate the mysids. This procedure was used at the end of each test and also on day 13 of the life cycle test. Day 13 was chosen because it preceded the expected release of juvenile mysids by the gravid females. Small dipping screens and pipettes were used to capture the mysids.

Analysis of variance with Duncan's multiple range test was employed to compare replicates. Significant differences ($\alpha = 0.05$) were attributed to the effects of the suspended particulates.

RESULTS AND DISCUSSION

Data from the 96-hour acute flow-through test suggest a trend for increased mortality of adult mysids with increased concentrations of suspended solids (Table 2). However, the differences observed are not statistically significant. Lack of significant mortality in 96 hours may indicate that longer tests are required to determine effects. With many toxicants, a 96-hour exposure underestimates effects observed only in long-term studies. Lack of mortality may have resulted from the use of adults to begin the test instead of juveniles, which were used in the life-cycle test.

Significant differences occurred in populations of mysids after exposure to suspended solids throughout a life cycle (Table 4). Effects were indicated by

Table 2. Survival of adult *Mysidopsis bahia* continuously exposed to concentrations of suspended solids for 96 hours. No significant differences occurred between control and experimental groups.

Concentration[a]	Number of adults at start Replicate 1	Replicate 2	Number of adults at 96 hours Replicate 1	Replicate 2
Control	10	10	9	10
0.1%	10	10	9	10
0.5%	10	10	8	8
2.0%	10	10	8	6

[a]Nominal concentration v/v.

fewer surviving adults (Table 3), fewer juveniles produced by surviving mysids, and fewer juveniles produced per female (Table 4). Therefore, the total number of mysids after 28 days was reduced by two factors acting in concert: (1) greater mortality of animals, *i.e.*, 32% survival of animals in the 1.02 g/liter treatment vs. 70% survival of controls, and (2) decreased production of juveniles by surviving females, *i.e.*, 19 in the 1.02 g/liter treatment vs. 102 in the controls. While it was unexpected in this study that the two factors would act together to reduce the population, similar observations have been made before. Nimmo *et al.* (1980) found that the insecticide, Dimilin, simultaneously affected survival and reproduction in mysids.

During the 28-day life-cycle test, the mysids were observed only once and that was on day 13. Numbers of surviving adults in the various concentrations of suspended solids were not significantly different from controls (Table 3), but mysids in the higher concentrations (1.02 g/liter and 0.23 g/liter) were smaller and usually less developed sexually than animals in the 0.045 g/liter suspension and controls. A similar effect also had been noted earlier when cadmium delayed the release of juveniles by at least 72 hours compared to controls (Nimmo *et al.*, 1978c).

Since turbidity reduced the ability to observe daily events in the test containers, it is difficult to explain the reasons for reduced population sizes. One

Table 3. Survival of *Mysidopsis bahia* continuously exposed to suspended solids for 28 days. Paired columns indicate the number of living individuals in each replicate.

Measured concentration	Number of animals by replicate: Start		Day 13		Day 28	
Control	25	25	19	20	15	20
0.045 g/liter	25	25	22	22	18	16
0.23 g/liter	25	25	14	21	13	14
1.02 g/liter	25	25	12	19	10	6

Table 4. Number of juveniles, number of juveniles per female mysid, and total population of mysids after 28-day exposure to suspended solids.

Measured concentration	Number of mysids by replicate					
	Total juveniles		Juveniles/female		Adults + juveniles	
Control	41	61	5.1	7.6	56	81
0.045 g/liter	35	56	3.2	5.1	53	72
0.23 g/liter	26	20	2.6	2.0	39	34
1.02 g/liter	13	6	2.6	1.2	23[a]	12[a]

[a]Significantly different from controls and other treatments ($\alpha = 0.05$) (ANOVA with Duncan's multiple range test for variable numbers).

factor may be the inability of the mysids to sense or obtain the *Artemia* because of confusion between food and suspended solids. In the laboratory, mysids feed on *Artemia* or other particulate foods by orienting to a current. When a mysid senses a food particle ahead, it moves to intercept the food, capturing the particle as the water current sweeps it by the mysid. Mysids will capture particles of almost any size; however, nonfood items are examined and quickly released. In the laboratory, copepods (*Calanus helgolandicus*) ingested particles of suspended "red mud" along with food (Paffenhofer, 1972). The author concluded that because of interference by the mud suspension, female copepods ingested reduced amounts of phytoplankton, which, in turn, hindered ovarian development.

Other problems encountered by the mysids could have included clogging of the animals' respiratory apparatus with debris, inability of the animals to orient in the water column, and interference with the elaborate mating behavior observed in the laboratory. Wilber (1971) suggested that turbidity may affect activity patterns and behavior of marine animals.

The particle concentrations used in this study were typical of those measured near dredging operations. Concentrations found in the water column in natural estuarine waters where dredging is not occurring have ranged from about 200 mg/liter to 400 mg/liter (Table 5). Those found in the water column during dredging operations often are greater than 1 g/liter, 10 g/liter being more typical; however, values sometimes exceed 100 g/liter (Table 6). Certainly, the

Table 5. Concentrations of suspended solids found in undisturbed estuarine waters.

Location	Concentrations (range)	Reference
Upper Chesapeake Bay	20-100 mg/liter	Biggs, 1970
Galveston Bay	200-400 mg/liter	Masch and Espey, 1967
Louisiana marshes	20-200 mg/liter	Mackin, 1961
Patuxent River	40- 80 mg/liter	Flemer *et al.*, 1970
Northern Chesapeake Bay	5-110 mg/liter	Schubel, 1968

Table 6. Concentrations of suspended solids reported in the water column during dredging operations.

Type of dredging/ location	Concentrations	Reference
Hopper dredge disposal, Carquinez Strait, San Francisco, Bay, California	< 0.2 g/liter in water column; 20 g/liter 1-2 m off bottom	Peddicord and McFarland, 1978
Shell dredging, Galveston Bay, Texas	10-175 g/liter depending on location	Masch and Espey, 1967
Shell dredging, Copano Bay, Texas	33-58 g/liter	Wilson, 1950
Shell dredging, Mobile Bay, Alabama	50-100 mg/liter at surface; 1- > 10 g/liter at bottom (concentration dependent on distance from dredge)	May, 1973

possibility exists that suspended particulate concentrations of 1 g/liter or more could occur, at least periodically, in the New York Bight. The projected volume of dredge spoil to be dumped in 1976 was 8.9 million m^3 (11.6 million yd^3), while the projected annual volume of raw and digested sludge was greater than 3.8 million m^3 (5 million yd^3) (Mueller *et al.*, 1976).

After an extensive review of the literature on dredging, Stern and Stickle (1978) concluded that the ecological effects of suspended solids in the water column were generally minimal and transitory. Peddicord and McFarland (1978) came to a similar conclusion after an extensive study in which they exposed juvenile and adult marine, estuarine, and freshwater fishes, and invertebrates to sediment slurries produced with a flow-through system.

The results of this study raise questions as to whether tests conducted in the past have been long enough to detect subtle changes in populations or whether enough studies have included larval stages. Relatively small quantities of suspended material resulted in decreased pumping rate by oysters (*Crassostrea virginica*). In water containing about 0.1 g silt/liter, there was an average reduction in pumping rate of 57%; at 1 g silt/liter, an 80% reduction occurred, and at 3-4 g silt/liter, a 94% reduction was observed (Loosanoff and Tommers, 1948).

Mysidopsis bahia is but a single representative of a myriad of primary consumers that occupy the water column. The various stages in its life cycle are important in estuarine food chains, and well over 50 species of fish have mysids as part of their diets. Stickney *et al.* (1974) found that three of four species of flounders captured mysids as a primary staple of their diets. It is obvious that, in some instances, the loss of mysids or any other related nektonic organisms could have serious ecological impact on the next trophic level.

CONCLUSIONS

Results of this study suggest that effects of continuous and long-term dredging in estuaries, producing concentrations of suspended particulates in excess of 1 g/liter, could reduce populations of planktonic or nektonic organisms. Judging from the literature, this is the first report of an aquatic species being used to assess effects of suspended particulates on an entire life cycle. Even though insufficient data exist to determine whether concentrations of suspended particulates used in this study were realistic simulations of those in the environment during dredging or disposal operations, the results raise questions as to whether tests conducted in the past have been designed correctly or have been of sufficient duration to detect changes in populations of planktonic or nektonic organisms.

REFERENCES

Anonymous. 1979. Test 6: *Mysidopsis bahia* life cycle. Federal Register 44(53): 16291.

Biggs, R.B. 1970. Geology and hydrology. *In:* Gross Physical and Biological Effects of Overboard Spoil Disposal in Upper Chesapeake Bay, University of Maryland, Natural Resources Institute Spec. Rep. 3. 6 pp.

Cronin, E.L., R.B. Biggs, D.A. Flemer, H.T. Pfitzenmeyer, F. Goodwin, Jr., W.L. Dovel, and D.E. Ritchie, Jr. 1970. Gross Physical and Biological Effects of Overboard Spoil Disposal in Upper Chesapeake Bay, University of Maryland, Natural Resources Institute Spec. Rep. 2. 66 pp.

Davis, H.C. 1960. Effects of turbidity-producing materials in sea water on eggs and larvae of the clam *(Venus (Mercenaria) mercenaria).* Biol. Bull. 118: 48-54.

Diaz, R.J. and D.F. Boesch. 1977. Impact of fluid mud dredged material on benthic communities of the tidal James River, Virginia. Tech. Rep. D-77-45, U.S. Army Engineer Waterways Experiment Station, CE, Vicksburg, MS. pp. 1-38.

Flemer, D.A., W.L. Dovel, H.T. Pfitzenmeyer, and D.E. Ritchie, Jr. 1968. Biological effects of spoil disposal in Chesapeake Bay. J. Sanitary Eng. Div. 94 (SA4): 683-706.

Flemer, D.A., D.H. Hamilton, C.W. Keefe, and J.A. Mihursky. 1970. The effects of thermal loading and water quality on estuarine primary production. University of Maryland, Natural Resources Institute, Chesapeake Biol. Lab. Ref. No. 71-6. 217 pp.

Harrison, W. 1967. Environmental effects of hydraulic dredging in estuaries. Alabama Mar. Res. Bull. 9: 1-85.

Loosanoff, V.L. 1961. Effects of turbidity on some larval and adult bivalves. Proc. Gulf Caribb. Fish. Inst. 14: 80-95.

Loosanoff, V.L. and F.D. Tommers. 1948. Effect of suspended silt and other substances on rate of feeding of oysters. Science 107: 69-70.

Mackin, J.G. 1961. Canal dredging and silting in Louisiana Bays. Publ. Inst. Mar. Sci. Univ. Tex. 7: 262-314.

Masch, F.D. and W.H. Epsey. 1967. Shell dredging—a factor in sedimentation in Galveston Bay. University of Texas, Center for Research in Water Resources Tech. Rep. No. 7. 168 pp.

May, E.B. 1973. Environmental effects of hydraulic dredging in estuaries. Alabama Mar. Res. Bull. 9: 1-85.

Mueller, J.A., J.S. Jeris, A.R. Anderson, and C.F. Hughes. 1976. Contaminant inputs to the New York Bight. NOAA Tech. Memo. ERL MESA-6. 347 pp.

Nimmo, D.R., L.H. Bahner, R.A. Rigby, J.M. Sheppard, and A.J. Wilson, Jr. 1977. *Mysidopsis bahia:* an estuarine species suitable for life-cycle toxicity tests to determine the effects of a pollutant. *In:* Aquatic Toxicology, F.L. Mayer and J.L. Hamelin (eds.), American Society for Testing and Materials Special Technical Publication 634, Philadelphia, PA. pp. 109-111.

Nimmo, D.R., T.L. Hamaker, J.C. Moore, and R.A. Wood. 1980. Acute and chronic effects of Dimilin on survival and reproduction of *Mysidopsis bahia. In:* Aquatic Toxicology, J.G. Eaton, P.R. Parrish, and A.C. Hendricks (eds.), American Society for Technical and Materials Special Technical Publication 707, Philadelphia, PA. pp. 366-376.

Nimmo, D.R., T.L. Hamaker, and C.A. Sommers. 1978a. Culturing the mysid (*Mysidopsis bahia*) in flowing sea water or a static system. *In:* Bioassay Procedures for the Ocean Disposal Permit Program. EPA-600/9-78-010.

Nimmo, D.R., T.L. Hamaker, and C.A. Sommers. 1978b. Entire life-cycle toxicity test using mysids (*Mysidopsis bahia*) in flowing water. *In:* Bioassay Procedures for the Ocean Disposal Permit Program. EPA-600/9-78-010.

Nimmo, D.R., R.A. Rigby, L.H. Bahner, and J.M. Sheppard. 1978c. The acute and chronic effects of cadmium on the estuarine mysid, *Mysidopsis bahia.* Bull. Environ. Contam. Toxicol. 19(1): 80-85.

Paffenhofer, G.A. 1972. The effects of suspended "red mud" on mortality, body weight, and growth of the marine planktonic copepod, *Calanus helgolandicus.* Water Air Soil Pollut. 1: 314-321.

Peddicord, R.K. 1977. Ecological evaluation of proposed discharge of dredged material into ocean waters (Implementation Manual for Sec. 103, P.L. 92-532). U.S. Army Engineer Waterways Experiment Station, CE, Vicksburg, MS.

Peddicord, R.K. and V.A. McFarland. 1978. Effects of suspended dredged material on aquatic animals. Tech. Rep. D-78-29, U.S. Army Engineer Waterways Experiment Station, CE, Vicksburg, MS. 102 pp.

Schubel, J.R. 1968. Suspended sediment of the northern Chesapeake Bay. The Johns Hopkins University, Chesapeake Bay Institute Tech. Rep. No. 35. 263 pp.

Stern, E.M. and W.B. Stickle. 1978. Effects of turbidity and suspended material in aquatic environments; literature review. Tech. Rep. D-78-21, U.S. Army Engineer Waterways Experiment Station, CE, Vicksburg, MS. 117 pp.

Stickney, R.R., G.L. Taylor, and R.W. Heard, III. 1974. Food habits of Georgia estuarine fishes. I. Four species of flounders (Pleuronectiformes: Bothidae). Fish. Bull. 72(2): 515-525.

Wilber, C.G. 1971. Turbidity. *In:* Marine Ecology, O. Kinne (ed.), Wiley-Interscience, New York. pp. 1181-1194.

Wilson, W.B. 1950. The effects of sedimentation due to dredging operations on oysters in Copano Bay, Texas. M.S. Thesis, Texas A&M College. 128 pp.

FATES AND BIOLOGICAL EFFECTS ON PLANKTON OF PARTICLE-SORBED PCB's IN COASTAL WATERS

H.B. O'Connors, Jr.,
C.D. Powers[1]
C.F. Wurster
K.D. Wyman[2]
G.M. Nau-Ritter
R.G. Rowland

Marine Sciences Research Center
State University of New York
Stony Brook, New York 11794

Abstract. The adsorption and desorption of polychlorinated biphenyls (PCB's) (Aroclor 1254) to and from organic and inorganic particulates characteristic of coastal waters were examined, as were the effects of PCB's on biological processes in natural phytoplankton assemblages and common estuarine copepods. Partitioning of PCB's in aqueous particle suspensions appeared related to the amount of organic matter present. Particle-desorbed PCB's reduced growth rate, carbon fixation, and chlorophyll *a* content of natural phytoplankton assemblages incubated for several days *in situ* and were more toxic than an equivalent amount of PCB's added directly to water. When a single dose of PCB's (1 or 10 μg/liter) were added directly to water containing a natural phytoplankton community, a dose-dependent suppression of cell biomass, photosynthesis, and chlorophyll *a* concentration was observed. Biomass within larger size classes, predominantly diatoms, was greatly reduced. PCB uptake by two copepod species (*Acartia clausi* and *A. tonsa*) from water was asymptotic, reaching maximum concentrations in 30 hours. Mortality of *A. tonsa* in water containing 1 μg PCB's/liter was greater than controls and was independent of the presence of PCB-contaminated food. However, in water containing 5, 10, or 20 μg PCB's/liter, mortality was significantly greater in the presence of phytoplankton that had accumulated PCB's.

INTRODUCTION

Coastal waters may be regarded as particle suspensions, consisting of an aqueous medium of variable salinity containing inorganic and organic particulates. Inorganic particles vary in mineral composition, concentration, and size. The bulk of suspended inorganic particles in slow-moving waters consists of alumino-silicate clays and fine silts, which are small (1-20 μm) and possess large sorptive surface areas per unit weight. Suspended organic matter consists of detrital particles and planktonic organisms occurring in seasonally varying quantities with sizes usually larger than those of the suspended inorganic particles.

PCB's introduced into coastal waters are partitioned and may remain dissolved in the aqueous medium, sorb onto suspended inorganic particles, or be removed

[1] Present address: Environmental Science Division, Oak Ridge National Laboratory, Oak Ridge, Tennessee 37830.

[2] Present address: Oceanographic Sciences Division, Brookhaven National Laboratory, Upton, New York 11973.

Contribution No. 258 of the Marine Sciences Research Center.

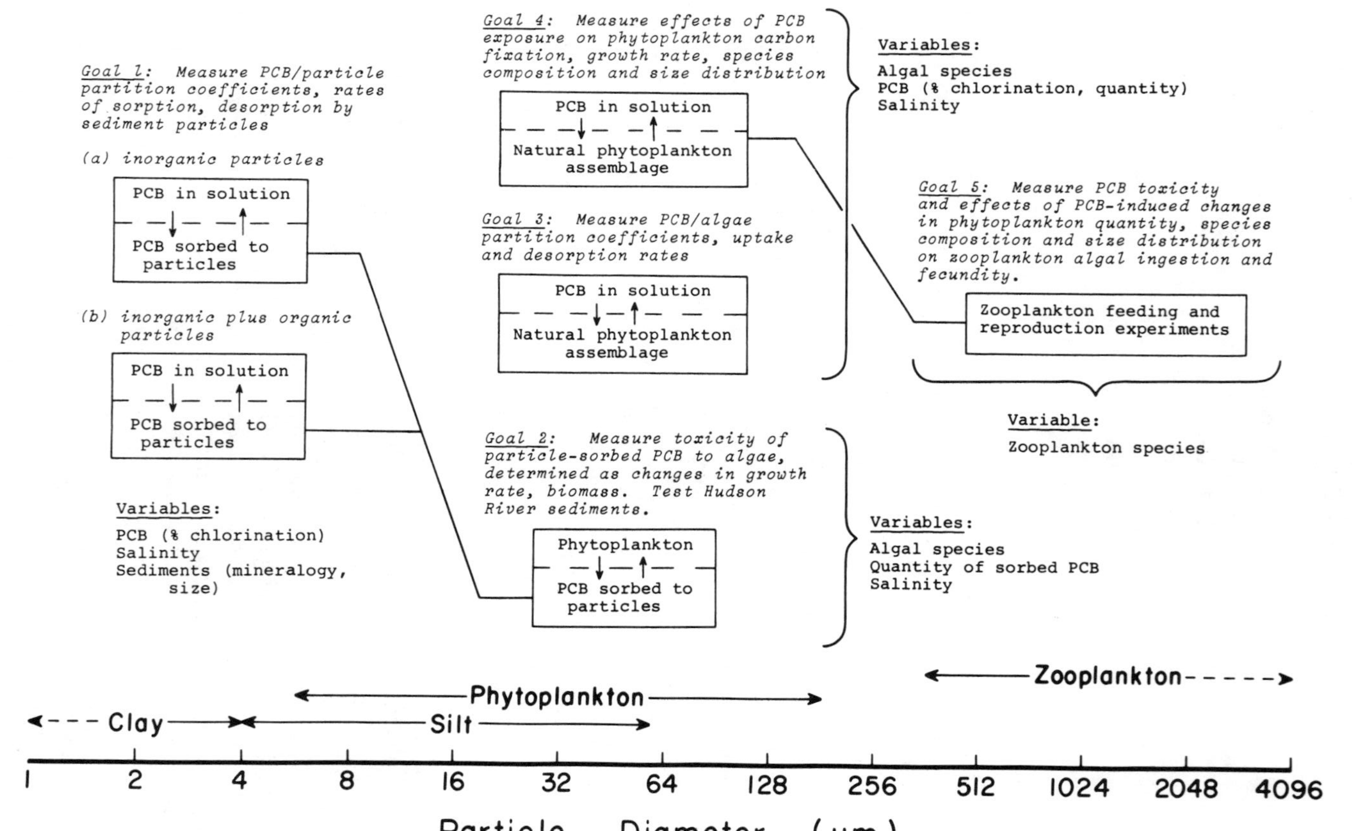

Figure 1. Research program designed to determine the partitioning and biological impacts of PCB's in coastal waters on simple food chains.

by detrital and living organic particulates (Figure 1). The particles may settle or remain in suspension. Most PCB's in water are associated with particulates.

Once associated with suspended particles, PCB's may remain there or, especially under conditions of reduced aqueous PCB concentration, may be released back into the aqueous medium and subjected to redistribution. Resuspension of fine sediments containing PCB's may provide a source of these chemicals to water and organisms that are relatively free of PCB's.

Biological impacts may include growth inhibition and changes in species composition in the phytoplankton community. Alterations in phytoplankton community composition may adversely affect herbivory, and PCB contamination of zooplankton from food and water may produce sublethal effects, such as reduced fecundity.

We report the results of several sets of experiments designed to measure the partitioning and biological impacts of PCB's in coastal waters, indicated as goals 1 through 5 in Figure 1. These data were obtained during the first of several years of experimentation required to accomplish the goals of our research programs.

MATERIALS AND METHODS

Phytoplankton

Collection and Culture

Natural assemblages of phytoplankton were collected from nearshore surface waters of central Long Island Sound. The water was passed through 132-μm nylon mesh to remove large zooplankton herbivores, and then divided into portions for appropriate treatment (PCB's dissolved in methanol or methanol alone) and incubation in dialysis bags floating in natural waters (Powers *et al.*, 1976; Powers *et al.*, 1977; O'Connors *et al.*, 1978).

Experiments requiring only two species of phytoplankton used laboratory cultures of *Skeletonema costatum* and *Thalassiosira pseudonana* grown in enriched, filtered Sound water.

Chlorophyll a Content

Aliquots of 1-10 ml each (depending upon cell density) were filtered through glass fiber filters. Each filter was macerated in acetone in a tissue grinder; the extract was filtered, and the chlorophyll *a* fluorescence of the filtrate was measured in a calibrated Turner Designs model 111 fluorometer (Strickland and Parsons, 1972).

Photosynthesis

Eighteen ml of water were dispensed into three screwcap tubes, two of them clear (light) and one black-taped (dark). Each tube contained 6 ml of sample and 0.1 ml (0.1 μCi) of ^{14}C-bicarbonate (New England Nuclear Corporation, Boston, Massachusetts). A blank consisted of 6 ml of cell-free, membrane-filtered (Millipore Corporation, Bedford, Massachusetts) Sound water with 0.1 ml of $H^{14}CO_3$. All tubes were incubated in natural light *in situ*. The contents were then filtered through 0.45 μm Millipore membranes that were placed in Aquasol 2 (New England Nuclear, Boston, Massachusetts) for determination of radioactivity by liquid scintillation counting (Mark II, Searle Analytic Inc., Des Plaines, Illinois).

Particle Concentration and Distribution

As previously described (Powers *et al.*, 1977), algae were identified by microscope, and particle suspensions were counted and sized using a Particle Data Electrozone Celloscope.

Zooplankton

Collection and Culture

Bottom-to-surface vertical tows were made in the same areas of Long Island Sound used for phytoplankton collections. Adult copepods were isolated and used within four hours of collection (*Temora* grazing experiments) or held overnight in filtered seawater to allow for clearance of the gut (*Acartia* PCB uptake studies).

Ingestion Rate

Groups of 75 *Temora longicornis* were placed in 500 ml each of Millipore-filtered Sound water containing natural phytoplankton assemblages or dilutions thereof. Algal assemblages dominated by species ranging in size from 5 μm to 40 μm were collected several times during spring and early summer (Table 1). Controls consisted of undiluted and diluted samples of the natural phytoplankton community with no copepods. Each beaker was emptied into a dialysis membrane bag and incubated in natural waters (Powers *et al.*, 1976) for 18-20 hours. A shield excluded nearly all sunlight.

Other experiments used bags in a plexiglas chamber (500-liter capacity) containing water from the original collection site. Agitated by streams of compressed air, the bags were incubated for 18-20 hours at 18°C in dim light.

The size distribution and concentration of suspended particles were measured (as under "Phytoplankton" above) in control cultures at the start of each experiment, and in control and grazed cultures at the conclusion of each experiment. Ingestion rates ($10^6 \mu m^3$ consumed/copepod/day) and mean food concentration ($10^6 \mu m^3$/ml) were computed using Frost's equations (1972).

PCB Uptake

Groups of 30 adult female *A. clausi* or *A. tonsa* were placed in 50-ml centrifuge tubes containing filtered seawater to which was added ^{14}C-PCB's in methanol (10 μg/liter) or pellets of Fithian illite or phytoplankton centrifuged from ^{14}C-PCB-treated particle suspensions following overnight exposure. Respective controls consisted of an equal volume of added methanol or an equivalent amount of uncontaminated illite or phytoplankton with methanol. Experiments under low light at 18°C were terminated after 48 hours. Analyses were made throughout all experiments to determine the distribution of

Table 1. Summary of phytoplankton and zooplankton data associated with 1977 and 1978 feeding experiments. I_{max}=maximum ingestion rate for the copepod, *Temora longicornis;* P_o = threshold food concentration. Columns are labeled with Roman numerals (I-IV) for easier reference in text.

Date of plankton collections and feeding experiments[1]	I Phytoplankton species present	II Diameter of particle volume mode (μm)	III I_{max} (10^6 μm^3/copepod day)	IV P_o (10^6 μm^3/ml)
20-24 June 1977 (a)	Numerous naked flagellates of the genus *Rhodomonas*; fewer cells and short chains of the diatoms, *Thalassionema nitzchoides* and *Skeletonema costatum*	4.8	20.6	3.5
12-13 July 1977 (b)	Numerous chains of *S. costatum*; fewer cells of the dinoflagellate, *Prorocentrum* sp., and small flagellates	11.5	36.2	1.1
6-7 July 1977 (c)	Cells and short chains of the diatom, *Rhizosolenia delicatula*; fewer numbers of small flagellates	14.1	43.8	2.2
8-9 June 1977 (d)	Cells and short chains of the diatom, *Leptocylindrus danicus*	22.8[2]	71.8	2.3
21-22 June 1978 (e)	Numerous chains of the diatom, *Cerataulina pelagica*; fewer cells and chains of *S. costatum* and *Asterionella japonica* (diatom)	37.9		

[1] Letter beneath collection dates indicates the corresponding particle volume size distribution in Figure 8.
[2] The average of 22.8 μm and 37.9 μm is 30.35 μm, which was used as the diameter of the particle volume mode for this combined data set.

^{14}C-PCB's in water, copepods, and particulates, and on centrifuge tube surfaces following methods described by Wyman and O'Connors (1980).

Mortality

Nonradioactive PCB's in methanol were injected into 200-ml aliquots of filtered seawater (some containing *S. costatum* and *T. pseudonana*) to yield initial PCB concentrations of 1, 5, 10, or 20 μg/liter; controls received an equivalent volume of methanol. Twenty *A. tonsa* were placed in each beaker, and the influence on survival of PCB's alone, or PCB's in the presence of contaminated food, was measured for 44 hours.

Fithian Illite (inorganic clay)

Source

Fithian illite clay was supplied by Dr. R. Dayal (Brookhaven National Laboratory), who obtained a 2-10 μm size fraction (4 μm mode) via gravity settling techniques from a bulk sample (Ward's Natural Sciences Establishment, Inc.). The mineral composition was confirmed by x-ray diffraction.

PCB Adsorption and Desorption

Centrifuge tubes containing an aqueous suspension of illite (450 mg/liter) were injected with 1 mg/liter of ^{14}C-PCB's. Large ^{14}C-PCB concentrations permitted measurement of the partitioning behavior of this chemical with relatively low replicate variability. Partition coefficients calculated from these data by the use of an adsorption isotherm equation (Steen *et al.*, 1978; Karickhoff *et al.*, 1979) were similar to coefficients calculated using data obtained from experiments in which the ^{14}C-PCB concentrations were lower. The proportion of ^{14}C-PCB's associated with particles, container surfaces, and water during the experiments was determined by liquid scintillation assay. Once PCB equilibrium was achieved between aqueous and solid phases, the ^{14}C-PCB-contaminated particles were centrifuged, the supernatants aspirated, and the pellets resuspended in uncontaminated filtered seawater. The redistribution of ^{14}C-PCB's desorbed from the particles was measured until a new equilibrium with the water particles and container surfaces was established.

The influence of organic matter, alone or in combination with illite particles, was examined. Simulating natural organic detritus, heat-killed diatoms (*T. pseudonana*) were used for 5% and 10% (dry wt) organic matter additions to illite suspensions.

To determine the effect of desorbed PCB's on natural phytoplankton, centrifuge tubes containing an aqueous suspension of illite (450 mg/liter) were injected with nonradioactive PCB's and placed overnight on a shaker table. Following centrifugation and aspiration of the supernatant, the PCB-loaded pellet was resuspended in filtered seawater and immediately added to 750 ml of a natural phytoplankton community. The samples were incubated in dialysis membrane bags in natural waters (Powers *et al.*, 1976). The effects on the variables described above (see "Phytoplankton") by PCB's desorbing from particles were recorded for three days.

PCB's

Source

Aroclor 1254 (Monsanto, St. Louis, Missouri), a mixture of tetra-, penta-, hexa-, and heptachlorobiphenyl isomers containing 54% chlorine by weight (Hutzinger *et al.*, 1974), was used in all PCB experiments. No attempt was made to distinguish the impacts and fates of individual isomers. ^{14}C-labeled Aroclor 1254 (31.3 μCi/m mol; lot 872-193) was obtained from New England Nuclear (Boston, Massachusetts). The ^{14}C label, locked within the biphenyl structure, was uniformly distributed amongst all the isomers (New England Nuclear, personal communication).

RESULTS

Particulate Adsorption and Desorption of PCB's and Their Impact on Natural Phytoplankton Communities

When added directly to water in closed glassware, 1 mg/liter of PCB's was partitioned among the water, suspended particles (450 mg/liter in all cases), and container surfaces. Equilibrium was reached within three hours (Figure 2), at which time more than half of the PCB's were associated with particulates, with 10% remaining in the water and 35% associated with container surfaces.

No statistically significant difference was detected in the amount of PCB's retained at equilibrium by inorganic particles (illite clay) alone or by inorganic particles plus 5% or 10% (dry wt) added organic matter. Neither was there a difference in the amount of PCB's remaining in the water. Larger quantities of PCB's were sorbed, however, by particles composed entirely of organic matter. Partition coefficients calculated from equilibrium concentrations of PCB's on the particles and in the water revealed a linear relationship with total organic content (Nau-Ritter, personal communication). The range in partition coefficients was 1.29×10^4 for inorganic clay to 4.93×10^4 for 100% organic detritus.

Resuspension of these PCB-contaminated particulates in PCB-free medium resulted in desorption of PCB's and their redistribution among particles, water, and container surfaces (Figure 3). After one and one-half hours no difference in the amount of PCB's retained by three particulate systems was observed. At three hours, equilibrium concentrations were achieved, and the organic content of the particles became significant in determining PCB adsorption. Particle assemblages with higher organic content showed greater PCB retention.

When PCB-contaminated particulates were added to natural phytoplankton assemblages contained in dialysis membrane bags and incubated in Flax Pond, desorbed PCB's reduced algal growth rates, carbon fixation, and chlorophyll *a* content. In communities exposed for three days to sediments bearing PCB

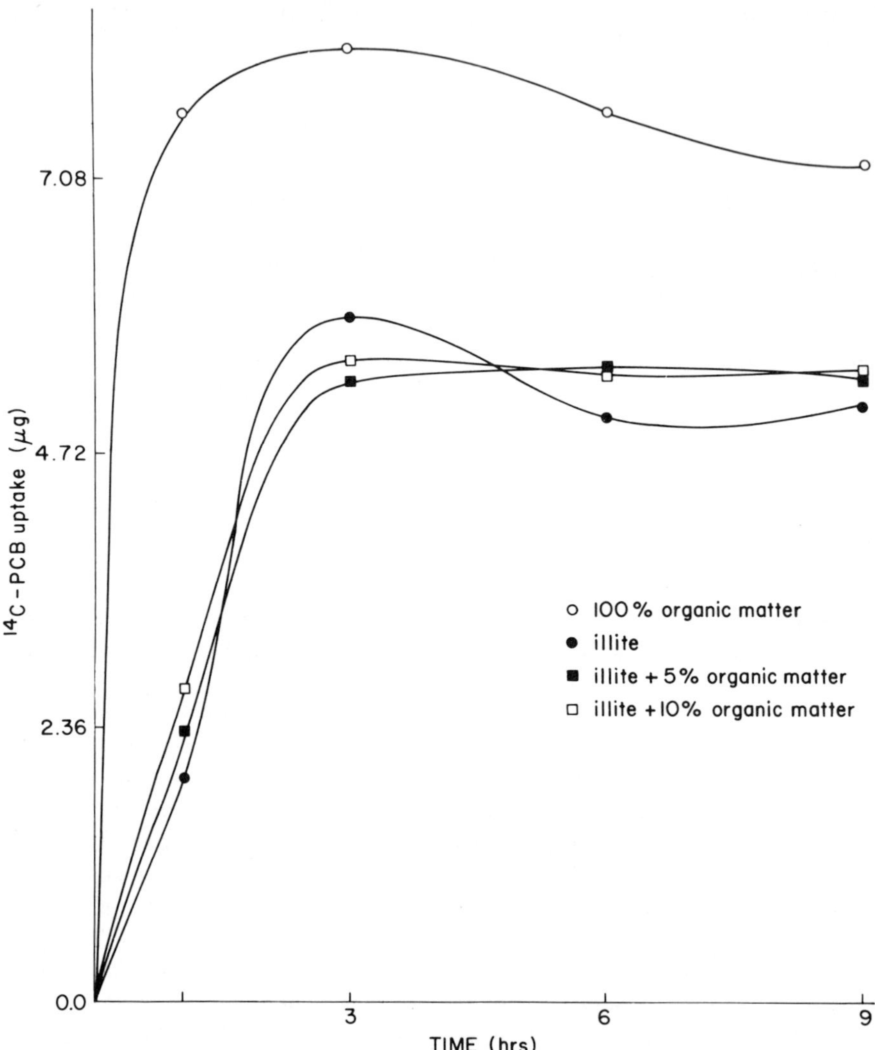

Figure 2. ^{14}C-PCB adsorption to inorganic illite, illite with added organic matter, and 100% organic matter. Particle concentration for all combinations of illite and organic matter was 450 mg/liter. Analysis of variance (Sokal and Rohlf, 1969) showed no significant difference between ^{14}C-PCB adsorption to illite, illite + 5% organic matter, and illite + 10% organic matter; the 100% organic matter adsorbed significantly more PCB's than the other particulates ($P < 0.001$).

concentrations (130 μg PCB sorbed to 90 mg illite/liter) in excess of those in the Hudson River, carbon-fixation was at least an order of magnitude lower than that observed in untreated communities (Figures 4, 5). This inhibition of photosynthesis was apparently an effect of PCB's rather than an effect of turbidity from suspended solids, since assemblages of phytoplankton in the presence of

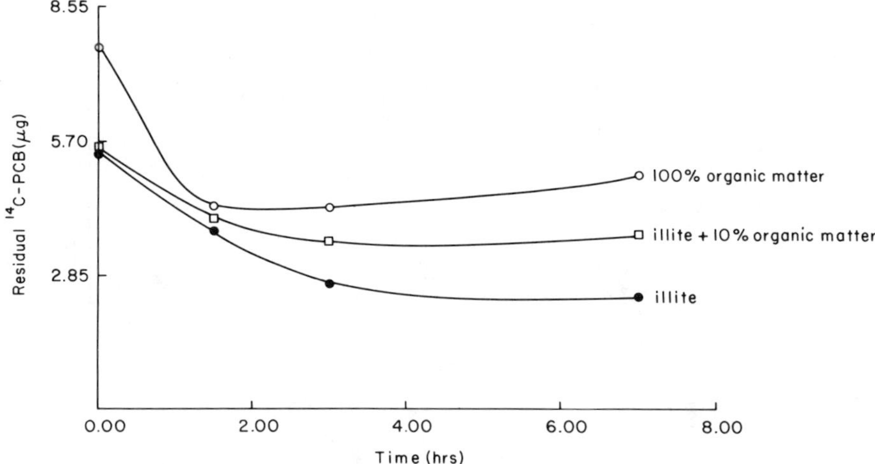

Figure 3. ^{14}C-PCB remaining on illite, illite + 10% organic matter, and 100% organic matter during time-course desorption into the water. A single classification analysis of variance (Sokal and Rohlf, 1969) showed no significant difference between ^{14}C-PCB contents of illite, illite + 10% organic matter, and 100% organic matter at 1.5 hours. A two-level nested analysis of variance (Sokal and Rohlf, 1969), however, showed that particles with organic matter had significantly ($P < 0.001$) higher ^{14}C-PCB quantities than inorganic particles after 1.5 hours.

PCB-free particles and those exposed to neither PCB's nor particles photosynthesized at the same rate. In Figure 4b, the increase in chlorophyll *a* concentration on day three in the presence of PCB's sorbed to illite was not accompanied by a corresponding increase in $^{14}C\text{-}HCO_3$ assimilation. Overcast weather conditions after day two may have reduced the photo-assimilation of carbon, but not the synthesis of chlorophyll *a* by the phytoplankton (Meeks, 1974).

Toxic effects on phytoplankton of PCB's desorbed from inorganic illite and from illite combined with 10% organic matter were not significantly different (Figure 5). The added 10% organic detritus simulated the organic component of natural sediments. This agrees with other reports in which sediment organic content played a minimal role in the sorption of PCB's to particles (Fulk *et al.*, 1975; Steen *et al.*, 1978).

Daily additions of particle and PCB concentrations comparable to those in the Hudson River estuary (10.0 μg PCB sorbed to 86 mg illite/liter) produced a more sustained reduction of photosynthesis and chlorophyll *a* than the same quantity of PCB's added directly to the water (Figure 6). After accounting for the inhibition of photosynthesis caused by turbidity from daily additions of sediments (Figure 6), single classification analyses of variance of each day's results revealed that photosynthesis in phytoplankton communities exposed to illite-sorbed PCB's was significantly ($P < 0.05$) lower by day two than in communities exposed to PCB's alone. This difference was not apparent on the first day, suggesting that the effects of daily PCB injections into the water

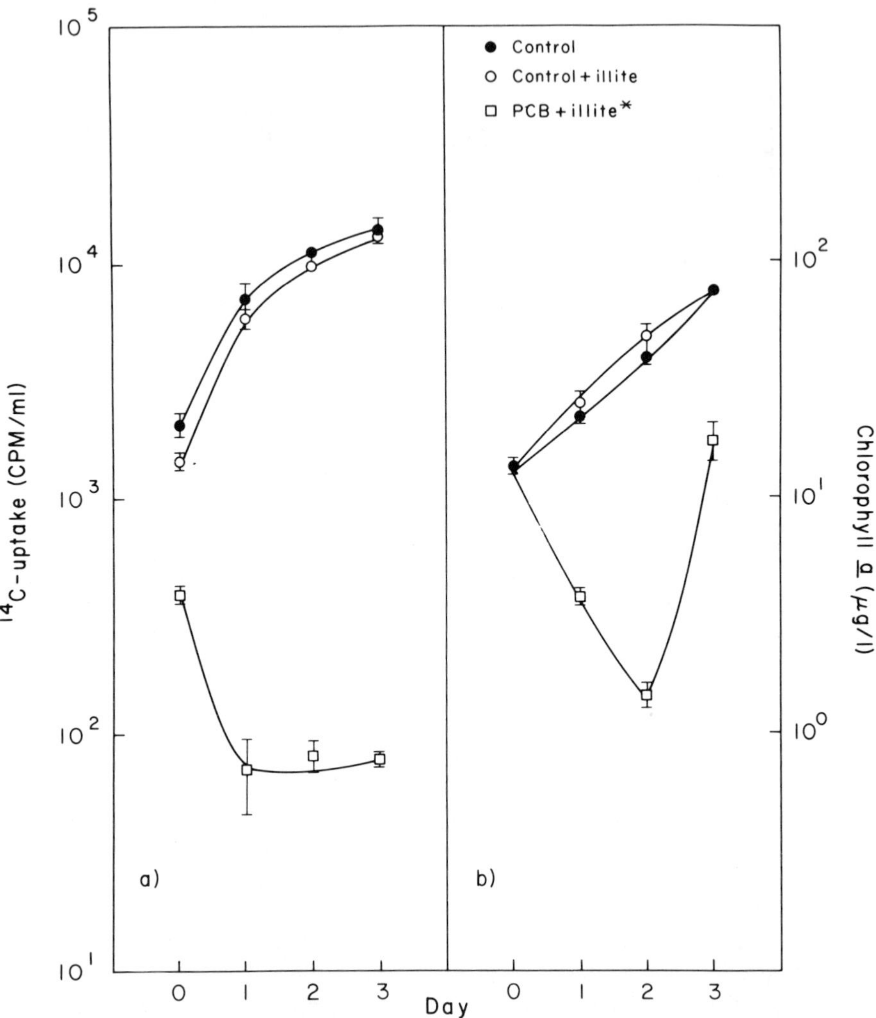

Figure 4. ^{14}C-fixation (a) and chlorophyll *a* content (b) of natural phytoplankton communities exposed to PCB's (130 μg/liter) associated with 90 mg of inorganic illite/liter. Experiment was begun 28 September 1978. Two-way analysis of variance (Sokal and Rohlf, 1969) showed that the time of exposure to PCB's (days), the concentration of PCB's, and the interaction term were all significant ($P < 0.01$), indicating that PCB sensitivity was highly dependent on time. No significant difference ($P > 0.05$) was found between control and control plus illite communities. An asterisk indicates that PCB's were added sorbed to particles. Bars represent standard errors of the means.

(*i.e.*, without particles) yielding PCB concentrations of 10 μg/liter may be comparable to those of PCB's desorbing from clay within the first 24 hours. As the experiment progressed, differences in ^{14}C-fixation and chlorophyll *a* content between PCB-exposed communities with and without clay particles were not

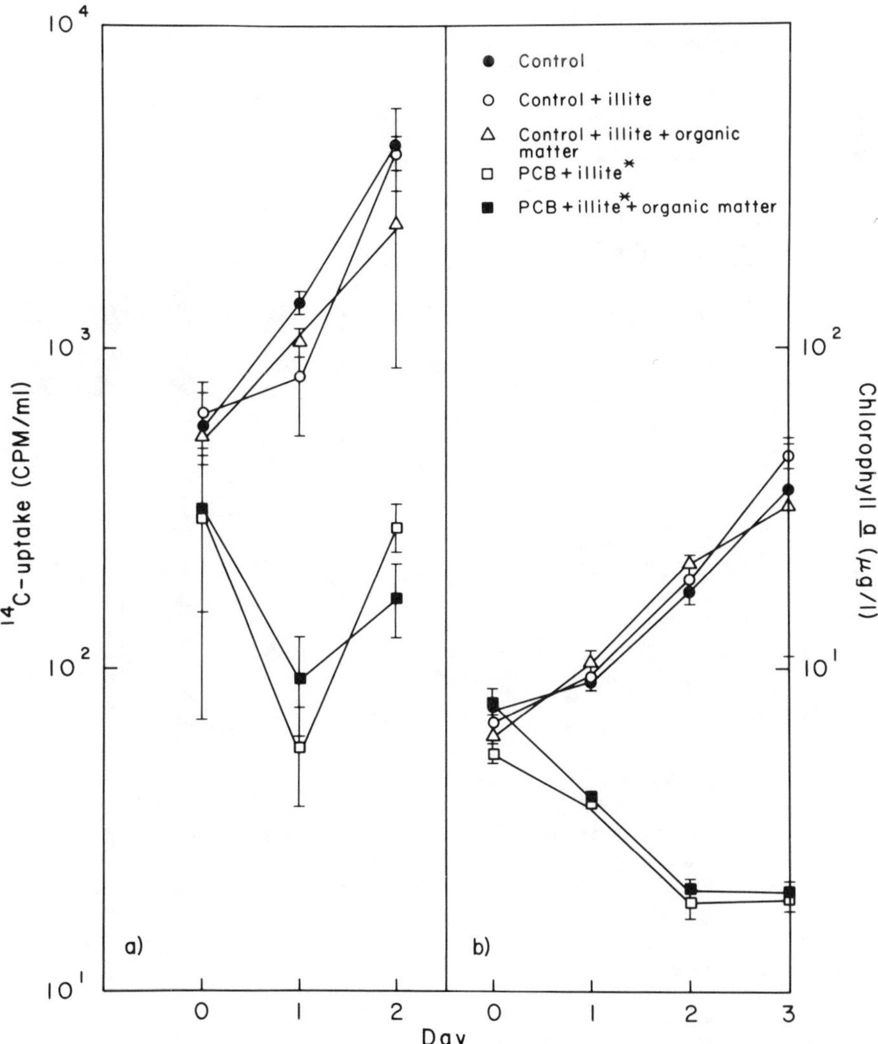

Figure 5. ^{14}C-fixation (a) and chlorophyll *a* content (b) of natural phytoplankton communities exposed to PCB's (130 μg/liter) associated with 90 mg/liter of inorganic illite and illite with 10% organic matter. Experiment was begun 1 November 1978. Two-way analyses of variance (Sokal and Rohlf, 1969) revealed no significant difference ($P > 0.05$) between the three control communities or between assemblages with PCB's plus illite and PCB's plus illite plus organic matter. An asterisk indicates that PCB's were added sorbed to particles. Bars represent standard errors of the means.

significant on day one ($P < 0.075$), but became significant on days two ($P < 0.05$) and three ($P < 0.001$). Presumably, greater inhibition in the presence of particle-associated PCB's was the result of the availability to the cells of higher concentrations of PCB's continuously desorbing from particles.

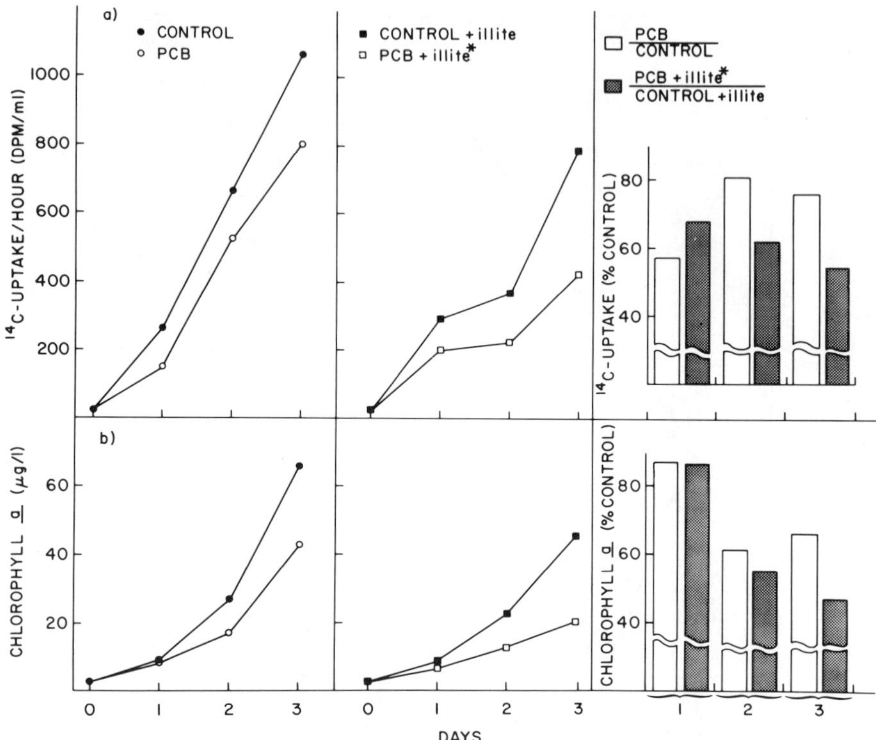

Figure 6. ^{14}C-fixation (a) and chlorophyll *a* content (b) of natural phytoplankton communities exposed to PCB's (10 μg/liter) added daily to the water or adsorbed to 86 mg illite/liter. Experiment was begun 17 July 1979. Since the introduction of illite each day increased the turbidity and reduced light available for photosynthesis, the effects from PCB's alone are expressed as percent of controls in the bar diagrams. Single classification analyses of variance on each day's results showed significant differences ($P < 0.05$) after day one between PCB added directly to the water and PCB sorbed onto illite. An asterisk indicates that PCB's were added sorbed to particles.

Effects of PCB's on Natural Phytoplankton Species and Size Composition

Natural springtime communities of estuarine phytoplankton were exposed to 1-10 μg PCB/liter under natural conditions within dialysis bags suspended in a tidally flushed arm of Long Island Sound (O'Connors *et al.,* 1978). Phytoplankton biomass and size were reduced by the PCB's (Figure 7). Suppression of algal biomass was dose-dependent; natural phytoplankton assemblages exposed to a single dose of PCB's at 10 μg/liter exhibited lower particle concentrations than controls and those receiving 1 μg/liter.

All size classes of algae were not equally affected. Treatment of the algal community with PCB's at 1 μg/liter suppressed the abundance of particles larger than 9 μm equivalent spherical diameter (ESD) for three days, but the

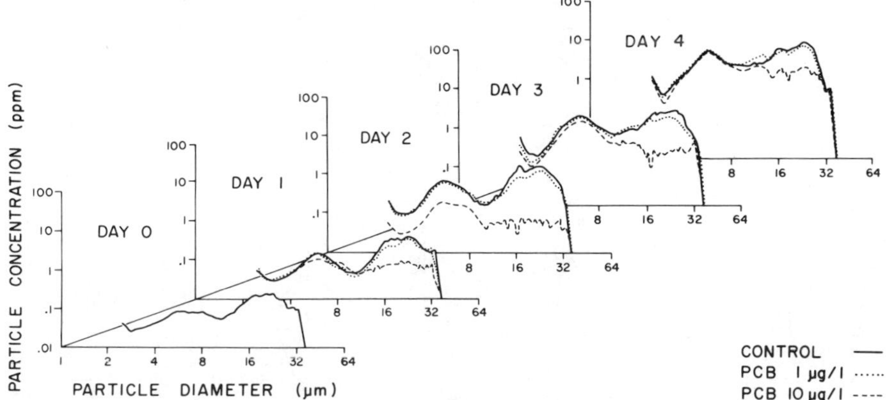

Figure 7. Particle concentration size distribution for natural phytoplankton community exposed to 1 and 10 μg/liter PCB's, 8-12 June 1976 (from O'Connors *et al.*, 1978).

abundance of smaller particles was unaffected (Figure 7). Treatment with 10 μg/liter suppressed the numbers of particles larger than 9 μm ESD by about an order of magnitude for 3 to 4 days, while concentrations of smaller particles recovered within three days.

Microscopic examination revealed that among particles larger than 9 μm ESD, the communities were dominated by the diatom, *Rhizosolenia fragilissima.* Other chain-forming diatoms present were *Skeletonema costatum, Thalassiosira decipiens, Thalassionema nitzchioides,* and *Leptocylindrus* sp. Particles smaller than 9 μm ESD consisted primarily of small phytoflagellates, with lesser numbers of pennate diatoms and a centric diatom resembling *T. pseudonana.* Suppression of particles larger than 9 μm ESD suggests that large diatoms are more sensitive to PCB's than smaller algae. PCB-induced suppression of diatom abundance in natural phytoplankton communities was observed by Biggs *et al.* (in press) and in the PCB sorption/desorption experiments reported in the previous section.

Phytoplankton Size and Zooplankton Herbivory

O'Connors *et al.* (1980) investigated the effects of reduced phytoplankton size on zooplankton herbivory in coastal waters. Naturally occurring ranges of cell sizes were found to be similar to those induced by exposure to PCB's, as reported above. The copepod, *Temora longicornis,* a ubiquitous estuarine herbivore, was fed natural phytoplankton assemblages collected on several occasions from the relatively PCB-free waters of central Long Island Sound. Collections varied in relative concentrations of small flagellates, larger dinoflagellates, and chain-forming diatoms in a manner similar to that observed for the PCB-induced changes (Table 1; Figure 8).

Generally, copepod feeding rates increased asymptotically with increasing

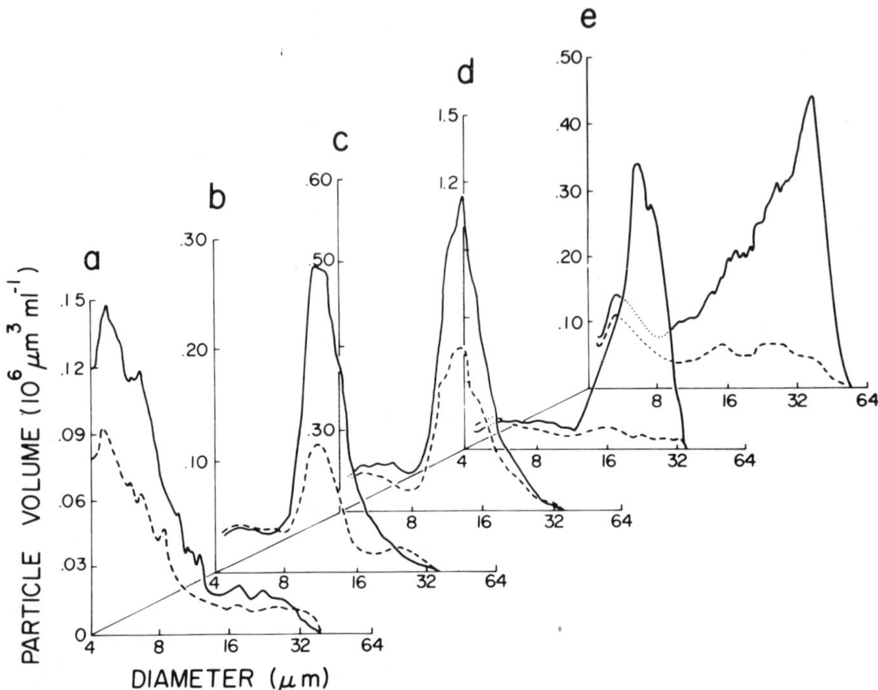

Figure 8. Particle volume size distribution for representative ungrazed (solid line) and grazed (dashed line) particle suspensions at the conclusion of grazing experiments run on 20-24 June 1977 (a), 12-13 July 1977 (b), 6-7 July 1977 (c), 8-9 June 1977 (d), and 21-22 June 1978 (e). The distributions are arranged in order of increasing frequency of larger particles. See Table 1, column (I), for species composition of phytoplankton.

phytoplankton concentration, reaching a maximum rate at some intermediate algal density (Figure 9a). When *Temora* were fed the several phytoplankton assemblages (each dominated by different-sized algae), the maximum ingestion rate increased linearly with particle diameter by a factor of 3.5. The modal diameter of the food particles in each of the assemblages increased from 5 µm (mostly microflagellates) to 30 µm (mostly large diatoms) (Figure 9b). Maximum copepod particle volume ingestion rates therefore were favored by large food particle size and reduced by small food size.

Short-Term PCB Uptake and Mortality

PCB uptake from contaminated water, sediments, and phytoplankton by *Acartia,* a common late spring and summer estuarine copepod, was measured, as was toxicity of PCB's to these animals.

Time-dependent uptake by *Acartia* from PCB-contaminated water and suspended inorganic sediments followed an asymptotic function, reaching equilibrium after approximately 30 hours exposure (Figures 10, 11a). *A. tonsa,*

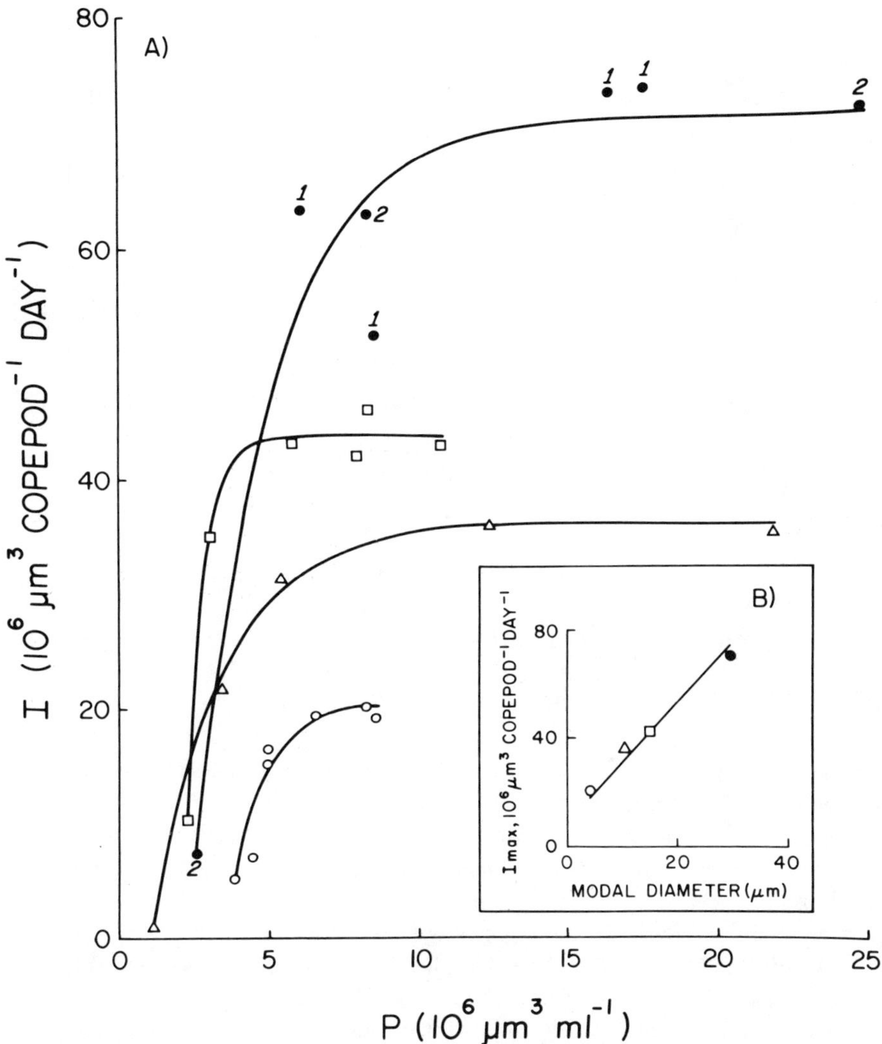

Figure 9. (a) *Temora longicornis* ingestion rates (I) as a function of food concentration (P) for experiments run on 20-24 June 1977 (○), 12-13 July 1977 (△), 6-7 July 1977 (□), 8-9 June 1977 (●[1]), and 21-22 June 1978 (●[2]). (b) I_{max} (from Table 1, column III) estimated from curves in Figure 9(a) as function of particle volume modal diameters (Table 1, column II). Symbols indicate dates as in Figure 9(a).

exposed to 10 µg PCB/liter of water, contained 28.82 x 10^{-4} µg PCB/copepod, while *A. clausi* contained 13.53 x 10^{-4} µg PCB/copepod at equilibrium (Figure 10). These concentrations correspond to weight-specific concentrations of 248 µg PCB/mg and 223 µg PCB/mg, respectively, given measured dry weights for *A. tonsa* of 11.6 µg and for *A. clausi* of 5.8 µg. *A. tonsa* achieved an equilibrium PCB concentration of 2.5 x 10^{-4} µg PCB/copepod (22 ng/mg dry wt) following

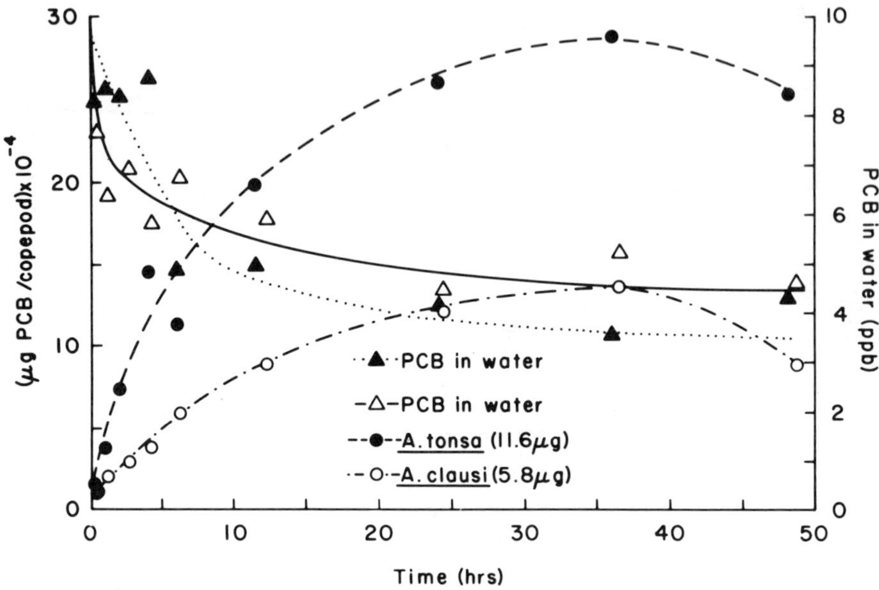

Figure 10. ^{14}C-PCB concentration in water (μg/liter or ppb) and copepods (*Acartia tonsa* and *A. clausi*) for 48-hour experiments examining uptake from PCB-contaminated water having an initial concentration of 10 μg/liter. Closed symbols are for experiments with *A. tonsa*. Copepod dry weights for *A. tonsa* and *A. clausi* are indicated in parentheses.

exposure to an illite clay suspension (100 mg/liter), upon which 20 ng PCB/mg dry wt initially had sorbed (Figures 11a, b).

Uptake by *A. tonsa,* fed a mixture of equal cell volumes of *Skeletonema costatum* and *Thalassiosira pseudonana* containing an initial PCB concentration of 80 ng/mg wet wt, was not asymptotic, despite water PCB concentrations similar to those measured in the previous experiments (Figures 12a,b, 13). After increasing rapidly for the first five hours when food was plentiful, PCB concentrations in the animals reached 7.02 x 10^{-4} μg PCB/copepod (61 ng/mg dry wt), then declined gradually following exhaustion of the contaminated food.

When *A. tonsa* was exposed to a PCB concentration of 1 μg/liter in water, significant mortality with time occurred over the 30-hour period, whereas no significant mortality was observed among control copepods (Figures 14a, b). Within the 1 μg/liter treatment (Figure 14b), no statistically significant difference in mortality between copepods exposed to contaminated water and copepods exposed to contaminated water plus phytoplankton (mixture of equal quantities of *S. costatum* and *T. pseudonana*) could be detected. However, a significant difference in mortality was observed between control copepods with phytoplankton and copepods exposed to the 1 μg/liter treatment with phytoplankton ($P < 0.001$). No such difference was seen between control and treatment beakers without phytoplankton ($P > 0.05$). Assuming 20% of the PCB's added was associated with the phytoplankton (Biggs *et al.,* in press), the

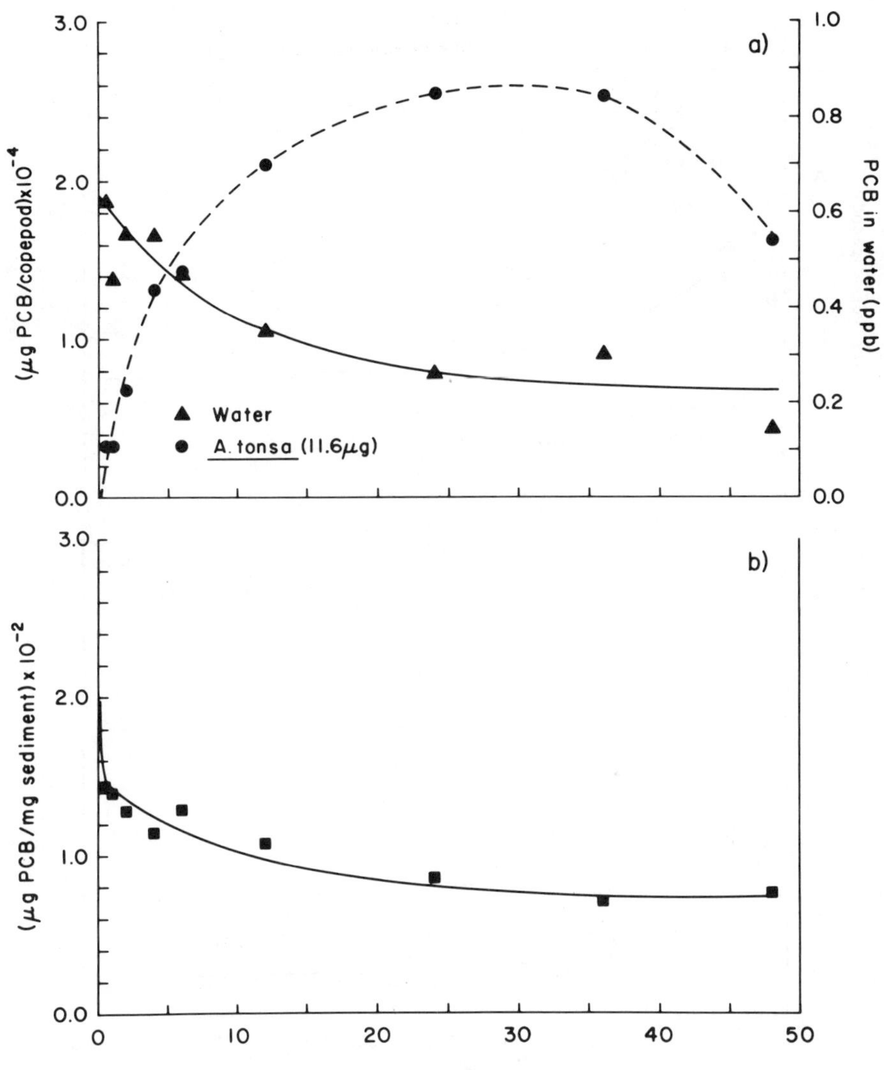

Figure 11. (a) ^{14}C-PCB concentration in water (μg/liter or ppb) and in copepods *(Acartia tonsa)* for a 48-hour experiment examining PCB uptake from contaminated Fithian illite. *A. tonsa* dry weight is indicated in parentheses. (b) ^{14}C-PCB concentration on Fithian illite during the experiment shown in Figure 11(a).

initial weight-specific PCB concentration on the phytoplankton was computed to be 14 ng/mg wet wt.

When *A. tonsa* was exposed to 5 μg/liter PCB, mortality among copepods with food was significantly greater than among those without food after 44 hours (Figure 14c). In this instance, initial phytoplankton PCB concentrations were calculated to be 70 ng/mg wet wt.

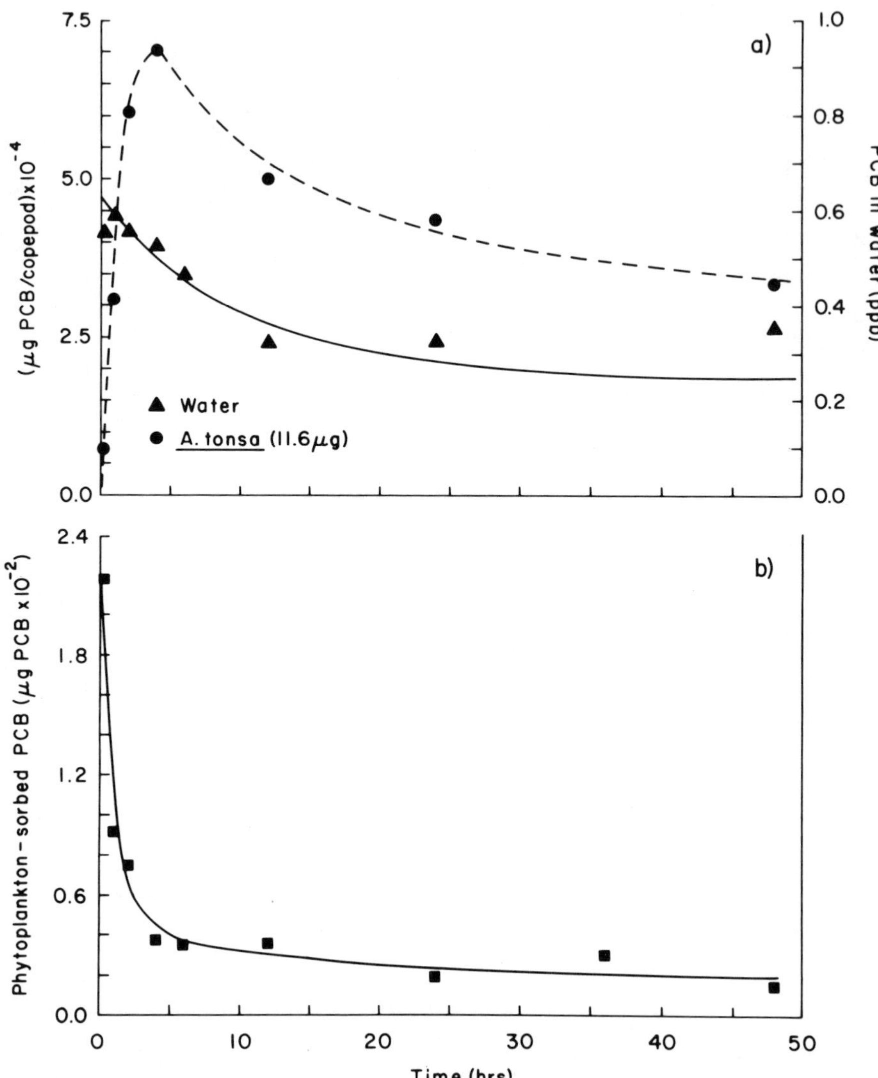

Figure 12. (a) ^{14}C-PCB concentration in water (μg/liter or ppb) and copepods (*Acartia tonsa*) for a 48-hour experiment examining PCB uptake from contaminated phytoplankton. *A. tonsa* dry weight is indicated in parentheses. (b) Phytoplankton-sorbed ^{14}C-PCB during the experiment shown in Figure 12(a). The phytoplankton were nearly all consumed after the first five hours.

At the 10 μg/liter PCB concentration, the presence of contaminated phytoplankton enhanced copepod mortality from the onset of the exposure, but after 44 hours all copepods had expired (Figure 14d). PCB concentrations on the phytoplankton were estimated to be 140 ng/mg wet wt initially.

At PCB concentrations of 20 μg/liter, all copepods died after 30 hours

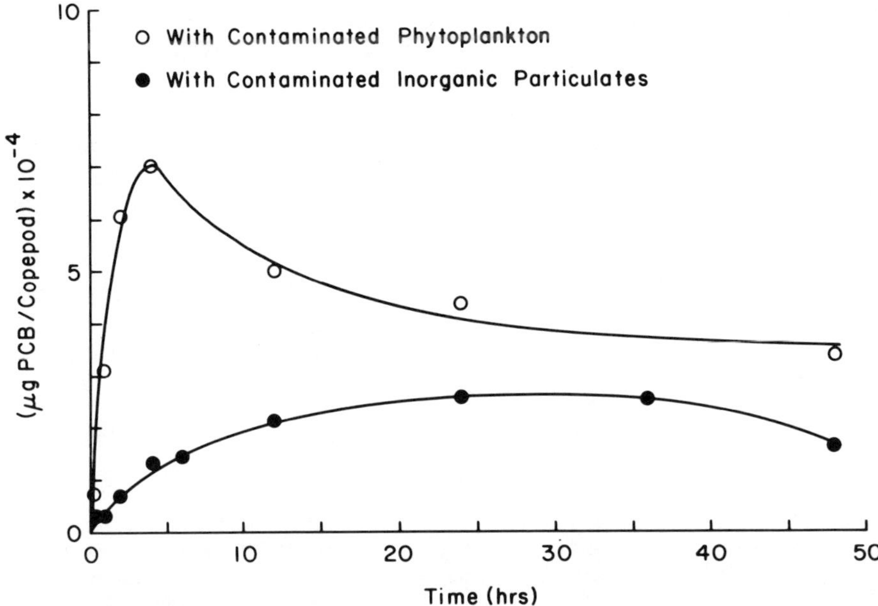

Figure 13. ^{14}C-PCB concentration in copepods (*Acartia tonsa*) with contaminated phytoplankton present and with contaminated illite present (with similar concentrations of PCB's in the water).

exposure (Figure 14e). PCB's in the water appeared to be great enough to cause significant mortality, regardless of whether contaminated phytoplankton was present. No significant difference could be detected in copepod mortality for fed and unfed animals; all copepods died after 30 hours exposure. Phytoplankton PCB concentrations were computed to be 280 ng/mg wet wt initially.

DISCUSSION

Heat-killed *Thalassiosira pseudonana* cultures were used as organic matter in the PCB-sorption experiments (Figure 2). Microscopic examination of the killed cultures showed that the 5-μm diameter cells had lysed, producing numerous particles less than 5 μm in size. Electronic size analysis revealed that the diameter of the illite clay was about 4 μm, larger than the size of the cell fragments. Equal weights of cell fragments and illite were used; therefore, the smaller, more numerous cell fragments possessed a greater surface area, which may account for the greater PCB sorptive capacity of the cell fragments alone.

While partitioning of PCB's among glass walls, water, and contaminated particles achieved equilibrium after three hours in a closed container, partitioning among contaminated particles, phytoplankton, and water in the dialysis membrane bags may not have achieved equilibrium during the four-day incubation. Biggs *et al.* (in press) found that contaminated natural particles

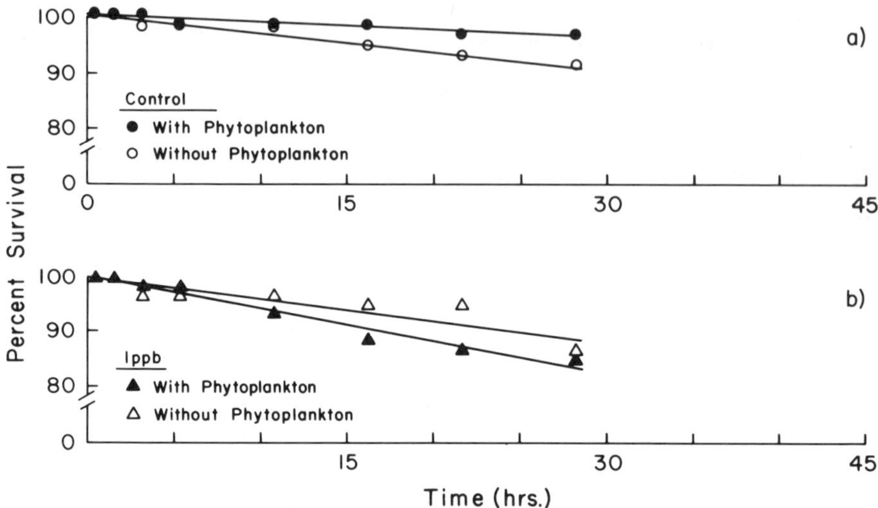

Figure 14. Above. (a) Percentage of survival among control copepods (*Acartia tonsa*) with and without phytoplankton. A two-way ANOVA (Sokal and Rohlf, 1969) showed no significant mortality with time or with the presence or absence of phytoplankton food ($P > 0.05$). (b) Percentage of survival among copepods (*A. tonsa*) exposed to 1 μg/liter (ppb) PCB's initially in the water with and without phytoplankton. A two-way ANOVA (Sokal and Rohlf, 1969) revealed no significant difference in mortality between copepods with or without food ($P > 0.05$); however, significant mortality occurred with time ($P < 0.001$). **Opposite.** (c) Percentage of survival among copepods (*A. tonsa*) exposed to 5 μg/liter (ppb) PCB's initially in the water with and without phytoplankton. A two-way ANOVA (Sokal and Rohlf, 1969) showed significant mortality with time and with presence or absence of food ($P < 0.001$). In addition, the interaction between time and presence or absence of food was significant ($P < 0.05$). (d) Percentage of survival among copepods (*A. tonsa*) exposed to 10 μg/liter (ppb) PCB's initially in the water with and without phytoplankton. A two-way ANOVA (Sokal and Rohlf, 1969) showed significant mortality with time, and showed that the difference in mortality between fed and unfed copepods was significant ($P < 0.001$). (e) Percentage of survival among copepods (*A. tonsa*) exposed to 20 μg/liter (ppb) PCB's initially in the water with and without phytoplankton. A two-way ANOVA (Sokal and Rohlf, 1969) showed significant mortality with time only ($P < 0.001$); no significant difference in mortality occurred between fed and unfed copepods ($P > 0.05$).

continued to desorb PCB's over a five-day period, losing about 80% of their sorbed PCB's while in a dialysis membrane bag incubated in natural waters (Figure 15).

PCB's, continuously desorbing from inorganic and organic particles, may cause more severe and prolonged toxicity to phytoplankton than PCB's dissolved in the water. Fine-grained clays carrying large quantities of PCB's are easily resuspended from the bottom sediments; hence, the residence time of PCB's in the water column is lengthened. With PCB concentrations as low as 10 μg/liter desorbing from sediments, significant changes in the production and biomass of natural phytoplankton communities may develop.

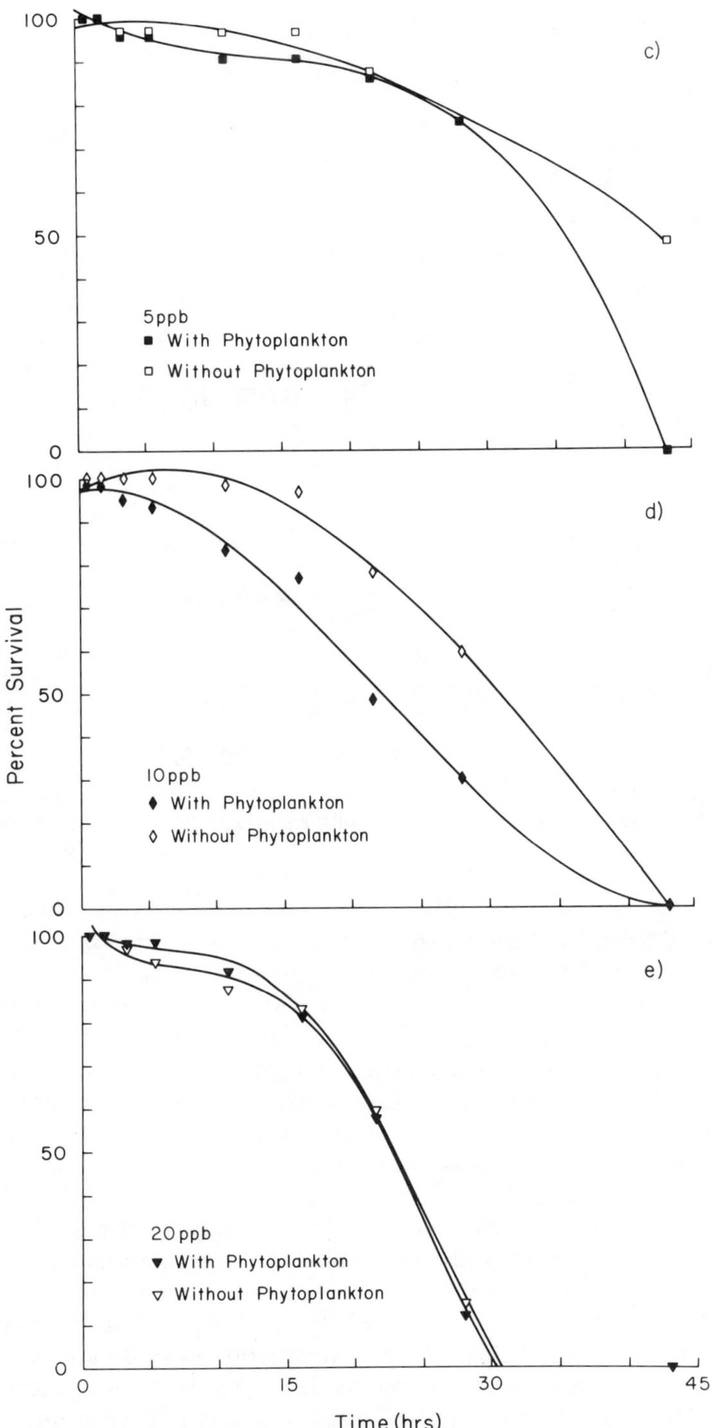
100
50
0
c)
5ppb
With Phytoplankton
Without Phytoplankton
100
50
0
d)
10ppb
With Phytoplankton
Without Phytoplankton
Percent Survival
100
50
0
e)
20ppb
With Phytoplankton
Without Phytoplankton
0
15
30
45
Time (hrs)

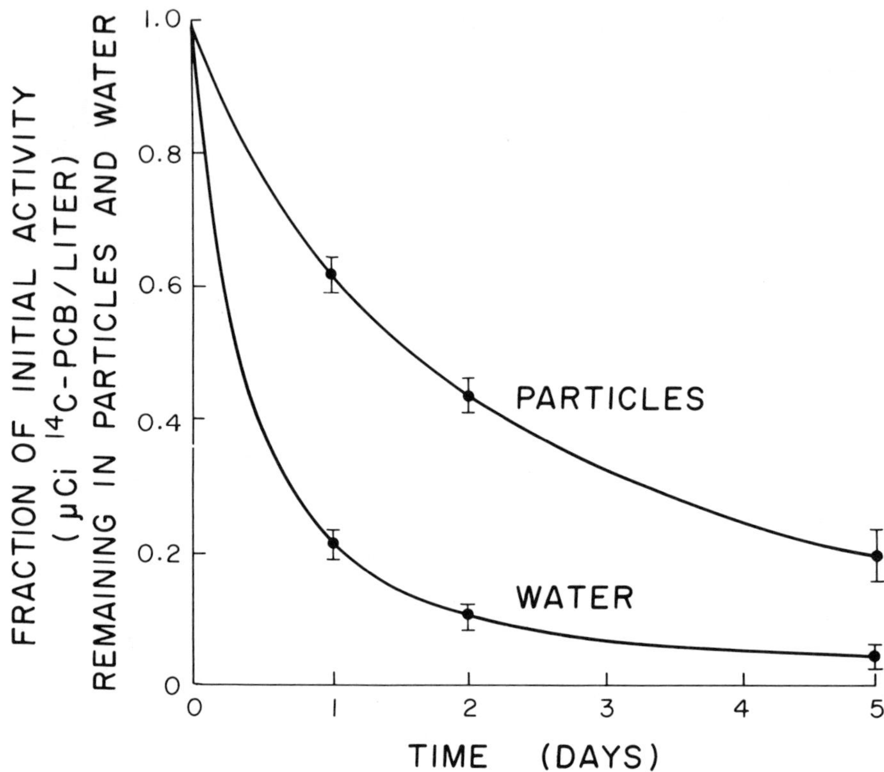

Figure 15. Loss of ^{14}C-PCB from dialysis cultures of estuarine phytoplankton incubated in Flax Pond. Bars represent standard errors of the means. (After Biggs *et al.*, 1980).

Our results showed that PCB's desorbed from inorganic illite clay and from illite with 10% organic matter (Figure 3) inhibited ^{14}C-fixation and chlorophyll *a* content in natural phytoplankton communities incubated in dialysis bags *in situ* (Figures 4, 5). Harding and Philips (1978) also observed a reduction in chlorophyll *a* content in axenic cultures of *Ditylum brightwelli* and *Lauderia borealis* containing PCB-contaminated microparticulates. They reported no significant difference, however, in ^{14}C-assimilation between control cultures and cultures with particles. Our results, by contrast, showed at least an order of magnitude difference on each day after day 0.

The concentration of PCB's associated with the single addition of 90 mg of clay particles to the phytoplankton community (Figures 4, 5) was 1.3×10^5 μg/g of sediment. Although atypical of most contaminated sediments, this level of PCB's approximates the reported concentration of 1.8×10^4 μg/g on Hudson River scum (Tofflemire *et al.*, 1979). Consistent with amounts detected in upper Hudson River sediments, the concentration of PCB's associated with the repeated daily additions of clay particles (Figure 6) was 100 μg/g of sediment. Nadeau and Davis (1976) listed sediment PCB concentrations of

6,700 μg/g at Ford Edward, New York, and Hetling *et al.* (1978) reported sediment values above 1,000 μg/g, with the majority of samples containing between 10 μg/g and 100 μg/g. Fine-textured sediments contained more PCB's than coarse fractions.

Exposure of natural phytoplankton communities to PCB's suppressed larger size classes of phytoplankton (O'Connors *et al.*, 1978; Biggs *et al.*, in press), which are food for herbivorous zooplankton in coastal waters. Results presented in Table 1, and Figures 7, 8, and 9 indicate that PCB-induced size reductions in natural phytoplankton could reduce markedly the volume of phytoplankton consumed by filter-feeding copepods.

At least two factors make an evaluation of the impact on copepod nutrition of changes in phytoplankton size in natural waters difficult. Small-sized phytoplankton in culture possess larger quantities of organic carbon per cell volume than large-sized species (Strathmann, 1967; Paffenhoeffer and Knowles, 1978). If the same is true of phytoplankton from natural waters, it is possible that *T. longicornis,* fed the four experimental food size series (Table 1), consumed nearly equal amounts of organic carbon, despite obvious differences in size-dependent volume ingestion rates, though laboratory experiments do not necessarily support this concept. While adults of two *Temora* species fed phytoplankton mixtures containing equal cell volumes of diatoms ranging in size from large to small consumed greater cell volumes of large diatoms, carbon equivalents of ingested cell volumes were not always equal to or lower than those observed when small diatoms were consumed (Table 5 in Paffenhoeffer and Knowles, 1978). Unfortunately, no method exists for measuring carbon per cell volume for individual phytoplankton species in a natural multispecies assemblage.

The second undetermined factor is the cost to copepods, in terms of energy expenditure, of filtering and ingesting small vs. large particles. Our results show, nevertheless, that reduction in phytoplankton size and biomass can decrease the volume of food consumed by *T. longicornis* in natural waters. Further, the size-dependent maximum volume ingestion rates suggest that pollutant-induced size reductions in the phytoplankton community (Greve and Parsons, 1977; O'Connors *et al.*, 1978) could result in a 50% reduction in the volume of phytoplankton eaten by these copepods.

Previous investigations concerning chlorinated hydrocarbons in the diet of aquatic organisms support the concept of equilibrium partitioning of these residues between seawater and lipid reserves within the biota. Field studies suggest a linear relation between PCB concentration in the lipid fraction of coastal marine zooplankton and that of the seawater in which they reside (Clayton *et al.*, 1977; Pavlou and Dexter, 1979). Further, increased chlorinated hydrocarbon levels have been observed in upper trophic levels, even though transfer of the residues through the food chain was not possible (Hamelink *et al.*, 1971). Indeed, the highest levels of PCB's in the biota often occur at the trophic level composed primarily of mixed plankton, thereby repudiating the concept of food chain biomagnification (Ware and Addison, 1973; Harvey *et al.*, 1974).

Exchange across body surfaces (*e.g.*, gills, integuments) and circulatory transport to internal lipid pools may be an important mechanism explaining these results. All of the aforementioned investigations, however, discount the importance of contaminated food as a mechanism for PCB accumulation and fail to discuss the implications of dietary inputs of PCB's on short-term uptake. Our experiments, while also lending credence to the concept of equilibrium partitioning, emphasize the importance of contaminated food in the diet of small estuarine copepods.

When adult *Acartia* are exposed to PCB-contaminated water or contaminated inorganic material, PCB uptake finds an equilibrium level and does not continue to increase with time. These uptake curves reflect a simultaneous accumulation and depuration of PCB residues by the copepods. Rarely, however, are aquatic organisms found in environments where food is not present. Figure 13 demonstrates the importance of this statement. The accumulation of PCB's from contaminated food is much greater than that from water alone, until the food supply is exhausted; then the levels in the copepods again approach equilibrium. Starved *Acartia* fill their guts within one hour following introduction of food (M. Dagg, personal communication). Therefore, only part of the PCB accumulation can be attributed to PCB-contaminated food in the gut. The mechanism of PCB accumulation in this instance must combine exchange across the body integument with assimilation of PCB residues from the contaminated food.

Several implications may be drawn from the increase in PCB accumulation due to ingestion of contaminated food. First, the presence of contaminated food allows equilibrium levels to be reached at a faster rate than would otherwise occur from contaminated water alone. Thus, small estuarine copepods may respond rapidly (in a few hours) to transient inputs of PCB residues.

Secondly, ingestion of PCB-contaminated food allows PCB residues to be packaged into fecal pellets, thereby providing a mechanism for transport of PCB's out of the water column and into the sediments. Such a mechanism has been used to explain the penetration of PCB residues into the deep ocean (Elder and Fowler, 1977).

Third, the presence of PCB-contaminated food within the gut of the copepods increases the PCB concentration available to predators of the next trophic level, such as many valuable fish species. While this may not be important with regard to food chain biomagnification, it may increase toxicity and reduce fecundity among upper trophic level organisms.

Finally, the increase in PCB accumulation because of ingestion of contaminated phytoplankton is sufficient to increase mortality and may reduce copepod fecundity. Though Morgan (1971) concluded that *Daphnia* fed PCB (Aroclor 1242)-contaminated algae survived better than those in PCB-contaminated water without algae, our experiments indicate a statistically significant increase in mortality among copepods exposed to contaminated food, compared to those without PCB's in their diet. This may be true at PCB concentrations observed in the water and phytoplankton at selected sites within the Hudson River estuary. Considering recent research regarding the transfer of PCB residues from microparticulates to marine phytoplankton (Harding and Phillips, 1978), the importance of PCB-contaminated phytoplankton in the mechanism of short-term uptake by small estuarine zooplankton cannot be overemphasized.

CONCLUSIONS

Particle-sorbed PCB's introduced into clean waters as resuspended PCB-contaminated sediments (*e.g.*, dredge spoils) will rapidly desorb into water and redistribute to uncontaminated particles, including plankton. Our results indicate significant amounts of PCB's will be redistributed within a period of a few hours.

PCB's sorbed by phytoplankton will inhibit growth and, within a period required by the algae to divide several times, may reduce the size composition of the community by altering species abundance. Such changes constitute significant alterations in food size for important zooplankton herbivores and may produce undesirable changes in the flow of biomass in coastal water food webs.

Zooplankton sorb PCB's from contaminated sediments via water and food. PCB accumulation and mortality in these animals are enhanced by the consumption of contaminated phytoplankton.

ACKNOWLEDGEMENTS

This research was supported by the New York State Department of Environmental Conservation, the New York Sea Grant Institute, the National Marine Fisheries Service, and the NOAA/MESA New York Bight Project.

REFERENCES

Biggs, D.C., C.D. Powers, R.G. Rowland, H.B. O'Connors, Jr., and C.F. Wurster. 1980. Uptake of polychlorinated biphenyls by natural phytoplankton assemblages: field and laboratory determination of ^{14}C-PCB particle-water index of sorption. Environ. Pollut. 22: 101-110.

Clayton, J.R., S.P. Pavlou, and N.F. Brietner. 1977. Polychlorinated biphenyls in coastal marine zooplankton: bioaccumulation by equilibrium partitioning. Environ. Sci. Technol. 11: 676-682.

Elder, D.L. and S.W. Fowler. 1977. Polychlorinated biphenyls: penetration into the deep ocean by zooplankton fecal pellet transport. Science 197: 459-461.

Frost, B.W. 1972. Effects of size and concentration of food particles on the feeding behavior of the marine copepod *Calanus finmarchicus*. Limnol. Oceanogr. 17: 805-815.

Fulk, R., D. Gruber, and R. Wullschleger. 1975. Laboratory studies of the release of pesticide and PCB materials to the water column during dredging and disposal operations. Contract Rep. D-75-6, U.S. Army Engineer Waterways Experiment Station, CE, Vicksburg, MS. 98 pp.

Greve, W. and T.R. Parsons. 1977. Photosynthesis and fish production: hypothetical effects of climatic change and pollution. Helgol. Wiss. Meeresunters. 30: 662-672.

Hamelink, J.L., R.C. Waybrant, and R.C. Ball. 1971. A proposal: exchange equilibria control the degree chlorinated hydrocarbons are biologically magnified in lentic environments. Trans. Am. Fish. Soc. 100: 207-214.

Harding, L.W., Jr. and J.H. Philips. 1978. Polychlorinated biphenyls: transfer from microparticulates to marine phytoplankton and effects on photosynthesis. Science 202: 1189-1192.

Harvey, G.R., H.P. Miklas, V.T. Bowen, and W.G. Steinhauer. 1974. Observations on the distribution of chlorinated hydrocarbons in Atlantic Ocean organisms. J. Mar. Res. 32: 103-118.

Hetling, L., E. Horn, and J. Tofflemire. 1978. Summary of Hudson River PCB Study Results. New York State Department of Environmental Conservation Tech. Pap. No. 51. 88 pp.

Hutzinger, O., S. Safe, and N. Zipko. 1978. The Chemistry of PCB's. CRC/Press, Cleveland, OH. 269 pp.

Karickoff, S., D. Brown, and T. Scott. 1979. Sorption of hydrophobic pollutants on natural sediments. Water Res. 13: 241-248.

Meeks, J. 1974. Chlorophylls. *In:* Algal Physiology and Biochemistry, Vol. 10, W. Stewart (ed.), University of California Press, Berkeley and Los Angeles. pp. 161-175.

Morgan, J.R. 1971. Effects of Aroclor 1242 (a polychlorinated biphenyl) and DDT on cultures of an alga, protozoan, daphnid, ostracod, and guppy. M.S. Thesis, University of Washington. 83 pp.

Nadeau, R.J. and R.A. Davis. 1976. Polychlorinated biphenyls in the Hudson River (Hudson Falls–Fort Edward, New York State). Bull. Environ. Contam. Toxicol. 16: 436-444.

O'Connors, H.B., Jr., D.C. Biggs, and D.V. Ninivaggi. 1980. Particle size dependent maximum grazing for *Temora longicornis* fed natural particle assemblages. Mar. Biol. 56: 65-70.

O'Connors, H.B., Jr., C.F. Wurster, C.D. Powers, D.C. Biggs, and R.G. Rowland. 1978. Polychlorinated biphenyls may alter marine trophic pathways by reducing phytoplankton size and production. Science 201: 737-739.

Paffenhoeffer, G.A. and S.C. Knowles. 1978. Feeding of marine planktonic copepods on mixed phytoplankton. Mar. Biol. 48: 143-152.

Pavlou, S.P. and R.N. Dexter. 1979. Distribution of polychlorinated biphenyls (PCB) in estuarine ecosystems. Testing the concept of equilibrium partitioning in the marine environment. Environ. Sci. Technol. 13: 65-70.

Powers, C.D., R.G. Rowland, H.B. O'Connors, Jr., and C.F. Wurster. 1977. Response to polychlorinated biphenyls of marine phytoplankton isolates cultured under natural conditions. Appl. Environ. Microbiol. 34: 760-764.

Powers, C.D., R.G. Rowland, and C.F. Wurster. 1976. Dialysis membrane chambers as a device for evaluating impacts of pollutants on plankton under natural conditions. Water Res. 10: 991-994.

Sokal, R.R. and F.J. Rohlf. 1969. Biometry. W.H. Freeman and Company, San Francisco. 776 pp.

Steen, W.C., D.F. Paris, and G.L. Baughman. 1978. Partitioning of selected polychlorinated biphenyls to natural sediments. Water Res. 12: 655-657.

Strathmann, R.R. 1967. Estimating the organic carbon content of phytoplankton from cell volume or plasma volume. Limnol. Oceanogr. 12: 411-418.

Strickland, J.D.H. and T.R. Parsons. 1972. A Practical Handbook of Seawater Analysis (2nd Edition). Fisheries Research Board of Canada, Ottawa. 310 pp.

Tofflemire, T.J., L.J. Hetling, and S.O. Quinn. 1979. PCB in the upper Hudson River. Sediment distribution, water interactions and dredging. New York State Department of Environmental Conservation Tech. Pap. No. 55. 68 pp.

Ware, D.M. and R.F. Addison. 1973. PCB residues in plankton from the Gulf of St. Lawrence. Nature 246: 519-521.

Wyman, K.D. and H.B. O'Connors, Jr. 1980. Implications of short term PCB uptake by small estuarine copepods (genus *Acartia*) from PCB contaminated water, inorganic sediments, and phytoplankton. Estuarine Coastal Mar. Sci. 11: 121-131.

FACTORS AFFECTING SHORT-TERM PCB AND DDT ACCUMULATION BY ZOOPLANKTON AND FISH FROM THE HUDSON ESTUARY

L.S. Peters[1]
J.M. O'Connor

New York University Medical Center
Institute of Environmental Medicine
A.J. Lanza Laboratories
Tuxedo, New York 10987

Abstract. Amphipods (*Gammarus daiberi*), mysid shrimp (*Neomysis americana*), and striped bass (*Morone saxatilis*) were exposed to ^{14}C-labeled PCB's and DDT in food and in water to determine contaminant accumulations under different conditions. These organisms accumulated both contaminants rapidly from the water. Levels in organisms of ~3-5 μg/g were observed during the first 48 hours of exposure in labeled water. PCB uptake from food and in the presence of suspended particulates was lower than that determined for direct water uptake. In general, bioconcentration factors for PCB's approximated 1×10^4 for food uptake, and ranged between 2×10^4 and 4×10^4 for water uptake. Organohalogen levels attained in test organisms during dietary experiments depended upon the amount of food present.

The clearance of DDT by *G. daiberi* was more rapid than the clearance of PCB's, although the rates of uptake for the two compounds were similar during 48-hour exposures. The rapidity and substantial degree of organohalogen accumulation by zooplankton suggest that zooplankton compete with fishes for direct accumulation of these compounds from water. However, the means by which foraging organisms serve to alter the transfer of organic contaminants is complex.

Bioaccumulation of chlorinated organics by estuarine organisms is dependent upon a variety of factors, including adsorption to particulates in the water column, amount of food and particulates present, and route of entry into the organism (*i.e.*, via direct water uptake or with food). Each of these factors is discussed with regard to contaminant levels common in the Hudson estuary and the coastal marine environment of the New York metropolitan region.

INTRODUCTION

Measurable levels of organic contaminants occur in the water, sediments, and biota of the New York Bight and adjacent waters (Kneip and Hernandez, 1969; Searl *et al.*, 1977; Tanacredi, 1977; Hetling *et al.*, 1978; O'Brien and Gere Engineers, Inc., 1979; Stainken and Rollwagen, 1979; O'Connor *et al.*, this volume). Three specific classes of organic compounds, the polychlorinated biphenyls (PCB's), chlorinated pesticides, and polynuclear aromatic hydrocarbons (PAH's), threaten the ecological integrity of the Bight ecosystem and/or human health and, therefore, should receive high priority for further study (O'Connor and Stanford, 1979).

Most studies of PCB's, pesticides, and PAH's in aquatic environments have established distributions in sediments, edible fishes, and shellfish, or their toxicity to organisms. Concentrations of all three classes that cause mortality of aquatic biota have been documented (Burdick *et al.*, 1964; Butler, 1969; Duke

[1] Present address: New York University, Center for Science and Technology Policy, 50 West 4th Street, New York, New York 10013.

et al., 1970; Kuhnhold, 1971; Nimmo *et al.*, 1971; Wells, 1972; Renzoni, 1973; Nimmo and Bahner, 1974; Dow, 1975). Survey and monitoring data have established that PCB residues in edible fishes and shellfish in estuarine and coastal marine areas are of particular concern (Foehrenbach *et al.*, 1971; Harvey *et al.*, 1974; Nadeau and Davis, 1976; Elder and Villeneuve, 1977; Hetling *et al.*, 1978; Wilson and Forester, 1978). Few studies, however, have measured these compounds in planktonic biota and assessed the role of zooplankton in transporting these contaminants in estuarine and coastal marine systems (Risebrough, 1971; Chiou *et al.*, 1977; O'Connors *et al.*, 1978; this volume). Harvey and Steinhauer (1975) concluded that the dynamics of PCB cycling in surface waters are dependent upon zooplankton activity. The relation between organohalogens and zooplankton has been discussed by O'Connors *et al.* (1978; this volume), but additional data are needed to develop models for assessing the hazards of organic contaminants in the aquatic environment.

We present here the results of experiments with zooplankton (*Neomysis americana*: Crustacea, Mysidaceae and *Gammarus daiberi*: Crustacea, Gammaridae) and fish (*Morone saxatilis*) that determine short-term rates of PCB and DDT uptake from water and food. Short-term, static bioaccumulation tests were designed to simulate the pulsed, non-repetitive exposure to organohalogens that occurs in aquatic environments when contaminants are discharged or resuspended by human activities, by periodic natural phenomena such as tidal scour, or by intermittent freshets, spills, and accidents.

Our studies emphasized PCB's and DDT because of the presence of relatively large quantities of PCB's and DDT in the Hudson-Bight system (Kneip and Hernandez, 1969; West *et al.*, 1976; Horn *et al.*, 1979; O'Connor *et al.*, this volume), and the defined need for studies on these halogenated organics (O'Connor and Stanford, 1979). Radiolabeled PCB and DDT compounds were employed to provide a rapid, reliable means for determining rates of organohalogen uptake in aquatic organisms (O'Connors *et al.*, 1978, this volume; Califano *et al.*, 1980).

MATERIALS AND METHODS

Test Organisms

Gammarus daiberi and *Neomysis americana* were collected from the Hudson River estuary and maintained in the laboratory in 38-liter aquaria with mixed algal cultures (*Chroococcus* sp., *Melosira* sp., and various phytoflagellates). *G. daiberi* adapted well to laboratory conditions and began reproducing. *N. americana* did not reproduce but remained active and fed for periods up to eight weeks.

Young striped bass (*Morone saxatilis*) were obtained from the Consolidated Edison hatchery facility at Buchanan, New York, or were seined from the Hudson River. Hatchery fish were ~17 mm, and wild fish were ~80 mm when used. All fish were held in 250-liter Min-o-Cool tanks and were fed brine shrimp (hatchery-reared fish) or chicken liver (wild fish).

Radiotracer Methodology

^{14}C-aroclor 1254 (specific activity, 0.096 mCi/mg) was purchased from New England Nuclear Corporation and ^{14}C-DDT (specific activity, 0.087 mCi/mg) from Amersham-Searle. Labeled compounds were added to water using acetone as a carrier (0.01% v/v). Although acetone has been identified as having a mild narcotic effect on *G. tigrinus,* its use as a carrier in preliminary experiments showed no effect on organohalogen uptake in test organisms (Cantilli, 1977). Detection of radiotracers was by liquid scintillation counting. For water samples, 1-5 ml of sample were placed in each counting vial; 15 ml scintillation fluor (4 g/liter POP, 0.3 g/liter POPOP dissolved in 2:1 toluene: triton-x-100) were added, and the sample was counted. ^{14}C levels in solids (*e.g.,* particles, organisms, tissues) were determined after drying and weighing the samples. All organohalogen concentrations in solids, except in one case (see Table 6), are reported in terms of dry weight. Samples were pulverized, wetted, treated with tissue solubilizer (Protosol), and counted after adding 15 ml scintillation fluor. A Nuclear Chicago Mark I LSC (Model 6860), calibrated with an Amersham primary standard and a set of quenched ^{14}C-standards, was used for counting. Sample counts first were converted for background and then for quench using the channel ratios method (Wang and Willis, 1965). Organohalogen concentrations were calculated from the specific activity of the labeled compound.

Uptake from Water

G. daiberi, N. americana, and *M. saxatilis* were used in water uptake studies with a single exposure format and two types of water: Hudson River water that had been passed through a 0.45-μm filter (FHRW), and Hudson River water that was unfiltered (HRW). Organisms were exposed to a nominal concentration of 1 ng/ml (0.5 ng/ml for 80-mm bass) ^{14}C-PCB or ^{14}C-DDT for 48 hours in glass vessels (200-ml capacity for amphipods and mysids; 950-ml capacity for 17-mm bass; 38-liter aquaria for 80-mm bass). The test compound was mixed thoroughly with the water before organisms were introduced. Each experiment was run in duplicate. Ten amphipods, mysids, or 17-mm bass were placed in each experimental vessel. Organisms from each vessel were divided between two counting vials. Six 80-mm bass were used per replicate; each large fish was counted individually.

Groups of organisms were removed from the test tanks at ten predetermined intervals (from 0.25 hours to 48 hours) and counted to determine ^{14}C-DDT or ^{14}C-PCB uptake. Whole body counts were made of *G. daiberi, N. americana,* and 17-mm *M. saxatilis.* Larger striped bass (*i.e.,* 80 mm) were dissected, and ^{14}C-PCB levels in epaxial muscle, gut, liver, and gills were determined.

Modified Uptake Studies

Modified uptake studies with *G. daiberi* and *M. saxatilis* examined the effects of specific environmental factors on ^{14}C-PCB uptake. Gammarids were exposed to ^{14}C-PCB in water in the presence of two food organisms: the rooted green plant, *Myriophyllum spicatum,* and the alga, *Chroococcus* sp. (Ginn, 1977; Poje, 1977). When *Myriophyllum* was used, *G. daiberi* were placed in test tanks containing a 2.5-cm length of the plant and water containing a calculated concentration of 1 ng/ml ^{14}C-PCB.

Gammarids were exposed to *Chroococcus* sp. under two conditions: (1) in a growing culture of *Chroococcus* to which 1 ng/ml ^{14}C-PCB had been added 24 hours earlier, and (2) in test tanks containing HRW and *Chroococcus* that had been exposed for one day to 1 ng/ml ^{14}C-PCB and washed free of radioactive water by centrifugation and rinsing with deionized water (3x). The *Chroococcus* in both of these experiments were from the same culture. *Chroococcus* concentrations in each test vessel were ~3 x 10^5 cells/ml. Gammarids were removed for counting at six predetermined intervals from 0.25 hours to 48 hours.

Each 17-mm striped bass was fed five ^{14}C-PCB labeled mysids (*N. americana*) that had been exposed to 1.0 ng/ml ^{14}C-PCB for 48 hours and washed in FHRW. Groups of bass were removed at predetermined intervals and counted to determine ^{14}C-PCB uptake from dietary intake.

Larger striped bass (80 mm) were fed five ^{14}C-PCB-labeled *G. daiberi* to determine the distribution of PCB's in the tissues of striped bass. Gammarids were labeled by 24-hour exposure to ^{14}C-PCB in FHRW, before being washed and fed to the striped bass. Groups of fish were removed at 24 hours and 48 hours after feeding and dissected. The gills, liver, entire gut, a subsample of muscle, and the remaining carcass were counted to determine PCB concentrations.

Striped bass (17 mm) were exposed to ^{14}C-PCB in water in a tank containing 2.5 cm of fine gravel or mud. The fish had free contact with the sediment surface. Exposures were in aquaria (100 liter) containing 80 liters of water and sediments collected by grab sampler from a region of the Hudson estuary not heavily contaminated by PCB's (<6 μg/g PCB). The mud was collected off Jones Point, New York and the gravel at Verplanck, New York. Fish were added to the tanks 24 hours after 1 ng/ml ^{14}C-PCB-labeled water had been mixed with the sediment.

Release Studies

Elimination of ^{14}C-PCB or ^{14}C-DDT by *G. daiberi* was determined. Gammarids were exposed, as described above, for 24 hours in HRW or FHRW. After 24 hours, one group of organisms was harvested to determine uptake. The remaining organisms were washed in FHRW to remove nonassimilated and nonabsorbed radiolabeled compounds and were placed in fresh, unlabeled FHRW containing *Myriophyllum* and *Chroococcus* as food. These gammarids were harvested at predetermined intervals, and ^{14}C activity was determined.

Retention of PCB's by 80-mm striped bass was studied after feeding them labeled *N. americana.* Whole-body PCB concentrations were determined 24 hours after feeding.

Each release study experiment was run in duplicate, as described above for studies testing bioaccumulation of organohalogens from water. Throughout all of the experiments described in this paper, water quality characteristics were: temperature, 22.0° ± 1°C; salinity, 3 ± 1 ‰; dissolved oxygen, ~8.0 mg/liter; and pH, 7.3.

RESULTS

Uptake from Water

Amphipods, mysids, and striped bass removed ^{14}C-PCB rapidly from water and showed significant body burdens beginning as early as 0.25 hours after exposure. Uptake by *Gammarus daiberi* exposed to ^{14}C-PCB in FHRW was examined in two separate experiments. After 48 hours, ^{14}C-PCB levels were similar, but values taken between 0.25 hours and 30 hours diverged (Figure 1a). The PCB concentration in gammarids exposed to ^{14}C-PCB in HRW for 48 hours was 5.5 ng/mg (Figure 1b), 36-45% more than that accumulated by *G. daiberi* exposed to FHRW. DDT uptake by *G. daiberi* in FHRW was similar in rate and final concentration to ^{14}C-PCB uptake (Table 1).

Bioaccumulation factors for gammarid uptake of PCB's from FHRW and HRW (^{14}C-PCB in organisms divided by ^{14}C-PCB measured in water) ranged from 1.7×10^4 to 3.8×10^4 (Table 2). For DDT, the bioaccumulation factor was 1.6×10^4.

Uptake of ^{14}C-PCB by *Neomysis americana* (Figure 1) was similar to that of *G. daiberi.* In FHRW, the mysid had 48-hour concentrations of 4.9 ng/mg. In HRW, the final concentration was 4.0 ng/mg. The shapes of the uptake curves were essentially identical. The bioaccumulation factor (48 hours) for *N. americana* uptake from FHRW was 3.5×10^4; the bioaccumulation factor from HRW was 1.7×10^4 (Table 2).

Morone saxatilis (17 mm) exposed to ^{14}C-PCB in HRW and FHRW had similar whole body concentrations after 24 hours and 48 hours exposure (Table 3). Final concentrations were 5.9 μg/g in fish exposed to PCB in HRW, and 4.96 μg/g in fish exposed to PCB in FHRW. Bioaccumulation factors were not calculated. Short-term accumulation of PCB and DDT from water by amphipods, mysids, and striped bass in all tested cases can be described by a power function with $r^2 > 0.92$.

Modified Uptake Studies

When unlabeled *Myriophyllum* was present, the rate of uptake of PCB's by *G. daiberi* was initially similar to the rate of uptake from FHRW or HRW. However, uptake after 12 hours was much lower than when no *Myriophyllum* was present. The final PCB concentration in *G. daiberi* was 2.4 ng/mg, and the final bioaccumulation factor was $\sim 1.0 \times 10^4$ (Tables 2, 4).

When ^{14}C-PCB labeled *Chroococcus* was the only source of PCB's, 48-hour uptake in *G. daiberi* was 0.17 ng/mg. When PCB's were available in the water and in algae, uptake after 48 hours reached 0.95 ng/mg (Table 4).

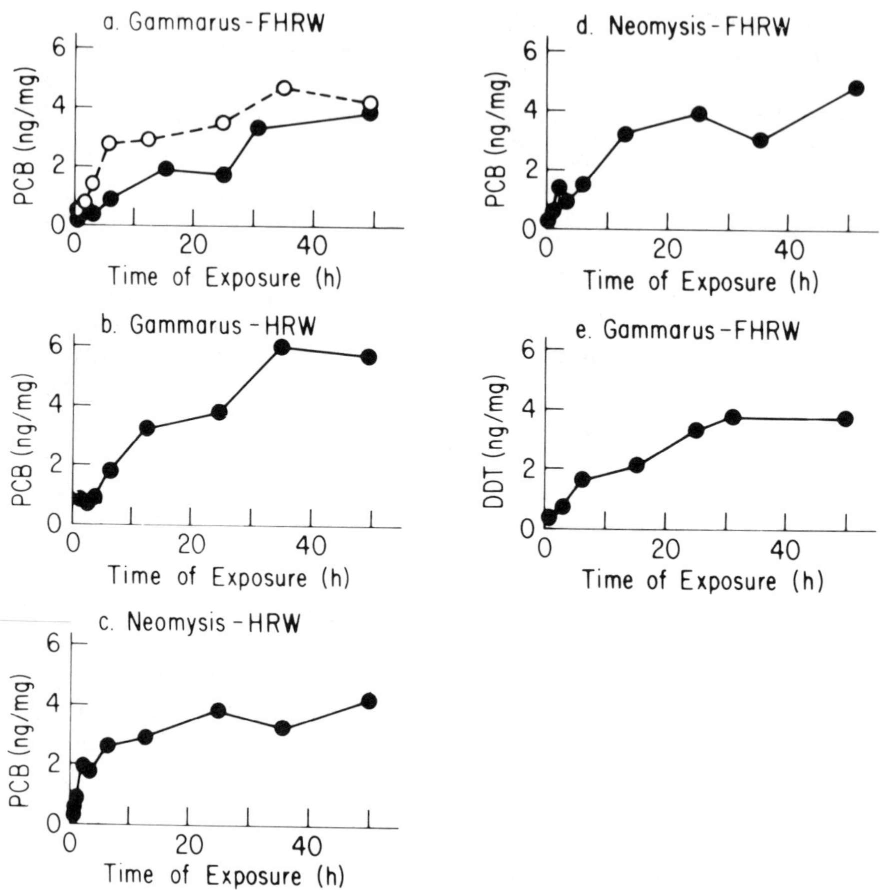

Figure 1. Uptake of PCB's and DDT for *Gammarus daiberi* and *Neomysis americana* exposed for 48 hours to contaminant concentrations of 1 ng/ml. (a) *G. daiberi* exposed to PCB's in filtered Hudson River water (FHRW); (b) *G. daiberi* exposed to PCB's in Hudson River water (HRW); (c) *N. americana* exposed to PCB's in HRW; (d) *N. americana* exposed to PCB's in FHRW; and (e) *G. daiberi* exposed to DDT in FHRW.

When 17-mm *M. saxatilis* were exposed to ^{14}C-PCB in systems containing Hudson River mud, 24-hour uptake values were lower than for when the fish were exposed to PCB's in water alone (Table 3). With a gravel substrate, uptake was essentially the same as that from FHRW. (Unfortunately, only a single data point is available.)

Striped bass (17 mm) fed ^{14}C-PCB labeled *N. americana* had body burdens of ~1 μg/g PCB approximately one hour after feeding (Table 5). After 24 hours, the whole body PCB concentration decreased to 58% and 71% of the initial body burden.

The distribution of ^{14}C-PCB in tissues of striped bass (80 mm) fed PCB-labeled *G. daiberi* showed greater concentrations in the liver and gut than in

Table 1. Concentrations of PCB's and DDT (ng/mg ± S.E.) in *Gammarus daiberi* following exposure in filtered Hudson River water (FHRW) or Hudson River water (HRW) for periods up to 48 hours. The PCB and DDT values are means of four replicates consisting of five organisms per sample. All exposures were to a concentration of 1.0 ng/ml ^{14}C-PCB or ^{14}C-DDT. Water values are in ng/ml ± S.E.

	Time, hours of exposure										
Exposure	0.25	0.50	1	2	3	6	12	15	24	30	34	48
PCB (FHRW)												
Gammarus	–	0.23 ±0.01	0.29 ±0.05	0.38 ±0.03	0.37 ±0.14	0.90 ±0.10	–	1.97 ±0.18	1.68 ±0.33	3.32 ±0.55	–	3.78 ±0.33
Water	–	0.66 ±0.08	0.36 ±0.05	0.45 ±0.01	0.42 ±0.09	0.42 ±0.01	–	0.28 ±0.01	0.19 ±0.03	0.19 ±0.02	–	0.10 ±0.02
PCB (FHRW)												
Gammarus	–	–	0.45 ±0.12	0.81 ±0.03	1.43 ±0.15	2.84 ±0.30	2.86 ±0.61	–	3.42 ±0.20	–	4.60 ±1.39	4.03 ±0.37
Water	–	–	2.87 ±0.14	0.39 ±0.14	0.28 ±0.02	0.23 ±0.03	0.21 ±0.03	–	0.23 ±0.03	–	0.20 ±0.01	0.24 ±0.02
PCB (HRW)												
Gammarus	0.29 ±0.01	0.36 ±0.01	0.86 ±0.13	0.81 ±0.11	0.86 ±0.08	1.84 ±1.36	3.24 ±0.36	–	3.72 ±0.52	–	5.92 ±0.30	5.47 ±0.89
Water	0.14 ±0.03	0.16 ±0.02	0.12 ±0.02	0.15 ±0.05	0.08 ±0.03	0.08 ±0.01	0.10 ±0.10	–	0.06 ±0.01	–	0.10 ±0.01	0.20 ±0.02
DDT (FHRW)												
Gammarus	–	0.27 ±0.02	–	–	0.60 ±0.09	1.60 ±0.47	–	2.14 ±0.23	3.30 ±0.22	3.70 ±0.34	–	3.70 ±0.09
Water	–	0.36 ±0.03	–	–	0.34 ±0.05	0.61 ±0.18	–	0.28 ±0.03	0.33 ±0.04	0.30 ±0.03	–	0.23 ±0.03

Table 2. 48-hour bioaccumulation factors for PCB and DDT uptake by *Gammarus daiberi* and *Neomysis americana*. PCB and DDT data are in ng/mg ± S.E. (organisms) and ng/ml ± S.E. (water) at the end of a 48-hour exposure. FHRW = filtered Hudson River water; HRW = Hudson River water.

Experiment	Uptake (48 hours)	Water (48 hours)	Bioaccumulation factor (x10^4)
PCB – *Gammarus* (FHRW)	3.78 ± 0.33	0.10 ± 0.02	3.8
PCB – *Gammarus* (FHRW)	4.03 ± 0.37	0.24 ± 0.02	1.7
PCB – *Gammarus* (HRW)	5.47 ± 0.89	0.20 ± 0.02	2.7
DDT – *Gammarus* (FHRW)	3.70 ± 0.09	0.23 ± 0.03	1.6
PCB – *Gammarus* with *Myriophyllum*	2.40 ± 0.16	0.24 ± 0.02	1.0
PCB – *Gammarus* in *Chroococcus* culture	0.95 ± 0.20	0.24 ± 0.01	0.4
PCB – *Gammarus* with labeled *Chroococcus*	0.17 ± 0.04	–	–
PCB – *Neomysis* (FHRW)	4.89 ± 1.75	0.14 ± 0.02	3.5
PCB – *Neomysis* (HRW)	4.00 ± 0.74	0.24 ± 0.02	1.7

muscle and gill tissue. The highest concentrations occurred in the liver. Gill concentrations were 21% and 11% of the liver 24 hours and 48 hours after feeding, respectively (Table 6). These percentages were essentially the same as those which occurred 24 hours and 48 hours following exposure to PCB's in water.

Following uptake from water, PCB's were initially eliminated more rapidly from gammarids than DDT (Table 7, Figure 2). After 72 hours, DDT release was rapid, attaining a final value, after 192 hours, of 1.1 ng/mg. The concentration of PCB's in *G. daiberi* after 192 hours was 1.9 ng/mg. The rate constants describing release of both DDT and PCB's after direct accumulation from HRW were smaller than those describing uptake. Assuming a simple negative exponential release of both DDT and PCB's, our data indicated that the biologi-

Table 3. PCB uptake by 17-mm *Morone saxatilis* (striped bass) following exposure to 1 ng/ml PCB in filtered Hudson River water (FHRW) and in Hudson River water (HRW) with and without sediment.

Time of exposure (hours)	PCB concentration (μg/g ± S.E.)			
	FHRW	HRW	HRW/mud	HRW/gravel
24	4.46 ± 0.23	5.10 ± 0.09	2.28 ± 0.17	–
48	4.96 ± 0.40	5.86 ± 0.32	–	4.64

Table 4. Uptake of PCB's by *Gammarus daiberi* in the presence of labeled food organisms (*Myriophyllum spicatum* and *Chroococcus* sp.). See text for description of tests. Data are in ng/ml ± S.E.

Time of exposure (hours)	*Myriophyllum*	*Chroococcus* Labeled but not washed	*Chroococcus* Labeled and washed
0.25	–	0.01 ± 0.01	0.01 ± 0.01
1	0.41 ± 0.11	0.04 ± 0.01	0.01 ± 0.01
2	0.56 ± 0.68	0.06 ± 0.01	–
3	1.28 ± 0.06	–	–
6	2.36 ± 0.29	–	–
12	2.28 ± 0.05	–	–
15	–	0.14 ± 0.01	0.09 ± 0.01
24	2.30 ± 0.15	0.54 ± 0.06	0.18 ± 0.05
34	2.40 ± 0.19	–	–
48	2.40 ± 0.16	0.95 ± 0.02	0.17 ± 0.04

cal half-life for release of DDT and PCB's from *G. daiberi* was $T_{1/2}$ = 5 days and $T_{1/2}$ = 12 days, respectively.

DISCUSSION

The common estuarine planktonic crustaceans, *Gammarus daiberi* and *Neomysis americana,* and young striped bass, *Morone saxatilis* (17 mm and 80 mm in length), rapidly accumulated PCB's and DDT from the water. The gammarids and striped bass also accumulated PCB's from food.

Rapid uptake rates and large (10^4) bioaccumulation factors for PCB and DDT uptake from water indicate that, for planktonic crustacea, as for fishes (Holden,

Table 5. Accumulation, retention, and elimination of PCB's by *Morone saxatilis* (17 mm) following feeding on five ^{14}C-Aroclor 1254 radiolabeled *Neomysis americana.*

	Replicate 1	Replicate 2
Accumulation		
Neomysis americana µg/g dry wt	5.87	6.05
Morone saxatilis µg/g dry wt ± S.E.	0.98 ± 0.10	1.00 ± 0.12
Retention after 24 hours		
Morone saxatilis µg/g dry wt ± S.E.	0.57 ± 0.06	0.71 ± 0.01
Percent retained	58.2	71.0

Table 6. Accumulation of PCB in tissues of *Morone saxatilis* (80 mm) after exposure to PCB's in food or in Hudson River water (0.5 ng/ml).

Tissue	Source of ^{14}C-Aroclor 1254 exposure	Mean concentration in fish tissue (ng/g wet wt) ± S.E. exposure period	
		24 hours	48 hours
Gill	Food (*Gammarus* spp.)	4 ± 2	7 ± 1
Liver		19 ± 2	64 ± 35
Gut		14 ± 1	15 ± 3
Muscle		1 ± 0.5	4 ± 2
Carcass*		2 ± 1	4 ± 2
Gill	Hudson River water	300 ± 60	600 ± 270
Liver		2900 ± 1200	2500 ± 340
Gut		660 ± 230	800 ± 270
Muscle		110 ± 19	150 ± 65
Carcass*		160 ± 40	240 ± 50

*Tissues remaining after removal of gills, liver, gut, and muscle.

1962; Grzenda *et al.*, 1970; Neely *et al.*, 1974; Defoe *et al.*, 1978), surface adsorption and partitioning into exposed tissues such as gills play a major role in uptake. However, it is unlikely that the concentration of PCB's in food and

Table 7. PCB and DDT concentrations remaining in *Gammarus daiberi* 24 hours to 192 hours after a single 24-hour exposure to 1 ng/ml of the contaminant in filtered Hudson River water. Data are given in ng/mg ± S.E. (organism), and ng/ml ± S.E. (water).

Time after initial exposure	PCB *Gammarus*	PCB Water	DDT *Gammarus*	DDT Water
24	3.10 ± 0.90	<0.01	2.70 ± 0.10	0.25 ± 0.03
48	2.5 ± 0.20	<0.01	3.0 ± 0.01	0.11 ± 0.03
72	–	–	3.50 ± 0.64	0.09 ± 0.01
96	–	–	2.40 ± 0.32	0.15 ± 0.01
144	2.30 ± 0.30	0.28 ± 0.01	–	–
192	1.90 ± 0.82	0.38 ± 0.01	1.10 ± 0.18	0.19 ± 0.01

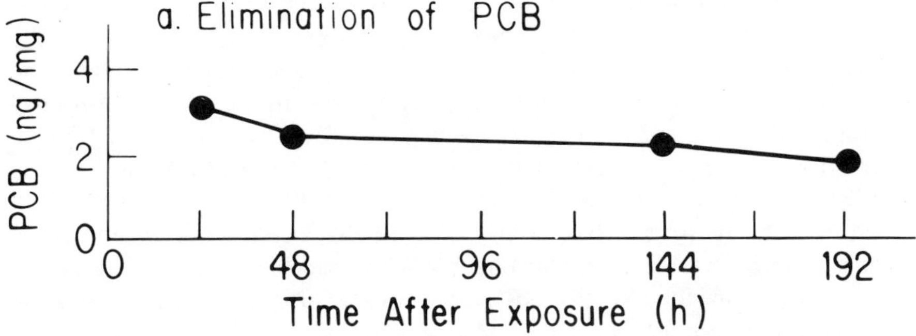

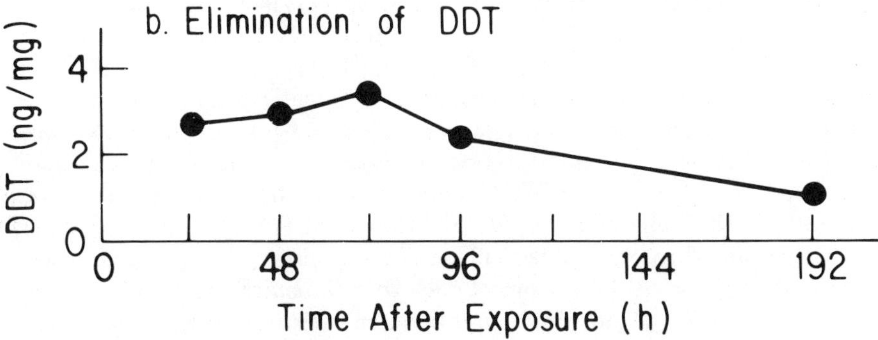

Figure 2. PCB concentrations in *Gammarus daiberi* following 24-hour exposure to (a) 1 ng/ml PCB's in filtered Hudson River water and (b) 1 ng/ml DDT in filtered Hudson River water.

potential for food chain transport can be predicted by knowing only water concentrations and octanol-water partition coefficients.

Harvey *et al.* (1974), Harvey and Steinhauer (1975), Chiou *et al.* (1977), and Hiraizumi *et al.* (1979) observed that particles scavenge PCB's and other chlorinated hydrocarbons from water and keep them in suspension. Adsorption concentrates organohalogens on particulates, and one of two events may occur. In highly contaminated water, PCB accumulation in filter-feeding plankton may be much higher than predicted from partition coefficients. In the Hudson River estuary, PCB's on particulates may be as high as 15 μg/g in normally suspended fine silts and clays (Hetling *et al.*, 1978). When estuarine copepods were fed ^{14}C-labeled particulates (clays or dead algae), PCB accumulation increased, as did copepod mortality (O'Connors *et al.*, this volume). Field studies showed that PCB concentrations in *Gammarus* and in microzooplankton in the Hudson are significantly correlated with suspended particles (O'Connor *et al.*, 1980).

Alternatively, when PCB concentrations are low (*e.g.*, coastal marine and open ocean environments; see MacLeod *et al.*, 1981), scavenging of PCB by suspended particles may reduce the quantities in the water and cause lower concentrations than predicted by partitioning.

Two aspects of our study confirm the above observations. Uptake by *G. daiberi* was least when ^{14}C-PCB-labeled *Chroococcus* were the only source of PCB's. When *G. daiberi* were exposed to labeled *Chroococcus* in labeled water, PCB uptake was greater than when *Chroococcus* were the only source of PCB's, but far less than when the PCB's were available in HRW or FHRW (Table 2). When striped bass were exposed to PCB's in sediment-water systems, uptake was least when fine particles, capable of adsorbing PCB's strongly, were included. Field observations of PCB concentrations in zooplankton (O'Connor *et al.*, 1980; MacLeod *et al.*, 1981) show variability among samples (up to factors of 10^4), suggesting that uptake is controlled by factors other than PCB concentration in water and partition coefficients.

Short-term exposure of planktonic organisms to organohalogens showed that uptake in gammarids, mysids, and larval striped bass (17 mm) was basically similar and conformed to power functions. Uptake rates depended upon contaminant availability, as demonstrated by the rapid, initial drop in organohalogen concentrations in water (*i.e.*, from 1.0 ng/ml to ~0.1-0.2 ng/ml). It is likely that uptake during the first few hours of exposure depended upon surface sorption of the contaminant, followed by partitioning to tissues. Harvey *et al.* (1974), Hiraizumi *et al.* (1979), and Horn *et al.* (1979) demonstrated that sorption occurs when organic particulates and organisms are exposed to chlorinated hydrocarbons. Although our studies with *G. daiberi, N. americana,* and 17-mm *M. saxatilis* did not distinguish between surface-sorbed and assimilated PCB's or DDT, the rapidity with which these organisms accumulated PCB's and DDT favors the conclusion that initial uptake values are the result of surface sorption.

At 24 hours, the tissues of 80-mm striped bass contained from 0.1 μg/g to 3.0 μg/g PCB (Table 6), showing that the organohalogen is incorporated into the tissues. Califano *et al.* (1980) showed that PCB uptake after 12 hours by striped bass larvae was nearly 80% of the total uptake for 48 hours. Subsequent experiments with larger fish showed that PCB concentrations in tissues (liver, muscle) continue to increase after the increase in "whole body" PCB concentration stops (O'Connor *et al.*, 1980). This suggests that substantial assimilation by the edible flesh may be delayed.

Feeding studies demonstrated that, while there is food-chain transport of organohalogens, PCB's are not biomagnified. In fact, with food present, PCB uptake by *G. daiberi* was less than from water, even when the total quantities of PCB available were similar. Elimination of PCB's by striped bass whose body burdens were obtained from food was more rapid than by fish that accumulated PCB's from the water. The same phenomenon has been observed in larger fishes (200-250 mm), including striped bass and white perch (*Morone americana;* O'Connor *et al.*, 1980).

Many workers have stated that PCB's are not biomagnified in aquatic food chains, although there is evidence for biomagnification in terrestrial food chains

(Risebrough, 1971). In coastal marine systems, PCB concentrations in the food of fishes and their likely predators are essentially the same, suggesting that body burdens are related more to partitioning and water uptake than to accumulation through food (Harvey *et al.*, 1974; Harvey and Steinhauer, 1975; O'Connor *et al.*, this volume). In fishes, the lack of biomagnification probably is related to the rapid elimination of PCB's derived from dietary sources, whereas PCB's accumulated from the water are retained for longer periods. In highly contaminated waters such as the Hudson River, the large quantities of PCB's available in the water, sediments, and food (Horn *et al.*, 1979) cause high body burdens in fishes (Nadeau and Davis, 1976; Spagnoli and Skinner, 1977; Hetling *et al.*, 1978; Horn *et al.*, 1979). However, the half-life of PCB's in Hudson River fishes is relatively short (≤2 years) when new inputs are low, and biomagnification is not observed in these populations (R. Sloan, New York State Department of Environmental Conservation, Albany, New York, personal communication).

An abundant literature on PCB's, DDT, and other organohalogens serves well to document the ubiquity of these compounds in water, sediments, and aquatic biota. The equally voluminous literature on the toxicity and potential for bioaccumulation of organohalogens provides insight into acceptable levels of these compounds in the environment. The impact of organohalogens on existing ecosystems, however, cannot be derived from available data. Further studies, similar in concept to those begun here, will be required to determine how environmental factors control the availability of organohalogen contaminants for biological accumulation. The results of such studies will determine how to manage future inputs of biodestructive compounds and the level of effort that must be directed toward rehabilitating contaminated systems.

ACKNOWLEDGEMENTS

The technical assistance of Richard J. Califano in the striped bass experiments and of Raquel Pachoa and Henry DeVegvar in the zooplankton experiments is appreciated. This work was supported in part by the New York State Department of Environmental Conservation, the National Institute of Environmental Health Sciences, Grant ES00260, and the NOAA MESA New York Bight Project, Grant 04-78-B01-16.

REFERENCES

Burdick, G.E., E.J. Harris, J.H. Dean, T.M. Wilker, J. Skea, and D. Colby. 1964. The accumulation of DDT in lake trout and the effect on reproduction. Trans. Am. Fish. Soc. 93: 127-136.

Butler, P.A. 1969. The significance of DDT residues in estuarine fauna. *In:* Chemical Fallout, M.W. Willer and G.G. Berg (eds.), Charles Thomas, Springfield, IL. pp. 205-220.

Califano, R.J., J.M. O'Connor, and L.S. Peters. 1980. Uptake, retention and elimination of PCB (Aroclor 1254) by larval striped bass, *Morone saxatilis.* Bull. Environ. Contam. Toxicol. 24: 467-472.

Cantilli, R. 1977. Effects of PCB on respiratory metabolism of the amphipod *Gammarus tigrinus.* M.S. Thesis, New York University. 66 pp.

Chiou, C.T., V. Freed, D. Schmedding, and R. Kohnert. 1977. Partition coefficient and bioaccumulation of selected organic chemicals. Environ. Sci. Technol. 11: 475-478.

Defoe, D.L., G.D. Veith, and R.W. Carlsen. 1978. Effects on Aroclor 1248 and 1260 on the fathead minnow (*Pimephales promelas*). J. Fish. Res. Board Can. 35: 997-1002.

Dow, R.L. 1975. Reduced growth and survival of clams transplanted to an oil spill site. Mar. Pollut. Bull. 6: 124-125.

Duke, T.W., J.I. Lowe, and A.J. Wilson, Jr. 1970. A polychlorinated biphenyl (Aroclor 1254) in the water, sediment, and biota of Escambia Bay, Florida. Bull. Environ. Contam. Toxicol. 5: 171-180.

Elder, D.L. and J.P. Villeneuve. 1977. Polychlorinated biphenyls in the Mediterranean Sea. Mar. Pollut. Bull. 8: 19-22.

Foehrenbach, J., G. Mahmood, and D. Sullivan. 1971. Chlorinated hydrocarbon residues in shellfish (Pelecypoda) from estuaries of Long Island, New York. Pestic. Monit. J. 5: 242-247.

Ginn, T.C. 1977. An ecological investigation of Hudson River macrozooplankton in the vicinity of a nuclear power plant. Ph.D. Thesis, New York University. 272 pp.

Grzenda, A.R., D.F. Paris, and W.J. Taylor. 1970. The uptake, metabolism and elimination of chlorinated residues by goldfish (*Carassius auratus*) fed a ^{14}C-DDT contaminated diet. Trans. Am. Fish. Soc. 99: 385-396.

Harvey, G.R., H.P. Miklas, V.T. Bowen, and W.G. Steinhauer. 1974. Observations on the distribution of chlorinated hydrocarbons in Atlantic Ocean organisms. J. Mar. Res. 32: 103-118.

Harvey, G.R. and W.G. Steinhauer. 1975. Biogeochemistry of PCB and DDT in the North Atlantic. *In:* Environmental Biochemistry, Vol. I, J.O. Nriagu (ed.), Ann Arbor Science Press, Ann Arbor, MI. pp. 203-221.

Hetling, L.E., E.G. Horn, and J. Tofflemire. 1978. Summary of Hudson River PCB study results. New York State Department of Environmental Conservation Tech. Rep. 51. 88 pp.

Hiraizumi, Y., M. Takahashi, and M. Nishimura. 1979. Adsorption of polychlorinated biphenyl onto seabed sediment, marine plankton and other adsorbing agents. Environ. Sci. Technol. 13: 580-584.

Holden, A.V. 1962. A study of the adsorption of ^{14}C-labelled DDT from water by fish. Ann. Appl. Biol. 50: 467-477.

Horn, E.G., L.J. Hetling, and J. Tofflemire. 1979. The problem of PCBs in the Hudson River system. *In:* Health Effects of Halogenated Aromatic Hydrocarbons, W.J. Nicholson and J.A. Moore (eds.), Ann. N.Y. Acad. Sci. 320: 591-609.

Kneip, T.J. and J.A. Hernandez. 1969. Pesticide residues in the Hudson River and biota. *In:* Development of a Biological Monitoring System and a Survey of Trace Metals, Radionuclides and Pesticides in the Lower Hudson River, M. Eisenbud and G.P. Howells (eds.), Rep. to New York State Department of Health, New York University Institute of Environmental Medicine, Tuxedo, NY. pp. 124-162.

Kuhnhold, W.W. 1971. The influence of crude oils on fish fry. Food and Agricultural Organization Fish. Rep. 99. 157 pp.

MacLeod, W.D., Jr., L.S. Ramos, A.J. Friedman, D.G. Burrows, P.G. Prohaska, D.L. Fisher, and D.W. Brown. 1981. Analysis of residual chlorinated hydrocarbons, aromatic hydrocarbons and related compounds in selected sources, sinks and biota of New York Bight. NOAA Tech. Memo. OMPA-6. 128 pp.

Nadeau, R.J. and R.A. Davis. 1976. Polychlorinated biphenyls in the Hudson

River (Hudson Falls-Fort Edward) New York. Bull. Environ. Contam. Toxicol. 16: 436-444.

Neeley, W.B., D. Branson, and G. Blau. 1974. Partition coefficient to measure bioconcentration potential of organic chemicals in fish. Environ. Sci. Technol. 8: 1113-1115.

Nimmo, D.R. and L.H. Bahner. 1974. Some physiological consequences of polychlorinated biphenyl- and salinity-stress in penaeid shrimp. *In:* Pollution and Physiology of Marine Organisms, F.J. Vernberg and W.B. Vernberg (eds.), Academic Press, New York. 427-443.

Nimmo, D., R. Blackman, A.J. Wilson, and J. Forester. 1971. Toxicity and distribution of Aroclor 1254 in the pink shrimp, *Penaeus duorarum*. Mar. Biol. 11: 191-197.

O'Brien and Gere Engineers, Inc. 1979. Bioaccumulation Studies on *Homarus americanus.* Rep. to U.S. Army Engineers District, New York. 52 pp.

O'Connor, J.M., J.B. Klotz, and T.J. Kneip. This volume. Sources, sinks and distribution of organic contaminants in the New York Bight ecosystem. pp. 631-653.

O'Connor, J.M., T.J. Kneip, and C.C. Lee. 1980. Biological Monitoring of PCBs in Hudson River Biota. Progress Report for 1979-80. Rep. of Office of Water Research, New York State Department of Environmental Conservation, Albany, NY. 37 pp.

O'Connor, J.S. and H.M. Stanford (eds.). 1979. Chemical Pollutants of the New York Bight; Priorities for Research. NOAA Special Report. 217 pp.

O'Connors, H.B., C.D. Powers, C.F Wurster, K.D. Wyman, G.M. Nau-Ritter, and R.G. Rowland. This volume. Fates and biological effects on plankton of particle-sorbed PCB's in coastal waters. pp. 423-449.

O'Connors, H.B., C.F. Wurster, C.D. Powers, D.C. Biggs, and R.G. Rowland. 1978. Polychlorinated biphenyls may alter marine trophic pathways by reducing phytoplankton size and production. Science 201: 727-739.

Poje, G.V. 1977. Growth and reproduction of *Gammarus tigrinus* Sexton, 1939. M.S. Thesis, New York University. 53 pp.

Renzoni, A. 1973. Influence of crude oil, derivatives, and dispersants on larvae. Mar. Pollut. Bull. 4: 9-13.

Risebrough, R.W. 1971. Chlorinated hydrocarbons. *In:* Impingement of Man on the Oceans, D.W. Hood (ed.), Wiley Interscience, New York. pp. 259-286.

Searl, T.D., H.L. Huffman, and J.P. Thomas. 1977. Extractable organics and nonvolatile hydrocarbons in New York harbor waters. *In:* Proceedings of Oil Spill Conference, American Petroleum Institute, Washington, DC. pp. 583-588.

Spagnoli, J. and L. Skinner. 1977. PCB's in fish from selected waters of New York State. Pestic. Monit. J. 11: 69-87.

Stainken, D. and J. Rollwagen. 1979. PCB residues in bivalves and sediments of Raritan Bay. Bull. Environ. Contam. Toxicol. 23: 690-697.

Tanacredi, J.T. 1977. Petroleum hydrocarbons from effluents: detection in marine environment. J. Water Pollut. Control Fed. 49: 216-226.

Wang, C. H. and D. L. Willis. 1965. Radiotracer Methodology in Biological Sciences. Prentice-Hall, Inc., Englewood Cliffs, NJ. 382 pp.

Wells, P.G. 1972. Influence of Venezuelan crude oil on lobster larvae. Mar. Pollut. Bull. 3: 105-106.

West, R.H., P.G. Hatcher, and D.K. Atwood. 1976. Polychlorinated Biphenyls and DDTs in Sediments and Sewage Sludge of the New York Bight. Final Rep. to NOAA/MESA New York Bight Project. 40 pp.

Wilson, A.J. and J. Forester. 1978. Persistence of Aroclor 1254 in a contaminated estuary. Bull. Environ. Contam. Toxicol. 19: 637-640.

CONTRIBUTED PAPERS ON THE EFFECTS OF NUTRIENTS AND CARBON LOADINGS

Opposite: The 26th Ward sewage treatment plant adjacent to Hendrix Creek and Jamaica Bay, Brooklyn, New York. The sewer systems of the New York-New Jersey metropolitan area represent primary sources of anthropogenic nutrients and carbon to the New York Bight and Hudson-Raritan estuary. (Courtesy New York Sea Grant Institute.)

THE IMPORTANCE OF NUTRIENTS TO PHYTOPLANKTON PRODUCTION IN NEW YORK HARBOR

J.J.A. McLaughlin
G.S. Kleppel[1]
M.P. Brown[2]
R.J. Ingram
W.B. Samuels[3]

Louis Calder Conservation and Ecology Study Center
Armonk, New York 10504

Abstract. A four-year study was conducted to identify the importance of nutrients in regulating phytoplankton growth and productivity in New York harbor.

While nitrogen (occurring primarily as ammonia in yearly concentrations of 25-88 μg-at/liter) and phosphorus appear to be present well in excess of concentrations capable of limiting phytoplankton growth, silicon delivery to the estuary and the New York Bight apex decreases sharply in May and June. The phytoplankton community at this time changes from diatom-dominated to chlorophyte- and dinoflagellate-dominated. Primary productivity reflects this change, initially decreasing between April and May, then increasing toward summer as the chlorophytes become dominant. Data from laboratory experiments suggest that diatom growth during the late spring could be restored by the addition of silicon, trace elements, and vitamins.

During peak photosynthesis periods (February-March, May-July), productivity seems to be regulated by light intensity, but during the fall, when productivity is low, unidentified constituents in the water appear to inhibit photosynthesis.

Water quality management strategies for the harbor must address all aspects of the system (planktonic, benthic, pelagic regimes) and consider the influence of the Bight apex and Long Island Sound as sources of biomass.

INTRODUCTION

Estuaries frequently have been used as repositories for nutrients and other bioactive materials resulting from domestic and industrial activities. Initially, the impact of these wastes on estuarine organisms may be minimal. In fact, some enrichment will enhance biological productivity. Further enrichment, however, may begin to alter certain food webs, as was the case in Moriches Bay, Long Island, New York, where nutrients from duck farm wastes altered the food stock of the hardshell clam, *Mercenaria mercenaria,* by replacing diatoms with less suitable food organisms. This caused deterioration of the fishery (Ryther, 1954). Eventually, continued enrichment, combined with the addition of toxic substances, may depress productivity and begin to impact adjacent coastal waters.

[1] Present address: Department of Biological Sciences, Allan Hancock Foundation, University of Southern California, University Park, Los Angeles, California 90007.

[2] Present address: Bureau of Water Resources, New York State Department of Environmental Conservation, 50 Wolfe Road, Albany, New York 12233.

[3] Present address: Environmental Quality Section, U.S. Geological Survey, Reston, Virginia 22070.

New York harbor is an estuary where the addition of domestic and industrial effluents results in an average biological oxygen demand (BOD) of over 30,000 metric tons per year (Hydroscience, 1975). Transport of part of this BOD out of the estuary periodically may affect productivity in waters of western Long Island Sound and the New York Bight apex (Bowman, 1977; Walsh *et al.*, 1978).

A four-year study of the effects of nutrient enrichment on the phytoplankton in New York harbor (Figure 1) has shown that growth and primary production are never controlled by nitrogen or phosphorus. However, other substances, such as silica, periodically may limit diatom growth, thus establishing conditions for the succession of planktonic populations. This paper presents a summary of these findings; additional details can be obtained in Ingram (1979), Kleppel (1979), and Samuels (1979).

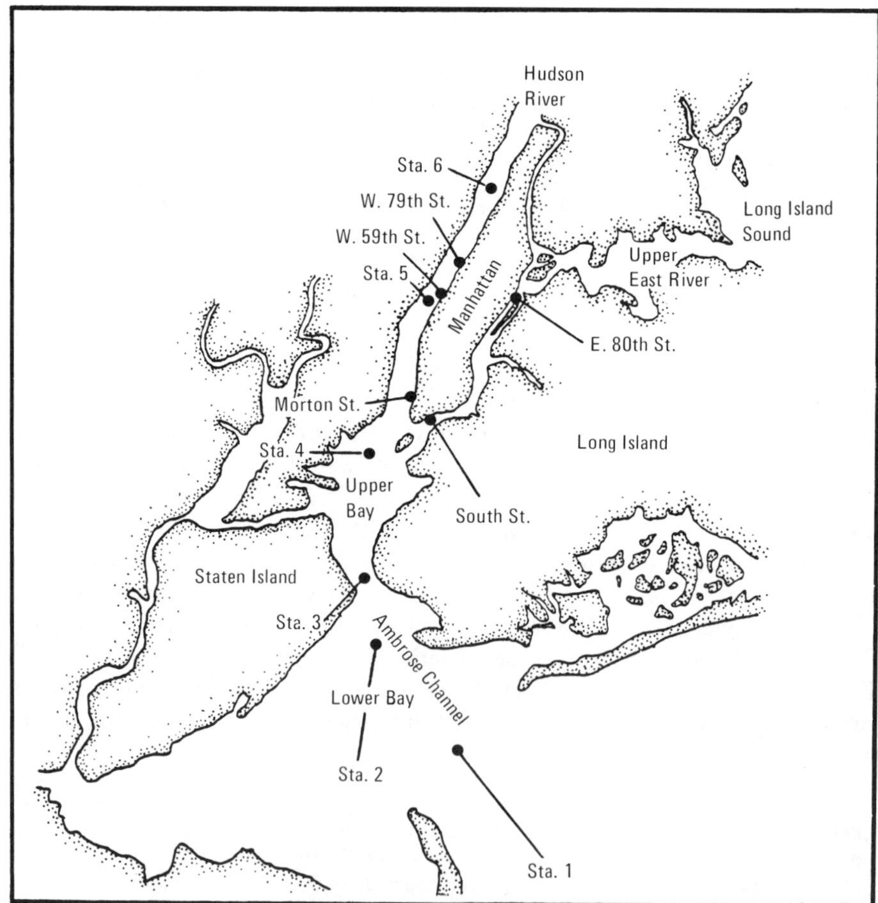

Figure 1. Station locations in New York harbor. Stations 1 through 6 were occupied from 1976 to 1978; all other stations were occupied from 1974 to 1978.

Table 1. Loading rates of inorganic nutrients to the lower Hudson River estuary from sewage effluent discharge.

	Metric tons per day			
Location	N	P	Si	Reference
Hudson estuary	169	17	–	Howells *et al.* (1970)
Hudson estuary	160	–	–	Garside *et al.* (1976)
New York harbor	140	–	–	Quirk, Lawler, and Matusky, Engineers (1970)
New York harbor	–	–	20	McLaughlin and Brown (1978)

NUTRIENTS IN NEW YORK HARBOR

Some recent estimates of the nutrient load delivered to the Hudson River estuary by wastewater discharge are presented in Table 1. The nitrogen (N) and phosphorus (P) input from sewage are higher than the national average, while the silicon (Si) load is about one eighth of the national average (Dunstan, 1975). More than 50% of the inorganic nitrogen present at any time occurs as ammonia-N (Ingram, 1979). This is added primarily in and above the Upper Bay (Figure 2) and represents a wastewater-N input from the New York City-New

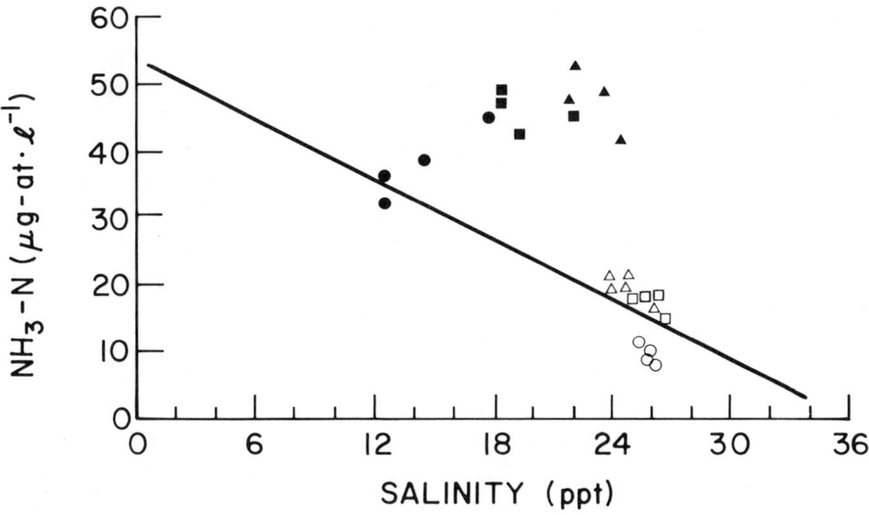

Figure 2. Relationship between salinity and NH_3-N concentration in the estuary. Data were collected in August 1977 from six stations in the Hudson River and indicate an ammonia source between Upper Bay and the George Washington Bridge. Symbols: Stations 1, 2, 3 (open circles, squares, triangles, respectively); Stations 4, 5, 6 (closed triangles, squares, circles, respectively).

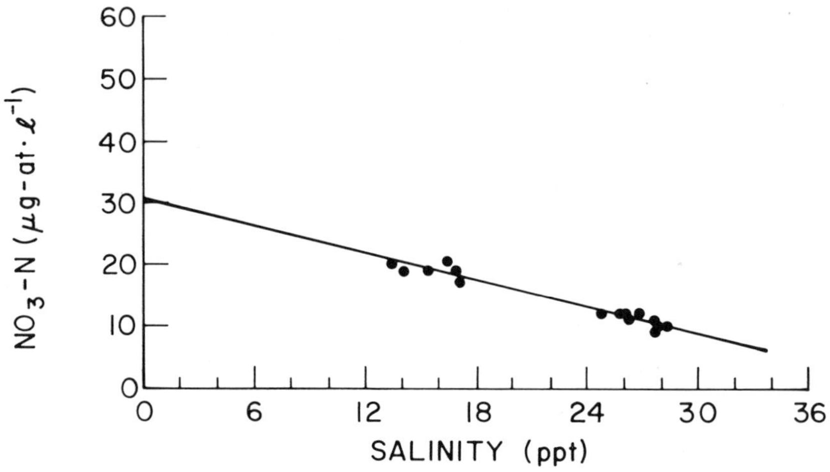

Figure 3. Relationship between salinity and NO_3-N concentrations. Data were collected in February 1977 from Stations 1 through 6 in the Hudson River.

Jersey metropolitan area of approximately 6.8 metric tons per day (Hammond, 1975). Ammonia-N concentrations in New York harbor, between the George Washington Bridge and the Rockway Point-Sandy Hook transect (Station 1), varied between 25 μg-at/liter and 88 μg-at/liter during 1975-1978.

Unlike ammonia, nitrate in the estuary behaves conservatively, decreasing linearly with salinity according to the relation,

$$NO_3\text{-N } (\mu\text{g-at/liter}) = 30 - 0.7 \text{ salinity (ppt)},$$

($r = 0.90$; $p < 0.01$). This regression explains about 81% of the variance in either parameter (Figure 3) and fits the general scheme for estuaries proposed by McCarthy *et al.* (1975) and Liss (1976).

Silicon delivery to New York harbor, like nitrate, is a function of freshwater flow. Figure 4 shows that the 20 metric tons per day of SiO_3-Si input by wastewater discharge contributes between 6% and 50% of the total load to Lower Bay. Clearly, urban wastewater is relatively more important as a source of SiO_3-Si during the summer, low-flow period than during the rest of the year.

Phosphorus in the harbor varied from 0.6 μg-at/liter to 17.1 μg-at/liter during our four-year study. Not surprisingly, concentrations always were highest in the summer, when freshwater flow was lowest (*i.e.*, mean value for *ortho*-P for June-September 1977 = 3.5 μg-at/liter; annual mean value for 1977 = 2.2 μg-at/liter). Phosphorous levels decreased exponentially as the flow rate increased. About half of the phosphorus exists as *ortho*-P. The example of low-flow period *ortho*-P distributions shown in Figure 5 suggests that, like ammonia, much of the *ortho*-P in the harbor is wastewater-derived.

Detailed descriptions of nutrient variability and dispersion in the Hudson River estuary are available (Hammond, 1975; Garside *et al.*, 1976; Malone, 1976; Ingram, 1979) and will not be considered further here.

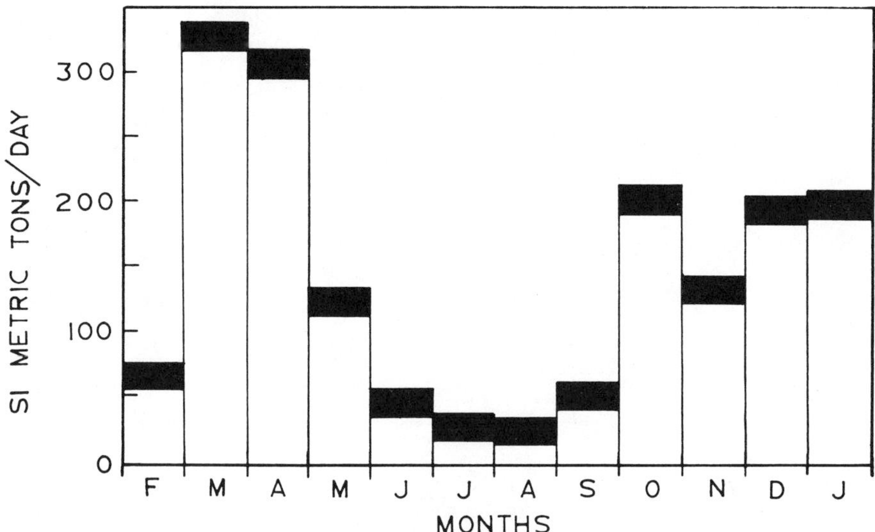

Figure 4. Silica delivery to New York harbor, February 1976 - January 1977. Dark areas represent approximately 20 metric tons of silicon per day added from sewage.

PHYTOPLANKTON AND NUTRIENTS

The phytoplankton populations in New York harbor that are present mainly as a result of advection from the New York Bight apex (Malone, 1977) utilize about 10% of the available nutrients. The remaining nutrients are discharged to the Bight apex (Garside *et al.*, 1976).

Measured values for primary productivity vary seasonally, ranging from 0.01 gC/m^2/day to 2.22 gC/m^2/day in Upper Bay (Malone, 1977). This is about twice as high as for the East River (0.007-1.13 gC/m^2/day) (Kleppel, 1979). Peak photosynthetic rates occur in the summer (July), and minimal rates occur in December in both the Upper Bay and the East River. A secondary peak is observed between February and April because of diatom biomass intrusions from the Bight apex and Long Island Sound. Unlike the summer peak, which largely results from high rates of photosynthesis by the chlorophyte, *Nannochloris atomus,* the diatom peak is a function of elevated biomass (Malone, 1977; Kleppel, 1979). Productivity, especially during peak periods, seems to be regulated principally by light and temperature (Malone, 1976, 1977).

The subtleties of the chemical regulation of phytoplankton production in polluted estuaries are poorly understood. Because of the chemical complexity of New York harbor water, it also is difficult to relate the responses of phytoplankton to individual or grouped constituents. For these reasons, the possible effects of bioactive materials were examined under controlled laboratory conditions. Details of the techniques used in these experiments are presented elsewhere (Kleppel, 1979; Samuels, 1979); however, a summary of the methodology and a discussion of major findings are presented below.

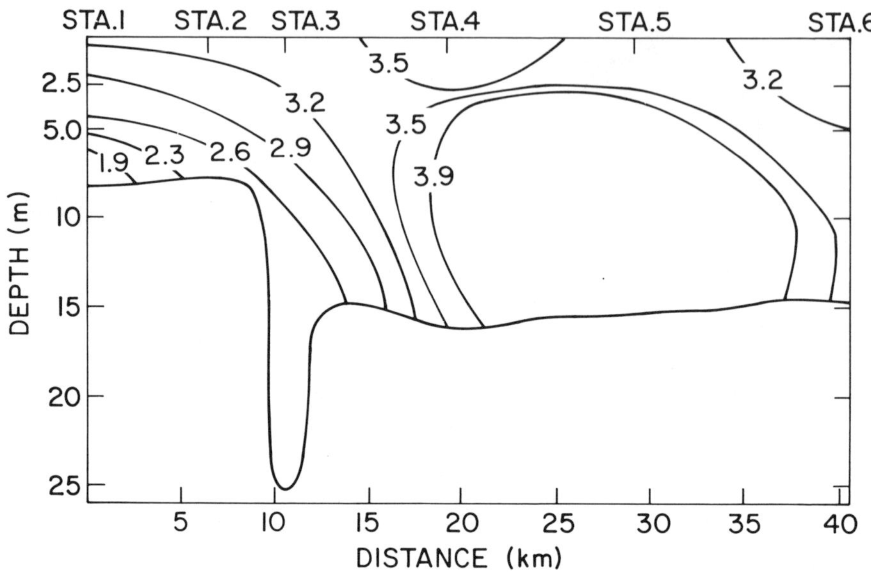

Figure 5. *Ortho*-P concentrations in the Hudson River of New York harbor. Data were collected in September 1977 and suggest a source between Upper Bay and the George Washington Bridge.

LABORATORY STUDIES OF DIATOM GROWTH AND PRODUCTIVITY

At the beginning of each month, water was collected from the East River, and subsamples were passed through a sterile filter (0.22 μm) (CS) or autoclaved (AUTO). *Skeletonema costatum* (Greville) Cleve, a common diatom in the estuary and one that is cultured axenically in our laboratory at optimal light intensities and temperatures, was inoculated into solutions containing 9 ml of CS or AUTO water and 1 ml of STP medium (Provasoli *et al.*, 1957).

Cultures either were counted (Palmer and Maloney, 1954) every other day for 14 days to assess growth rate or used in ^{14}C-uptake experiments five days after inoculation. ^{14}C-uptake in CS cultures qualitatively simulated the results of *in situ* measurements of ^{14}C-uptake made at the end of each month in the East River (Figure 6a,b). The AUTO cultures showed an increase in ^{14}C-uptake relative to uptake in CS and *in situ* situations in water collected during the fall (Figure 6c). However, in the spring (the end of April for *in situ* measurements and the beginning of May for laboratory studies) autoclaving failed to reverse the decrease in ^{14}C-uptake. Apparently, the factors causing primary productivity to be low in the fall differ from those responsible for low productivity in the spring.

The results of growth rate experiments support Malone's (1977) hypothesis that light and temperature limitation causes low winter photosynthetic rates. When these limits were removed in the laboratory, both photosynthesis and growth were enhanced during the winter. In water collected during late spring,

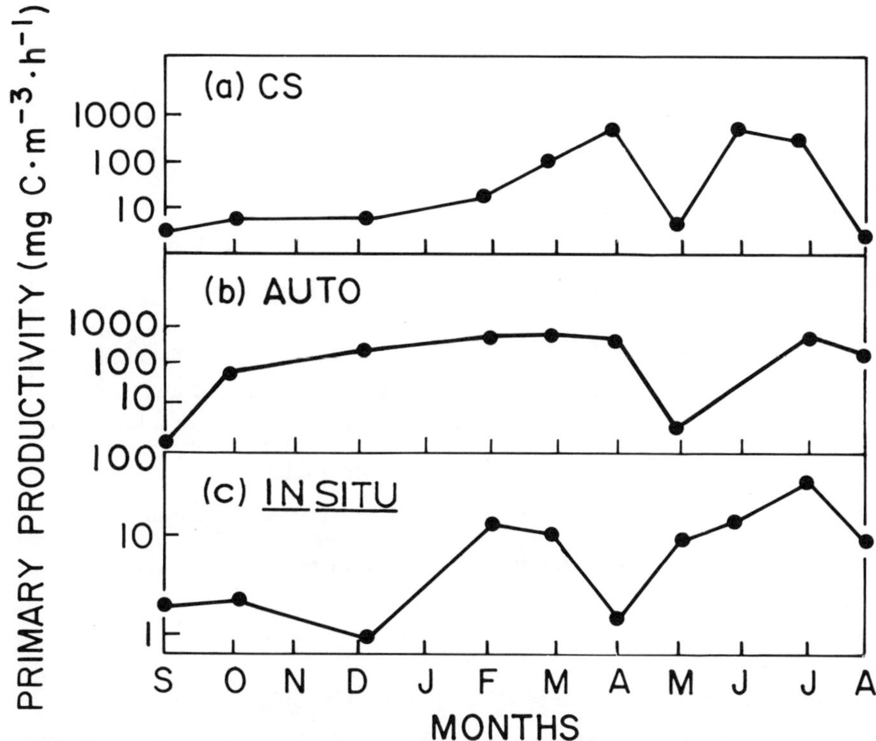

Figure 6. Primary productivity measured using laboratory cultures of *Skeletonema costatum* in (a) sterile filtered (CS) water and (b) autoclaved (AUTO) water from the East River. *In situ* productivity in the East River is indicated in (c). Measurements were made from 1977 to 1978.

growth, like ^{14}C-uptake, was low and did not recover when light and temperature were optimal or when the water was autoclaved. However, it was possible to stimulate growth by adding SiO_3-Si, trace metals, or vitamin B_{12} (Figure 7) (cf. Samuels, 1979).

The bioassay results indicate that SiO_3-Si may become a limiting nutrient during the spring. Silicon delivery to the estuary between March and May decreases by about 80% (Figure 4), and although diatom populations may not become silicon-depleted in the estuary, populations in the Bight apex from which the estuarine flora is derived probably do become depleted. It is hypothesized that in April-May, when the dilution rate decreases (U.S. Geological Survey, 1976, 1977, 1978) and light penetration increases (the time when phytoplankton growth in the harbor might be expected to improve), SiO_3-Si becomes less available. It is possible that the addition of vitamins or trace metals permits diatoms to utilize more efficiently any silicon that is present. Dunstan (1975) proposed that trace substances periodically may limit growth and production in nutrient-saturated estuaries.

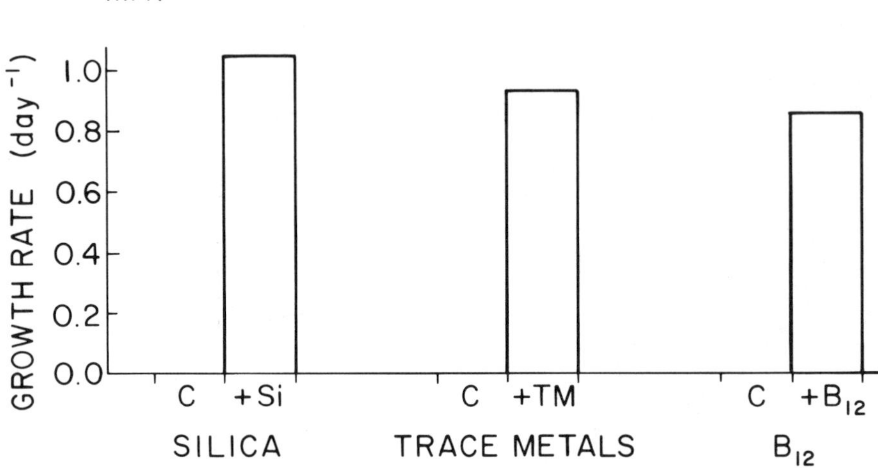

Figure 7. Growth rates of *Skeletonema costatum* in water collected from the East River in May 1977 (C), and in the same water enriched with either silica (+Si; 35 μg-at/liter), trace metals (+TM; PII metals mix of Provasoli *et al.*, 1957), or vitamin B_{12} (+B_{12}; 3 μg/liter).

FURTHER RESEARCH AND MANAGEMENT PERSPECTIVES

Quite clearly, the role of silicon and trace nutrients in regulating productivity and plankton species composition in New York harbor merits further study. The recent presentation by Malone and Garside (1980) on the possibility of silicon limitation in the Bight apex represents a good start. Additional work mandates the inclusion of waters beyond the harbor (such as Long Island Sound).

It has been proposed by some (Segar, personal communication) that advanced wastewater treatment should be implemented to remove all nitrogen from sewage effluent. If such a goal were feasible, removal of 80-90% of the nitrogen now present in the harbor would result in nitrogen limitation to phytoplankton (Garside *et al.*, 1976). Furthermore, the relationship between nitrogen and silicon would be altered, *i.e.*, silicon, which is delivered from the Hudson River would become increasingly available and might cause alterations in the composition of the phytoplankton community (Officer and Ryther, 1980). Greater silicon availability principally would impact the diatoms, which might replace chlorophytes as the dominant component of the spring-summer flora. It is unlikely, however, that improved sewage treatment can remove 80-90% of the inorganic nitrogen in the estuary since as much as 50% of the total dissolved inorganic nitrogen is delivered by the Hudson River. However, if the input of ammonium were reduced significantly, the relationship between nitrogen and silicon might be altered during the period of low river flow (late spring-summer). This also is the time when succession from diatoms to chlorophytes occurs in the estuary.

A question arises about the benefit of a diatom-dominated versus a chlorophyte-dominated system. The answer to this still is unsettled, but it has been suggested that diatoms form the basis for some food webs that are commercially important to man (Ryther, 1954; Officer and Ryther, 1980). These include food webs that support the hard shell clam and oyster. The potential for an improved shellfishery in the lower Hudson estuary exists. However, establishing such a fishery will require more than simply altering the food supply; it will require reduction of BOD and a general improvement of benthic sediment conditions. Furthermore, in considering food web alteration, the importance of detritus as a food source must be considered. The nutritional aspects of detritus are poorly understood at present, and whether specific types of detrital material in the estuary are more or less nutritious than phytoplankton needs to be determined.

When considering improvements to fisheries, it also is necessary to consider primary productivity. Productivity in the New York harbor is limited by light penetration to a varying extent throughout the year. Simple models suggest that small improvements in light penetration can cause large increases in production rates. However, enhanced light penetration will require more than improved sewage treatment. Malone (this volume) has suggested that secondary treatment will result in minimal changes in light penetration, since a large portion of the total particulate load is delivered from north of Manhattan Island. To increase light penetration, sophisticated runoff control seems necessary. The feasibility of this remains questionable. If such an improvement is made, the possibility exists that accumulations of particulate organic carbon could occur in the harbor, especially during the low river flow period. These ultimately could increase BOD levels.

The need to reduce BOD produced by unutilized phytoplankton biomass should be studied in light of the potential for food web engineering mentioned above. That is, if phytoplankton biomass fits into a scheme whereby it is removed as food for harvested secondary production, then BOD problems might be reduced. Such a strategy, however, calls for a commitment to a total systems approach for reclaiming New York harbor and must be preceded by considerable applied, multidisciplinary research.

ACKNOWLEDGEMENTS

We thank R.L. Whalen for performing silica analyses. This work was supported, in part, by U.S. EPA Grant R803370030.

REFERENCES

Bowman, M.J. 1977. Nutrient distributions and transport in Long Island Sound. Estuarine Coastal Mar. Sci. 5: 531-548.

Dunstan, W.M. 1975. Problems of measuring and predicting influence of effluents on marine phytoplankton. Environ. Sci. Technol. 9: 635-638.

Garside, C., T.C. Malone, O.A. Roels, and B.A. Sharfstein. 1976. An evaluation of sewage derived nutrients and their influence on the Hudson estuary and New York Bight. Estuarine Coastal Mar. Sci. 4: 281-289.

Hammond, D.E. 1975. Dissolved gases and kinetic processes in the Hudson River estuary Ph.D. Thesis, Columbia University. 161 pp.

Howells, G.P., T.J. Kneip, and M. Eisenbud. 1970. Water quality in industrial areas: profile of a river. Environ. Sci. Technol. 4: 26-35.

Hydroscience, Inc. 1975. Development of Steady State Water Quality Model for New York Harbor. Report to the Interstate Sanitation Commission, Hydroscience, Inc., Mahwah, NJ. 363 pp.

Ingram, R.J. 1979. Selected phytoplankton nutrients in the lower Hudson estuary. Ph.D. Thesis, Fordham University. 153 pp.

Kleppel, G.S. 1979. Aspects of primary productivity in the East River of New York harbor. Ph.D. Thesis, Fordham University. 153 pp.

Liss, P.S. 1976. Conservative and non-conservative behavior of dissolved constituents during estuarine mixing. *In:* Estuarine Chemistry, J.D. Burton and P.S. Liss (eds.), Academic Press, London. pp. 93-130.

McCarthy, J.J., W.R. Taylor, and J.L. Taft. 1975. The dynamics of nitrogen and phosphorus cycling in the open waters of the Chesapeake Bay. *In:* Marine Chemistry in the Coastal Environment, T.M. Church (ed.), ACS Symposium Ser. No. 18. pp. 664-681.

McLaughlin, J.J.A. and M.P. Brown. 1978. Examination of processes associated with phytoplankton production in New York harbor waters. Rep. to Hazen and Sawyer, Inc., Fordham University Contrib. 567. 31 pp.

Malone, T.C. 1976. Phytoplankton productivity in the apex of the New York Bight: September 1973 to August 1974. NOAA Tech. Memo. ERL MESA-5. 100 pp.

Malone, T.C. 1977. Environmental regulation of phytoplankton productivity in the lower Hudson estuary. Estuarine Coastal Mar. Sci. 5: 157-171.

Malone, T.C. This volume. Factors influencing the fate of sewage-derived nutrients in the lower Hudson estuary and New York Bight. pp. 389-400.

Malone, T.C. and C. Garside. 1980. Evidence of silicon limited diatom growth in the plume of the Hudson River. Abstr. of Papers Submitted for the Second Winter Meeting, Am. Soc. Limnol. Oceanogr., Inc., January 3-February 4, 1980.

Officer, C.B. and J.H. Ryther. 1980. The possible importance of silicon in marine eutrophication. Mar. Ecol. Prog. Ser. 3: 83-91.

Palmer, C.M. and T.E. Maloney. 1954. A new counting slide for nannoplankton. Am. Soc. Limnol. Oceanogr. Spec. Publ. 21: 1-7.

Provasoli, L., J.J.A. McLaughlin, and M.R. Droop. 1957. The development of artificial media for marine algae. Arch. Mikrobiol. 25: S.392-428.

Quirk, Lawler, and Matusky, Engineers. 1970. Hudson River Water Quality and Waste Assimilative Capacity Study, Final Report to New York State Department of Environmental Conservation, Pearl River, NY.

Ryther, J.H. 1954. The ecology of plankton blooms in Moriches Bay and Great South Bay, Long Island, New York. Biol. Bull. 106: 198-209.

Samuels, W.B. 1979. Bioassay of New York harbor water using the marine diatom *Skeletonema costatum* (Greville) Cleve as test organism. Ph.D. Thesis, Fordham University. 157 pp.

United States Geological Survey. 1976. Water Resources Data for New York, Vol. 1. U.S. Geological Survey Water-Data Report NY-76-1.

United States Geological Survey. 1977. Water Resources Data for New York, Vol. 1. U.S. Geological Survey Water-Data Report NY-77-1.

United States Geological Survey. 1978. Water Resources Data for New York, Vol. 1. U.S. Geological Survey Water-Data Report NY-78-1.

Walsh, J.J., T.E. Whitledge, F.W. Barnevik, C.D. Wirick, S.O. Howe, W.E. Esaias, and J.T. Scott. 1978. Wind events and food chain dynamics within the New York Bight. Limnol. Oceanogr. 23: 659-683.

ENVIRONMENTAL EFFECTS OF SEWAGE SLUDGE AT THE PHILADELPHIA DUMPING SITE

Donald W. Lear[1]
Marria L. O'Malley[2]

Environmental Protection Agency
Region III Field Office
Annapolis, Maryland 21401

William C. Muir
George Pence

Environmental Protection Agency
Region III Headquarters
6th and Walnut Streets
Philadelphia, Pennsylvania 19106

Abstract. The Philadelphia sewage sludge disposal site is located about 70 km east of Ocean City, Maryland. Sludge has been dumped at this site since 1973, and the environmental effects of the practice have been under detailed observation since that time. The site showed no evidence of departures from typical, temperate mid-shelf conditions prior to the onset of sludge dumping. Since then, trends of environmental modification and degradation have appeared in areas contaminated by sludge-derived materials. The changes seem limited to the ocean bottom environment and include increased concentrations of metals in organisms, elevated metals and total organic carbon levels in sediments, changes in abundance of some species, apparent mortalities of molluscan shellfishes, unique occurrences of sewage bacteria, and appearance of pathological conditions in endemic crustaceans.

INTRODUCTION

The Philadelphia sewage sludge dumping site (Figure 1) is a 172-km^2 area located approximately 70 km east of Ocean City, Maryland, at 38°23′N, 74°15′W. The site lies over continental shelf waters 40-60 m deep. The bottom in the vicinity of the dumping area is comprised primarily of sands and exhibits a gently rolling topography of low, parallel ridges separated by shallow depressions trending in a northeast-southwest direction. Seasonal water mass characteristics and stratification patterns observed at the site appear to be typical of the Middle Atlantic Bight (Palmer and Lear, 1973; Lear *et al.*, 1974; Lear and Pesch, 1975) and are in agreement with the regional setting summarized by Beardsley *et al.* (1976).

Regulation of dumping at the Philadelphia site is administered by the U. S. Environmental Protection Agency (EPA) Region III, Philadelphia, Pennsylvania. The history of waste disposal at this location was summarized by Muir *et al.* (1978). This site has been in use since May 1973 when EPA Region III

[1,2]Present address: EPA-NOAA Oceanographic Unit, Route 2, Box 385, Stevensville, Maryland 21666.

Presented to Domestic Wastes Workshop.

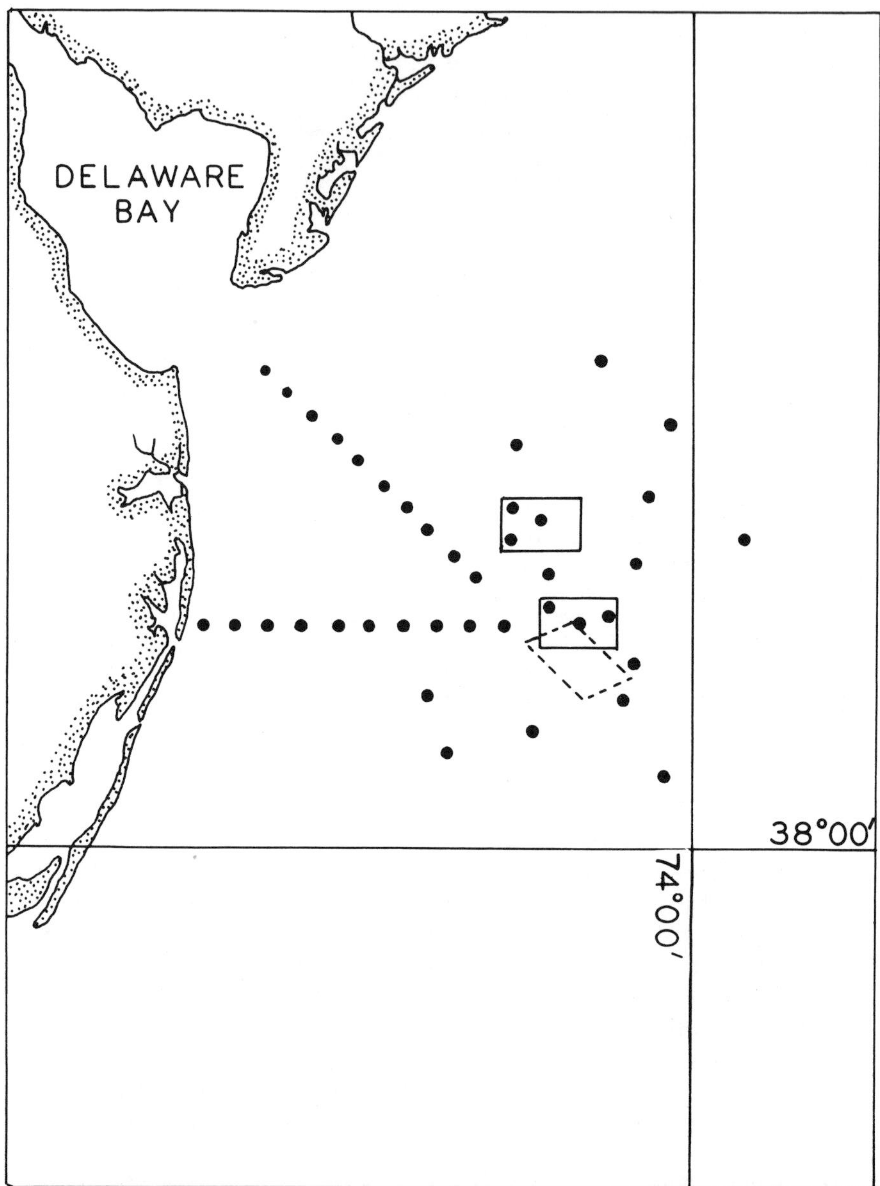

Figure 1. Site location map showing the industrial acid waste dumpsite (upper rectangle) and the sewage sludge dumpsite (lower rectangle) off the Maryland-Delaware coast. Isolated dots are selected stations in the "far-field" grid; the "near-field" intensive grid is represented by the dashed line parallelogram in and south of the sludge site.

designated the location as the disposal site for sludge from the cities of Philadelphia, Pennsylvania and Camden, New Jersey. Prior to that time sludge from the city of Philadelphia was discharged at a shallower site near the mouth

Table 1. Total quantity (wet wt) of sewage sludge dumped at the Philadelphia sewage site since 1973. Figures given for 1979 and 1980 are estimates.

Year	Sludge discharged ($kg \times 10^6$)
1973	222
1974	700
1975	544
1976	544
1977	481
1978	381
1979	240
1980	136
1981	0

of Delaware Bay. Between 1973 and 1977 more than 2.8×10^9 kg of sludge from the two cities was released. Dumping of sludge from Camden was phased out at the 70-km site during 1978. Sludge from Philadelphia was dumped until November 25, 1980, after which the practice was halted. The amounts of sludge released at the site each year since 1973, together with projections of amounts to be dumped until 1981, are summarized in Table 1.

A second ocean disposal site is located 9 km northwest of the sewage sludge release site (Figure 1). This dumpsite is designated for industrial acid iron waste disposal and was active from 1968 to 1978. Wastes dumped at this site were from a titanium ore extraction process and contained significant amounts of metals, especially vanadium and titanium (Figure 2).

During spring 1973, EPA Region III initiated an intensive field research and monitoring program to measure the short-term and long-term environmental effects of sludge dumping at the Philadelphia site. The program continues, and in October 1979 it became the responsibility of the National Oceanic and Atmospheric Administration (NOAA) National Ocean Survey, Ocean Dumping and Monitoring Division. The purpose of this paper is to summarize the results and findings of the program to date.

METHODS

The primary objectives of the study were to determine the fate of disposed materials in the environment and the effects of these materials on benthic populations in the area. Monitoring at the sludge site began in May 1973, prior to dumping of sewage sludge. A large area (6,860 km^2) sampling grid (Figure 1) was established to determine baseline conditions at this location. Sampling of a "near-field" grid (1.8 km x 5.6 km, Figure 1) was established adjacent and south of the sewage sludge release area to investigate apparent accumulations of sludge components and departures from ambient conditions. In 1978 additional stations were added based on the mesoscale topography, to sample more

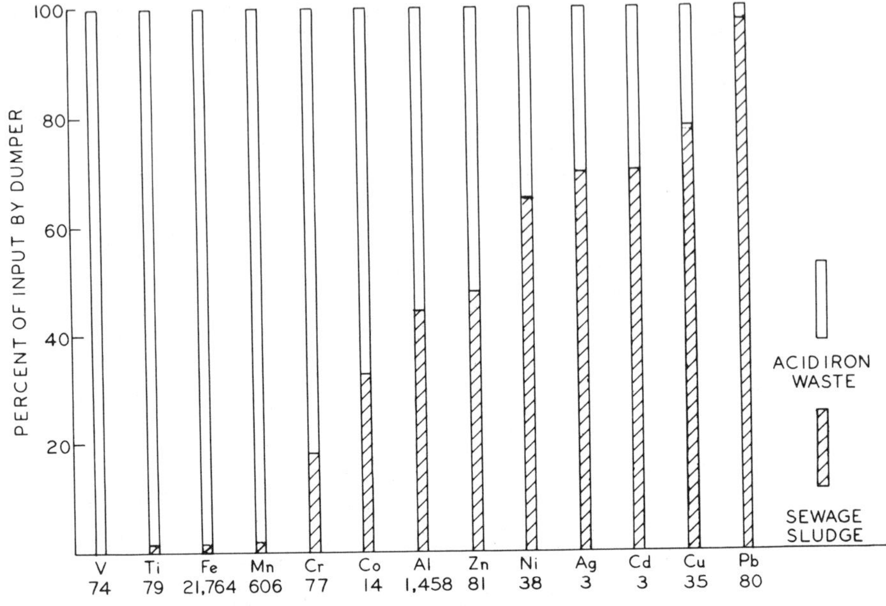

Figure 2. Graphical representation by percent of metals input from acid-iron waste and sewage sludge disposal through 1974. Loadings of each metal, in metric tons, indicated along the abscissa.

definitively contrasting ridge and swale habitats. The study area was confined to the mid-shelf environment between the 36-m and 63-m isobaths to minimize zonation effects of benthic species.

The assessment has included physical oceanographic characterizations, barge plume studies, hydraulic models, bathymetry, sedimentological studies, faunal surveys, metals analyses of sediments and organisms, population dynamics assessments, and pathological comparisons of animals near and remote from the sludge release site. Field sampling and laboratory methodology are described in detail by Palmer and Lear (1973), Lear (1974), Lear *et al.* (1974, 1977, in press), and Lear and Pesch (1975).

Technical participants in the program included personnel from the Annapolis EPA laboratory and other EPA units, scientists from universities, consulting firms, and federal agencies including the U. S. Coast Guard, NOAA, and the U. S. Public Health Service, Food and Drug Administration.

RESULTS

The baseline survey of the area in 1973 revealed no aberrant conditions. Water quality determinations for nutrients, dissolved oxygen, and pH were well within limits expected for a temperate, non-stressed, mid-shelf environment. Zooplankton and phytoplankton populations were typical of temperate coastal waters. Coliform and fecal coliform bacteria were not present in significant quantities in the water or sediment sampled, nor was there evidence of fin rot or external diseases in the vertebrates collected. The benthic community was characteristic of a clean, firm, sandy mid-shelf habitat. The benthic fauna was diverse and abundant and included at least 120 invertebrate species from eight phyla. Results of the baseline survey are reported in detail in Palmer and Lear (1973). The findings of the site characterization effort collectively indicated the disposal area was unpolluted and largely free from terrestrial influences at the advent of dumping in 1973 (Palmer and Lear, 1973).

In 1975 several evidences of pollution were found in the benthic environment at and adjacent to the sludge dumping site. Components of the sludge discharged at the Philadelphia site appeared to collect on the ocean bottom, for the most part in regions to the southeast, south, and southwest of the dumpsite, rather than primarily within the boundaries of the site (Lear *et al.*, 1977). Bottom drifter studies indicated prevailing net water movement in the area was in those directions. Tidal and wind-induced currents were apparently of sufficient strength to sweep most released sludge outside the boundaries of the release site. Rather than accumulating on the bottom in a uniform layer, the sludge-derived materials collected in multiple, separate deposits in shallow topographic depressions running through the area (Lear *et al.*, in press). The exact mechanisms of sludge dispersal and the overall spatial distribution of sludge-derived materials in the region remain only partially determined at this time.

The dumped sewage sludge contains large quantities of metals, several of which are known to have detrimental biological effects in elevated concentrations. These include copper, chromium, lead, silver, cobalt, nickel, cadmium, and zinc (Figure 2). It has been demonstrated that more than 85% of the copper, silver, lead, nickel, and cadmium entering the dumping area is introduced in sludge. These metals can be used as tags for locating sludge accumulations (Pesch *et al.*, 1977).

An example of the accumulations of materials is provided by the near-field distribution of lead in 1975 (Figure 3). Each data point represents the means of three replicate samples. The data were analyzed by the Duncan multiple range test. Hatched areas contain significantly higher concentrations of lead at the 0.05 level. These areas correspond well with mesoscale depressions. Other metals examined showed similar distributions (Lear *et al.*, 1977).

Less conservative, but nonetheless significant when dealing with sewage sludge, was total organic carbon (Figure 4). Concentrations of total organic carbon (TOC) were significantly higher at the 0.05 level in the hatched areas. TOC distributions confirmed subjective organoleptic observations of odor and sediment texture (Lear *et al.*, 1977) characteristic of sewage sludge. These observations also coincided well with metals distributions and topography (Lear *et al.*, 1977).

The response of the benthic infauna is illustrated by the distributions of the archiannelid, *Protodrilus*, and the polychaetes, *Lumbrinereis impatiens* and *L.*

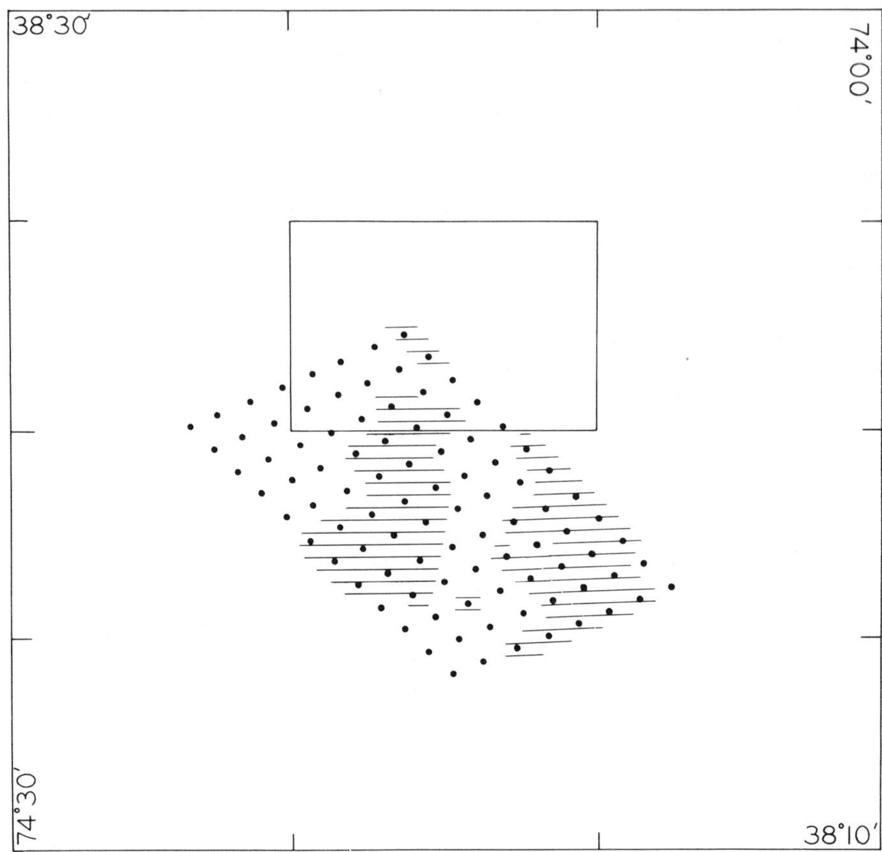

Figure 3. Distribution of lead in "near-field" grid, December 1975. Hatched areas signify statistically higher concentrations of lead, 6.6-9.7 ppm (dry wt) at the 0.05 level. Concentrations of lead ranged from 1.4 ppm to 6.1 ppm (dry wt) in the non-hatched areas.

acuta. Protodrilus is a ubiquitous taxon in this area and appeared to be indifferent to topography or accumulated wastes. It occurred in relatively uniform abundance at all stations. The two lumbrinereids, however, exhibited contrasting distributions. *L. acuta* occurred in the cleaner habitats (Figure 5), while *L. impatiens* showed a great degree of pollution tolerance (Figure 6). Further details of metals distributions and benthic infaunal distributions, as well as data from two later cruises, are presented by Lear *et al.* (1977).

Another consequence of municipal sewage sludge disposal has been the bioaccumulation of metals in harvestable shellfishes. Several studies have demonstrated statistically the incidence of high levels of metals in sea scallops, *Placopecten magellanicus,* from within and adjacent to the dumpsite. Silver, copper, and nickel, metals found in sewage sludge, were measured in scallop tissues. Dry weight concentrations of 2-10 ppm silver, 7.5-12.5 ppm copper, and 6-15 ppm nickel were found in scallops from areas impacted by dumping

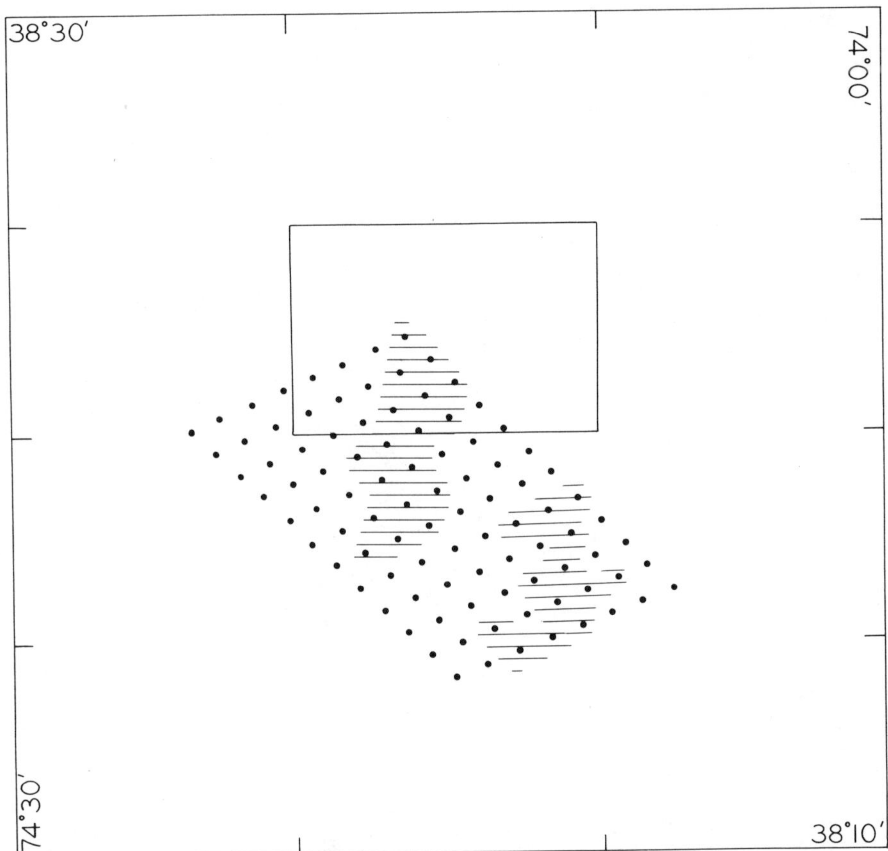

Figure 4. Distribution of total organic carbon (TOC) in "near-field" grid, December 1975. Hatched areas signify statistically higher concentrations of TOC, 1,136-2,396 ppm (dry wt) at the 0.05 level. Concentrations of TOC in the non-hatched areas ranged from 380 ppm to 1,080 ppm (dry wt).

activity, compared with concentrations of 0.4-1.0 ppm silver, 4-7 ppm copper, and 1-4 ppm nickel in specimens from unimpacted areas (Pesch, 1976; Pesch *et al.*, 1977; Reynolds, 1977). These authors also examined several other metals.

The mahogany clam, *Arctica islandica,* occupied the entire study area and was particularly common in waters between 36 m and 54 m. Over the duration of the site survey, elevated mortalities of this animal were observed in several areas contaminated by dumping activities. The incidence of mortalities did not appear to be seasonal (Lear *et al.*, in press). Mortality was measured in terms of the number of "clappers" (pairs of valves with intact hinge ligaments) sampled at collection stations over the duration of the field surveys. Evidence of greater incidence of mortality appeared at two collecting stations inside the boundaries of the acid waste and sewage sludge dumpsites and at two stations outside the dumping areas, one located 37 km east of the release sites. In all cases, the mortality appeared linked to dumping by high local sediment concentrations of

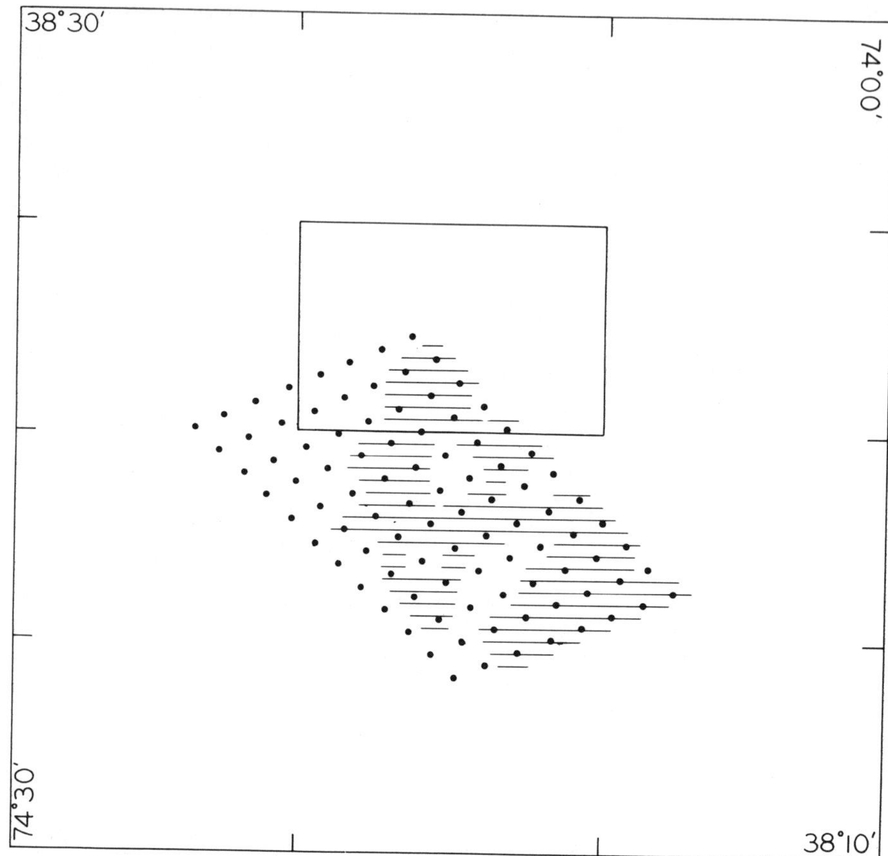

Figure 5. Distribution of the polychaete, *Lumbrinereis impatiens,* in "near-field" grid during December 1975. The mean number of *L. impatiens* in hatched areas was greater than one for three replicates per station; non-hatched areas contained no *L. impatiens.*

metals. At the station 37 km east of the release area, the contaminating metals included lead and chromium (Lear *et al.*, in press). At this time, the relative importance of sewage sludge disposal vs. acid waste disposal in bringing about these mortalities is unclear.

Dermal disorders similar to those known from fishes collected at sludge discharge sites in the New York and California areas (Murchelano and Ziskowski, 1976, this volume; Mearns and Sherwood, 1977; Sherwood, this volume) were not detected in fishes from the disposal area. However, pathological disorders have appeared in bottom-dwelling crabs (*Cancer irroratus*). The disorders included necrotic lesions of the exoskeleton and melanistic accumulations in the gills (Sawyer, this volume).

The incidence of sewage indicator bacteria (coliforms, fecal coliforms) was examined over the study area, with additional transects to the mouth of Delaware Bay and to Ocean City, Maryland (Figure 7). These data showed the

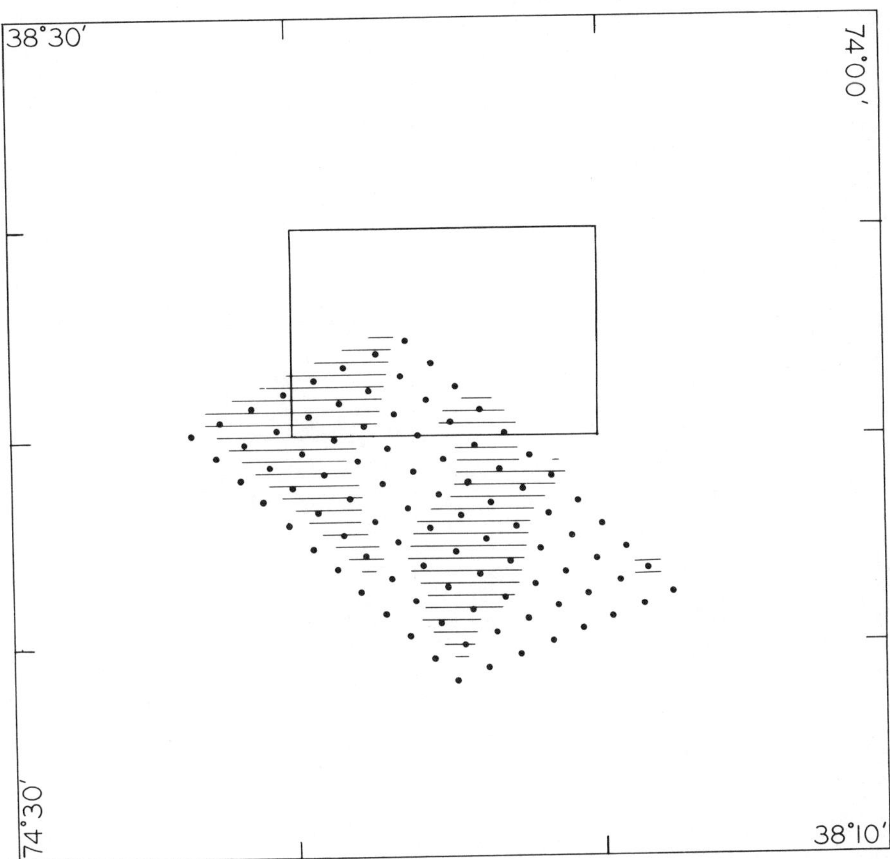

Figure 6. Distribution of the polychaete, *Lumbrinereis acuta,* in "near-field" grid during December 1975. The mean number of *L. acuta* in hatched areas was greater than one for three replicates per station; non-hatched areas contained no *L. acuta.*

highest concentrations of indicator bacteria in and near the sludge release site. The contamination patterns were consistent with the known hydrography and hydraulics of the system. The discrete distribution of these organisms in and around the release site gives additional credence to the observed increases in metals and TOC associated with the sludge site, compared to ambient mid-shelf levels (Lear *et al.*, in press).

DISCUSSION

Documented responses of marine ecosystems to waste disposal have shown significant alterations and degradation (Pearce *et al.*, 1976; Word, 1979; Boesch, this volume; Mearns and Word, this volume; Steimle *et al.*, this volume). At the Philadelphia dumpsite, a mid-shelf site with relatively small loadings and an active hydraulic regime, ecosystem changes might be expected to be more subtle and variable with time than at more heavily loaded sites such as in the New York Bight.

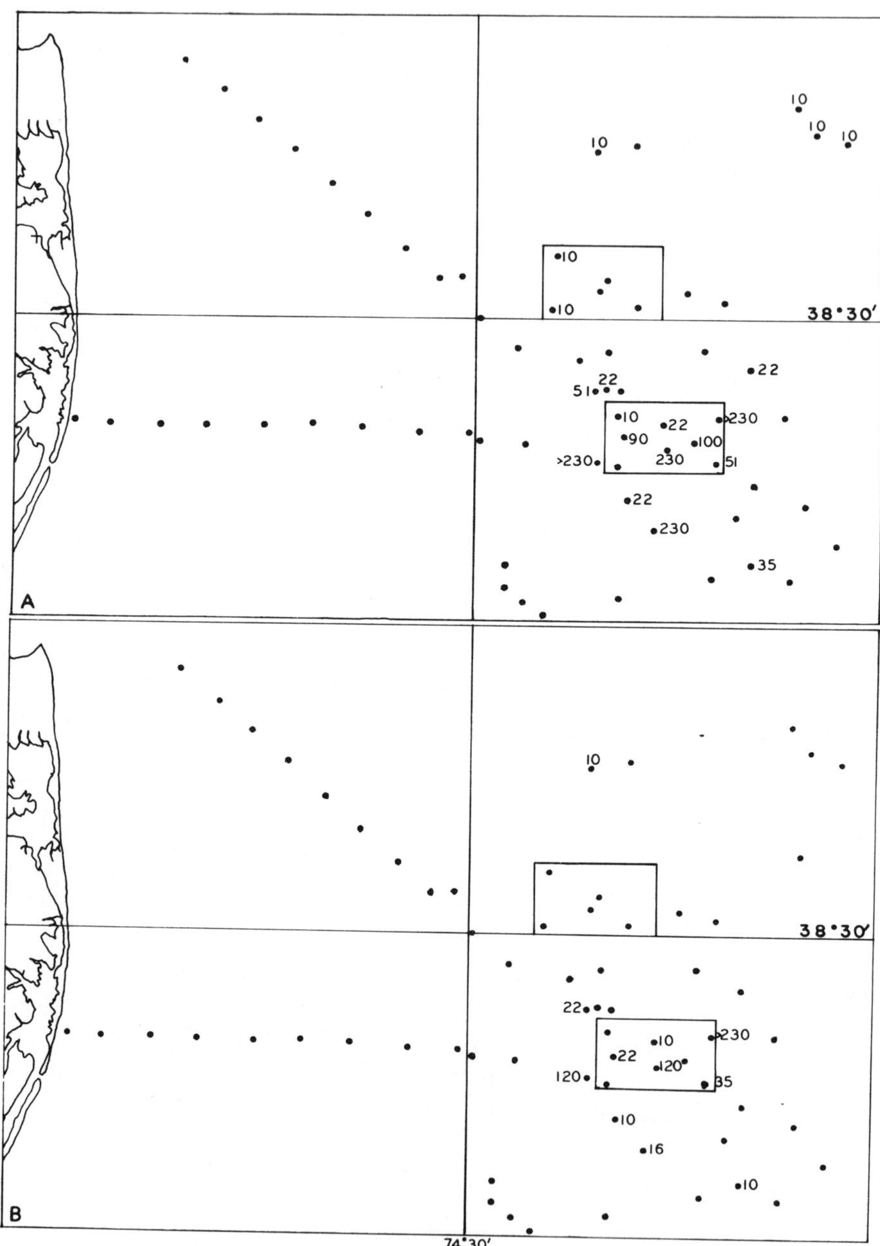

Figure 7. Distribution of coliform bacteria (A) and fecal coliform bacteria (B), April 1978, in and around dumpsites (MPN/gm).

The results of this study indicate a modification of ocean bottom environmental quality at and around the sewage sludge disposal site. The trend apparently is linked to sludge dumping but in part also may result from the release of industrial acid wastes prior to 1978 at the dumping area northwest of the sludge site (Muir *et al.*, 1978). The localized areas of high concentrations of metals, TOC, and sewage bacteria that have been identified in and near the sludge release site are good indicators of the fate of sewage sludge materials in this ecosystem. Although it is difficult to predict the exact response of the benthic community occupying the dumping area to such environmental stress, it is likely that the composition of the community will change as pollution-tolerant taxa are encouraged, pollution-sensitive forms are diminished, and opportunists are recruited. One of the benthic species examined, the mahogany clam, appeared to have been subjected to physiological stress as a result of dumping, as evidenced by the large number of "clappers" in the vicinity of the sludge site. The apparent mortalities were observed to coincide with areas of high metals accumulations (Lear *et al.*, in press).

These studies, which are continuing, indicate a transition in the monitored area from a "typical" or "normal" mid-shelf habitat to one with localized areas with measurable changes. Studies in stressed ecosystems such as the New York Bight and the California outfall areas provide useful guides and indices against which to monitor such transitions.

During the course of the study reported in this paper, the data and conclusions were examined critically for use in adversary ocean dumping permit hearings conducted by U. S. EPA, Region III. Comments on the hearing data that differed from the conclusions presented in this paper were published by Guarino *et al.* (1975, 1977, 1979). These comments, however, assumed deposition of ocean dumped materials within the release site or southwest of the site, in accordance with a constant southwesterly drift. These assumptions did not adequately describe actual transport mechanisms or the role of mesoscale topography in determining the fate of materials deposited in the benthic environment. In addition, data collected since February 1977 were not considered. Nevertheless, the comments were useful for modifying sampling rationales for later investigations.

Sludge dumping at the Philadelphia site ceased in 1980, and for this reason a unique opportunity exists to examine the recovery as well as the degradation of the benthic environment. It is our intent to continue field work at the site for at least three years to observe the recovery phase. The work initiated during this period will focus primarily on the geochemical, geological, and biological changes that appear in the ocean bottom environment.

ACKNOWLEDGEMENTS

The U.S. Environmental Protection Agency, Region III, Philadelphia, Pennsylvania, initiated the study and was primarily responsible for the aspects reported here. However, without the cooperation of many individuals and institutions this work would not have been possible. Personnel at the EPA Annapolis Field Office made the metals and TOC determinations. The U. S. Public Health Service, Food and Drug Administration laboratory at Davisville,

Rhode Island, under the direction of Captain James Verber, performed the shipboard bacteriological analyses. The tedious task of identifying and counting the benthic infauna was done by Marine Research, Inc., Falmouth, Massachusetts. Dr. Thomas K. Sawyer, NOAA National Marine Fisheries Service, Oxford, Maryland, was responsible for the crustacean histopathological studies. Determinations of metals in shellfishes were done by Dr. G. Pesch and Mr. Bruce Reynolds at the EPA, Narragansett, Rhode Island laboratory. The U. S. Coast Guard provided ship time and navigational expertise aboard the U. S. Coast Guard Cutter ALERT, Cape May, New Jersey. This program is continuing under the aegis of the NOAA National Ocean Survey, Ocean Dumping and Monitoring Division, Rockville, Maryland.

REFERENCES

Beardsley, R.C., W.C. Boicourt, and D.V. Hansen. 1976. Physical oceanography of the Middle Atlantic Bight. *In:* Middle Atlantic Continental Shelf and the New York Bight, M. Grant Gross (ed.), Am. Soc. Limnol. Oceanogr. Spec. Symp. 2:20-34.

Boesch, D.F. This volume. Ecosystem consequences of alterations of benthic community structure and function in the New York Bight region. pp. 543-568.

Guarino, C.F., M.D. Nelson, and S.S. Almeida. 1979. Ocean dispersal as an ultimate disposal method. J. Water Pollut. Control Fed. 51:773-782.

Guarino, C.F., M.D. Nelson, and S. Townsend. 1977. Philadelphia sludge disposal in coastal waters. J. Water Pollut. Control Fed. 49:737-744.

Guarino, C.F., M.D. Nelson, S.A. Townsend, T.E. Wilson, and E.F. Ballotti. 1975. Land and sea solids management alternatives in Philadelphia. J. Water Pollut. Control Fed. 47:2551-2564.

Lear, D.W. 1974. Supplemental Report: Environmental Survey of Two Interim Dumpsites, Middle Atlantic Bight. EPA Report, EPA Region III, Philadelphia, PA. 53 pp.

Lear, D.W., M.L. O'Malley, and S.K. Smith. 1977. Effects of Ocean Dumping Activity, Mid-Atlantic Bight—1976; Interim Report. EPA Document 4031-77-029. 168 pp.

Lear, D.W., M.L. O'Malley, and S.K. Smith. In press. Effects of ocean dumping on a temperate mid-shelf environment. *In:* Ocean Dumping of Industrial Wastes, B.H. Ketchum, D.R. Kester, and P.K. Park (eds.), Plenum Press, New York.

Lear, D.W. and G.G. Pesch. 1975. Effects of Ocean Disposal Activities on Mid-Continental Shelf Environment off Delaware and Maryland. EPA Document 903/9-75-015. 223 pp.

Lear, D.W., S.K. Smith, and M.L. O'Malley. 1974. Environmental Survey of Two Interim Dumpsites, Middle Atlantic Bight. EPA Document 903/9-74-010a. 141 pp.

Mearns, A.J. and M.J. Sherwood. 1977. Changes in the prevalence of fin erosion off Los Angeles and Orange counties. *In:* Coastal Water Research Project Annual Report for the Year 1977, Southern California Coastal Water Research Project, El Segundo, CA. pp. 143-145.

Mearns, A.J. and J.Q. Word. This volume. Forecasting effects of sewage solids on marine benthic communities. pp. 495-512.

Muir, W.C., G.M. Horowitz, and G.D. Pence. 1978. Ocean disposal in the Mid-Atlantic Bight. Proc. Mar. Tech. Symp. 1978:1-6.

Murchelano, R.A. and J. Ziskowski. 1976. Fin rot disease studies in the New York Bight. *In:* Middle Atlantic Continental Shelf and New York Bight, M. Grant Gross (ed.), Am. Soc. Limnol. Oceanogr. Spec. Symp. 2:329-336.

Murchelano, R.A. and J. Ziskowski. This volume. Fin rot disease in the New York Bight (1973-1977). pp. 347-358.

Palmer, H.D. and D.W. Lear. 1973. Environmental Survey of an Interim Ocean Dumpsite, Middle Atlantic Bight. EPA Document 903/9-73-001-A. 134 pp.

Pearce, J.B., J.V. Caracciolo, M.B. Halsey, and L.H. Rogers. 1976. Temporal and spatial distributions of benthic macroinvertebrates in the New York Bight. *In:* Middle Atlantic Continental Shelf and the New York Bight, M. Grant Gross (ed.), Am. Soc. Limnol. Oceanogr. Spec. Symp. 2:394-403.

Pesch, G.G. 1976. Testimony at EPA Region III Ocean Dumping Permit Hearing for the City of Philadelphia. Georgetown, DE. 28 April, 1976.

Pesch, G.G., B. Reynolds, and P. Rogerson. 1977. Trace metals in scallops from within and around two ocean disposal sites. Mar. Pollut. Bull. 8:224-228.

Reynolds, B.H. 1979. Trace Metals Monitoring at Two Ocean Disposal Sites. EPA Research Report 600/3-79-037. 63 pp.

Sawyer, T.K. This volume. Distribution and seasonal incidence of "black gill" in the rock crab, *Cancer irroratus.* pp. 199-211.

Sherwood, M.J. This volume. Fin erosion, liver condition, and trace contaminant exposure in fishes from three coastal regions. pp. 359-377.

Steimle, F.J. Caracciolo, and J.B. Pearce. This volume. Impacts of dumping on New York Bight apex benthos. pp. 213-223.

Word, J.Q. 1979. The infaunal trophic index. *In:* Annual Report 1978 of the Southern California Water Research Project. Southern California Coastal Water Research Project, El Segundo, CA. pp. 19-39.

FORECASTING EFFECTS OF SEWAGE SOLIDS ON MARINE BENTHIC COMMUNITIES

Alan J. Mearns[1]
Jack Q. Word[2]

Southern California Coastal Water Research Project
646 West Pacific Coast Highway
Long Beach, California 90806

Abstract. Solids from marine municipal discharges settle to the sea bottom where they cause major but potentially reversible changes in the biomass and trophic structure of macrobenthic communities. Analysis of data from macrobenthic surveys around five southern California coastal discharge sites indicates that there are significant exponential relationships between the mass emission rate of solids and the size of altered macrobenthic communities. Size was expressed both in terms of biomass (metric tons, mt, wet wt) and area (km^2) occupied by communities dominated by surface and subsurface deposit-feeding organisms.

The relationships may be useful in forecasting changes in the size and structure of benthic communities in response to changes in solids emissions and in providing guidance for the design and execution of responsive monitoring surveys.

INTRODUCTION

Municipal wastewaters contain a mixture of gross and trace constituents that, in the absence of toxic concentrations of pollutants, can increase the abundance of a number of benthic invertebrate species and decrease the abundance of others (Smith, 1974; Word *et al.*, 1977; Young and Young, this volume). At present, it is possible to forecast which benthic species will increase or decrease in response to a particular kind of discharge or input (Word *et al.*, 1977; Pearson and Rosenberg, 1978; Rhoads *et al.*, 1978). It also has been demonstrated that recovery to normal benthic community structure can take place when discharge is terminated (Smith, 1974; Pearson and Rosenberg, 1978).

While helpful, this kind of information is not sufficient for dischargers and regulatory agencies to assess the total ecological costs and benefits of alternative treatment strategies. How much marine life is affected? Will a proposed change in emissions change the total environmental impact by 50%, 75%, or 90%? Needed are predictive relationships between the amount of sewage discharged and the magnitude of biological changes resulting from the discharge. Relationships should be expressed in terms that are easy to understand and that have an impact on decision makers. We believe such relationships exist, and we demonstrate here several such relationships derived from assessments of conditions at a suite of southern California deepwater municipal sewage discharges.

[1] Present address: Pacific Office, Office of Marine Pollution Assessment, National Oceanic and Atmospheric Administration, 7600 Sand Point Way, N.E., Seattle, Washington 98115.
[2] Present address: College of Fisheries, University of Washington, Seattle, Washington 98195.

BACKGROUND

The five discharges and discharge areas cited in this report (Figure 1) are especially suited for examining quantitative relationships between pollutant inputs and biological responses. Each of the five has one or more outfalls that have been discharging relatively constant amounts of sewage for one or more decades (Schafer, 1977, 1979). Each discharges subtidally at similar depths (20-60 m) and into similar hydrographic and faunal regimes. All discharge sites also are subjected to predominantly longshore currents that are variable but have similar energy spectra (Hendricks, 1977). Although the contaminant inputs of individual discharges may differ from one another by as much as two orders of magnitude, depending on the contaminant, each discharge emits relatively constant amounts of bioactive materials, especially suspended solids, which contain much of the biochemical oxygen demand (BOD), chemical oxygen demand (COD), nitrogen, and phosphorus (Schafer, 1977, 1979).

In all effluents most of the trace and gross constituents (metals, chlorinated hydrocarbons) are associated with or attached to particulates. These particulates, or suspended solids, are discharged at rates ranging from 1,000 to over 100,000 metric tons per year (mt/yr). Both field and laboratory measurements indicate that between 10% and 20% of the particulates settle to the sea floor at most sites (Herring and Abati, 1979). The particulates create elliptical zones of contamination centered around the outfalls, which cover from several to nearly 200 square kilometers (Young *et al.*, 1976; Eganhouse *et al.*, 1978; Young *et al.*, 1978; Bascom, 1979).

During the past decade there have been frequent benthic invertebrate surveys around each discharge site and at distant control areas (Smith, 1974; Mearns and Greene, 1976; Smith and Greene, 1976; Word *et al.*, 1977; Bascom, 1979; Word, 1979b; Mearns, 1981). Increased biomass, decreased species diversity, and a shift from suspension- to deposit-feeding benthic communities are among the responses common to these wastewater discharges (Mearns and Greene, 1976; Word *et al.*, 1977). Word (1979b) described this shift in terms of an Infaunal Trophic Index (ITI), which is described below. The changes are similar to those observed in areas where increased inputs of organic carbon and nitrogen into surface sediments have led to the increases in the production of more tolerant deposit-feeding species at the expense of more sensitive detrital-feeding species (Smith *et al.*, 1973; Pearson and Rosenberg, 1978; Young and Young, this volume).

Initial comparisons of benthic communities at the California sites revealed that the increase in average biomass (expressed as g/m^2), decrease in average diversity (H′), and area of the sea floor (expressed in km^2) over which significant changes occurred were approximately proportional to sewage flow and mass emission rate of suspended solids but were poorly correlated with concentrations of suspended solids or other contaminants (Mearns and Greene, 1976; Mearns and Young, 1978; Bascom *et al.*, 1979). On the basis of these observations, mathematical formulations were developed linking solids mass emission rates

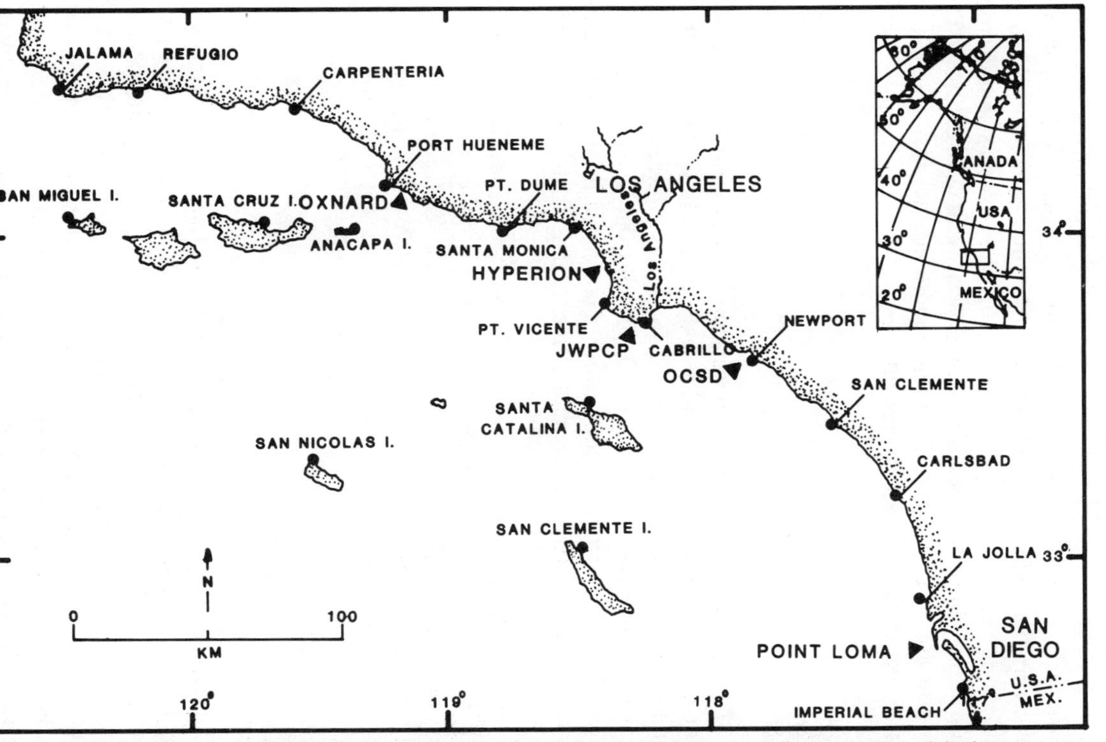

Figure 1. The Southern California Bight and coastal zone showing locations of five major municipal wastewater treatment plants (▲). Dischargers, discharge depths, and flows: Oxnard–20 m, 42.4 x 10^6 liters/day, and 11-km (7-mi) sludge outfall–90 m, 17.4 x 10^6 liters/day; Joint Water Pollution Control Plant (JWPCP). County of Los Angeles–60 m, 1,268 x 10^6 liters/day; Orange County Sanitation Districts (OCSD)–60 m, 681 x 10^6 liters/day; Point Loma Treatment Plant, City of San Diego–60 m, 439 x 10^6 liters/day.

(independent variable) and the size of the impacted area or mass of organisms affected (dependent variable). This was done because it was felt that more attention would be drawn to a response involving tens or hundreds of square kilometers and impacting hundreds or thousands of tons of organisms, than to responses involving the traditionally-expressed biomass (g/m^2) or information theory species diversity (H or H′) averaged over a set of stations. Thus, in the following analysis, commonly used sample variables such as biomass are presented in terms that reflect the magnitude of benthic community changes (*i.e.*, metric tons of excess standing crop, area occupied by excess standing crop, or area occupied by abnormal abundances of deposit-feeding organisms). The relationships also are used to forecast benthic conditions that may result from alterations in sewage inputs or waste treatment.

METHODS

Inputs

Each of the five major southern California dischargers measures flow and gross effluent constituents from daily and weekly composites. Schafer (1977, 1979) computed annual mass emission rates from these measurements. For suspended solids, these values in 1977 were: 1,500 mt from Oxnard, 20,500 mt for Point Loma, 33,300 mt from Orange County, a combined total of 78,800 mt from Hyperion Treatment Plant sludge and effluent, and 102,000 mt from the Joint Water Pollution Control Plant (JWPCP).

For this report, data on suspended solids mass emission rates for 1977 were compared to ITI's computed using 1977 and 1978 benthic survey data. In addition, averaged 1975-1977 suspended solids mass emission rates were compared to average benthic standing crops from 1975-1977 surveys. This procedure was required because the total suite of data for computing both biomass and other indices of community structure during 1975-1977 was not taken in the same surveys.

Benthic Surveys

Since 1973 each discharge site was sampled by at least four and by as many as 14 synoptic benthic macrofaunal grab surveys, with two to four replicate samples taken at from 10 to 44 stations, depending on the survey site (Mearns and Greene, 1976). Various devices (from 0.04-m^2 Shipek grabs to 0.16-m^2 Van Veen grabs) and several screen sizes were used in the surveys. An overall summary of the samples inspected for this report is presented in Table 1.

Differences resulting from heterogeneous collection methods were studied (Word, 1976; Word *et al.*, 1976a), and abundance and biomass data were adjusted to comparable units (densities or biomass per square meter). More recent sample collections were made with uniform procedures (0.1-m^2 Van Veen, 1.0-mm screen). Techniques for identification, enumeration, and weighing also were standardized among all present monitoring and investigative groups

Table 1. Summary of monitoring and research benthic surveys that provided data used in this study. Agency abbreviations: AHF–Allan Hancock Foundation, University of Southern California; JWPCP–Joint Water Pollution Control Plant, County of Los Angeles; LA–City of Los Angeles; OCSD–Orange County Sanitation District; PLTP–Point Loma Treatment Plant, City of San Diego; SCCWRP–Southern California Coastal Water Research Project. Frequencies: A–annual; Q–quarterly; S–semiannual.

Site of survey	Agency	Survey period	Approximate grid size (km^2)	Number of stations	Number of replicates	Frequency	Screen size (mm)	Sampling device	Reference
Monitoring surveys									
Oxnard	City	1974	27	19	3	Q	0.5	Shipek	M.B.C., Inc., 1975
		1976-1978	1	7	3	Q	0.5	Shipek	Unpublished data
Santa Monica Bay	LA	1971-1977	197	22	4	S	1.0	Shipek	Unpublished data
		1978-1979	197	22	1	S	1.0	Van Veen	Unpublished data
Palos Verdes	JWPCP	1973-1979	62	40	4	S	1.0	Shipek	Unpublished data
		1979	62	18	4	S	1.0	Van Veen	Unpublished data
Orange County	OCSD	1974-1976	20	18	3	Q	1.0	Forest-Petersen	Unpublished data
		1976-1979	20	17	2	A	1.0	Van Veen	Unpublished data
Point Loma	PLTP	1973-1975	30	18	3	S	1.0	Petersen	Unpublished data
		1976-1979	30	17	4	Q	1.0	Van Veen	Unpublished data
Research surveys									
60-m control	SCCWRP	Apr.-Aug. 1977	–	70	1	Once	1.0	Van Veen	Word and Mearns, 1979
Synoptic shelf	SCCWRP	Feb.-Dec. 1978	1,800	310	1	Once	1.0	Van Veen	Bascom, 1979
Orange County	SCCWRP	Mar. and Aug. 1975	35	35	≥2	Twice	1.0 and 5.0	Van Veen	Greene, 1976
Grab comparison	SCCWRP	Dec. 1974 and Mar. 1975	–	4	60-80	Once	1.0, 0.7, and 0.5	8 types	Word, 1976
Entire coastal shelf	AHF	1956-1958	3,900	176	1	Once	0.7	Orange peel	Allen Hancock Foundation, 1965

(Word *et al.*, 1976b; Word, 1977). This provided an internally consistent taxonomic data base.

Charts showing contoured data from areal surveys are presented by Bascom (1979) for Santa Monica Bay, Palos Verdes, Orange County, and Point Loma; by M.B.C., Inc. (1975) for Oxnard; and by Greene (1976) for Orange County.

Estimating the Magnitude or Size of Affected Populations

Three variables were chosen from the surveys to estimate the overall size of affected benthic communities. First, the overall excess standing crop (mt) of infauna around each of the five major discharge sites shown in Figure 1 was estimated. Values of benthic biomass (g/m^2) for discrete single surveys were plotted and contoured to reveal the size and shape of regions of enhanced biomass (*i.e.*, above control values). The total area of each contour was measured by planimetry, and the products of average biomass (g/m^2) and area (m^2) were summed. The difference between the sum for each outfall area and that for expected total biomass, based on average values for adjacent control transects, was taken as an estimate of the excess standing crop (mt, wet wt) for each site and survey.

Contours of benthic biomass also were used to estimate the total area (km^2) occupied by the excess standing crop. This was used as a second, but not necessarily independent, measure of the size (km^2) of the affected community.

The third variable chosen was the area (km^2) occupied by infaunal communities dominated by deposit-feeding organisms, as determined by the ITI. The ITI is a numerical description of the distribution of four groups of selected macrofaunal invertebrate species in a grab sample. The ITI is calculated as follows:

$$\mathrm{ITI} = 100 - \left[33.\bar{3}\left(\frac{0N_1 + 1N_2 + 2N_3 + 3N_4}{N_1 + N_2 + N_3 + N_4}\right)\right],$$

where N_i = the number of individuals in group "i". The groups contain species that largely are distinguished by feeding categories, as shown in Table 2, and described in more detail by Word (1979a,b). Values of the ITI range between 0 (dominated by subsurface deposit-feeding invertebrates) and 100 (dominated by suspension-feeding invertebrates). Recent studies (Bascom *et al.*, 1979; Word, 1979b) confirm that the ITI is inversely correlated with the concentration of organic matter in surface sediments (which, in waste discharge areas, is related to the mass emission rate of wastewater organic matter and suspended solids). Species presently used in the ITI (Table 2) account for at least 60% of the individuals collected in southern California macrobenthos samples taken on the open coastal shelf in fine sediments at depths between 20 m and 200 m. Additional attributes and the logic behind development and application of the ITI are described elsewhere (Word, 1979a,b).

Table 2. Groups of species considered in calculating the Infaunal Trophic Index. The wastewater discharge factor affecting the abundance of each group appears to be the deposition or accumulation of organic particulate material (Word, 1979a, b). UI = unidentified form.

Group and description	Species	
I. Suspension feeders primarily; dominant in control or background areas; decrease in abundance with increasing proximity to wastewater discharges; 19 species, 7 taxa.	*Amphiodia (Amphispina) urtica* *Amphiodia (Amphispina) digitata* *Amphiodia psara* *Amphiodia occidentalis* *Amphiodia* spp. *Ampelisca pacifica* *Ampelisca hancocki* *Ampelisca brevisimulata* *Ampelisca macrocephala* *Ampelisca cristata*	*Paraphoxus bicuspidatus* *Metaphoxus frequens* *Heterophoxus oculatus* *Ampelisca* sp. *Paraphoxus* sp. *Metaphoxus* sp. *Heterophoxus* sp. *Sthenelenella uniformis* *Phoronis* sp.
II. Surface detritus feeders; may be abundant in control areas but are not dominant; increase in abundance in areas slightly affected by wastewater discharges; 14 species, 7 taxa.	*Photis brevipes* *Photis californica* *Photis* spp. *Euphilomedes producta* *Euphilomedes carcharodonta* *Euphilomedes longiseta*	*Mediomastus* spp. *Myriochele gracilis* *Myriochele* sp. *Axinopsida serricata* *Mysella pedroana* *Mysella tumida*
III. Surface deposit feeders primarily; often present but never abundant in areas moderately affected by wastewater discharges; 4 species, 4 taxa.	*Parvilucina tenuisculpta* *Macoma carlottensis* *Bittium* spp. *Spiochaetopterus costarum*	
IV. Subsurface deposit feeders; rare at control sites; most abundant in areas heavily affected by wastewater discharges; 10 species, 8 taxa.	*Armandia bioculata* *Schistomeringos longicornis* *Schistomeringos* sp. *Ophryotrocha* sp. Dorvilleidae, UI[1]	*Capitella capitata* Tubificidae, UI *Solemya panamensis* *Solemya* sp. Stenothoidae, UI

[1] Undescribed species, D. Montagne, Los Angeles County Sanitation Districts, Whittier, California (personal communication).

ITI values were calculated from over 300 Van Veen grab samples taken around the five major urban outfall discharge sites (Bascom, 1979) and at distant control sites (Word, 1979b; Word and Mearns, 1979). Distant areas unaffected by municipal waste discharges had ITI values that ranged from 83 to 100 (*i.e.*, dominated by suspension-feeding invertebrates). ITI values between 60 and 80 occurred near outfall and non-outfall sites and indicated a fauna still numerically dominated by detrital-feeding invertebrates. Values below 60 (*i.e.*, fauna dominated by deposit-feeding invertebrate species) occurred almost exclusively near outfall sites. Areas with ITI values between 30 and 60 (dominated by surface deposit-feeding infauna) were considered "changed", and areas with ITI values below 30 (dominated by subsurface deposit-feeding infauna) were considered "degraded" (Bascom *et al.,* 1979). The sizes of affected areas within each site were measured by planimetry.

Correlation with Inputs

Once computed, estimates of the size of affected areas and excess standing crops were plotted against suspended solids mass emission rates. Correlations and bestfit lines were computed using linear regression techniques on log transformed data.

RESULTS

During 1977 there was a 100-fold range of nutrient, BOD, and suspended solids mass emission rates among the five major discharges (*i.e.,* suspended solids ranging from 1,500 mt/yr at Oxnard to 102,000 mt/yr at Palos Verdes) (Table 3). Previous data (Schafer, 1977) indicate these emissions have been relatively constant during the past decade.

Contours of sediment volatile solids, an indicator of sediment organic content (Mitchell and Schafer, 1975), benthic infaunal biomass, and ITI indicate that the size and shape of affected bottom areas vary among the discharge sites. For example, at Point Loma, volatile solids increased from a background of 2.8% to a high value of 5.6% near the outfall, while benthic biomass increased from 70 g/m^2 to 120 g/m^2, and the ITI dropped from 93.5 to 33. Overall, a total of 5.5 km^2 exhibited elevated biomass, and 4 km^2 were characterized by ITI values below 60 (Table 4). In contrast, peak values for volatile solids and benthic

Table 3. Total mass emission rates and concentrations of various constituents in southern California discharges, 1977 (Schafer, 1979). ND = no data.

	JWPCP	Hyperion[1]		Orange County	Point Loma	Oxnard
		8 km	11 km			
Flow						
Total (liters/day x 10^6)	1,268	1,207	17.4	681	439	42.4
Total suspended solids						
Total (mt/yr)	102,000	27,300	51,500	33,300	20,500	1,500
Concentration (mg/liter)	220	62	8,100	132	128	98
Five-day BOD						
Total (mt/yr)	102,000	63,900	ND	47,900	26,700	3,900
Concentration (mg/liter)	220	145	ND	197	167	258
Ammonia nitrogen						
Total (mt/yr)	18,000	7,590	1,650	9,830	3,810	317
Concentration (mg/liter)	39	17.2	259	36	23.8	20.5

[1] Two outfalls, one 8 km (5 mi) with a combined effluent of primary and secondary wastes and the second 11 km (7 mi) with waste-activated sludge, discharge within 3 km of each other in Santa Monica Bay.

biomass at Palos Verdes reached 27% and 879 g/m^2, respectively, and the ITI reached a minimum of 0.3. Overall, an area >94 km^2 was characterized by ITI's below 60 (Table 4). Conditions in Santa Monica Bay and off Orange County were intermediate (Mearns and Greene, 1976; Bascom *et al.*, 1979). At Oxnard, the smallest discharge site, there were no stations with ITI's below 60, and an area of less than 0.01 km^2 was affected (Table 4).

It is apparent in Table 4 that both excess standing crops and areal extent of enhanced biomass and depressed ITI values generally vary with suspended solids mass emission rates. As a first approximation, the relationships between suspended solids mass emission rates (J_e) and benthic effects are exponential and can be described by linear regressions of $\log_{10}$ transformed data (Figures 2,3,4).

FORECASTING

The strong correlations between solids mass emission rates and the magnitude of benthic response (Figures 2, 3, 4) are assumed to be indicative of a cause-effect relationship. Thus, it should be possible to forecast changes in the size and

Table 4. Summary of estimates of excess standing crops (mt, wet wt) and areas (km^2) occupied by excess standing crop and Infaunal Trophic Indices (ITI's) characteristic of deposit-feeding communities at five southern California waste discharge sites. MER = mass emission rate. Values computed from results of macrobenthic surveys taken between 1975 and 1977 for excess standing crop and its area, and in 1978 (Bascom, 1979) for ITI's ⩽ 60 and ⩽ 30.

	Oxnard	Point Loma	Orange County	Santa Monica Bay	Palos Verdes
Suspended solids MER (mt/yr) 1975	2,181	18,725	33,396	110,180	130,966
1977	1,500	20,500	33,300	78,800	102,000
Excess standing crop (mt) 1975-1977	17	150	455	5,625	11,760
Area (km^2) occupied by excess standing crop	1	5	5	30	45
Area (km^2) occupied by benthic assemblages with ITI					
⩽ 60	0 (<0.01)[1]	⩽ 4.0	10.8	> 57.7	> 94
⩽ 30	0 (<0.01)[1]	⩽ 0.16[2]	0.25[3]	3.3	8.4

[1] Not detected; arbitrary assignment.

[2] One station within 0.4 km of outfall had value < 30 on several occasions. Limit of detection based on station density in grid.

[3] One station within 0.5 km of diffuser had value < 30 on one occasion. Limit of detection based on station density in grid.

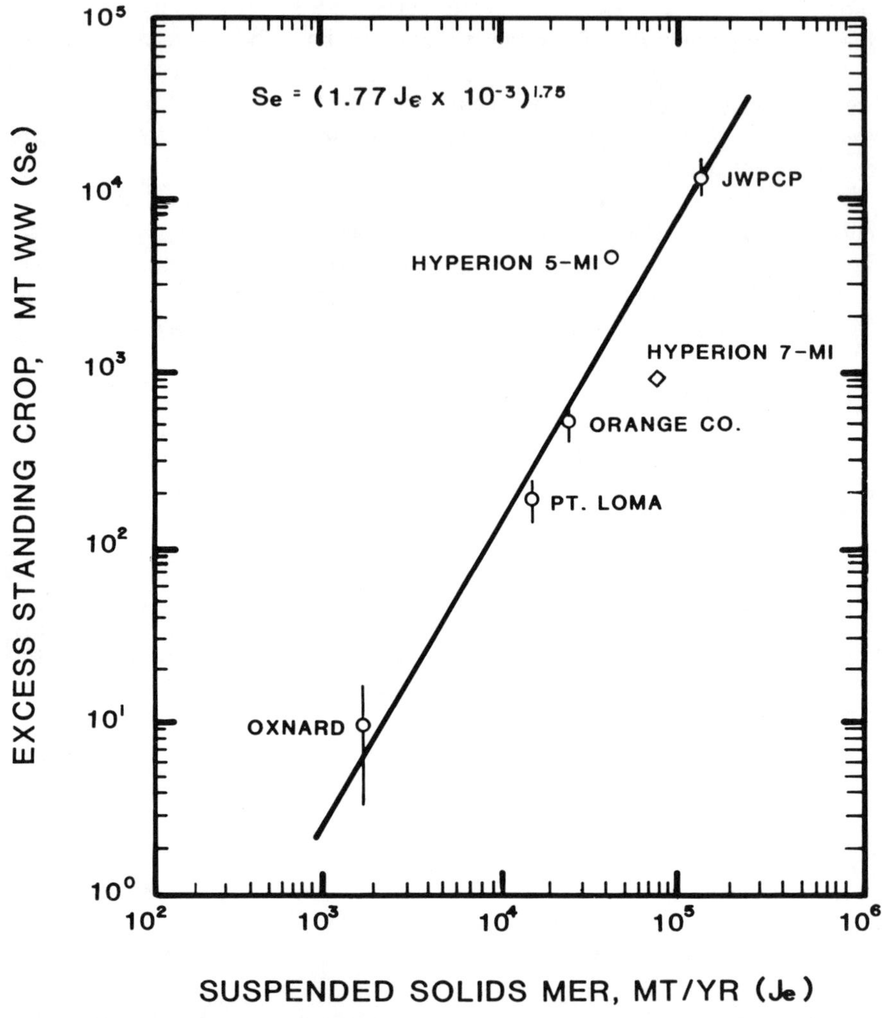

Figure 2. Relation between suspended solids mass emission rates (J_e, mt/yr) and total excess standing crop of benthic infauna (S_e, mt wet wt) surrounding six outfalls on the southern California mainland shelf. Hyperion 8-km (5-mi) effluent and 11-km (7-mi) sludge outfalls plotted separately. Circles and diamond indicate average excess standing crop; bars indicate range of standing crops for each location.

general composition of these benthic macrofaunal communities that would accompany increases or decreases in the mass emission rates of solids. Some aspects of making such forecasts are considered below.

Characteristics of the Relationships

Inspection of exponents in the equations of Figures 2, 3, and 4 indicates that each community variable has a different response to increases or decreases in

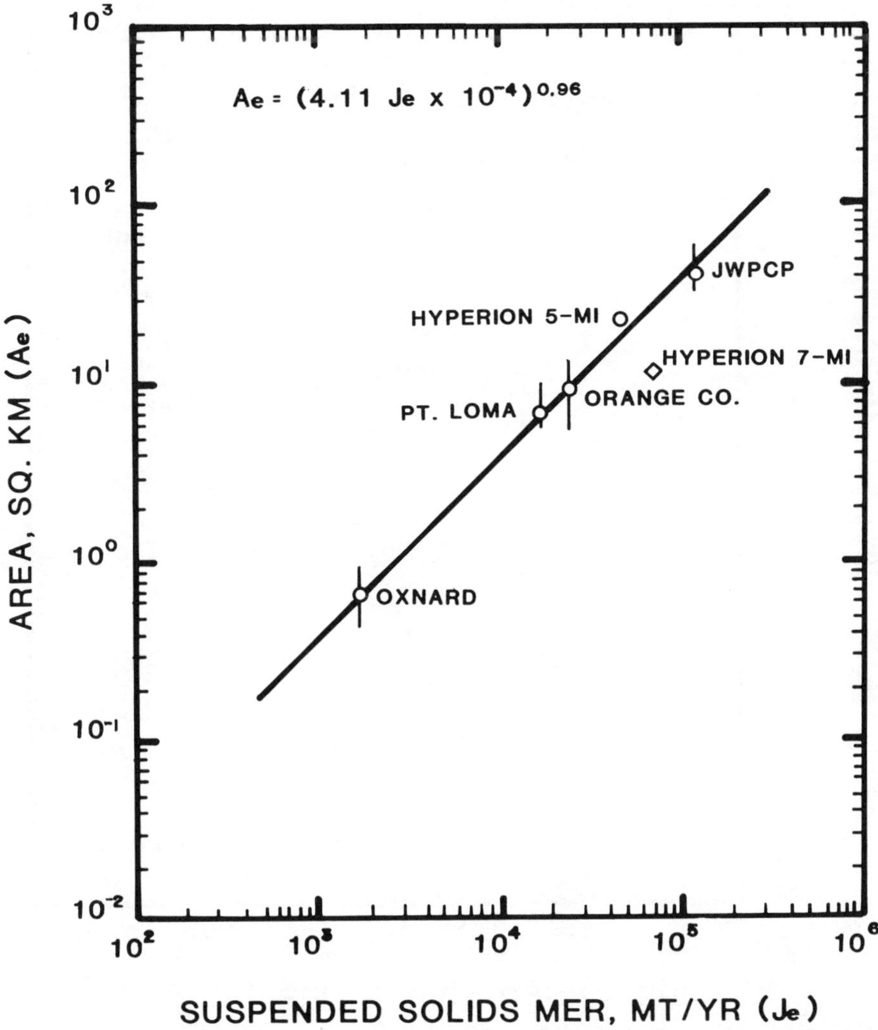

Figure 3. Relation between suspended solids mass emission rates (J_e, mt/yr) and size of areas (km^2) occupied by total excess standing crop (A_e) of infauna surrounding six outfalls on the southern California mainland shelf. Circles and diamond indicate average of area occupied by excess standing crops; bars indicate ranges for each location.

suspended solids emission. For example, back calculations (Table 5) indicate that doubling suspended solids emissions from 25,000 mt/yr to 50,000 mt/yr (typical range of values for moderate outfalls) will increase (1) the excess standing crop by a factor of 3.4 (from 759 mt to 2,553 mt), (2) the area occupied by that standing crop by a factor of 2 (from 9.4 km^2 to 18.2 km^2), (3) the area occupied by both surface and subsurface deposit-feeding infauna (characterized by ITI's $\leqslant 60$) by a factor of 4.5 (from 5 km^2 to 22 km^2), and (4) the area

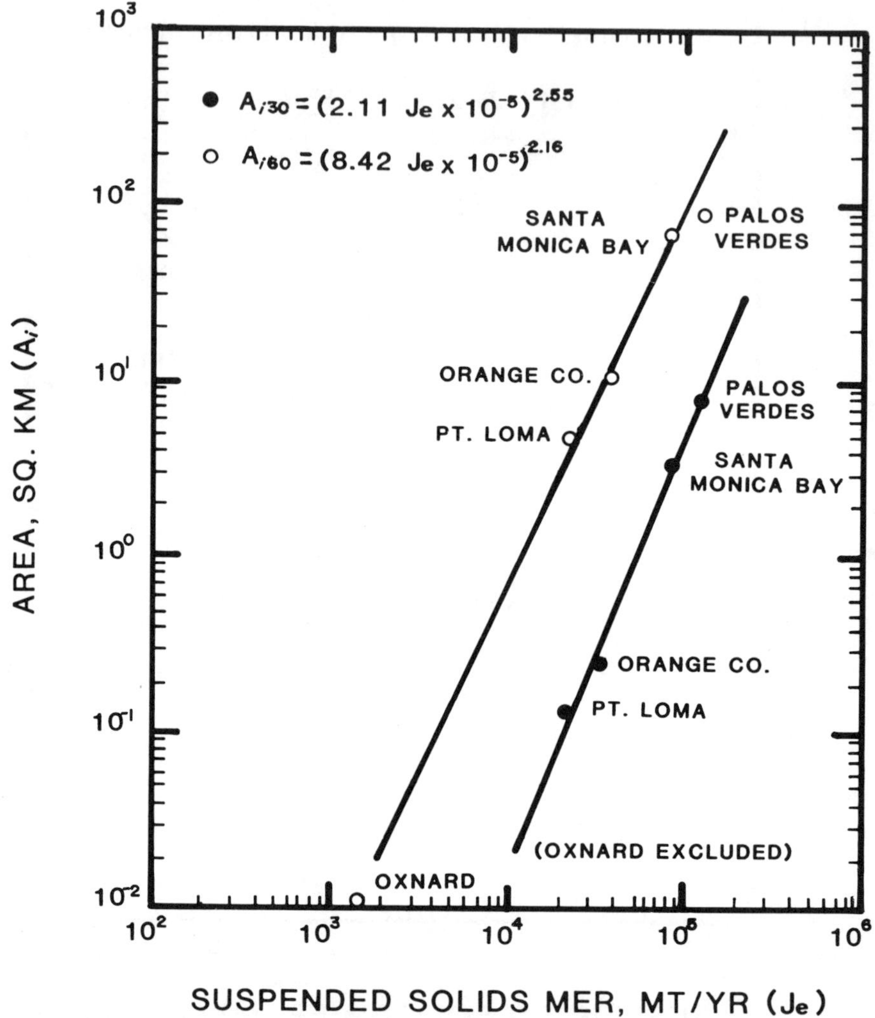

Figure 4. Relation between suspended solids mass emission rates (J_e, mt/yr) and size of areas (A_i, km^2) occupied by macrobenthic fauna dominated by subsurface deposit-feeding organisms ($i \leqslant 30$) and by surface and subsurface deposit-feeding organisms ($i \leqslant 60$); calculations based on the Infaunal Trophic Index (ITI, see text). Data points represent values from surveys reported by Bascom (1979). There were no samples with ITI's $\leqslant 30$ at Oxnard.

occupied only by subsurface deposit feeders (such as *Capitella capitata*) by a factor of 5.9 (from 0.02 km^2 to 1.14 km^2).

Forecast for a Southern California Ocean Outfall

In 1978 an intergovernmental agency asked the Coastal Water Research Project to forecast ecological benefits of proposed reductions of specific wastewater constituents in several southern California effluents. The largest

Table 5. Calculations of macrobenthic excess standing crops (S_e), areas occupied by excess standing crops (A_e), and areas occupied by macrobenthic communities dominated by both surface and subsurface deposit-feeding organisms (A_{i60}) or only by subsurface deposit-feeding organisms (A_{i30}) for a wide range of annual suspended solids mass emission rates (MER).

Suspended solids MER (mt/yr)	S_e (mt)	Macrobenthos response A_e (km²)	A_{i60} (km²)	A_{i30} (km²)
10,000	153	3.9	0.7	0.018
25,000	759	9.4	5.0	0.0195
50,000	2,553	18.2	22.4	1.141
70,000	4,600	25.1	46.0	2.69
100,000	8,588	35.4	100.4	6.68
165,560[1]	20,753	57	299	24.2

[1] New York Bight sewage sludge (Mueller *et al.*, 1976).

reduction was to occur in the discharge off Palos Verdes where the 1975 suspended solids emission of 131,000 mt/yr was to be reduced by 85% to 19,900 mt/yr. In fact, this discharger will soon implement secondary treatment of 756 x 10^6 liters/day (200 mgd) of a total flow of about 1,323 x 10^6 liters/day (350 mgd). Actual effluent quality forecasts are not yet available, but in the interim an 85% removal of suspended solids for the 756 x 10^6 liters/day (200 mgd) flow is assumed. This would produce a combined flow of 43,000 mt/yr of suspended solids. In addition, solids emission from this discharge already has decreased to about 104,000 mt/yr in 1979.

Using the equations described in Figures 2, 3, and 4, predictions were made of infaunal benthic conditions that might prevail with the reduced emissions. As shown in Table 6, the 1979 emissions are forecast to produce slight (12% to 22%) decreases in the areas dominated by subsurface deposit-feeding infauna (*i.e.*, degraded, A_{i30}), the area occupied by excess standing crop (A_e), and the weight of the excess standing crop (S_e). However, no measurable change in area dominated by combined surface and subsurface deposit-feeding infauna (A_{i60}) is forecast. This lack of change results, in part, from a survey grid that did not completely encompass the affected area (*i.e.*, >97 km²).

In contrast, implementation of the facilities now under construction (*i.e.*, 756 x 10^6 liters/day secondary treatment) could result in reduced effects on the benthos, ranging from a 64% decrease in area occupied by excess standing crop (A_e) to a 91% decrease in area dominated by subsurface deposit-feeding infauna (to 0.8 km²). The net effect would be a total bottom area of 16 km² that would harbor about 1,990 mt of excess infauna of abnormal composition. Within this impacted area, less than 1 km² would be dominated by subsurface deposit-feeding infauna (Group IV and some Group III species, Table 2). Full treatment, as proposed by the intergovernmental task force, would reduce further the total affected area to about 7.5 km², within which would occur about 510 mt of

Table 6. Forecast of changes in infaunal benthos parameters off Palos Verdes in 1979, when partial secondary treatment is implemented and if full secondary treatment is implemented. Percent reduction from 1975 conditions are given. See Table 5 for abbreviations.

	1975	1979[1]	Future	
			Partial treatment[2]	Full treatment[3]
Solids, MER, mt/yr	131,000	104,000 (21%)	43,300 (67%)	19,900 (85%)
S_e (mt)	11,760	9,200 (22%)	1,985 (83%)	510 (96%)
A_e (km^2)	46	37 (18%)	16 (64%)	7.5 (83%)
A_{i60} (km^2)	<94	109 (112%)[4]	16 (83%)	3.1 (97%)
A_{i30} (km^2)	8	7 (12%)	1 (91%)	0.1 (99%)

[1] Actual measured input with forecasts of infaunal benthos parameters

[2] Assuming a combination of 567 x 10^6 liters/day (150 mgd) of existing primary effluent with 33,300 mt/yr suspended solids plus 756 x 10^6 liters/day (200 mgd) of secondary effluent with 85% removal of suspended solids (10,000 mt/yr discharged).

[3] Under conditions assigned by intergovernmental task force (Mearns, 1978).

[4] No detectable change; 1975 survey underestimated area affected.

excess infaunal biomass, a 3.1-km^2 area of changed infaunal community structure, and a very small, possibly undetectable, 0.11-km^2 area dominated by subsurface deposit-feeding organisms (Table 6).

It would be tempting to conduct an incremental analysis of energy and treatment cost vs. magnitude of ecological change suggested by these forecasts; however, not all of the data required for such an analysis are in hand at this time. Nevertheless, these forecasts may have an intermediate benefit to the discharger and to regulatory agencies in that they represent hypotheses that could be tested by revised monitoring programs.

Application to the New York Bight

It is not certain whether the relationships described above will be applicable to the New York Bight or other coastal areas where sewage solids are discharged or dumped. Based on a mass emission (dumping) rate of 165,560 mt/yr for barged sewage solids into the apex (computed from Mueller *et al.*, 1976), there may be an excess standing crop of infauna of approximately 21,000 mt occupying an area of 57 km^2. In addition, an area of about 300 km^2 may be dominated by surface and subsurface deposit-feeding infauna, within which 24 km^2 may be dominated by subsurface deposit-feeding infauna (Table 5). To our knowledge, biomass has not been determined from apex benthic surveys, but Boesch (this volume) refers to an area on the order of 10-15 km^2 that supports a depauperate fauna and is surrounded by a larger area of 240 km^2, characterized as an "enhanced transitional area." In addition, our inspection of data in Steimle *et al.* (this volume) suggests that an area of 150 km^2 is nearly devoid of ampeliscid amphipods. These organisms are suspension feeders, and in southern

California are among the first to disappear from areas impacted by sewage outfalls. These organisms occur in samples characterized by ITI > 60 but are rare where ITI $\leqslant 60$. These estimates of measurable change in benthic communities are well within a factor of two projections, presented in Table 5, suggesting that the techniques described in this paper might be applicable to the New York Bight and should be explored further.

Uncertainties and Limitations

Applying the described approach to southern California and other locations, such as the New York Bight, requires a careful reexamination of each assumption and the inclusion of other information, such as corrections for differences in the settling rates of suspended solids (Herring and Abati, 1979). There also are a number of uncertainties and problems with the present approach; consequently, we recommend that relationships and equations be tailored to individual situations. The assumption that mass emission rates are relatively constant has been valid for several discharges but not for others (*e.g.*, those at Palos Verdes where discharges have been decreasing). More importantly, none of the benthic surveys were designed for estimating total population size, and only a few (Bascom, 1979) were designed for clearly resolving the size and shape of affected bottom areas. New surveys should be conducted to estimate more accurately both parameters, and new equations should be computed. This study also included an outfall (Oxnard) that was significantly shallower than the rest (20 m vs. 60 m). Even if the solids settling characteristics of this effluent were identical to those from the other outfalls, the increased current and wave action inshore could lead to alterations in the size of the affected area and to an underestimate of the impact of the Oxnard suspended solids mass emission (1,500 mt/yr) on benthic communities, were it discharged in deeper water.

This analysis also assumes that the relation between benthic biomass and suspended solids results from the nutritive (organic carbon) content of the solids, that all effluent solids have the same assimilative capacity, and that production and cropping (mortality) rates are proportional to emissions at all sites. We know, in fact, that solids from waste-activated sludge discharged via the Hyperion 11-km (7-mile) outfall have a different organic (BOD) content than the solids remaining in the 8-km (5-mile) effluent. As an extreme example, it would be inappropriate to compare effects of suspended solids of dredged material with those from the same discharge of sewage solids. In addition, we know from stomach content analysis that the infauna is indeed cropped by fishes, but at unknown rates. Finally, we also assume that toxic effects are of second order importance. We think this is true for most of the California study sites except Palos Verdes, where high concentrations of H_2S and DDT have prevailed over a large area (Smith and Greene, 1976) within which low biomass occurs.

Despite these problems and uncertainties, further development of untraditional approaches, such as this, is needed to resolve real management problems and to provide a more quantitative approach for projecting the full magnitude of effects of discharges on benthic communities.

ACKNOWLEDGEMENTS

This work was part of a team effort to forecast physical, chemical, and biological changes at several southern California waste discharge sites, and we thank our staff and associates for their help. We also thank laboratory supervisors at the Hyperion Treatment Plant, the County Sanitation Districts of Los Angeles, the County Sanitation Districts of Orange County, the Point Loma Treatment Plant (City of San Diego), and the Oxnard Treatment Plant (County of Ventura), for use of their staff and data. We also thank Dr. Garry Mayer, Office of Marine Pollution Assessment, Northeast Office, for his helpful review. This work was supported, in part, by an agreement with the Los Angeles and Orange County Regional Sludge Management Agency (LAOMA) and, in part, by Grant R801152 from the U.S. Environmental Protection Agency, Corvallis, Oregon.

REFERENCES

Allan Hancock Foundation. 1965. An oceanographic and biological survey of the southern California mainland shelf. Appendix: data. State of California Resources Agency, State Water Quality Control Board, Publication 27. 232 pp.

Bascom, W. 1979. Life in the bottom: San Pedro and Santa Monica Bays, Palos Verdes and Point Loma Peninsulas. *In:* Coastal Water Research Annual Report 1978, Southern California Coastal Water Research Project, El Segundo, CA. pp. 57-80.

Bascom, W., A.J. Mearns, and J.Q. Word. 1979. Establishing boundaries between normal, changed and degraded areas. *In:* Coastal Water Research Project Annual Report 1978, Southern California Coastal Water Research Project, El Segundo, CA. pp. 81-94.

Boesch, D.F. This volume. Ecosystem consequences of alterations of benthic community structure and function in the New York Bight region. pp. 543-568.

Eganhouse, R.P., D.R. Young, and J.N. Johnson. 1978. Geochemistry of mercury in Palos Verdes sediments. Environ. Sci. Technol. 12(10): 1151-1157.

Greene, C.S. 1976. Response and recovery of the benthos at Orange County. *In:* Coastal Water Research Project Annual Report, 1976, Southern California Coastal Water Research Project, El Segundo, CA. pp. 197-203.

Hendricks, T.J. 1977. Coastal currents. *In:* Coastal Water Research Project Annual Report, 1977, Southern California Coastal Water Research Project, El Segundo, CA. pp. 53-62.

Herring, J.R. and A.L. Abati. 1979. Effluent particle dispersion. *In:* Coastal Water Research Project Annual Report, 1978, Southern California Coastal Water Research Project, El Segundo, CA. pp. 113-125.

M.B.C., Inc. 1975. Predischarge Receiving Water Monitoring Study. Final Summary Report. Submitted to City of Oxnard, Department of Public Works, Oxnard, CA.

Mearns, A.J. 1978. Changes in the biomass of benthic infauna. Testimony before the Subcommittee on Water Resources, Committee on Public Works and Transportation, U.S. House of Representatives, 95th Congress, Second Session. May 24-25, 1978. Washington, DC. pp. 172-179.

Mearns, A.J. 1981. Effects of municipal discharges on open coastal ecosystems. *In:* Studies in Marine Environmental Pollution, R.A. Geyer (ed.), Elsevier Press, Amsterdam. pp. 25-66.

Mearns, A.J. and C.S. Greene. 1976. Comparison of the benthos at several wastewater discharge sites. *In:* Coastal Water Research Project Annual Report, 1976, Southern California Coastal Water Research Project, El Segundo, CA. pp. 211-216.

Mearns, A.J. and D.R. Young. 1978. Impact of nearshore development on open coastal resources. *In:* The Urban Harbor Environment, J.N. Baskin, M.D. Dailey, S.N. Murray, and E. Segal (eds.), Southern California Ocean Studies Consortium, Harbor Plaza, Long Beach, CA. pp. 23-47.

Mitchell, F.K. and H.A. Schafer. 1975. Effects of ocean sludge disposal. *In:* Coastal Water Research Project Annual Report, 1975, Southern California Coastal Water Research Project, El Segundo, CA. pp. 153-162.

Mueller, J.A., J.S. Jeris, A.R. Anderson, and C.F. Hughes. 1976. Contaminant inputs to the New York Bight. NOAA Tech. Memo. ERL MESA-6. 347 pp.

Pearson, T.H. and R. Rosenberg. 1978. Macrobenthic succession in relation to organic enrichment and pollution of the marine environment. Oceanogr. Mar. Biol. Annu. Rev. 16: 229-310.

Rhoads, D.C., P.L. McCall, and J.Y. Yingst. 1978. Disturbance and production on the estuarine seafloor. Am. Sci. 66: 557-586.

Schafer, H.A. 1977. Characteristics of municipal wastewater discharges, 1977. *In:* Coastal Water Research Project Annual Report, 1978, Southern California Coastal Water Research Project, El Segundo, CA. pp. 19-23.

Schafer, H.A. 1979. Characteristics of municipal wastewater discharges, 1977. *In:* Coastal Water Research Project Annual Report, 1978, Southern California Coastal Water Research Project, El Segundo, CA. pp. 97-101.

Smith, G.B. 1974. Some effects of sewage discharge to the marine environment. Ph.D. Thesis, Scripps Institution of Oceanography. 334 pp.

Smith, K.L., Jr., G.T. Rowe, and J.A. Nichols. 1973. Benthic community respiration near the Woods Hole sewage outfall. Estuarine Coastal Mar. Sci. 1: 65-70.

Smith, R.W. and C.S. Greene. 1976. Biological communities near submarine outfall. J. Water Pollut. Control Fed. 48(8): 1894-1912.

Steimle, J., J. Caracciolo and J.B. Pearce. This volume. Impacts of dumping on New York Bight apex benthos. pp. 213-223.

Word, J.Q. 1976. Biological comparison of grab sampling devices. *In:* Coastal Water Research Project Annual Report, 1976, Southern California Coastal Water Research Project, El Segundo, CA. pp. 189-194.

Word, J.Q. 1977. Taxonomic standardization developments. *In:* Coastal Water Research Project Annual Report, 1977, Southern California Coastal Water Research Project, El Segundo, CA. pp. 85-87.

Word, J.Q. 1979a. Marine environmental assessment aided by the use of the Infaunal Trophic Index. *In:* Proceedings of the Ecological Damage Assessment Conference, J. Siva, E. Birchard, K. Guziak, D. Chamberlain, and G. Cox (eds.), Society of Petroleum Industrial Biologists, American Petroleum Institute, Washington, DC. pp. 41-58.

Word, J.Q. 1979b. The Infaunal Trophic Index. *In:* Coastal Water Research Project Annual Report 1978, Southern California Coastal Water Research Project, El Segundo, CA. pp. 19-39.

Word, J.Q., T.J. Kawling, and A.J. Mearns. 1976a. A Comparative Field Study of Benthic Sampling Devices Used in Southern California Benthic Surveys. Task Report, Grant R801152, U.S. Environmental Protection Agency, Corvallis, OR. 79 pp.

Word, J.Q. and A.J. Mearns. 1979. 60-meter control survey off southern California. Southern California Coastal Water Research Project Tech. Memo. 229. 50 pp.

Word, J.Q., A.J. Mearns, and B.L. Myers. 1977. Animals that are indicators of marine pollution. *In:* Coastal Water Research Project Annual Report, 1977, Southern California Coastal Water Research Project, El Segundo, CA. pp. 199-206.

Word, J.Q., B.L. Myers, and L. Harris, 1976b. Taxonomic standardization program. *In:* Coastal Water Research Project Annual Report, 1976, Southern California Coastal Water Research Project, El Segundo, CA. pp. 195-196.

Young, D.R., T. K. Jan, and T.C. Heesen, 1978. Cycling of trace metals and chlorinated hydrocarbon wastes in the Southern California Bight. *In:* Estuarine Interactions, M.L. Wiley (ed.), Academic Press, New York. pp. 481-496.

Young, D.R., D.J. McDermott, and T.C. Heesen. 1976. DDT in sediments and organisms around Southern California outfalls. J. Water Pollut. Control Fed. 48(8): 1919-1928.

Young, M.W. and D.K. Young. This volume. Marine macrobenthos as indicators of environmental stress. pp. 527-539.

IN SITU PHOSPHATE REGENERATION IN MIDDLE ATLANTIC COASTAL WATERS

Jonathan H. Sharp

College of Marine Studies
University of Delaware
Lewes, Delaware 19958

Abstract. Concern exists over man's disposal of bioactive nutrients into the marine environment. These nutrients can stimulate microbial plant production that, if not consumed in marine food chains, can cause serious oxygen demands. To understand the impact of man's input of nutrients it is necessary first to gain a better understanding of natural nutrient cycles. This paper attempts to quantify natural phosphate regeneration in bottom waters of three regions in Middle Atlantic coastal waters and to evaluate the contributions that *in situ* nutrient regeneration could make toward supporting microbial plant production. It is striking that the apex of the New York Bight, with high production, receives a much smaller proportion of its phosphate supply from *in situ* regeneration than adjacent New Jersey and Long Island coastal waters. The approach taken here is the first of its kind for quantifying and comparing coastal nutrient regeneration.

INTRODUCTION

The New York Bight receives appreciable amounts of nutrients from terrigenous sources, and these sources are considered to be primarily anthropogenic (Gross, 1976). Nitrogen and phosphorus are the major growth controlling factors for phytoplankton in this coastal region (Ryther and Dunstan, 1971; Malone, 1977). These nutrients enter the water either in the form of inorganic ions (*i.e.*, ammonium, nitrate, nitrite, and phosphate) or as organic matter that eventually is converted to inorganic ions. Nutrients also are recycled in these waters by natural processes. To understand the importance of external nutrient inputs, we must be able to quantify *in situ* processes.

Nutrients are taken up within the photic zone by phytoplankton and incorporated into particulate organic matter. Some dissolved organic nitrogen and phosphorus may be excreted or leaked by the phytoplankton, but the majority is potentially available as particulate matter introduced into the marine food web through herbivores. Subsequent feeding losses and direct organic excretion by zooplankton and nekton generate dissolved and particulate detrital organic matter. The detrital organic matter eventually is oxidized, primarily by bacterial action, to return inorganic nutrients to the system. Bacterial nutrient regeneration takes place throughout the water column as well as at the water-sediment interface. In addition to bacterial regeneration of inorganic nutrients, zooplankton and nekton also regenerate inorganic nutrients via excretion. For stratified oceanic surface waters, biogenic regeneration of nutrients within the photic zone via excretion has been estimated to be the major nutrient source for phytoplankton (Eppley *et al.*, 1973; McCarthy and Goldman, 1979). In coastal waters, excretion provides approximately 25% of the phytoplankton need (Fournier *et al.*, 1977; Smith and Whitledge, 1977). Walsh *et al.* (1978) attempted a partial nitrogen budget for the New York Bight, but their evaluation, like most others, encountered difficulties quantifying microbial nutrient regeneration.

It is difficult to estimate directly microbial nutrient regeneration in the photic zone since the uptake of nutrients by phytoplankton occurs simultaneously with microbial production. A recent attempt was made by Harrison (1978) using an isotope dilution technique, but it is somewhat indirect and requires discrete sampling. In and below the photic zone, quantitative estimates are marred by the artificiality of containing samples during measurement and by heterogeneity in discrete sampling. An attempt has been made by Sharp and Church (1981) to study nutrient dynamics in Middle Atlantic coastal bottom waters with integration over six months and tens to hundreds of kilometers. With this attempt, an average nutrient regeneration rate has been calculated for isolated bottom waters. A rate also has been calculated for nitrogen flux across the thermocline, and an approximate rate has been estimated by extrapolation for photic zone microbial nutrient regeneration. The basic principles of the Sharp and Church study are given in the following section. This paper is a partial extension of that study with comparison of phosphate regeneration in three regions of Middle Atlantic coastal waters.

COASTAL NUTRIENT REGENERATION

The study of Sharp and Church (1981) was done in coastal waters off New Jersey and Delaware (Figure 1). Central shelf stations were sampled in winter and throughout the summer of 1977 (six cruises). Measurements revealed that waters were isothermal with essentially homogeneous chemistry in March. From April to September they showed progressive density stratification with a strong pycnocline developing around 12-18 m and intensifying throughout the summer; the bottom depth was 36-40 m. Bottom water was fairly well-mixed as a water mass and showed decreasing dissolved oxygen and increasing dissolved inorganic carbon, nitrogen, and phosphorus over the summer. The surface layer also was well-mixed. Values were higher for dissolved oxygen, slightly lower for dissolved inorganic carbon, slightly lower for phosphate, and considerably lower for inorganic nitrogen. Values for these parameters remained relatively constant over the study period. Surface waters also were higher in dissolved organic carbon than were bottom waters.

The lack of strong or consistent north to south chemical gradients along the 150-km transect suggested there was no net chemical input over time along the longshore axis. The lack of net chemical transfer to bottom waters along the onshore axis also was assumed on the basis of the isolation of this water mass as seen in a composite temperature-salinity (TS) diagram (Sharp and Church, 1981). This is a critical point since intrusion of offshore waters was postulated by Riley (1967) and demonstrated by Wright (1976) for adjacent coastal waters. Upwelling also has been shown for the Long Island coast (Walsh *et al.*, 1978). Sharp and Church (1981) demonstrated that chemical changes in bottom waters did not indicate intrusion of nutrient-rich deep water. Instead, bottom

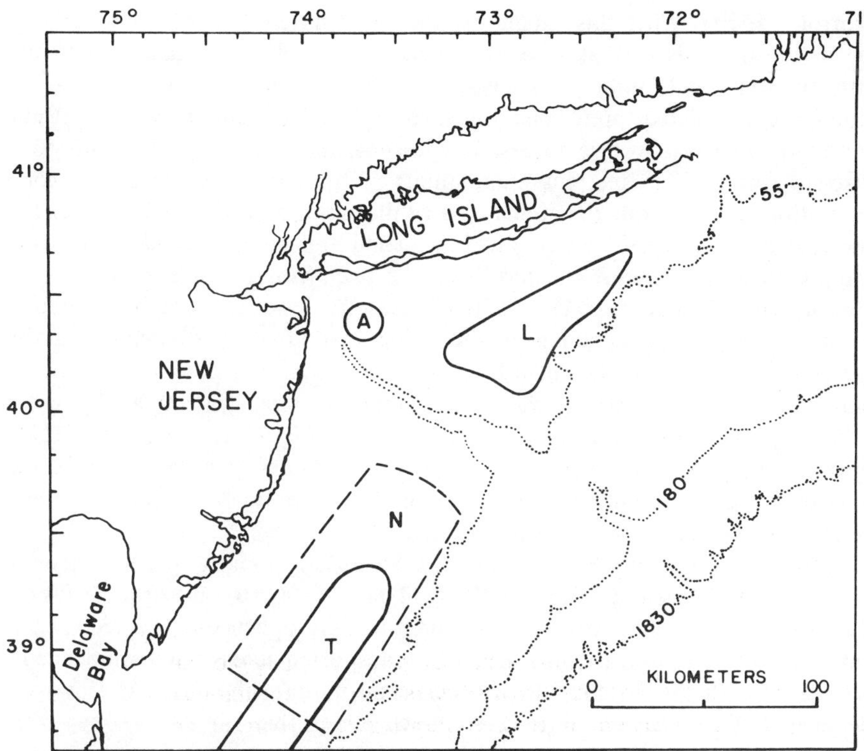

Figure 1. Sampling areas for nutrient regeneration estimates. T = data from Sharp and Church (1981); N, A, and L = XWCC cruise area; N = New Jersey coast stations, A = apex station, and L = Long Island coast stations. Contours are shown in meters.

waters were considered to remain fairly well isolated and the oxygen decrease and nutrient increase to result primarily from *in situ* oxidation of organic matter, rather than horizontal diffusion or advection. The absolute isolation of these waters could be examined elsewhere. Here it only is necessary to state that nutrient regeneration by *in situ* processes is chemically consistent and can be modeled.

The apparent oxygen utilization (AOU) model of Redfield *et al.* (1963) was used to examine nutrient-organic-oxygen dynamics in the bottom waters. The AOU model gives a stoichiometry (the Redfield ratio) for carbon, nitrogen, phosphorus, and oxygen in the oxidation of marine organic matter to produce carbon dioxide, nitrate, and phosphate. It can be used if one can measure temporal changes in the dissolved nutrients (N and P) and can estimate oxygen utilization and production of dissolved inorganic carbon. These estimates are made by comparing measured values with theoretical ones based upon atmospheric saturation of oxygen and carbon dioxide at ambient temperature, salinity, and pH. This approach was used originally for oxygen (Redfield *et al.*, 1963); the extension for carbon dioxide relies on basic chemical oceanography

(Skirrow, 1975). With this model, Sharp and Church (1981) found a good correlation of AOU with apparent production of dissolved inorganic carbon and with the temporal buildup of phosphate. The model also was verified by comparing calculated preformed values for phosphate and dissolved organic carbon to observed surface values. The seasonal increase in dissolved inorganic nitrogen (nitrate, nitrite, plus ammonium) of the bottom water gave a good correlation to AOU, but the stoichiometry did not match the AOU model. The discrepancy and progressively negative calculated preformed nitrogen values were interpreted as evidence for selective loss of nitrogen from the bottom waters.

Sharp and Church (1981) concluded that the rate of phosphate buildup was indicative of the general regeneration process and that the discrepancy in the nitrogen buildup could be quantified as a nitrogen flux across the seasonal thermocline. The bottom water regeneration rate was 0.0044 μ moles P/liter/day. If nitrogen were generated at the same rate, the Redfield ratio (16/1) times the phosphate regeneration rate gives its kinetics. The latter minus the calculated nitrogen buildup gave a loss of 0.02 μ moles N/liter/day that was interpreted as a flux across the thermocline.

Using the nitrogen flux, an average measured primary productivity of 1.75 μ moles C/liter/day, and the Redfield ratio (106:16), nitrogen diffusion could supply 8% of the nitrogen requirement for phytoplankton. It is recognized that oxygen and phosphate also would be transported across the thermocline in accordance with the same diffusivity coefficient used for the nitrogen. However, the deep:surface nitrogen ratio is so much larger than those for oxygen or phosphate (about one to two orders of magnitude greater) that the net nitrogen flux would be much more profound. Nonetheless, the calculated flux is considered to be underestimated slightly.

In the Sharp and Church study, the bottom and surface layers were of about equal thickness; the average water depth was about 40 m, while the pycnocline was near the middle of the water column (12-18 m). Thomas *et al.* (1976) estimated water column respiration rates for these waters and found the surface layers, which correspond approximately with the photic zone, to have respiration rates equal to or greater than the bottom layers. Hence, conservatively, microbial nutrient regeneration in the surface layer should be of the same magnitude as in the bottom layer. Extrapolation of the estimated bottom water nutrient regeneration rate (0.0044 μ moles P/liter/day) to surface waters results in the conservative estimate that surface water microbial nutrient regeneration could supply 27% of the nitrogen requirement of the phytoplankton. Photic zone microbial regeneration thus could supply 27% of the nitrogen requirement, and nitrogen diffusion from bottom waters could supply 8%, for a total of 35% of the photic zone nitrogen requirement.

METHODS

The attempt here is to estimate nutrient regeneration for three separate regions of Middle Atlantic coastal waters. Nitrogen is the most limiting nutrient in these waters, as was stated above. However, in a situation similar to Riley's classical food chain model (Riley, 1963), modeling must be done with phosphate since adequate total inorganic nitrogen data are not available. Phosphate regeneration rates should be similar to nitrogen rates for the several months studied (Golterman, 1964; Grill and Richards, 1964). That particulate carbon, nitrogen, and phosphorus ratios for estuarine and coastal waters in the Middle Atlantic area are close to the classical Redfield ratios (Sharp, unpublished data) further validates this approach. Model C/N ratios also have been shown for other coastal waters (Eppley *et al.*, 1977), which is in contrast to the oceanic picture where C/N ratios are appreciably higher than the Redfield ratio (Goldman *et al.*, 1979; Sharp *et al.*, 1980).

Data used for the current analysis came from the NOAA-MESA Expanded Water Column Characterization (XWCC) cruises of April-October 1977 (Hazelworth *et al.*, 1978a,b,c,d,e). These data are from the same time period as those of Sharp and Church (1981), with partial spatial overlap. It was not possible to make AOU calculations since the listed oxygen electrode data gave an inconsistent seasonal trend and were suspect in their accuracy because of questionable electrode calibration.

Analyses were performed on XWCC data from stations in the broad central shelf region. This region appears to have consistent stratification throughout the summer, but the inner and outer peripheries show irregular thermal structure as the result of frontal processes (Boicourt and Hacker, 1976; Halliwell and Mooers, 1979). Partial breakdown of the stratification occurred by the October sampling (cruise XWCC 16), so data from the other four cruises (XWCC 12-15, late April to early August) were used. Stations chosen were (1) in the central shelf region (bottom depth of 30-60 m), (2) distant from bottom topographic anomalies (*e.g.*, the Hudson Canyon), and (3) consistently stratified (with a seasonally progressing pycnocline and nutricline). Data from three regions were used for calculations (Figure 1). While data from only one sampling location could be used from the apex area, data from seven and ten locations were available for the Long Island and New Jersey coasts, respectively. Tables 1-3 list the XWCC station numbers.

In calculating phosphate regeneration rates, bottom waters were assumed to be isolated from the surface for the summer period. Increases within the bottom waters thus should reflect *in situ* microbial processes. Bottom water samples were defined as those showing density and phosphate values indicative of being below the pycnocline and nutricline. To calculate regeneration rates, sampling dates were used as independent variables (first date sampled on cruise XWCC 12 was set as 0), and least squares regressions were calculated with phosphate concentrations as the dependent variables. The resulting regression slopes are reported in units of micromoles phosphate-phosphorus per liter per day (μ moles P/liter/day).

RESULTS AND DISCUSSION

Tables 1-3 present the data and calculated phosphate regeneration rates for the three regions. Several suspiciously low phosphate values were excluded from

Table 1. Phosphate concentration data used for regeneration rate calculations for apex stations. Cruise designations (*e.g.*, XWCC 12) are explained in methods section.

Cruise number	Date	Station	Depth (m)	Day	T (°C)	S (‰)	PO_4 (μ moles P/liter)
XWCC-12	4/29/77	13	20	0	6.79	32.69	0.51
		13	35	0	4.66	32.95	0.70
XWCC-13	6/1/77	13	31	33	6.14	32.68	0.92
		13	39	33	4.78	32.89	1.02
XWCC-14	6/27/77	13	38	59	8.26	32.90	1.26
XWCC-15	8/1/77	13	20	93	11.13	32.51	2.56
		13	30	93	10.35	32.64	2.78
		13	37	93	10.14	32.67	2.83

Slope = 0.0235 μ moles P/liter/day; n = 8, r = 0.96.

the regressions (6 out of 110), as indicated in Tables 2 and 3. To illustrate further the separate nature of the three regions, a composite TS diagram is given for all samples used in the calculations (Figure 2). From the envelopes enclosing the regions and averages of temperatures and salinities for each region on each date, it appears that the apex "water mass" is fresher throughout the season than either of the others, and that the Long Island "water mass" initially is colder and fresher, and later is warmer and fresher than the New Jersey "water mass." These observations, plus appreciably different phosphate concentrations on the final sampling, justify treating phosphate regeneration in the three regions separately. The regeneration rates given at the bottom of Tables 1-3 show that phosphate production in the apex region is more than double that of the New Jersey coast region, which in turn is slightly less than that in the Long Island coast region.

The bottom water regeneration rates are applied to surface waters, as was discussed above (also see below). To assess the impact of nutrient regeneration (assuming that the limiting nutrient, nitrogen, is regenerated at its stoichiometrically equivalent rate), photic zone productivity must be used. Approximate values for the apex and the New Jersey coast were taken from August 1976 measurements of Thomas *et al.* (1979). The areal measurements (mg C/m^2/day) were converted to integral values by dividing by the pycnocline depth (assuming it to be similar in depth to the photic zone), to give values expressed in μg C/liter/day. These in turn were converted to phosphate equivalents by:

$$\mu \text{ moles P/liter/day} = [(\mu\text{g C/liter/day}) \div 12] \div 106,$$

with 12 being the atomic weight of carbon and 106 being the C/P Redfield ratio. Apex station numbers 34, 41, and 109, and New Jersey coast stations

Table 2. Phosphate data used for regeneration rate calculations for Long Island coast. Asterisk beside phosphate data indicates three points from XWCC-13 excluded from regression (see text).

Cruise number	Date	Station	Depth (m)	Day	T (°C)	S (‰)	PO_4 (μ moles P/liter)
XWCC-12	4/30/77-5/2/77	27	34	0	3.89	33.18	0.48
		28	42	1	4.03	33.51	0.69
		62	30	2	2.82	33.36	0.43
		62	40	2	2.81	33.37	0.56
		62	46	2	2.79	33.36	0.58
		66	30	1	3.04	33.40	0.60
		66	41	1	3.03	33.40	0.58
		68	30	1	3.11	33.45	0.56
		68	48	1	3.10	33.46	0.63
		69	39	1	3.23	33.22	0.63
		113	47	1	3.97	33.55	0.74
XWCC-13	6/1/77-6/3/77	27	(not sampled below 25 m)				
		28	42	33	5.25	33.20	0.48
		62	30	34	6.45	33.28	0.49
		62	40	34	6.30	33.28	0.53
		62	47	34	6.07	33.31	0*
		66	30	34	6.67	33.11	0.14*
		66	43	34	6.63	33.11	0.25*
		68	33	33	6.11	33.37	0.46
		68	46	33	5.67	33.37	0.46
		69	30	34	6.33	33.18	0.48
		69	40	34	6.28	33.18	0.48
		113	40	33	4.70	33.28	0.44
		113	48	33	4.66	33.28	0.46
XWCC-14	6/77-7/77	(Long Island coastal stations were not sampled.)					
XWCC-15	8/7/77-8/8/77	27	34	99	9.77	32.58	1.89
		28	30	99	9.18	32.70	2.43
		28	41	99	9.12	32.70	2.43
		62	30	98	9.60	32.67	1.36
		62	40	98	9.54	32.91	1.69
		62	46	98	9.43	32.91	1.96
		66	31	98	9.25	32.90	1.13
		66	42	98	9.24	32.96	1.67
		68	40	99	9.07	33.07	1.64
		68	48	99	9.07	33.09	1.80
		69	30	99	9.48	32.77	1.67
		69	40	99	8.97	32.89	2.01
		113	40	99	8.41	33.02	2.24
		113	47	99	8.38	33.02	2.27

Slope = 0.0145 μ moles/liter/day; n = 34, r = 0.88.

Table 3. Phosphate data for regeneration rate calculations for New Jersey coast. Asterisk beside phosphate data indicates three points excluded from regression (see text.)

Cruise number	Date	Station	Depth (m)	Day	T (°C)	S (‰)	PO_4 (μ moles P/liter)
XWCC-12	5/3/77-	123	20	0	5.15	33.72	0.48
	5/5/77	123	35	0	5.13	33.72	0.48
		124	30	0	5.15	33.90	0.60
		124	36	0	5.15	33.90	0.51
		132	20	1	5.00	33.74	0.48
		132	30	1	4.99	33.74	0.28
		133	20	2	5.84	33.98	0.30
		133	30	2	5.82	33.96	0.31
		133	41	2	5.82	33.95	0.28
		134	40	2	5.38	33.98	0.49
		134	52	2	5.36	33.99	0.51
		142	21	1	4.35	33.53	0.34
		142	31	1	4.32	33.54	0.45
		143	20	2	5.23	33.88	0.37
		143	36	2	5.10	33.88	0.37
		144	30	1	5.94	34.10	0.40
		144	40	1	5.94	34.10	0.37
		153	30	1	4.72	33.74	0.45
		154	30	1	4.97	33.98	0.45
		154	40	1	4.97	33.97	0.52
XWCC-13	6/5/77-	123	29	33	5.88	33.22	0*
	6/7/77	123	33	33	5.84	33.21	0.45
		124	29	33	5.56	33.40	0.50
		124	38	33	5.38	33.41	0.47
		132	20	33	7.88	33.01	0.75
		132	30	33	7.43	33.01	0.79
		133	30	33	6.21	33.14	0.82
		133	45	33	6.11	33.22	0.82
		134	40	33	4.75	33.49	0.80
		134	54	33	4.75	33.49	0.75
		142	32	34	7.61	33.20	0.44
		143	30	34	6.68	33.45	0.37
		144	30	34	5.81	33.60	0.31
		144	44	34	5.65	33.51	0.33
		153	31	34	6.66	33.41	0.48
		154	30	35	6.51	33.73	0.63
		154	44	35	6.46	33.76	0.63
XWCC-14	6/29/77-	123	30	58	6.96	32.98	0.87
	6/30/77	123	33	58	6.88	32.98	0.84
		124	30	58	7.57	33.33	0.60
		124	35	58	7.56	33.32	0.60
		132	33	58	7.63	33.05	0.59

Table 3. *Continued*

Cruise number	Date	Station	Depth (m)	Day	T (°C)	S (‰)	PO4 (μ moles P/liter)
XWCC-14	6/29/77-6/30/77	133	30	58	6.59	33.17	0.72
		133	45	58	6.50	33.19	0.72
		134	53	58	6.62	33.31	0.73
		142	30	57	6.94	32.96	0.70
		143	30	58	6.42	33.17	0.56
		143	36	58	6.36	33.19	0.14*
		144	30	58	5.84	33.22	0.58
		144	40	58	5.83	33.20	0.56
		144	44	58	5.82	33.22	0.57
		153	32	57	7.61	33.02	0.59
		154	30	57	6.12	33.35	0.57
		154	45	57	6.11	33.38	0.58
XWCC-15	8/3/77-8/6/77	123	30	92	9.52	32.85	1.62
		123	35	92	9.50	32.85	1.46
		124	30	95	9.19	33.04	0.12*
		124	38	95	8.99	33.05	1.39
		132	20	92	10.09	32.86	1.48
		132	34	92	9.67	33.00	1.54
		133	30	92	8.97	33.11	1.26
		133	43	92	8.98	33.11	1.25
		134	50	94	8.48	33.09	1.44
		142	32	93	9.48	32.95	1.68
		143	37	92	8.47	33.01	1.86
		144	30	94	8.35	33.35	1.32
		144	44	94	8.31	33.26	1.27
		153	33	93	8.76	33.03	1.57
		154	30	93	8.51	33.33	1.42
		154	40	93	8.51	33.32	1.49
		154	46	93	8.51	33.32	1.47

Slope = 0.0106 μ moles P/litter/day; n = 68, r = 0.86

207, 217, and 227 (Thomas *et al.*, 1979) were averaged for the two areas since they are close in location to stations used for the regeneration calculation. An approximate value for the Long Island coast was calculated similarly from Figures 2 and 5 of Walsh *et al.* (1978). Table 4 gives these phosphate equivalents as phosphorus requirements and compares them to regeneration rates. The phosphorus requirement and phosphate regeneration rate for the New Jersey coast are higher than those presented by Sharp and Church (1981). Although the productivity value of Sharp and Church is similar to that used here, the

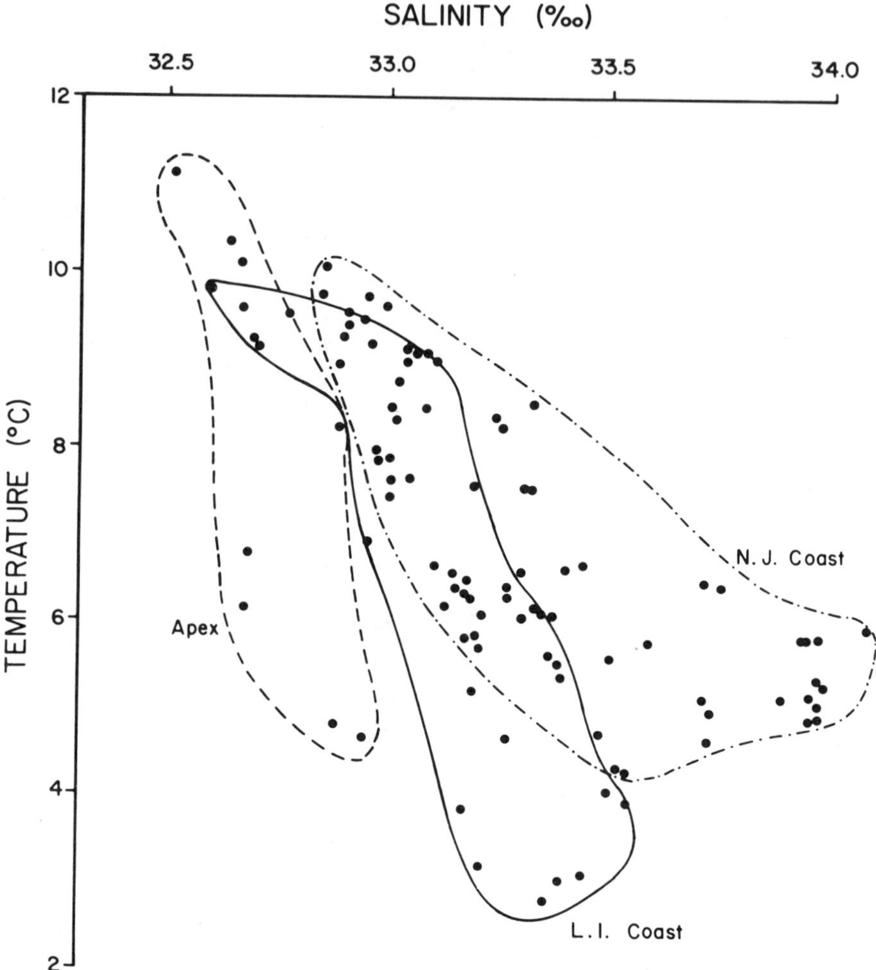

Figure 2. Composite temperature-salinity (TS) diagram for all samples (April-August 1977) used in phosphate regeneration calculations. Envelopes enclose samples from each region.

sampling areas were slightly different, with this study extending further north and further across the shelf. The time coverage of the present study also was shorter, and the data set for late August shows phosphate values considerably higher than those reported for September by Sharp and Church. Nonetheless, the contribution to productivity is of the same magnitude, and it is possible to consider the value of Sharp and Church with those here for the Long Island and New Jersey coasts as being in the range of 27-45%.

Although the apex nutrient regeneration rate is considerably higher than those of the New Jersey and Long Island coasts, the contribution of nutrient regeneration to productivity is much lower. This paper has assumed that surface nutrient regeneration should be similar in magnitude to that calculated for

Table 4. Comparisons of productivity (mg C/m^2/day), phosphorus requirement (μ moles P/liter/day), phosphate regeneration rate (μ moles P/liter/day), and proportion of phosphorus requirement met by *in situ* regeneration (%) for the New York Bight apex, Long Island coast, and New Jersey coast. Productivities for apex and New Jersey coast are from Thomas *et al.* (1979), while that for Long Island Coast is from Walsh *et al.* (1978).

Apex	Productivity	P requirement	P regeneration	%
Apex	6,515	0.33	0.0235	7
Long Island coast	900	0.035	0.0145	41
New Jersey coast	650	0.024	0.0106	44

bottom waters. Respiration data for the apex are provided by Thomas *et al.* (1976). Using values from their Figure 7 for apex stations 34, 51, and 109, shallow water respiration rates appear to be approximately double those of coastal bottom waters. Even if the nutrient regeneration rate for the apex were twice that indicated in Table 4, it still would represent a comparatively small contribution to the productivity ($< 15\%$). A significant outcome of this exercise is that the nutrient regeneration in the apex appears to supply a smaller proportion of the phytoplankton requirement than in the two central shelf regions.

Further comparisons may be made between present estimates of nutrient regeneration and the water column respiration rates of Thomas *et al.* (1976). The average respiration rate for apex bottom water in August 1975 was 0.58 ml O_2/liter/day (Thomas *et al.*, 1976: stations 34, 51, and 109). The phosphate regeneration rate presented in Table 4 of this paper can be translated into oxygen by:

$$\text{ml } O_2/\text{liter/day} = [(\mu \text{ moles P/liter/day})(32)(138)] \div 1.43 \times 10^3,$$

with the molecular weight of $O_2 = 32$, the Redfield O/P = 138, 1.43 mg O_2/liter = 1 ml O_2/liter, and 10^3 μg = 1 mg. With an average phosphate regeneration of 0.0235 μ moles P/liter, the apex oxygen consumption would equal 0.073 ml/liter/day, which is 12.5% of the respiration rate measured by Thomas *et al.* This appears reasonable since respiration includes metazoan, phytoplankton, and bacterial activity, and thus does not reflect exclusively nutrient regeneration processes. Additionally, the respiration estimates are from short-term (six-hour) enclosed samples from one date (August), while the regeneration rates are integrated over four months for non-enclosed bottom waters (from which some loss obviously occurred). Furthermore, respiration rates were estimated in 1975, while regeneration rates were calculated for 1977. That the two estimates are of the same order of magnitude, and that the nutrient regeneration rate gives a smaller oxygen demand confirms that both approaches to understanding the physiology of these bottom waters are basically valid.

The major point of this paper is that *in situ* nutrient regeneration appears to supply a significant portion of the phytoplankton demand in the Long Island

and New Jersey central shelf regions, but not in the apex of the New York Bight. A somewhat novel approach has been taken to quantify this difference. If direct nutrient excretion by zooplankton (about 25% of nutrient requirement:Fournier *et al.*, 1977; Smith and Whitledge, 1977) is added to the microbial regeneration, 50-70% of the phytoplankton nutrient requirement can be accounted for by *in situ* processes in the central shelf areas. In these areas, a small additional nutrient input probably is supplied to the photic zone by diffusion from bottom waters (about 10%; Sharp and Church, 1981). The remaining 20-40% of required nutrients presumably are derived from offshore (Walsh *et al.*, 1978). The apex area may receive a proportional amount of nutrients from direct zooplankton excretion, a proportional or slightly larger amount from diffusion from the bottom waters, and a smaller amount from offshore. The remaining required nutrients, which for the apex are probably in excess of 50%, must come from estuarine outflow.

REFERENCES

Boicourt, W.C. and P.W. Hacker. 1976. Circulation on the Atlantic continental shelf of the United States, Cape May to Cape Hatteras. Mem. Soc. R. Sci. Liege, 6th Series 10: 187-200.

Eppley, R.W., W.G. Harrison, S.W. Chisholm, and E. Stewart. 1977. Particulate organic matter in surface waters off Southern California and its relationship to phytoplankton. J. Mar. Res. 35: 671-696.

Eppley, R.W., E.H. Renger, E.L. Venrick, and M.M. Mullin. 1973. A study of plankton dynamics and nutrient cycling in the central gyre of the north Pacific Ocean. Limnol. Oceanogr. 18: 534-551.

Fournier, R.O., J. Marra, R. Bohrer, and M. Van Det. 1977. Plankton dynamics and nutrient enrichment of the Scotian Shelf. J. Fish. Res. Board Can. 34: 1004-1018.

Goldman, J.C., J.J. McCarthy, and D.G. Peavey. 1979. Growth rate influence on the chemical composition of phytoplankton in oceanic waters. Nature 279: 210-215.

Golterman, H.L. 1964. Mineralization of algae under sterile conditions or by bacterial breakdown. Int. Ver. Theor. Angew. Limnol. Verh. 15: 544-548.

Grill, E.V. and F.A. Richards. 1964. Nutrient regeneration from phytoplankton decomposing in seawater. J. Mar. Res. 22: 51-69.

Gross, M. (ed.). 1976. Middle Atlantic Continental Shelf and the New York Bight. Am. Soc. Limnol. Oceanogr. Spec. Symp. 2. 441 pp.

Halliwell, G.R. and C.N.K. Mooers. 1979. The space-time structure and variability of the shelf water/slope water and Gulf Stream surface temperature fronts and associated warm-core eddies. J. Geophys. Res. 84: 7707-7725.

Harrison, W.G. 1978. Experimental measurements of nitrogen remineralization in coastal waters. Limnol. Oceanogr. 23: 684-694.

Hazelworth, J.B., A. Herman, T.M. Niedrauer, and P.G. Hanson. 1978a. MESA New York Bight Project Expanded Water Dolumn Characterization Cruise (XWCC-12), NOAA Ship GEORGE B. KELEZ, April-May 1977. NOAA Data Report ERL MESA-33. 176 pp.

Hazelworth, J.B., T.M. Niedrauer, and P.G. Hanson. 1978b. MESA New York Bight Project Expanded Water Column Characterization Cruise (XWCC-13),

NOAA Ship GEORGE B. KELEZ, May-June 1977. NOAA Data Report ERL MESA-34. 215 pp.

Hazelworth, J.B., T.M. Niedrauer, and P.G. Hanson. 1978c. MESA New York Bight Project Expanded Water Column Characterization Cruise (XWCC-14), NOAA Ship GEORGE B. KELEZ, June-July 1977. NOAA Data Report ERL MESA-37. 117 pp.

Hazelworth, J.B., T.M. Niedrauer, and P.G. Hanson. 1978d. MESA New York Bight Project Expanded Water Column Characterization Cruise (XWCC-15), NOAA Ship GEORGE B. KELEZ, August 1977. NOAA Data Report ERL MESA-38. 192 pp.

Hazelworth, J.B., T.M. Niedrauer, and P.G. Hanson. 1978e. MESA New York Bight Project Expanded Water Column Characterization Cruise (XWCC-16), NOAA Ship GEORGE B. KELEZ, October 1977. NOAA Data Report ERL MESA-39. 122 pp.

McCarthy, J.J. and J.C. Goldman. 1979. Nitrogenous nutrition of marine phytoplankton in nutrient-depleted waters. Science 203: 670-672.

Malone, T.C. 1977. Environmental regulation of phytoplankton productivity in the lower Hudson estuary. Estuarine Coastal Mar. Sci. 5: 157-171.

Redfield, A.C., B.H. Ketchum, and F.A. Richards. 1963. The influence of organisms on the composition of sea water. *In:* The Sea, Volume II, M.N. Hill (ed.), Interscience, New York. pp. 26-77.

Riley, G.A. 1963. A mathematical model of regional variations in plankton. Limnol. Oceanogr. 10 (Suppl.): R202-R215.

Riley, G.A. 1967. Mathematical model of nutrient conditions in coastal waters. Bull. Bingham Oceanogr. Collect. 19: 72-80.

Ryther, J.H. and W.M. Dunstan. 1971. Nitrogen, phosphorus, and eutrophication in the coastal marine environment. Science 171: 1008-1013.

Sharp, J.H. and T.M. Church. 1981. Biochemical modeling in coastal waters of the Middle Atlantic States. Limnol. Oceanogr. 26: 843-854.

Sharp, J.H., M.J. Perry, E.H. Renger, and R.W. Eppley. 1980. Phytoplankton rate processes in the oligotrophic waters of the central north Pacific Ocean. J. Plankton Res. 2:335-353.

Skirrow, G. 1975. The dissolved gases-carbon dioxide. *In:* Chemical Oceanography, Second Edition, Volume 2, J.P. Riley and G. Skirrow (eds.), Academic Press, New York. pp. 1-192.

Smith, S.L. and T.E. Whitledge. 1977. The role of zooplankton in the regeneration of nitrogen in a coastal upwelling system off northwest Africa. Deep-Sea Res. 24: 49-56.

Thomas, J.P., J.E. O'Reilly, A.F.J. Draxler, J.A. Babinchak, C.N. Robertson, W.C. Phoel, R.I. Waldhauer, C.A. Evans, A. Matte, M.S. Cohn, M.F. Nitkowski, and S. Dudley. 1979. Biological processes: production and respiration. *In:* Oxygen Depletion and Associated Benthic Mortalities in New York Bight, 1976, R.L. Swanson and C.J. Sindermann, U.S. Department of Commerce, NOAA. pp. 231-261.

Thomas, J.P., W.C. Pheol, F.W. Steimle, J.E. O'Reilly, and C.A. Evans. 1976. Seabed oxygen consumption-New York Bight apex. *In:* Middle Atlantic Continental Shelf and the New York Bight, M.G. Gross (ed.), Am. Soc. Limnol. Oceanogr. pp. 354-639.

Walsh, J.J., T.E. Whitledge, F.W. Barvenik, C.D. Wirick, S.O. Howe, W.E. Esaias, and J.T. Scott. 1978. Wind events and food chain dynamics within the New York Bight. Limnol. Oceanogr. 23: 659-683.

Wright, W.R. 1976. The limits of shelf water south of Cape Cod, 1941-1972. J. Mar. Res. 34: 1-14.

MARINE MACROBENTHOS AS INDICATORS OF ENVIRONMENTAL STRESS

Martha W. Young

U.S. Fish and Wildlife Service
National Coastal Ecosystems Team
NASA/Slidell Computer Complex
Slidell, Louisiana 70458

David K. Young

Naval Ocean Research and
Development Activity
Oceanography Division
NSTL Station, Mississippi 39529

Abstract. Certain assumptions are inherent in the use of marine macrobenthos in impact assessment: (1) species densities are relatively persistent within the time frame of sampling, (2) species densities are reflective of and are quantitative indicators of environmental disturbance, and (3) species that are most dense or contribute the most to overall biomass are the most "important" trophically. These assumptions are examined in the light of field experiments in the Indian River estuary, Florida.

Three selected species (the mollusc, *Diastoma varium;* the amphipod, *Grandidierella bonnieroides;* and the polychaete, *Capitella capitata capitata*) are discussed in terms of their responses to predator exclusion cages with and without the addition of processed sewage sludge. *C. capitata capitata* showed the most dramatic response to organic enrichment by increasing densities in a manner corresponding to that reported in many field observations. Densities of *G. bonnieroides* initially increased inside cages, compared with uncaged areas, suggesting its trophic value as food for fishes. *D. varium* densities did not differ with caging or with organic enrichment.

The value of field experimental studies is emphasized as a means for clarifying results of environmental impact studies through a knowledge of the life histories and natural population fluctuations of individual species.

INTRODUCTION

The macrobenthos often are selected from among other animal groups for assessment of environmental impact of man's activities because they are relatively large in size (usually defined operationally as invertebrates retained on a 1-mm mesh screen), diverse, numerous, relatively stationary, and often trophically important (*i.e.,* in transfer of energy in the food web from primary producers to higher level consumers). In addition, specimens are fairly easily collected, processed, and identified taxonomically. Certain inherent assumptions are that (1) species densities are relatively persistent within the time frame of the sampling period, (2) species densities reflect and are quantitative indicators of environmental disturbance, and (3) species that are most dense or contribute the most to overall biomass are trophically most "important." That these (among other mutually related assumptions) are not necessarily correct is recognized by many benthic ecologists but often are not taken into account when the marine macrobenthos is utilized to assess projected or ongoing environmental impact.

The purpose of this paper is to examine these assumptions with regard to population densities of three taxonomically different macrobenthic species that were selected from a "community" of 197 species sampled over a two-year period of time. Factors regulating densities of these species were determined from field experiments using predator exclusion cages and uncaged treatments, both with and without organic enrichment by periodic addition of processed sewage sludge. The selected species were the gastropod mollusc, *Diastoma varium;* the amphipod crustacean, *Grandidierella bonnieroides;* and the polychaete worm, *Capitella capitata capitata.* These three species occur in benthic samples taken over wide geographic areas and are representative of a range of response to environmental disturbance.

METHODS

The study site in the Indian River, a bar-built estuary on the east central coast of Florida, was established at a shallow-water, subtidal location in well-sorted sediment (88.0% sand, 9.6% silt, 2.4% clay) with extensive cover of the seagrass, *Halodule wrightii.* Four replicate rectangular cores of seagrass and sediment (15 cm x 15 cm x 20 cm deep = 4,500 cm^3) were sampled from the study site from September 1974 to September 1975 at monthly intervals (except during January and August 1975), and from October 1975 to October 1976 at two-month intervals. The latter sampling intervals coincided with sampling periods for the field experiments described below. Each experimental treatment was administered on a 4-m square area. Uncaged areas were staked off in a manner similar to the caged areas, but they were not surrounded by hardware cloth. Cages were constructed of 12-mm mesh hardware cloth, and were 2 m on a side and 2 m high. Each cage penetrated the sediment to 5-cm depth and extended above highest tidal stages.

For the purpose of the paper, implications of results from four of 12 previously reported experimental treatments (Young and Young, 1978) will be considered: (1) caged–to inhibit natural predator effects on the macrobenthos; (2) uncaged–to serve as a control of cage effects; (3) caged Milorganite[1]–to test effects of monthly applications of 567 g/m^2 (5 lb/4 m^2) of processed sewage sludge upon the macrobenthos and to inhibit effects of natural predators; and (4) uncaged Milorganite–to test effects of processed sewage sludge (567 g/m^2) additions in the absence of a cage. Statistical analyses of results from these experimental treatments are described in Young and Young (1978).

RESULTS AND DISCUSSION

The assumption that densities of macrobenthos are relatively persistent within the time frame of sampling is basic to many assessment studies that

[1] Milorganite is the trade name of processed sewage sludge in a granular form from Milwaukee, Wisconsin. The most common use of this product is as a high organic nitrogen fertilizer. The minimum guaranteed analysis is as follows: total nitrogen (6.00%), available phosphoric acid (4.00%), Fe_2O_3 (7.5%), ZnO (0.10%), and SO_3 (2.5%).

attempt to generalize from samples taken over a given time interval. The interval frequently is too short a period to warrant such an assumption (Hedgpeth, 1978). Our data from two years of sampling in the Indian River estuary, Florida, show a range of variability of species densities (Figure 1) consistent with data from longer studies, such as those of Segerstrale (1960), Buchanan *et al.* (1974), and Boesch *et al.* (1976). These demonstrate that annual fluctuations of numerically abundant species are the rule rather than the exception for coastal marine and estuarine macrobenthos.

The three species selected for discussion in this paper have certain similarities with the groupings of Buchanan *et al.* (1974). The mollusc, *Diastoma varium,* is a "volatile" species, displaying high productivity with widely fluctuating densities. The amphipod, *Grandidierella bonnieroides,* is a more "conservative" species, having persistently low, relatively unfluctuating densities over long periods of time. The polychaete, *Capitella capitata capitata,* is widely accepted as an "opportunistic" species, rapidly increasing in density with disturbance, but maintaining fairly uniform densities (within an order of magnitude) under natural circumstances. The rationale for these groupings is examined for each species in relation to results from the caging experiments, as follows.

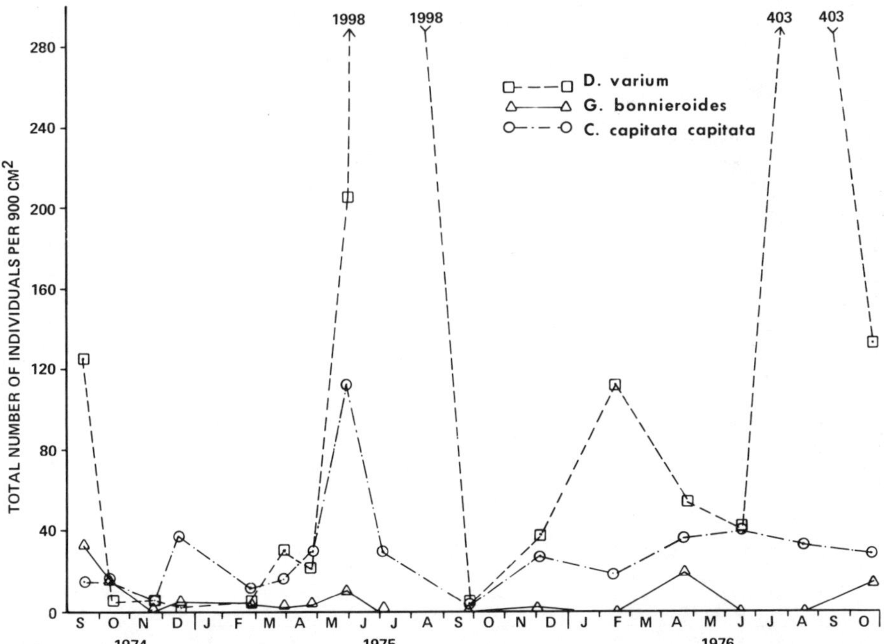

Figure 1. Total number of individuals per 900 cm^2 for *Diastoma varium* (open squares and dashed line), *Grandidierella bonnieroides* (open triangles and solid line), and *Capitella capitata capitata* (open circles and broken line) from the Indian River study site from September (S) 1974 to October (O) 1976.

Diastoma varium (Pfeiffer, 1840)

Diastoma (=*Bittium*) *varium,* a gastropod mollusc in the family Cerithiidae, commonly is found among seagrasses from Maryland to Florida and from Texas to Brazil (Abbott, 1974). Adams and Angelovic (1970) demonstrated in laboratory experiments that *D. varium* feeds upon epiphytes and seagrass detritus. Marsh (1973) reported this species as the numerically dominant epifaunal taxon collected from the seagrass, *Zostera marina,* in the York River of the Chesapeake Bay.

Densities of *D. varium* were not affected by caging or by the addition of Milorganite (Figure 2). An earlier experiment at the same study site with caged and uncaged treatments resulted in densities of *D. varium* that were not significantly different inside and outside cages over a seven-month period, but that differed significantly with time (Buzas, 1978).

High uncaged densities of *D. varium* might lead to unrealistic conclusions regarding the trophic importance of these animals. For example, Petersen (1918), upon collecting up to 100,000 specimens/m^2 of an ecologically similar snail among seagrass in Danish waters, concluded that they were "of considerable importance as fish food [p. 41]." The high densities of these snails, presumably the result of one generation per year, were collected during the

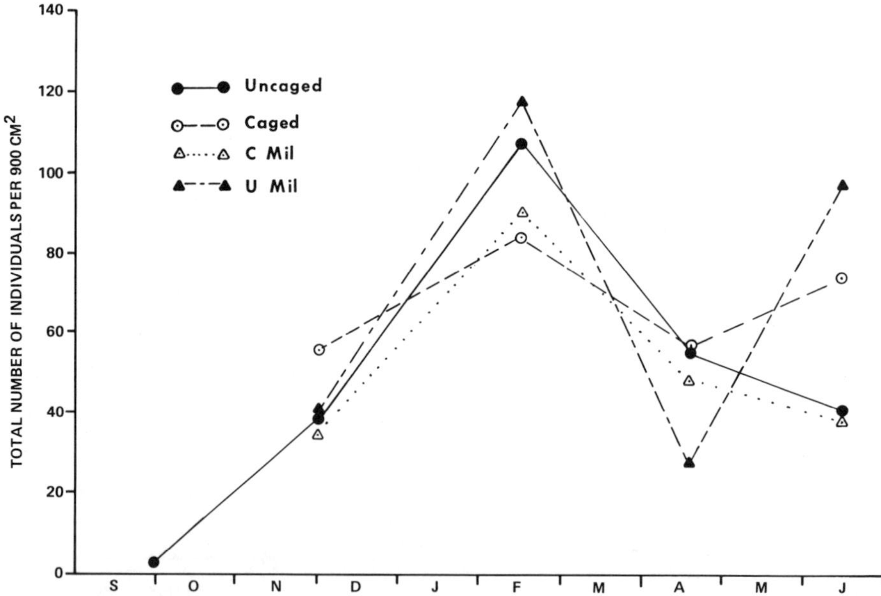

Figure 2. Total number of individuals per 900 cm^2 for *Diastoma varium* for uncaged treatment (solid circles and solid line), caged treatment (open circles and dashed line), caged Milorganite treatment (open triangles and dotted line), and uncaged Milorganite treatment (solid triangles and broken line) from September (S) 1975 to June (J) 1976.

period from spring to autumn. Rasmussen (1973) later reevaluated Petersen's conclusions, suggesting that the trophic importance of these snails (*Rissoa* sp.) was "greatly over-estimated [p. 429]" precisely because specimens were collected in such high densities. Rasmussen believed that Petersen did not consider crustaceans (gammarids, isopods, and mysids) as important ecologically as the gastropods; but Petersen (1918:p. 42) stated:

> From investigation of stomach content, the crustacea appear to play a more important part as fish food than do the *Rissoa,* despite their comparatively smaller stock; it is possible, however, that the stock is renewed twice a year or more often, and that the annual production is thus really larger than would appear from the size of the stock at any time.

Grandidierella bonnieroides (Stephenson, 1948)

Grandidierella bonnieroides is a gammaridean amphipod in the family Corophiidae. The range of this species extends southward along the coast of the United States in seagrass habitats from North Carolina and the east coast of Florida, throughout the Caribbean, and off Madagascar (Zimmerman, 1978). Zimmerman *et al.* (1979) have performed laboratory experiments indicating that although *G. bonnieroides* prefers seagrass detritus as a source of food, it occasionally eats epiphytic algae.

The pioneer field experimental study of Blegvad (1928) demonstrated that macrobenthic species that became more dense inside cages were not necessarily those that were collected abundantly from the adjacent uncaged sea floor. In our study when *G. bonnieroides* was caged, an initial increase in density resulted (Figure 3). This increase corresponded to the season when juvenile fish predators, small enough to pass through the 12-mm cage mesh, were absent from the Indian River estuary (G. Gilmore, personal communication). The impact of predation by small fishes upon subtidal amphipods was believed by Zimmerman (1978) to be the major regulator of species densities in Puerto Rican waters. He reported a predation rate of 17,480 amphipods (17.2 g dry, 68.4 g wet) per m^2/yr, with an estimated turnover rate of the amphipod population of once every three months.

Capitella capitata capitata Warren, 1976

Capitella capitata capitata, described by Eisig (1887) and Hartman (1947) as *Capitella capitata,* is a burrowing deposit feeder of cosmopolitan distribution. This species is found in temperate latitudes off the coast of the United States and northern Europe, as well as in the Mediterranean and southward along South America and Africa to the Antarctic. Many studies of the macrobenthos do not distinguish among subspecies of *C. capitata.* Three distinctly different morphological forms of the genus *Capitella* routinely were found in our samples that correspond to descriptions in the taxonomic literature of *C. capitata*

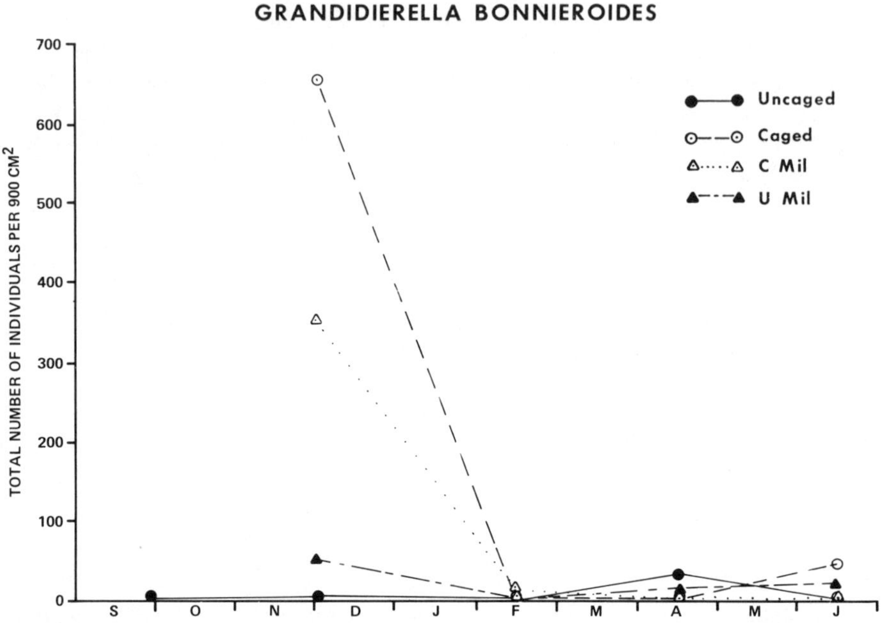

Figure 3. Total number of individuals per 900 cm² for *Grandidierella bonnieroides* from uncaged treatment (solid circles and solid line), caged treatment (open circles and dashed line), caged Milorganite treatment (open triangles and dotted line), and uncaged Milorganite treatment (solid triangles and broken line) from September (S) 1975 to June (J) 1976.

capitata, C. capitata ovincola, and *C. capitata tripartita* (Warren, 1976). We report here only the data for *C. capitata capitata.* Ecological observations about *C. capitata* that follow apply also to *C. capitata capitata.*

In our field experiments, densities of *C. capitata capitata* increased greatly following the addition of Milorganite to the cage (Figure 4). Specimens of *C. capitata capitata* also were more robust (defined as the width of the thoracic region at the level of the fifth setiger) from Milorganite-treated plots than from those without Milorganite (Young and Young, 1978).

Much has been written about the use of *C. capitata* as an indicator of pollution (*e.g.,* Wilhelmi, 1916; Reish, 1957a; Felice, 1959; Warren, 1977), and other authors (Reish, 1957b; Rosenberg, 1972; Anger, 1975; Gray, 1976) further specified that high densities of this polychaete are associated with organic enrichment.

Pearson (1975) noted that as enrichment levels increased in a Scottish sea loch polluted by pulp and paper mill effluent, total numbers of species decreased while densities of pollution-tolerant species remained high. Pearson reported *C. capitata* as the most abundant species at the sampling station nearest to the point of effluent discharge during those years when the mean effluent load and biological oxygen demand were greatest.

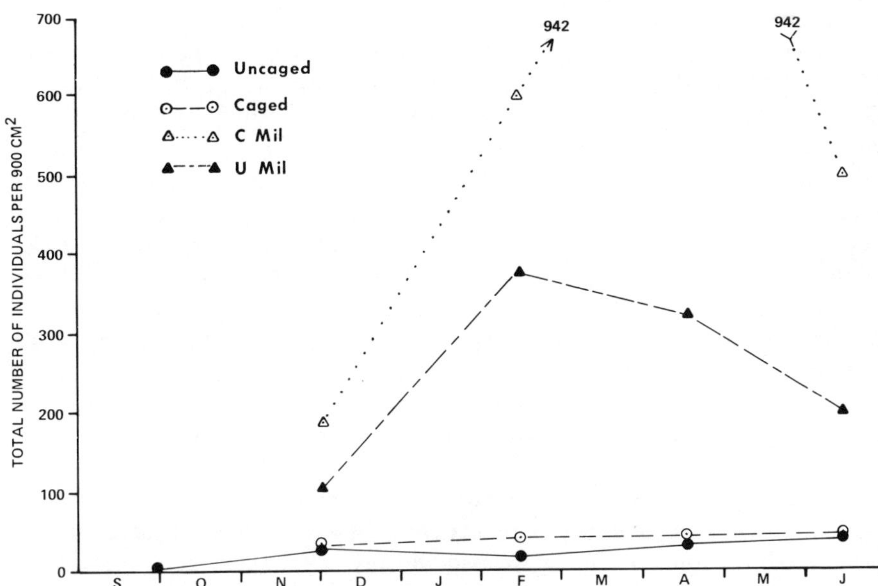

Figure 4. Total number of individuals per 900 cm² for *Capitella capitata capitata* from uncaged treatment (solid circles and solid line), caged treatment (open circles and dashed line), caged Milorganite treatment (open triangles and dotted line), and uncaged Milorganite treatment (solid triangles and broken line) from September (S) 1975 to June (J) 1976.

Smith *et al.* (1973) indicated that organic overload results in the numerical dominance of *C. capitata* and other macrobenthic species that can utilize increased organic matter and tolerate low levels of dissolved oxygen. Warren (1977) contended that *C. capitata* is no more resistant to stress factors resulting from polluted conditions (*e.g.*, high concentrations of hydrogen sulfide: Theede *et al.*, 1969; lowered levels of dissolved oxygen: Muus, 1967; Reish, 1970; Mangum and Van Winkle, 1973; Wells and Warren, 1975; extreme variations of temperature: Warren, 1977; and extreme variations in salinity: Estcourt, 1967) than many macrobenthic species that are not usually found in polluted environments. These environmental factors, however, may be important in determining the presence or absence of other species in polluted areas–including predators (McNulty, 1970) that may be "caged out" or excluded by stressful conditions (Young and Young, 1977).

Grassle and Grassle (1974) described the opportunistic aspects of *C. capitata* life history, suggesting that this species is composed of a number of sibling species that are able to exploit various types of disturbed environments because of their inherent genetic variability. *C. capitata* reproduces the year around, completes its life cycle rapidly (30-40 days), produces large numbers of eggs per female, enjoys brood protection of larvae, and, in some populations, is entirely benthonic, avoiding the hazards of a pelagic phase. Populations of *C. capitata* typically increase rapidly following disturbance or the exploitation of an open environment; then numbers decline because of the organism's poor competitive ability, perhaps resulting from a shortage of the "most labile food components" (Grassle and Grassle, 1974: p. 276). Anger (1975) indicated that species such as *C. capitata* may, in fact, compete for space more successfully than other species when the food supply is increased by sewage particles.

Sanders *et al.* (1972) speculated that *C. capitata* is scarce or absent where the macrobenthos is diverse, because of competition and/or predation. Whether or not *C. capitata* is a "poor" or "good" competitor is still an open question because of contradictory evidence. For example, data from Cape Cod Bay (Young and Rhoads, 1971) show high densities of *C. capitata* (10,600/m^2) at the station with the highest numbers of species, highest numbers of individuals, and highest biomass–where *C. capitata* clearly was "competing" successfully.

How does *C. capitata* readily take advantage of organic enrichment? Laboratory observations (K. Eckelbarger, personal communication) showed that particles of processed sewage sludge are eaten directly as well as used in tube-building. Warren (1977) contended that this species digests microorganisms associated with decomposing plant material. Henriksson (1969) suggested that increases in *C. capitata* are correlated directly with numbers of bacteria in polluted environments. Whether *C. capitata* utilizes enriched organic matter directly or indirectly via associated microorganisms is a question that cannot be addressed adequately here.

Our field experimental results (Young and Young, 1978) support the direct correlation between densities and biomass (measured as robustness) of *C. capitata capitata* and high concentrations of organic matter, as observed by numerous workers. These results also provide the first field experimental evidence in support of the hypothesis of Tenore (1977) that this species is nitrogen-limited in most natural environments. Measurements by others (Blegvad, 1932; Chen, 1970; Anger, 1975; Orlob and O'Leary, 1977) have shown increased production of macrobenthos in sewage-enriched areas of the sea floor where dissolved oxygen is not depleted. Anger (1975: p. 420) pointed out that certain macrobenthos that are preadapted to process sewage particles provide "valuable biomass for the food web" when these abundant food organisms are eaten.

Petersen (1918) clearly understood the ecological distinction between production, measured as standing crop, and productivity, measured as total production over a period of time. In addition, he recognized that certain macrobenthic species were "doubtless devoured by crabs [p. 41]" and that the crabs, in turn, were eaten by cod. Not stated by Petersen, however, was the corollary that, if left uneaten by crabs, these macrobenthic species would not be utilized efficiently (in terms of energy transfer) in the food web. In the New York Bight, Pearce (1972a: p. 78) reported that cancroid crabs "frequently constitute from 90 to 100 percent of the diet of some of the larger ground fish." That such crabs utilize other macrobenthic species as food seems apparent, but their role as "translators" of energy (Young and Young, 1978), as well as bioaccumulators of certain toxins, remains largely unassessed. Without question, the high incidence of necrotic lesions of cancroid crabs from sewage sludge disposal areas in the New York Bight (Pearce, 1972b) is a cause for concern and must be balanced with any consideration of enhanced macrobenthic production by organic enrichment.

IMPLICATIONS

Field experiments, either occurring in relation to natural or man-induced events, provide the opportunity to study large scale perturbations of natural systems. Dimon and Dauer (1977) reported recolonization in Tampa Bay, Florida by opportunistic polychaetes in the first month after a red tide, followed by recolonization by molluscs and amphipods, with the establishment of an equilibrium level of species by month 11 after defaunation. Pioneering species settling in trays of defaunated mud in Long Island Sound reached maximum abundance in 10 to 50 days (McCall, 1977). Rhoads *et al.* (1978) cited the value of comparing the Long Island Sound experimental field results with data from dredged material disposal sites in the same area. In a Scottish sea loch, a doubling of macrobenthic production in two to three years followed intentional fertilization (Raymont, 1949), which, in turn, resulted in a great increase in the growth rate of flounders (Gross, 1949).

Many species of macrobenthos may be unavailable to predators because of deep-burrowing habits, or because they are unattractive as food items. Buchanan and Warwick (1974) suggested that certain species, such as the burrowing anemone, *Cerianthus lloydi,* contribute considerably to overall benthic biomass but little to benthic productivity. Mills (1975) observed that certain long-lived sessile epifauna (*e.g.,* crinoids, tunicates, and sponges) are indigestible because a large portion of their body weight consists of inedible structural material. Such benthic animals, while they may be the largest individuals available, are not selected by fish predators and, as a result, contribute little to fish production.

Pearson and Rosenberg (1976:p. 79) suggested that succession from echinoderms to large populations of infaunal annelids, in response to increasing organic waste, might result in "an enhanced food supply for bottom-feeding fish." Pearce *et al.* (1976:p. 401) raised essentially the same question, "Does the augmentation of pollution-resistant, opportunistic, detritus feeders result in increased biomass available as food to the finfish?"

The laboratory experiments of Trevallion and Ansel (1967), in which the bivalve, *Tellina tenuis,* was grown in enriched seawater, were not able to demonstrate the transfer of the resulting increased production of *Tellina* to adult plaice. While individual experiments add only small pieces to the puzzle, the answers to the questions raised by Mills (1975) appear to lie in further manipulative field and laboratory experiments. Hargrave and Levings (1975:p. 2273) stated "Perturbation, either by enhancement or disturbance . . . is the most direct way in which to consider the underlying mechanisms that determine pathways and rates of processes in ecological systems."

Allowing that one-point-in-time comparisons of macrobenthic communities are largely meaningless because of the unpredictable nature of densities of the component species, sampling for assessment purposes must be geared to the early weeks of environmental stress to take recruitment and rapid development of opportunists into account. When an unpredicted environmental stress occurs, either from a natural disturbance or as a result of man's activities, speed and completeness of sampling the macrobenthos must be emphasized (*e.g.*, Sanders *et al.*, 1972 for the West Falmouth oil spill; Boesch *et al.*, 1976 for the aftereffects of Hurricane Agnes). Sampling should follow at longer intervals, as frequently as is logistically possible, to establish long-term population trends. Adequate controls in unperturbed areas should be established to avoid misinterpretation of natural variability as a response to the stress under consideration.

Time limitations of sampling and problems of spatial variability often are compounded because observed environmental variables tend to be those that are easiest to measure rather than those that are the most valuable for interpretation of data. Recognition of the importance of predation as a density regulator must accompany evaluation of carefully selected physical and chemical variables.

Field experimentation is necessary to insure accurate assessments of the effects of limited variables upon macrobenthic densities at any given time. While field experiments such as caging have limitations (Virnstein, 1978), at least the number of variables is controllable to some degree, and cause-and-effect relationships are more discernable than in field monitoring and assessment studies. Although practical constraints of impact assessment studies (*e.g.*, time, manpower, economics) may not permit manipulative field experiments, the results of such experiments may be extrapolated to influence interpretation of field studies that do not include controlled manipulations (Rhoads *et al.*, 1978).

The macrobenthos cannot be viewed as responding to stress as an entity unto itself. Animals that are closely related taxonomically may not react to a perturbation in the same way (Young and Young, 1978). An in-depth understanding of life histories (*e.g.*, Grassle and Grassle, 1974 and Warren, 1977 for capitellids) and ecological observations of each species in its natural environment through field experimentation are needed.

The statement of Howmiller and Scott (1977:p. 810) for freshwater macrobenthos is equally applicable for the use of marine macrobenthos in environmental assessment studies:

> The full potential of benthic studies for environmental assessment will be realized only through studies that consider data on community structure at the species level in combination with existing information on the ecological attributes of the constituent species.

ACKNOWLEDGEMENTS

We thank two anonymous reviewers for their helpful suggestions and Ms. R. Mauffray for typing the manuscript. This study was funded by the Harbor

Branch Foundation, Inc. and by a grant from the Atlantic Foundation to the Smithsonian Institution. Support for preparation of the manuscript was provided by Naval Ocean Research and Development Activity and Mississippi State University Research Center.

REFERENCES

Abbott, R.T. 1974. American Seashells. The Marine Mollusca of the Atlantic and Pacific Coasts of North America. Van Nostrand Reinhold Co., New York. 663 pp.

Adams, S.M. and J.W. Angelovic. 1970. Assimilation of detritus and its associated bacteria by three species of estuarine animals. Chesapeake Sci. 11: 249-254.

Anger, K. 1975. On the influence of sewage pollution on inshore benthic communities in the South of Kiel Bay. Helgol. Wiss. Meeresunters. 27: 408-438.

Blegvad, H. 1928. Quantitative investigations of bottom invertebrates in the Limfjord 1910-1927 with special reference to plaice food. Rep. Danish Biol. Sta. 34: 33-52.

Blegvad, H. 1932. I. Investigations of the bottom fauna at outfalls and drains in the Sound. Rep. Danish Biol. Sta. 37: 1-20.

Boesch, D.F., M.L. Wass, and R.W. Virnstein. 1976. The dynamics of estuarine benthic communities. *In:* Estuarine Processes, Vol. I, M. Wiley (ed.), Academic Press, New York. pp. 177-196.

Buchanan, J.B., P.F. Kingston, and M. Sheader. 1974. Long-term population trends of the benthic macrofauna in the offshore mud of the Northumberland coast. J. Mar. Biol. Assoc. U.K. 54: 785-795.

Buchanan, J.B. and R.M. Warwick. 1974. An estimate of benthic macrofaunal production in the offshore mud of the Northumberland coast. J. Mar. Biol. Assoc. U.K. 54: 197-222.

Buzas, M.A. 1978. Community unity? Patterns in molluscs and foraminifera. *In:* Estuarine Interactions, M. Wiley (ed.), Academic Press, New York. pp. 173-190.

Chen, C.W. 1970. Effects of San Diego's wastewater discharge on the ocean environment. J. Water Pollut. Control Fed. 42: 1458-1467.

Eisig, H. 1887. Monographie der Capitelliden des Golfes von Neapel. *In:* Fauna und Flora des Golfes von Neapel, Monogr. 906 pp.

Estcourt, I.N. 1967. Ecology of benthic polychaetes in the Heathcote estuary, New Zealand. N.Z. J. Mar. Freshwater Res. 1: 371-394.

Felice, F.P. 1959. The effect of wastes on the distribution of bottom invertebrates in the San Francisco Bay estuary. Wasmann J. Biol. 17: 1-17.

Grassle, J.F. and J.P. Grassle. 1974. Opportunistic life histories and genetic systems in marine benthic polychaetes. J. Mar. Res. 32: 253-284.

Gray, J.S. 1976. The fauna of the polluted River Tees estuary. Estuarine Coastal Mar. Sci. 4: 653-676.

Gross, F. 1949. Further observations on fish growth in fertilized sea loch (Loch Craiglin). J. Mar. Biol. Assoc. U.K. 28: 1-8.

Hargrave, B.T. and C.D. Levings. 1975. A prognosis for Canadian benthic biological oceanography. J. Fish. Res. Board Can. 32: 2268-2275.

Hartman, O. 1947. Polychaetous annelids. Part 7. Capitellidae. Allan Hancock Pacific Expeditions 10: 391-481.

Hedgpeth, J.W. 1978. As blind men see the elephant: the dilemma of marine ecosystem research. *In:* Estuarine Interactions, M. Wiley (ed.), Academic Press, New York. pp. 3-15.

Henriksson, R. 1969. Influence of pollution on the bottom fauna of the Sound (Öresund). Oikos 20: 507-523.

Howmiller, J.P. and M.A. Scott. 1977. An environmental index based on relative abundance of oligochaete species. J. Water Pollut. Control Fed. 49: 809-815.

McCall, P.L. 1977. Community patterns and adaptive strategies of the infaunal benthos of Long Island Sound. J. Mar. Res. 35: 221-266.

McNulty, J.K. 1970. Effects of abatement of domestic sewage pollution on the benthos, volumes of zooplankton, and the fouling organisms of Biscayne Bay, Florida. Stud. Trop. Oceanogr. (Miami) 9: 1-107.

Mangum, C. and W. Van Winkle. 1973. Responses of aquatic invertebrates to declining oxygen conditions. Am. Zool. 13: 529-541.

Marsh, G.A. 1973. The *Zostera* epifaunal community in the New York River, Virginia. Chesapeake Sci. 14: 87-97.

Mills, E.L. 1975. Benthic organisms and the structure of marine ecosystems. J. Fish. Res. Board Can. 32: 1657-1663.

Muus, B.J. 1967. The fauna of Danish estuaries and lagoons. Medd. Dan. Fisk. Havunders. N.S. 5: 1-316.

Orlob, G.T. and D.A. O'Leary. 1977. Impact on marine benthos of wastewater discharge. J. Environ. Eng. Div., EE2. pp. 307-320.

Pearce, J.B. 1972a. Invertebrate resources; available forms and potentials. *In:* Resources of the World's Oceans, H. Frey (ed.), New York Institute of Ocean Resources, Inc., New York. pp. 75-90.

Pearce, J.B. 1972b. The effects of solid waste disposal on benthic communities in the New York Bight. *In:* Marine Pollution and Sea Life, M. Ruivo (ed.), Fishing News (books) Ltd., Surrey-London, England. pp. 404-411.

Pearce, J.B., J.V. Caracciolo, M.B. Halsey, and L.H. Rogers. 1976. Temporal and spatial distributions of benthic macroinvertebrates in the New York Bight. *In:* Middle Atlantic Continental Shelf and the New York Bight, M. Gross (ed.), Am. Soc. Limnol. Oceanogr. Spec. Symp. 2: 394-403.

Pearson, T.H. 1975. The benthic ecology of Loch Linnhe and Loch Eil, a sea-loch system on the west coast of Scotland. IV. Changes in the benthic fauna attributable to organic enrichment. J. Exp. Mar. Biol. Ecol. 20: 1-41.

Pearson, T.H. and R. Rosenberg. 1976. A comparative study of the effects on the marine environment of wastes from cellulose industries in Scotland and Sweden. Ambio 5: 77-79.

Petersen, C.G.J. 1918. The sea bottom and its production of fish food. A survey of the work done in connection with valuation of the Danish waters from 1883-1917. Rep. Danish Biol. Sta. 25: 1-82.

Rasmussen, E. 1973. Systematics and ecology of the Isefjord marine fauna (Denmark). With a survey of the eelgrass (*Zostera*) vegetation and its communities. Ophelia 11: 1-507.

Raymont, J.E.G. 1949. Further observations on changes in the bottom fauna of a fertilized sea loch. J. Mar. Biol. Assoc. U.K. 28: 9-19.

Reish, D.J. 1957a. Effect of pollution on marine life. Industrial Wastes September-October 2: 114-118.

Reish, D.J. 1957b. The relationship of the polychaetous annelid *Capitella capitata* (Fabricius) to waste discharged of biological origin. *In:* Biological Problems in Water Pollution, C. Tarzwell (ed.), U.S. Public Health Service. pp. 195-200.

Reish, D.J. 1970. The effects of varying concentrations of nutrients, chlorinity and dissolved oxygen on polychaetous annelids. Water Res. 4: 721-735.

Rhoads, D.C., P.L. McCall, and J.Y. Yingst. 1978. Disturbance and production on the estuarine seafloor. Am. Sci. 66: 577-586.

Rosenberg, R. 1972. Benthic faunal recovery in the Swedish fjord following the closure of a sulphite pulp mill. Oikos 23: 92-108.

Sanders, H.L., J.F. Grassle, and G.R. Hampson. 1972. The West Falmouth oil spill. I. Biology. Woods Hole Oceanogr. Inst. Tech. Rep. WHOI 72-20. 48 pp.

Segerstrale, S.G. 1960. Fluctuations in the abundance of benthic animals in the Baltic Sea. Commentat. Biol. Soc. Sci. Fenn. 23: 3-19.

Simon, J.L. and D.M. Dauer. 1977. Reestablishment of a benthic community following natural defaunation. *In:* Ecology of Marine Benthos, B.C. Coull (ed.), University of South Carolina Press, Columbia, SC. pp. 139-154.

Smith, K.L., Jr., G.T. Rowe, and J.A. Nichols. 1973. Benthic community respiration near the Woods Hole sewage outfall. Estuarine Coastal Mar. Sci. 1: 65-70.

Tenore, K.R. 1977. Growth of *Capitella capitata* cultured on various levels of detritus derived from different sources. Limnol. Oceanogr. 22: 936-941.

Theede, H., A. Ponat, K. Hiroki, and C. Schlieper. 1969. Studies on the resistance of marine bottom invertebrates to oxygen-deficiency and hydrogen sulphide. Mar. Biol. 2: 325-337.

Trevallion, A. and A. Ansell. 1967. Studies on *Tellina tenuis* Da Costa. II. Preliminary experiments in enriched sea water. J. Exp. Mar. Biol. Ecol. 1: 257-270.

Virnstein, R.W. 1978. Predator caging experiments in soft sediments: caution advised. *In:* Estuarine Interactions, M. Wiley (ed.), Academic Press, New York. pp. 261-273.

Warren, L.M. 1976. A review of the genus *Capitella* (Polychaeta: Capitellidae). J. Zool. London 180: 195-209.

Warren, L.M. 1977. The ecology of *Capitella capitata* in British waters. J. Mar. Biol. Assoc. U.K. 57: 151-159.

Wells, R.M.G. and L.M. Warren. 1975. The function of the cellular hemoglobins in *Capitella capitata* (Fabricius) and *Notomastus latericeus* Sars (Capitellidae: Polychaeta). Comp. Biochem. Physiol. 51A: 737-740.

Wilhelmi, J. 1916. Übersicht über die biologische Beurteilung des Wassers. Ges. Naturf. Freunde Berlin: 297-306.

Young, D.K. and D.C. Rhoads. 1971. Animal-sediment relations in Cape Cod Bay, Massachusetts. I. A transect study. Mar. Biol. 2: 242-254.

Young, D.K. and M.W. Young. 1977. Community structure of the macrobenthos associated with seagrass of the Indian River estuary, Florida. *In:* Ecology of Marine Benthos, B.C. Coull (ed.), University of South Carolina Press, Columbia, SC. pp. 359-381.

Young, D.K. and M.W. Young. 1978. Regulation of species densities of seagrass-associated macrobenthos: evidence from field experiments in the Indian River estuary, Florida. J. Mar. Res. 36: 569-593.

Zimmerman, R.J. 1978. The feeding habits and trophic position of dominant gammaridean amphipods in a Caribbean seagrass community. Ph.D. Thesis, University of Puerto Rico. 85 pp.

Zimmerman, R.J., R. Gibson, and J. Harrington. 1979. Herbivory and detritivory among gammaridean amphipods from a Florida seagrass community. Mar. Biol. 54: 41-47.

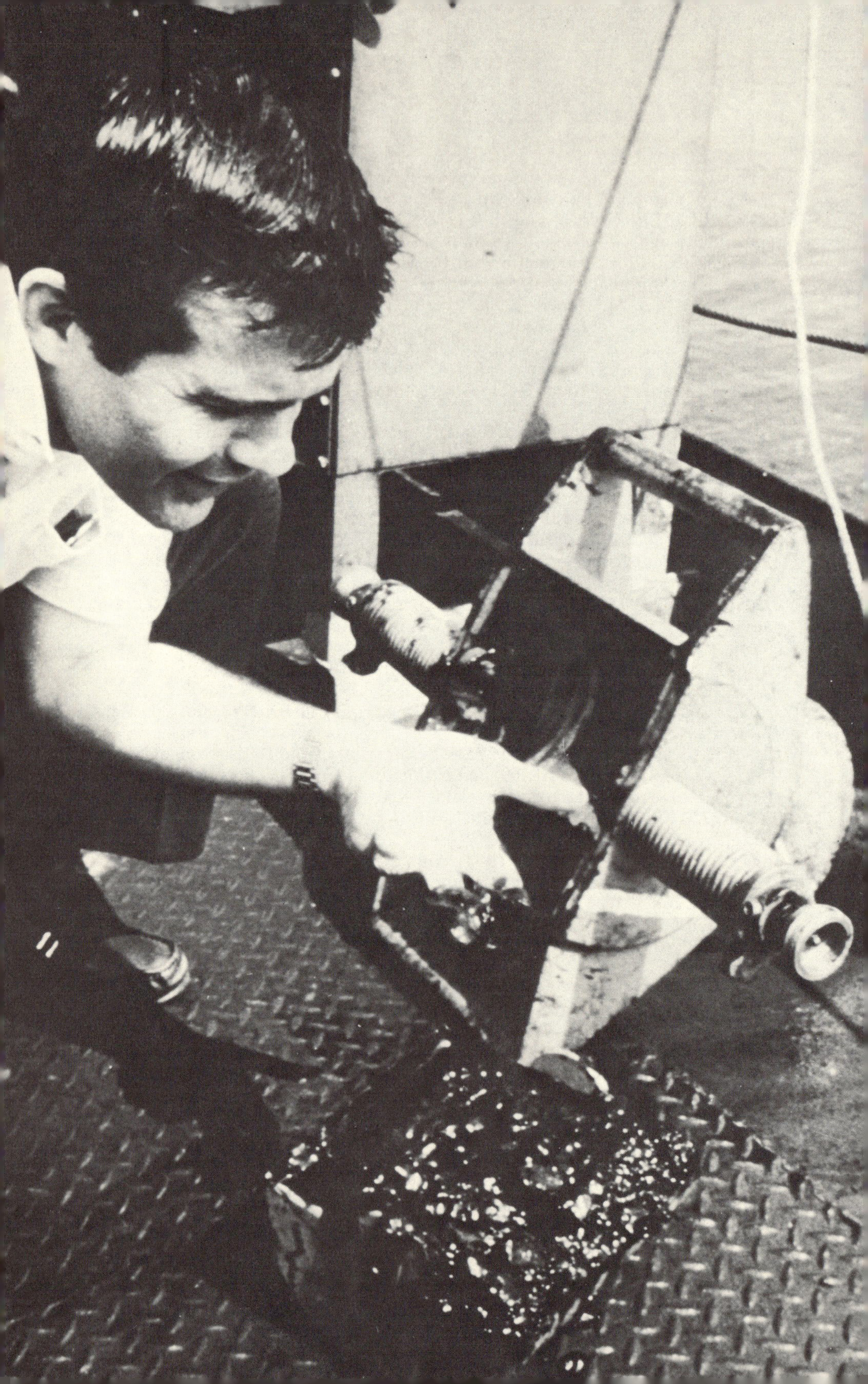

CONTRIBUTED PAPERS ON THE EFFECTS OF TOXIC SUBSTANCES

Opposite: Grab sample of highly contaminated sediments from off Shooters Island, Staten Island, New York. Sediments from this portion of the Hudson-Raritan estuary look and smell oily and have been shown to contain elevated levels of petroleum hydrocarbons and polychlorinated biphenyls. (Courtesy R.L. Swanson, NOAA Office of Marine Pollution Assessment.)

ECOSYSTEM CONSEQUENCES OF ALTERATIONS OF BENTHIC COMMUNITY STRUCTURE AND FUNCTION IN THE NEW YORK BIGHT REGION

Donald F. Boesch[1]

Virginia Institute of Marine Science
Gloucester Point, Virginia 23062

Abstract. Structure and function are divorced in most studies of the ecology of benthos. Furthermore, the interaction of the benthos with the remainder of the ecosystem is seldom considered. Data on the structure and function of benthos in the New York Bight region are drawn together, and inferences are made about the role of benthos in these ecosystems and the consequences of human alterations of benthic communities.

Alterations to the macrobenthic communities in the Hudson-Raritan estuary, the Christiaensen Basin, and upper Hudson Shelf Valley have resulted in reductions in populations of species that are predominant food items for demersal fishes and invertebrates. The Christiaensen Basin supports a dense and productive benthic community, but little of this productivity appears to be transferred to high trophic levels. Energy flow through the seabed has been increased due to waste additions, but this may result in localized depletion of oxygen in bottom waters during the summer when water masses are stratified. Nutrients regenerated from the seabed do not appear to be influential in supporting the high primary productivity in the Bight apex, and the seabed may actually be an important nutrient sink. Benthic biological processes affecting toxicants concentrated in sediments may determine their availability to benthic and demersal organisms, but are probably not responsible for significant reintroduction of these toxicants into the water column, except as they affect resuspension of bottom sediments.

INTRODUCTION

In considerations of the ecological effects of human activities in coastal environments, attention is typically focused on the seabed and its inhabitants, the benthos. The reasons for this special attention are several. Activities such as channel dredging, dredged material dumping, sand mining, and filling may directly effect major modifications of benthic habitats. Furthermore, nutrient or organic enrichment and toxic discharges generally have repercussions on the seabed, following sedimentation of particle-borne pollutants. Such contaminants may accumulate and persist in bottom sediments, while concentrations in the overlying water remain low. Benthic organisms are predominantly sedentary, and many are relatively long-lived. They must, therefore, tolerate the variable environment, unlike both nektonic organisms, which can avoid stressful conditions, and planktonic organisms, which can recover from induced alterations dynamically.

Consequently, there is a long and extensive history of study of the response of the benthos to various forms of pollution. The response of the macrobenthos to organic enrichment has been particularly well-characterized (Pearson and

[1] Present address: Louisiana Universities Marine Consortium, Star Route Box 541, Chauvin, Louisiana 70344.

Contribution No. 935, Virginia Institute of Marine Science and School of Marine Science, College of William and Mary.

Rosenberg, 1978), and several conceptual models have been developed to explain the generally observed trends in community structure in relation to environmental stress (Pearson and Rosenberg, 1978; Gray, 1979; Boesch and Rosenberg, 1981).

Attention was directed to the benthos relatively early in the investigation of ecological effects of human activities in the New York Bight and Hudson-Raritan estuary. Dean and Haskin (1964) and McGrath (1974) studied macrobenthos in the Raritan estuary. Since 1968 extensive studies have been conducted in the New York Bight apex, largely to assess the effects of ocean dumping (Rowe, 1971; National Marine Fisheries Service, 1972a,b; Pearce, 1972; Pearce *et al.*, 1976; Botton, 1979). These studies mainly have involved describing the distribution and abundance of benthic animals and various attributes derived from these distributions, such as species diversity and biomass, in relation to geographic, hydrographic, and sedimentary properties. Because of the complex interaction of environmental factors influencing animal distribution, it has been difficult to relate the observed biological patterns to specific activities, such as sewage sludge dumping (although there have been oversimplified attempts to do so), and virtually impossible to relate alterations to specific toxic agents. Despite criticisms of the conclusions drawn from some investigations of New York Bight benthos (Pararas-Carayannis, 1973), certain effects of waste disposal activities on the macrobenthos seem to be supported by ample evidence. These will be reviewed later in this paper. The vexing problem remains, however, in interpreting the implications of these modifications to the economies of both nature and man. This paper examines this last issue.

Deducing the ecosystem consequences of alterations of benthic communities is not a well-worn path. As Mills (1975) pointed out, knowledge of the relationship between benthos and marine ecosystems is woefully inadequate—the subject has been largely avoided by benthic ecologists, despite the fact that their discipline originated as a branch of fisheries research. Benthic ecological research has diverged in three directions: community structure studies emphasizing species composition, community functional studies emphasizing energy and materials flow, and fisheries studies. As Hargrave and Levings (1975) lamented, practitioners of these three approaches seldom communicate or collaborate.

In this perspective, I will attempt to weave interrelationships between the community structure of the benthos and the two most important modes of interaction with the broader ecosystem: (1) energy flow and biogeochemical cycling, and (2) trophic support of nektonic predators. It is only in the context of such understanding that sound evaluation can be made of the natural resource costs and benefits of human activities to the benthos.

OBSERVED ALTERATIONS OF THE BENTHOS

Modifications resulting from human activities in the New York Bight region undoubtedly have been greatest in the inshore habitats of the Raritan and Hudson estuarine systems and smaller coastal embayments, which are ironically less well-known than those in the offshore New York Bight apex. McGrath

(1974) reported an impoverished macrobenthic community in the muddy sediments of western Raritan Bay and Sandy Hook Bay that he attributed to depressed dissolved oxygen levels and that McDonough (1976) related to high trace metal (Cr, Cu, Ni, Pb, and Zn) concentrations in sediments. In the lower New York Bay, the sandy sediments of this better flushed habitat supported a denser and more diverse benthic community somewhat similar to that of the shallow continental shelf. Dean and Haskin (1964) described recovery of freshwater and brackish water benthos in the upper Raritan estuary following improvement of sewage waste treatment.

In summary, the macrobenthos of the Hudson-Raritan estuarine system apparently has been altered grossly from its natural state because of both the discharge of toxic and oxygen demanding wastes and physical modifications of habitats, including channel dredging, landfilling, and altered patterns of sedimentation (Gross, 1976). No "functional" studies of energy or material flow have been published on estuarine benthic communities of the region, and little is known of the trophic coupling of the benthos and nektonic predators.

The seabed of the New York Bight apex is influenced primarily by continental shelf waters, which are higher in salinity and less variable in temperature than inshore waters. Nevertheless, the Bight apex is influenced by estuarine discharges of sediments, organic material, nutrients, and toxicants, all of which have repercussions in the benthos. Furthermore, the direct discharges of sewage sludge, dredge spoil, and other wastes affect the seabed. Critical to understanding the benthic ecology of the Bight apex is knowledge of the complex physiography and associated sedimentary patterns of the region. The Hudson Shelf Valley, extending from the harbor entrance to the Hudson Canyon at the edge of the shelf, is the dominant physiographic feature. The terminus of the Shelf Valley was largely filled with solid wastes by the 1930's, and, just to the south, the Valley has been partially occluded by more recent disposal of dredged materials (Williams, 1979). As presently expressed in surface features, the Shelf Valley terminates in the broad Christiaensen Basin, 26-38 m deep. The Basin is the site of deposition of much of the fine-grained sediments introduced into the Bight apex and is surrounded by relatively clean, sandy sediments (Figure 1) off the New Jersey and Long Island shores and on Cholera Bank to the east. The muddy sediments of the Basin and Hudson Shelf Valley have much higher concentrations of organic carbon (Hatcher and Keister, 1976) and trace metals (Carmody *et al.*, 1973) than do surrounding areas and apparently serve as a major sink for many particle-associated compounds of biological significance. Strong currents move both up and down the Shelf Valley in response to meteorological forcing, eroding and redistributing these fine sediments toward the Long Island shoreline or down the Hudson Shelf Valley (Swift *et al.*, 1976; Freeland *et al.*, 1979).

The muddy fine sand of the Christiaensen Basin and the Hudson Shelf Valley support a qualitatively different benthic biota, with naturally high diversity and biomass, than the surrounding sand bottoms (Rowe, 1971; Pearce *et al.*, 1976). Several assessments of the impact of ocean dumping activities on benthos in the

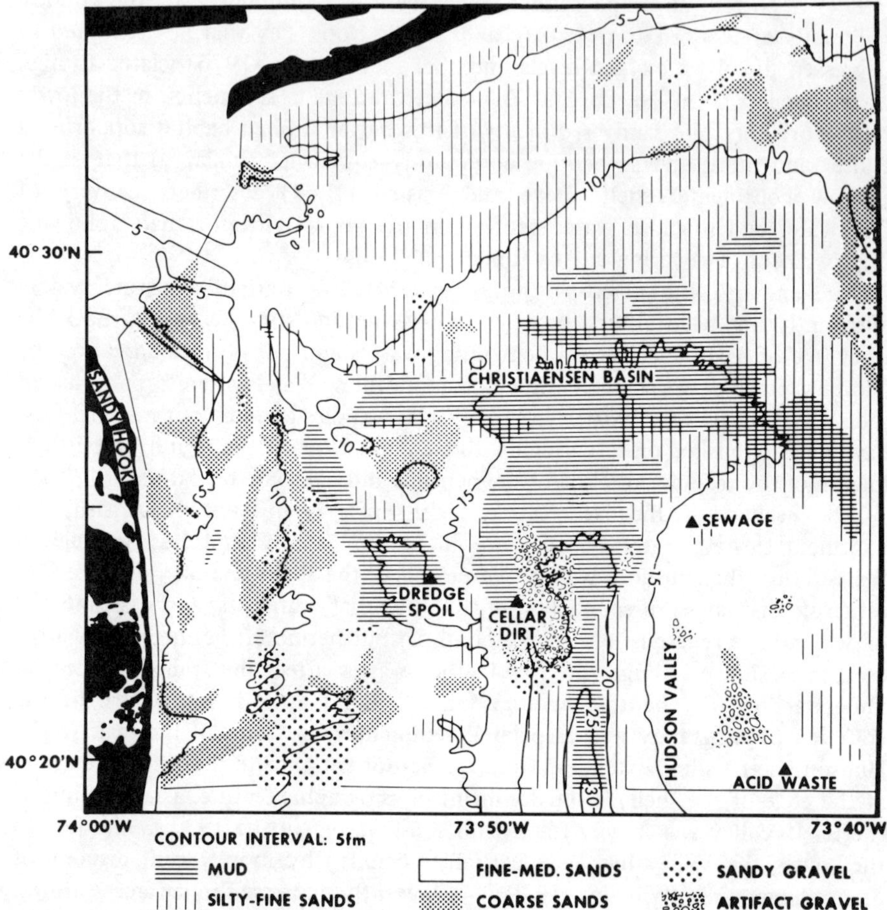

Figure 1. Distribution of sediment types in the New York Bight apex (Freeland *et al.*, 1976). Note the distribution of muddy sediments in the Christiaensen Basin and upper Hudson Shelf Valley.

area have been misled by comparisons of communities at sites within the Basin containing high contaminant levels with "control" communities inhabiting sandy sediments with low contaminant levels (*e.g.*, McDonough, 1976). Such comparisons should be made on a within-habitat basis (*i.e.*, habitats similar in sediment properties, but with varying contaminant levels, should be compared). For the Christiaensen Basin this presents a real dilemma, as it is a unique environment in the otherwise sandy Middle Atlantic shelf. The most realistic comparisons probably are made along a gradient from the Christiaensen Basin down the Hudson Shelf Valley, along which sediment properties remain relatively constant, but contaminant levels (such as trace metals and organic carbon) decline rapidly away from their source (Figure 2).

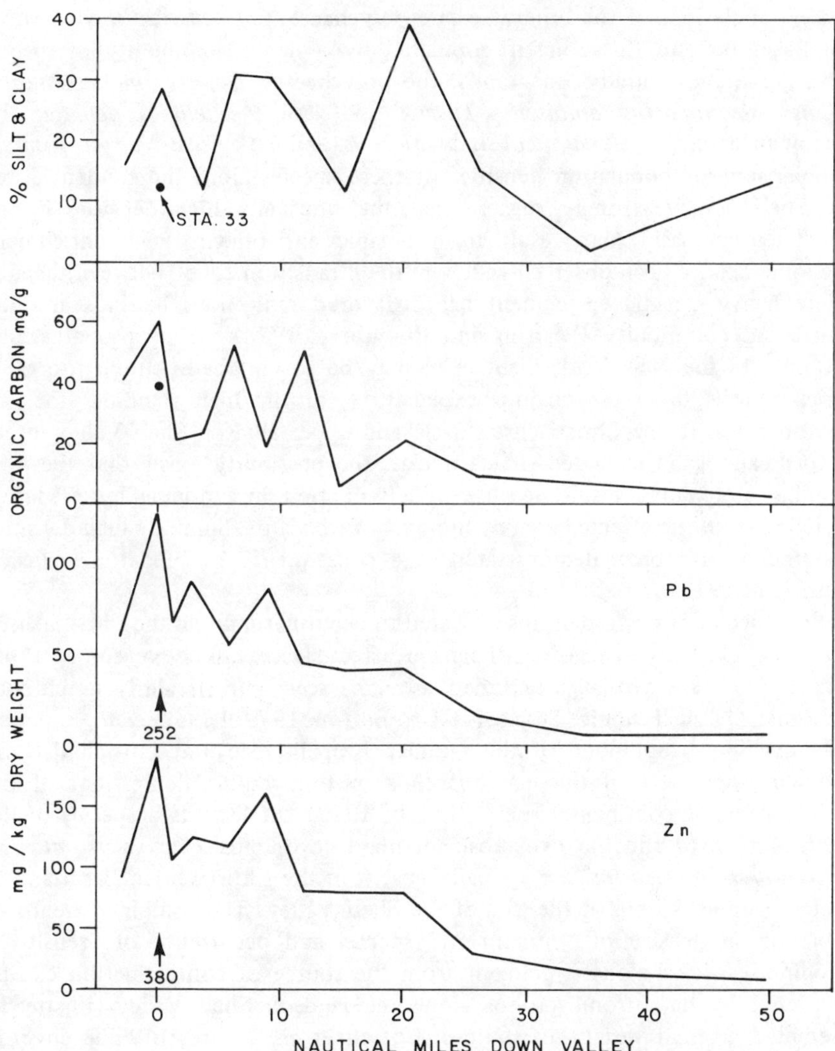

Figure 2. Percent silt and clay, total organic carbon concentration, and concentrations of lead and zinc in sediments collected along the axis of the Hudson Shelf Valley. Values for station 33, located near the margin of the Christiaensen Basin in the vicinity of the sewage sludge dumpsite, are indicated by dots or arrows (Boesch *et al.*, in press).

Pearce (1972), Pearce *et al.* (1976), and Steimle *et al.* (this volume) found a depauperate macrobenthic community in muddy sediments near the sewage sludge and dredged material disposal sites. In our detailed analysis of all existing data (Boesch *et al.*, in press), we have found this most affected zone typically exists in a 10-15 km^2 area west of the sewage sludge disposal site on the margin of the Basin. The area supports few macrobenthic species and is characterized by dense populations of the otherwise rare polychaete, *Capitella capitata.* Most of the Basin beyond this zone if populated by a dense community of species characteristic of muddy fine sands: the polychaetes, *Nephtys incisa, Pherusa affinis, Mediomastus ambiseta, Tharyx* sp., and *Asabellides oculata;* the burrowing anemone, *Ceriantheopsis americanus;* and the bivalve, *Nucula proxima.* Comparisons of population densities of these species along the gradient down the Shelf Valley strongly suggest that the unusually high densities in the Christiaensen Basin may result from organic enrichment. Such enrichment phenomena have been observed widely in the "transition zone" between the area where heavy organic enrichment has disfavored indigenous benthos and the unaffected community (Pearson and Rosenberg, 1978). An area greater than 240 km^2 in the New York Bight apex may be so enriched. Although organic enrichment is the most obvious explanation for the high standing stock of macrobenthos in the Christiaensen Basin and upper Hudson Shelf Valley, other factors cannot be excluded. In particular, the possibility exists that the high standing stock may result because of a reduction in cropping by predators excluded from the affected area by hypoxia or toxicants. Similarly high densities of infauna have been demonstrated as a consequence of predator exclusion (Virnstein, 1977).

Few species of macrobenthos expected to occur naturally in the Christiaensen Basin are excluded, except in the most affected areas. Notable among those animals that are virtually excluded are crustaceans, particularly amphipods (National Marine Fisheries Service, 1972a; Botton, 1979; Boesch *et al.*, in press). Tube-dwelling amphipods of the families Ampeliscidae and Corophiidae are dominant members of the macrobenthos of topographic depressions of the Middle Atlantic continental shelf (Boesch, 1979). At least two species of the genus *Ampelisca* and the two most abundant corophiids, *Unciola irrorata* and *Erichthonius rubricornis,* are virtually absent in the Christiaensen Basin and at least the upper 15 km of the axis of the Shelf Valley. The combined trends of reduction in density of "enrichment" species and occurrence of "sensitive" amphipod species along a gradient from the source of contamination can be illustrated by data from stations along the Hudson Shelf Valley (Figure 3), assembled from a variety of sources (Boesch *et al.,* in press). While environmental factors related to increasing depth (*e.g.,* temperature) may be partially responsible for the biotic patterns witnessed, sediments are reasonably similar at the stations represented, and the patterns closely reflect the down-valley trends in sediment trace metals and total organic carbon (Figure 2).

Less information is available concerning the functional ecology of the benthos of the New York Bight. Seabed oxygen consumption, which is mainly attributable to microbial metabolic activities, has been studied by Smith *et al.* (1974), Rowe *et al.* (1976), and Thomas *et al.* (1976). Comparisons of community metabolism (oxygen consumption rates) among the habitats of the Bight apex are complex because of the pervasive effect of temperature on metabolic rates. The benthos of shallow sandy bottoms consumes relatively little oxygen during the winter, but may demonstrate higher consumption rates than

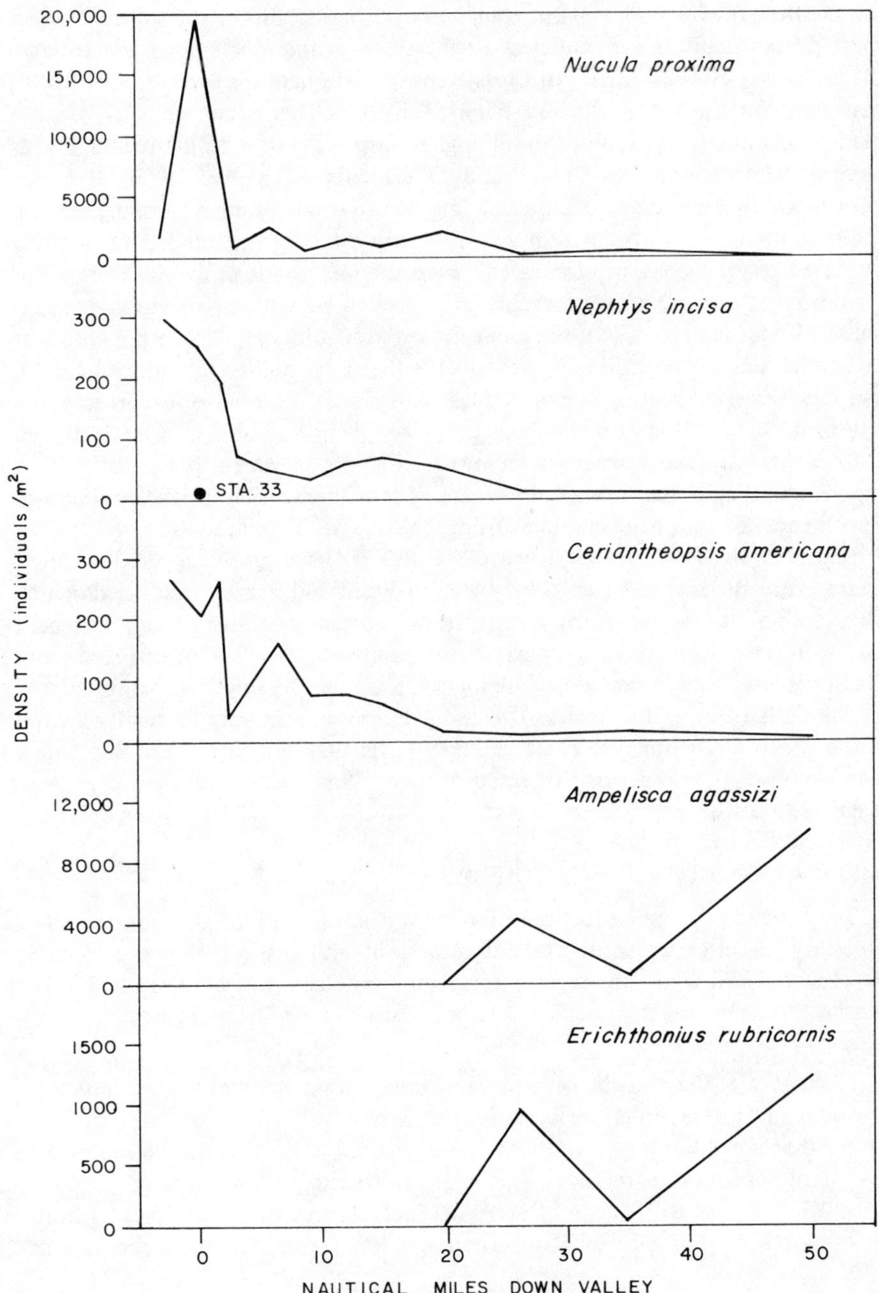

Figure 3. Patterns of abundance of macrobenthic species apparently favored or disfavored by organic enrichment along the axis of the Hudson Shelf Valley (Boesch *et al.*, in press). Abundance of *Nephtys incisa* at station 33 is indicated by a dot. All of the other species considered in this figure were not found at station 33.

the benthos of the muddy Basin-Shelf Valley complex during the summer, when these deeper habitats are insulated by the seasonal thermocline. On the average, the oxygen consumption rate of Christiaensen Basin benthos is about 50% higher than that on the sandy Cholera Bank (Table 1). This is not unusual because muddy sediments typically exhibit higher organic carbon content and higher oxygen consumption rates than sandy sediments (Hargrave, 1973). It is not known whether increased organic loading due to sewage sludge dumping, dredge spoil dumping, or nutrient stimulation of *in situ* production has resulted in increased oxygen consumption rates, but some increase is probable. The rate of oxidation of organic matter reaching the seabed in the Christiaensen Basin is limited by temperature, oxygen, and nutrient availability; thus, it is possible that carbon loading of the sediments exceeds the biotic capability for mineralization, and much organic matter is more or less permanently buried. Furthermore, the rate of decomposition of organic matter by sediment microbes may be decreased if toxicants also accumulate in sediments. Chervin *et al.* (in press) found that supplemental cadmium inhibited the rate of remineralization of labeled urea and glycolic acid by sediment microbes from the New York Bight apex.

Smith *et al.* (1974) and Rowe *et al.* (1976) have also compared nutrient fluxes from the seabed in the Christiaensen Basin and nearby sandy sediments. They found that while nitrogen flux from the sandy sediments approximated that estimated from oxygen demand measurements, the flux of nitrogen from Christiaensen Basin sediments, while higher, was considerably less than predicted on the basis of oxygen demand. This they attributed primarily to denitrification in the anaerobic sediments. Thus, the highly organic sediments of the Bight apex may serve as a greater sink for fixed nitrogen than was the case prior to large scale waste discharges.

IMPLICATIONS TO THE ECOSYSTEMS

To evaluate the consequences of observed alterations of the benthos, it is necessary to integrate structure and function within the benthos, as well as to evaluate the significance of interactions between the benthos and other ecosystem components. Specific questions that will be addressed are:

1. What are the effects of altered macrobenthic community structure on trophic interactions, particularly with fishes?
2. What are the effects of alterations of macrobenthos on the flux of energy and important materials (*e.g.*, oxygen, nutrients, toxicants)?
3. What is the significance of benthic processes and alterations thereof in the functioning of the Hudson-Raritan and New York Bight apex ecosystems?

Trophic Coupling

The benthos has been considered important to ecosystem productivity and human interests principally in terms of supplying food for bottom-feeding fishes and epibenthic invertebrates, such as crabs, lobsters, and shrimp. The importance

Table 1. Summary of estimates of benthic community metabolism and organic carbon budgets for the New York Bight apex.

Benthic metabolism (g 0_2/m^2/day)

	Thomas *et al.* (1976)	Smith *et al.* (1974)	Rowe *et al.* (1976)	Garside and Malone (1977)
Bight apex	0.31			1.86
Christiaensen Basin	0.55	1.24	1.09	
Dredge spoil site	0.62			
Cholera Bank	0.32	0.81		

Carbon budget for Bight apex (1,250 km^2)

	Organic carbon (metric tons C/day)		
	Segar and Berberian (1976)	Thomas *et al.* (1976)	Garside and Malone (1977)
Sources			
Primary production	1,146 (40%)	1,266 (37%)	1,300 (77%)
Estuarine inputs	1,073 (37%)	1,501 (44%)	280 (17%)
Sewage sludge	110 (4%)	235 (7%)	110 (7%)
Dredged materials	540 (19%)	380 (11%)	
Total	2,869	3,382	1,690
Sinks			
Water column respiration		1,730-5,646 (92-98%)	810 (48%)
Benthic respiration		145 (2-8%)	880 (52%)
Sedimentation and export		?	?
Total		1,875-5,719	1,690

of the benthos to fishery resources in the New York Bight region is made apparent by considering those resource species that feed primarily on benthic prey or themselves live on the seabed (Table 2).

Although a wide variety of benthos are important to fish and invertebrate predators, species grouped as macrobenthos are generally considered most important. Meiobenthos may constitute an important food resource for post-larval and juvenile fishes, but a lack of knowledge about trophic importance and the limited data on the effects of waste discharges on the meiobenthos (Tietjen, 1980) precludes further discussion of the implications of such effects. Thus, this discussion will focus exclusively on macrobenthos.

Obviously, if an area of seabed were devoid of macrobenthic life, the value of this area as a feeding ground for demersal fishes would be virtually nil. More difficult to evaluate are the effects of benthic community composition, abundance, and productivity on predator populations. Can the predators utilize any available macrobenthos, thus making alterations in macrobenthic com-

Table 2. Fishery resource species of the New York Bight region that live on the seabed or depend on benthic organisms as an important food source (McHugh, 1977).

Continental shelf	
Species that live on seabed	
Surf clam	*Spisula solidissima*
Sea scallop	*Placopecten magellanicus*
American lobster	*Homarus americanus*
Rock crab	*Cancer irroratus*
Fishes that feed on benthic prey	
Red hake	*Urophycis chuss*
White hake	*Urophycis tenuis*
Silver hake	*Merluccius bilinearis*
Haddock	*Melanogrammus aeglefinus*
Scup	*Stenotomus chrysops*
Black sea bass	*Centropristis striata*
Summer flounder	*Paralichthys dentatus*
Winter flounder	*Pseudopleuronectes americanus*
Yellowtail flounder	*Limanda ferruginea*
Tilefish	*Lopholatilus chamaeleonticeps*
Estuaries and nearshore waters	
Species that live on seabed	
American oyster	*Crassostrea virginica*
Blue mussel	*Mytilus edulis*
Hard clam	*Mercenaria mercenaria*
Soft clam	*Mya arenaria*
Bay scallop	*Argopecten irradians*
Conch	*Busycon* spp.
Species that feed on benthic prey	
Blue crab	*Callinectes sapidus*
Atlantic sturgeon	*Acipenser oxyrhynchus*
Tautog	*Tautoga onitis*
Weakfish	*Cynoscion regalis*
Atlantic croaker	*Micropogon udulatus*
Spot	*Leiostomus xanthurus*
Northern kingfish	*Menticirrhus saxatilis*
White perch	*Morone americana*
Striped bass	*Morone saxatilis*
Northern puffer	*Sphaeroides maculatus*

munity structure of little consequence? Does a high standing crop of benthos make more food available for the predators? Does greater production of the benthos yield greater production of demersal predators? And, finally, are the demersal predators so limited by other factors (recruitment success, higher predators, or fishing pressure) that partial reduction of their trophic resources will not affect their total population?

Although there have been few studies that have compared the consumption of demersal predators with the composition of the macrobenthic community, the available information from fish-food habit studies indicates a recurring pattern wherein certain components of the benthos are consumed in excess of their proportional representation in the community, whereas others important in the benthos are seldom consumed. This so-called "selectivity" or "electivity" may be as much a result of prey availability as purposeful selection by the predator, with prey avoidance, size, and living position in the sediment all important in determining which components of the benthos are cropped. Disproportionate cropping was illustrated in Arntz's (1978) studies of the food habits of cod (*Gadus morhua*) and dab (*Limanda limanda*) in the western Baltic (Figure 4). The proportion of biomass of molluscs consumed was far less than in the benthic community, and polychaetes were about equally represented in fish stomachs and the benthos. However, crustaceans and, in the case of the dab, echinoderms were disproportionately represented in fish stomachs. Arntz further showed "selective" cropping within higher prey taxa. For example, both fishes showed high electivity for the polychaete, *Harmothoe sarsi,* while other polychaetes were inconsistently selected or consistently rare in stomachs.

Within the New York Bight region, Sedberry and Musick (1979) compared the food habits of nine dominant bottom-feeding fishes to the macrobenthic community within two regions of the outer continental shelf (40-90 m) off New Jersey. Two species, the silver hake (*Merluccius bilinearis*) and the goosefish (*Lophius americanus*) preyed primarily on nekton or other fishes. All other species had as their principal prey benthic crustaceans, particularly amphipods (*Unciola, Erichthonius, Byblis,* and *Ampelisca*), crabs (*Cancer*), and shrimp (*Crangon* and *Dichelopandalus*) (Table 3). Only the little skate, which can forage deeply in the sediment, fed on molluscs to any appreciable extent, and only the Gulf Stream flounder appreciably preyed on polychaetes.

Food habits of fishes generally change as the predator grows. Typically, post-larvae prey on plankton or meiofauna, juveniles and smaller adults on amphipods and other peracaridan crustaceans, and larger individuals on decapod crustaceans and other fishes (Figure 5). Thus, the small crustacean prey are particularly important for juvenile stocks, which are both less motile and more dependent on specific prey.

In the New York Bight apex, most of those species of macrobenthos apparently advantaged by waste discharges are not known to be important prey for fishes in the western North Atlantic (Table 4). Only the polychaete, *Nephtys incisa,* has been reported with any frequency (McEachran *et al.,* 1976). The other species are not generally consumed, either because of their small size (*e.g., Mediomastus*), predator avoidance behavior (*Ceriantheopsis*), burrowing life style (*Tharyx* and *Pherusa?*), or thick shells (*Nucula*). On the other hand, those species largely excluded from the upper Valley and the Christiaensen Basin, but abundant in similar sediments of the Hudson Shelf Valley, are among the most important prey species known for demersal fishes in the New York Bight region (*i.e.,* tubicolous amphipods and rock crabs). Only scant data exist on the food

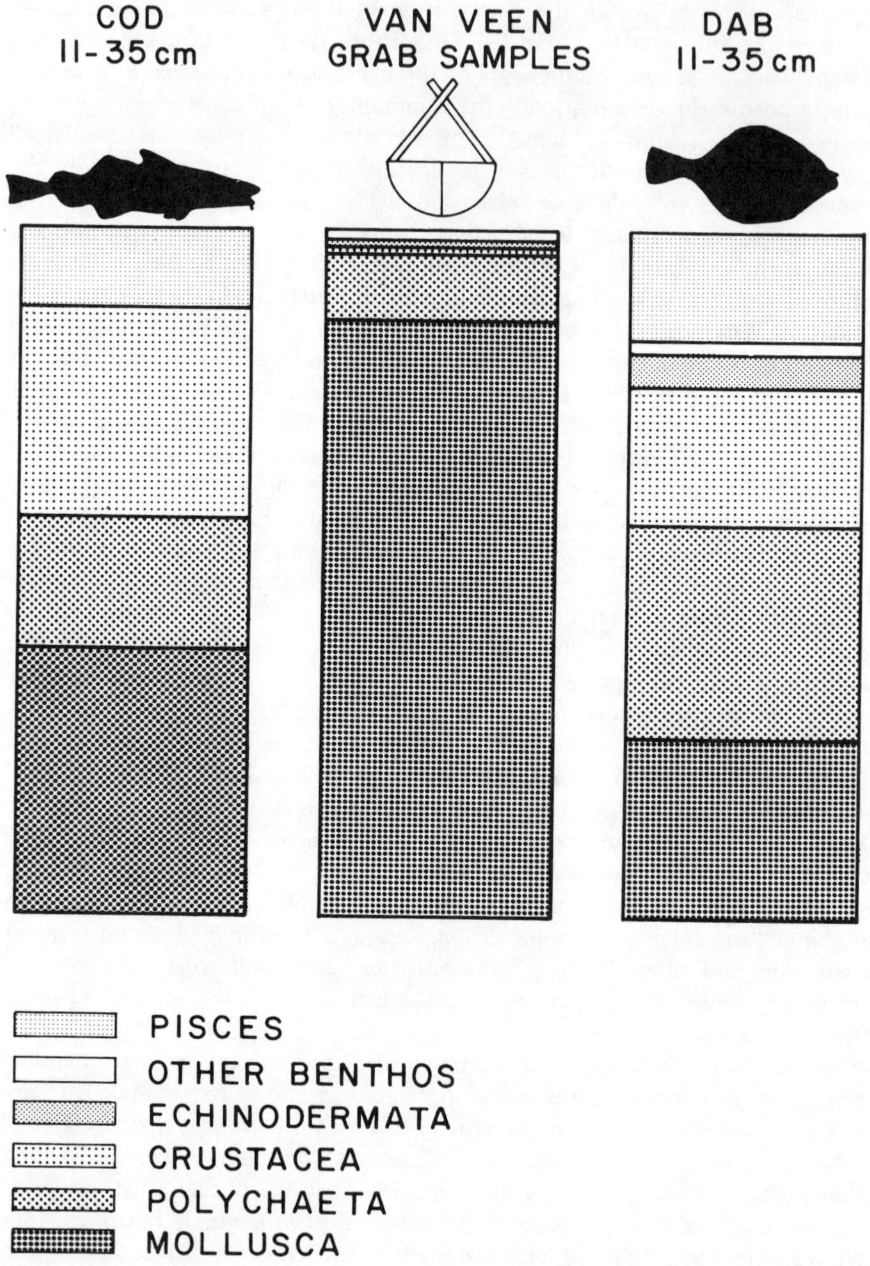

Figure 4. Comparison of the biomass of major benthic taxa consumed by cod and dab to that in the benthic community in Kiel Bay, western Baltic Sea (Arntz, 1978).

Table 3. Principal benthic prey of bottom-feeding fishes most abundant in otter trawl catches on the outer continental shelf off New Jersey. Prey are listed in order of relative importance based on frequency, abundance, and volume in fish stomachs (Sedberry and Musick, 1979).

Predator	Prey species
Raja erinacea (little skate)	*Cancer irroratus* (crab)
	Unciola irrorata (amphipod)
	Erichthonius rubricornis (amphipod)
	Byblis serrata (amphipod)
	Ampelisca vadorum (amphipod)
	Diastylis spp. (cumacean)
	Ensis directus (bivalve)
	Crangon septemspinosa (shrimp)
Urophycis chuss (red hake)	*Unciola irrorata*
	Erichthonius rubricornis
	Cancer irroratus
	Dichelopandalus leptoceras (shrimp)
	Byblis serrata
Urophycis regius (spotted hake)	*Cancer irroratus*
	Crangon septemspinosa
	Dichelopandalus leptoceras
	Unciola irrorata
	Citharichthys arctifrons (fish)
Macrozoarces americanus (ocean pout)	*Erichthonius rubricornis*
	Unciola irrorata
	Cancer irroratus
Stenotomus chrysops (scup)	*Erichthonius rubricornis*
	Unciola irrorata
	Chone infundibuliformis (polychaete)
	Ampelisca spp.
Citharichthys arctifrons (Gulf Stream flounder)	*Unciola irrorata*
	Erichthonius rubricornis
	Byblis serrata
	Ampelisca vadorum
	Euchone sp. (polychaete)
	Chone infundibuliformis
	Clymenura sp. (polychaete)
Hippogolossina oblonga (fourspot flounder)	*Cancer irroratus*
	Unciola irrorata
	Dichelopandalus leptoceras
	Crangon septemspinosa
	Erichthonius rubricornis
	Byblis serrata

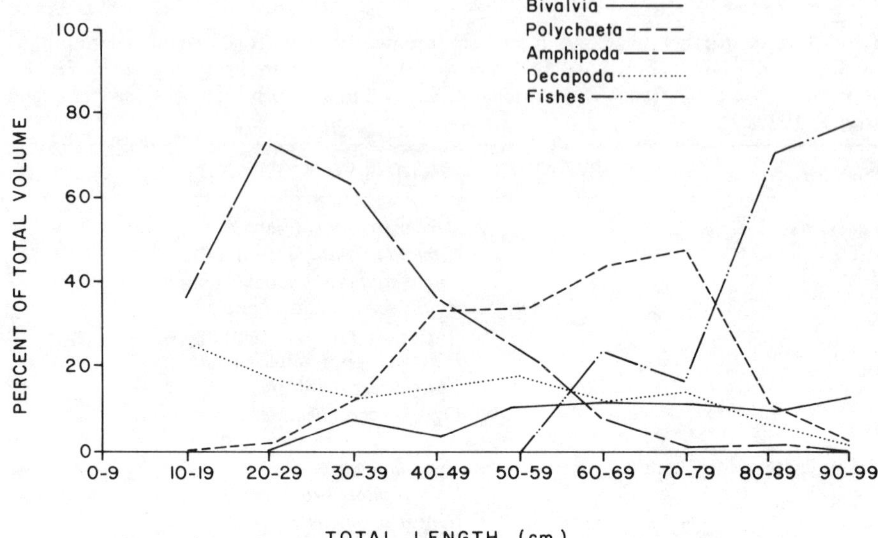

Figure 5. Change in food habits of the skate, *Raja ocellata*, with size (McEachran *et al.*, 1976).

habits of fishes in the New York Bight apex, and none of these data are based on fishes collected from the Christiaensen Basin. Gut contents of demersal fishes (*Urophysis chuss, Merluccius bilinearis, Limanda ferruginia,* and *Pseudopleuronectes americanus*) collected on Cholera Bank were dominated by crustaceans, although polychaetes, including flabelligerids (probably *Pherusa affinis*), were also frequently represented (National Marine Fisheries Service, 1972b). Despite this lack of specific information on the food habits of fishes in altered habitats of the Bight apex, it is reasonable to conclude on the basis of the literature that the trophic resource "potential" of a substantial portion of the Bight apex has been reduced. Although the demersal predators may to some degree switch to feed on the abundant polychaete and bivalve populations in the "enriched zone" in lieu of the "preferred" crustaceans, these prey, if available, should be more difficult for the predators to harvest.

In areas of the southern California shelf where the benthos has been greatly altered by sewage additions, Word (1979) reported that the Dover sole (*Microstomus pacificus*), a generalist feeder, consumed prey in similar proportions to their importance in the benthic community. The Dover sole was thus able to "switch" prey in the altered habitats. The yellow chin sculpin (*Icelinus quadriseriatus*), on the other hand, normally feeds on small epibenthic crustaceans and did not switch prey in areas where the crustaceans were excluded; as a consequence, the sculpin was not abundant in these altered habitats. In the most altered habitats off the Palos Verdes Peninsula, Allen (1975) noted that bottom-feeding fishes comprised a very small proportion of the fish fauna, compared to

Table 4. Important species of macrobenthos whose distribution or abundance apparently has been affected by waste discharges into the New York Bight apex.

Species	Importance as prey for fishes	Feeding type and effect on sedimentary processes
Species advantaged in most heavily contaminated sediments		
Capitella capitata	Unimportant	Deposit-feeding, surface tube dweller; pelletizes surface sediments
Species advantaged in the Christiaensen Basin		
Ceriantheopsis americanus	Unimportant	Surface contact predator; forms heavy mucous tube that penetrates deeply into the sediment, but is not ventilated
Nephtys incisa	Moderately important, skates, flatfishes, and gadoids	Motile predator or deposit feeder; ventilates temporary burrows and mixes surface sediments
Pherusa affinis	Occasionally important	Surface deposit-feeding, motile burrower; active in mixing sediments
Mediomastus ambiseta	Unimportant	Subsurface deposit-feeding, shallow dweller
Tharyx spp.	Unimportant	Surface deposit-feeding, motile burrower; possibly active in mixing sediments
Asabellides oculata	Unimportant	Surface deposit-feeding, tube dweller; possibly ventilates tube
Nucula proxima	Unimportant	Subsurface motile deposit feeder; mixes surface sediments
Species disadvantaged in the Christiaensen Basin		
Ampelisca spp.	Occasionally important, especially for flatfishes	Seston-surface deposit-feeding, shallow infaunal tube dweller; stabilizes sediments
Erichthonius rubricornis	Very important for a wide variety of fishes	Seston-feeding, epifaunal tube dweller; stabilizes sediments
Unciola irrorata	Very important for a wide variety of fishes	Surface deposit-feeding shallow infaunal tube dweller; stabilizes sediments
Cancer irroratus	Very important for a wide variety of fishes	Motile epibenthic predator; causes small scale disturbances in surface sediments

Santa Monica and San Pedro Bays, and that most of the fishes off Palos Verdes were water column feeders.

It is not likely that the higher standing crops and apparently higher production of macrobenthos in the "enriched zone" result in more food available and higher yields to the demersal nekton. As stated earlier, high standing crops of macrobenthos may result from a lack of predator cropping (Virnstein, 1977).

It is not possible yet to evaluate the significance of benthic changes to Bight fisheries resources, because other factors controlling populations of demersal predators are poorly understood. Complete lack of success of Bight apex populations may be inconsequential to the total Middle Atlantic Bight stocks, but of considerable importance to local populations. The domain of concern realistically will vary with the prey species, depending on its motility and migratory habits.

Community Structure and Energy and Materials Flow

Energy flow (reflected by respiration) and mineralization of nutrients in the benthos are processes quantitatively dominated by microbes. For example, the proportion of seabed oxygen consumption attributable to bacterial respiration is typically 80% or more (Hargrave, 1973). However, the importance of larger benthic organisms in regulating these processes is becoming increasingly realized (Aller, 1978). Many macrobenthic animals inhabiting fine-grained sediments ventilate burrows or tubes to obtain oxygen from the overlying water, since the pore water in the surrounding sediment is often anoxic. Other animals increase sediment porosity by burrowing or pelletizing sediment through deposit-feeding (Rhoads, 1974). Both ventilation and enhanced porosity have the effect of increasing the availability of oxygen and sulfate to sediment microbes, thus stimulating microbial respiration and mineralization. Furthermore, the activities of macrobenthos greatly increase the surface area of the interface of oxidizing and reducing (anoxic) regions of the sediment. This interface or redox potential discontinuity is an active site of geochemical reactions and exchange. Thus, the activities of macrobenthos substantially increase the flux of dissolved materials, including toxicants such as trace metals, between sediment and overlying water relative to molecular diffusion (Aller, 1978; McCaffrey *et al.*, 1980).

Profiles of sediment interactions for macrobenthic species advantaged or disadvantaged by waste discharges in the New York Bight apex are given in Table 4. In the most heavily contaminated sediments dominated by *Capitella capitata,* the macrobenthos is probably relatively ineffective in enhancing energy and materials flux. *Capitella* builds tubes near the sediment surface and accomplishes little sediment mixing (Rhoads *et al.*, 1977; Pearson and Rosenberg, 1978). The remainder of the macrobenthos is depauperate. In the "enriched zone," however, dense populations of species capable of enhancing sediment-water exchange are present. Tube and burrow ventilation (*Nephtys* and, possibly, *Asabellides*), sediment pelletization (*Nucula* and the deposit-feeding polychaetes), bioturbation by burrowers, and vertical transport of sediment (*Mediomastus*) are accomplished. This suggests that the macrobenthos may significantly enhance microbial metabolism, nutrient flux into the water column, sediment resuspension, and, possibly, mobilization of toxicants in the Christiaensen Basin. The crustacean species excluded from the Basin, on the other hand, probably do not significantly enhance materials exchange because they live at the sediment surface and tend to reduce sediment resuspension and, hence, geochemical exchange.

Although the inferred stimulatory effect of macrobenthos on biogeochemical exchange in the "enriched zone" must be weighed against possible inhibitory effects of toxicants on microbes (Chervin *et al.*, in press), the image developed here is analogous to a biological waste treatment system containing oligochaetes and aquatic insects that enhance decomposition of organic wastes. Waste discharges in the Bight apex have caused development of a community that has a greater capacity to oxidize organic matter than the indigenous community, but is less suitable for support of higher trophic levels. Both functions are of value to man. Waste assimilation is considered an appropriate use of the marine environment by many (Standing Committee on the Disposal of Sewage Sludge, 1978), and few can argue that the Bight apex is not effectively assimilating substantial amounts of wastes. The concern as to whether such a use is judicious is based on the relative value of impaired uses (*e.g.*, fishery resources, public health concerns) and the fact that the Bight apex is not a closed system. We have little knowledge and no control of effects transcending the immediate apex.

The macrobenthos of the Hudson-Raritan estuarine system is more depauperate than that found in the "enriched zone" of the Bight apex (McGrath, 1974). Suspension feeders and small surface deposit feeders dominate, indicating that the macrobenthos in this heavily stressed estuarine system plays a less important role in geochemical exchange than in the Christiaensen Basin. Hypoxic stress, toxicants, and habitat modifications may have so altered the communities that only small opportunistic species ineffective in enhancing geochemical exchange (Rhoads *et al.*, 1977) persist. The overall capacity of the estuarine system to assimilate wastes thus would be reduced.

Significance of Benthic Processes

The flux of biologically important materials through the seabed must be considered in light of their overall importance in the larger ecosystem. What is the role of benthic metabolism in depleting the dissolved oxygen in bottom waters? What is the importance of the seabed in regenerating or removing nutrients? What role do bottom sediments and biological processes therein play in the fate of toxicants? And, ultimately, have these functions been significantly affected by waste discharges?

The consumption of dissolved oxygen by the seabed results directly from the aerobic respiration of benthic organisms, principally microorganisms, and indirectly from the oxidation of reduced byproducts of anaerobic metabolism of sediment microbes. Benthic community metabolism (seabed oxygen uptake) has been directly measured in at least three separate studies in the Bight apex (Table 1); Garside and Malone (1977) indirectly estimated mean benthic respiration based on the relationship between surface oxygen flux and particulate organic carbon (Table 1). Estimates vary widely from a rate averaged over the Bight apex and throughout the year of 0.31 g O_2/m^2/day by Thomas *et al.* (1976) to 1.86 g O_2/m^2/day from the O_2-POC flux estimates of Garside and Malone (1977). The *in situ* measurements of Smith *et al.* (1974) and Rowe *et al.* (1976) are at least twice as high as those of Thomas *et al.* (1976), but there is agreement that benthic metabolism averages about 50% higher in the Christiaensen Basin than in the sandy habitat of Cholera Bank.

The dissimilar estimates of respiration obviously yield somewhat different perceptions of the importance of the benthos in the carbon or oxygen budgets for the Bight apex, but the relative differences are further exaggerated by disparate estimates of water column respiration (Table 1). Based on curiously high estimates of water column respiration, Thomas *et al.* (1976) concluded that benthic respiration is not more than 8% of total oxygen consumption and therefore of little consequence in oxygen depletion phenomena. According to Thomas *et al.*, the respiration rate of the bottom meter of the water column exceeds the respiration of the seabed. On the other hand, Garside and Malone (1977) estimated that water column respiration is two to seven times less than reported by Thomas *et al.*; consequently, they concluded that benthic respiration accounts for over half of the total oxidation of organic carbon in the Bight apex.

Since bottom water oxygen depletion is known primarily from the deeper areas of the Bight apex, below the seasonal pycnocline at about 30 m, it seems most appropriate to evaluate the role of seabed oxygen consumption in the oxygen budget of the sub-pycnocline water mass, which averages less than 10 m thick over the Christiaensen Basin and the upper Hudson Shelf Valley. Using only direct measurements of benthic respiration for summer stratified conditions of about 30 ml O_2/m^2/hr (*ca* 1 g O_2/m^2/day) (Thomas *et al.*, 1976), the proportion of total respiration below the pycnocline attributable to the seabed is about 15% using the water column respiration estimates of Thomas *et al.* (1976) and 50% using that of Garside and Malone (1977). If, hypothetically, organic loading by wastes has increased benthic respiration in the Christiaensen Basin by 50% (the degree to which respiration in the Basin is higher than surrounding bottoms), this may represent roughly a 5-17% increase in oxygen demand below the summer pycnocline, exclusive of the potentially enhanced respiration in the water column.

Unaccounted for in the published carbon budgets for the New York Bight apex (Table 1) are burial in sediments (primarily in the Christiaensen Basin and Hudson Shelf Valley) and export out of the region. Dredged material may contribute as much as 19% of the total organic carbon load of the Bight apex (Segar and Berberian, 1976). Comparisons of historical bathymetric surveys (Freeland *et al.*, 1976) have indicated that most of the dredged material previously dumped can be accounted for in deposits. On the other hand, organic and toxicant-rich fine particles are preferentially dispersed, compared to the coarser components (MESA, 1978), and the organic matter in dispersed sediments may be more susceptible to oxidation. Cochran and Aller (1979) concluded on the basis of ^{234}Th profiles in sediments near the dredged materials dumpsite that the accumulation of mud is episodic and probably rapid when it occurs. This could be attributable either to massive deposition at the time of dredged material disposal or periodic redistribution of the dredged material deposits. The presence of *Nucula proxima* greater than one year old suggested to Cochran and Aller that heavy deposition had not taken place within a year. The presence of laminations in the sediment that had not been bioturbated except in the upper 4 cm indicates that, although episodic redistribution of sediments is possible, long periods of bottom sediment stability must exist, constituting an indefinite sink for organic carbon. Furthermore, a larger portion of the organic pool in dredged material should be relatively refractory compared to other carbon sources. Thus, although some of the organic carbon contained in the dredged material is oxidized, the permanence and volume of the dredged material deposits and the nature of the organic carbon suggest that much of the organic matter contained is buried and not mineralized.

All authors agree that *in situ* primary production and estuarine inputs (including sediment discharges) dominate the organic carbon budget of the apex. Presumably, the fate of these inputs is diffuse, and they serve as fuel for water column and seabed respiration. Sedimentation of particulate organic carbon from these sources is probably greater in the depositional basin of the Bight apex, as inferred simply on the basis of the distribution of fine-grained sediments. The infusion of organic matter from sewage sludge disposal, although of considerably smaller magnitude, is more intensely localized and thus appears to be predominantly responsible for the high organic content of muds in the Christiaensen Basin (Hatcher and Keister, 1976). This organic material is not immediately buried, as with dredged material, and it is less refractory than that in dredged material. The rate of decomposition of this sedimented organic material is limited by the low temperatures of the basin (Thomas *et al.*, 1976) and the availability of oxygen, sulfate, and inorganic nutrients. It is probable that most of the sewage sludge organics are decomposed, although burial may be significant.

Nutrient regeneration also may be an important coupling between the benthos and the pelagic part of marine ecosystems. The relative importance of benthic regeneration of nutrients in continental shelf waters is being debated (Carpenter and McCarthy, 1978; Rowe, 1978). Benthic regeneration apparently is quite important in some bays and estuaries (Nixon *et al.*, 1975), but on the outer continental shelf, upwelling of slope water is the dominant source of inorganic nutrients (Carpenter and McCarthy, 1978). Rowe and Smith (1977) and Rowe (1978) have argued that benthic regeneration is important on the inner continental shelf in the New York Bight.

Nitrogen is usually the limiting nutrient element to coastal phytoplankton

(Ryther and Dunstan, 1971). Ammonium is the principal form of dissolved inorganic nitrogen (DIN) regenerated from the benthos and is also the form preferentially utilized by phytoplankton (MacIsaac and Dugdale, 1969). DIN regeneration rates of 24 mg N/m^2/day in the Christiaensen Basin and 35 mg N/m^2/day in sandy sediments along New Jersey and south of Long Island have been measured (Rowe *et al.*, 1976; Rowe, 1978). Assuming an overall average of 30 mg N/m^2/day, benthic regeneration could supply about 13-17% of the DIN required for annual average primary productivity of over 1 g C/m^2/day (Malone, 1976). However, Malone (1976) concluded that the estuarine contribution of DIN exceeds phytoplankton demand throughout the year, except during the summer. During the summer (June-August) there is more DIN assimilated in the estuary and consequently less exported. Furthermore, very high rates of primary productivity (up to 6 g C/m^2/day) rapidly deplete available DIN, and productivity must be sustained by autochthonous DIN regeneration. However, the vertical flux of DIN during the summer in the Bight apex must be slow because of the seasonal thermohaline stratification, as evidenced by water column ammonium profiles (Rowe *et al.*, 1976). Surface mixed-layer concentrations of DIN are reduced as a result of high phytoplankton assimilation, but the proportion of ammonium in the DIN pool is increased. This suggests that pelagic regeneration must be primarily responsible for supplying phytoplankton DIN demand during the summer. Benthic regeneration of nutrients from particulate wastes (sewage sludge and dredge spoil) probably does not, therefore, contribute significantly to the high primary productivity of the New York Bight apex or the Hudson-Raritan estuary. The nutrient dynamics of these ecosystems are apparently dominated by sewage-derived nutrients entering the estuary (Garside *et al.*, 1976; Malone, this volume).

The seabed may, on the other hand, serve as a significant nutrient sink. Some nutrients are buried in sedimentary environments in the estuary or in the Bight apex. Although large quantities of NH_4^+ are released where dredged material is dumped (MESA, 1978), it is likely, for the same reasons discussed earlier, that a large portion of the nutrients in dredged material (a source estimated to contribute over 20% of the total nitrogen supplied to the Bight apex [Segar and Berberian, 1976]) is not released to the water column. Denitrification by anaerobic sediment microbes resulting in the production of N_2 also may be an important nitrogen sink (Nixon, in press), especially in muddy sediments such as the Christiaensen Basin. Rowe *et al.* (1976) found that the regeneration of DIN in the Christiaensen Basin was 25% less than that estimated to be mineralized on the basis of oxygen demand measurements. The low ratio of N:P regenerated from the benthos (Rowe *et al.*, 1976) probably results from loss of fixed nitrogen and may be responsible for the importance of nitrogen as a major limiting nutrient in coastal ecosystems (Nixon, in press). The organic enrichment of the Christiaensen Basin by waste discharges probably has resulted in conditions more conducive to denitrification (*i.e.*, reducing conditions), although this rate of loss of fixed nitrogen is almost certainly less than the enhanced flux of organic nitrogen.

Potentially toxic organic and inorganic compounds are concentrated in the finer sediments of the Hudson-Raritan estuary and the New York Bight apex. Most of the trace metals and radionuclides accommodated in bottom sediments probably are not available to the biota, as they exist in strongly adsorbed or insoluble form. Experimental evidence suggests that uptake of trace metals and radionuclides by deposit feeders is greater from the aqueous phase than from ingested particles (Beasley and Fowler, 1976). The elevated concentrations of trace metals in some benthic animals living in areas of contaminated sediments (Young *et al.,* 1978) may result from higher concentrations of these pollutants in interstitial waters. However, sediment-borne organic toxicants, such as PCB's, are relatively more bio-available, and uptake from ingested sediments may predominate over that from water (Fowler *et al.,* 1978). The importance of exchange of inorganic contaminants between the aqueous medium and organism underlies the apparent absence of food chain magnification for inorganic contaminants. On the other hand, the uptake of organic toxicants from food is of greater importance; thus, food chain magnification is characteristic for such organic compounds as DDT and PCB's (Young and Mearns, 1979).

Physical and biochemical activities of benthic organisms may affect the flux of toxic materials between water and sediment. Complex redox reactions, organic complexing, and metabolic transformations (*e.g.,* catabolic breakdown, methylation of trace metals) may affect availability and toxicity. Furthermore, benthic animals may facilitate release of toxicants from sediments by reworking or ventilating sediments (Bryan, 1976; Aller, 1978).

Waste disposal in the Bight generally has increased the prevalence of anoxic habitats in sediments because of organic enrichment. Trace metals tend to be bound as insoluble sulfides in anoxic sediments, suggesting that the sedimentary basins serve as important sinks of these toxicants. However, activities of benthic animals may serve to remobilize sediment-bound toxicants well after abatement of discharges (Bryan, 1976). The distribution of dissolved and suspended particulate trace metals in the Bight apex shows no obvious concentration gradients above the seabed, but rather reflects the importance of estuarine sources (Segar and Cantillo, 1976). Significant remobilization of sedimented toxicants is not apparent, except perhaps as associated with resuspended sediments, but lack of specific knowledge precludes definitive conclusions regarding the role of the benthos in the fate of important toxicants.

Animals that live in or on the sediment, however, must be exposed to higher concentrations of many toxicants than if they lived off the seabed. Contact with sediments bearing high concentrations of toxicants is thought to be a factor in pathologic maladies, such as gill and shell disease in crustaceans (Young and Pearce, 1975) and fin erosion disease in demersal fishes (Murchelano and Ziskowski, 1976; Sherwood, 1979, this volume).

CONCLUSIONS

Alterations to the structure and function of benthic communities in the New York Bight region have been documented or are strongly suggested by available data. The consequences of these alterations are much less clear, but it appears that the ability of the Hudson-Raritan estuary and the Bight apex to sustain living resources harvested by man has been impaired.

Benthic communities that have developed in the sedimentary basins in the Bight apex serve the useful function of decomposing organic matter disposed in the Bight or produced through nutrient stimulation of phytoplankton. Although it is improbable that the enhanced oxygen consumption of the seabed is responsible for widespread oxygen depletion, such as occurred in the summer of 1976 (Malone, 1978), it may contribute to more localized hypoxia below the pycnocline in the Bight apex and in the estuary. More frequent or prolonged anoxia in these regions may diminish the efficiency of the decomposition process by exterminating infaunal animals. This would further exacerbate oxygen depletion by allowing a buildup of sedimented organics, punctuated by periodically high rates of seabed oxygen consumption.

Regeneration from the seabed probably is not an important source of nutrients for phytoplankton in the estuary or Bight apex because of the dominance of allochthonous nutrient sources. Rather, the bottom sediments may serve as a considerable nutrient sink through burial and denitrification. The benthos thus may influence the rate of export of nutrients onto the continental shelf beyond the immediate Bight apex. Less is known about the role benthic organisms play in the transfer of toxicants, such as trace metals and synthetic organics. Benthic biological activities may affect the availability of the toxicants to animals and microbes that live on or near the seabed, but it is unlikely that under the present conditions they are significantly influential in the transfer of toxicants back into the water column.

ACKNOWLEDGEMENTS

Research support was provided by the NOAA Marine EcoSystems Analysis Program (MESA) New York Bight Project (Grant No. 04-78-B01-12). Some of the results presented were obtained under contract support from the Bureau of Land Management (Contract No. AA550-CT6-62). I am grateful for comments on an early manuscript by G.F. Mayer, R.J. Diaz, M.A. Bowen, L.C. Schaffner, and three anonymous reviewers; they greatly helped sharpen my arguments.

REFERENCES

Allen, M.J. 1975. Regional variation in the structure of fish communities. *In:* Annual Report of the Southern California Coastal Water Research Project, Southern California Coastal Water Research Project, El Segundo, CA. pp. 99-102.

Aller, R.C. 1978. The effects of animal-sediment interactions on geochemical processes near the sediment-water interface. *In:* Estuarine Processes, M.L. Wiley (ed.), Academic Press, New York. pp. 151-171.

Arntz, W.E. 1978. The "upper part" of the benthic food web: the role of macrobenthos in the western Baltic. Rapp. P.-V. Reun. Cons. Int. Explor. Mer 173: 85-100.

Beasley, T.M. and S.Q. Fowler. 1976. Plutonium and americium: uptake from contaminated sediments by the polychaete *Nereis diversicolor.* Mar. Biol. 38: 95-100.

Boesch, D.F. 1979. Benthic ecological studies: macrobenthos. Middle Atlantic outer continental shelf environmental studies. Virginia Inst. Mar. Sci. Spec. Sci. Rep. 194. 301 pp.

Boesch, D.F., M.A. Bowen, L.C. Schaffner, and F. Steimle. In press. The macrobenthos of the New York Bight apex: effects of waste disposal. *In:* The Macrobenthos of the New York Bight Region, NOAA/OMPA Tech. Rep.

Boesch, D.F. and R. Rosenberg. 1981. Response to stress in marine benthic communities. *In:* Stress Effects on Natural Ecosystems, G.M. Barrett and R. Rosenberg (eds.), John Wiley and Sons, New York. pp. 179-200.

Botton, M.L. 1979. Effects of sewage sludge on the benthic invertebrate community of the inshore New York Bight. Estuarine Coastal Mar. Sci. 8: 169-180.

Bryan, G.W. 1976. Heavy metal contamination in the sea. *In:* Marine Pollution, R. Johnston (ed.), Academic Press, London. pp. 185-302.

Carmody, D.J., J.B. Pearce, and W.E. Yasso. 1973. Trace metals in sediments of New York Bight. Mar. Pollut. Bull. 4: 132-135.

Carpenter, E.J. and J.J. McCarthy. 1978. Benthic nutrient regeneration and high rate of primary production in continental shelf waters. Nature 274: 188-189.

Chervin, M.B., C. Garside, C.D. Litchfield, T.C. Malone, and J.P. Thomas. In press. Synoptic investigation of nutrient cycling in the plume of the Hudson and Raritan Rivers. NOAA/OMPA Tech. Rep.

Cochran, J.K. and R.C. Aller. 1979. Particle reworking in sediments from New York Bight apex: evidence from $^{234}Th/^{238}U$ disequilibrium. Estuarine Coastal Mar. Sci. 9: 739-747.

Dean, D. and H. Haskin. 1964. Benthic repopulation of the Raritan River estuary following pollution abatement. Limnol. Oceanogr. 9: 551-563.

Fowler, S.W., G.G. Polikarpov, D.L. Elder, P. Parsi, and J.P. Villeneuve. 1978. Polychlorinated biphenyls: accumulation from contaminated sediments and water by the polychaete *Nereis diversicolor.* Mar. Biol. 48: 303-309.

Freeland, G.L., D.J.P. Swift, W.L. Stubblefield, and A.E. Con. 1976. Surficial sediments of the NOAA-MESA study areas in the New York Bight. Am. Soc. Limnol. Oceanogr. Spec. Symp. 2: 90-101.

Freeland, G.L., D.J.P. Swift, and R.A. Young. 1979. Mud deposits near the New York Bight dumpsite: origin and behavior. *In:* Ocean Dumping and Marine Pollution, H.O. Palmer and M.G. Gross (eds.), Dowden, Hutchinson & Ross, Stroudsburg, PA. pp. 73-95.

Garside, C. and T.C. Malone. 1977. Monthly oxygen and carbon budgets of the New York Bight apex. Estuarine Coastal Mar. Sci. 6: 93-104.

Garside, C., T.C. Malone, O.A. Roels, and B.C. Sharfstein. 1976. An evaluation of sewage-derived nutrients and their influences on the Hudson estuary and New York Bight. Estuarine Coastal Mar. Sci. 4: 281-289.

Gray, J.S. 1979. Pollution-induced changes in populations. *In:* The Assessment of Sublethal Effects of Pollutants in the Sea, H.A. Cole (ed.), Royal Society, London. pp. 545-561.

Gross, M.G. 1976. Sources of urban wastes. Am. Soc. Limnol. Oceanogr. Spec. Symp. 2: 150-161.

Hargrave, B. 1973. Coupling carbon flows through some pelagic benthic communities. J. Fish Res. Board Can. 30: 1317-1326.

Hargrave, B.T. and C.D. Levings. 1975. A prognosis for Canadian benthic biological oceanography. J. Fish. Res. Board Can. 32: 2268-2275.

Hatcher, P.G. and L.E. Keister. 1976. Carbohydrates and organic carbon in New York Bight sediments as possible indicators of sewage contamination. Am. Soc. Limnol. Oceanogr. Spec. Symp. 2: 240-248.

McCaffrey, R.J., A.C. Myers, E. Davey, G. Morrison, M.L. Bender, N.A. Luedtke, J.D. Cullen, P. Froelich, and G. Klinkhammer. 1980. Benthic fluxes of nutrients and manganese in Narragansett Bay, Rhode Island. Limnol. Oceanogr. 25: 31-44.

McDonough, K.B. 1976. A benthic index of environmental quality for the New York Bight apex and Raritan Bay. M.S. Thesis, State University of New York, Stony Brook. 89 pp.

McEachran, J.D., D.F. Boesch, and J.A. Musick. 1976. Food division within two sympatric species-pairs of skates (Pisces:Rajidae). Mar. Biol. 35: 301-317.

McGrath, R. 1974. Benthic macrofaunal census of Raritan Bay—preliminary results, Pap. 24. *In:* Hudson River Ecology, Third Symposium on Hudson River Ecology, March 22-23, 1973, Hudson River Environmental Society, Inc. 40 pp.

McHugh, J.L. 1977. Fisheries and fishery resources of New York Bight. NOAA Tech. Rep. NMFS Circ. 401. 50 pp.

MacIsaac, J.J. and R.C. Dugdale. 1969. The kinetics of nitrate and ammonium uptake by natural populations of marine phytoplankton. Deep-Sea Res. 16: 445-457.

Malone, T.C. 1976. Phytoplankton productivity in the apex of the New York Bight: environmental regulation of productivity/chlorophyll *a.* Am. Soc. Limnol. Oceanogr. Spec. Symp. 2:360-272.

Malone, T.C. 1978. The 1976 *Ceratium tripos* bloom in the New York Bight: causes and consequences. NOAA Tech. Rep. NMFS Circ. 410. 14 pp.

MESA. 1978. MESA New York Bight Project Annual Report for Fiscal Year 1977. NOAA Special Report. 131 pp.

Mills, E.L. 1975. Benthic organisms and the structure of marine ecosystems. J. Fish. Res. Board Can. 32: 1657-1663.

Murchelano, R.A. and J. Ziskowski. 1976. Fin rot disease studies in the New York Bight. Am. Soc. Limnol. Oceanogr. Spec. Symp. 2: 329-339.

National Marine Fisheries Service. 1972a. The Effects of Waste Disposal in the New York Bight. Section 2. Benthic Studies. National Marine Fisheries Service, Sandy Hook Laboratory, Highlands, NJ. 277 pp.

National Marine Fisheries Service. 1972b. The Effects of Waste Disposal in the New York Bight. Section 4. Finfish Studies. National Marine Fisheries Service, Sandy Hook Laboratory, Highlands, NJ. 24 pp.

Nixon, S.W. In press. Remineralization and nutrient cycling in coastal marine ecosystems. *In:* International Symposium on Nutrient Enrichment in Estuaries, B.J. Neilson and L.E. Cronin (eds.), Humana Press, Clifton, NJ.

Nixon, S.W., C.A. Oviatt, and S.S. Hale. 1975. Nitrogen regeneration and the metabolism of coastal marine bottom communities. *In:* Role of Terrestrial and Aquatic Organisms in Decomposition Processes, J. Anderson and A. Macfayden (eds.), Blackwell, Oxford, England. pp. 269-283.

Pararas-Carayannis, G. 1973. Ocean dumping in the New York Bight: an assessment of environmental studies. U.S. Coastal Eng. Res. Center Tech. Memo. 39. 159 pp.

Pearce, J. 1972. The effects of waste disposal on benthic communities in the New York Bight. *In:* Marine Pollution and Sea Life, M. Ruvio (ed.), Fishing News (books) Ltd., London. pp. 404-411.

Pearce, J.B., J.V. Caracciolo, M.B. Halsey, and L.H. Rogers. 1976. Temporal and spatial distributions of benthic macroinvertebrates in the New York Bight. Am. Soc. Limnol. Oceanogr. Spec. Symp. 2: 394-403.

Pearson, T.H. and R. Rosenberg. 1978. Macrobenthic succession in relation to organic enrichment and pollution of the marine environment. Oceanogr. Mar. Biol. Annu. Rev. 16: 229-311.

Rhoads, D.R. 1974. Organism-sediment relations on the muddy sea floor. Oceanogr. Mar. Biol. Annu. Rev. 12: 263-300.

Rhoads, D.R., R.C. Aller, and M.B. Goldhaber. 1977. The influence of colonizing benthos on physical properties and chemical diagnosis of the estuarine seafloor. *In:* Ecology of Marine Benthos, B.C. Coull (ed.), University of South Carolina Press, Columbia, SC. pp. 113-138.

Rowe, G.T. 1971. The effects of pollution on the dynamics of the benthos of New York Bight. Thalassia Jugosl. 7: 353-359.

Rowe, G.T. 1978. Benthic nutrient regeneration and rate of primary production in continental shelf waters–reply. Nature 274: 189-190.

Rowe, G.T. and K.L. Smith. 1977. Benthic pelagic coupling in the Mid-Atlantic Bight. *In:* Ecology of Marine Benthos, B.C. Coull (ed.), University of South Carolina Press, Columbia, SC. pp. 55-66.

Rowe, G.T., K.L. Smith, and C.H. Clifford. 1976. Benthic-pelagic coupling in the New York Bight. Am. Soc. Limnol. Oceanogr. Spec. Symp. 2: 370-376.

Ryther, J.H. and W.M. Dunstan. 1971. Nitrogen, phosphorous and eutrophication in the coastal marine environment. Science 171: 1008-1013.

Sedberry, G.R. and J.A. Musick. 1979. Food habits of dominant demersal fishes of the outer continental shelf, Middle Atlantic Bight. Virginia Inst. Mar. Sci. Spec. Rep. 197. 80 pp.

Segar, D.A. and G.A. Berberian. 1976. Oxygen depletion in the New York Bight apex: causes and consequences. Am. Soc. Limnol. Oceanogr. Spec. Symp. 2: 220-239.

Segar, D.A. and A.Y. Cantillo. 1976. Trace metals in the New York Bight. Am. Soc. Limnol. Oceanogr. Spec. Symp. 2: 171-198.

Sherwood, M.J. This volume. Fin erosion, liver condition, and trace contaminant exposure in fishes from three coastal regions. pp. 359-377.
Coastal Water Research Project, El Segundo, CA. pp. 203-221.

Sherwood, Marjorie J. This volume. Fin erosion, liver condition, and trace contaminant exposure in fishes from three coastal regions. pp.

Smith, K.L., G.T. Rowe, and C.H. Clifford. 1974. Sediment oxygen demand in an upwelling and outwelling area. Tethys 6: 223-230.

Standing Committee on the Disposal of Sewage Sludge. 1978. Sewage Sludge Disposal Data and Reviews of Disposal to Sea. U.K. Department of the Environment, London. 49 pp.

Steimle, F., J. Caracciolo, and J.B. Pearce. This volume. Impacts of dumping on New York Bight apex benthos. pp. 213-223.

Swift, D.J.P., G.L. Freeland, P.E. Gadd, G. Han, J.W. Lavelle, and W.L. Stubblefield. 1976. Morphologic evolution and coastal sand transport, New York-New Jersey shelf. Am. Soc. Limnol. Oceanogr. Spec. Symp. 2: 69-89.

Thomas, J.P., W.C. Phoel, F.W. Steimle, J.E. O'Reilly, and C.A. Evans. 1976. Seabed oxygen consumption–New York Bight apex. Am. Soc. Limnol. Oceanogr. Spec. Symp. 2: 354-369.

Tietjen, J. 1980. Population structure and species composition of the free living nematodes inhabiting sands of the New York Bight apex. Estuarine Coastal Mar. Sci. 10: 61-73.

Virnstein, R.W. 1977. The importance of predation by crabs and fishes on benthic infauna in Chesapeake Bay. Ecology 58: 1199-1217.

Williams, S.J. 1979. Geologic effects of ocean dumping on the New York Bight inner shelf. *In:* Ocean Dumping and Marine Pollution, H.D. Palmer and M.G. Gross (eds.), Dowden, Hutchinson & Ross, Stroudsburg, PA. pp. 51-72.

Word, J.Q. 1979. The infaunal index, a relative abundance measure of the benthic infauna and how it may be applied to fish food habits studies. *In:* Gutshop '78, S.J. Lipovsky and C.A. Simenstad (eds.), Univ. Washington Sea Grant Publ. WSG-WO-79-1. pp. 37-42.

Young, D.R., T.-K. Jan, and T.C. Heesen. 1978. Cycling of trace metals and chlorinated hydrocarbon wastes in the Southern California Bight. *In:* Estuarine Interactions, M.L. Wiley (ed.), Academic Press, New York. pp. 481-496.

Young, D.R. and A.J. Mearns. 1979. Pollutant flow through food webs. *In:* Annual Report, 1978 of the Southern California Coastal Water Research Project, Southern California Coastal Water Research Project, El Segundo, CA. pp. 185-202.

Young, J.S. and J.B. Pearce. 1975. Shell disease in crabs and lobsters from the New York Bight. Mar. Pollut. Bull. 4: 70.

IN SITU MONITORING OF THE EFFECTS OF WATER QUALITY ON BENTHIC DETRITAL DECOMPOSITION

John J. Lee
Carmine A. Mastropaolo
Marie E. McEnery
John H. Tietjen
Judith R. Garrison

Department of Biology
City College of New York
Convent Avenue at 139th Street
New York, New York 10031

Abstract. Detrital decomposition is an important process that contributes to the fertility of seas, particularly in estuarine and coastal waters. The process involves a complex community of microorganisms and small animals, which intereact in a manner similar to that observed in forest litter and composts. Plastic chambers to measure the decomposition rates of *Spartina alterniflora* (a marsh grass that is a major contributor of detritus on United States East and Gulf coasts) were placed subtidally at five sites in the northeast: Towd Point, Southampton, New York; the effluent quarry of the Millstone power plant on Niantic Bay, Connecticut (Long Island Sound); a control site near the Millstone effluent quarry; Winsor Cove, Cataumet, Massachusetts (the site of an oil spill); and Sippewissett marsh, Falmouth, Massachusetts (a control site for Winsor Cove). Stations were visited monthly. Several techniques were used to measure rates of decomposition and growth of sediment microbial and animal populations. A long-term method involving the capture of mineralized $^{14}CO_2$ is described. This appears to be a promising measure of the integrated effects of pollution stress on important but slow environmental processes. The procedures may be more environmentally meaningful than LD_{50}'s or similar short-term measures of transient stress.

INTRODUCTION

Two principal food webs are characteristic of highly productive East and Gulf coast estuarine marine ecosystems. One, the plankton grazing web, is based on phytoplankton; the other, the detrital food web, is based on the production and decomposition of seagrasses and cord grasses that grow in shallow waters and wetlands bordering the sea.

Only a small quantity of sea- and cord grasses is grazed directly by marine organisms; the bulk of the material first is processed by microorganisms and small animals. Marine systems are not unique in this, the process being similar to litter decomposition in grasslands and forest soils. As much as two thirds of the annual production of forest leaves enter the litter decomposition cycle. Comparable estimates in the marine environment suggest that approximately 90% of the annual production of sea- and cord grasses is channeled into the detrital food web (Mann, 1972).

The detrital marine food web, particularly in northeast and middle Atlantic states, has been impaired by rapid destruction and removal of salt marshes, and pollution-related perturbation of detritus-based benthic production. Although many factors have contributed to the sharp decline of commercial fish landings in the New York marine district over the last 20 years (Jensen, 1977), sharp reductions in the quantity and quality of habitats suitable for supporting the detrital food web must have been a major contributing factor.

How can the impact of stresses on remaining detrital habitats be judged? Assessment of impact implies an ability to judge objectively the quality of a habitat and its suitability to support detrital decomposition. Unfortunately, assay methods for this task are not yet available (Tietjen, this volume). We accepted the challenge to test the feasibility of an *in situ* microcosm approach because the problem seemed technically soluble by extending to the field laboratory techniques already in use for studying detrital decomposition (Lee *et al.*, 1975; Tenore *et al.*, 1977).

Detrital decomposition is one of the most complex benthic marine processes. Diverse species of microorganisms, microflora, and small animals (protozoa and meiofauna) are involved in the decomposition of detritus and transformation of plant structural materials into compounds utilizable by animals at higher trophic levels (Lee, 1980). A number of methods can be used to estimate detrital decomposition and the activity of communities associated with the process. The community can be visualized by scanning electron microscopy (SEM) and the living biomass estimated by its ATP content. Since nitrogen is assimilated and fixed during decomposition, the carbon/nitrogen (C/N) ratio should become smaller. Measurements of this ratio over time represent a third potential means of gauging decomposition. Finally, radionuclide tracers can be used to make direct estimates of detrital decomposition. Unlabeled and ^{14}C-labeled detritus can be mixed together in a known ratio, and the amount of ^{14}C released as CO_2 can be used to trace the rate of detrital decomposition. Each of these techniques has been investigated in the course of this study.

MATERIALS AND METHODS

Flow-through reaction vessel/respirometers were built of plexiglass to test the effects of water quality on detrital decomposition, under conditions as close to ambient as practical. The reaction vessels consisted of two parts. The lower portion was a solid box (Figures 1a, b) that held the sediment and detritus being tested. The upper portion had 9.5-mm (3/8-in.) holes covered outside with nylon window screening and insite with nylon and fiberglass Millipore filters (NRWP 14250; 1 μm pore). Plastic shields surrounded the reaction vessels to deflect horizontal waves and to prevent damage from browsing animals. During periods when respired $^{14}CO_2$ was trapped, the caps of the reaction vessels were removed and replaced with water-tight, glass-enclosed vessels containing filter paper saturated with 40% KOH (Figures 1a, 2a). Measurements of $^{14}CO_2$ release were made at low tide during the day and lasted from two to three hours.

Reaction vessels deployed in the field were cleaned weekly to reduce the accumulation of aufwuchs and to prevent the plugging of reaction vessel pores. Plastic surfaces surrounding the filters were scrubbed with commercial plastic pads designed for cleaning cooking pots. At the same time, filter surfaces were brushed gently with a 1.25-cm paint brush. At the termination of each experiment, the reaction vessel was removed from its holder under water and placed in a plastic bucket of seawater. This prevented drying and changing the characteristics of the small aufwuchs community that accumulated on the vessels, despite the efforts described above. It also prevented damage to the membranes by unbalanced water pressure during transport to the laboratory. Upon return to the laboratory, ^{24}Na, used as a tracer to measure efflux rates, was mixed thoroughly with the seawater in the reaction vessels. The seawater in the buckets surrounding the vessels was sampled at five-minute intervals. Radioactivity was measured by γ-spectrometry. The rate of seawater flux through the reaction vessels was at least 12 times the vessel volume per hour in cases with the most fouling; it was much higher (> 20 times) in most cases.

Sediment in the reaction vessels was thoroughly washed beach sand, with approximately the same grain size distribution as found in Southampton, New York salt marshes (Matera and Lee, 1972). This was enriched with 50 g or 4% (by weight) ground dried *Spartina alterniflora.* The *Spartina* was collected in winter (December) and, after it was washed and dried, was ground in a Wiley Mill to particles of approximately 120 μm. Labeled *Spartina* was grown from seedlings (from Environmental Concern, St. Michaels, Maryland) in a transparent plexiglas cabinet, under a $^{14}CO_2$ atmosphere (Ferguson and Williams, 1974). After the *S. alterniflora* reached a height of approximately 0.8 m, the photoperiod in the chamber was shortened and the temperature lowered to 15°C during the dark cycle to induce senescence before harvesting. After harvest, the leaves and stems were placed in gently flowing freshwater for two weeks, dried in glass trays in a hood for two months, ground in a Wiley Mill, and assayed for activity.

The reaction vessels and their holders were placed in sublittoral zones at five experimental locations: (1) the Millstone atomic power plant effluent quarry in Waterford, Connecticut; (2) a control site at the nearby New England Environmental Laboratory at the Millstone site; (3) Winsor Cove on Buzzards Bay, Bourne, Massachusetts; (4) Greater Sippewissett Marsh on Buzzards Bay, Falmouth, Massachusetts; and (5) Towd Point on North Sea Harbor, an embayment of greater Peconic Bay at the eastern end of Long Island, New York (a control site). Several factors entered into the choice of experimental sites. Since all sites had restricted access, vandalism was not a problem. The Millstone atomic power plant effluent quarry was chosen because it offered the potential for studying the combined effects of elevated temperatures and trace metals on detrital decomposition. Winsor Cove was chosen because it was the site of a well-studied but minor No. 2 fuel oil spill.

At mean low tide, the vessels were covered by approximately 1 m of water. Depending upon ambient conditions and lunar cycles, they were covered by 2-3 m of water at high tide.

At each site the reaction vessels were divided into two groups: those inoculated with microflora alone, and those inoculated with meiofauna, microflora, and microfauna. Three replicate reaction vessels were deployed at each station. With one exception, both groups were inoculated with populations

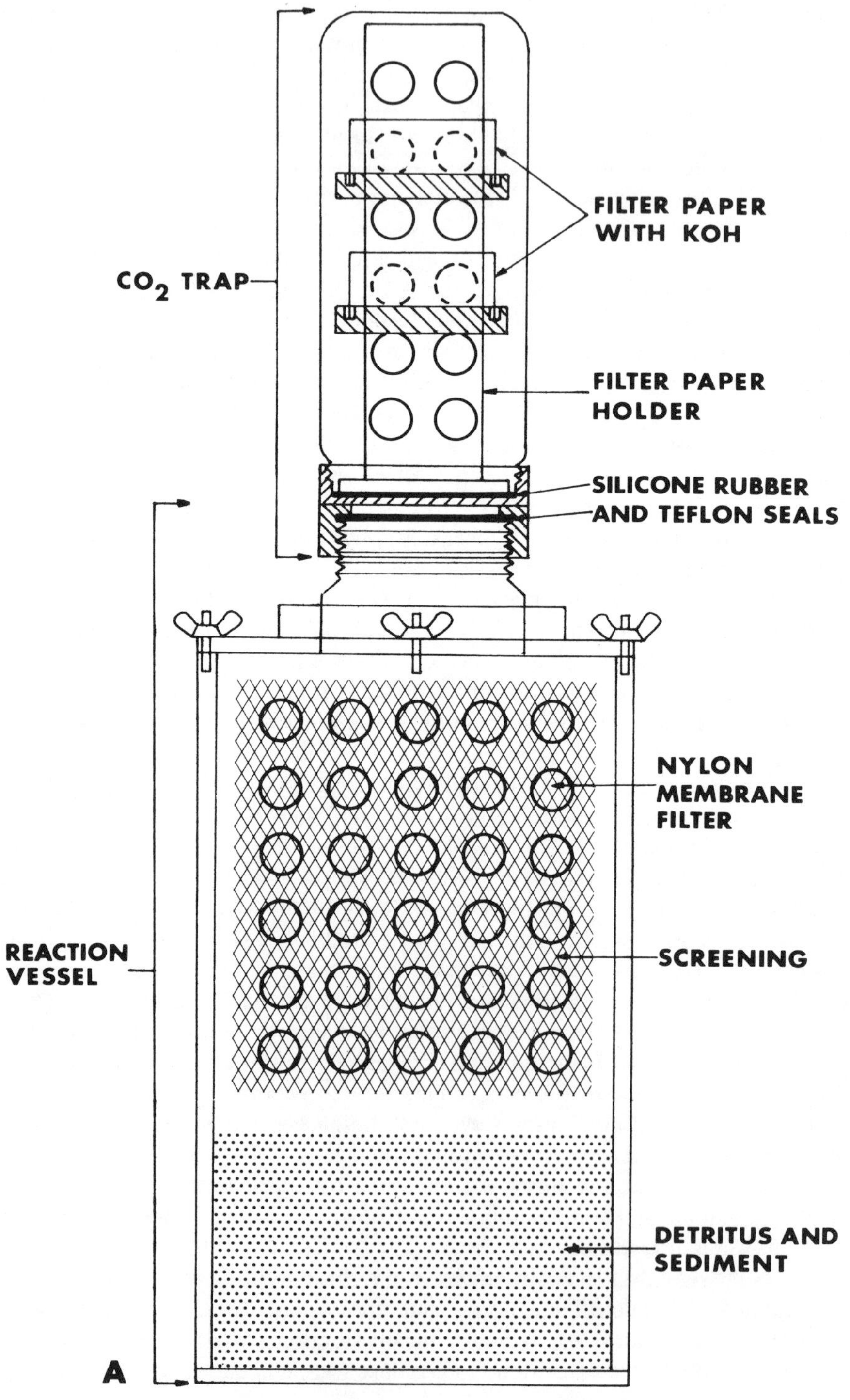
FILTER PAPER
WITH KOH
CO2 TRAP
FILTER PAPER
HOLDER
SILICONE RUBBER
AND TEFLON SEALS
NYLON
MEMBRANE
FILTER
REACTION
VESSEL
SCREENING
DETRITUS AND
SEDIMENT
A

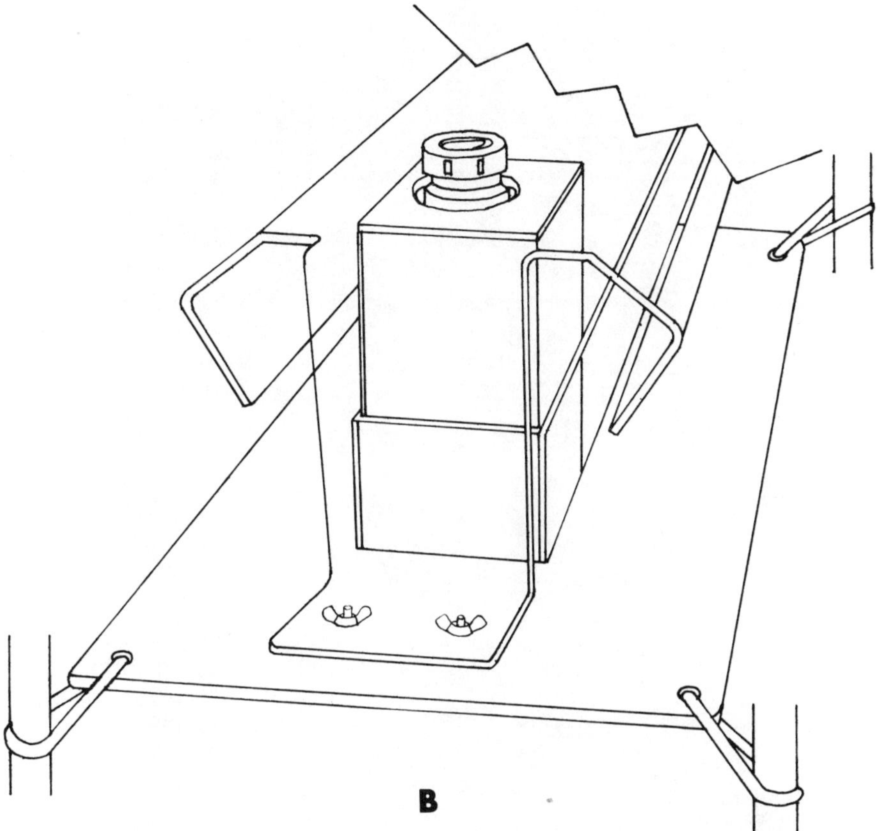

Figure 1. Opposite. (a) Diagram of an *in situ* detritus incubator/respirometer vessel. **Above.** (b) Diagram of the apparatus that holds the vessel in place on the sea floor. The detritus incubator is constructed of plexiglass. The respirometer housing is made of glass. The filter paper holders are constructed of Teflon.

from indigenous detritus rafts. Winsor Cove, which still had oil-saturated sediments and lacked an indigenous detrital decomposing community, was the exception. The inoculum for this site was collected from an adjacent, unimpacted cove.

One group of vessels at each site was inoculated with a detrital fraction containing meiofauna, microflora, and microphytes. This fraction was prepared at each site by gently sieving locally collected detritus through a 0.5-mm screen. A 1.50 g aliquot of the detrital fraction was diluted with Millipore SC (8 μm) filtered seawater until newspaper print was easily read through the graduated cylinder used for dilutions. This represented a solution whose optical density was approximately equal to a 1 x 10^7/ml nutrient broth culture of *Escherichia coli.* A 70-ml inoculum of the solution was added to each of the meiofauna-microfauna-microflora experimental vessels. Samples of the inoculum

EXPERIMENT IN
PROGRESS
DO NOT DISTURB
A
B
C

were fixed in a Rose Bengal-formalin mixture for later determination of the meiofaunal species composition.

The remainder of the gently sieved detrital fraction from each site was used as the source for microflora. These organisms were dislodged from the detritus by vigorous agitation and separated from larger particles and meiofauna by filtration through membrane filters (8-μm pore). The filtrate containing the microflora was diluted with filtered (8 μm) seawater to the same approximate optical density as the fraction that contained the meiofauna, microflora, and microphytes before being used as inoculum for the microflora experimental vessels.

Each site was visited once a month; samples were taken aseptically from the vessels with sterile pipettes, and rates of $^{14}CO_2$ liberation were measured from reaction vessels containing decomposing *Spartina*. In the field, each sample of sediment, organisms, and pore water was divided into three aliquots that were frozen on dry ice for C/N analysis, extracted for 30 minutes in cold (4°C) 35% perchloric acid containing 17 mM EDTA before freezing on dry ice for ATP analysis, or fixed in 4% seawater-buffered glutaraldehyde for later examination by SEM. ATP in frozen perchlorate extracts was neutralized with 0.7 ml of 2.6 N KOH to pH 7.2-7.4 and measured using a JRB luminescent biometer and a purified, lyophilized, luciferin-luciferase preparation (E.I. DuPont de Nemours Company) in MOPS buffer (pH 7.4) (Falkowski, 1977). Precision was facilitated by the automated rapid injection and switching system of Chapman *et al.* (1971).

Material for particulate carbon and nitrogen analysis was filtered onto precombusted (two hours at 200°C) Gelman (cat. #61641) glass fiber filters. The filters were dried, weighed, and analyzed on a Model 185 Hewlett-Packard C-H-N analyzer (Sharp, 1974). Acetanilide was used as a standard.

Filter papers with captured $^{14}CO_2$ were folded carefully in the field, placed in small polycarbonate capsules, and frozen on dry ice. The capsules were counted for activity after oxidation to CO_2 using an Intertechnique Sample Oxidizer (Model IN 4101) (Peterson *et al.*, 1969). Radioactivity in the samples was measured with a Beckman β liquid scintillation counter (LS 250). ^{14}C standards (Amersham-Searle CFR 101) were oxidized with each batch of 40 samples to calculate and compensate for recovery losses and alkaline quench in counting.

Samples for SEM examination were dehydrated through a graded series of ethyl alcohol, placed in iso-amyl acetate, and transferred to lens paper for easier manipulation during critical point drying and mounting on stubs. The stubs were coated with an Au-Pb mixture in a vacuum evaporator and examined in a Cambridge Stereoscan SEM (Model 50-10).

Figure 2. Opposite Reaction vessels at lowest tide with respirometer in place (a). Edge of the marsh at Winsor Cove where an oil spill killed most of the *Spartina* six years ago; erosion and crumbling of the marsh peat are obvious in foreground (b). Close-up picture of marsh peat showing mussel repopulation of the littoral zone (c).

RESULTS AND DISCUSSION

Bearing in mind that samples in this feasibility study were limited to two or three replicates at each study site, and that there are stochastic aspects to the development of isolated natural communities, it was surprising to find generally good reproducibility among replicate reaction vessels incubated at the same test site (*e.g.*, Figure 4). The small number of replicates at each site and time necessitated plotting the data as arithmetic means.

Of the four methods used to estimate detrital decomposition and the communities associated with the process, the C/N ratio seemed the least useful because the amounts of fixed nitrogen remained low in proportion to the carbon in undecomposed detritus. Vessels with meiofauna, microflora, and microfauna incubated at Southampton had the lowest initial C/N ratios (approximately 20:1) and remained so for the duration of the experiment (Figure 3). Vessels at the Millstone quarry and Winsor Cove sites inoculated with similar complex mixtures dropped from initial levels of approximately 36:1 to lower values in August or September (Figure 3). The C/N ratios of vessels with microflora at

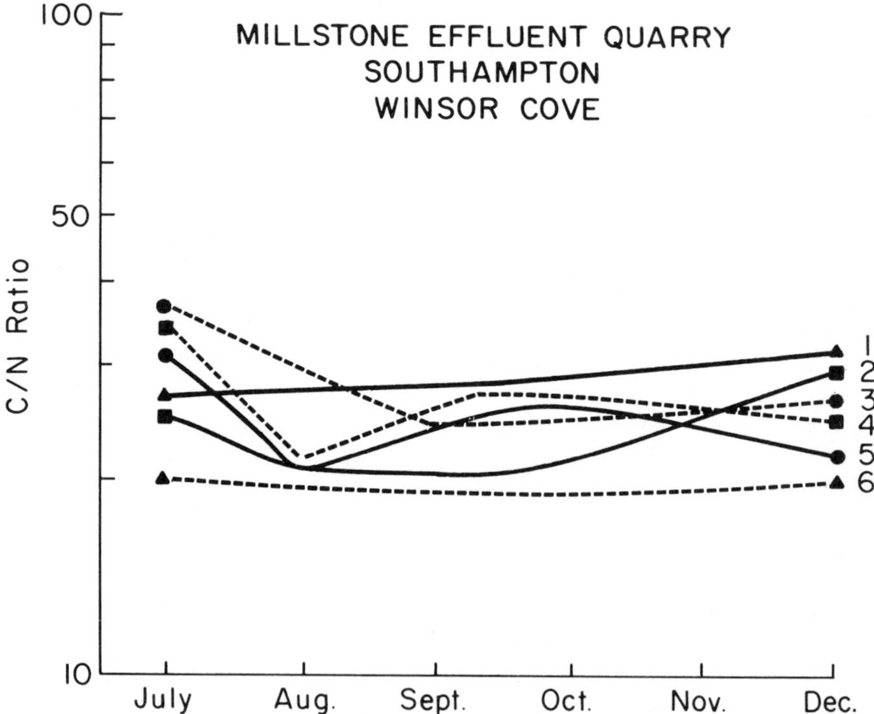

Figure 3. Changes in carbon/nitrogen (C/N) ratio during incubation. Data plotted, starting after the first month of incubation, are averages of three vessels. Curves 1 and 6 from Southampton (▲), curves 2 and 4 from Winsor Cove (■), and curves 3 and 5 from Millstone effluent quarry (●). Vessels represented by solid lines (curves 1, 2, and 5) inoculated with microflora alone; vessels represented by dashed lines (curves 3, 4, and 6) inoculated with microflora, microfauna, and meiofauna.

Southampton rose slightly during incubation. At Millstone, vessels with microflora alone fluctuated but generally fell from slightly above 30:1 to near 20:1 in December. The C/N ratios in the other vessels with microflora alone gradually rose during incubation.

On the other hand, total ATP content per gram detritus seemed a promising method for estimating the size of the living community associated with detrital decomposition. While there was generally good reproducibility among replicate flasks (*e.g.*, Figure 4), there were occasional major deviations in samples from vessels containing meiofauna. For example, September values for two of the samples incubated with meiofauna at Southampton were similar, while that for the third was four times greater (Figure 5). We believe the variance is readily explained. Meiofauna contain considerable ATP (Sikora *et al.*, 1977; Yingst, 1978) and are heterogeneously distributed. Yingst's data and our own suggest considerable variance in ATP measurements may be expected from small volume samples containing meiofauna because of the distribution of these organisms. Neglecting this problem for the moment, total ATP gradually increased in all vessels during incubation (Figures 4, 5). With the exception of vessels from the Millstone effluent quarry, after six months, total ATP per gram detritus was greater in vessels containing meiofauna than in corresponding vessels with only microflora. The slowest growing communities were those incubated at Southampton where, after six months, ATP values for samples from vessels with only microflora (~10 ng ATP/g detritus) were 6-13% of those observed in vessels with only microflora incubated at Winsor Cove (160 ng ATP/g detritus) or those observed at both Millstone sites (~80 ng ATP/g detritus).

As expected, because detrital carbon is mineralized during decomposition, rates of $^{14}CO_2$ release from detritus declined during incubation (Figures 6, 7). Also contributing to the declines are changes in ambient temperature. During incubation, temperatures at the various sites increased from an average 14°C in June to 25°C in August, before falling to lows of 4°C (range 2.5°-6.5°C) in December. Rates of mineralization also were affected by changes in the ratio of less refractory to more refractory detritus. The less refractory fraction is decomposed more rapidly during initial stages of microbial attack, leaving more refractory components behind (Lee *et al.*, 1975).

At two of the sites, Winsor Cove and Southampton, there were considerable differences in the amount of $^{14}CO_2$ released from reaction vessels with and without meiofauna (Figure 6). After one month, rates of mineralized carbon respired at Southampton averaged approximately 5.5 mg C/hour in sediments in reaction vessels with meiofauna, compared to approximately 0.78 mg C/hour in sediments with microflora alone. At the same sites in September, after three months of incubation when the water was still fairly warm (15°-17°C) at both sites, rates of mineralization generally declined but still ranged between 0.2-3.3 mg C/hour. Rates observed in vessels with microflora alone from these two locations differed somewhat from trends seen in vessels with meiofauna. Mineralization rates in vessels from Southampton rose until September before declining; those from Winsor Cove declined until September and then rose

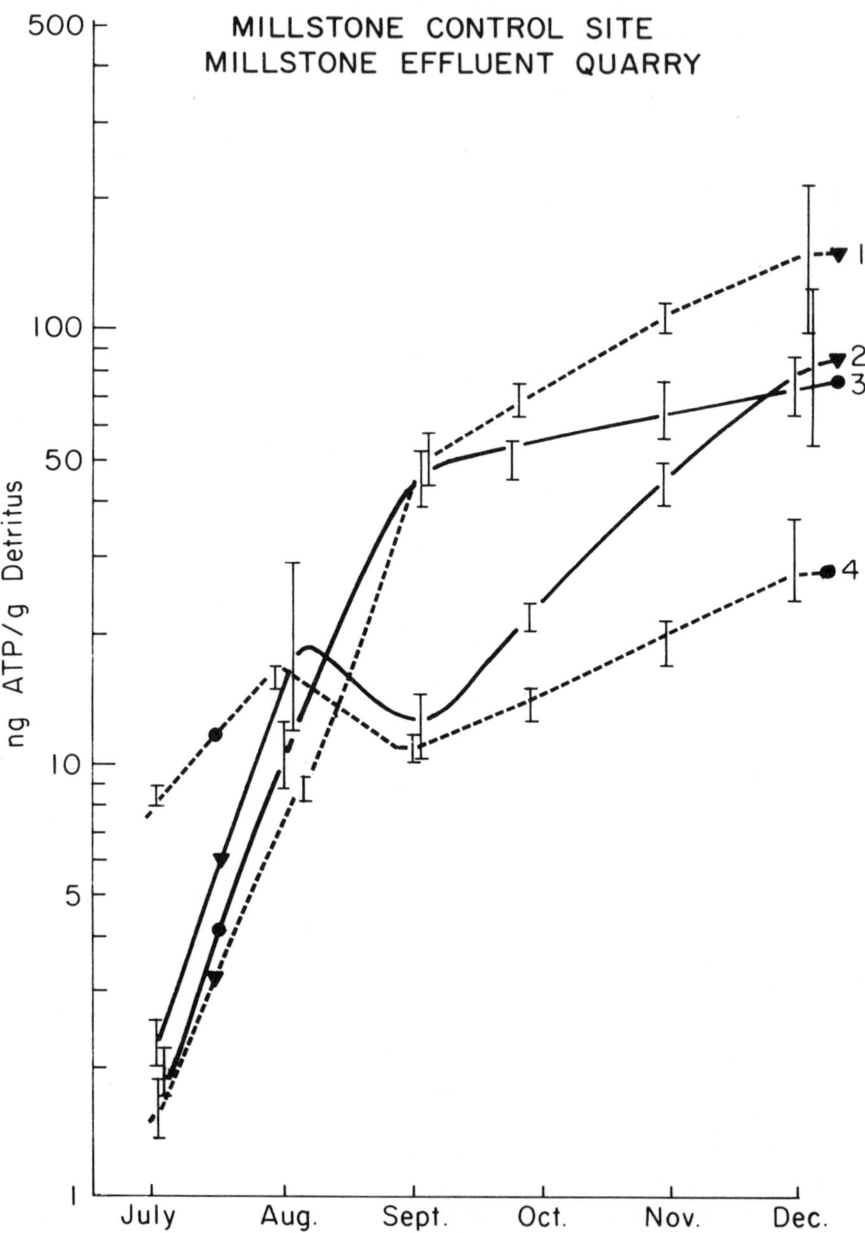

Figure 4. ATP per gram of detritus in vessels incubated in the Millstone control site and effluent quarry. Data plotted, starting after first month of incubation, are averages of three vessels. Bars indicate the range of data points. Curves 1 and 2 from Millstone control site (▾) and curves 3 and 4 from Millstone effluent quarry (•); symbols are used merely to identify individual curves rather than to locate data points. Vessels represented by dashed lines (curves 1 and 4) were inoculated with microflora, microfauna, and meiofauna; vessels represented by solid lines (curves 2 and 3) were inoculated only with microflora.

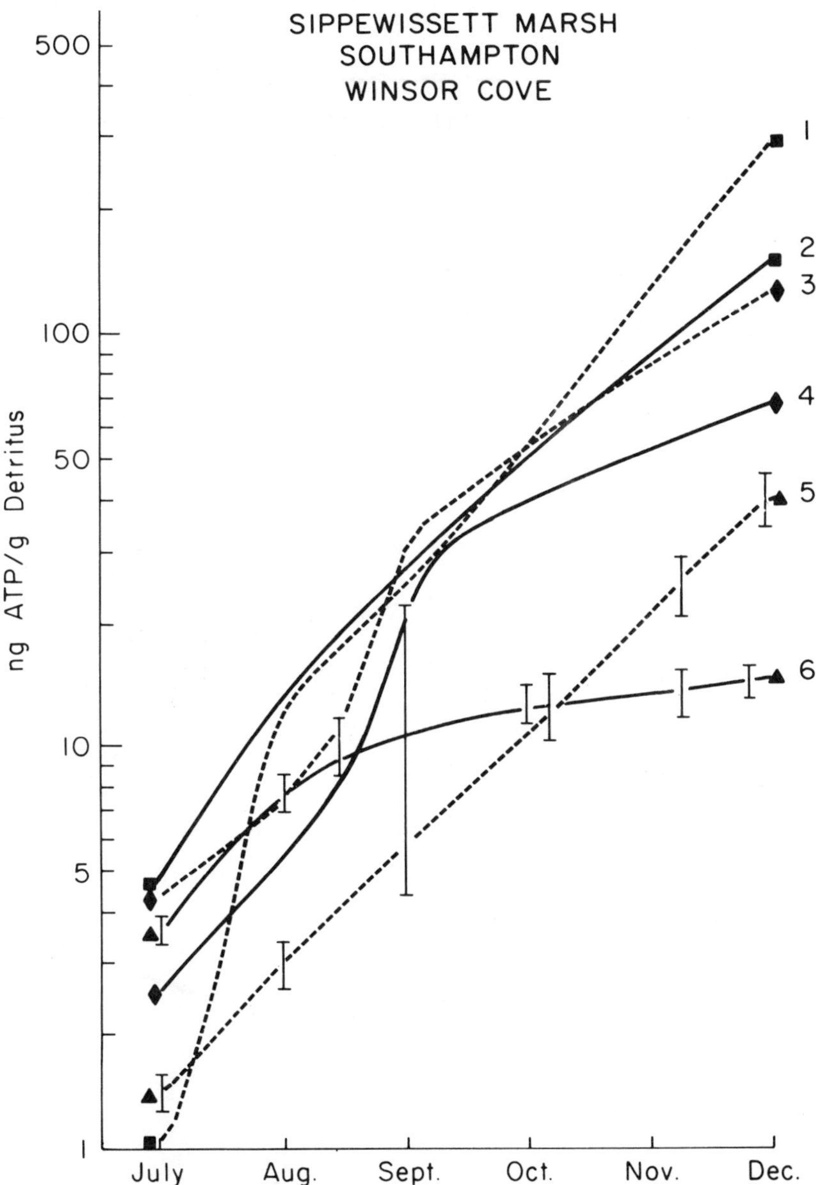

Figure 5. ATP per gram of detritus in vessels incubated at Sippewissett marsh, Winsor Cove, and Southampton. Data plotted after first month of incubation. Curves 1 and 2 from Winsor Cove (■), curves 3 and 4 from Sippewissett marsh (♦), and curves 5 and 6 from Southampton (▲); symbols are used merely to identify individual curves rather than to locate data points. Data from Southampton are averages of three points; bars represent range of data points. Data from Sippewissett Marsh and Winsor Cove represent individual points. Vessels represented by dashed lines (curves 1, 3, and 5) were inoculated with microflora, microfauna, and meiofauna. Vessels represented by solid lines (curves 2, 4, and 6) were inoculated only with microflora.

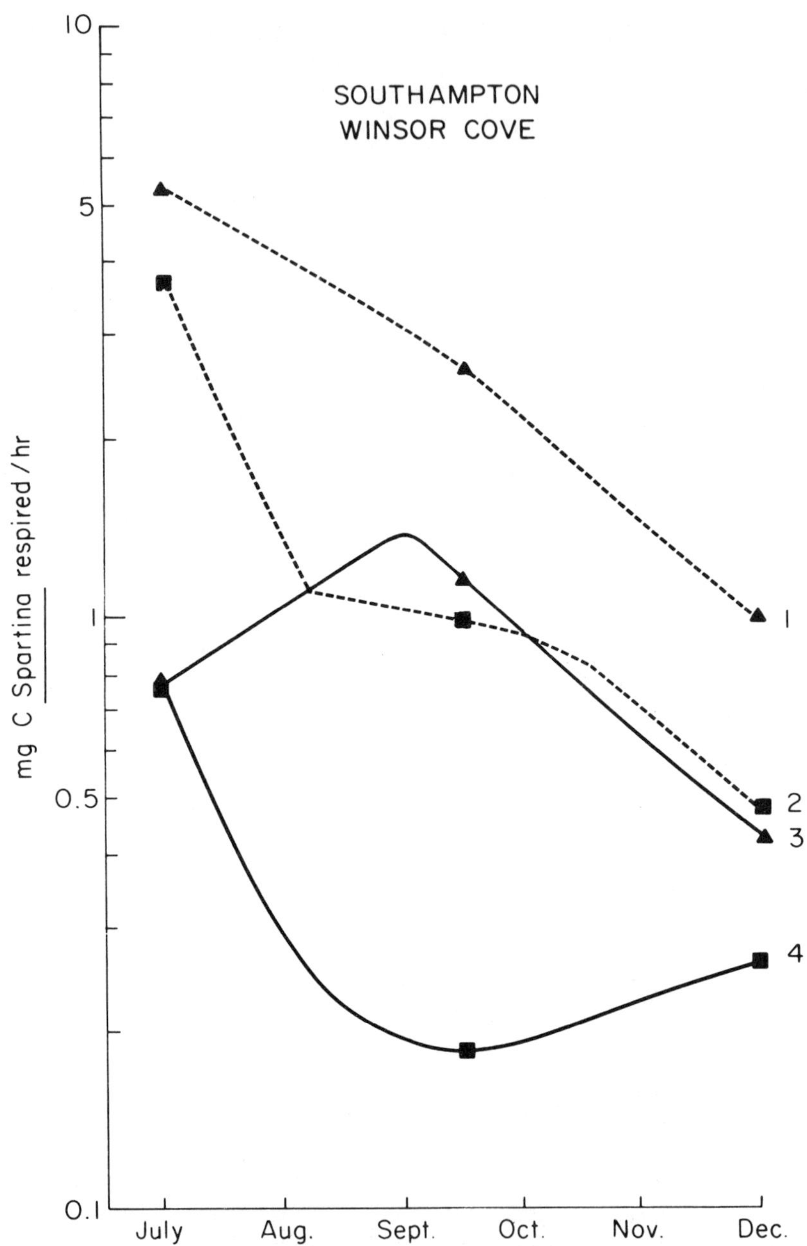

Figure 6. Rate of detrital carbon released as CO_2 per hour in vessels at Southampton (▲, curves 1 and 3) and Winsor Cove (■, curves 2 and 4); symbols are used merely to identify individual curves rather than to locate data points. Data calculated from measurements of $^{14}CO_2$ tracer captured in respirometers. Data are plotted as averages of six subsamples in each of three vessels. Data plotted after first month of incubation. Vessels represented by dashed lines (curves 1 and 2) were inoculated with microflora, microfauna, and meiofauna; vessels represented by solid lines (curves 3 and 4) were inoculated only with microflora.

slightly until the end of the experimental period. At the Sippewissett and Millstone control sites the mineralization rates in vessels with and without meiofauna were much closer and generally followed the same trends found at the other two sites (Figure 7). The rates of mineralization in the reaction vessels within the Millstone effluent quarry were on the average 30% lower than those at nearby control sites (Figure 7). This may indicate that the effluent conditions are less conducive to detrital decomposition.

SEM observations of samples from the reaction vessels suggested great variation in the microbial colonization of the *Spartina* fragments (Figure 8). Some particles were heavily colonized after only one month of incubation, whereas others in the same reaction vessel were sparsely colonized after five months of incubation. In general, there was a great variety of morphological types of microbes and more clumping of particles into small aggregates in vessels with meiofauna that in those without. SEM observations offer promise of yielding much useful information and should be pursued in future studies.

Of the four variables measured in the reaction vessels (changes in C/N ratio, buildup of detrital-based microbial and meiofaunal communities as observed by ATP concentrations or SEM, and CO_2 evolution), only two seem reasonably practical as means of monitoring the effects of pollutant stress on detrital decomposition. There was little change in C/N ratio because of the presence of much undecomposed cellulose relative to the fraction of nitrogen that can be fixed and acquired from the water column by the biota. Too little is known about the stochastic aspects of microbial cellulose degradation in shallow coastal waters and the conditions favoring some flora over others to permit the use of SEM or selective growth media in determining when detrital communities function at reduced rates under pollution stress. On the other hand, total ATP (energy charge) and release of mineralized carbon (as traced with $^{14}CO_2$) are reasonably simple and promising methods for measuring the effects of various types of stress on detrital decomposing systems in nearshore areas. The ^{14}C tracer method and the total ATP method measure overlapping but not identical aspects of benthic community function. The ^{14}C tracer measures the rate of detrital decomposition and reflects the availability of benthically-derived nutrients and carbon to food webs linked to this process. Total ATP reflects the number of organisms linked wholly or partially to detrital decomposition and to other benthic processes (*e.g.*, algal primary production, chemolithotrophy, photolithotrophy).

In unstressed communities there probably is a relationship between total microbial, microphytic, and small animal biomass (estimated by the ATP method) and detrital decomposition (measured by the tracer method). The relationship probably is skewed in communities in which large inputs of biodegradable substrates (*e.g.*, oil, sewage sludge) promote the growth of large microbial communities not necessarily linked to desirable food webs.

Although the data base for this preliminary study was small (two or three replicates at each site, two variables: microflora alone or microflora plus microfauna and meiofauna), the similarity of results from replicates and from

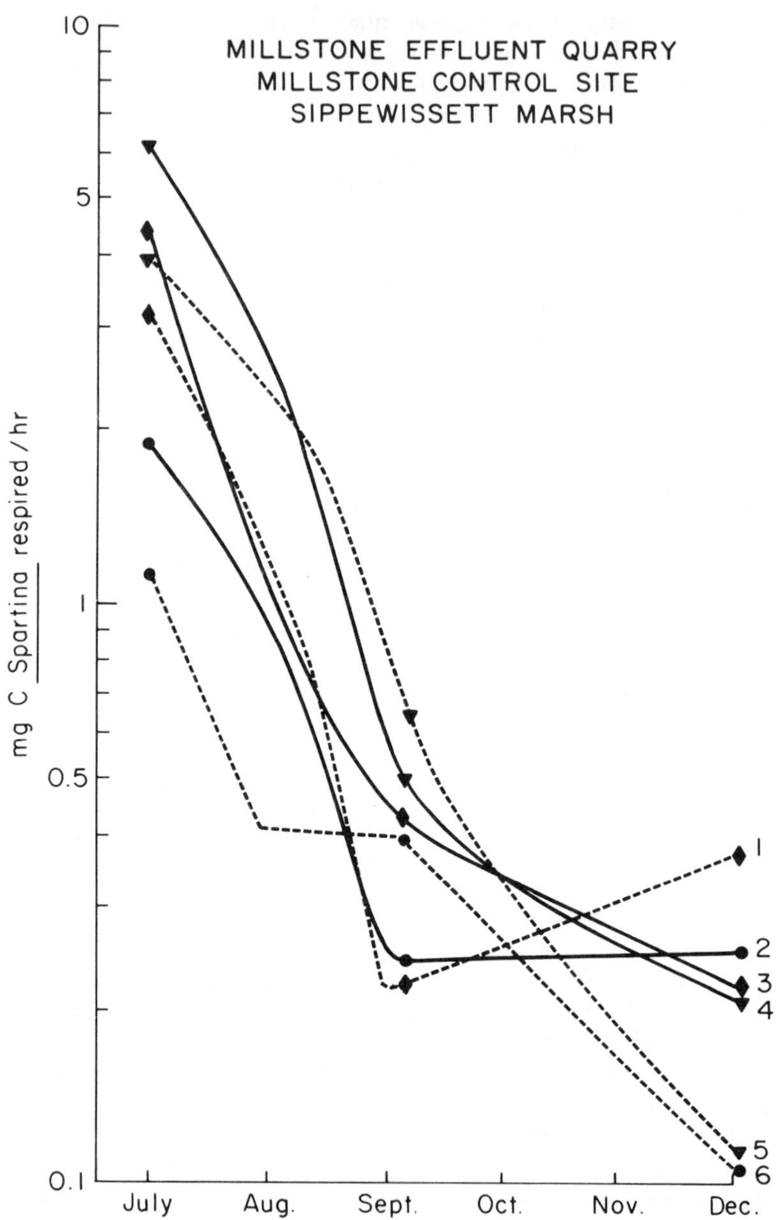

Figure 7. Rate of detrital carbon released as CO_2 per hour in vessels incubated at Sippewissett marsh (♦, curves 1 and 3), Millstone effluent quarry (•, curves 2 and 6), and Millstone control site (▾, curves 4 and 5); symbols are used to identify individual curves rather than to locate data points. Data calculated from measurements of $^{14}CO_2$ tracer captured in respirometers. Data are plotted as averages of six subsamples in each of three vessels. Data plotted after first month of incubation. Vessels represented by dashed lines (curves 1, 5, and 6) were inoculated with microflora, microfauna, and meiofauna; vessels represented by solid lines (curves 2, 3, and 4) were inoculated only with microflora.

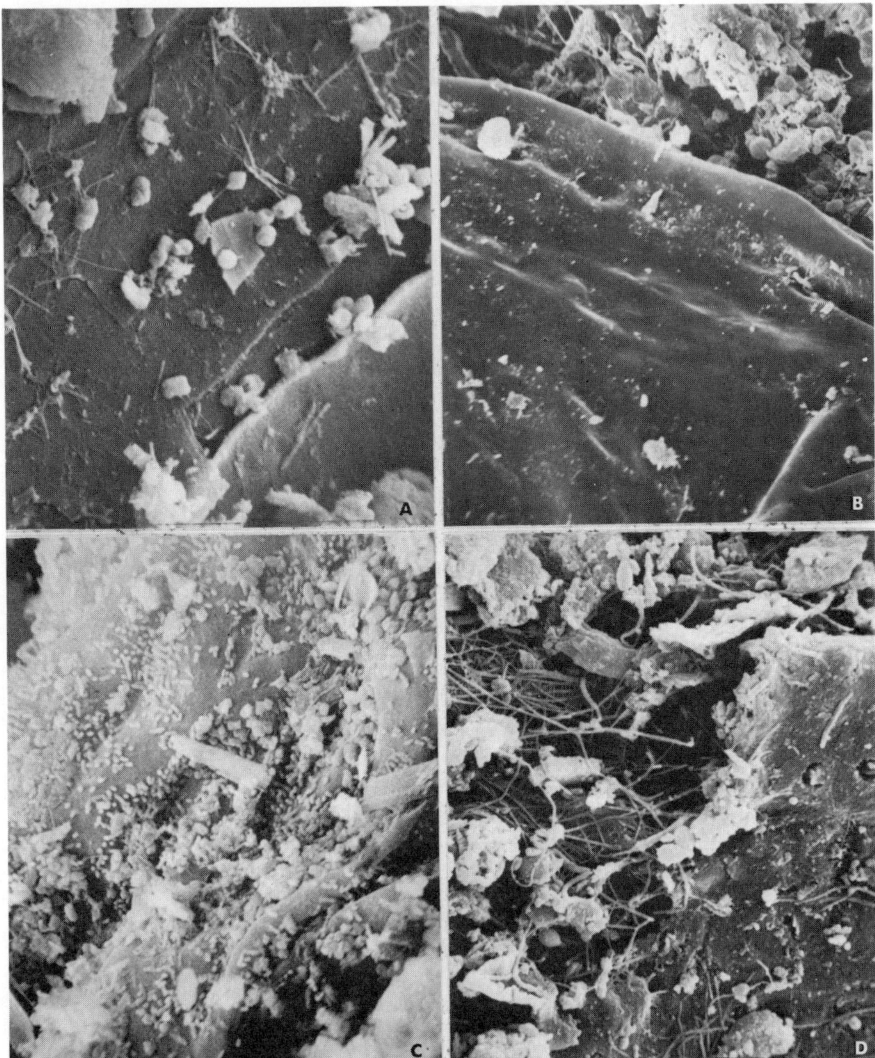

Figure 8. Scanning electron microscopy (SEM) photographs of representative microbial populations on the surfaces of *Spartina* fragments from reaction vessels. (a) Typical colonization on the surface of *Spartina* after one month incubation (x2,800); (b) extensive microbial colonization on one fragment (upper third) and no colonization on an adjacent fragment (lower two thirds); (c) extensive colonization by bacillus and cocci on fragments incubated for several months (x2,800); and (d) extensive mycelial-like growth of filamentous bacteria and flexibacteria (x2,750).

sites geographically separated was surprising. To establish a more realistic idea of the "normal" ranges of initial rates of detrital colonization and decomposition, the data base should be expanded. This is particularly true because of the stochastic aspects of the decomposition processes. While a uniformly assembled "natural" substrate was used, it would be more realistic to use native substrates for studies of this type. Nowhere was this point made more clearly than at Winsor Cove. Mussels have returned to this oil-impacted site and are feeding on available plankton (Figure 2c). However, after five years, *Spartina alterniflora* still is receding along the shoreline (Hampson and Moul, 1978), and meiofauna rarely are found in the oil-soaked sediment, which oozes oil as one walks on it (Figure 2b). While our study shows that water quality at the site is satisfactory for detrital decomposition, such decomposition probably is not occurring.

One approach used earlier in laboratory flow-through chambers with detrital decomposition systems was to withdraw, freeze, and thaw the native substrate and then "spike" it with ^{14}C-labeled ground *Spartina* detritus (Tenore *et al.*, 1977). There are few technical obstacles to doing this, and, in fact, it might be advantageous to incubate native "spiked" and artificially constructed substrates side by side as controls. Although of more academic than pragmatic interest, experiments relating $^{14}CO_2$ release during decomposition to total mineralized carbon, and longer experiments relating the cropping of microbial and meiofaunal assemblages to rates of *in situ* decomposition may be worthwhile.

The *in situ* incubation system described in this paper, which permits the capture of mineralized $^{14}CO_2$, shows considerable promise as a practical means to monitor the effects of water and sediment qualities on detrital decomposition in nearshore benthic habitats. However, considerable effort still is needed to develop the assay for routine applications.

ACKNOWLEDGEMENTS

Many people aided us while we were doing this study. Dr. Howard Sanders and Mr. George Hampson of the Woods Hole Oceanographic Institution (WHOI) assisted in setting up the experiments at Winsor Cove, Massachusetts. Drs. John Teal of WHOI and Ivan Valiella of the Boston University Marine Program helped us set up in the Greater Sippewissett marsh in Falmouth, Massachusetts. Dr. William Renfro and Mr. Paul Jacobson of the North East Nuclear Energy Company provided laboratory space and helped us set up experiments and controls at the Millstone Atomic Power Plant at Niantic, Connecticut. Mr. Tom Rewinski of the Southampton Town Board of Trustees granted us permission to place the experiment in the Towd Point Sanctuary, Southampton, New York.

The ATP analysis was done under the tutelage and in the laboratory of Dr. Paul Falkowski, Department of Oceanography, Brookhaven National Laboratory. The C/N measurements were made in the Biology Department, Lamont-Doherty Geological Observatory of Columbia University. We are grateful to Dr. Thomas Malone and Ms. Elizabeth Cosper for instructing us in the use of the equipment and making it available to us.

The study was supported by DOE Contract (11-1) 3395. Support from NSF grant OCE 78-09962 to complete the project also is gratefully acknowledged.

REFERENCES

Chapman, A.G., L. Fall, and D. Atkinson. 1971. Adenylate energy charge in *Escherichia coli* during growth and starvation. J. Bacteriol. 108: 1072-1086.

Falkowski, P. 1977. The adenylate energy charge in marine phytoplankton: the effect of temperature on the physiological state of *Skeletonema costatum* (Grev.) Cleve. J. Exp. Mar. Biol. Ecol. 27: 37-45.

Ferguson, R.L. and R.B. Williams. 1974. A growth chamber for the production of ^{14}C labeled salt marsh plants and its application to smooth cordgrass, *Spartina alterniflora* Loisel. J. Exp. Mar. Biol. Ecol. 4: 251-259.

Hampson, G. and E. Moul. 1978. No. 2 fuel oil spill in Bourne, Massachusetts: immediate assessment of the effects of marine invertebrates and a three year study of growth and recovery of a salt marsh. J. Fish. Res. Board Can. 35: 731-744.

Jensen, A.C. 1977. New York's marine fisheries: changing needs in a changing environment. N.Y. Fish Game J. 24: 99-128.

Lee, J.J. 1980. A conceptual model of marine detrital decomposition and the organisms associated with the process. *In:* Advances in Aquatic Microbiology, Vol. 2, H. Jannasch and M.R. Droop (eds.), Academic Press, London. pp. 259-299.

Lee, J.J., K. Tenore, J. Tietjen, and C. Mastropaolo. 1975. An experimental approach toward understanding the role of meiofauna in a detritus based marine food web. *In:* Radioecology and Energy Resources, C.E. Cushing, Jr. (ed.), Dowden, Hutchinson & Ross, Inc., Stroudsburg, PA. pp. 140-147.

Mann, K.H. 1972. Macrophyte production and detritus food chains in coastal areas. *In:* IBP-UNESCO Symposium on Detritus and Its Ecological Role in Aquatic Ecosystems (Supplement 29), Member Institute of Italian Hydrobiology, Pallanza, Italy pp. 353-383.

Matera, N.J. and J.J. Lee. 1972. Environmental factors affecting the standing crop of foraminifera in sublittoral and psammolittoral communities of a Long Island salt marsh. Mar. Biol. 14: 89-103.

Peterson, J., F. Wagner, S. Siegel, and W. Nixon. 1969. A system for convenient combustion preparation of titrated biological samples for scintillation analysis. Anal. Biochem. 31: 189-203.

Sharp, J.H. 1974. Improved analysis for "particulate" organic carbon and nitrogen from sea water. Limnol. Oceanogr. 19: 984-992.

Sikora, J.P., W.B. Sikora, C. Erkenbrecher, and B. Coull. 1977. Significance of ATP, carbon and caloric content of meiobenthic nematodes in partitioning benthic biomass. Mar. Biol. 44: 7-14.

Tenore, K., J. Tietjen, and J. Lee. 1977. Effect of meiofauna on incorporation of aged eelgrass, *Zostera marina* detritus by the polychaete *Nephthys incisa.* J. Fish. Res. Board Can. 34: 536-567.

Tietjen, J. H. This volume. Potential roles of nematodes in polluted ecosystems and the impact of pollution on meiofauna. pp. 225-233.

Yingst, J.Y. 1978. Patterns of micro- and meiofaunal abundance in marine sediments, measured with the adenosine triphosphate assay. Mar. Biol. 47: 41-54.

INFLUENCE OF CADMIUM ON THE MICROBIAL POPULATIONS AND PROCESSES IN SEDIMENTS FROM THE NEW YORK BIGHT APEX

C.D. Litchfield[1]
M.A. Devanas[2]
J. Zindulis[3]
M. Meskill
J. Freedman[4]
C. McClean

Center for Coastal and Environmental Studies
Marine Sciences Division
Rutgers–The State University
New Brunswick, New Jersey 08903

Abstract. The effects of cadmium on the ecology of bacteria from selected New York Bight apex sediments were studied during a two-year period. Cadmium chloride supplementation to media and mineralization experiments indicated a reduction in the numbers of aerobic heterotrophic bacteria but either no effect or slight stimulation of the rates of decomposition of urea and glycolic acid. This pattern was observed repeatedly during the sampling period and indicated that a rather constant percentage of cadmium-resistant bacteria in sediments accounts for the majority of the microbial mineralization capacity. Microbial strains demonstrating resistance to cadmium on initial isolation were cultured, purified, and identified, and their antibiotic patterns were established. Seventy-two percent of the isolates were resistant to multiple classes of antibiotics, indicating a widespread distribution of both cadmium resistance and antibiotic resistance in non-enteric sedimentary bacteria. Additional studies with isolate A-4-20 confirmed that both the uptake of amino acids and their incorporation into protein by cadmium-resistant bacteria were unaffected by the presence of cadmium at concentrations tested during field studies.

INTRODUCTION

Although cadmium is of environmental concern because of its extreme toxicity (Friberg *et al.*, 1971; Fleischer *et al.*, 1974), relatively few studies have examined the effects of cadmium on microorganisms. Bacteria play two primary roles in the aquatic ecosystem: they serve as primary decomposers and converters of poor quality organic matter into biomass, and as food sources for higher trophic levels. Cadmium may affect both of these functions because it could inhibit the degradative-conversion activities of bacteria or become concentrated by bacterial cells and so proceed through the food web.

During the past several years, we have examined the effects of cadmium on mineralization capacities of sedimentary bacteria from the New York Bight apex. This area is heavily polluted with the metal (Carmody *et al.*, 1973), and it

[1] Present address: E.I. du Pont de Nemours and Company, Haskell Laboratory, Wilmington, Delaware 19898.
[2] Present address: Department of Microbiology, University of Georgia, Athens, Georgia 30602.
[3] Present address: Department of Biology, Rensselaer Polytechnic Institute, Troy, New York 12181.
[4] Present address: Albert Einstein College of Medicine, New York, New York 10461.

was of interest to determine whether major microbial processes were inhibited, or whether microbial processes continued apace, despite the pollution of this region. Questions relating to the level of cadmium resistance of the microflora, the distribution of this resistance throughout the microbial population, and the mechanisms by which bacteria survived and functioned in the presence of elevated levels of cadmium also were examined. This paper presents the results of studies on the specific effects of cadmium additions on microbial metabolic processes in sediments.

MATERIALS AND METHODS

Sampling Locations

Four stations in the vicinity of the New York Bight apex were selected for study (Figure 1). Three of the stations were chosen from the MESA-SYMAP sampling grid (see Thomas *et al.*, 1976: Figure 1). Station 34 lies between the sewage sludge and dredged materials dumpsites in the Christiaensen Basin, a natural topographic low in the apex. Station 109 is located adjacent to the dredged materials and cellar dirt disposal sites, and Station 51 is at the acid

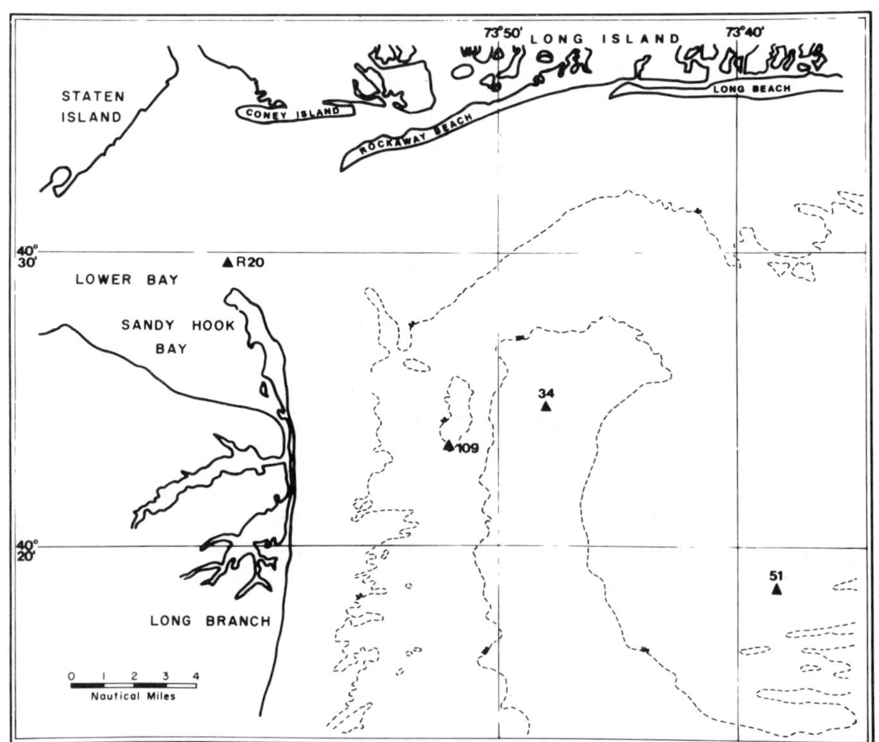

Figure 1. Sample locations for sediment stations in the New York Bight apex.

waste disposal site. The fourth station is located at navigation buoy R-20, at the mouth of the Hudson-Raritan estuary system.

Sample Treatment

Samples were obtained with a Shipek or Smith-McIntyre grab and were processed immediately according to previously described procedures for pH, redox potential (Eh), and temperature (Litchfield *et al.*, 1976). Additional portions were taken for the determintation of percent moisture and for later analyses of total organic carbon and nitrogen (Perkin-Elmer CHN Analyzer).

Microbiological Procedures

Sediment samples composed of unmixed surface layers (0-3 cm) were transferred aseptically to a dilution bottle containing sterile artificial seawater (ASW), at a salinity of 36 °/₀₀ and pH of 7.4. The exact amounts removed depended on the number and types of analyses to be performed but generally averaged 200-250 g wet sediment. This approximate 1:10 dilution became the source for all subsequent microbiological and mineralization studies.

Aliquots from the above dilution were taken for direct counting of bacteria via the acridine orange epifluorescent microscopy method (AODC) (Hobbie *et al.*, 1977); ATP analysis using the cold sulfuric acid extraction procedure of Karl and LaRock (1975), as modified by Zindulis and Litchfield to include substitution of glycine as the buffer and the presence of ^{14}C-ATP throughout the extraction procedure for more accurate calculation of percent recovery (Zindulis and Litchfield, in press); and surface-spread plating of serial dilutions onto low nutrient heterotrophic medium (ESWA) (Litchfield *et al.*, 1975), urea + glycerol medium (urea), and 0.1% glycolic acid added to $(NH_4)_2SO_4$ medium (glycolate) (Litchfield *et al.*, 1976). These three basic media also were supplemented with 1 ppm or 50 ppm (final concentration) cadmium as cadmium chloride. There were four replicates for each ATP analysis, five for each dilution used for surface spread plating on each medium, and two for each direct count. Incubation conditions and enumeration followed published procedures (Litchfield *et al.*, 1976). All data were normalized to a gram dry weight basis.

Heterotrophic Mineralization

Heterotrophic mineralization was determined on aliquots of the sediment-ASW dilution prepared for microbiological analyses. Within one hour of collection, mineralization studies were performed according to previously described methods (Litchfield *et al.*, 1979a) and used ^{14}C-urea or ^{14}C-glycolic acid as the substrates. Replicate sets of incubation bottles contained a concentrated cadmium chloride solution designed to yield a final level of 1 ppm Cd. Respired $^{14}CO_2$ from the various test conditions was counted via liquid scintillation, averaged, and corrected to disintegrations per minute (dpm); calculations of decomposition rates were completed as described previously (Litchfield *et al.*, 1979a). All values were normalized to a gram dry weight basis.

Statistical Analyses

Statistical analyses for Spearman Rank correlation coefficients were accomplished using the SAS package (Barr *et al.*, 1976) at the Center for Computer Information Services, Rutgers–The State University. Student's *t*-tests and calculations of the ATP data were performed on a Monroe Model 1860 Programmable Calculator.

Antibiotic Testing

Pure cultures were grown in enriched seawater broth (ESWB). Log phase cells were washed in ASW and resuspended; 0.1 ml of the suspension was surface-spread plated onto dry ESWA media. After 30-60 min to allow for drying, combinations of the following commercially prepared antibiotic discs (BBL, Cockeysville, Maryland) were placed on each plate: sulfachloropyridazine (0.25 μg), penicillin G (10 U), bacitracin (10 U), tetracycline (30 U), streptomycin (10 U), novobiocin (30 U), neomycin (30 U), nalidixic acid (30 μg), kanamycin (5 μg), and chloramphenicol (30 μg). Five to six discs were placed on each plate. Each test was repeated three times, and the resulting zones of inhibition for each antibiotic were averaged.

Growth Studies

The effects of cadmium on specific growth rates (Meynell and Meynell, 1965) of isolate A-4-20 were tested by the addition of 0 ppm, 1 ppm, 10 ppm, 50 ppm, 75 ppm, and 100 ppm Cd as cadmium chloride to replicate tubes of ESWB: proteose peptone #3, 1.0 g; yeast extract (Difco), 1.0 g; sodium glycerophosphate, 0.05 g; and filtered ASW (36 ‰) (Rila Marine Mix, Utility Chemical Company, Teaneck, New Jersey), 1,000 ml, pH 7.4 ± 0.2. Microbial growth was determined by turbidity at 600 nm in an Hitachi Perkin-Elmer Model 139 Spectrophotometer and correlated with cellular protein concentrations by the Coomasie blue protein determination technique of Bradford (1976) as well as by the numbers of viable cells determined from plate counts of inocula. The Bradford method was selected because salt does not interfere with color development until a concentration of about 10% NaCl is reached (unpublished data).

Whole cell uptake of ^{14}C-leucine (240 mCi/mM, ICN) and the resultant $^{14}CO_2$ respired or amino acid incorporation into protein were determined as for the heterotrophic mineralization studies with the exception that cells were resuspended in an HEPES/ASW buffer (10.0 mM N-2-hydroxyethylpiperazine-N′-2 ethane sulfonic acid) (Calbiochem).

RESULTS

Physical/Chemical Studies

Selected physical, chemical, and microbiological characteristics of sediments from the four study sites are listed in Table 1. The pH values declined steadily

Table 1. Summary of physical, chemical, and microbiological characteristics of sediments at four reference stations in New York Bight apex, May 1977 to March 1978. Dash (–) indicates data not taken.

Station	Sampling date	Depth overlying water (m)	pH	Eh (mV)	Percent moisture	Total organic carbon (mg)[a]	Total nitrogen (mg)[a]	ATP (ng)[b]	Log direct count[b]	Log CFU ESWA[b]
R-20	May 1977	7.6	7.4	470	35	8.95	0.788	40.0	–	6.78
	July	8.5	7.9	140	34	8.13	0.603	148.0	8.03	5.22
	November	5.0	6.5	335	33	11.82	0.635	251.0	8.29	6.49
	March 1978	8.5	4.5	45	39	11.15	0.632	56.9	8.36	6.64
109	May 1977	38.1	7.3	635	36	16.05	1.546	0.0	–	5.03
	July	20.1	7.2	410	46	19.16	1.532	995.0	8.41	6.39
	November	24.0	7.3	215	16	14.99	0.768	247.0	8.69	5.86
	March 1978	17.8	4.2	85	22	1.09	0.046	–	6.32	3.71
51	May 1977	24.4	7.6	110	21	0.73	0.076	0.5	–	4.04
	July	21.9	7.3	85	19	0.59	0.067	96.6	7.60	4.23
	November	20.0	5.9	245	21	0.76	0.058	7.0	6.14	3.60
	March 1978	24.8	8.0	110	22	0.91	0.073	7.4	7.23	4.39
34	May 1977	32.0	7.4	480	34	11.51	1.054	26.1	–	4.26
	July	31.7	7.5	470	49	17.69	1.578	510.0	7.74	5.74
	November	30.0	6.1	205	59	23.30	1.933	218.0	8.27	5.70
	March 1978	34.1	7.5	–	53	22.12	1.843	26.1	7.94	5.88

[a]All relevant data normalized to a per gram dry weight basis and constitute averages of at least two determinations.

[b]All data are normalized to a per gram dry weight basis.

during the sampling period at Stations R-20 and 109 and paralleled a decrease in the Eh at these stations. This coordinated decline, however, appears to have had no effect on the total AODC or colony-forming units (CFU) from these samples ($r = 0.03$ and 0.26, respectively, for AODC and 0.11 and 0.10, respectively, for CFU according to Spearman rank correlational analysis). Similarly, there was no apparent relationship between the log AODC and the total organic carbon (TOC) ($r = 0.54$) or the total organic nitrogen (TON) ($r = 0.54$). Similar values were obtained for the relationship between the log of the total aerobic CFU and TOC or TON. A high level of correlation was obtained between the log AODC and the log total aerobic CFU ($r = 0.83$), which had a probability value of $p \leqslant 0.001$. This was true for individual stations and for individual sampling periods. Although AODC values are higher than the enumerated CFU, the consistency of correlation between these two parameters demonstrates that they both reflect the microbial population in these sediments.

The percent moisture values vary markedly at the dredge spoils (Station 109) and Christiaensen Basin (Station 34) sites, indicating the continual input of different types of material. Despite this, there is a high level of correlation ($p \leqslant 0.001$) between the percent moisture content of these sediments and TOC and TON ($r = 0.80$ and 0.81, respectively). The ATP concentrations were correlated with the TOC, TON, log AODC, and log CFU at the $p \leqslant 0.05$ level.

Colony-Forming Units and Cadmium

Seasonal change in the number of CFU on the urea medium is shown in Figure 2. In general, the CFU on the urea medium supplemented with 1 ppm Cd parallel the CFU obtained on unsupplemented urea medium, particularly at Station R-20, although at a reduced level. Counts obtained from the other stations show similar but not as striking trends compared to the unsupplemented urea-based medium. In 20 of the 27 possible cases, there were statistically significant reductions in the colony-forming units obtained from the urea media supplemented with 1 ppm or 50 ppm Cd. The 1 σ (standard deviation) levels averaged $\log_{10}$ 0.2 CFU. Previous studies (Litchfield *et al.*, 1979a) demonstrated that at least a 0.5 log difference was necessary before statistically significant differences could be noted.

Essentially similar recoveries of CFU were obtained on glycolate and glycolate plus 1 ppm Cd (Figure 3). As expected, there was reduced growth on the glycolate plus 50 ppm Cd media. However, in March 1978, the CFU values from the 50 ppm Cd-supplemented media were equivalent or slightly higher than those obtained on the other media. Only a small percentage of the microbial population appears capable of growth in the presence of cadmium. Regardless of the isolation medium, mean CFU values for all stations were 30% for the urea medium and 26% for the glycolate medium when each medium was supplemented at 1 ppm Cd. When 50 ppm Cd was added, CFU values dropped to 8% and 10%, respectively. This lends support to Stevenson's hypothesis that in natural systems there exists a predominantly resting or zymogenous bacterial population (Stevenson, 1978).

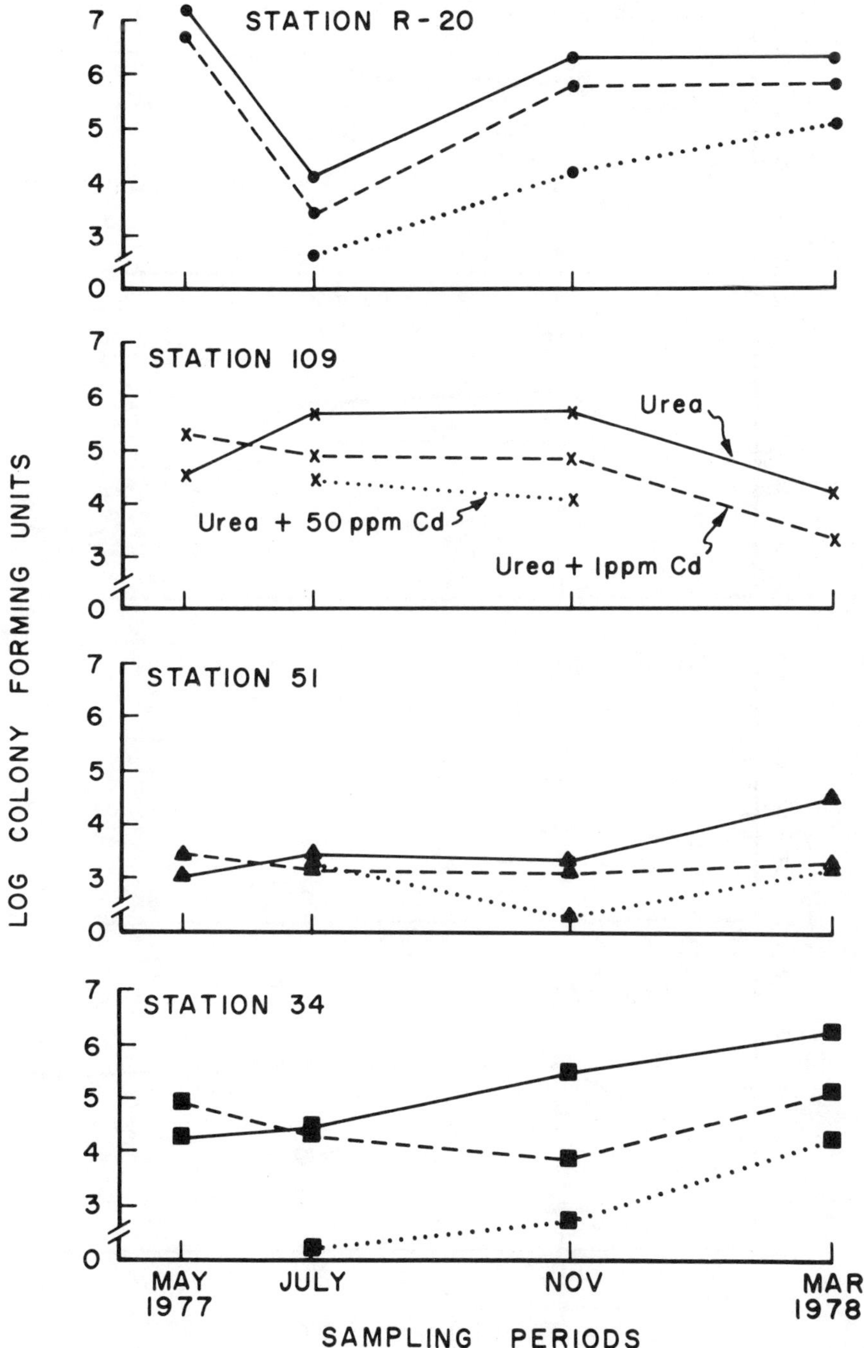

Figure 2. Seasonal variations of colony-forming units on urea (—), urea + 1 ppm Cd (– – –), and urea + 50 ppm Cd (. . .) media.

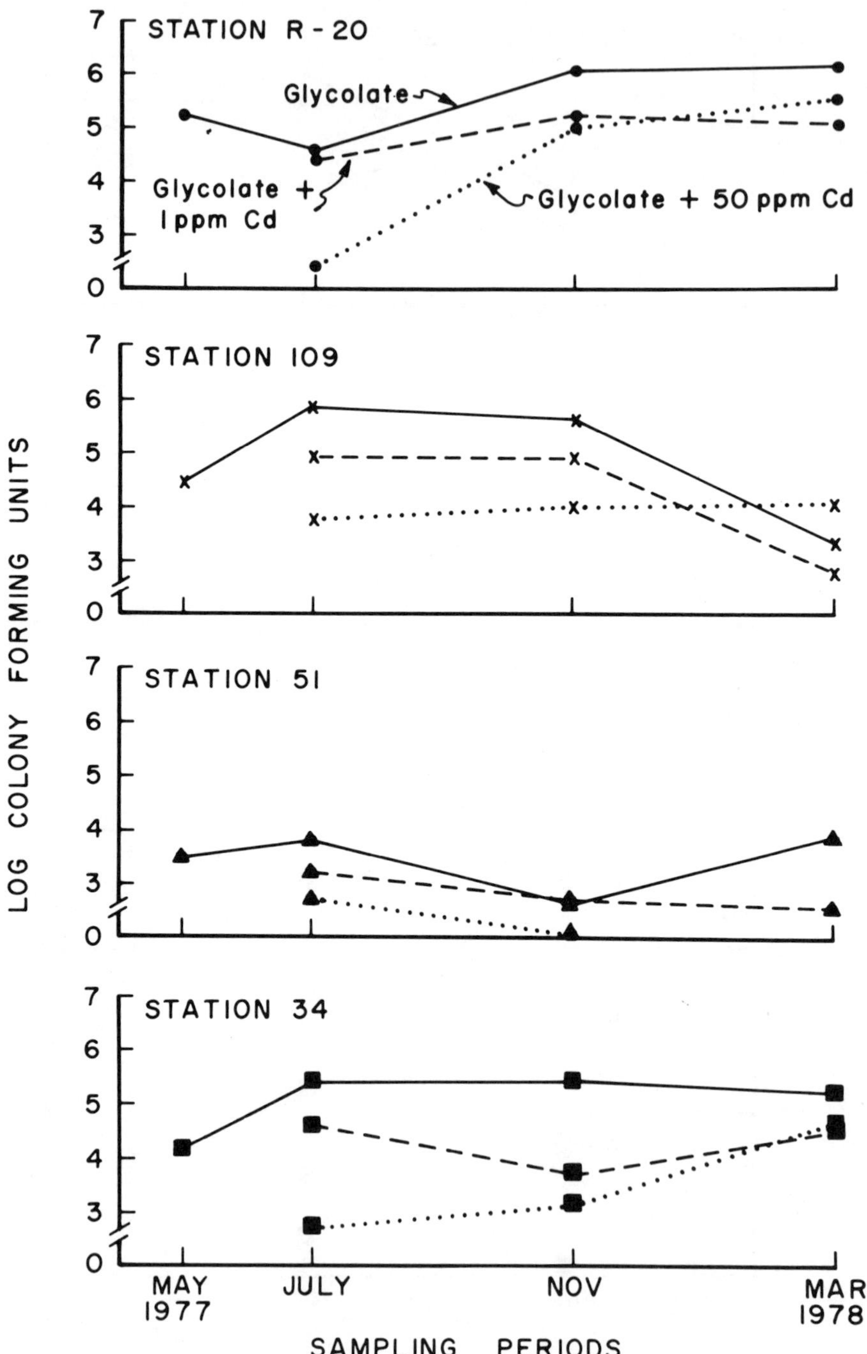

Figure 3. Seasonal variations of colony-forming units on glycolate (—), glycolate + 1 ppm Cd (– – –), and glycolate + 50 ppm Cd (. . .) media.

Table 2. Effect of added cadmium (1 ppm) on the turnover times of urea as determined from the saturation kinetic model (T_t) or first-order kinetic model (T_d) of Wright and Hobbie (1966). Dash (–) indicates no comparison possible because of mixed kinetic data.

Station	Cruise date	No Cd		1 ppm Cd		Percent change[a]	Gross effect on decomposition
		T_t	T_d	T_t	T_d		
R-20	May 1977	60.3		74.6		23	Inhibited
	July	68.3			115.0	–	–
	November	84.1		2.8		-96	Stimulated
	March 1978	10.9		12.4		13	Inhibited
109	May 1977	12.2		15.0		22	Inhibited
	July	59.5		78.5		31	Inhibited
	November	2.0		0.8		-60	Stimulated
	March 1978		763.3	166.8		–	–
51	May 1977	91.7			237.3	–	–
	July		225.0	21.2		–	–
	November	160.9		136.9		-14	Stimulated
	March 1978	4.7		191.5		3974	Inhibited
34	May 1977	46.7		91.3		95	Inhibited
	July	46.6		46.1		-1	Stimulated
	November	98.0		34.3		-65	Stimulated
	March 1978	41.1		39.0		-5	Stimulated

[a]A positive percent change indicates longer turnover times with cadmium, while a negative percent change indicates shortened turnover times.

Kinetic Analyses: Turnover Times

In an attempt to define more precisely the effects of cadmium on the decomposer role of the sedimentary microflora, 1 ppm $CdCl_2$ was added to replicate sets of mineralization bottles used to estimate the rates of glycolic acid and urea decomposition. The resultant turnover times (T_t and T_d) are shown in Tables 2 and 3. The T_t values were calculated from incubation conditions that conformed to the saturation kinetic model; the T_d values were derived from first-order plots of substrate decomposition rates (Wright and Hobbie, 1966). The shortest turnover times on urea occurred during March at Stations R-20, 51, and 34, while at Station 109, November T_t rates were the most rapid. When cadmium was present, except for Station 51, either November or March was the period when urea was most rapidly turned over. *In situ* urea levels were not elevated during the fall-winter sampling period (Litchfield *et al.*, 1979b). In general, the uptake rates for urea decomposition conformed to the saturation kinetic model and indicated less variability in the sources of urea metabolism. Also the T_t's generally were within a two- to four-fold range for each treatment, which was shown previously to be statistically insignificant when employing our present methods (Litchfield *et al.*, 1979a).

Table 3. Effect of added cadmium (1 ppm) on the turnover times of glycolate as determined from the saturation kinetic model (T_t) or first-order kinetic model (T_d) of Wright and Hobbie (1966). ND indicates comparisons not possible because glycolate + 1 ppm Cd was not used during May 1977 cruise; dash (–) indicates no comparison possible because of mixed kinetic data.

		No Cd		1 ppm Cd			
Station	Cruise date	T_t	T_d	T_t	T_d	Percent change[a]	Gross effect on decomposition
20	May 1977	0.91				ND	
	July	0.01		0.01		0	No change
	November	64.4		46.7		-27	Stimulated
	March 1978		12.7		40.0	214	Inhibited
109	May 1977	0.1				ND	
	July		19.1	9.0		–	–
	November		0.3	0.7		–	–
	March 1978	111.5			32.3	–	–
51	May 1977		28.7			ND	
	July	14.7		10.9		-25	Stimulated
	November	14.7			643.8	–	–
	March 1978	89.4		41.2		-53	Stimulated
34	May 1977	0.6				ND	
	July	2.9		1.5		-48	Stimulated
	November	2.5			2.2	–	–
	March 1978	15.0			22.1	–	–

[a]A positive percent change indicates longer turnover times with cadmium, while a negative percent change indicates shortened turnover times.

The predominant effect of cadmium addition to glycolate was a slight stimulation of the rate of decomposition; however, this was not statistically significant ($p \leqslant 0.01$). The most rapid turnover times for glycolic acid decomposition generally occurred during July or November, depending on the station (Table 3). Because it is assumed that algal utilization is depicted more accurately by first order kinetics (Wright and Hobbie, 1966), the mixed kinetic responses implied that, unlike urea, different organisms and/or trophic levels were metabolizing glycolic acid during the different seasons.

Cadmium had little effect on the mode of substrate utilization, whether it was predominantly extracellular (urea) or intracellular (glycolic acid). This is in contrast to studies of the inhibition of glucose uptake by mixed microbial populations in activated sludge, which showed approximately 20% inhibition (Cenci and Morrozzi, 1977).

Seasonal Changes in Carbon Mineralization

The decomposer role of the microbial flora also may be evaluated in terms of the amount of carbon mineralized. The relative amounts of urea-carbon utilized

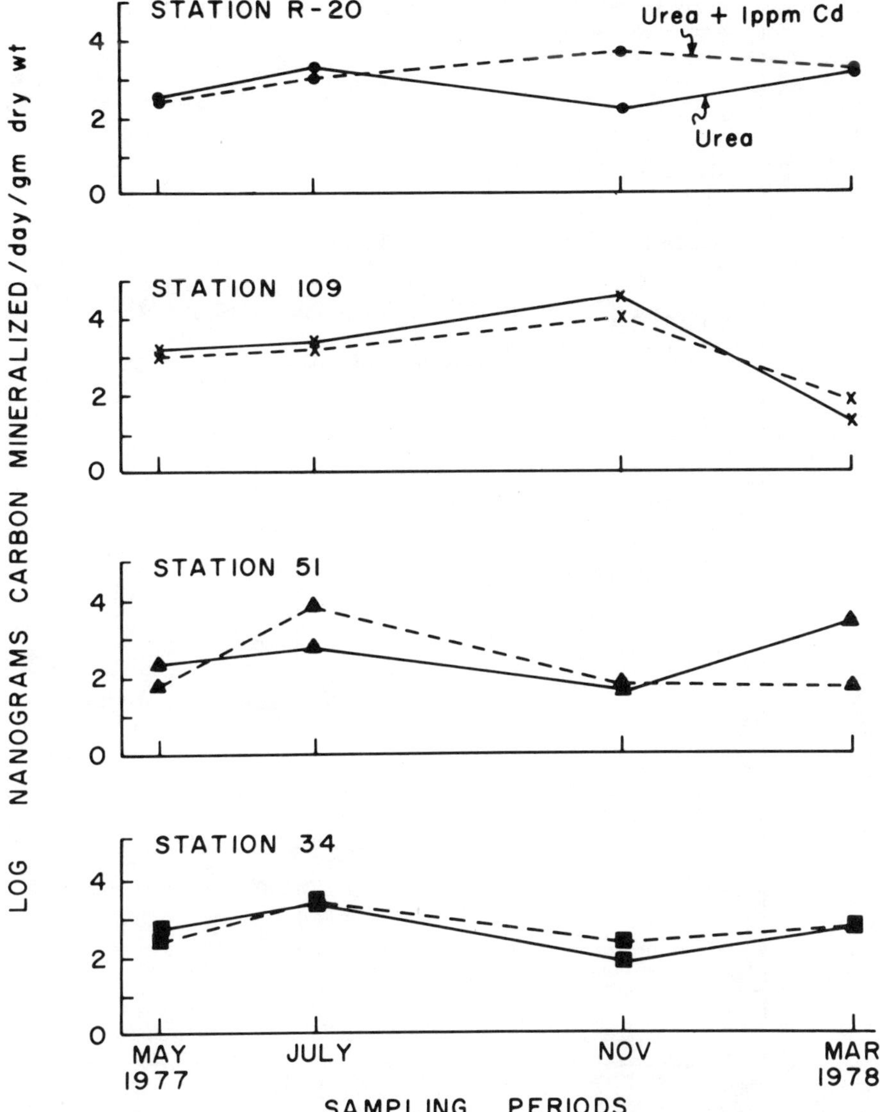

Figure 4. Seasonal variation in nanograms of carbon mineralized by sedimentary microorganisms from the New York Bight apex with urea as substrate (—) and urea + 1 ppm Cd (– – –).

on a 24-hour basis are shown in Figure 4. Similar calculations for glycolic acid are shown in Figure 5. Again, the presence of additional cadmium, which resulted in a decrease in the numbers of CFU on the different media, did not significantly affect the amounts of glycolic acid-C or urea-C utilized by the microflora of these sediments (see previous section). The combination of these

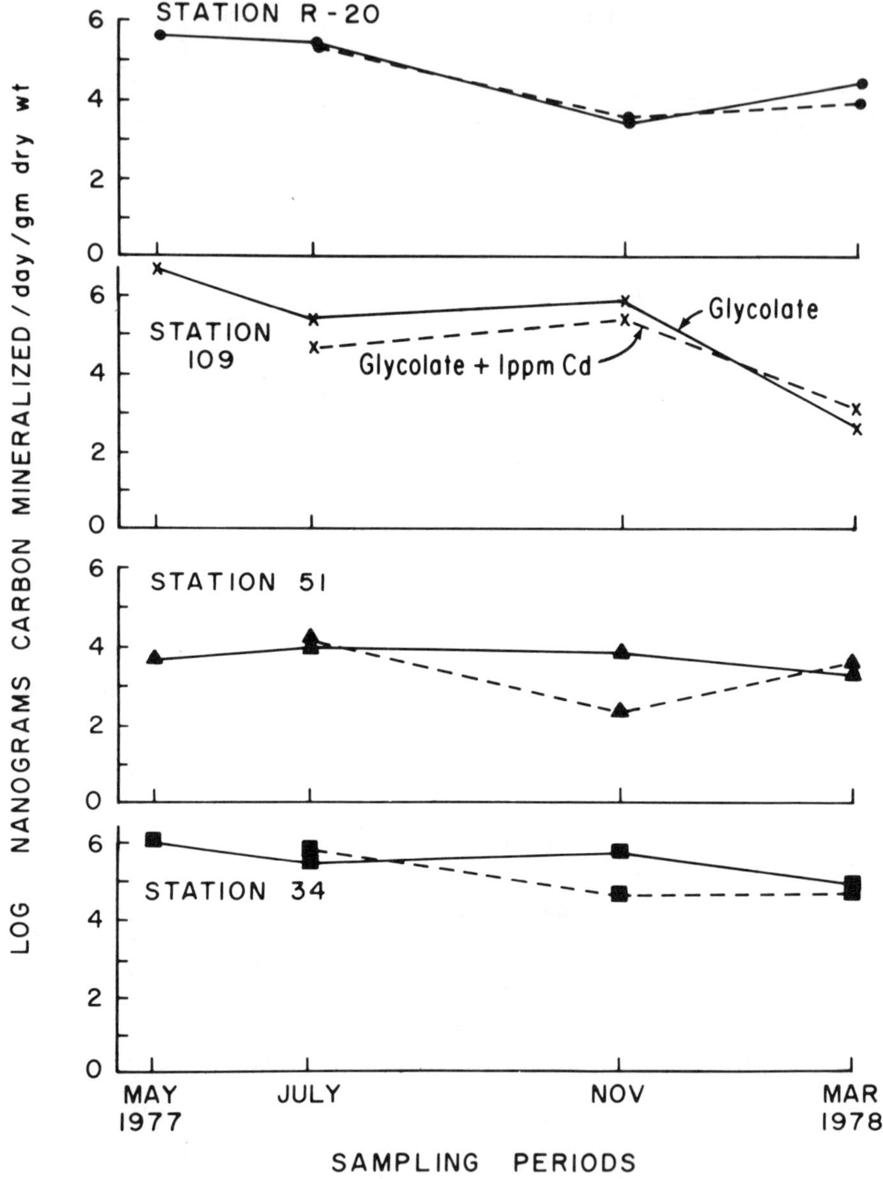

Figure 5. Seasonal variation in nanograms of carbon mineralized by sedimentary microorganisms from the New York Bight apex with glycolic acid as substrate (—) and glycolic acid + 1 ppm Cd (– – –).

data lends further support to the hypothesis that, except for a few isolated situations, at any given time only a small portion of the microbial flora was actively metabolizing in the Bight's cadmium-polluted sediments.

Table 4. Patterns of antibiotic resistance among the cadmium-resistant isolates from the sediments in the New York Bight apex and the pure culture of the marine coryneform, A-4-20. NT = not tested; R = resistant, and S = sensitive.

		Sedimentary bacteria		
Type of drug or antibiotic action	Number of antibiotics tested of that type	Number of microbial strains tested	Number of strains resistant to one or more antibiotics of that type	A-4-20 (single isolate)
Sulfa drug	1	51	16	NT
Cell wall synthesis	3	53	32	R
P-site in protein synthesis	1	53	19	R
Binds to 30S ribosomes	3	53	28	R
DNA-cross linking	2	53	19	NT
A-site in protein synthesis	1	53	19	S

Antibiotic Resistance

Resistance to heavy metals has been linked to antibiotic resistance, and this, in turn, has been shown to be plasmid-mediated (Chopra, 1971; Smith and Novick, 1972; Sizemore and Colwell, 1977). Consequently, it was of interest to determine whether the ability of bacteria to grow in the presence of additional cadmium would be reflected by a resistance to antibiotics. Isolates resistant to 1 ppm Cd or greater were selected from the four reference stations. These isolates were tested against 11 antibiotics. The observed antibiotic resistance patterns are summarized in Table 4.

Resistance to classes of antibiotics, rather than individual antibiotics, was examined to eliminate bias caused by testing similar antibiotics that might be expected to give similar results (*e.g.*, ampicillin and penicillin). Calculations based on the multiple resistance patterns summarized in Table 4 showed that 72% of the cadmium-resistant isolates also were resistant to two or more classes of antibiotics. Of these antibiotic-resistant isolates, 33% were gram positive, 52% were gram negative, and 15% were yeast. This is a higher percentage of gram negative bacteria than would be expected on the basis of studies by Babich and Stotzky (1977). These data also show that cadmium-resistance and multiple antibiotic resistance can occur in environmental isolates other than *Bacillus* (Timoney *et al.*, 1978) and the enterics (Koditschek and Guyre, 1974).

Pure Culture Studies

The antibiotic resistance data for A-4-20 are recorded separately on Table 4, because this coryneform bacterium exhibited resistance to three major classes of antibiotics–those that inhibit cell wall synthesis (penicillin), those that inhibit protein synthesis at P and A sites (erythromycin and chloramphenicol, respectively), and those that inhibit binding to 30-S ribosomes (streptomycin, kanamycin, and neomycin)–and is capable of growing in highly elevated concentrations of cadmium. In fact, it consistently demonstrated a higher specific growth rate at 75 ppm Cd than at 1 ppm Cd (Table 5).

Protein Synthesis

When A-4-20 was tested after growth in both unsupplemented and cadmium-supplemented media, there was considerable change in the rate of ^{14}C-leucine uptake (Table 6). These data suggest that growth in the presence of cadmium may result in cellular changes that "protect" the cell from additional cadmium levels and may stimulate bacterial activity. When cadmium chloride was present during the assay, but the organisms had not been exposed previously to this compound, there was an inhibition of both carbon dioxide respiration and leucine incorporation into protein (22% and 15% inhibition, respectively). Prior exposure to 75 ppm Cd did not affect carbon dioxide respiration but appeared to stimulate incorporation by 35%. The combination of inducible resistance to cadmium, as demonstrated by uptake and incorporation rates, and multiple antibiotic resistance patterns are presumptive evidence for the existence of plasmid-mediated resistance factor(s) in this bacterium. Preliminary data indicate that a plasmid is present in this organism (Litchfield *et al.*, 1980). Studies currently are underway to elucidate further the exact mechanism of cadmium resistance in this bacterium.

DISCUSSION

The effects of environmental pollutants on the abundances of bacteria have been well-documented. The effects of these same pollutants on microbial environmental processes have not been studied so extensively. To date, this is the first report on the direct effects of cadmium on the decomposer role of bacteria in marine sediments. The CFU data indicated that when cadmium concentrations in media are elevated to environmental levels occurring in the apex, a 0.5-1.0 order of magnitude reduction occurred in the numbers of

Table 5. Influence of cadmium concentrations on the growth rates of the marine coryneform, A-4-20. Hours per doubling calculated according to Meynell and Meynell (1965).

Cadmium present (ppm)	Specific growth rate
0	9.80
1	6.62
10	7.14
50	7.30
75	7.69
100	Toxic

culturable aerobic heterotrophic bacteria. However, such concentrations of cadmium generally had little or no effect on the metabolism of the two test substrates, urea and glycolic acid, which normally are present in these sediments. The only exceptions to this were the statistically significant ($p \geqslant 0.01$) increases at Station 51 (July) and at Station R-20 (November) (Figure 4). These cannot be explained until more information is available regarding the type(s), locations, and inhibitors of marine bacterial urease; this is currently under investigation. Regardless of the ionic form of the supplemental cadmium, the important point to note is the decrease in the numbers of culturable bacteria concomitant with the lack of a significant effect of cadmium on metabolic activities. Thus, it is likely that the metabolically active bacteria are the cadmium-resistant strains.

Recently, there has been an increased interest in the non-chromosomal gene pool represented by plasmids. That plasmids may serve as a mechanism for the dispersion of beneficial phenotypic characteristics in the environment has been a widely discussed possibility (Sizemore and Colwell, 1977; Litchfield *et al.*, 1980). The high correlation between patterns of multiple antibiotic resistance and heavy metal resistance prompted Colwell and co-workers to examine their mercury-resistant cultures (Allen *et al.*, 1977). They found presumptive evidence that these two phenomena were related to specific plasmids, a hypothesis that was confirmed by later studies (Sizemore and Colwell, 1977). Gram-positive bacteria, which comprise from 8% to 31% of all our sediment isolates, have been studied extensively for plasmids (Smith and Novick, 1972), so their presence in these isolates is to be expected. The remaining portions of the bacteria isolated from New York Bight sediments are gram-negative bacteria. Koditschek and Guyre (1974) and Babinchak *et al.* (1977) presented evidence suggesting the widespread occurrence of antibiotic resistance in coliforms isolated from New York Bight apex sediments, but other gram-negative genera have not been examined extensively. More importantly, enterics accounted for less than 20% of

Table 6. Effect of cadmium on ^{14}C-leucine uptake, respiration, and incorporation by whole cells of the coryneform bacterium, A-4-20. Washed log-phase whole cells were used in all experiments. Data are compared on a dpm/mg protein basis with log-phase ESWB grown cells as control. Incubation conditions followed those described in the Materials and Methods section. NT = not tested.

Culture condition	Cadmium in assay system (ppm)	Percent change		
		Uptake	$^{14}CO_2$ respired	^{14}C-leucine incorporated
ESWB	0	0	0	0
ESWB + 75 ppm Cd	0	0	0	+8%
ESWB	75	NT	−22%	-15%
ESWB + 75 ppm Cd	75	0	0	+35%

the gram negative strains ever isolated (data not presented). Regardless, 72% of all the marine bacteria isolated from cadmium-supplemented media during this study were resistant to two or more classes of antibiotics. The probability, therefore, is great that a large number of these strains will prove to contain plasmids during continued study of this question.

There are several strategies a cell could adopt to overcome toxic metallic environments. The toxic material could be prevented from entering the cytoplasm by at least two mechanisms: the excretion of a specific metal chelator, or the excretion of a material for the entrapment of the metal. This latter mechanism recently has been demonstrated for *Klebsiella aerogenes,* which synthesizes an extracellular polysaccharide capsule that either may entrap or weakly bind copper or cadmium, and thus effectively prevents these elements from interfering with the cell (Bitton and Freihofer, 1978). How widespread this polysaccharide is among the bacteria described here is an open question. Examples of specific metal chelators, however, are widespread in mammalian physiology where metallothioneins have long been recognized as a major detoxification mechanism (Friberg *et al.,* 1971). Recently, Gauthier and Flatau (1977) reported the presence of metallothioneins in marine vibrios during growth in the presence of cadmium. Preliminary studies in our laboratory indicate tht A-4-20 similarly produces a metallothionein-like protein along with a plasmid (Devanas *et al.,* in press). These two factors, then, may explain the ability of this coryneform to grow at elevated cadmium concentrations.

The maintenance of a metabolically active microbial population under conditions of heavy metal pollution may be explained by several genetic characteristics of the microflora: the production of extracellular polysaccharides, the synthesis of a metallothionein-like chelator, or some other adaptive mechanism. Most important, the gene pool for these special syntheses could reside in plasmids, for which presumptive evidence is presented in this paper.

ENVIRONMENTAL IMPLICATIONS

1. Long-term exposure to cadmium has resulted in a cadmium-resistant microbial population that is the metabolically active microbial population in New York Bight apex sediments.
2. Concident resistance to cadmium and multiple types of antibiotics also has been noted in bacterial strains from the apex.
3. A plasmid-mediated gene pool is the most likely genetic mechanism for multiple resistance. This mechanism could explain the wide distributions of resistant bacteria from New York Bight apex sediments.

ACKNOWLEDGEMENTS

This research was supported by NOAA MESA New York Bight Project Grants 04-06-022-44020 and 04-07-022-44013 and by the Center for Coastal and Environmental Studies, Rutgers—The State University. The authors express their appreciation to the officers and crews of the NOAA ship GEORGE B. KELEZ,

R/V RUTGERS, R/V ONRUST, R/V ALBATROSS and R/V CAPE HENLOPEN for their assistance in the collection of the samples.

REFERENCES

Allen, D.A., B. Austin, and R.R. Colwell. 1977. Antibiotic resistance patterns of metal-tolerant bacteria isolated from an estuary. Antimicrob. Agents Chemother. 12: 545-547.

Babich, H. and G. Stotzky. 1977. Sensitivity of various bacteria, including actinomycetes and fungi to cadmium and the influence of pH on sensitivity. Appl. Environ. Microbiol. 33:681-695.

Babinchak, J.A., J.T. Graikoski, S. Dudley, and M.F. Nitkowski. 1977. Distribution of fecal coliforms in bottom sediments from the New York Bight. Marine Pollut. Bull. 8: 150-153.

Barr, A.J., J.H. Goodnight, J.P. Sall, and J.T. Helwig. 1976. A User's Guide to SAS 76. SAS Institute, Raleigh, NC. 329 pp.

Bitton, G. and V. Freihofer. 1978. Influence of extracellular polysaccharides on the toxicity of copper and cadmium toward *Klebsiella aerogenes.* Microbial Ecol. 4: 119-125.

Bradford, M.M. 1976. A rapid and quantitative method for the quantitation of microgram quantities of protein utilizing the principle of protein-dye binding. Anal. Biochem. 73: 248-254.

Carmody, D., J. Pearce, and W.E. Yasso. 1973. Trace metals in sediments of New York Bight. Marine Pollut. Bull. 4: 132-135.

Cenci, G. and G. Morrozzi. 1977. Evaluation of the toxic effect of Cd^{2+} and $Cd(CN_4)^{2-}$ ions on the growth of mixed microbial population of activated sludges. Sci. Total Environ. 7: 131-143.

Chopra, I. 1971. Decreased uptake of cadmium by a resistant strain of *Staphylococcus aureus.* J. Gen. Microbiol. 63: 265-267.

Devanas, M.A., C.D. Litchfield, C. McClean, and J. Gianni. 1981. Concidence of cadmium and antibiotic resistance in New York Bight apex benthic micro-organisms. Mar. Pollut. Bull. 11:264-269.

Fleischer, M., A.F. Sarofin, D.W. Fassett, P. Hammond, H.T. Shacklette, I.C.T. Nisbet, and S. Epstein. 1974. Environmental impact of cadmium: a review by the panel on hazardous trace substances. Environ. Health. Perspect., May 1974.

Friberg, L., M. Piscator, and G.F. Nordberg. 1971. Cadmium in the Environment, CRC Press, Cleveland, OH. 166 pp.

Gauthier, M. and G. Flatau. 1977. Concentration et mode de fixation der cadmium per an *Vibrio marin.* Comptes Rendus Acad. Sci., Paris, Ser. D. 285: 817-820.

Hobbie, J.E., R.J. Daley, and S. Jasper. 1977. Use of nuclepore filters for counting bacteria by epifluorescence microscopy. Appl. Environ. Microbiol. 33: 1225-1228.

Karl, D.M. and P.A. LaRock. 1975. Adenosine triphosphate measurements in soil and marine sediments. J. Fish. Res. Board Can. 32: 599-607.

Koditschek, L.K. and P. Guyre. 1974. Antimicrobial-resistant coliforms in New York Bight. Marine Pollut. Bull. 5: 71-74.

Litchfield, C.D., M.A. Devanas, C. McClean, J. Gianni, and S.M. O'Connell. 1980. Correspondence between plasmids and bacterial survival in heavy metal polluted sediments. Abstr. Annu. Meet. Am. Soc. Microbiol. 1980: N. 107.

Litchfield, C.D., M.A. Devanas, J. Zindulis, C.E. Carty, J.P. Nakas, and E.L. Martin. 1979a. Application of the ^{14}C-organic mineralization technique to marine sediments. *In:* Methodology for Biomass Determinations and Microbial Activities in Sediments, C.D. Litchfield and P.L. Seyfried (eds.), American Society for Testing and Materials Special Technical Publication 673, Philadelphia. pp. 78-94.

Litchfield, C.D., H.M. Edenborn, M. Meskill, and J. Zindulis. 1979b. Microbial role in the urea cycle in the New York Bight. Abstr. Annu. Meet. Am. Soc. Microbiol. 1979: N. 63.

Litchfield, C.D., J.P. Nakas, and R.H. Vreeland. 1976. Bacterial flux in some New Jersey estuarine sediments. Am. Soc. Limnol. Oceanogr. Spec. Symp. 2: 340-353.

Litchfield, C.D., J.B. Rake, J. Zindulis, R.T. Watanabe, and D.J. Stein. 1975. Optimization of procedures for the recovery of heterotrophic bacteria from marine sediments. Microbial Ecol. 1: 219-233.

Meynell, G.C. and E. Meynell. 1965. Theory and Practice in Experimental Bacteriology, Cambridge University Press, London. 288 pp.

Sizemore, R.K. and R.R. Colwell. 1977. Plasmids carried by antibiotic-resistant marine bacteria. Antimicrob. Agents Chemother. 12: 373-382.

Smith, K. and R.P. Novick. 1972. Genetic studies on plasmid-linked cadmium resistance in *Staphylococcus auereus.* J. Bacteriol. 112: 761-772.

Stevenson, L.H. 1978. A case for bacterial dormancy in aquatic systems. Microbial Ecol. 4: 127-133.

Thomas, J.P., W.C. Phoel, F.W. Steimle, J.E. O'Reilly, and C.A. Evans. 1976. Seabed oxygen consumption–the New York Bight apex. Am. Soc. Limnol. Oceanogr. Spec. Symp. 2: 354-369.

Timoney, J.F., J. Port, J. Giles, and J. Spanier. 1978. Heavy-metal and antibiotic resistance in the bacterial flora of sediments of New York Bight. Appl. Environ. Microbiol. 36: 465-472.

Wright, R.T. and J.E. Hobbie. 1966. Use of glucose and acetate by bacteria and algae in aquatic ecosystems. Ecology 47: 447-464.

Zindulis, J. and C.D. Litchfield. In press. ATP methology for marine sediments. NOAA OMPA Tech. Memo.

LONG-TERM VARIABILITY IN COASTAL SYSTEMS: BACKGROUND NOISE AND ENVIRONMENTAL STRESS

Robert J. Livingston

Department of Biological Science
Florida State University
Tallahassee, Florida 32306

Abstract. One of the principal problems of field impact studies in coastal systems is the difficulty of differentiating anthropogenic effects from natural background variability. Our long-term analysis of two bay systems in the northern Gulf of Mexico indicates that various organisms at different levels of biological organization respond in different ways to stress with differential lags of recovery, uncoupled biomass and species richness indices, and species-specific responses to habitat alteration. If equilibrium and resilience of coastal areas are to be evaluated, it will be necessary to conduct long-term, multidisciplinary studies in disparate systems to account for the natural variability and sensitivity of such areas to disturbances. Without such an approach, resource management will remain limited and arbitrary.

INTRODUCTION

If we can assume that certain questions concerning the environmental impact of a given pollutant can be answered in the field, it becomes necessary to differentiate such effects from natural (background) variability or noise. Coastal biological systems frequently are controlled to varying degrees by key climatic features and land runoff patterns. The spatial heterogeneity and temporal variability of dynamic coastal systems often are overlooked in attempts to establish causal relationships. This, in turn, tends to complicate management decisions, especially with regard to the long-term implications of the interactions of complex sets of factors within a given system. While there is a range of experimental and field approaches to this problem, the fact that there is no single approach to explain actual field impact reflects the present confused state of the environmental sciences. Variability remains misunderstood and even ignored by scientists and managers who are searching for easy answers to complex questions.

There are various levels of temporal cycling in coastal systems. Periods may range from minutes to tens of years, although the most obvious temporal cycles involve diurnal, lunar, or seasonal periods. As pointed out by Coull and Fleeger (1977), most "long-term" studies of benthic community dynamics usually approximate one year, while longer studies often are done on a quarterly basis. The failure of quarterly sampling to establish trends or levels of variation has been demonstrated by Livingston (1976). For a given system, the frequency and duration of sampling should be determined by the range of system variability and the hypothesis to be tested. While inadequate sampling can lead to over-generalization, contradiction, and confusion, a field program designed without adequate background information or a clear objective often becomes tangled in circular reasoning, with a resultant lack of scientific resolution. Hence, meaningless "baseline" studies have literally flooded the ecological literature without advancing our understanding of system function.

The literature concerning long-term cycles in coastal systems is diverse and inconclusive. There are descriptive analyses of changes in phytoplankton (Reid, 1977), benthic macrophytes (Bayley *et al.,* 1978), meroplankton (Williams, 1972), zooplankton (Longhurst *et al.,* 1972), meiobenthic copepods (Coull and Fleeger, 1977), and fishes (McErlean *et al.,* 1973). In most cases, causative factors are not identified. This could be because biological systems do not necessarily have linear relationships to key physical or chemical variables (Livingston *et al.,* 1978; Meeter and Livingston, 1978). System-specific variability should be an integral part of impact evaluation. The relationship of natural catastrophic change and long-term system response is simply too poorly understood for such evaluation. Storms and hurricanes have varied effects on coastal areas. Biological alteration may be associated with heavy rainfall and runoff, as noted by Boesch *et al.* (1976) for soft-bottom communities in the lower Chesapeake Bay. This factor also is mentioned by Saloman and Naughton (1977) concerning benthic fauna on beaches in the northern Gulf of Mexico. Temperature changes, especially winter lows, often affect coastal systems (Snelson and Bradley, 1978). However, such effects are seldom documented beyond two years, even though there is ample, though indirect, evidence from commercial fisheries data that long-term population changes may be determined by specific climatic features (Sutcliffe *et al.,* 1977; Meeter *et al.,* 1979; Van Winkle *et al.,* 1979). Fisheries data often are affected by other factors, such as fishing effort or socioeconomic conditions (Clark and Brown, 1977; Meeter *et al.,* 1979), and thus are no substitute for long-term, multidisciplinary analyses of representative coastal systems.

The basis of problems in biomonitoring studies lies in the natural spatial and temporal variability of aquatic systems. The extent of spatial variability depends to some degree on the specific interest of a given investigator. This interest can range from a few microns (in microbial studies) to global phenomena. Basic ecological functions of coastal systems vary considerably from place to place. Thus, there are problems when small-scale studies are expanded to broad generalizations since there is really little empirical basis for such extrapolation. Comparative data between heterogeneous systems simply are not available, and thus the basis for extrapolation remains undetermined. The complications of spatial variability of aquatic systems and the development of adequate sampling methods to answer specific questions have been reviewed elsewhere (Michael *et al.,* 1977; Green, 1979) and will not be discussed here. It is sufficient to say that statistical tools are available for adequate sampling approaches to the spatial question if the research objective is reasonably well-drawn.

Temporal variability is somewhat more complex since, at any given instant, a natural system represents a composite series of different temporal sequences of varying periods superimposed over one another as the result of an almost infinite number of cause-and-effect reactions. The problem of finding order of causality is heightened by the fact that these overlapping cycles may differ along habitat or microhabitat gradients and at different levels of biological organization. Consequently, the term "background noise" becomes a euphemism for our inability to determine the temporal or sequential modes of system function. Modelling efforts often assume that systems are in a state of equilibrium, without defining the actual extent of temporal variability. Terms such as stability, resilience, and diversity are used to give a theoretical framework to what is essentially a lack of consistent observations of organisms under field conditions. This situation has led to the improbable point where hypothesis testing is ignored, terminology becomes "nonoperational" (Peters, 1977), and the nature of environmental stress becomes part of a tautology that leads from some form of theoretical hunch to predictive models.

Since 1971 our research group has carried out continuous, integrated field and laboratory studies in two bay systems in the northeast Gulf of Mexico in an attempt to understand spatial and temporal variability in diverse assemblages of coastal species. This paper is an attempt to outline the problems associated with impact analysis in a coastal system, with particular emphasis on the nature and importance of background noise and natural (long-term) variability.

Area of Study

The area of study is shown in Figure 1. The Apalachicola Bay system is a shallow, river-dominated barrier island estuary that is physically controlled by the Apalachicola River, a major drainage in the southeastern United States. The ecological energy of the Bay is based on periodic phytoplankton blooms and river-derived inputs of allochthonous detritus and dissolved organic matter (Livingston *et al.*, 1976a). High turbidity and sedimentation maintain mud flat

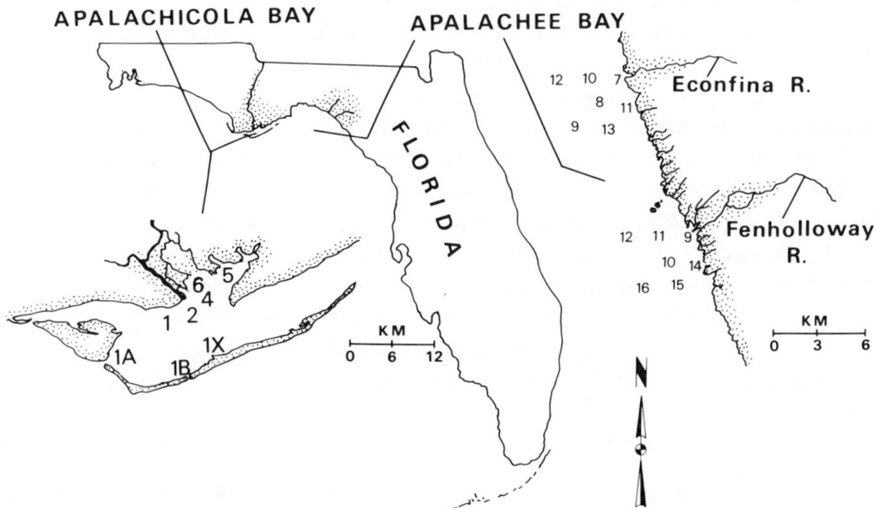

Figure 1. Area of study in the northeastern Gulf of Mexico showing permanent sampling stations in the Apalachicola and Apalachee Bay systems.

and oyster reef development, and the Apalachicola estuary is a major nursery for various commercially important species, such as shrimp, oysters, blue crabs, and finfishes. In Apalachee Bay, on the other hand, benthic macrophytes provide the main source of production and microhabitat availability (Livingston, 1974). This relatively shallow bay system receives input from a series of small rivers, and offshore areas are characterized by low turbidity and color. Except for pulp mill effluents that were discharged into Apalachee Bay prior to 1975 (see below), the basic habitat characteristics (river flow rate, temporal runoff patterns, physiography of the receiving basin, water quality characteristics) were similar for the Econfina and Fenholloway estuaries. Thus, the premise for an essentially experimental approach to a field impact analysis was established prior to habitat stratification and the development of a program of study.

For the past nine years, a comparison has been made between the unpolluted Econfina drainage system and the Fenholloway system, which has been contaminated by kraft pulp mill effluents from 1954 to 1974. The impact analysis was carried out from 1971 to 1973. In 1974 a pollution control system was initiated that eliminated considerable portions of the kraft mill effluents. Because of the general lack of other sources of pollution in Apalachee Bay, the recovery of the Fenholloway system, relative to the Econfina baseline or control, was considered to be an appropriate follow-up to previous impact analyses. In this way, the long-term comparison of the two Apalachee Bay estuaries served as the basis for a broader comparative study of the clear, macrophyte-dominated Apalachee Bay with the turbid, phytoplankton-dominated Apalachicola estuarine system.

METHODS

In both the Apalachicola and Apalachee Bay systems, the same quantitative field methods have been used to measure continuous changes in various physical, chemical, and biological parameters. These include water temperature, salinity, pH, dissolved oxygen, color, turbidity, and light penetration (Secchi readings) (Livingston, 1976; Livingston *et al.,* 1976b). Dissolved nutrients (ammonia, silicate, nitrite, nitrate, orthophosphate) have been analyzed along with micro- and macro-particulate matter, chlorophyll *a*, particulate organic carbon (POC), and total organic carbon (TOC) (Livingston *et al.,* 1976a). Long-term changes in organochlorine compounds also have been documented (Livingston *et al.,* 1978). In the Apalachicola estuary, the emphasis was on phytoplankton productivity and allochthonous organic matter. The benthic macrophytes were intensively studied in Apalachee Bay (Livingston *et al.,* 1976b; Zimmerman and Livingston, 1976a,b, 1979). Details concerning the various sampling methods for abiotic and biotic factors are available in the above references.

RESULTS AND DISCUSSION

Long-term Trends

A preliminary overview of long-term trends in the Apalachicola system is given in Figures 2 and 3. Over a six-year period of study, there were several

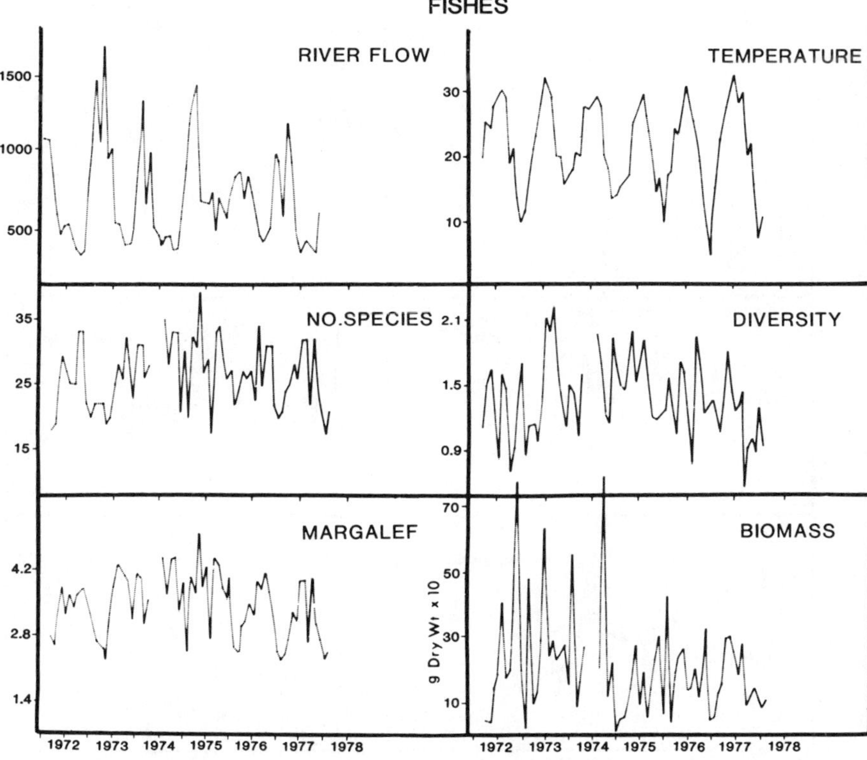

Figure 2. Apalachicola River flow (m^3/sec), temperature (°C), and various indices (number of species, Shannon Diversity Index, Margalef Index, dry weight biomass) for fishes taken with repetitive otter trawl tows at permanent sampling stations in the Apalachicola Bay system. The data were taken monthly from March 1972 through February 1978 and are cumulative totals of 70 minutes of trawling per month.

major climatic events that tended to influence the system. During the winter of 1973, there was a major river flooding episode that was reflected in the salinity and color characteristics of inshore and offshore Apalachicola Bay. This evidently resulted in prolonged reductions in abundance, species richness, and diversity of both the fish and invertebrate assemblages. During the succeeding two years (1974-1975), all such measurements tended to increase in time. Fish biomass and invertebrate numbers peaked in 1974, after which there was a steady decline in most of the fish indices. The invertebrate indices tended to peak in 1976; following the cold winter of 1976-1977, there were major changes in the invertebrate assemblages. In both instances, the fish and invertebrate assemblages were influenced by runoff and winter (low) temperatures. The important point is that, despite fairly regular seasonal fluctuations, there were regular progressions or changes in species assemblages that were directly affected by specific climatic shocks to the system. Such changes were the result of the

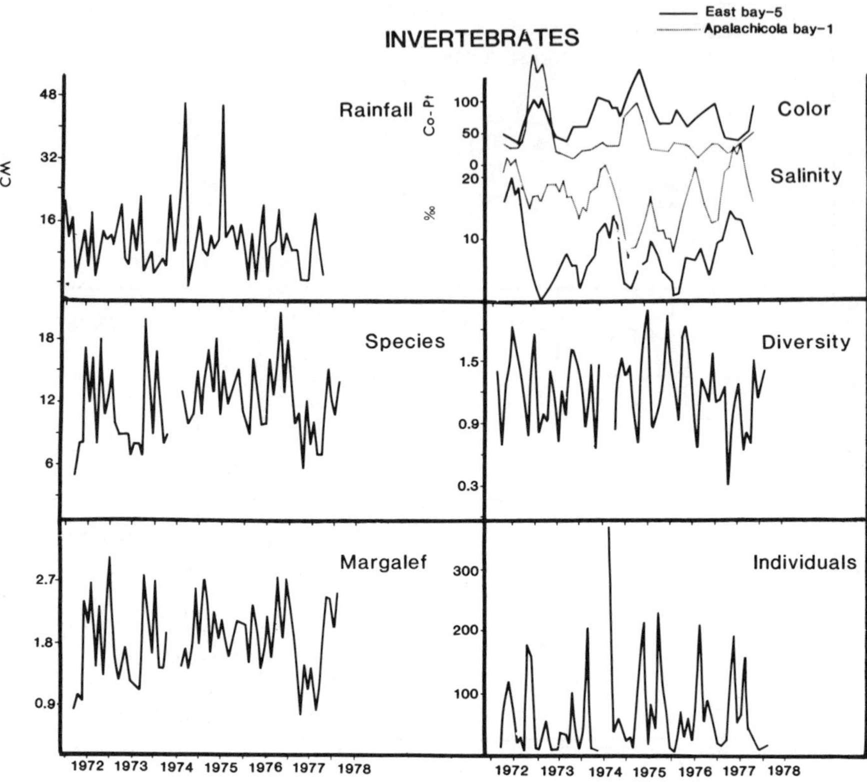

Figure 3. Apalachicola total monthly rainfall (cm), salinity (ppt), water color (APHA Pt-Co units), and various indices (number of species, Shannon Diversity Index, Margalef Index, number of individuals) for invertebrates taken with repetitive otter trawl tows at permanent sampling stations in the Apalachicola Bay system. The data were taken monthly from March 1972 through May 1978 and are cumulative totals of 70 minutes of trawling per month.

reactions of various populations to natural habitat alterations. In addition, there was an indication of long-term cycles that could have important implications in the use of species assemblages in impact analysis. Meeter *et al.* (1979) found that Apalachicola River flow peaks follow five- to seven-year cycles. The interaction of such runoff cycles and longer-term periods of minimum temperatures as controlling features of this coastal system is still under study. It is obvious, however, that there are regular progressions of community changes that, though somewhat variable from one group (fishes) to another (invertebrates), may reflect influences of physical events or shocks that occurred months or even years before. Annual variability is real, and the system tends to show patterned responses of successive biological changes that are controlled by recurrent combinations of climatic change.

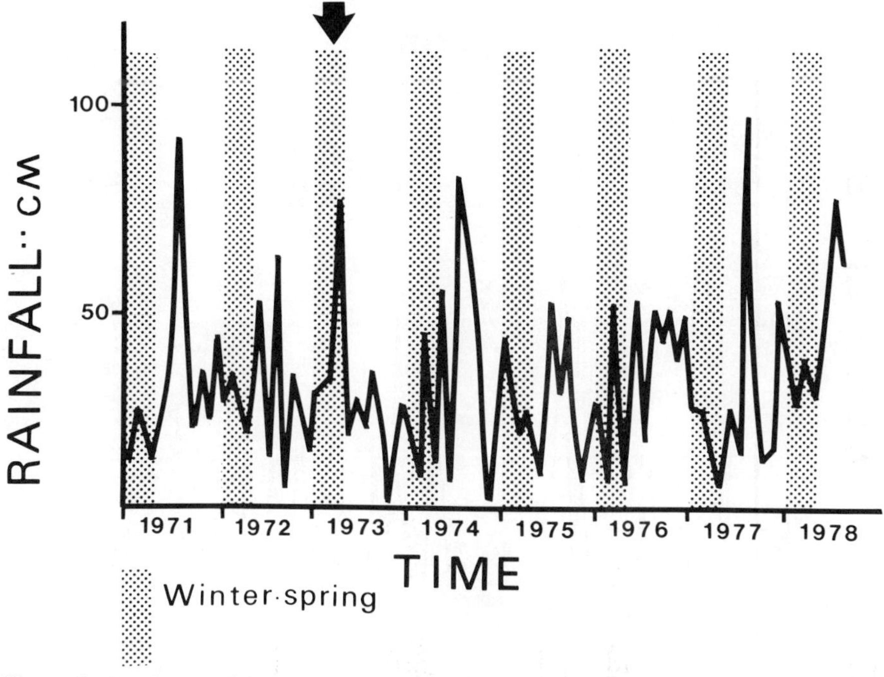

Figure 4. Total monthly rainfall (cm) in the Econfina-Fenholloway drainage area from January 1971 through September 1978. Stippled bars indicate winter-spring periods; arrow indicates period of 1973 major rainfall.

Whereas the Apalachicola estuary is dominated by drainage features of an extensive (Piedmont) system, Apalachee Bay is largely influenced by local (Florida) climatic features. Rainfall patterns in north Florida are usually seasonal with major peaks occurring during summer months. During 1973, however, this pattern changed, and there was a major rainfall during the winter-early spring period (Figure 4). In the summer, the associated wetlands usually buffer such rainfall to the extent that inshore coastal systems reflect only minimal variation in water quality functions such as salinity and color. Reduced evapotranspiration rates in winter allow increased runoff, when compared to the runoff from rainfall events of similar magnitude during summer periods. The heavy rainfall in the spring of 1973 seems to have had a major influence on coastal water quality (Figure 5), with seven-year low salinities and high color levels observed in the Econfina system. Such water quality conditions were the result of the timing and extent of local climatic events and wetland functions.

During the early spring of 1973, die-offs of benthic macrophytes were noted throughout the offshore Econfina system. This phenomenon was correlated with the 1973 spring rainfall. Macrophyte biomass was considerably lower in subsequent years (Figure 6). The dominant species, turtlegrass (*Thalassia testudinum*), was particularly affected by the spring runoff. There was a

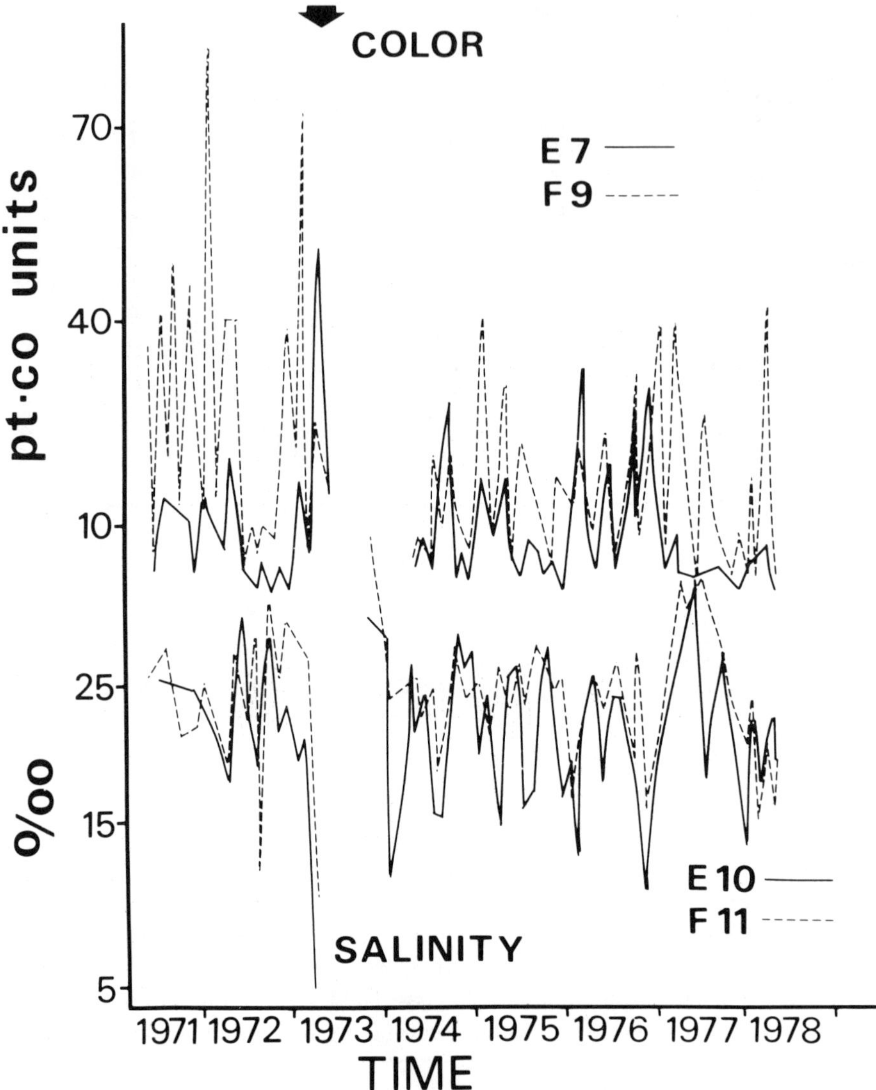

Figure 5. Salinity (parts per thousand) and water color (APHA Pt-Co units) taken monthly at stations E7 and F9 from June 1971 through March 1978. Arrow indicates period of 1973 major rainfall.

recovery of this species during the next four years. These trends were not reflected in long-term changes in the number of macrophyte species (Figure 7), however, that peaked in the Econfina system during the 1974-1975 period at the same time that standing crops reached seven-year lows. Fish species richness followed this trend closely (Figure 7), with peaks in the Econfina species numbers during the summer of 1974. In the Fenholloway system, this index

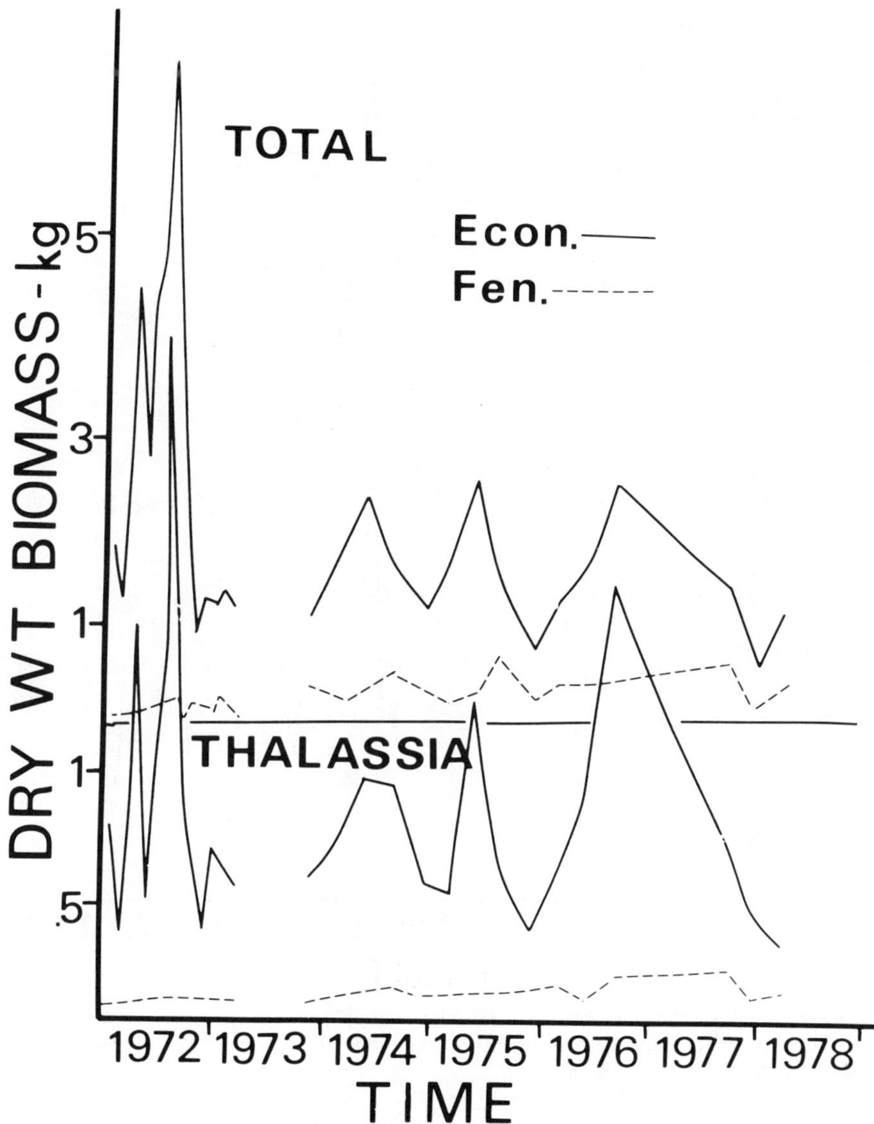

Figure 6. Total dry weight biomass of benthic macrophytes taken at seven stations in the Econfina and Fenholloway drainage areas. Figures represent multiple samples extrapolated to biomass per square meter at each station. Also shown are total biomass figures for the dominant seagrass, *Thalassia testudinum.*

tended to peak in 1975. Following installation of the pollution-control program in 1974, there was a recovery of species richness compared to control stations (Figure 7). The total numbers of fishes and numbers of the dominant pinfish (*Lagodon rhomboides*) tended to follow these trends (Figure 8). Thus, as in the

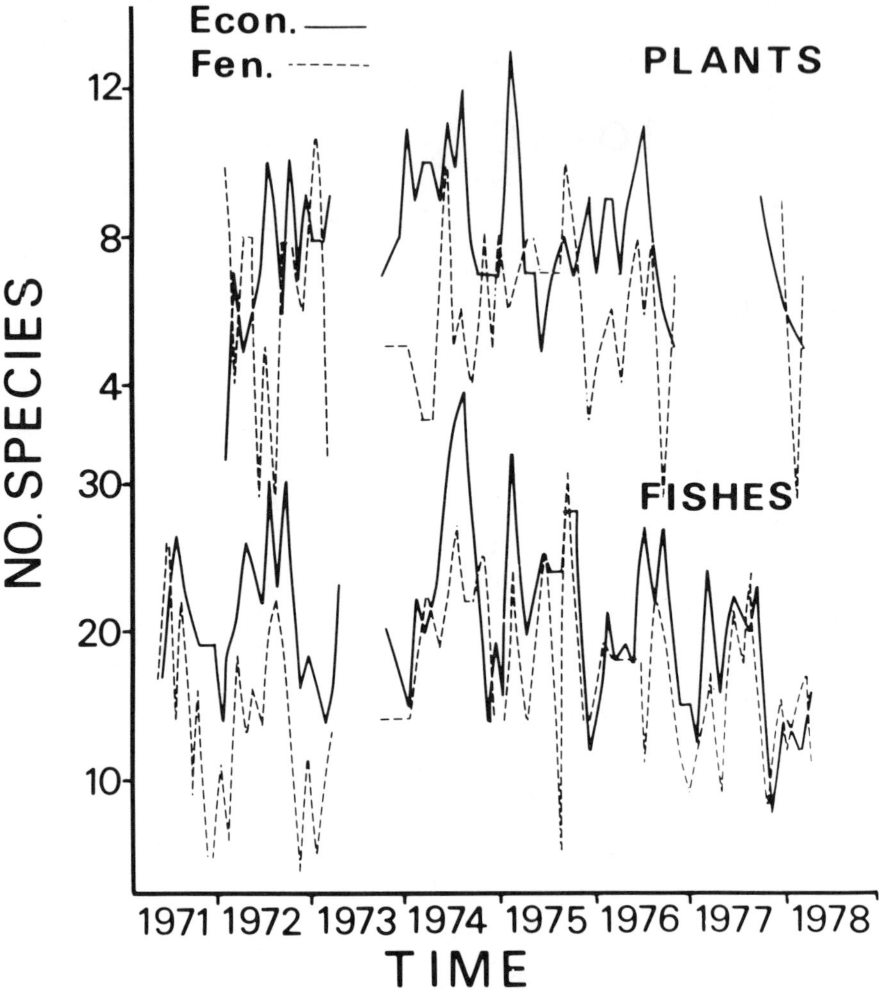

Figure 7. Total numbers of species of benthic macrophytes (gm/m² at E8 and F10) and fishes (21 two-minute trawl tows at E7, E8, E10, and F9, F10, F11) taken monthly from June 1971 through May 1978.

Apalachicola system, specific climatic shocks led to biotic trends at various levels of biological organization. The recovery of a formerly polluted system could only be evaluated relative to long-term trends in the control area. Indices commonly used to evaluate coastal communities often are uncoupled relative to various independent climatological forcing events. Thus, macrophyte biomass recovery was not associated with species richness factors.

Causal Relationships and Stability

The establishment of causal toxic influence on biological components of coastal systems is difficult because of the complexity of multiple interactions

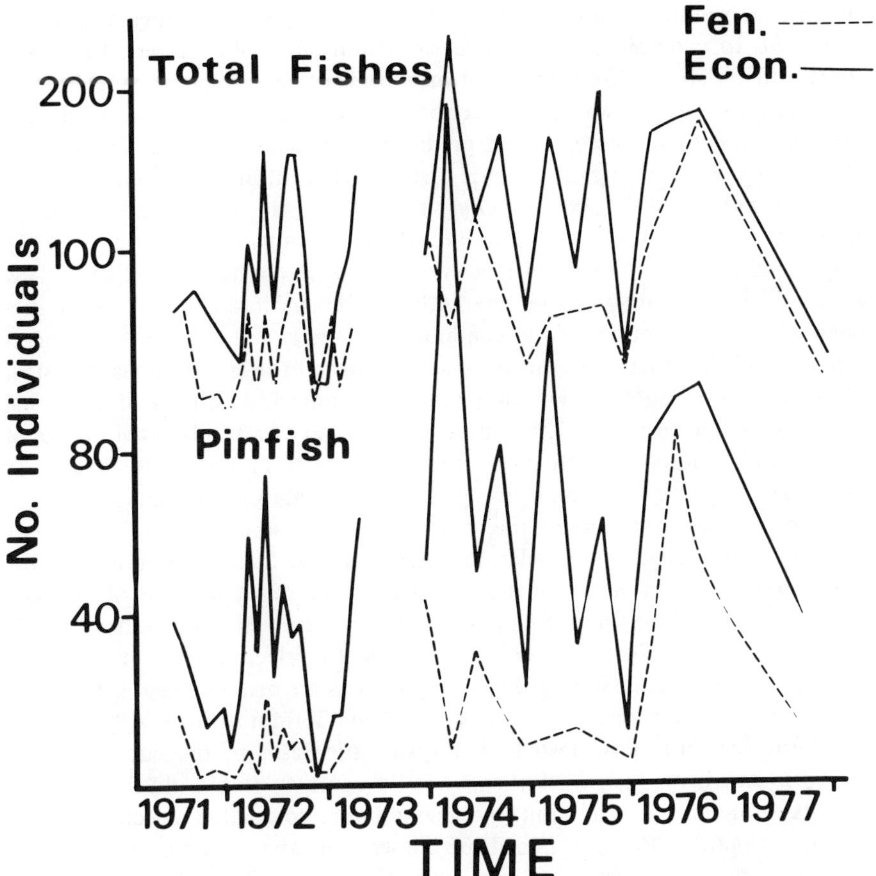

Figure 8. Total numbers of individuals of fishes taken by repetitive (49) otter trawl tows at seven permanent stations in the Econfina and Fenholloway offshore systems from March 1971 through September 1977. Also shown are total numbers of individuals of the dominant pinfish, *Lagodon rhomboides.*

and the lack of adequate controls. There are constant fluctuations in key climatic features, microhabitat distributions, water quality functions, and biological interactions (*i.e.,* predator-prey relationships, competition). Aside from the obvious impact in acutely affected areas, there has been relatively little evidence of single-factor cause and response at the system level (Livingston *et al.,* 1974). Background variation is perhaps the most vexing problem in the identification of toxicant effects at this level.

There has been considerable discussion concerning the concept of stability (return to normal after periods of stress or perturbation) as it relates to populations and communities. Orians (1974) recognized various aspects of stability, and Harrison (1979) attempted to clarify the ambiguity of the term by breaking it down to four mathematical formulations. Resistance (the ability of the system to return to normal; Webster *et al.,* 1975), persistence (the ability of the system to maintain population levels within acceptable ranges; Lewontin, 1969), and variability (frequency and magnitude of changing functions) have all been used in various ways to describe system-level phenomena. There are numerous models that describe and predict functional mechanisms, and this has led to even greater heights of semantic diversity (Lewontin, 1969; Holling, 1973; May, 1973; Orians, 1974; Webster *et al.*, 1975). The principal misunderstandings are derived, in part, from poor hypothesis formulation (Dayton, 1979), over-generalization from limited (specific) studies or points of view, and a lack of comprehensive data from representative systems. McNaughton (1977), in a review of the problems related to community diversity and stability, considers it "fascinating" that there is so much modeling and so little empirical testing. Most authors simply overlook the fact that without an empirical estimate of the actual variability of a given system, the impact of stress and catastrophic environmental perturbation cannot be evaluated. This will vary from system to system, as determined by particular combinations of biotic and abiotic cycling.

Wiens (1978) questioned the assumption of stochastic influences of environmental change on population variability. He focused on the need for long-term population data (relative to generation time) because of possible temporal variability in the overlapping of resource utilization traits. In addition, the relationship of periodic ecological stress to selection remains virtually unknown. There is a possibility that if populations are maintained at suboptimal levels by predation pressure there may be an equilibrium of sorts, although this hypothesis has not been tested in a systematic fashion. Indeed, the rocky intertidal paradigm (*i.e.,* disturbance by predation or by physical factors prevents competitive exclusion by maintaining competing species at low densities; Connell, 1961; Paine, 1966) does not appear applicable to soft sediment benthic or grassbed communities (Peterson, 1979). There have been recent challenges to the application of traditional concepts of succession (Odum, 1953) to fouling communities (Sutherland and Karlson, 1977). The equilibrium state of coral reef fish assemblages has been questioned (Talbot *et al.,* 1978). However, these studies are based on limited or arbitrary periods of study, with too little attention to environmental variability and the natural history of individual populations for a serious examination of the relationship of biological equilibrium to causal factors. Such studies only emphasize the need for more comprehensive and consistent observations regarding biological variability in marine systems before the temporal structure of population and community interactions can be subjected to broad generalizations.

SUMMARY OF CONCLUSIONS

This paper represents a preliminary analysis of one portion of a nine-year study of two coastal systems in the Gulf of Mexico.

1. Spatially diverse coastal systems tend to follow temporal sequences of changes in species composition. Cyclic phenomena have varying periodic-

ities at different levels of biological organization. Such periods are superimposed over one another to form complex, long-term patterns of community response that reflect catastrophic and prolonged changes in key climatic features, such as rainfall, river flow, and temperature.

2. Various groups of organisms (*i.e.*, benthic plants, fishes, invertebrates) respond to stress and periodic environmental shocks in different ways. Differential lags, uncoupled biomass and species richness indices, and spectral changes in long-term (supra-annual) trends of biological functions have been identified.
3. Traditional concepts of stability and resilience of aquatic systems have been based on inadequate data and faulty assumptions. Natural variability in coastal systems is considerable, in terms of both amplitude and temporal periodicity. Impact analysis should be designed for specific trends of key physicochemical variables. Multidimensional variability is still poorly understood in most coastal systems. Unverified models based on linear relationships of specific populations or groups of populations are often poor predictors of long-term trends because of the high natural variability of coastal systems. Impact studies should limit the scope of analysis and hypothesis testing to system-specific periods.
4. Assumptions regarding community stability and the ability of species assemblages to withstand stress are often invalid without adequate estimation of the relationship of equilibrium factors to natural, long-term biological variability.
5. Demonstrated increases and reductions of key biological indices are often part of longer-term cycles that are characteristic of the natural variability of diverse coastal systems. Impact evaluations should account for such changes.
6. Coastal management can benefit from coordinated, long-term interdisciplinary studies since water quality standards simply cannot be applied indiscriminately to functionally disparate systems. Without estimates of background variability and the differential sensitivity of component subsystems, management prerogatives will remain limited and arbitrary.

ACKNOWLEDGEMENTS

The author acknowledges G.C. Woodsum for his help with the computer programs used for this project. Portions of the data collection were funded by NOAA Office of Sea Grant, Department of Commerce, under Grant No. 04-3-158-43. Data analysis was supported by EPA Program Element No. 1 BA025 under Grant No. R-803339. Matching funds came from the people of Franklin County, Florida.

REFERENCES

Bayley, S., V.D. Stotts, P.F. Springer, and J. Steenis. 1978. Changes in submerged aquatic macrophyte populations at the head of Chesapeake Bay, 1958-1975. Estuaries 1: 171-182.

Boesch, D.F., R.J. Diaz, and R.K. Virnstein. 1976. Effects of tropical storm Agnes on soft-bottom macrobenthic communities of the James and York estuaries and the lower Chesapeake Bay. Chesapeake Sci. 17: 246-259.

Clark, S.H. and B.E. Brown. 1977. Changes in biomass of finfishes and squids from the Gulf of Maine to Cape Hatteras, 1963-74, as determined from research vessel survey data. Fish. Bull. 75: 1-22.

Connell, J.H. 1961. Effects of competition, predation by *Thais lapillus,* and other factors on natural populations of the barnacle *Balanus balanoides.* Ecol. Monogr. 31: 61-104.

Coull, B.C. and J.W. Fleeger. 1977. Long-term temporal variation and community dynamics of meiobenthic copepods. Ecology 58: 1136-1143.

Dayton, P. 1979. Ecology: a science and a religion. *In:* Ecological Processes in Coastal and Marine Systems, R.J. Livingston (ed.), Plenum Press, New York. pp. 3-18.

Green, R.H. 1979. Sampling Design and Statistical Methods for Environmental Scientists. John Wiley & Sons, New York. 257 pp.

Harrison, G.W. 1979. Stability under environmental stress: resistance, resilience, persistence, and variability. Am. Nat. 113: 659-669.

Holling, C.S. 1973. Resilience and stability in ecological systems. Annu. Rev. Ecol. Syst. 5: 1-24.

Lewontin, R.C. 1969. The meaning of stability. *In:* Diversity and Stability in Ecological Systems, G.M. Woodwell and H.H. Smith (eds.), Academic Press, New York. pp. 13-24.

Livingston, R.J. 1974. The Ecological Impact of Pulp Mill Effluents on Aquatic Flora and Fauna in North Florida: Comparison of a Polluted Drainage System (Fenholloway) with an Unpolluted One (Econfina). Florida Coastal Coordinating Council. 87 pp.

Livingston, R.J. 1976. Time as factor in environmental sampling programs: diurnal and seasonal fluctuations of estuarine and coastal populations and communities. *In:* Symposium on the Biological Monitoring of Water Ecosystems, J. Cairns, Jr. (ed.), Am. Soc. Test. Mater. Spec. Tech. Publ. 607: 212-234.

Livingston, R.J., R.L. Iverson, R.H. Estabrook, V.E. Keys, and J. Taylor, Jr. 1974. Major features of the Apalachicola Bay system: physiography, biota, and resource management. Fla. Sci. 37: 245-271.

Livingston, R.J., R.L. Iverson, and D.C. White. 1976a. Energy Relationships and the Productivity of Apalachicola Bay. Final Report, Florida Sea Grant Tech. Paper: 437 pp.

Livingston, R.J., R.S. Loyd, and M.S. Zimmerman. 1976b. Determination of adequate sample size for collections of benthic macrophytes in polluted and unpolluted coastal areas. Bull. Mar. Sci. 26: 569-575.

Livingston, R.J., N. Thompson, and D. Meeter. 1978. Long-term variation of organochlorine residues and assemblages of epibenthic organisms in a shallow north Florida (U.S.A.) estuary. Mar. Biol. 46: 355-372.

Longhurst, A., M. Colebrook, J. Gulland, R. Le Brasseur, C. Lorenzen, and P. Smith. 1972. The instability of ocean populations. New Sci. 1: 500-502.

McErlean, A.J., S.G. O'Connor, J.A. Mihursky, and C.I. Gibson. 1973. Abundance, diversity, and seasonal patterns of estuarine populations. Estuarine Coastal Mar. Sci. 1: 19-36.

McNaughton, S.J. 1977. Diversity and stability of ecological communities: empiricism in ecology. Am. Nat. 111: 515-525.

May, R.M. 1973. Stability and Complexity in Model Ecosystems. Princeton University Press, Princeton, NJ. 265 pp.

Meeter, D.A. and R.J. Livingston. 1978. Statistical methods applied to a four-year multivariate study of a Florida estuarine system. *In:* Biological Data in Water Pollution Assessment: Quantitative and Statistical Analyses, J. Cairns, K. Dickson, and R.J. Livingston (eds.), Am. Soc. Test. Mater. Spec. Tech. Publ. 652: 53-67.

Meeter, D.A., R.J. Livingston, and G.C. Woodsum. 1979. Short- and long-term hydrological cycles of the Apalachicola drainage system with application to Gulf coastal populations. *In:* Ecological Processes in Coastal and Marine Systems, R.J. Livingston (ed.), Plenum Press, New York. pp. 315-338.

Michael, A.D., D. Boesch, C. Hershner, R.J. Livingston, K. Roos, and D. Straughan. 1977. Benthos. *In:* Oil Spill Studies: Strategies and Techniques, Publication No. 4286, American Petroleum Institute, Washington, DC. pp. 57-75.

Odum, E. 1953. Fundamentals of Ecology. Saunders, Philadelphia. 244 pp.

Orians, G.H. 1974. Diversity, stability, and maturity in natural ecosystems. *In:* Unifying Concepts in Ecology, W.H. van Dobben and R.H. Lowe-McConnell (eds.), W. Junk, Hague, Netherlands. pp. 139-150.

Paine, R.J. 1966. Food web complexity and species diversity. Am. Nat. 100: 65-75.

Peters, R.H. 1977. The unpredictable problems of tropho-dynamics. Environ. Biol. Fish. 2: 97-101.

Peterson, C.H. 1979. Predation, competitive exclusion, and diversity in the soft-sediment benthic communities of estuaries and lagoons. *In:* Ecological Processes in Coastal and Marine Systems, R.J. Livingston (ed.), Plenum Press, New York. pp. 233-264.

Reid, P.C. 1977. Continuous plankton records: changes in the composition and abundance of the phytoplankton of the northeastern Atlantic Ocean and North Sea, 1958-1974. Mar. Biol. 40: 337-339.

Saloman, C.H. and S.P. Naughton. 1977. Effect of Hurricane Eloise on the benthic fauna of Panama City Beach, Florida, U.S.A. Mar. Biol. 42: 357-363.

Snelson, F.F. and W.K. Bradley, Jr. 1978. Mortality of fishes due to cold on the east coast of Florida, January, 1977. Fla. Sci. 41: 1-12.

Sutcliffe, W.H., Jr., K. Drinkwater, and B.S. Muir. 1977. Correlations of fish catch and environmental factors in the Gulf of Maine. J. Fish. Res. Board Can. 34: 19-30.

Sutherland, J.P. and R.H. Karlson. 1977. Development and stability of the fouling community at Beaufort, North Carolina. Ecol. Monogr. 47: 425-446.

Talbot, F.H., B.C. Russell, and G.R.V. Anderson. 1978. Coral reef fish communities: unstable, high diversity systems? Ecol. Monogr. 48: 425-440.

Van Winkle, W., B.L. Kirk, and B.W. Rust. 1979. Periodicities in Atlantic coast striped bass (*Morone saxatilis*) commercial fisheries data. J. Fish. Res. Board Can. 36: 54-62.

Webster, J.R., J.B. Waide, and B.C. Patten. 1975. Nutrient recycling and the stability of ecosystems. *In:* Mineral Cycling in Southeastern Ecosystems, F.G. Howell (ed.), Atomic Energy Commission Symp. Ser. Conf. 70-0513. pp. 1-27.

Wiens, J.A. 1978. On competition and variable environments. Am. Sci. 65: 590-597.

Williams, A.B. 1972. A ten year study of meroplankton in North Carolina estuaries: mysid shrimps. Chesapeake Sci. 13: 254-262.

Zimmerman, M.S. and R.J. Livingston. 1976a. Seasonality and physico-chemical ranges of benthic macrophytes from a north Florida estuary (Apalachee Bay). Contrib. Mar. Sci. Univ. Tex. 20: 34-45.

Zimmerman, M.S. and R.J. Livingston. 1976b. The effects of kraft mill effluents on benthic macrophyte assemblages in a shallow bay system (Apalachee Bay, North Florida, U.S.A.). Mar. Biol. 34: 297-312.

Zimmerman, M.S. and R.J. Livingston. 1979. Dominance and distribution of benthic macrophyte assemblages in a north Florida estuary (Apalachee Bay, Florida). Bull. Mar. Sci. 29: 27-40.

THE POTENTIAL CONTRIBUTION OF PETROLEUM HYDROCARBONS TO CHANGES IN BENTHIC COMMUNITIES OF THE NEW YORK BIGHT

Allan D. Michael

Taxon, Inc.
50 Grove Street
Salem, Massachusetts 01970

Abstract. Contaminant levels in the lower Hudson River estuary and New York Bight have affected the distribution and abundance of benthic invertebrates. One of the major contaminants, petroleum hydrocarbons, includes persistent and toxic components such as polyaromatic hydrocarbons. Input levels, mostly from non-point sources, are high and difficult to control. Since a large number of contaminants are present in the New York Bight, it is difficult, if not impossible, to identify individual contaminants that have caused changes in benthic communities. However, oil spills have occurred where levels of other contaminants were extremely low and where the effects on benthos can be traced directly to hydrocarbons. These effects include reductions in the numbers of species and individuals and increases in the number of opportunistic species. Changes in benthic community patterns observed in parts of the New York Bight and the Hudson River estuary are similar to those caused by oil spills and theoretically could be attributed solely to petroleum hydrocarbons.

INTRODUCTION

Heavy human population densities and industrial activity in the New York-New Jersey area have introduced substantial amounts of a wide range of contaminants into receiving waters. The realization that input cannot continue indefinitely without some understanding of the resulting problems and without control provided impetus for the NOAA MESA New York Bight program. While MESA researchers have tried to attribute ecological changes to contaminant levels and excess nutrient loadings, there are virtually no situations where a single contaminant or nutrient is involved. Further, there is little background data against which to assess potential or realized changes. The task of assessing damage after the fact, when a wide range of potential causes is involved, seems impossible. However, management decisions, obviously needed in the case of the New York Bight, must be based on the best available evidence. The approach used here is to assemble information on input rates and to use data from the literature to assess the probable relative importance of petroleum hydrocarbons as a cause of changes that have taken place in benthic communities.

INPUTS

Mueller *et al.* (1976) summarized contaminant inputs to the New York Bight by evaluating dumping, municipal and industrial waste disposal activities, runoff, and atmospheric inputs. Included in the list of materials were nutrients, such as phosphorus, ammonia, and organic nitrogen, and several broad classifications such as suspended solids and biological oxygen demand. These broad categories do not identify specific contaminants but often indicate the presence of contaminants. In addition, Mueller *et al.* (1976) reported on eight substances that normally are considered contaminants: oil and grease, lead, mercury, iron, cadmium, copper, chromium, and zinc (Table 1).

It has been estimated that 50% to 70% of oil and grease wastes are petroleum hydrocarbons (National Academy of Science, 1975). Assuming a figure of 60%, the daily input of petroleum hydrocarbons discharged directly into the New York Bight (*i.e.*, by dredged materials and sewage sludge dumping) is approximately 275 metric tons. However, the actual mass load reaching the Bight will be greater because of the contribution of oil and grease discharged into the New York harbor and transported to the Bight-proper through surface and water column dispersive processes. The daily direct input of petroleum hydrocarbons to the Bight is approximately 40 times that of lead and over 4,500 times that of mercury. The only metal approaching the total input of hydrocarbons to the Bight is iron, with a daily release of 190 metric tons. However, unlike metals, petroleum hydrocarbons dissolve, evaporate, and degrade, leading to considerable losses from the system. Despite this, hydrocarbon levels of 300-6,000 μg/g occur in sediments in the New York Bight apex (MESA, 1976; Farrington and Tripp, 1977). Farrington and Tripp (1977) measured levels up to 3,000 μg/g dry wt total hydrocarbons and argued that fossil fuels contributed the bulk of this total. MESA (1976) reported levels up to 6,530 ppm in sediments at the dredged materials dumpsite and 80 ppm at two proposed alternate sewage sludge disposal sites approximately 64 km offshore. More recently, Anderson (this volume) reported polynuclear aromatic hydrocarbon concentrations in the parts per million range from sediments slated for dredging in the Hudson-Raritan estuary.

The sources of hydrocarbon inputs to the New York Bight are summarized in Figure 1. All estimates, except those from the atmosphere, are from Mueller *et al.* (1976) and have been reduced to reflect petroleum hydrocarbon inputs, as described above (National Academy of Science, 1975). The estimate of atmospheric inputs is from Gibson *et al.* (1979).

Table 1. Total mass into the New York Bight. Recalculated from the data of Mueller *et al.* (1976) excluding transect zone loadings. Loadings as metric tons/day.

Parameter	Total mass load	Load contribution by location Direct Bight	New Jersey	Long Island
Oil and grease	454.1	370.6	78.3	5.2
Lead	6.8	6.7	0.06	0.03
Mercury	0.05	0.03	0.02	0.002
Iron	193.7	188.6	4.6	0.5
Cadmium	2.01	2.0	0.01	0.002
Copper	7.6	7.5	0.1	0.03
Chromium	2.8	2.6	0.03	0.2
Zinc	15.9	15.5	0.3	0.1

No value is provided in Figure 1 for estuarine inputs to the Bight (*i.e.*, transect zone contributions of Mueller *et al.*, 1976) because it has not been possible to distinguish quantitatively the fraction of hydrocarbons that sink to the bottom and become incorporated in estuarine sediments from that which is flushed out of the estuary into the Bight. Mueller *et al.'s* calculations (1976) assume that all inputs to the estuary enter the Bight. This clearly is an over-estimation.

The Mueller *et al.* (1976) data further are complicated by the direct Bight category of inputs (Figure 1), which includes materials dredged from the estuary and dumped at the dredged materials dumpsite in the apex. Since much of the contaminant burden of dredged materials is of estuarine or riverine origin, the

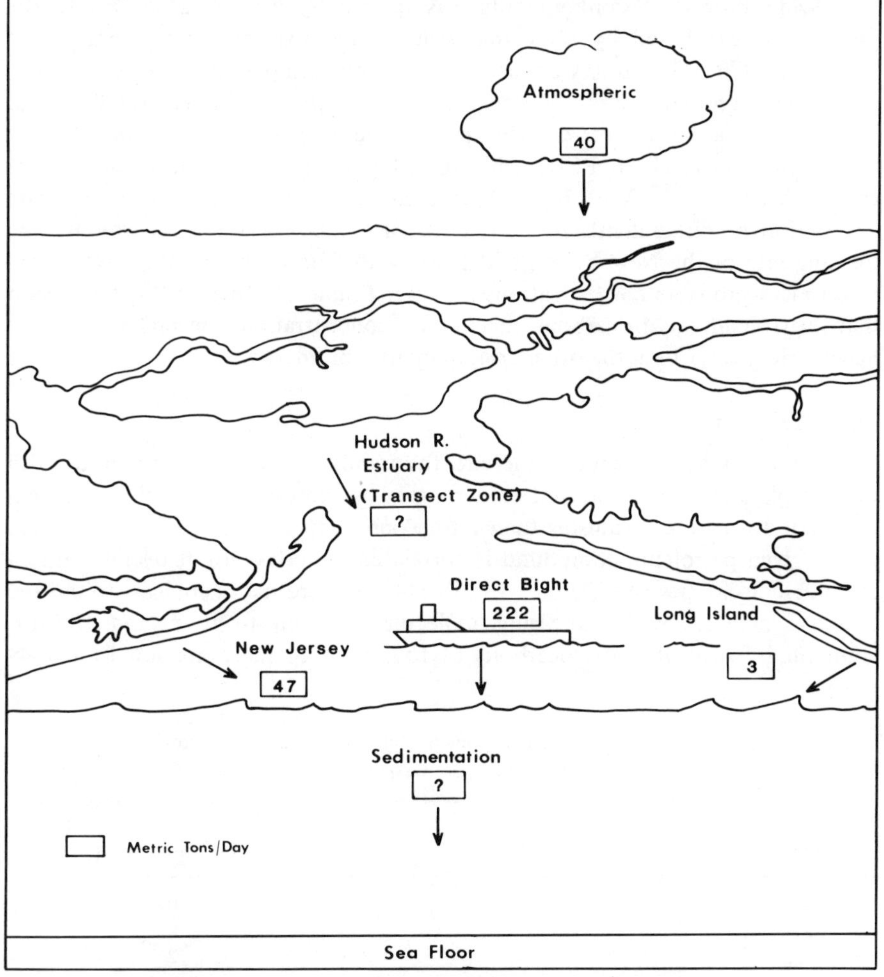

Figure 1. Summary of petroleum hydrocarbons inputs (metric tons/day) to the New York Bight. See text for explanation.

contaminants in dredged materials are, in effect, accounted for twice, *i.e.*, in the direct Bight statistics and in the transect zone statistics. By omitting the latter data from Figure 1, a more conservative estimate is achieved, and a potential double accounting of petroleum hydrocarbon inputs is avoided.

Nevertheless, the Mueller *et al.* (1976) data illustrate that the Hudson River estuary, either directly or indirectly, is the principal source of petroleum hydrocarbons entering the New York Bight. Within the Hudson estuary, urban runoff and municipal wastewaters contribute approximately 40% and 20% of the total, respectively. Barge dumping, the largest input shown on Figure 1, and municipal wastewaters are discharges that can be controlled. It would be difficult, if not impossible, to reduce significantly the contribution made by urban runoff.

A wide variety of contaminants was ranked by a panel according to the probable threat posed by these materials to the environment (O'Connor and Stanford, 1979). Of primary concern were chlorinated pesticides, lead, mercury, polynuclear aromatic hydrocarbons, polychlorinated biphenyls (PCB's), and plutonium. Carpenter *et al.* (in press) found insufficient comprehensive information on levels of PCB's, chlorinated pesticides, and plutonium for the apex area of the New York Bight. They used mass balance equations to estimate levels of five contaminants in New York Bight apex waters. The volume and flushing rate of the New York Bight apex, with data on input levels, were used to estimate probable concentrations in water (Table 2). These estimates suggest that petroleum hydrocarbons occur at concentrations several orders of magnitude greater than the other contaminants considered.

TOXICITY

"Petroleum hydrocarbons" include thousands of organic compounds with varying physical, chemical, and toxicological properties. Crude oil generally is less toxic to marine organisms than refined products. Acute toxicity associated with a given petroleum compound is correlated directly with its overall content of soluble aromatics (Neff *et al.*, 1976). The monoaromatics are least toxic, and toxicity increases with increasing molecular size up to four- and five-ring compounds. Aromatic hydrocarbons of four rings or more are not as acutely

Table 2. Probable concentrations of five contaminants in waters of New York Bight apex (from Carpenter *et al.*, in press).

Pollutant	Concentration (ppm)
Mercury	0.02
Cadmium	0.5
Lead	2.6
PCB's	0.002
Petroleum hydrocarbons	100

toxic, possibly because of their large molecular size and low solubility. The toxicity of di- and tri-aromatic compounds appears to increase with alkylation. Methylphenanthrene is the most toxic hydrocarbon evaluated to date (Anderson, 1977). Naphthalene and its alkyl derivatives are toxic to marine organisms in the 0.1-1.0 ppm range. Because polynuclear aromatic hydrocarbons have low solubility, they persist in the marine environment in association with suspended solids and sediments.

It is impossible to identify levels of crude oil or fuel oil necessary to produce acute toxicity because of the variety and differing proportions of compounds in various fuel oils. The variety of inputs to the New York Bight ensures that a broad spectrum of petroleum hydrocarbons exists there. While individual hydrocarbons may not be the most toxic of various possible contaminants, their relative concentrations suggest that they are a serious problem.

Up to 5,400 μg/g dry wt of hydrocarbons have been reported from sediments in the Providence River, and concentrations of 300 μg/g have been found in sediments in Narragansett Bay (Van Vleet and Quinn, 1978). These petroleum-like hydrocarbons, which resemble a mixture of No. 2 fuel oil, No. 6 fuel oil, and used crankcase oil, are introduced to the River and Bay largely by chronic input from a municipal wastewater treatment plant. Although crude oil is shipped through the New York Bight, oil spills and operational discharges are a small fraction of the total input (Munson *et al.*, 1978). It is probable that the "average" composition of petroleum hydrocarbons entering the Bight is similar to that described by Van Vleet and Quinn (1978), *i.e.*, a partially- or well-aged fuel or lubricating oil. Tanacredi (1977) demonstrated the presence of automotive waste crankcase oil components in treated wastewater effluents of Jamaica Bay, New York.

COMMUNITY CHANGES INDUCED BY PETROLEUM HYDROCARBONS

In intertidal and subtidal benthic communities, changes caused by petroleum hydrocarbon contamination have been documented frequently (Sanders *et al.*, 1972; Clark *et al.*, 1975; Michael *et al.*, 1975; Schultz and Tebo, 1975; Straughan, 1977; Thomas, 1977). Changes range from mass mortalities of virtually all species, when heavily oiled with fresh No. 2 fuel oil, to slight alterations in community structure. Such alterations include the loss of rare, sensitive species or an increase in the proportion of opportunistic species (Michael *et al.*, 1975). Michael (1977) suggested that increasing amounts of hydrocarbon contamination tend to push macrobenthic communities back along a successional sequence. With light contamination, an increased number of species and individuals may occur. Higher levels of contamination lead to a reduction in the number of species, to an extreme where one might find monocultures of opportunistic species such as the polychaete, *Capitella capitata.*

At the site of the West Falmouth oil spill Michael *et al.* (1975) found slight changes in macrobenthic communities with low levels (20-50 μg/g) of aged (four years) fuel oil. These changes included a slight reduction in the total number of species and a higher percentage of opportunistic species than in control areas. At higher oil levels (several hundred micrograms per gram), the numbers of species and individuals were reduced more markedly. Other workers also have found reduced numbers of species associated with the presence of petroleum hydrocarbons (Clark *et al.*, 1975; Schultz and Tebo, 1975; Straughan, 1977). Some of these spills occurred in areas removed from population centers, where the environments can be considered pristine. At West Falmouth, there are no industrial discharges, and sediment samples analyzed for fecal coliforms showed no evidence of human contamination. Thus, it is safe to attribute the effects on the biota to petroleum hydrocarbons alone. The METULA spill in the Strait of Magellan, the ARROW spill in Chedabucto Bay, Nova Scotia, and the GENERAL M.C. MEGIS spill on the northwest coast of Washington are other examples of spills that occurred in areas not otherwise contaminated by industrial or municipal wastes.

At the University of Rhode Island Marine Ecosystems Research Laboratory, changes in benthic communities resulted after oil was introduced into experimental ecosystems in large cylindrical tanks (Elmgren *et al.*, in press). These tanks were supplied with water from Narragansett Bay, and No. 2 fuel oil was added to maintain levels of 75-300 ppb in the water. During five and one-half months, oil in the sediment reached a maximum of 1,000 μg/g dry wt but was more typically 400-500 μg/g (Grassle, personal communication). The abundance of macrofauna began to drop soon after oil was added. At the end of the experiment the oiled tanks had less than one tenth of the macrofaunal biomass found in controls. Suspension feeding animals virtually were excluded from oiled tanks.

CHANGES IN NEW YORK BIGHT MACROBENTHIC COMMUNITIES

Most obvious changes in macrobenthos of the New York Bight have occurred in the Christiaensen Basin and in an area just west of the sewage sludge disposal site on the margin of the basin (Rowe, 1971; Pearce *et al.*, 1976; Boesch, this volume; Steimle *et al.*, this volume). The latter area appears to be most impacted and supports few macrobenthic species in addition to dense populations of *Capitella capitata*, a species that is rare in other areas of the Bight. Other areas of the Basin support a dense benthic community characteristic of muddy sands. Dominant are the polychaetes, *Nephtys incisa, Pherusa affinis, Mediomastus ambiseta, Tharyx* sp., and *Asabellides oculata*, the burrowing anemone, *Cerianteopsis americanus*, and the bivalve, *Nucula proxima.*

Boesch (this volume) suggests that organic enrichment or the exclusion of predators by hypoxia or contaminants causes high standing crops of *Capitella* in the Christiaensen Basin. A few species, notably amphipods that occur elsewhere in the region, are absent in the heavily affected areas. These include two species of *Ampelisca* and the corophiid, *Unciola irrorata*, both dominant members of the macrobenthos in many areas of the Middle Atlantic continental shelf (Boesch, 1979).

McGrath (1974) reported greatly depressed macrobenthic communities in Raritan Bay. The average number of species per sample (0.1 m^2) was four, and the average number of individuals was only eleven. A striking feature was the absence of the amphipod genus *Ampelisca,* which ten years earlier was a dominant (Dean and Haskin, 1964). McGrath reported seeing an oily film in the grab samples and an apparent increase in oil content of sediments toward the Arthur Kill industrial complex. While this clearly is not an example of cause and effect, because of the presence of other contaminants and the lack of quantitative data on hydrocarbons, the presence of petroleum hydrocarbons in amounts that produce slicks is sufficient in itself to explain the elimination of *Ampelisca* (Sanders *et al.*, 1972).

DISCUSSION

Very high levels of petroleum hydrocarbons are found in the Christiaensen Basin and at the New York sewage sludge and dredged materials disposal sites (MESA, 1976; Farrington and Tripp, 1977). It is reasonable to expect that levels of 3,000-6,000 ppm of degraded fuel oils in sediments can produce changes in benthic communities, such as those seen at the sewage sludge dumpsite. This is not to suggest that other contaminants are not important, or that they do not contribute to the changes seen. Information on virtually all the major contaminants, including hydrocarbons, is scant, and it is virtually impossible to identify clearly the effects of each. Yet, information on input levels suggests that petroleum hydrocarbons are, by volume, the largest individual source of toxic contaminants. Although a significant proportion of this material is degraded rapidly, levels found in sediments at the dumpsites are much higher than those of any other contaminants that have been measured. Laboratory and field studies clearly have established that some petroleum hydrocarbons are toxic to marine organisms. Changes in macrobenthic communities of the nature and extent found in the New York Bight have been related to petroleum hydrocarbons in other studies where other contaminants were not involved.

Pearson and Rosenberg (1978) described modifications to benthic communities caused by organic enrichment. In some cases these changes were similar to those described here for benthic communities in the New York Bight apex and to others reported for oil spills. Pearson and Rosenberg (1978) argued that data were too scarce to indicate whether general community changes along an organic enrichment gradient were also applicable to pollution by contaminants such as heavy metals and chlorinated hydrocarbons. They admitted, however, that in many of their case studies, the sediments might have been contaminated by pollutants other than organic wastes. They further suggested that pollution is not likely to favor any particular species. Results from bioassays and oil spill studies do not support the latter suggestion. Sanders *et al.* (1972), Grassle and Grassle (1974), and Michael *et al.* (1975) documented differing survival rates of various species after exposure to No. 2 fuel oil. Population shifts clearly were related to the effect of oil rather than to organic enrichment. If changes in the benthos occurred because oil provided a carbon source, one might expect an increased abundance of fauna with increased levels of oil. To the contrary, higher oil levels were related to lower faunal abundances. There is no evidence that benthic community changes induced by oil spills are a result of organic enrichment *per se.*

It is obvious that no single factor is responsible for all the changes observed in benthic communities of the New York Bight. Boesch's suggestion in this volume that changes result from a combination of factors, including organic enrichment, hypoxia, toxicants, and reduced predation, seems most reasonable. However, contamination by petroleum hydrocarbons must be considered a major factor.

REFERENCES

Anderson, J.W. 1977. Response to lethal levels of petroleum hydrocarbons: Are they sensitive indicators and do they correlate with tissue contamination? *In:* Proceedings of the Symposium on the Fate and Effects of Petroleum Hydrocarbons in Marine Ecosystems and Organisms, November 10-12, 1976, NOAA, Seattle, WA. pp. 95-114.

Anderson, J.W. This volume. The transport of petroleum hydrocarbons from sediments to benthos and the potential effects. pp. 165-179.

Boesch, D.F. 1979. The distribution and dynamics of macrobenthos on the Middle Atlantic continental shelf. Virginia Inst. Mar. Sci. Spec. Sci. Rep. 194. 301 pp.

Boesch, D.F. This volume. Ecosystem consequences of alterations of benthic community structure and function in the New York Bight region. pp. 543-568.

Carpenter, J.H., P.W. Anderson, A. Calabrese, K.S. Kamlet, T. Kneip, and A. Michael. In press. Implications of current and possible future contaminant levels in the apex area of the New York Bight. NOAA OMPA Tech. Rep.

Clark, R.C., Jr., J.S. Finley, B.G. Patten, and Edward E. DeNike. 1975. Long-term chemical and biological effects of a persistent oil spill following the grounding of the GENERAL M.C. MEIGS. *In:* Proceedings of 1975 Conference on Prevention and Control of Oil Pollution, American Petroleum Institute, U.S. Environmental Protection Agency, and U.S. Coast Guard, Washington, DC. pp. 479-487.

Dean, D. and H. Haskin. 1964. Benthic repopulation of the Raritan River estuary following pollution abatement. Limnol. Oceanogr. 9: 551-563.

Elmgren, R., G.A. Vargo, J.F. Grassle, J.P. Grassle, D.R. Heinle, G. Langlois, and S.L. Vargo. In press. Trophic interactions in experimental marine ecosystems perturbed by oil. International Conference for Exploration of the Sea.

Farrington, J.W. and B.W. Tripp. 1977. Hydrocarbons in western North Atlantic surface sediments. Geochim. Cosmochim. Acta 41: 1627-1641.

Gibson, C.E., L. Ciaccio, B. Ketchum, and A.D. Michael. 1979. Petroleum hydrocarbons. *In:* Chemical Pollutants of the New York Bight; Priorities for Research, J.S. O'Connor and H.M. Stanford (eds.), NOAA Special Report. pp. 20-32.

Grassle, J.F. and J.P. Grassle. 1974. Opportunistic life histories and genetic systems in marine polychaetes. J. Mar. Res. 32(2): 253-284.

McGrath, R. 1974. Benthic macrofaunal census of Raritan Bay-preliminary results, Pap. 24. *In:* Hudson River Ecology, Third Symposium on Hudson River Ecology, March 22-23, 1973, Hudson River Environmental Society, Inc. 40 pp.

MESA. 1976. Evaluation of proposed sewage sludge dump site areas in the New York Bight. NOAA Tech. Memo. ERL MESA-11. 212 pp.

Michael, A.D. 1977. The effects of petroleum hydrocarbons on marine populations and communities. *In:* Proceedings of the Symposium on the Fate and Effects of Petroleum Hydrocarbons in Marine Ecosystems and Organisms, NOAA, Seattle, WA. pp. 129-137.

Michael, A.D., C.R. Van Raalte, and L.S. Brown. 1975. Long-term effects of an oil spill at West Falmouth, Massachusetts. *In:* Proceedings of 1975 Conference on Prevention and Control of Oil Pollution, American Petroleum Institute, U.S. Environmental Protection Agency, and U.S. Coast Guard, Washington, DC. pp. 573-582.

Mueller, J.A., J.S. Jeris, A.R. Anderson, and G.F. Hughes. 1976. Contaminant inputs to the New York Bight. NOAA Tech. Memo. ERL MESA-6. 347 pp.

Munson, J.S., R.D. Moskowitz, M.D. Rowe, A.G. Tingle, and D.A. Dicterle. 1978. The benefits of a reduction in operational discharges to the New York Bight. A Report Energy Policy Analysis Division, National Center for the Analysis of Energy Systems, Brookhaven National Laboratory, Upton, NY. 46 pp.

National Academy of Science. 1975. Petroleum in the marine environment. Workshop on Inputs, Fates and the Effects of Petroleum in the Marine Environment, May 21-25, National Academy of Science, Washington, DC. 107 pp.

Neff, J.M., J.W. Anderson, B.A. Cox, R.B. Laughlin, Jr., S.S. Rossi, and H.E. Tatem. 1976. Effects of petroleum on survival, respiration and growth of marine animals. *In:* Sources, Effects and Sinks of Hydrocarbons in the Aquatic Environment, American Institute of Biological Sciences, Arlington, VA. pp. 515-539.

O'Connor, J.S. and H.M. Stanford (eds.). 1979. Chemical Pollutants of the New York Bight; Priorities for Research. NOAA Special Report. 217 pp.

Pearce, J.B., J.V. Caracciolo, M.B. Halsey, and L.H. Rogers. 1976. Temporal and spatial distributions of benthic macroinvertebrates in the New York Bight. Am. Soc. Limnol. Oceanogr. Spec. Symp. 2: 394-403.

Pearson, T.H. and R. Rosenberg. 1978. Macrobenthic succession in relation to organic enrichment and pollution of the marine environment. Oceanogr. Mar. Biol. Annu. Rev. 16: 229-311.

Rowe, G.T. 1971. The effects of pollution on the dynamics of the benthos of New York Bight. Thalassia Jugosl. 7: 353-359.

Sanders, H.L., J.F. Grassle, and G.R. Hampson. 1972. The West Falmouth oil spill. Part I. Biology. Woods Hole Oceanogr. Inst. Tech. Rep. WHOI-72-20. 23 pp.

Schultz, D. and L.B. Tebo, Jr. 1975. Boone Creek oil spill. *In:* Proceedings of 1975 Conference on Prevention and Control of Oil Pollution, American Petroleum Institute, U.S. Environmental Protection Agency, and U.S. Coast Guard, Washington, DC. pp. 583-588.

Steimle, F.J. Caracciolo, and J.B. Pearce. This volume. Impacts of dumping on New York Bight apex benthos. pp. 213-223.

Straughan, D. 1977. Biological survey of intertidal areas in the Straits of Magellan in January, 1975, five months after the METULA Oil Spill. *In:* Fate and Effects of Petroleum Hydrocarbons in Marine Organisms and Ecosystems, D.A. Wolfe (ed.), Pergamon Press, Oxford, England. pp. 247-260.

Tanacredi, J.T. 1977. Petroleum hydrocarbons from effluents: detection in marine environment. J. Water Pollut. Control Fed. 49: 216-226.

Thomas, M.L. 1977. Long term biological effects of bunker C oil in the intertidal zone. *In:* Fate and Effects of Petroleum Hydrocarbons in Marine Organisms and Ecosystems, D.A. Wolfe (ed.), Pergamon Press, Oxford, England. pp. 238-245.

Van Vleet, E.S. and J.G. Quinn. 1978. Contributions of chronic petroleum inputs to Narragansett Bay and Rhode Island Sound sediments. J. Fish. Res. Board Can. 35(5): 536-543.

SOURCES, SINKS, AND DISTRIBUTION OF ORGANIC CONTAMINANTS IN THE NEW YORK BIGHT ECOSYSTEM

J.M. O'Connor
J.B. Klotz[1]
Theo. J. Kneip

New York University Medical Center
Institute of Environmental Medicine
A.J. Lanza Laboratories
Long Meadow Road
Tuxedo Park, New York 10987

Abstract. Data are given on the concentrations of chlorinated pesticides, polychlorinated biphenyls (PCB's), and polynuclear aromatic hydrocarbons (PAH's) in the sediments, water, and biota of the Hudson River, New York harbor, and New York Bight. Major sources of organic contaminants to the estuarine/coastal system are, for pesticides, Hudson River transport and municipal wastewater; for PCB's, Hudson River transport, atmospheric fallout, and surface runoff; and for PAH's, atmospheric fallout, surface runoff, and oil spills. Major sinks for all contaminants are the harbor, the Lower Bay, and the Bight.

Of the three classes of compounds, the most common contaminants in marine and estuarine biota are the PCB's. Pesticides and PAH's, although present in many organisms, occur at relatively low levels. Except for PCB's in striped bass from the Hudson River, no instances of contaminants exceeding U.S. Food and Drug Administration (FDA) Action Levels were recorded. The implications of these data for ecosystem management and reclamation are discussed.

INTRODUCTION

Chlorinated pesticides, polychlorinated biphenyls (PCB's), and polynuclear aromatic hydrocarbons (PAH's) have been identified as "major perceived threats [to humans and aquatic biota] that require continued study [O'Connor and Stanford, 1979: Table 1]." These classes of compounds have been detected in water, suspended sediments, deposited sediments, and biota from the Hudson River, the Hudson-Raritan estuary, and the New York Bight. PCB loadings have closed the striped bass commercial fishery and have generated concern about PCB distribution and cycling in the environment (Horn *et al.,* 1979; Thomann and St. John, 1979), and about potential ecological impacts on marine systems (O'Connors *et al.,* 1979; Peters and O'Connor, this volume).

Detailed studies of organic contaminants in marine and coastal waters have been restricted. Research to account for sources, sinks, and distribution factors has been confined primarily to PCB's (McDermott *et al.,* 1976; Munson *et al.,* 1976; Southern California Coastal Water Research Project, 1977; Horn *et al.,* 1979; Thomann and St. John, 1979), but some studies on PAH's have emerged (Pavlou *et al.,* 1973; MacLeod *et al.,* 1977; Brown *et al.,* 1979). As more attention is given by regulatory agencies to management of organic contaminants in marine waters and to reclamation of resources contaminated by potentially toxic compounds, a greater number of studies will emerge.

[1]Present address: Department of Environmental Sciences, Columbia University, School of Public Health, New York, New York 10004.

Table 1. Pesticides detected in water, sediments, and biota of the Hudson River, the Hudson estuary, and the New York Bight region.

1, 2-Dibromo-3-chloropropane	Heptachlor epoxide
Hexachlorobenzene	α-Chlordane
α-Benzenehexachloride	Trans-nonachlor
Lindane	Endrin
Heptachlor	Mirex[a]
Aldrin	DDT (and metabolites)
Dieldrin	

[a]Mirex detected in a single sample (MacLeod *et al.*, in press) may represent laboratory-induced artifact. Identity of the compound was verified by GC-MS analysis.

This paper describes the sources, sinks, and distribution of organic contaminants in the New York Bight and contiguous waters (Figure 1). For compounds such as DDT and PCB's, limited time series measurements are available. For DDT and metabolites, data are available from 1967 to present. PCB data are available from 1973 to 1978. Data on PAH's are available only from sampling carried out in 1978 and 1979 (MacLeod *et al.*, 1981; O'Connor *et al.*, in press a,b).

CHLORINATED PESTICIDES

The pesticides of concern in the New York Bight region have been defined as aldrin/dieldrin, chlordane and metabolites, DDT and metabolites, endosulfan and metabolites, endrin and metabolites, heptochlor and metabolites, hexachlorocyclohexanes, and toxaphene (O'Connor and Stanford, 1979). The chlorinated pesticides identified in the water, sediments, and biota of the Hudson River and the New York Bight region are listed in Table 1; other pesticides (*e.g.*, organophosphates and carbamates) have not been studied in any detail in the Hudson River, the Hudson estuary, or the Bight region.

Sources

Pesticides in the Hudson River and the Bight region derive from agricultural applications in the Hudson/Mohawk and Raritan watersheds, and in the predominantly agricultural zones of eastern Long Island. An undetermined quantity of pesticides is used in the metropolitan region for control of cockroaches and other vermin. Except for data on regional usage of aldrin, chlordane, and toxaphene (Young *et al.*, 1979), distribution and use patterns for chlorinated pesticides in the New York Bight region are not available. Estimates for the use of chlordane; aldrin, and toxaphene in the northeastern United States in 1972 were 206, 62, and 42 kg, respectively. Usage in areas bordering the Bight region must be assumed to be a fraction of these figures.

Chlorinated pesticides other then DDT and dieldrin appear only sporadically in samples of water, sediments, and biota. Kneip and Hernandez (1969) showed that few compounds other than DDT and dieldrin were regularly present in

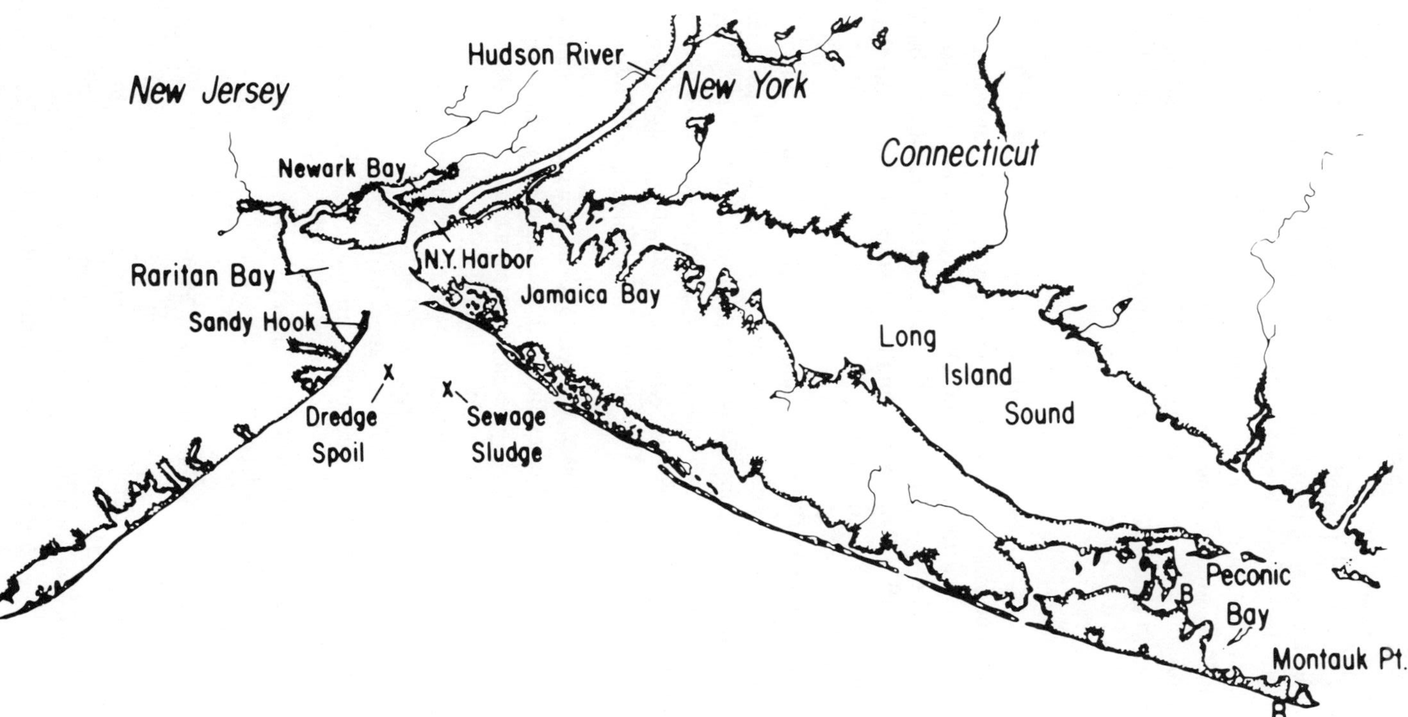

Figure 1. The Hudson-Raritan estuary and New York Bight study area.

plankton, sediments, and fishes from the Hudson River; the most abundant were heptachlor and heptachlor epoxide. In New York harbor, the Interstate Sanitation Commission (1977) reported detectable values only for a-benzene-hexachloride. Other chlorinated pesticides, if present, were below detection limits. Chlorinated pesticides in sediments, biota, and sewage sludge samples from the Bight region are dominated by DDT and dieldrin (MacLeod *et al.*, 1981; O'Connor *et al.*, in press a, b). Sources of chlorinated pesticides other than DDT and dieldrin are minimal and are not widespread in the region. The presence of substantial concentrations of DDT and trans-nonachlor in sewage sludge suggests a significant source (possibly vermin control) for these compounds within the city proper.

Sinks

Environmental concentrations of chlorinated pesticides are represented in this report by values for dieldrin and total DDT (consisting of the added values, within a sample, for o,p'-DDE; p,p'-DDE; o,p'-DDD; p,p'-DDD; o,p'-DDT; p,p'-DDT; and/or TDE). Data from a number of stations throughout the Bight and Hudson River are given in Table 2.

Dieldrin values for sediments at two stations in the Hudson River above New York City (km 66 and km 21) were similar (15-16 ng/g, dry wt) (Kneip and Hernandez, 1969). Concentrations in sediments taken downstream, near the George Washington Bridge (km 15), were significantly lower (1 ng/g, dry wt). Dieldrin was not detected in sediments from Lower Bay and from the outer Bight, or in sewage sludge samples (O'Connor *et al.*, in press a); however, dieldrin was present in sediments from the Christiaensen Basin and from the sewage sludge dumpsite (Table 2).

The data suggest that the New York metropolitan area is a significant source of DDT to the sediments of Lower Bay and the Bight apex. In independent studies (West *et al.*, 1976; O'Connor *et al.*, in press a), DDT concentrations in sewage sludge were found to range from ~800 ng/g to ~1,100 ng/g, dry wt. Total DDT in sediments at the sludge dumpsite and in nearby regions of the Christiaensen Basin ranged from ~50 ng/g to 160 ng/g, dry wt.

DDT was present in sediments from the upper harbor region and from various shipping channels at concentrations ranging from non-detectable to ~20 ng/g, dry wt (O'Connor *et al.*, in press a). Average total DDT for Newtown Creek, Gowanus Canal, and Pierhead Channel sediment samples was 7 ng/g, dry wt. Total DDT in sediment samples from the dredged materials dumpsite was 9 ng/g, dry wt (West *et al.*, 1976).

The source of DDT to the sediments of New York harbor and Lower Bay is a combination of Hudson River discharge, municipal wastewater effluents, and combined sewer overflow discharge. The high concentrations of total DDT in sewage sludge from the New York City area suggest that the many raw and treated sewage discharges to the Hudson River, Upper Bay, and Lower Bay may contribute significant quantities of DDT to the system. Data from a sampling

Table 2. Concentrations of total DDT and dieldrin (ng/g, dry wt) in sediments from the Hudson River estuary and New York Bight, and in sewage sludge. N.D. indicates compound not detected; N.S. indicates not sampled or compound not analyzed for.

Material Location or number	Dieldrin	Total DDT[a]
Sediment		
Hudson River (km 66)[b]	15	10
Hudson River (km 21)[b]	16	15
Hudson River (km 15)[c]	1	10
Lower Bay[b]	N.D.	34
Sandy Hook transect[d]	N.S.	0.4
Christiaensen Basin[c]	7	160
Dredged materials dumpsite[d]	N.S.	9
Sewage sludge dumpsite[d]	N.S.	120
Sewage sludge dumpsite[c]	3	47
Outer Bight[c]	N.D.	N.D.
Sewage sludge		
Sample 1[c]	N.D.	1,094
Sample 2[d]	N.S.	775[e]

[a]Total DDT = DDT and metabolites as listed in text.

[b]From Kneip and Hernandez (1969).

[c]From O'Connor *et al.* (in press a).

[d]From West *et al.* (1976).

[e]Mean of average values from four different sewage plants.

program designed to differentiate DDT input in sewage, storm-water runoff, and Hudson River discharge are required to evaluate the sources effectively.

Sewage sludge from the New York metropolitan area is a significant sink of DDT and, along with dredged materials, serves as a source of DDT to the Bight apex. Using total DDT concentrations in sewage sludge from West *et al.*, (1976) and O'Connor *et al.*, (in press a), and average DDT concentrations in river and harbor sediments (MacLeod *et al.*, 1981; O'Connor *et al.*, in press a), one may calculate the mean annual mass loads of DDT present in the Bight apex as a result of sewage sludge dumping and dredged materials dumping. If one assumes a loading factor of 4,047 x 10^3 metric tons (4,461 x 10^3 tons) (wet) of sewage sludge per year and 7,845 x $10^3 m^3$ (10,261) yd^3) (wet) of dredge spoil per year as average values derived from figures for 1974 through 1978 (P. Anderson, U.S. Environmental Protection Agency, personal communication; D. Suszkowski, U.S. Army Engineers, New York District, personal communication), yearly inputs of DDT from sewage sludge may range from ~106 kg/yr to 231 kg/yr whereas inputs of DDT from dredged materials may be less than half of that (65 kg/yr) (Table 3). Errors may occur in estimates of contaminant concentrations, solids content of sludge or dredge spoils, volumes dumped, or other factors, all of which may affect calculations of DDT loading. It should be kept in mind that only now are comparable surveys to determine DDT levels in sewage sludge,

Table 3. Calculated annual mass load of total DDT dumped in the Bight apex from materials dredged from New York harbor and Lower Bay, and in sewage sludge. Volumes of dredged materials for 1974-1978 from D. Suszkowski (personal communication). Volumes of sludge dumped for 1974-1978 from P. Anderson (personal communication).

Source	Mean amount dumped, 1974-1978	Annual DDT[a] load (kg)
Dredged materials[b]	7,845 x 10^3 m^3 (wet)	65
Sewage sludge[c]	4,047 x 10^6 kg (wet)	106
Sewage sludge[d]	4,047 x 10^6 kg (wet)	231

[a]Total DDT = DDT and metabolites as listed in text.

[b]Mean DDT concentration of 16.5 ng/g, dry wt; solids content ~50% (MacLeod *et al.*, 1981).

[c]Mean DDT concentration of 775 ng/g, dry wt; solids content ~ 5% (West *et al.*, 1976).

[d]Mean DDT concentration of 1,094 ng/g, dry wt; solids content ~ 5% (O'Connor *et al.*, in press a).

dredged materials, and harbor sediments being conducted. The estimates given here of annual DDT input to the Bight from sewage sludge dumping and from disposal of dredged sediments will vary. However, the important point is that DDT-contaminated sewage sludge, despite its low volume relative to dredged material, represents a significant, if not major, source of DDT to the Bight ecosystem.

Estimates of total DDT in Hudson River discharge have not been made, and the quantity of DDT in the River discharge relative to the DDT in dumped sewage sludge and dredge spoil is unclear. Total DDT concentrations in sediments along the Jersey shore, in the presumed track of the Hudson River plume, were higher than in all other Bight sediments, except those near the dredged materials dumpsite and the sewage sludge dumpsite (West *et al.*, 1976). The relative impact of Hudson River discharge on total DDT in the apex may be substantial and should be studied further.

Distribution

Total DDT and dieldrin in the biota of the New York Bight region span a wide range of concentrations and are dependent upon species, habitat, and environmental concentration of the compounds being measured. Among striped bass (*Morone saxatilis*) captured in the Hudson estuary, in New York harbor, and at Montauk Point (Table 4), flesh concentrations of DDT were higher in fish from the estuary than in fish from the ocean by a factor of between four and ten. Whether the fish from the Montauk Point area could have spent time in the estuary previously is important in assessing the significance of pesticide concentrations. Data from sediments in the outer Bight (Table 2) indicate no major sinks of dieldrin or DDT in offshore areas.

Fishes and shellfishes from regions with relatively high sediment concen-

Table 4. Concentrations of dieldrin and total DDT (ng/g, wet wt) in the flesh of various fishes and shellfishes from the New York Bight region. N.D. indicates compound not detected; N.S. indicates compound not analyzed for.

Species (location)	Dieldrin	Total DDT[a]
Striped bass[b] *(Morone saxatilis)* (km 15; Hudson River)	200	390
Striped bass[c] *(M. saxatilis)* (km 66; Hudson River)	N.D.	930
Striped bass[b] *(M. saxatilis)* (Montauk Point)	1	95
Winter flounder[b] *(Pseudopleuronectes americanus)* (Raritan Bay)	4	9
Winter flounder[d] *(P. americanus)* (Great Bay)	N.S.	14
Winter flounder[b] *(P. americanus)* (Christiaensen Basin)	N.D.	20
Blue crab[c] *(Callinectes sapidus)* (km 66; Hudson River)	730	1,340
Lobster[b] *(Homarus americanus)* (Raritan Bay)	N.D.	26
Lobster[b] *(H. americanus)* (Ambrose Light)	N.D.	7
Blue mussel[b] *(Mytilus edulis)* (Sandy Hook)	N.D.	65
Blue mussel[b] *(M. edulis)* (Coney Island)	1	32
Surf clam[b] *(Spisula solidissima)* (Shark River)	N.D.	12
Surf clam[b] *(S. solidissima)* (Rockaway Beach)	N.D.	4

[a]Total DDT = DDT and metabolites as listed in text.
[b]Data from O'Connor *et al.* (in press b).
[c]Data from Kneip and Hernandez (1969).
[d]Data from Sherwood *et al.* (1978).

trations of total DDT (*e.g.*, Christiaensen Basin; Table 2) showed no magnification of total DDT relative to sediment values (Table 4). For example, the flesh of winter flounder (*Pseudopleuronectes americanus*) from the Christiaensen Basin contained 20 ng/g, wet wt total DDT, while sediments from the same region contained 160 ng/g, dry wt total DDT (80 ng/g, wet wt). The flesh of lobsters (*Homarus americanus*) from Raritan Bay had total DDT concentrations of 26 ng/g, wet wt. Blue mussels (*Mytilus edulis*), assumed to be accurate indicators of contamination (Goldberg *et al.*, 1978), also had relatively low concentrations of total DDT–65 ng/g, wet wt at Sandy Hook and 32 ng/g, wet wt at Coney Island (Table 4). However, these data are too few and too scattered to permit conclusions about DDT levels in biota from the New York Bight region. Complementary data presented in Tables 2 and 4 were not always the result of simultaneous sampling or sampling at identical locations.

In contrast, blue crabs (*Callinectes sapidus*) tested by Kneip and Hernandez (1969) in the Hudson River showed very high concentrations of dieldrin and total DDT (Table 4). Concentrations of dieldrin (730 ng/g, wet wt) were very close to the FDA action limit (1 μg/g, wet wt; U.S. Food and Drug Administration, 1978); total DDT values were well above the FDA limit of 1 μg/g, wet wt. More data are needed to determine whether the high values for these pesticides are a continuing problem in Hudson River blue crabs.

POLYCHLORINATED BIPHENYLS (PCB's)

PCB's were first noted in natural systems by Jensen (1966), who identified them as interfering compounds in gas chromatographic analyses for chlorinated pesticides. Since then, many surveys have established PCB's as essentially ubiquitous in the environment (Risebrough, 1971; Wasserman *et al.*, 1979). Data available are insufficient to state which of the 209 possible PCB isomers are toxic and environmentally hazardous; however, most regulatory actions aimed at controlling PCB levels in food or water (U.S. Food and Drug Administration, 1978; U.S. Environmental Protection Agency, 1980a) consider the critical value to be total PCB's, rather than concentrations of individual isomers or classes of PCB compounds. PCB's in New York waters have been the subject of intensive study since 1976, when the New York State Departments of Environmental Conservation and Health initiated a study to document the distribution, transport, and bioaccumulation of PCB's in the Hudson River (Hetling *et al.*, 1978, 1979; Horn *et al.*, 1979; Thomann and St. John, 1979) and in shellfishes and fishes from other parts of the state (Spagnoli and Skinner, 1977).

Sources

PCB's enter the aquatic environment in industrial and manufacturing discharges, from atmospheric fallout (wet and dry), in effluent from sewage treatment plants, in municipal wastewater discharges, and in the leachate from landfills (U.S. Environmental Protection Agency, 1980a). In the New York region, the single largest point source of PCB's is the now-dispersed discharge from capacitor manufacturing plants located at Ft. Edward and Hudson Falls, New York, approximately 250 km upstream from Manhattan. Horn *et al.* (1979) state that 200,000-300,000 kg of PCB's are present in the sediments of the upper Hudson. Lawler, Matusky, and Skelly Engineers (1978), Hetling *et al.*

(1978), and Horn *et al.* (1979) state that PCB discharge to the lower Hudson depends upon freshwater discharge at the Green Island Dam (Troy, New York). At average flows (425-482 m^3/sec; 15,000-17,000 cfs), approximately 3,100 kg of PCB's per year enter the lower Hudson from upstream sources.

Essentially no data exist to estimate the quantities of PCB's transported through the tidal Hudson to New York City. Bopp (1979) showed, in general, that PCB concentrations in Hudson River sediments decline from Albany (km 248) to New York harbor (km 0-10) (Table 5). A similar downstream gradient occurs in plankton PCB concentrations (O'Connor and Hernandez, unpublished data); however, noticeable increases of PCB's in plankton occur in the vicinity of Poughkeepsie, New York (km 135). Given the large volume of tidal flow in the lower Hudson, the high degree of mixing in the system (Abood, 1974), and the tendency for PCB's to adsorb to inorganic, organic, and living particulates (Hiraizumi *et al.*, 1979), the probability is high that a large proportion of PCB's entering the system is transported to the harbor region. More data are needed, however, on rates of PCB deposition in areas of high siltation, and on the extent to which PCB's may be lost from the system by volatilization (Horn *et al.*, 1979) and by bacterial decomposition.

Mancini *et al.* (in press) have estimated the mass of PCB's discharged to the lower Hudson and to the Hudson-Raritan estuary from sources other than upper Hudson discharge. Their data suggest that major sources are sewage treatment plant discharges (200-1,000 kg/yr) and atmospheric fallout (~1,200 kg/yr). Urban runoff and raw sewage by-passes may account for an additional 165 kg/yr. Added together, sources below the Green Island Dam contribute nearly as much PCB to the harbor/Bay system as the upper Hudson (~2,500 kg/yr vs. ~3,100 kg/yr from the upper Hudson). This has not been considered in ongoing studies of PCB's in the Hudson system.

Concentrations of PCB's in sediments throughout the Bight region and in sewage sludge (Table 5) range from ~0.4 ng/g to 9 μg/g, dry wt. PCB concentrations in the vicinity of the sewage sludge dumpsite and dredged materials dumpsite, and in the Christiaensen Basin ranged from 0.4 μg/g to 2 μg/g, dry wt. Values at the Sandy Hook transect and in the outer Bight were 0.9 ng/g and 0.4 ng/g, dry wt, respectively. Estimates of PCB's in sewage sludge (West and Hatcher, 1980; O'Connor *et al.*, in press a) were 3 μg/g and 9 μg/g, dry wt.

The mass load of PCB's to the Bight from ocean dumping of dredged materials and sewage sludge may be estimated from the average PCB content of sediments sampled from the harbor and Bay, and the average PCB concentration of sewage sludge. Data on PCB concentrations in the New York harbor region are still being gathered. Recent data from MacLeod *et al.*, (1981) and results from unpublished survey work conducted for the New York District, U.S. Army Engineers (N. Rubinstein, U.S. EPA Gulf Breeze Environmental Research Laboratory, personal communication) show PCB concentrations in sediments from the New York harbor region ranging from about 3 μg/g (Newark Bay) to 0.001 μg/g, dry wt (Lower Bay). The average PCB content of sediments from the

Table 5. Total PCB concentrations (μg/g, dry wt) in sediments from the Hudson River and New York Bight region, and in sewage sludge.

Material Location or number	Total PCB's (μg/g dry wt)
Sediment	
Hudson River (km 240)[a]	60
Hudson River (km 15)[b]	0.5
Hudson River (N.Y. harbor)[a]	6
Pierhead Channel[b]	1
Gowanus Canal[b]	2
Newtown Creek[b]	0.4
Raritan Bay[c]	0.2
Sandy Hook transect[d]	0.0009
Christiaensen Basin[b]	0.7
Dredged materials dumpsite[d]	0.4
Sewage sludge dumpsite[d]	2
Outer Bight[b]	0.0004
Sewage sludge	
Sample 1[b]	9
Sample 2[d]	3

[a]Data from Horn *et al.* (1979).

[b]Data from O'Connor *et al.* (in press a).

[c]Data from Stainken and Rollwagen (1979).

[d]Data from West *et al.* (1976).

New York metropolitan region was estimated as 0.9 μg PCB per gram dry sediment, on the basis of data from Stainken and Rollwagen (1979), Bopp (1979), MacLeod *et al.*, (1981), and various unpublished sources. Using this estimate, the calculated annual PCB loading to the Bight from dredged materials is 3,500 kg. For sewage sludge, values range from ~600 kg to >2,000 kg, depending upon PCB concentrations in sludge (Table 6). PCB values in sediments may differ depending upon the source of sediments (Table 5). PCB estimates in sewage sludge (Table 5) vary by a factor of three; estimates for Bight sediments may vary by a factor of 6,000 (Bopp, 1979; MacLeod *et al.*, 1981). At the extreme of possible ranges for calculated values, PCB inputs to the Bight from sewage sludge and dredged materials dumping may overlap. Actual differences in loading factors can only be determined from analysis of PCB levels in materials dumped into the system. The sources of dredged sediments and sewage sludge are the critical factors determining the magnitude of annual PCB loadings to the Bight apex from ocean dumping.

Sinks

The major identifiable inputs for PCB's to the Bight region are, in order of importance: Hudson River sediments, sewage sludge, the atmosphere, and New York harbor sediments. Data available at present are insufficient to determine

Table 6. Calculated annual mass loads of total PCB's dumped in the New York Bight apex in materials dredged from New York harbor and Lower Bay, and in sewage sludge. Volumes of dredged materials for 1974-1978 from D. Suszkowski (personal communication). Volumes of sludge dumped for 1974-1978 from P. Anderson (personal communication).

Source	Mean volume dumped, 1974-1978	Annual PCB load (kg)
Dredged materials[a]	7,845 x 10^3 m^3	3,500
Sewage sludge[b]	4,047 x 10^6 kg (wet)	600
Sewage sludge[c]	4,047 x 10^6 kg (wet)	2,000

[a]Mean PCB concentration of 1.68 μg/g, dry wt; solids content ~50% (MacLeod *et al.*,1981).
[b]Mean PCB concentration of 3 μg/g, dry wt; solids content ~ 5% (West *et al.*, 1976).
[c]Mean PCB concentration of 9 μg/g, dry wt; solids content ~ 5% (O'Connor *et al.*, in press a).

the mobility of PCB's in any of these compartments; however, discharge from the upper Hudson, in combination with atmospheric fallout and combined stormwater and wastewater discharges from large urban areas (*e.g.*, Albany, Poughkeepsie, Kingston, and Newburgh) probably represent the major contributors of PCB's to the New York Bight. Further research on PCB's in stormwater runoff, municipal wastewater, atmospheric fallout, and suspended particulate matter in the tidal Hudson will be required to quantify the PCB input to the Bight from these diverse PCB sources, and to estimate the extent to which the mobility of PCB's in the compartments is affected by seasonally variable phenomena such as rainfall, tidal exchange, sedimentation, and fallout.

Distribution

Heavy PCB contamination of biota in the Bight region is confined primarily to the tidal Hudson River and Hudson River estuary (Table 7). Data on PCB's in Hudson River fishes have been summarized in Nadeau and Davis (1976), Spagnoli and Skinner (1977), Hetling *et al.* (1978), Horn *et al.* (1979), and Thomann and St. John (1979). PCB data for a variety of planktonic organisms, bivalves, and various crustacea are available in Spagnoli and Skinner (1977), Stainken and Rollwagen (1979), O'Brien and Gere Engineers, Inc. (1979), and O'Connor *et al.* (in press b). Data presented here are confined to species for which comparative PCB data are available from several points in the Hudson River/Hudson River estuary/New York Bight region.

Striped bass (*Morone saxatilis*) within the tidal Hudson River contained the greatest total PCB concentration, ranging from 9μg/g to 16 μg/g, wet wt. Striped bass from open ocean environments contained less than 1 μg/g, wet wt. Open ocean fish were not classified as to river of origin; no conclusion can be made as to past history of PCB exposure in either the Hudson or other coastal systems.

Flesh of the winter flounder (*Pseudopleuronectes americanus*) contained ⩽0.1 μg/g PCB's, wet wt, regardless of the location of capture of the fish (Table

Table 7. PCB concentrations (μg/g, wet wt) in flesh of fishes and shellfishes from the Hudson River and New York Bight region. Data given as total PCB's or as component PCB classes. N.D. indicates Aroclor not detected; * indicates Aroclor not quantified; and TR indicates trace amounts present, but not quantifiable by techniques used.

Species (Location)	Aroclor 1016	Aroclor 1254	Total PCB
Striped bass[a] *(Morone saxatilis)* (km 248; Hudson River)	N.D.	15	15
Striped bass[b] *(M. saxatilis)* km 122; Hudson River)	N.D.	16	16
Striped bass[b] *(M. saxatilis)* (km 44; Hudson River)	N.D.	9	9
Striped bass[c] *(M. saxatilis)* (km 30; Hudson River)	*	*	12
Striped bass[c] *(M. saxatilis)* (Montauk Point)	*	*	0.6
Striped bass[a] *(M. saxatilis)* ("Atlantic Ocean")	N.D.	0.6	0.6
Winter flounder[c] *(Pseudopleuronectes americanus)* (Raritan Bay)	*	*	0.07
Winter flounder[c] *(P. americanus)* (Christiaensen Basin)	*	*	0.1
Winter flounder[a] *(P. americanus)* ("Atlantic Ocean")	N.D.	0.09	0.09
Winter flounder[a] *(P. americanus)* ("Atlantic Ocean")	N.D.	0.1	0.1
Mackerel[a] *(Scomber scombrus)* ("Atlantic Ocean")	0.5	N.D.	0.5
Mackerel[c] *(S. scombrus)* (Bight apex)	*	*	0.3
Lobster[c] *(Homarus americanus)* (Raritan Bay)	*	*	0.3
Lobster[e] *(H. americanus)* (Dredged materials dumpsite)	*	*	0.2
Lobster[c] *(H. americanus)* (Ambrose)	*	*	0.03
Blue mussel[a] *(Mytilus edulis)* ("Atlantic Ocean")	N.D.	TR	–

Table 7. Continued.

Species (Location)	Aroclor 1016	Aroclor 1254	Total PCB
Blue mussel[c] *(M. edulis)* (Sandy Hook)	*	*	0.2
Blue mussel[c] *(M. edulis)* (Coney Island)	*	*	0.4
Hard clam[a] *(Mercenaria mercenaria)* ("Atlantic Ocean")	N.D.	N.D.	0.2
Hard clam[d] *(M. mercenaria)* (Raritan Bay)	*	*	0.1

[a]Spagnoli and Skinner (1977).
[b]Hetling *et al.* (1978).
[c]O'Connor *et al.* (in press b).
[d]Stainken and Rollwagen (1979).
[e]O'Brien and Gere Engineers, Inc. (1979).

7). For Atlantic mackerel (*Scomber scombrus*), American lobster (*Homarus americanus*), blue mussels (*Mytilus edulis*), and hard clams (*Mercenaria mercenaria*), PCB concentrations in the flesh were no greater than 0.5 μg/g, wet wt.

No consistent relationship could be established, even for sedentary species, between PCB concentrations in flesh and the site of capture. The rather consistent values for PCB's in all organisms except striped bass suggest that PCB body burdens are related primarily to direct water uptake at rates dependent upon the partition coefficient of PCB's between water and flesh (Neely *et al.*, 1974; Chiou *et al.*, 1977) not supplemented with additional uptake via the food chain. Recent data on the tissue distribution of PCB's in fishes and lobsters (Cahn *et al.*, 1977; O'Connor *et al.*, in press b) demonstrate that species such as striped bass, winter flounder, bluefin tuna (*Thunnus thynnus*), and lobsters may have vastly elevated levels of PCB's in the liver or hepatopancreas despite the presence of low levels in the flesh. This may reflect exposure to PCB's in the diet; because of selective tissue accumulation, such exposure may not be reflected by PCB concentrations in the muscle.

POLYNUCLEAR AROMATIC HYDROCARBONS

The polynuclear aromatic hydrocarbons derive primarily from petroleum and petroleum combustion products, from combustion of coal, wood, and coke, and as leachates from slag piles of coal and coke ash. O'Connor and Stanford (1979) identified PAH's as compounds of concern in the Bight. Ciaccio (1979) summarized the characteristics of PAH's found in crude oil and provided a general discussion of pertinent data regarding the terminology of PAH's and related substances. PAH's are of significance primarily because many PAH compounds or their metabolites are potent carcinogens (Dipple, 1976; U.S. Environmental Protection Agency, 1980b).

LaFlamme and Hites (1978) reviewed the global distribution of polycyclic aromatic hydrocarbons in recent sediments and determined that qualitative patterns are similar for most regions studied. PAH abundances increased in proximity to urban centers and were related primarily to combustion products.

The ecological and environmental health impacts associated with PAH's in natural systems, including toxicity, bioaccumulation potential, carcinogenicity, and environmental persistence, are correlated with numbers of rings and substitutions (Dipple, 1976; Roesijadi *et al.,* 1978; U.S. Environmental Protection Agency, 1980b; Anderson, this volume). In general, it has been shown that, although PAH's may be rapidly taken up by marine organisms (Lee *et al.,* 1972; Lee, 1977; Lee and Anderson, 1977; Roesijadi *et al.,* 1978), tissue clearance rates resulting from excretion or metabolism may be rapid (Lee and Anderson, 1977; Malins, 1977).

Sources

Several authors have reported on abundance and distribution of hydrocarbons in the New York Bight region. For the most part, these data relate to extractable organics and nonvolatile hydrocarbons, with little attention paid to qualitative analysis for specific PAH's. Monaghan *et al.* (1974) determined that Bight surface waters contained more petroleum hydrocarbons than did water from a depth of 10 m. Tanacredi (1977) and Searl *et al.* (1977) studied hydrocarbons in Jamaica Bay and New York harbor, respectively, and determined that extractable organics were more concentrated near sewage treatment plant effluents than in waters associated with docking areas and shipping channels. Tanacredi's data (1977) showed a wide range of values for total extractable hydrocarbons from treated effluents discharged into the Jamaica Bay region. Values ranged from 6.07 mg/liter discharged by the Rockaway Treatment plant to 21.8 mg/liter discharged in effluents from the 26th Ward Treatment Plant. Hydrocarbon values within various New York harbor regions were, in all cases, higher than those at the Sandy Hook/Rockaway transect (Searl *et al,* 1977). Based upon correlation of UV-fluorescence emission spectra and GC profiles between water samples, effluent samples, and "waste crankcase oil," Tanacredi (1977) concluded that appreciable quantities of hydrocarbons in the Jamaica Bay system are attributable to waste automotive petroleum products.

O'Connor *et al.* (in press a,b) reported concentrations of 15 PAH's in sediments and biota from the New York harbor region. Data for selected compounds and total PAH's detected are given in Table 8. PAH's were most concentrated in the sediments of Newtown Creek, Gowanus Canal, and Lower

Table 8. Selected PAH and total PAH concentrations (ng/g, dry wt) in sediments from the Hudson-Raritan estuary and New York Bight region, and in sewage sludge. Total PAH is the total polynuclear aromatic hydrocarbons detected as target compounds by MacLeod *et al.* (in press) (see O'Connor *et al.,* in press a for additional details). N.D. indicates compound not detected.

Material Location	Naphthalene	Phenanthrene	Anthracene	Benz (a) anthracene	Total PAH
SEDIMENT					
Hudson-Raritan estuary					
15 km north of the Battery	60	120	60	330	2,000
Pierhead Channel	200	300	200	500	3,200
Gowanus Canal	100	1,000	500	3,000	16,400
Newtown Creek	120,000	14,600	9,600	5,600	182,000
Lower Bay	100	600	300	2,000	9,900
New York Bight region					
Christiaensen Basin	800	500	300	1,000	6,000
Sewage sludge dumpsite	80	70	40	200	1,100
Outer Bight	0.6	3	N.D.	3	22
SEWAGE SLUDGE	2,200	4,400	1,100	1,100	20,400

Bay, and in sewage sludge. PAH values for sediments from the Hudson River above New York harbor, the sewage sludge dumpsite, and the outer Bight were lower. These trends in PAH concentration reflect closely the trends in extractable organics data presented by Searl *et al.* (1977).

The available data, though heterogeneous as to method of detection and concentrations reported, suggest that PAH's in the New York harbor region derive from three major sources: sewage treatment plant effluents, combined sewer overflows, and oil spillage in the vicinity of major shipping areas (Gibson *et al.,* 1979). Unfortunately, the distribution of sampling sites is inadequate to document precisely the PAH's in the harbor. Tanacredi (1977) and O'Connor *et al.* (in press a) provide evidence that a significant fraction of the hydrocarbons in the harbor region derive from automotive petroleum products. Further data on sources are unavailable; more study is needed on the distribution and abundance of unsubstituted and alkyl-substituted compounds before probable sources can be assigned for PAH's in the metropolitan region (LaFlamme and Hites, 1978).

Dumping of sewage sludge and dredged materials into the Bight apex represents a significant source of PAH's to the Bight ecosystem (Table 9). PAH's associated with dredged materials were estimated at an annual mean load of 167,000 kg; PAH's in sewage sludge were estimated at an annual mean load of 4,100 kg, based on the average volumes of dredged materials and sewage sludge dumpted at sea from 1974 through 1978.

Sinks

Major sinks for PAH's exist in the sediments of Lower Bay as well as in the sediments of open water areas and docks and slips of New York harbor and Raritan Bay. Although not thoroughly investigated, preliminary data show that PAH burdens in the sediments of oil tanker lightering areas may be high (E. Sutro, New Jersey Department of Environmental Protection, personal communication). Concentrations of PAH's in the water column cannot be estimated because data are lacking. Although not directly equatable to PAH concentrations, data on concentrations of extractable hydrocarbons may provide insight into likely trends of PAH concentrations in the water column. Tanacredi (1977) and Searl *et al.* (1977) have shown that surface waters of the harbor region contain concentrations of extractable hydrocarbons ranging from approximately 70 μg/liter at the entrance to the Lower Bay to ~180 μg/liter in the Kill van Kull/Arthur Kill region. In Jamaica Bay, values in surface waters ranged from ~500 μg/liter to more than 2,000 μg/liter.

Sediments from the Hudson River contained relatively low concentrations of PAH's, in comparison to sewage sludge and sediments from New York harbor and Lower Bay. It is unlikely, therefore, that the heavy burdens of PAH in the harbor region are derived from Hudson River flows. Oil and grease data from wastewater mass load estimates (Mueller *et al.*, 1976) showed that materials discharged to the East River and Upper Bay together are more than double the oil and grease input to the Hudson (100.7 metric tons/day vs. 44.2 metric tons/day). If one assumes that oil and grease values are in some fashion related to PAH's, then it can be concluded that discharges from the metropolitan region exceed those from the Hudson River. Furthermore, wastewater loads do not account for stormwater runoff or combined sewer overflow loads. Oil and grease loads in combined sewer overflows to Newtown Creek during storm events, for example, have been measured as high as 5,000 mg/liter (Interstate Sanitation Commission, 1973). The quantities of oil discharged in lightering tankers and discharged from ships during normal operations in ports or at anchor have not been estimated.

In addition to petroleum-related sources of PAH's, the PAH's derived from fallout on the metropolitan region contain a variety of hydrocarbons (Mueller *et al.*, 1976) and PAH's, *e.g.*, benzo(a)pyrene and chrysene (Daisey *et al.*, 1979), which are deposited and eventually discharged into surface waters by surface runoff. Estimates of mass loads present in the atmospheric sink are not available presently.

Table 9. Calculated annual mass load of total PAH to the New York Bight in materials to be dredged and in sewage sludge. Data from O'Connor *et al.* (in press a). Volumes of dredged materials for 1974-1978 from D. Suszkowski (personal communication). Volumes of sludge dumped for 1974-1978 from P. Anderson (personal communication).

Source	Mean volume dumped, 1974-1978	Annual PAH load (kg)
Dredged materials[a]	7,845 x 10^3 m^3 (wet)	167,000
Sewage sludge[b]	4,047 x 10^6 kg (wet)	4,100

[a]Mean total PAH concentration of 43 μg/g, dry wt; solids content ~ 50%.
[b]PAH concentration of 20 μg/g, dry wt.

Distribution

PAH concentrations in biota of the Bight region were highest in blue mussels (*Mytilus edulis*) and lobsters (*Homarus americanus*) and lowest in fishes (Table 10). At this point, correlations between water column or sediment concentrations of PAH's and tissue concentrations in organisms cannot be established because the data are insufficient. O'Connor *et al.* (in press b) noted that in lobsters and winter flounder (*Pseudopleuronectes americanus*), PAH's in the hepatopancreas and liver, respectively, were higher than in the edible flesh. Fishes and crustaceans take PAH's up rapidly but also are able to clear the compounds rapidly (Lee and Anderson, 1977; Roesijadi *et al.*, 1978).

CONCLUSIONS AND IMPLICATIONS

The Hudson River, Hudson-Raritan estuary, and New York Bight contain substantial quantities of chlorinated pesticides, PCB's, and PAH's. The sources vary for each class of compound. For pesticides, as exemplifed by dieldrin and DDT, the major source to the estuary is Hudson River flow. The major source to the Bight is from sewage sludge and dredged material dumping operations. A major proportion of the pesticides derives from use in the urban metropolitan region.

PCB's enter the Hudson River estuary primarily by downstream transport from industrial and municipal sources, as well as from stormwater runoff. Other than upstream industrial sources of PCB's, the major source is atmospheric fallout. Major PCB sinks are harbor sediments and the atmosphere. Dumping of dredged materials contributes large quantities of PCB's to the Bight apex each year.

The most important source of PAH's to the system is the New York metropolitan region. Oil spillage, sewer overflows, municipal wastewater discharges, and atmospheric fallout contribute PAH's to harbor sediments. Dumping of dredge spoils in the Bight apex contributes most of the PAH's to the coastal Bight ecosystem. Of the three classes of compounds considered in this

Table 10. Selected PAH and total PAH concentrations (ng/g, wet wt) in flesh of various fishes and shellfishes from the Hudson River and New York Bight region. Total PAH is the total polynuclear aromatic hydrocarbons detected as target compounds by MacLeod *et al.* (1981) (see O'Connor *et al.*, in press b for additional details). Sample locations are as indicated. N.D. indicates compound not detected.

Species (location)	Naphthalene	Phenanthrene	Anthracene	Biphenyl	Total PAH
Atlantic mackerel *(Scomber scombrus)* (New York Bight apex)	N.D.	10	N.D.	N.D.	10
Winter flounder *(Pseudopleuronectes americanus)* (Christianesen Basin)	2	N.D.	N.D.	6	8
Winter flounder *(P. americanus)* (Raritan Bay)	2	1	N.D.	N.D.	5
Striped bass *(Morone saxatilis)* (Montauk Point)	7	N.D.	N.D.	N.D.	19
Striped bass *(M. saxatilis)* (Hudson River)	4	N.D.	N.D.	4	8
Lobster *(Homarus americanus)* (New York Bight)	7	N.D.	N.D.	N.D.	7
Lobster *(H. americanus)* (Raritan Bay)	5	5	N.D.	N.D.	25
Lobster *(H. americanus)* (Raritan Bay)	7	N.D.	N.D.	N.D.	77
Blue mussel *(Mytilus edulis)* (Sandy Hook)	6	6	N.D.	4	250
Blue mussel *(M. edulis)* (Shark River)	20	10	1	40	120

paper, the PAH's represent the majority of organic contaminants discharged to the Bight apex each year.

The biota of the Bight region accumulate pesticides, PCB's, and PAH's in significant quantities. The only instances of contaminant accumulation above

FDA action limits for edible flesh occurred for PCB's in striped bass from the Hudson River. The ecological impact of organic contaminants in the Hudson estuary and Bight ecosystem has not been determined.

Special significance should be attached to the following factors related to organic contamination in the Bight region:

1. For PCB's, sources other than the upper Hudson River are of great significance in determining sediment loads of PCB's, rate and magnitude of PCB accumulation in biota, and mass transport of PCB's to the harbor, the Lower Bay, and the Bight. Plans for mitigating the Hudson River PCB problem by dredging upriver "hot spots" should be reassessed with regard to the probable beneficial impact on downstream sediments and fisheries.

2. Significant sources of chlorinated pesticides exist in the New York metropolitan region and, presumably, in other urban areas of the drainage basin. It is likely that expansion and upgrading of sewage treatment facilities in the New York metropolitan region will result in a reduction of direct discharges of chlorinated pesticides to New York harbor, Raritan Bay, and Lower Bay. Eventually, such reductions in discharge from combined sewer systems will result in lower concentrations of DDT and other pesticides in the sediments of New York harbor. However, in order to bring about substantial reductions in pesticide loadings to the Bight ecosystem, it would be necessary to cease ocean dumping of sewage sludge and dredged materials. Whether this is a realistic option, considering the relatively small quantities of contaminants in question, the virtual absence of upland disposal sites, and the essential nature of the Port of New York, can be decided only with much more research and much more public debate on the problem.

3. The majority of organic contaminants in the Bight region are PAH's derived from anthropogenic sources in the urban metropolitan region. Control of oil spills and treatment of stormwater runoff could reduce significantly the input of PAH's to the system.

4. Contamination of biota with pesticides, PCB's, and PAH's is essentially ubiquitous in the system, but is at relatively low levels. Whatever the effects of these materials, they represent only a portion of the overall impact on estuarine and coastal marine ecosystems caused by other activities such as dredging, filling, shipping, shoreline development, nutrient enrichment, and overfishing. At present, the Federal government and agencies in virtually every coastal state are funding studies to determine how the various factors affect living marine resources.

ACKNOWLEDGEMENTS

This work was performed as part of the Marine EcoSystems Analysis (MESA) New York Bight Project (NOAA Contract 04-78-B01-16) and was supported in part by a grant from the New York State Department of Environmental Conservation PCB Estuary Study, and the National Institutes of Environmental Health Sciences Grant ES00260 to the New York University Medical Center,

Institute of Environmental Sciences. We thank Dr. W. MacLeod and Mr. L. Scott Ramos for contribution of data and comments on interpretation. Other original data were provided by Drs. L.S. Peters and K. Salamon. The skills of Ms. Eleanor Clemm and Ms. Toni Moore in preparing many drafts of the manuscript and tables are gratefully acknowledged.

REFERENCES

Abood, K.A. 1974. Circulation in the Hudson estuary. Ann. N.Y. Acad. Sci. 250: 29-111.

Anderson, J.W. This volume. The transport of petroleum hydrocarbons from sediments to benthos and the potential effects. pp. 165-179.

Bopp, R.F. 1979. The geochemistry of polychlorinated biphenyls in the Hudson River. Ph.D. Dissertation, Columbia University. 191 pp.

Brown, D.W., A.J. Friedman, D.G. Burrows, G.R. Snyder, B.G. Patten, W.E. Ames, L.S. Ramos, P.G. Prohaska, D.G. Gennaro, D.D. Dungan, M.Y. Uyeda, and W.D. MacLeod, Jr. 1979. Investigation of petroleum in the marine environs of the Strait of Juan de Fuca and northern Puget Sound. EPA 600/7-79-164. U.S. Office of Energy, Minerals and Industry, Office of Research and Development, U.S. Environmental Protection Agency, Washington, DC. 107 pp.

Cahn, P.H., J. Foehrenbach, and W. Guggino. 1977. PCB levels in certain organs of some feral fish from New York State. *In:* Physiological Responses of Marine Biota to Pollutants, F.J. Vernberg, A. Calabrese, F.P. Thurberg, and W.B. Vernberg (eds.), Academic Press, New York. pp. 51-61.

Chiou, C.T., V. Freed, D. Schmedding, and R. Kohnert. 1977. Partition coefficient and bioaccumulation of selected organic chemicals. Environ. Sci. Technol. 11:475-478.

Ciaccio, L.L. 1979. Polyaromatic hydrocarbons in petroleum crudes, petroleum products and ocean water. *In:* Chemical Pollutants of the New York Bight; Priorities for Research, J.S. O'Connor and H.M. Stanford (eds.), NOAA Special Report, pp. 151-159.

Daisey, J.M., M.A. Leyko, M.T. Kleinman, and E. Hoffman. 1979. The nature of the organic fraction of the New York City summer aerosol. *In:* The New York Summer Aerosol Study, 1976, T.J. Kneip and M. Lippmann (eds.), Ann. N.Y. Acad. Sci. 322: 125-141.

Dipple, A. 1976. Polynuclear aromatic hydrocarbons. *In:* Chemical Carcinogens, C.E. Searle (ed.), ACS Monograph 173, American Chemical Society, Washington, DC. pp. 245-314.

Gibson, C.I., L. Ciaccio, B. Ketchum, and A. Michael. 1979. Petroleum hydrocarbons subpanel report. *In:* Chemical Pollutants of the New York Bight; Priorities for Research, J.S. O'Connor and H.M. Stanford (eds.), NOAA Special Report. pp. 20-32.

Goldberg, E.D., V.T. Bowen, J.W. Farrington, G. Harvey, J.H. Martin, P.L. Parker, R.W. Risebrough, W. Robertson, E. Schneider, and E. Gamble. 1978. The mussel watch. Environ. Conserv. 5: 101-125.

Hetling, L.J., E.G. Horn, and T.J. Tofflemire. 1978. Summary of Hudson River PCB study results. New York State Department of Environmental Conservation, Tech. Rep. 51. 88 pp.

Hetling, L.J., T.J. Tofflemire, E.G. Horn, R. Thomas, and R. Mt. Pleasant. 1979. The Hudson River PCB problem: management alternatives. *In:* Health Effects of Halogenated Aromatic Hydrocarbons, W.J. Nicholson and J.A. Moore (eds.), Ann. N.Y. Acad. Sci. 320: 630-650.

Hiraizumi, Y., M. Takahashi, and M. Nishimura. 1979. Adsorption of polychlorinated biphenyl onto seabed sediment, marine plankton and other adsorbing agents. Environ. Sci. Technol. 13: 580-584.

Horn, E.G., L.J. Hetling, and T.J. Tofflemire. 1979. The problem of PCB's in the Hudson River system. *In:* Health Effects of Halogenated Aromatic Hydrocarbons, W.J. Nicholson and J.A. Moore (eds.), Ann. N.Y. Acad. Sci. 320: 591-609.

Interstate Sanitation Commission. 1973. Combined sewer overflow study for the Hudson River Conference. EPA-RZ-73-152. Office of Research and Monitoring, U.S. Environmental Protection Agency, Washington, DC. 287 pp.

Interstate Sanitation Commission. 1977. Report of the Interstate Sanitation Commission on the Water Pollution Control Activities and the Interstate Air Pollution Program. 1977. Interstate Sanitation Commission, New York. 136 pp.

Jensen, S. 1966. Report of a new chemical hazard. New Sci. 32: 612.

Kneip, T.J. and J.A. Hernandez. 1969. Pesticide residues in the Hudson River and biota. *In:* Development of a Biological Monitoring System and a Survey of Trace Metals, Radionuclides and Pesticides in the Lower Hudson River, M. Eisenbud and G.P. Howells (eds.), Report to New York State Department of Health. New York University Institute of Environmental Medicine, Tuxedo Park, NY. pp. 124-162.

LaFlamme, R.E. and R.A. Hites. 1978. The global distribution of polycyclic aromatic hydrocarbons in recent sediments. Geochim. Geophys. Acta 42: 289-303.

Lawler, Matusky, and Skelly Engineers. 1978. Upper Hudson River PCB No Action Alternative Study. Final Report to New York State Department of Environmental Conservation. Lawler, Matusky, and Skelly Engineers, Pearl River, NY.

Lee, R.F. 1977. Fate of petroleum components in estuarine waters of the southeastern United States. *In:* Conference on Prevention and Control of Oil Pollution, American Petroleum Institute, Washington, DC. pp. 86-102.

Lee, R.F. and J.W. Anderson. 1977. Fate and effect of napthalenes; Controlled Ecosystem Pollution Experiment. Bull. Mar. Sci. 21: 127.

Lee, R.F., R. Sauerherber, and J.B. Bobbs. 1972. Uptake, metabolism and discharge of polycyclic aromatic hydrocarbons by marine fish. Mar. Biol. 17: 201-208.

McDermott, D.J., D.R. Young, and T.C. Heesen. 1976. PCB contamination of southern California marine organisms. *In:* EPA, 1976, National Conference on Polychlorinated Biphenyls. EPA 560/6-75-004. U.S. Environmental Protection Agency, Washington DC. pp. 209-217.

MacLeod, W.D., Jr., D.W. Brown, R. Jenkins, L.S. Ramos, and V. Henry. 1977. A pilot study on the design of a petroleum hydrocarbon baseline investigation for northern Puget Sound and Strait of Juan de Fuca. NOAA Tech. Memo. ERL MESA-8. 53 pp.

MacLeod, W.D., Jr., L.S. Ramos, A.J. Friedman, D.G. Burrows, P.G. Prohaska, D.L. Fisher, and D.W. Brown. 1981. Analysis of residual chlorinated hydrocarbons, aromatic hydrocarbons and related compounds in selected sources, sinks and biota of New York Bight. NOAA Tech. Memo. OMPA-6. 128 pp.

Malins, D.C. 1977. Metabolism of aromatic hydrocarbons in marine organisms. *In:* Aquatic Pollutants and Biologic Effects with Emphasis on Neoplasia, H.F. Kraybill, C. Dawe, J. Harshbarger, and R. Tardiff (eds.), Ann. N.Y. Acad. Sci. 298: 482-496.

Mancini, J.L., T.W. Bober, J.C. Goldman, J.A. Mueller, E.P. Myers, A.I. Mytelka, and J.R. Roswell. In press. Urban loads to the New York Bight. NOAA Special Report.

Monaghan, P.H., D. Brandon, R. Brown, T. Searl, and J.J. Elliot. 1974. Measurements in Atlantic, Mediterranean, Gulf of Mexico and Persian Gulf. *In:* Measurement and Interpretation of Non-volatile Hydrocarbons in the Ocean. Rep. AID.IDJ.B. 74 EPR. 4EX.4, Maritime Administration, U.S. Department of Commerce, Washington, DC. 73 pp.

Mueller, J.A., J.S. Jeris, A.R. Anderson, and C.F. Hughes. 1976. Contaminant inputs to the New York Bight. NOAA Tech. Memo. ERL MESA-6. 347 pp.

Munson, T.O., H. Palmer, and J. Forns. 1976. Transport of chlorinated hydrocarbons in the upper Chesapeake Bay. *In:* EPA, 1976, National Conference on Polychlorinated Biphenyls. EPA 560/6-75-004. U.S. Environmental Protection Agency. Washington, DC. pp. 218-229.

Nadeau, R.J. and R.A. Davis. 1976. Polychlorinated biphenyls in the Hudson River (Hudson Falls-Fort Edward) New York. Bull. Environ. Contam. Toxicol. 16: 436-444.

Neely, W.B., D. Branson, and G. Blau. 1974. Partition coefficient to measure bioconcentration potential of organic chemicals in fish. Environ. Sci. Technol. 8: 1113-1115.

O'Brien and Gere Engineers, Inc. 1979. Bioaccumulation Studies on *Homarus americanus.* Report to U.S. Army Engineers District. O'Brien and Gere Engineers, Inc., Syracuse, NY. 52 pp.

O'Connor, J.M., T.J. Kneip, R.A. Greig, F.P. Thurberg, D.M. Goodrich, and H.M. Stanford. In press a. Organic contaminants in the New York Bight ecosystem. Environ. Pollut.

O'Connor, J.M., T.J. Kneip, R.A. Greig, F.P. Thurberg, D.M. Goodrich, and H.M. Stanford. In press b. Organic contaminants in biota of the New York Bight. Environ. Pollut.

O'Connor, J.S. and H.M. Stanford (eds.), 1979. Chemical Pollutants of the New York Bight; Priorities for Research. NOAA Special Report. 217 pp.

O'Connors, H.B., C.F. Wurster, C.D. Powers, D.C. Biggs, and R.G. Rowland. 1979. Polychlorinated biphenyls may alter marine trophic pathways by reducing phytoplankton size and production. Science 201: 737-739.

Pavlou, S.P., K. Krogslund, R. Dexter, and J. Clayton. 1973. Data report. R/V ONAR cruise 434, 450, 469, 502. Synthetic organics in Puget Sound; hydrographic, chemical and biological measurements. Spec. Rep. 54, Department of Oceanography, University of Washington, Seattle, WA. 178 pp.

Peters, L.S. and J.M. O'Connor. This volume. Factors affecting short-term PCB and DDT accumulation by zooplankton and fish from the Hudson estuary. pp. 451-465.

Risebrough, R.W. 1971. Chlorinated hydrocarbons. *In:* Impingement of Man on the Oceans, D.W. Hood (ed.), Wiley Interscience, New York. pp. 259-286.

Roesijadi, G., J.W. Anderson, and J. Blaylock. 1978. Uptake of hydrocarbons from marine sediments contaminated by Prudhoe Bay crude oil: influence of feeding type of test species and availability of polycyclic aromatic hydrocarbons. J.Fish. Res. Board Can. 35: 608-614.

Searl, T.D., H.L. Huffman, and J.P. Thomas. 1977. Extractable organics and nonvolatile hydrocarbons in the New York harbor waters. Proceedings of Oil Spill Conference, American Petroleum Institute, Washington, DC. pp. 583-588.

Southern California Coastal Water Research Project. 1977. Annual Report of the Southern California Coastal Water Research Project, Southern California Coastal Water Research Project, El Segundo, CA. 253 pp.

Sherwood, M.J., A. Mearns, D. Young, B. McCain, R.A. Murchelano, G. Alexander, T. Heesen, and Tsu-Kai Jan. 1978. A Comparison of Trace Contaminants in Diseased Fishes from Three Areas. Report to NOAA MESA New York Bight Project. Southern California Coastal Water Research Project. El Segundo, CA. 116 pp.

Spagnoli, J. and L. Skinner. 1977. PCB's in fish from selected waters of New York State. Pestic. Monit. J. 11: 69-87.

Stainken, D. and J. Rollwagen. 1979. PCB residues in bivalves and sediments of Raritan Bay. Bull. Environ. Contam. Toxicol. 23: 690-697.

Tanacredi, J.T. 1977. Petroleum hydrocarbons from effluents: detection in marine environment. J. Water Pollut. Control Fed. 49: 216-226.

Thomann, R.V. and J. St. John. 1979. The fate of PCB's in the Hudson River ecosystem. *In:* Health Effects of Halogenated Aromatic Hydrocarbons, W.J. Nicholson and J.A. Moore (eds.), Ann. N.Y. Acad. Sci. 320: 610-629.

U.S. Environmental Protection Agency. 1980a. Ambient Water Quality Criteria for Polychlorinated Biphenyls. EPA 440/5-80-068. Office of Water Regulation and Standards, U.S. Environmental Protection Agency, Washington, DC. 191 pp.

U.S. Environmental Protection Agency. 1980b. Proposed Ambient Water Quality Criteria for Polynuclear Aromatic Hydrocarbons. EPA 440/5-80-069. Office of Water Regulation and Standards, U.S. Environmental Protection Agency, Washington, DC. 155 pp.

U.S. Food and Drug Administration. 1978. Criteria and Standards for Contaminants in Foodstuffs. U.S. Food and Drug Administration, Washington, DC. 4 pp.

Wasserman, M., D. Wassermann, S. Cucos, and H. Miller. 1979. World PCB's map: storage and effects in man and his biologic environment in the 1970's. *In:* Health Effects of Halogenated Aromatic Hydrocarbons, W.J. Nicholson and J.A. Moore (eds.), Ann. N.Y. Acad. Sci. 320: 69-124.

West, R.H. and P.G. Hatcher. 1980. Polychlorinated biphenyls in sewage sludge and sediments of the New York Bight. Mar. Pollut. Bull. 11: 126-129.

West, R.H., P.G. Hatcher, and D.K. Atwood. 1976. Polychlorinated Biphenyls and DDT's in Sediments and Sewage Sludge of the New York Bight. Final Report to NOAA MESA New York Bight Project. NOAA Atlantic Oceanographic and Meteorological Laboratory, Miami, FL. 40 pp.

Young, D., G. Bowes, P. Moskowitz, and C. Wurster. 1979. Halogenated hydrocarbons sub-panel report. *In:* Chemical Pollutants of the New York Bight: Priorities for Research. J.S. O'Connor and H.M. Stanford (eds.), NOAA Special Report. pp. 33-43.

PERSPECTIVES ON METALS IN NEW YORK BIGHT ORGANISMS: FACTORS CONTROLLING ACCUMULATION AND BODY BURDENS

Joseph M. O'Connor

New York University Medical Center
Institute of Environmental Medicine
Sterling Forest
Tuxedo Park, New York 10987

Joseph W. Rachlin

Department of Biological Sciences
Lehman College of CUNY
Bedford Park Boulevard
Bronx, New York 10468

Abstract. Despite the vast array of data on metals bioconcentration factors (BCF's) and levels of toxic metals in marine organisms, the data generally are not applicable to impact assessment. This is because adequate data are not available for defining metal species in seawater and marine sediments, and because laboratory-derived data for BCF's and toxic effects generally relate to metals species not measured in field studies.

Metals partitioning from food and sediment and the potential for metals regulation in marine organisms were evaluated with the use of selected data from the New York Bight and other coastal systems. It was concluded from these data that marine organisms regulate essential and nonessential metals, although regulation by various species differed. Metals regulation is accomplished by several mechanisms, ranging from selective uptake and selective elimination to sequestration by metal-binding proteins. A general scheme for internal metals regulation is provided, along with recommendations for future field and laboratory research.

INTRODUCTION

Coastal marine ecosystems such as the New York Bight receive large quantities of metals derived from a variety of sources. These include waterborne metals from the weathering of soils and bedrock, deposition of metals from the atmosphere, and waste discharges from domestic and industrial effluents (Gross, 1970; Stumm and Morgan, 1970; Sherk, 1972; Mueller *et al.,* 1976; Beijer and Jernelov, 1979; Goldberg, 1979). The transport and fate of metals are, in general, controlled by sorption processes in deposited and suspended sediments, as well as by reactions with organic material in the water column (Stumm and Morgan, 1970; Callahan *et al.,* 1979).

While it is safe to assume that the waters of the Bight and the Bight apex, like other coastal systems, are in a steady state with metals derived from crustal materials (National Academy of Sciences, 1975), alteration of the steady state by man may contaminate the system and degrade the quality of living marine resources (Gross, 1970; National Academy of Sciences, 1975; Hall *et al.,* 1978; Carpenter *et al.,* in press). The metals balance of sediments, biota, and the water column has been altered by a variety of human activities (Carmody *et al.,* 1973; Segar and Cantillo, 1976; Greig *et al.,* 1977a,b; Klinkhammer, 1977; Hall *et al.,* 1978). Mueller *et al.* (1976) estimated loading rates of metals to the New York

Bight and Bight apex from various sources and determined that a large portion of the potentially toxic cadmium (Cd), copper (Cu), lead (Pb), mercury (Hg), and zinc (Zn) comes from ocean dumping of dredged materials and sewage sludge. Three of these metals (lead, mercury, and cadmium) have been cited as threats to the Bight ecosystem (O'Connor and Stanford, 1979). Mercury contamination is a problem within the Hackensack Meadowlands of New Jersey, which drain into the Raritan River/Raritan Bay system (Lipsky *et al.*, 1980); Kneip and O'Connor (1980) have identified potentially hazardous levels of cadmium in blue crabs (*Callinectes sapidus*) from the Hudson estuary. While instances of increased concentrations of toxic metals have been detected in shellfishes from the New York Bight apex (surf clam, *Spisula solidissima;* oyster, *Crassostrea virginica;* Greig *et al.*, 1977b; Wenzloff *et al.*, 1979; Carpenter *et al.*, in press), the overall significance of increased levels of metals in organisms in the Bight has not been determined (Hall *et al.*, 1978; Saila and Segar, 1979).

Hall *et al.* (1978) determined that a trend toward increased metals concentrations exists for finfishes and shellfishes from the inner Bight and western Long Island Sound. Whether this trend will continue and eventually manifest itself as an ecological problem or as a public health concern has been the subject of several evaluative and summary documents (Gross, 1970; Goldberg, 1979; O'Connor and Stanford, 1979; Carpenter *et al.*, in press; O'Connor, in press). This paper examines aspects of metals contamination in marine organisms and points out relationships between environmental metals concentrations and levels of metal in tissues from various marine plants and animals. The thesis presented by this paper is that metals accumulation in marine organisms reflects organismal requirements for trace metals; adaptations at the biochemical level make it possible for organisms to acclimatize to increased environmental and tissue levels of essential and nonessential metals. This may occur without significant changes in survival, growth, or reproduction provided that environmental metal concentrations remain below incipient lethal concentrations for the population.

The data base for this paper is derived from many laboratories and was developed over a period of at least ten years of sampling and analysis. During this time, procedures for environmental metals analysis underwent a rapid evolution aimed at improving methods of extraction, recognizing sources of errors, and producing greater standardization among laboratories. The data used in calculations for this paper have not been screened for quality according to present criteria, and individual data points are not considered to be representative of absolute values. In fact, calculations are independent of most analytical errors, other than extraction efficiency. The indices, ratios, and factors cited are considered to be representative of trends likely to exist in marine organisms and in the marine environment.

ACCUMULATION AND CONCENTRATION OF METALS

Body burdens of metals in aquatic organisms are the result of direct uptake from water, uptake from food, and, especially in the case of algae, surface sorption resulting from physical adsorption, partitioning to cellular material, or chemisorption by ligands (Davies, 1978; Hoover, 1978; Coombs, 1979; see also review by Neff *et al.,* 1978). Metals burdens in different tissues may depend upon multiple routes of uptake and distribution (Hazen, 1981). In this discussion the accumulation of metals will be treated only with regard to total concentrations in the medium. Despite the oft-mentioned need for data on chemical species of metals in marine systems and their influence on accumulation and effects (Lowman *et al.,* 1971; Hoover, 1978; Neff *et al.,* 1978; Callahan *et al.,* 1979; Allen *et al.,* 1980; Carpenter *et al.,* in press), such data are unlikely to be widely available soon. Thus, it is necessary to examine metal concentrations in marine organisms through bioconcentration and bioaccumulation from total metals available in environmental compartments.

Bioconcentration and bioaccumulation factors are simple ratios that reflect the partitioning of metals to biota and should not be confused with rates of uptake. Here "bioconcentration" is treated as the concentration of a metal in an organism relative to the total concentration of the same metal in water. The bioconcentration factor (BCF) implies that body burdens of metals in an organism are determined by a constant that reflects the steady state between metals uptake and elimination. BCF's from environmental data preclude statements regarding rates and routes of metal uptake; however, laboratory studies with metals indicate that uptake rates from water approximate a logarithmic function related to volume or weight/volume ratio of the organism (Fromm and Stokes, 1962; Zubarik and O'Connor, 1978; Amiard and Amiard-Triquet, 1979; Sato *et al.,* 1980).

Bioaccumulation has been defined as the uptake of a contaminant from all sources (*i.e.,* direct uptake from water, from food, and from other metals sources available to the organism). Recently, some authors have approached metals uptake in shellfishes as a special case, determined possibly by the metal concentration of the sediment (Atwood *et al.,* 1979), and have calculated partition coefficients (K) based on the ratio of shellfish body burden to sedimentary metal concentration. Such an approach may be simplistic, but it does reflect more closely the relationship between metals sources and benthic organisms than does the BCF.

Many publications have summarized metal BCF values for marine organisms. Gathered here (Tables 1-5) are BCF data for the three metals of greatest concern in the Bight and Bight apex (cadmium, mercury, and lead; O'Connor and Stanford, 1979) as well as data for copper and zinc, two metals that represent essential trace elements for marine organisms and that also are potential toxicants.

BCF values obtained from the literature and calculated from available data are highly variable; they do not lend themselves readily to interpretation because of variation in BCF's among species tested. This point alone suggests that BCF data must be used with caution and must not be applied to problems of contamination in the marine environment without recognizing the limitations of the data. However, if BCF data are viewed as possible indicators of physiological processes associated with uptake, some inferences can be drawn on the bioaccumulation of individual metals. The first is that within groups of marine organisms, the nonessential metals (cadmium, mercury, and lead) have higher potentials for maximum bioconcentration than essential metals (copper and zinc). Whether this potential for bioconcentration of nonessential metals can be assigned any significance with regard to ecological impact cannot be ascertained. It is consistent with the notion that organisms have developed mechanisms to regulate essential metals, and may not regulate nonessential metals to the same degree.

The value of the BCF as an indicator of ecological impact is questionable. No systematic analysis of metals bioaccumulation and effects has been undertaken; instead, most authors have concluded that bioconcentration and bioaccumulation imply effects at the ecosystem level. Not only is bioconcentration not a necessary condition for ecological impact of metals, but also it has been argued that accumulation and effect of metals on marine organisms may be inversely

Table 1. Bioconcentration factors (BCF's) for cadmium in marine organisms. The data are from a variety of sources; only those BCF values calculated from Greig *et al.* (1977a) and Hall *et al.* (1978) are specific to New York Bight organisms.

Biological group	BCF
Algae	11-20[1]
Phytoplankton	350-6,000[1]
Marine plants	1,000[2]
Algae	670[3]
Herbivorous zooplankton	80-10,000[1]
Predaceous zooplankton	300-10,000[1]
Zooplankton	3,800[4]
Benthic crustacea	37-700[3]
Marine invertebrates	250,000[2]
Shellfishes	100,000-2,000,000[1]
Crassostrea virginica	149-2,600[3]
Mya arenaria	160[3]
M. arenaria	150[6]
Mytilus edulis	45[3]
Scallop	168[3]
Mercenaria mercenaria	109[3]
Fishes	10[1]
Fundulus heteroclitus	48[5]
Marine fishes	3,000[2]
Fishes	120-1,000[7]

[1] Bowen *et al.*, 1971.
[2] Callahan *et al.*, 1979.
[3] U.S. Environmental Protection Agency, 1980a.
[4] Calculated from Greig *et al.*, 1977a.
[5] Eisler, 1971.
[6] Calculated from Eisler, 1977.
[7] Calculated from Hall *et al.*, 1978.

related (National Research Council Ocean Affairs Board, 1971). It is necessary to consider only three examples to illustrate this point. Among the algae, recent research (see review in Davies, 1978; Allen *et al.,* 1980) has shown that metals uptake is related to simple ion exchange phenomena at the cell surface. However, the surface charge on algae and the presence of certain ionizing groups (*e.g.,* glycoproteins, polysaccharides) can affect metals binding and diffusion into the cell. The net charge on the cell surface can be related to pH, salinity, stage of growth, and, possibly, nutritional state of the organism (Hayward, 1969; Cossa, 1976; Davies, 1978; Allen *et al.,* 1980; Jensen and Rachlin, personal communication). Thus, for essential metals as well as for nonessential metals, BCF values in algae may reflect the organisms' ability to concentrate the element relative to variables that are either uncontrolled or, as yet, unknown. Similarly, the presence of metals in algae may indicate past history of the availability of metal binding sites in proteins (Motohashi and Tsuchida, 1974), events that are at least partially independent of metals contamination in the environment.

Benthic organisms also present strong evidence against the use of BCF as an indicator of pollution. Blue crabs (*Callinectes sapidus*) from the Hudson River estuary, for example, show much higher cadmium BCF's in the mainstem Hudson than in certain cadmium-contaminated backwaters, such as Foundry Cove (Kneip and Hazen, 1979; Kneip and O'Connor, 1980). This results from

Table 2. Bioconcentration factors (BCF's) for mercury in marine organisms.

	Biological group	BCF
Inorganic mercury	Marine plants	1,000[1]
	Algae	800-7,400[2]
	Algae	49,000-1,100,000[3]
	Marine invertebrates	100,000[1]
	Copepods	14,000-230,000[3]
	Amphipods	1,300-1,500[3]
	Crangon septemspinosa	18,000[3]
	Homarus americanus	129[2]
	Crassostrea virginica	2,800-10,000[2]
	Fishes	1,670[1]
	Morone saxatilis larvae	1,300-7,600[3]
	Fishes	6,000-200,000[4]
Organic mercury	Algae	12,000-390,000[3]
	Copepods	9,800-360,000[3]
	Amphipods	2,600-4,900[3]
	Salt water decapods	1,150[2]
	Salt water molluscs	30,000[2]
	C. virginica	40,000[2]
	M. saxatilis larvae	5,000-10,000[3]

[1] Callahan *et al.,* 1979.
[2] U.S. Environmental Protection Agency, 1980d.
[3] Zubarik and O'Connor, 1978.
[4] Calculated from Greig *et al.,* unpublished data and Segar and Cantillo, 1976.

the ability of crabs to store significant quantities of cadmium in the hepatopancreas, regulating tissue levels at a consistent concentration regardless of the quantities of cadmium available. Under such circumstances, BCF's appear larger for crabs in non-contaminated areas than in those that may be in highly contaminated areas. Erroneous conclusions regarding potential ecological effects easily could be drawn from such data.

Cutshall and co-workers (Cutshall *et al.*, 1977, 1978) showed that for marine fishes, body burdens of mercury, cadmium, and zinc are not necessarily related to metals concentrations in the water, but instead may be related to latitude, size of the organisms, or the metals content of the diet. Based upon their studies of mercury in pacific hake (*Merluccius productus*), Cutshall *et al.* (1978) concluded that, without species-specific data on the metabolism of metals and food habits, it is impossible to conclude that fishes are indicators of water pollution patterns.

The variability in BCF values for metals (Tables 1-5) and the realization that their value as an index of ecological impact is suspect would dictate that studies of metals distribution and potential effects in organisms might better be directed toward food chain studies and research on physiological factors affecting metals metabolism. The information in historical data files applicable to such an approach is inadequate; however, one might consider determining the ratio of

Table 3. Bioconcentration factors (BCF's) for lead in marine organisms.

Biological group	BCF
Marine plants	200[1]
Algae	900-1,000[2]
Algae (sessile)	8,000-20,000[3]
Phytoplankton	< 1,000-3,000,000[3]
Phytoplankton	40,000[4]
Herbivorous zooplankton	3,000-2,000,000[3]
Predaceous zooplankton	200-60,000[3]
Zooplankton	3,000[4]
Zooplankton	600,000[5]
Marine invertebrates	200[1]
Crassostrea virginica	68-1,400[2]
Mya arenaria	112[2]
Mytilus edulis	800-2,500[2]
Fishes	60[1]
Fishes	2,300[6]

[1] Callahan *et al.*, 1979.
[2] U.S. Environmental Protection Agency, 1980c.
[3] Bowen *et al.*, 1971.
[4] Lowman *et al.*, 1971.
[5] Calculated from Greig *et al.*, 1977a and Segar and Cantillo, 1976.
[6] Calculated from Greig *et al.*, unpublished data and Segar and Cantillo, 1976.

metals in specific organisms and their food. This could provide some insight on the magnitude of metals flow in a natural system. From average metals data for fishes, shellfishes, and zooplankton in the New York Bight region and from information on probable dietary constituents for the different species (Bigelow and Schroeder, 1953; Greig *et al.*, 1977a,b; Hall *et al.*, 1978), estimates have been made of the proportional concentrations of metals in various planktivorous and predatory fishes, and the metals concentrations in their food organisms (Table 6).

The proportion of metals in the organism to that in food may be viewed as a "partition coefficient" (K; Atwood *et al.*, 1979), which represents the existing concentration of metals in a consumer as determined by the concentration in food and several variables, including assimilation efficiency of the metal across the gut, total quantity of metal ingested, and rate of metals elimination. The approach used by Atwood *et al.* was specifically for cadmium in shellfishes. The same approach is valid for other organisms and their food resources, since similar principles of ingestion and assimilation apply whether the metal in question be cadmium, copper, lead, mercury, or zinc, and whether the organism be a finfish, mollusc, or crustacean.

Table 4. Bioconcentration factors (BCF's) for copper in marine organisms. The data are from a variety of sources; only those BCF values calculated from Greig *et al.* (1977a) are specific to New York Bight organisms.

Biological group	BCF
Phytoplankton	30,000[1]
Marine plants	1,000[2]
Algae	74-617[3]
Zooplankton	6,000[1]
Zooplankton	710-940[4]
Marine invertebrates	1,670[3]
Molluscs	30,000[2]
Mya arenaria	11,000[5]
M. arenaria	3,300[4]
Scallops	3,000-4,000[3]
Crassostrea virginica	20,000-28,000[3]
Mytilus edulis	90-208[3]
Polychaetes	200-2,500[3]
Fishes	667[2]
Fishes	80-3,700[6]

[1] Lowman *et al.*, 1971.
[2] Callahan *et al.*, 1979.
[3] U.S. Environmental Protection Agency, 1980b.
[4] Calculated from Greig *et al.*, 1977a.
[5] Calculated from Eisler, 1977.
[6] Calculated from Greig, unpublished data.

Table 5. Bioconcentration factors (BCF's) for zinc in marine organisms. The data are from a variety of sources; only those BCF values calculated from Greig *et al.* (1977a) and Hall *et al.* (1978) are specific to New York Bight organisms.

Biological group	BCF
Algae (sessile)	80-3,000[1]
Phytoplankton	200-1,300[1]
Phytoplankton	26,000[2]
Marine plants	80-2,500[3]
Algae	50,000[4]
Algae	1,800-10,000[5]
Herbivorous zooplankton	125-500[1]
Predaceous zooplankton	50[1]
Zooplankton	8,000[2]
Amphipods	400[5]
Zooplankton	6,500[7]
Molluscs	2,100-300,000[3]
Crassostrea virginica	24,000[4]
C. virginica	27,000[5]
Mytilus edulis	100-500[5]
Shellfishes	1,400-10,000[1]
Crustacea	1,700-15,000[3]
Carcinus maenas	5,000-9,000[5]
Cancer irroratus	4,000[6]
Homarus americanus	1,860[6]
Fishes	280-20,000[1]
Fishes	280-15,000[3]
Marine fishes	400-2,000[4]
Fishes	100-1,000[8]

[1] Bowen *et al.*, 1971.
[2] Lowman *et al.*, 1971.
[3] Jinks and Eisenbud, 1972.
[4] Callahan *et al.*, 1979.
[5] U.S. Environmental Protection Agency, 1980e.
[6] Calculated from Hall *et al.*, 1978.
[7] Calculated from Greig *et al.*, 1977a.
[8] Calculated from Wenzloff *et al.*, 1979.

The partition coefficients (K_f) estimated here demonstrate low K_f's for planktivorous species and generally greater K_f's for predatory species (Table 6). Except for concentrations of lead found in bluefish, partitioning of metals between food organisms and consumers was less than 1.0, which indicates some level of metals regulation, even for the nonessential metals. The apparent potential for metal regulation was highest among the planktivorous forms (alewife, shad, and Atlantic herring). Interestingly, the metals concentrations found in the planktivorous herrings by Hall *et al.* (1978) were, in some cases,

Table 6. Partition coefficients (K_f) between metals in food organisms and selected commercial fishes from the New York Bight region. Data for zooplankton, as food for alewife, herring, and shad, are from Greig *et al.* (1977b). All other data are from Hall *et al.* (1978).

Fish	Cd	Cu	Pb	Zn
Alewife (*Alosa pseudoharengus*)	0.05	0.04	0.04	0.005
Atlantic herring (*Clupea harengus harengus*)	0.07	0.06	0.05	0.007
American shad (*A. sapidissima*)	0.02	0.05	0.02	0.002
Bluefish (*Pomatomus saltatrix*)	0.56	0.19	1.11	0.73
Striped bass (*Morone saxtilis*)	0.89	0.20	0.78	0.22
Winter flounder (*Pseudopleuronectes americanus*)	0.33	0.05	0.59	0.29

significantly greater than in the larger, predatory species such as bluefish, striped bass, and winter flounder. This infers that regulation of metals among fishes may occur even when the body burden is relatively high.

Assuming that the partition coefficients are representative of steady-state conditions, one may derive from them initial estimates of regulatory potential. For cadmium and lead, both nonessential metals, regulatory potential is high among the planktivores and comparatively low among the predators. Although the predaceous species appear to regulate copper and zinc more effectively, partition coefficients for these metals are greater than in the planktivorous species. Bluefish and striped bass regulated cadmium with a ratio greater than 0.5 of that in presumed food organisms. For all three predaceous species, the proportion of lead is greater than 0.5 of that in the food organisms. While this is not necessarily suggestive of a contamination problem or rapid bioaccumulation, it may suggest that cadmium and lead would be the first to accumulate to harmful levels in predatory fishes.

Also calculated are sediment-organism partition coefficients (K_s) for benthic organisms from the New York Bight and Long Island Sound dumpsites (Greig *et al.*, 1977b; Wenzloff *et al.*, 1979), and for benthos from Gulf coast embayments (Guthrie *et al.*, 1979). At the dumpsites (Table 7), partition coefficients were below 0.5 for all species-metals comparisons and were lower for the bivalve molluscs than for the rock crab or the channeled whelk. All species, however, had metals concentrations significantly below those in the surrounding sediment, which suggests regulation of metals concentration by virtue of low uptake rates or rapid rates of removal, despite previous data indicating that bivalve molluscs tend to be poor regulators of heavy metals (Goldberg *et al.*, 1978).

The benthos from the Gulf coast (Table 8), though exposed to less contaminated sediments, showed a broader range of K_s's. In general, copper, cadmium, and mercury were regulated, but K_s for zinc in oysters, barnacles, and polychaetes showed accumulation to levels greater than background concentrations. It may be that the capacity of benthos to regulate metals increases with increased background either because of induction of metabolic metal regulatory

Table 7. Partition coefficients (K_S) between metals in sediments and benthic organisms from New York Bight and Long Island Sound dumpsites. Data are from Greig *et al.* (1977b) and Wenzloff *et al.* (1979).

Organism (site)	Cu	Pb	Zn
Rock crab, *Cancer irroratus* (Long Island Sound)	0.15	0.07	0.16
Rock crab, *C. irroratus* (New York Bight)	0.17	–	0.33
Surf clam, *Spisula solidissima* (New York Bight)	0.03	–	0.19
Ocean quahog, *Arctica islandica* (New York Bight)	0.06	0.01	0.09
Channeled whelk, *Busycon canaliculatum* (Long Island Sound)	0.22	0.01	0.19

processes or by adaptation in the population, which results in enhanced regulatory potential.

If benthic organisms are capable of regulating metals efficiently, partition coefficients should vary in response to changing environmental metals concentrations. Neff *et al.* (1978) found in laboratory studies that sediment metals concentrations did not correlate with metals in exposed organisms, which suggests that K_s was variable. This would enable an organism to maintain metals in the body at relatively constant levels, despite variable sources of metals in sediments. On the other hand, the geographic trend of metals in benthos on the Atlantic coast corresponds with a general trend in sediment metals loads, indicating that partition coefficients may be fixed for some organisms, and that increases in metals loads in the environment will result in increased metals in organisms (Atwood *et al.*, 1979; Wenzloff *et al.*, 1979).

These data point out the need for additional research in several critical areas. First, and most important, is the question of whether metals in food or sediments are the major routes of transport from the environment to organisms. When the relative importance of uptake from water and uptake from food is determined, environmental assessments may be made more easily, based upon direct estimates of metals available to fishes and benthos and their ability to regulate accumulation.

Second, it is essential to know how metals are stored in organisms and whether values for body burdens or concentrations can be correlated with the potential for environmental impact. Recent studies of cadmium uptake in fishes (Hazen, personal communication) demonstrate that cadmium distribution in organs differs depending upon whether the metal was accumulated directly from

Table 8. Partition coefficients (K_S) between metals in sediments and various benthic organisms from two Texas Bays. Data are from Guthrie *et al.* (1979).

Organism	Cd	Cu	Zn	Hg
Barnacle	0.63	0.10	24.18	0.57
Crab (*Callinectes sapidus*)	0.07	0.02	0.13	0.01
Clam (*Rangia*)	0.63	0.15	0.48	0.22
Polychaete (*Nereis*)	0.93	0.29	1.53	0.29
Oyster (*Crassostrea virginica*)	0.26	0.27	3.86	0.14

water or from food. If metals are stored in certain organs in organic matrices and rendered inactive biochemically, concentration may reflect regulation, not contamination. If metals exist in an active form able to affect metabolic processes, even low levels may be critical.

Third, it is essential to know whether partition coefficients between food organisms and fishes or between sediments and benthos are constant over a range of concentrations.

BIOCHEMISTRY OF METALS IN MARINE ORGANISMS

Earlier it was stated that the ability to predict acute and chronic effects of toxic metals on marine organisms was dependent upon better data regarding the speciation of metals in marine waters. Ionization, sorption, and chelation of metals in water and sediments all may affect availability for uptake by organisms and the eventual expression of environmental effect. Thus, estimates of bioconcentration factors published for various metals in various organisms partly reflect bioavailability. For the metals included in this discussion, BCF's range over four to five orders of magnitude.

Other significant factors in bioconcentration of metals operate within the organism and affect uptake from water and food, storage and/or excretion in the organism, and potential for acute or chronic effects at the level of the individual organism.

The primary site for physiological regulation of any metal includes the site of uptake (either the gill surface or the gut epithelium). No reports were found for the efficiency of metals assimilation across respiratory epithelia in marine organisms. Efficiency of uptake across the gut epithelium has been studied for cadmium in green crabs (*Carcinus maenas*) and was found to be approximately 10% (Jennings and Rainbow, 1979). This is essentially the same assimilation efficiency found for cadmium in mammals (Friberg *et al.*, 1979). Table 9 presents assimilation efficiencies for various metals in vertebrates. Given the similarity of enzymatic systems among organisms, it is reasonable to assume that respiratory and gut assimilation efficiencies at least would be similar in mammals and in marine organisms. From these data one may conclude that, of the nonessential metals, the major problem in contamination of marine organisms

Table 9. Percent of applied dose of metals absorbed by organisms exposed via inhalation and food.

Site of uptake	Cd[1]	Cu[2]	Pb[3]	Inorg. Hg[4]	Org. Hg[4]	Zn[5]
Resp. epithelium	25-50	–	30	–	80	10-90
Gut	5-10	50	10	10	90	10-90

[1] Friberg *et al.*, 1979; Jennings and Rainbow, 1979.
[2] Piscator, 1979; no data available for inhalation uptake.
[3] Tsuchiya, 1979.
[4] Berlin, 1979; no data available for inorganic (mercuric) uptake by inhalation.
[5] Elinder and Piscator, 1979; uptake variable and dependent upon many factors.

would be organic mercury. Cadmium, lead, and inorganic mercury all are absorbed with very low efficiencies, although studies have shown that the uptake for all of these metals may be affected by an organism's nutritional status (Camner *et al.*, 1979).

A variety of mathematical and conceptual models have been applied to uptake and loss kinetics of metals in marine organisms. In general, uptake/loss data approximate first order or quasi-first order kinetics descriptive of single-compartment models (Cutshall, 1974; Callahan *et al.*, 1979). However, such approaches shed little light on mechanisms of metals metabolism within the organism (*i.e.*, if metals are accumulated from the environment, do they necessarily exert a physiological impact that can affect the organism negatively?). In this regard, it is essential to recall, from an earlier section of this paper, the relationships established for regulation of metals by marine organisms. There it was noted that, despite the availability of substantial quantities of metals in food material, substantial regulation of metals may occur, even for nonessential metals such as cadmium, lead, and mercury.

Much attention currently is being given to the role of metalloproteins as regulators of metal metabolism and detoxification in all organisms. One class of metal binding proteins, the metallothioneins, has been identified in organisms from algae through mammals (Kojima and Kägi, 1978) and are thought to play a major role in the metabolism of cadmium, copper, mercury, zinc, and other metals. While the physiological function of metallothioneins is not fully understood, several facts are clear regarding their relationship to essential and nonessential metals at both physiologic and potentially toxic levels: (1) metallothioneins are capable of binding significant quantities of metals, primarily in the liver and kidney of vertebrates and in analogous organs of marine invertebrates (Kojima and Kägi, 1978; Ridlington and Fowler, 1979; Frankenne *et al.*, 1980); (2) levels of metallothioneins can be increased or induced by concentrations of metals in the environment (Kojima and Kägi,

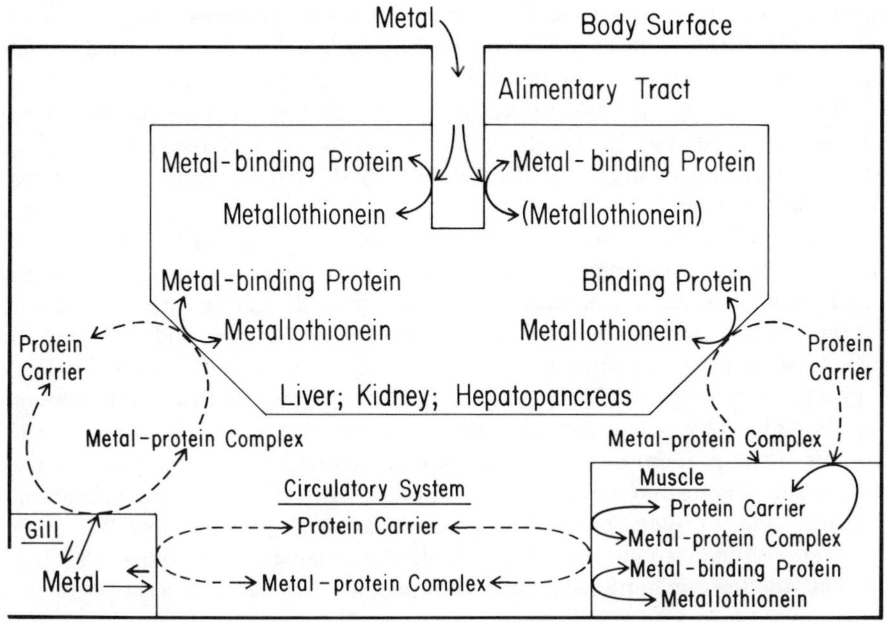

Figure 1. Schematic model for behavior of metals in marine organisms. Generalized and adapted from Jennings *et al.*, 1979.

1978; Jennings *et al.*, 1979; Olafson *et al.*, 1979); and (3) metallothioneins have been implicated in mechanisms of metals detoxification (Piscator, 1964; Jennings *et al.*, 1979; Vallee, 1979; Webb, 1979).

A general scheme for the way metal-binding proteins function in metabolism and detoxication has been put forth by Jennings *et al.* (1979). This fits well with available data on tissue distributions of metals in marine organisms and major sites for synthesis and activity of metallothioneins. A modified and generalized version of the Jennings model (Figure 1) is presented here for aquatic organisms.

In the model, transport across the gut is envisioned as diffusional; transport across the gill is envisioned as ligand-assisted diffusion, as proposed by Coombs (1979). This is consistent with observations that diffusion is the primary mode of metal entry into aquatic plants (Davies, 1978) and into aquatic invertebrates and fishes (Hazen, 1981). Metals assimilated by the gut are transported rapidly to the liver or analogous organs (*e.g.*, hepatopancreas) and become bound to metalloproteins (metallothionein or related compounds). In the liver, metalloprotein (*e.g.*, metallothionein) production is inducible so that, under most circumstances, a sufficient quantity becomes available for binding of any metals present. Once proteins are bound in a metal-metalloprotein complex, sequestration is dependent upon continual metalloprotein synthesis because metalloproteins, such as metallothioneins, have a relatively short half-life (7-14 days). As metalloprotein breakdown occurs, newly synthesized material binds the released metal and retains the metal in the tissue. If metals are excessive, they cannot be bound effectively and they enter the circulation, usually as ligand-bound complexes, and may be accumulated in muscle (Jennings *et al.*, 1979).

Metals crossing the gill surface enter the circulation, become bound to proteins in the blood, and are distributed to all tissues, including muscle, liver, kidney, or analogous organs. In liver and kidney tissues, the metals are bound to metallothioneins and sequestered.

Evidence that the concept proposed by Jennings *et al.* (1979) is accurate comes from an increasing body of field and laboratory data on metals distribution in tissues. The data demonstrate that all marine organisms tested contain metallothionein-like proteins in appropriate organs, and that typical metal distributions in marine organisms show higher metal levels in those organs typically containing metallothioneins or similar proteins. The gradual buildup of certain metals (*e.g.*, cadmium, mercury) in marine organisms with age may occur because metal-protein complexes may not be excreted from organs such as the crustacean hepatopancreas or fish kidney. With prolonged sequestration, redistribution of metal-protein complexes may take place as metalloprotein production is induced in other tissues, especially somatic muscle tissue. This may account for the apparent accumulation of mercury in large, long-lived predatory fishes, such as sharks and billfishes.

SUMMARY AND IMPLICATIONS

Despite the vast literature on the distribution of potentially toxic metals in coastal marine systems and estuaries, most authors agree that the available data cannot be applied to assessments of present or future ecological impact. Until more data become available on the species of metals in water and sediments, it will remain difficult to estimate acute or chronic metals effects on marine organisms, or the rate at which metals accumulate in marine food chains.

Bioconcentration factors for metals in marine organisms range over four to five orders of magnitude and differ according to the species of metals in the environment, the organism in question, and aspects of metals metabolism that cannot be ascertained from BCF data. Since the distribution of metals species in water is usually unknown, BCF data are essentially uninterpretable; laboratory studies add no information to the BCF, since most studies are carried out with ionic metal species and do not provide a realistic model for metals in natural systems.

Ratios of metals in organisms compared to their food or to sediments (partition coefficients) demonstrate that marine organisms regulate metals either at the site of uptake or by internal physiological mechanisms. Such partition coefficients are available from very few studies; however, data estimated for benthic invertebrates, planktivorous fishes, and predatory fishes suggest that the metals most likely to be accumulated to high levels are zinc (for benthic invertebrates), cadmium (for fishes and benthic invertebrates), and lead (for fishes).

Despite the existence of high partition coefficients for some metals in some organisms, metals accumulation alone cannot be assumed to represent an adverse ecological impact. Marine organisms employ a variety of strategies for sequestering and detoxifying metals, including the isolation of excess essential and nonessential metals in metallothionein and other metal-binding proteins. Under conditions of relatively low-level metal contamination, or with biological regulation of metals transport in the food web, binding and isolation of metals in nontoxic matrices may explain the apparently high levels of metals seen in some organisms, coupled with an apparent lack of ecological impact.

The apparent absence of ecological impact from metals in the New York Bight region is not suggestive of an unlimited capacity for the system to accept additional metals burdens. Calculated partition coefficients point out that zinc, cadmium, organic mercury, and lead may not be regulated well by marine organisms, and that excessive loads of these metals may lead to excessive accumulations, and possible acute or chronic effects.

If one assumes that rates of metals loading to the Bight region will not increase in future years, it is likely that organisms in the Bight will maintain their ability to regulate metals effectively and will suffer no negative impact from the metals present or introduced into the system. It should be noted, however, that different assumptions would lead to different conclusions. In localized regions where metals concentrations are at or near toxic levels, one cannot deny that impacts already are severe. However, factoring local effects into system-wide impact assessments requires great caution and extensive study employing techniques other than general surveys.

Surveys to determine metals in organisms, in water, and in sediments provide data useful for calculating metals budgets, estimating body burdens, and identifying species or locations in which concentrations of metals may accumulate. New research is required to relate environmental metals data to real or potential ecological impact. The most critical areas for new research are: (1) developing techniques for identifying the various forms of metals in sediments and in the water column; (2) determining the forms of metals stored in marine organisms; (3) determining the specificity of various forms of metals for sequestration in different tissues; and (4) establishing in the laboratory the metals-regulating capacity for finfishes and shellfishes so that critical tissue burdens may be established and monitored.

ACKNOWLEDGEMENTS

This report was prepared as part of a research contract from the U.S. Department of Commerce, NOAA/MESA New York Bight Project, No. 04-78-B01-16, and was supported in part by NIEHS Grant ES00260 to the New York University Medical Center, Institute of Environmental Medicine. Dr. Theo. J. Kneip provided much valuable input to the manuscript, as did Messrs. A. Wiedow and R. Hazen. Ms. Toni Moore prepared the manuscript.

REFERENCES

Allen, H.E., R.H. Hall, and T.D. Bresbin. 1980. Metal speciation. Effects on aquatic toxicity. Environ. Sci. Technol. 14: 441-443.

Amiard, J.C. and C. Amiard-Triquet. 1979. Distribution of cobalt-60 in a mollusc, a crustacean and a freshwater teleost: variations as a function of the source of pollution and during elimination. Environ. Pollut. 20: 199-213.

Atwood, D., D. Brown, V. Cabelli, J. Farrington, C. Carside. G. Han, D. Hansen, G. Harvey, K. Kamlet, J.S. O'Connor, L. Swanson, D. Swift, J. Thomas, J. Walsh, and T. Whitledge. 1979. The New York Bight. *In:* Assimilative Capacity of United States Coastal Waters for Pollutants, E.D. Goldberg (ed.), Proceedings of a Workshop, Working Paper No. 1, Federal Plan for Ocean Pollution Research Development and Monitoring, NOAA ERL, Boulder, CO. pp. 148-178.

Beijer, K. and A. Jernelov. 1979. Sources, transport and transformation of metals in the environment. *In:* Handbook on the Toxicology of Metals, L. Friberg, G. Nordberg, and V. Vouk (eds.), Elsevier, New York. pp. 47-64.

Berlin, M. 1979. Mercury. *In:* Handbook on the Toxicology of Metals, L. Friberg, G. Nordberg, and V. Vouk (eds.), Elsevier, New York. pp. 503-530.

Bigelow, H.B. and W.C. Schroeder. 1953. Fishes of the Gulf of Maine. U.S. Fish. Wildl. Serv. Fish Bull. 74. 577 pp.

Bowen, V.T., J.S. Olsen, C.L. Osterberg, and J. Ravera. 1971. Ecological interactions of marine radioactivity. *In:* Radioactivity in the Marine Environment, A.H. Seymour (ed.), National Academy of Sciences, Washington, DC. pp. 200-222.

Callahan, M.A., M. Slimak, N. Gabel, I. May, C. Fowler, J. Freed, P. Jennings, R. Durfie, F. Whitmore, B. Maestri, W. Mabey, B. Holt, and C. Gould. 1979. Water Related Environmental Fate of 129 Priority Pollutants, Vol. I. EPA 440/4-79-029B. U.S. Environmental Protection Agency, Office of Water Planning and Standards, Washington, DC. 514 pp.

Camner, P., T.W. Clarkson, and G.F. Nordberg. 1979. Routes of exposure, dose and metabolism of metals. *In:* Handbook on the Toxicology of Metals, L. Friberg, G. Nordberg, and V. Vouk (eds.), Elsevier, New York. pp. 65-98.

Carmody, D.J., J.B. Pearce, and W.E. Yasso. 1973. Trace metals in sediments of New York Bight. Mar. Pollut. Bull. 4: 132-135.

Carpenter, J.H., P.W. Anderson, A. Calabrese, K.S. Kamlet, T.J. Kneip, and A. Michael. In press. Implications of Current and Possible Future Contaminant Levels in the Apex Area of the New York Bight. NOAA Special Report.

Coombs, Th. L. 1979. Cadmium in aquatic organisms. *In:* The Chemistry, Biochemistry and Biology of Cadmium. Topics in Environmental Health, Vol. 2, M. Webb (ed.), Elsevier, New York. pp. 93-141.

Cossa, D. 1976. Sorption du cadmium par une population de la diatomée *Phaeodactylum tricornutum* en culture. Mar. Biol. 34: 163-167.

Cutshall, N. 1974. Turnover of zinc-65 in oysters. Health Phys. 26: 327-331.

Cutshall, N.H., J.R. Naidu, and W.G. Pearcy. 1977. Zinc and cadmium in the Pacific hake *Merluccius productus* off the western U.S. coast. Mar. Biol. 44: 195-201.

Cutshall, N.H., J.R. Naidu, and W.G. Pearcy. 1978. Mercury concentrations in Pacific hake *Merluccius productus* (Ayres), as a function of length and latitude. Science 200: 1489-1491.

Davies, A.G. 1978. Pollution studies with marine plankton. Part II. Heavy metals. Adv. Mar. Biol. 15: 381-508.

Eisler, R. 1971. Cadmium poisoning in *Fundulus heteroclitus* (Pisces, Cyprinodontidae) and other marine organisms. J. Fish. Res. Board Can. 28: 1225-1234.

Eisler, R. 1977. Acute toxicities of selected heavy metals on the softshell clam, *Mya arenaria.* Bull. Environ. Contam. Toxicol. 17: 137-144.

Elinder, C.G. and M. Piscator. 1979. Zinc. *In:* Handbook on the Toxicology of Metals, L. Friberg, G. Nordberg, and V. Vouk (eds.), Elsevier, New York. pp. 675-686.

Frankenne, F., F. Noel-Lambot, and A. Disteche. 1980. Isolation and characterization of metallothioneins from cadmium-loaded mussel, *Mytilus edulis.* Comp. Biochem. Physiol. 66(C): 179-182.

Friberg, L., T. Kjellstrom, G. Nordberg, and M. Piscator. 1979. Cadmium. *In:* Handbook on the Toxicology of Metals, L. Friberg, G. Nordberg, and V. Vouk (eds.), Elsevier, New York. pp. 355-382.

Fromm, P.O. and R.M. Stokes. 1962. Assimilation and metabolism of chromium by trout. J. Water Pollut. Control Fed. 1962: 1151-1156.

Goldberg, E.D. 1979. Assimilative Capacity of U.S. Coastal Waters for Pollutants. Working Paper #1, Federal Plan for Ocean Pollution. U.S. Department of Commerce, NOAA ERL, Boulder, CO. 284 pp.

Goldberg, E.D., V.T. Bowen, J.W. Farrington, G. Harvey, J.H. Martin, P.L. Parker, R.W. Risebrough, W. Robertson, E. Schneider, and E. Gamble. 1978. The mussel watch. Environ. Conserv. 5: 101-125.

Greig, R.A., A. Adams, and D.R. Wenzloff. 1977a. Trace metal content of plankton and zooplankton from the New York Bight and Long Island Sound. Bull. Environ. Contam. Toxicol. 18: 3-8.

Greig, R.A., D.R. Wenzloff, A. Adams, B. Nelson, and C. Shelpuk. 1977b. Trace metals in organisms from ocean disposal sites of the middle-eastern United States. Arch. Environ. Contam. Toxicol. 6: 395-409.

Gross, M.G. 1970. Preliminary analyses of urban wastes, New York metropolitan region. Marine Sciences Research Center, State University of New York, Stony Brook, Tech. Rep. 7. 13 pp.

Guthrie, R.K., E.M. Davis, D.S. Cherry, and H.E. Murray. 1979. Biomagnification of heavy metals by organisms in a marine microcosm. Bull. Environ. Contam. Toxicol. 21: 53-61.

Hall, R.G., G.M. Meaburn, S. Chang, R. Greig, and J.B. Pearce. 1978. Retrieval and Interpretation of Data on Trace Metals in Fish and Shellfish from N.Y. Bight. Rep. to NOAA MESA New York Bight Program. 52 pp.

Hayward, J. 1969. Studies on the growth of *Phaeodactylum tricornutum.* V. The relationship to iron, manganese and zinc. J. Mar. Biol. Assoc. U.K. 49: 439-446.

Hazen, R. 1981. Cadmium in the Foundry Cove ecosystem. Ph.D. Thesis, New York University. 188 pp.

Hoover, T.B. 1978. Inorganic Species in Water: Ecological Significance and Analytical Needs. U.S. Environmental Protection Agency 600/3-78-064, ERL, Athens, GA. 100 pp.

Jennings, J.R. and P.S. Rainbow. 1979. Studies on the uptake of cadmium by the crab *Carcinus maenas* in the laboratory. I. Accumulation from seawater and a food-source. Mar. Biol. 50: 131-139.

Jennings, J.R., P.S. Rainbow, and A.G. Scott. 1979. Studies on the uptake of cadmium by the crab *Carcinus maenas* in the laboratory. II. Preliminary investigation of cadmium binding proteins. Mar. Biol. 50: 141-149.

Jinks, S. and M. Eisenbud. 1972. Concentration factors in the marine environment. Radiation Data and Reports, U.S. Environmental Protection Agency, Office of Radiation Programs 13: 243-247.

Klinkhammer, G. 1977. The distribution and partitioning of some trace metals in the Hudson River estuary. M.S. Thesis, University of Rhode Island Graduate School of Oceanography. 125 pp.

Kneip, T.J. and R.E. Hazen. 1979. Cadmium in an Aquatic Ecosystem. Final

Report to National Science Foundation from New York University Medical Center, Institute of Environmental Medicine, Tuxedo Park, NY. 217 pp.

Kneip, T.J. and J.M. O'Connor. 1980. Cadmium in Foundry Cove Crabs: Health Hazard Assessment. Final Report to Health Research Council, New York State Health Planning Commission, Albany. NY. 20 pp.

Kojima, Y. and H.R. Kägi. 1978. Metallothionein. Trends Biochem. Sci. 3: 90-93.

Lipsky, D., R.J. Reed, and R. Harkov. 1980. Mercury Levels in Berry's Creek. State of New Jersey, Department of Environmental Protection, Trenton, NJ. 25 pp.

Lowman, F.G., T. Rice, and F. Richards. 1971. Accumulation and redistribution of radionuclides by marine organisms. *In:* Radioactivity in the Marine Environment, A.H. Seymour (ed.), National Academy of Sciences, Washington, DC. pp. 161-199.

Motohashi, K. and T. Tsuchida. 1974. Uptake of cadmium by pure cultured diatom *Skeletonema costatum.* Bull. Plankton Soc. Japan 21: 55-59.

Mueller, J.A., J.S. Jeris, A.R. Anderson, and C.F. Hughes. 1976. Contaminant inputs to the New York Bight. NOAA Tech. Memo. ERL MESA-6. 347 pp.

National Academy of Sciences. 1975. Assessing Potential Ocean Pollutants. Report of the Panel on Assessing Potential Ocean Pollutants. U.S. National Academy of Sciences, Washington, DC. 438 pp.

National Research Council Ocean Affairs Board. 1971. Marine Environmental Quality—Suggested Research Programs for Understanding Man's Effect on the Oceans. National Academy of Sciences, Washington, DC. 107 pp.

Neff, J.W., R.S. Foster, and J.F. Sloney. 1978. Availability of sediment-adsorbed heavy metals to benthos with particular emphasis on deposit-feeding infauna. Tech. Rep. D-78-42, U.S. Army Engineer Waterways Experiment Station, Vicksburg, MS. 286 pp.

O'Connor, J.M. In press. Environmental Health: Chemical Contaminants. Monograph 32, New York Bight Atlas, New York Sea Grant Institute, Albany, NY.

O'Connor, J.S. and H. M. Stanford. 1979. Chemical Pollutants of the New York Bight; Priorities for Research. NOAA Special Report. 217 pp.

Olafson, R.W., A. Kearns, and R.G. Sim. 1979. Heavy metal induction of metallothionein synthesis in the hepatopancreas of the crab *Scylla serrata.* Comp. Biochem. Physiol. 62(B): 417-424.

Piscator, M. 1964. Cd in the kidneys of normal beings and the isolation of metallothionein from the liver of rabbits exposed to Cd. Nord. Hyg. Tidshr. 45: 76-82.

Piscator, M. 1979. Copper. *In:* Handbook on the Toxicology of Metals, L. Friberg, G. Nordberg, and V. Vouk (eds.), Elsevier, New York. pp. 411-420.

Ridlington, J.W. and B.A. Fowler. 1979. Isolation and partial characterization of a cadmium binding protein from the American oyster (*Crassostrea virginica*). Chem.-Biol. Interactions 25: 127-138.

Saila, S. and D. Segar. 1979. Metals sub-panel report. *In:* Chemical Pollutants of the New York Bight; Priorities for Research, J.S. O'Connor and H.M. Stanford (eds.), NOAA Special Report. pp. 10-19.

Sato, T., Y. Ose, and T. Sakai. 1980. Toxicological effect of selenium on fish. Environ. Pollut. (series A) 21: 217-224.

Segar, D. and A. Cantillo. 1976. Trace metals in the New York Bight. *In:* Middle Atlantic Continental Shelf and the New York Bight, M.G. Gross (ed.), Am. Soc. Limnol. Oceanogr. Spec. Symp. 2: 171-198.

Sherk, J.A., Jr. 1972. Current status of the knowledge of the biological effects of suspended and deposited sediments in Chesapeake Bay. Chesapeake Sci. 13: 5137-5144.

Stumm, W. and J.J. Morgan. 1970. Aquatic Chemistry. Wiley Interscience, New York. 583 pp.

Tsuchiya, K. 1979. Lead. *In:* Handbook on the Toxicology of Metals, L. Friberg, G. Nordberg, and V. Vouk (eds.), Elsevier, New York. pp. 451-484.

U.S. Environmental Protection Agency. 1980a. Ambient Water Quality Criteria for Cadmium. EPA 440/5-80-025. EPA Office of Water Regulation and Standards, Washington, DC.

U.S. Environmental Protection Agency. 1980b. Ambient Water Quality Criteria for Copper. EPA 440/5-80-036. EPA Office of Water Quality Regulation and Standards, Washington, DC.

U.S. Environmental Protection Agency. 1980c. Ambient Water Quality Criteria for Lead. EPA 440/5-80-057. EPA Office of Water Quality Regulation and Standards, Washington, DC.

U.S. Environmental Protection Agency. 1980d. Ambient Water Quality Criteria for Mercury. EPA 440/5-80-058. EPA Office of Water Regulation and Standards, Washington, DC.

U.S. Environmental Protection Agency. 1980e. Ambient Water Quality Criteria for Zinc. EPA 440/5-80-079. EPA Office of Water Quality Regulation and Standards, Washington, DC.

Vallee, B.L. 1979. Metallothionein: historical review and perspectives. *In:* Metallothionein: Proceedings of 1st International Meeting on Metallothionein and Other Low Molecular Weight Metal Binding Proteins, J.H.R. Kägi and M. Nordberg (eds.), Kirkhauser Verlag, Boston. pp. 19-41.

Webb, M. 1979. Functions of hepatic and renal metallothioneins in the control of the metabolism of cadmium and certain other bivalent cations. *In:* Proceedings of 1st International Meeting on Metallothionein and Other Low Molecular Weight Metal Binding Proteins, J.H.R. Kägi and M. Nordberg (eds.), Kirkhauser Verlag, Boston. pp. 313-321.

Wenzloff, D.R., R.A. Greig, A.S. Merril, and J.W. Rope. 1979. A survey of heavy metals in the surf clam, *Spisula solidissima* and the ocean quahog, *Arctica islandica* of the mid-Atlantic coast of the United States. Fish. Bull. 77: 280-285.

Zubarik, L.S. and J.M. O'Connor. 1978. A radioisotopic study of mercury uptake by Hudson River biota. *In:* Energy and Environmental Stress in Aquatic Ecosystems, J.H. Thorpe and J.W. Gibbons (eds.), DOE Symposium Series 48: 273-289.

CONSEQUENCES OF INDUSTRIAL WASTE DISPOSAL AT THE 106-MILE OCEAN WASTE DISPOSAL SITE

Thomas P. O'Connor[1]
P. Kilho Park[2]

National Oceanic and Atmospheric Administration
National Ocean Survey
Ocean Dumping and Monitoring Division
Rockville, Maryland 20852

Abstract. Industrial wastes are dumped beyond the edge of the continental shelf at the 106-Mile Ocean Waste Disposal Site. These wastes are mainly liquids that, in some cases, form a precipitate floc after mixing with seawater. Rather than descending to the sea floor, wastes are distributed primarily as plumes within the upper mixed layer. The method of dumping yields a plume with a maximum initial concentration of about 200 ppm waste. Within about four hours, the maximum concentration is reduced to approximately 10 ppm with a waste plume width of about 1 km. Further dilution and plume growth proceed slowly, leading to the conclusion that storm events may be necessary to dilute extensively the waste. The average flow through the dumpsite seems to be sufficient, when compared to the frequency of dumping, for each dump to be an independent event so that later dumps do not occur in previously created plumes. It is estimated that, on the average, the plumes move toward the southwest along the contour of the continental shelf.

Laboratory studies have shown that oceanic strains of phytoplankton are less tolerant of waste than coastal strains of the same species, and that there is considerable variation in response to waste among species. Although no tested phytoplankton species have been found to be affected at 10 ppm waste, it is possible that within a waste plume there will be a readjustment of phytoplankton communities favoring more resilient species. Tests of this are proceeding, as are tests on the sublethal responses of zooplankton to wastes, histopathological and chemical analyses of field-collected organisms, studies of the stability of waste-derived organic compounds, and studies of the partitioning of waste constituents between dissolved and particulate phases.

INTRODUCTION

The 106-Mile Ocean Waste Disposal Site, formerly Deep Water Dumpsite 106 (38°40'N to 39°00'N and 72°00'W to 72°30'W), is 196 km (106 nmi) southeast of the entrance to New York harbor and about 185 km east of Cape May, New Jersey (Figure 1). It is an area of approximately 1,400 km^2, beyond the edge of the continental shelf, overlying depths of 1,800-2,700 m. Since 1972 it has been a site for regulated disposal of industrial and some municipal wastes. During 1978, for example, it received about 800 x 10^6 liters of waste. It and a site north of Puerto Rico, which received about 350 x 10^6 liters of waste in 1978, are the only two deep ocean sites being used for U.S. Environmental Protection Agency (EPA)-regulated disposal of wastes (Anderson and Dewling, 1981).

The objective of the National Oceanic and Atmospheric Administration (NOAA) Ocean Dumping Program is to discern the consequences of using the

[1,2]Present address: National Oceanic and Atmospheric Administration, Office of Marine Pollution Assessment, Ocean Dumping Program, Rockville, Maryland 20852.
Presented to Industrial Wastes Workshop.

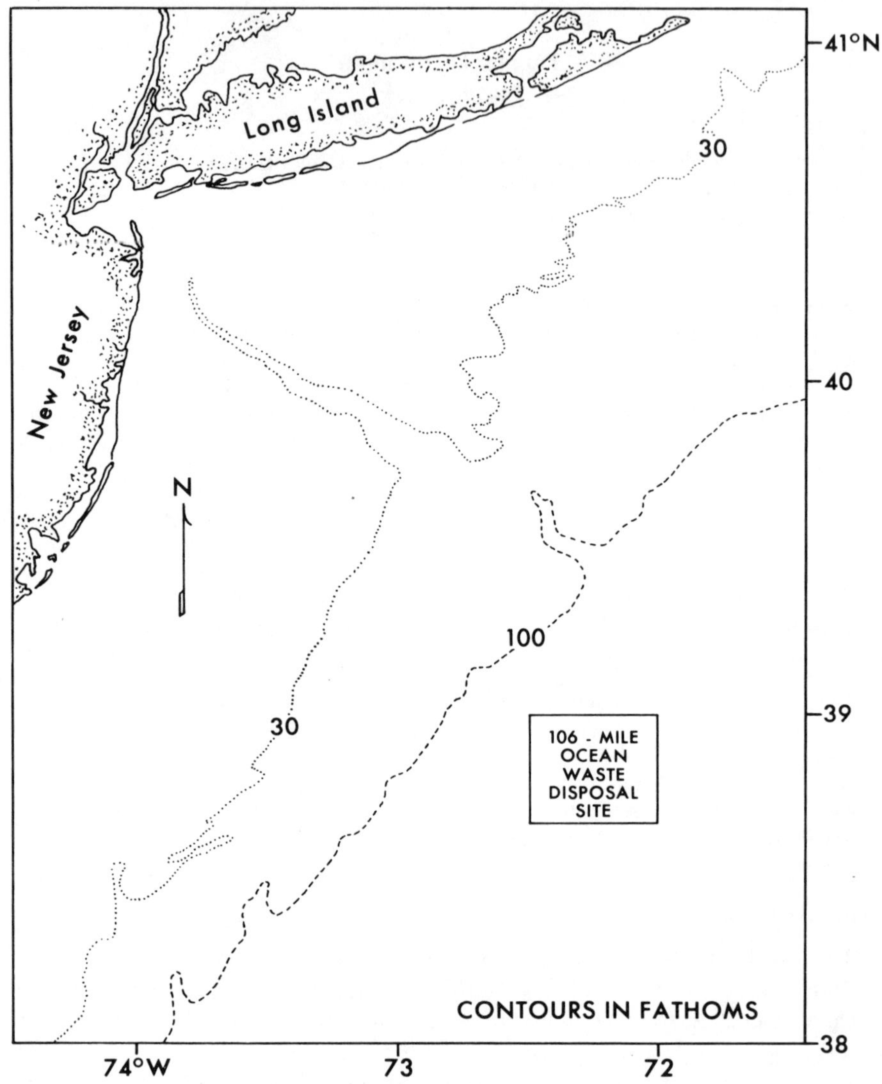

Figure 1. Location of 106-Mile Site.

ocean for waste disposal. In dealing with deep ocean industrial sites, the Program centers on defining the dispersal of waste in terms of its location and concentration in the ocean as a function of time after dumping, the chemical interaction of waste with the marine environment, and, most importantly, the response of the marine ecosystem to the introduction of contaminants. In this paper we discuss some results from ongoing studies on the chemistry, dispersion, advection, and biological effects of wastes dumped at the 106-Mile Site.

WASTE CHARACTERISTICS

Of the 800 x 10^6 liters dumped at the 106-Mile Site in 1978, about 400 x 10^6 liters came from the DuPont plant in Edge Moor, Delaware; 200 x 10^6 liters from the DuPont-Grasselli plant in Linden, New Jersey; 120 x 10^6 liters from the American Cyanamid Warner plant, also in Linden; and the remainder from four smaller sources, one of which was Camden, New Jersey municipal sewage sludge. The NOAA program has considered primarily the three major wastes.

DuPont-Edge Moor waste, derived from titanium dioxide production, is a highly acid solution of ferric chloride (2-4 M HC1 and 0.5-1 M Fe) containing chromium and vanadium at about 100 parts per million (ppm); copper, zinc, nickel, and lead in the 10-100 ppm range; and cadmium at less than 1 ppm. At the rate the waste is dumped, dilution and reaction with the oceanic carbonate system quickly neutralize the acid so there is little effect on oceanic pH. Because of neutralization, the iron precipitates as a hydrous iron floc that adsorbs other waste components (Kester *et al.,* 1981). Current studies with the waste include investigations of the size range of precipitates and changes in particle size that may occur with time. The availability of these particles and adsorbed components to zooplankton and other filter-feeding organisms is affected by particle size.

DuPont-Grasselli waste is an alkaline solution (pH 12-13) of sodium sulfate with small amounts of dissolved metals and 0.5-1.0% by weight organic carbon derived from the manufacture of N,O dimethyl hydroxylamine (DMHA) and anisole (methyphenyl ether). Organic compounds in the waste include methyl sulfate, methanol, phenol, and small amounts of product or unknown compounds. When high concentrations of this material are mixed with seawater, as at the outlet port of a dumping barge, the high alkalinity causes magnesium to precipitate from seawater as magnesium hydroxide (Kester *et al.,* 1978). The precipitate theoretically is unstable at normal oceanic pH of about 8, which prevails almost immediately behind a dumping barge. Nevertheless, particles persist and have been observed acoustically in the ocean after dumping events (Orr and Hess, 1978).

American Cyanamid waste is a slightly acidic (pH 5) solution containing 1-4% organic carbon. Of the three major wastes, it is the most difficult to characterize because it is derived from an array of plant operations. The major waste sources are production of organophosphorus pesticides and water treatment chemicals, with additional wastes from the production of intermediates, surfactants, and chemicals used in the rubber, mining, and paper industries. Unlike the other wastes, no significant particulate phase is associated with the material.

DISPERSION

Measurements of dispersion are basically analyses of the distribution of specific waste characteristics in the ocean after a dumping operation. Given the large initial dilutions that occur, useful characteristics are constituents that are highly concentrated in or are totally unique to the waste. Dispersion measurements have been based on iron analyses, dye that has been added to waste, acoustic sensing of particulate phases that form upon dumping, color changes on the sea surface as seen from satellites or aircraft, and, most recently, gas-chromatographic analysis for organic compounds that are unique to waste.

Wastes are delivered to the 106-Mile Site in approximately 4 x 10^6 liter lots, usually by barge, though occasionally by tanker. By regulation, each dump occurs within a given quadrant of the site, and the quadrant for each waste is changed quarterly. Wastes are discharged by gravity from moving barges or by pumping from moving tankers at a rate not to exceed an EPA-regulated maximum that varies from 70,000 liters/km to 120,000 liters/km. As a result of this procedure, approximately 200 U-shaped, 45 km-long ribbons of waste were laid down within the 106-Mile Site during 1978. Critical questions are: what are the concentrations of waste in the ocean? where in the vertical structure of the ocean does the waste lie? how wide are the waste plumes? and is there any accumulation or additive effect due to repeated dumping events?

Water column stability, acoustically-sensed particle distribution, and iron concentrations measured after dumping events are depicted for summer conditions in Figures 2 through 4, and for winter conditions in Figures 5 through 7. These data indicate that wastes are distributed between the sea surface and the pycnocline. The depth of this mixed layer is shallower in summer than in winter. There is a permanent pycnocline at the 106-Mile Site at about 150-200 m, and between approximately May and September a seasonal pycnocline forms at 10-30 m, as a result of sea surface warming (Ingham *et al.*, 1977). That waste particles are confined to the mixed layer is indicated from high frequency (200 KHz) acoustic sensing (Figures 3, 6; Orr *et al.*, 1980). Chemical data (Figures 4, 7) do not confirm the presence of a vertical limit to waste penetration because in neither case were samples available from below the pycnocline (Kester *et al.*, 1981; Mukherji and Kester, 1980). However, more recent summer data (Mukherji and Kester, 1980) and older information (EG&G, 1977a) collected without ancillary acoustic sensing show, by iron analysis, that waste is distributed primarily within the mixed layer. The presence of a permanent pycnocline in deep water should preclude accumulation of waste on the sea floor. This contrasts with the situation over the continental shelf where, although a seasonal pycnocline is present during the warmer months, the mixed layer during colder months extends to the sea floor.

In principle, waste components should descend below the pycnocline over long periods of time. Dissolved waste constituents and particles could be incorporated with biological detritus or fecal material, which would enhance sinking. However, the greater the time required for sinking, the greater the opportunity for waste dilution by horizontal dispersion, and the lesser the likelihood for waste detection in bottom sediment. Inorganic analyses of sediments (Grieg and Wenzloff, 1977) have not shown increased heavy metal concentrations. Unlike inorganic analyses, where dumping-induced changes must be sought in the context of natural levels of heavy metal concentrations, analyses for waste-specific organic compounds should yield clear evidence of

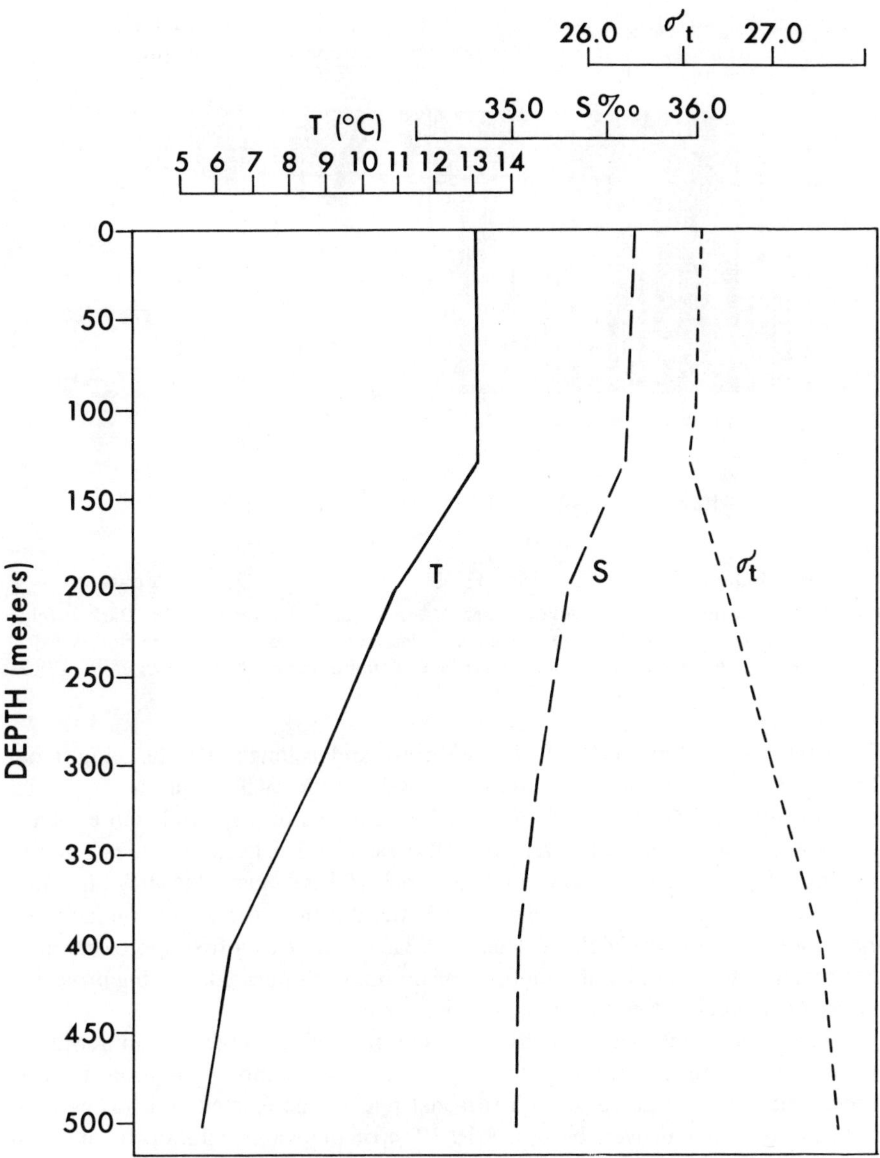

Figure 2. Temperature (T), salinity (S), and density (σ_t) profile from February 1978 (Orr *et al.*, 1980).

waste on the sea floor. Such analyses have not indicated the presence of waste (G.R. Harvey, personal communication).

Waste dispersion, then, is a matter of lateral spreading with restricted vertical migration from a plume initially created by a dumping event. Csanady (1981) indicated extensive dilution created by barge-generated turbulence. Basically, it can be expected that wastes will be mixed into volumes that are about 2.5 times

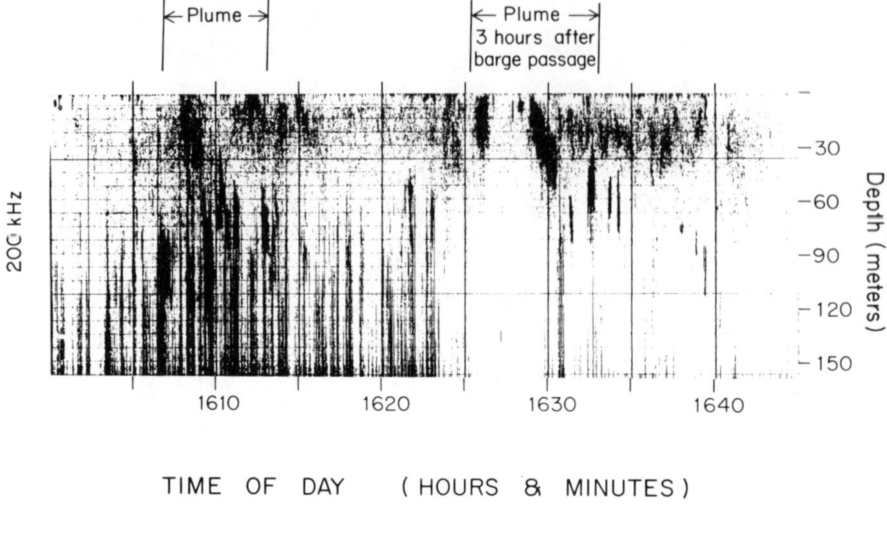

Figure 3. Acoustic signal of DuPont-Edge Moor waste in February 1978. Dark patches extending to about 100 m are due to backscatter from hydrous iron oxide floc. Vertical lines appearing throughout record are an artifact of transducer movement (Orr *et at.*, 1980).

the barge width, three times the barge's draft, and as long as the dumping track. Thus, 4×10^6 liters of waste, dumped over 45 km, will result in an initial dilution factor of 5,000. This is about 1.5 to 2 orders of magnitude more than is achieved by a well-designed sewage outfall (Csanady, 1981) and requires a dumping time of only five to six hours at 250 cm/sec (5 kts). Waste densities vary from 1.02 g/cm^3 to 1.6 g/cm^3, but given the initial dilution, the effective gravity of the wastes become negligible (Csanady, 1981). To at least a first approximation, dilution is determined by the method of dumping and oceanic mixing processes and is independent of waste characteristics.

Some dispersion measurements are summarized in Figure 8 in terms of maximum measured concentrations converted to minimum dilution factors. These data were obtained when a seasonal pycnocline existed and wastes were vertically distributed over the upper 10-30 m of the water column. In any such series of measurements one cannot be sure that maximum measured concentrations are, in fact, the maximum concentrations present. However, taken as a whole, these data indicate that (1) dilution factors within one hour after a dump are about 6,000-8,000, in conformance with the above mentioned mechanism, and (2) subsequent dilution may be slow. In two cases where wastes were followed for about a day, dilutions at the end of the experiment were less than ten times those measured about four hours after the dump. It can be concluded that maximum waste concentrations in the range of 5-50 ppm persist for some time. Lower concentrations prevail under winter conditions because

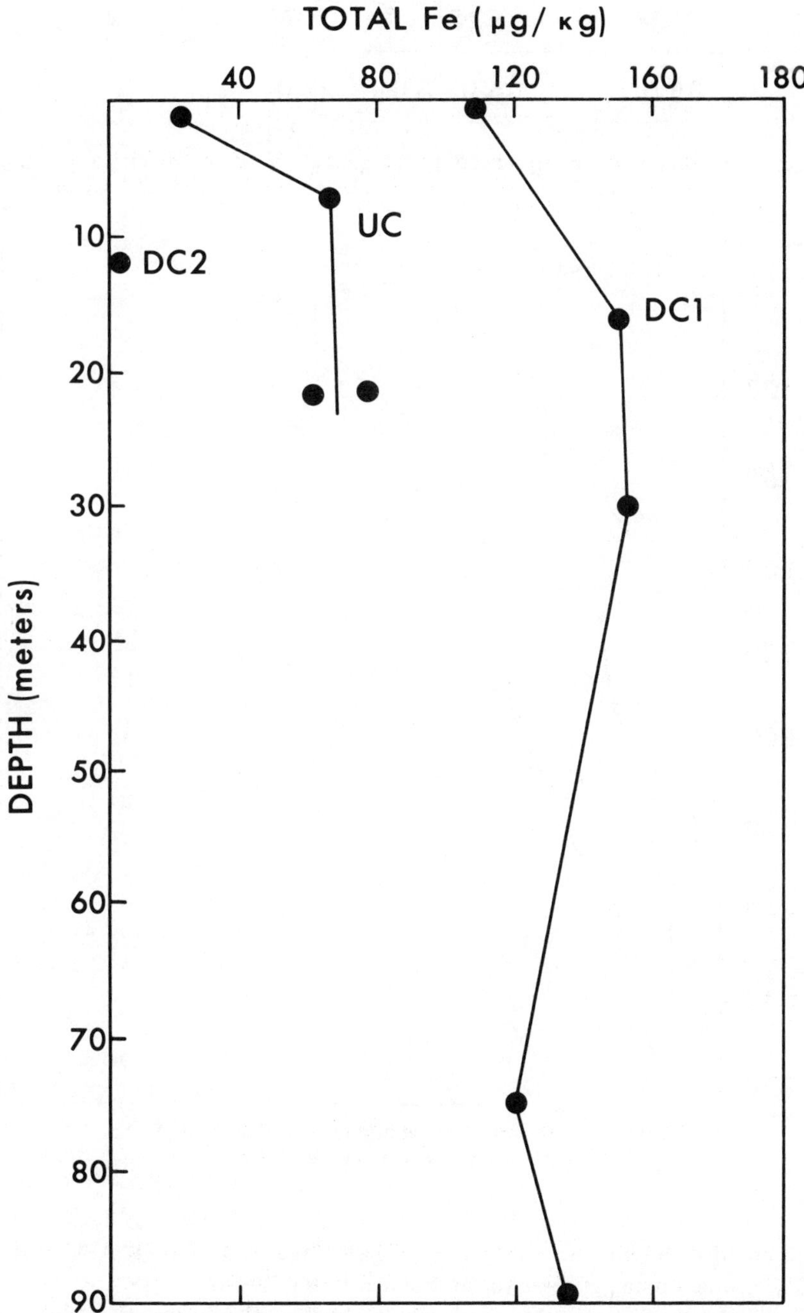

Figure 4. Iron concentration at four hours post-dump in February 1978. Concentrations before dump or away from plume ranged from 0.2 g/kg to 0.9 g/kg. The second downcast (DC2) was made immediately after initial downcasts (DC1) and upcasts (UC) (Mukherji and Kester, 1980).

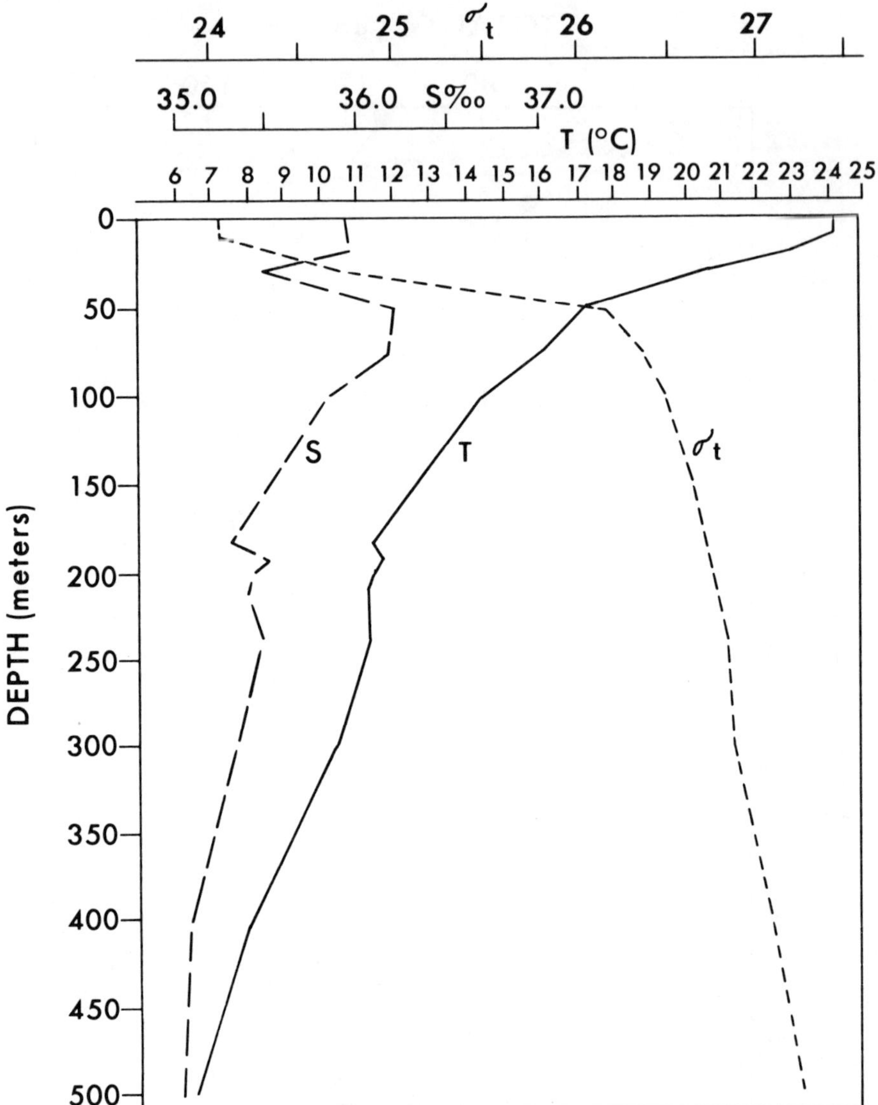

Figure 5. Temperature (T), salinity (S), and density (σ_t) profile from July 1977 (Orr *et al.*, 1980).

wastes are distributed more deeply (compare Figures 4 and 7). Also, during storms, oceanic mixing appears to cause much faster dilution.

Waste dilution is correlated with plume width. The image of a plume of DuPont-Edge Moor waste taken from the LANDSAT II satellite about 12 hours after the plume was created indicates a width of 750 m (Figure 9). Plume widths measured from shipboard can be misleading because without closely spaced

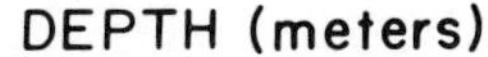

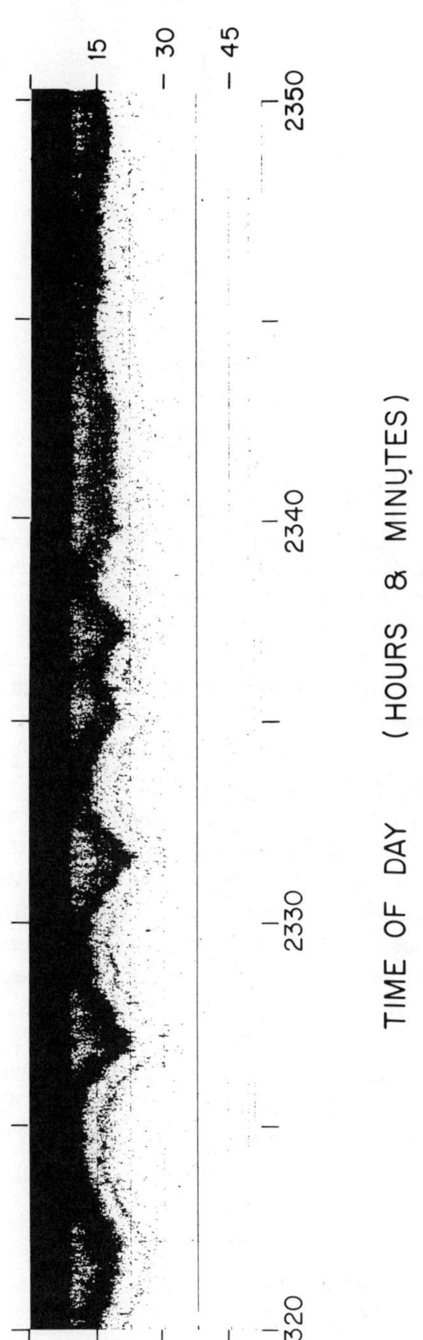

ALBATROSS IV 26 JULY 1977 Edge Moor Waste

riding an internal wave

8 hours after barge passage

Figure 6. Acoustic record of DuPont-Edge Moor waste in July 1977 (Orr *et al.*, 1980).

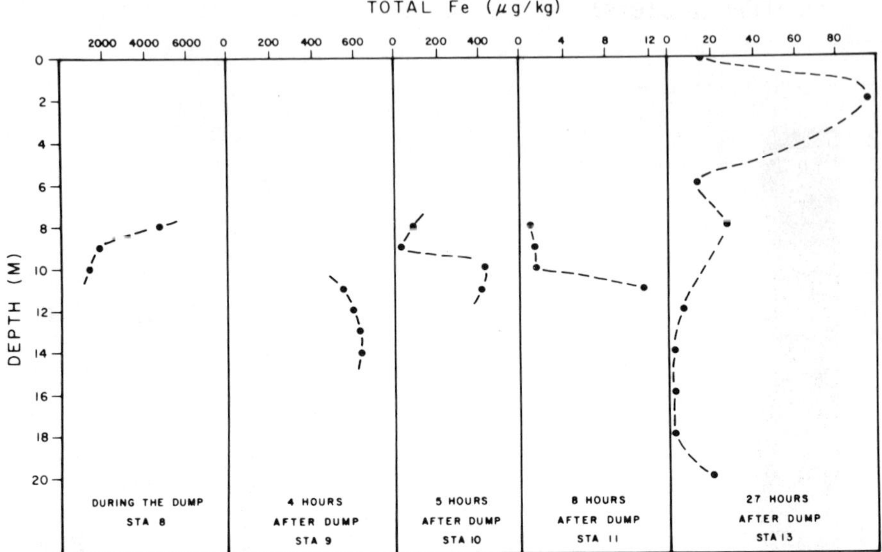

Figure 7. Iron concentrations in DuPont-Edge Moor plume in July 1977. Background concentrations in mixed layer ranged from 0.2 μg/kg to 0.6 μg/kg. Samples obtained within plume were not obtained at exactly the same times, so apparently vertical structure may include horizontal variations in concentration (Kester *et al.*, 1981).

navigational fixes and frequent crossings of one part of a plume, oblique crossings cannot be separated from those at right angles to the plume. Shipboard estimates made over a period of 2-24 hours after a dump may range from a few hundred meters to one or two thousand meters (Orr *et al.*, 1980).

An indication of plume width is provided in Figure 4. Samples for iron analysis were taken by lowering a rosette sampler at what seemed by acoustic data to be the plume center. Five samples were taken while the rosette was descending, four were taken on the ascent, and one bottle (which failed to trip below the pycnocline on the original descent) was filled at 12 m on the second descent. The ship drifted about 180 m in the time required to take these samples, yet the iron concentration near 10 m decreased from 150 μg/liter to 3 μg/liter.

Csanady (1981) considered waste dispersion during the summer, where vertical waste distribution due to initial mixing coincides approximately to the depth limit of the mixed layer. Under these circumstances, dilution is basically a one-dimensional process whereby:

$$D_t = D_o L_o^{-1} \sigma_t \sqrt{2\pi}$$

with: D_t = dilution factor at time, t, after a dump;
D_o = dilution factor from initial mixing;
L_o = initial width of plume;
σ_t = standard deviation of plume width at time, t.

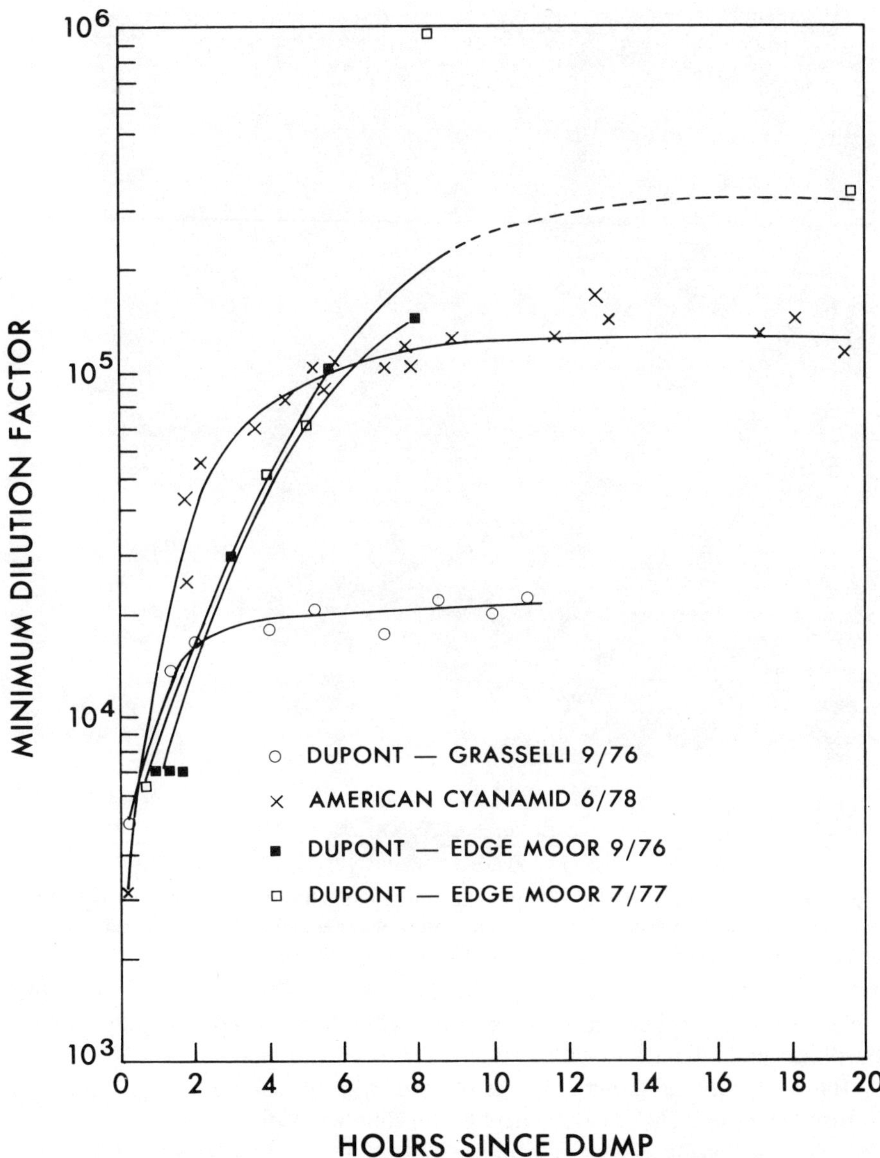

Figure 8. Measured minimum dilution factors, wind speeds, and sea heights. DuPont-Edge Moor, September 1976: 10-12 m/sec and 1-2 m (EG&G, 1977a). DuPont-Edge Moor, July 1977: 10-15 m/sec, 1-2 m; curve shown as dashed line beyond eight hours where data are sparce (Kester *et al.*, 1981). DuPont-Grasselli, September 1976: 4-5 m/sec, calm sea (EG&G, 1977b). American Cyanamid, June 1978: 5 m/sec, 1m or less, except during first two hours (NOAA, unpublished data).

The horizontal waste distribution is assumed to be Gaussian, so that a plume width of $4.3\sigma_t$ incorporates 97% of the total distribution and is a width over

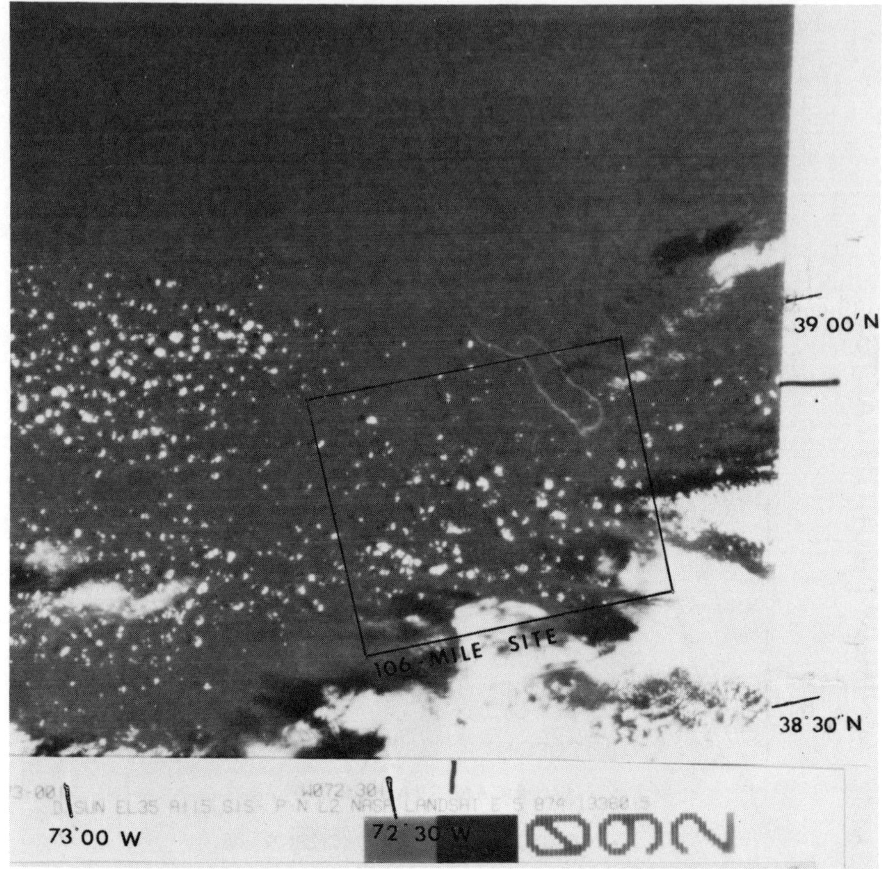

Figure 9. LANDSAT II image of 106-Mile Site (outlined) and surrounding area at 1015 EDT, September 9, 1977. A dump of DuPont Edge-Moor waste began at 2050 EDT, September 8, in the NE quadrant of the site (courtesy of C. Ohlhurst, NASA).

which the waste concentrations rise from 10% of the maximum value to the maximum and fall back down to 10%.

Rates of plume growth are associated with oceanic mixing, which for illustrative purposes is described here by six simple models:

I	$\sigma_t = \sqrt{2\,Kt}$	with	$K = 0.01\ L^{4/3}\ cm^2/sec$;
II	same as I	with	$K = 0.01\ L^{1.15}\ cm^2/sec$;
III	same as I	with	$K = 10^3\ cm^2/sec$;
IV	same as I	with	$K = 10^4\ cm^2/sec$;
V	$\sigma_t = wt$	with	$w = 0.2$ cm/sec;
VI	same as V	with	$w = 2.0$ cm/sec.

In models I and II the eddy diffusion coefficient, K, is assumed to increase as the plume widens and is progressively mixed by larger and larger eddies. The exponent 4/3 in model I is from the classical "4/3 law," while the value 1.15 in model II was chosen by Okubo (1971) to fit best a collection of diffusion experiment results, where L varied from 10^2 m to 10^5 m. Model III employs a constant eddy diffusion coefficient of 10^3 cm^2/sec, which Csanady (1973) recommended as a useful approximation for describing initial plume behavior in the ocean. The value 10^4 cm^2/sec in model IV was estimated by Ichiye *et al.* (1981) from drogue behavior during a waste tracking experiment. The last two models invoke a linear dependence of plume width on time, which could be the consequence of a shear in the vertical distribution of waste (Csanady, 1981). The coefficients 0.2 and 2.0 cover a range of oceanic conditions from quiescence to those in the presence of strong wind.

Minimum dilution factors determined by these models for 20 hours following a dump, assuming a uniform vertical waste distribution over 15 m, are shown in Figure 10. Comparison of these curves with Figure 8 indicates that models III, IV, or V are reasonable approximations in that they predict rather slow changes in concentration and waste levels during the 4-20 hour post-dump period in the 5-50 ppm range. The same models yield values for plume width as a function of time that, as seen in Figure 11 for models III, IV, and V, indicate widths in the 600-1,200 m range at 20 hours after a dump.

There seem to be sufficient data to conclude that crude models that describe the formation of narrow plumes with maximum waste concentrations in the 5-50 ppm range are relevant to waste dispersion over the first day after a dump. A most important question is: for how long do these models apply? The result of extending model III over long times is shown in Table 1. At one month, the model predicts a maximum concentration of 2 ppm and a plume width of 3,000 m. Csanady (1981) suggested that plumes disperse slowly and are narrow until a storm occurs. Under storm conditions, the diffusion coefficient can be expected to increase and horizontal shear will split the plume. Diffusion in both horizontal directions, rather than merely perpendicular to the plume, will decrease concentrations and increase the area over which waste is spread.

ADVECTION

The rate of flow through the dumpsite relative to the frequency of dumping events determines the likelihood that individual plumes will, to some extent, coalesce with one another. If plumes interact, the level of contamination to which organisms are exposed will increase. Therefore, estimates of the residence time of water within the dumpsite and trajectories of water upon leaving the site are needed to assess the consequences of using the site for waste dumping. These estimates differ with the presence or absence of warm-core Gulf Stream eddies at the site.

Eddies form in the northwest Atlantic when a portion of the Gulf Stream encircles a parcel of Sargasso Sea water from its southern boundary and breaks away from the main stream as a clockwise circling eddy of about 50 km in radius. These eddies move roughly along the contour of the continental shelf,

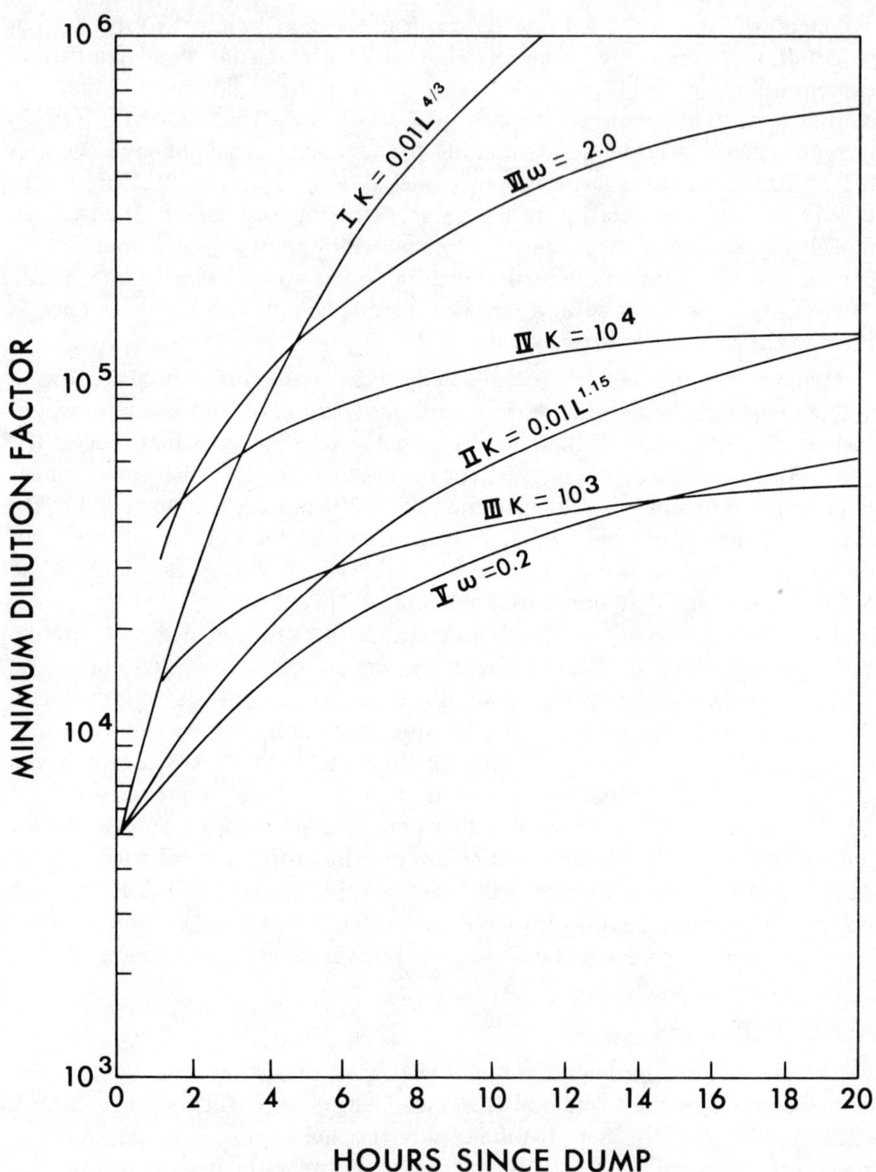

Figure 10. Minimum dilution factors as calculated for models I-VI (see text) assuming vertically uniform waste distributions over upper 15 m of water column.

which near the 106-Mile Site implies a trajectory toward the southwest (Bisagni, 1976). Mizenko and Chamberlain (in press) report that the 106-Mile Site was completely or partially within an eddy about 20% of the time during 1974, 1975, and 1976, and 70% of the time during 1977.

Waste plumes within an eddy move clockwise with the eddy water at speeds

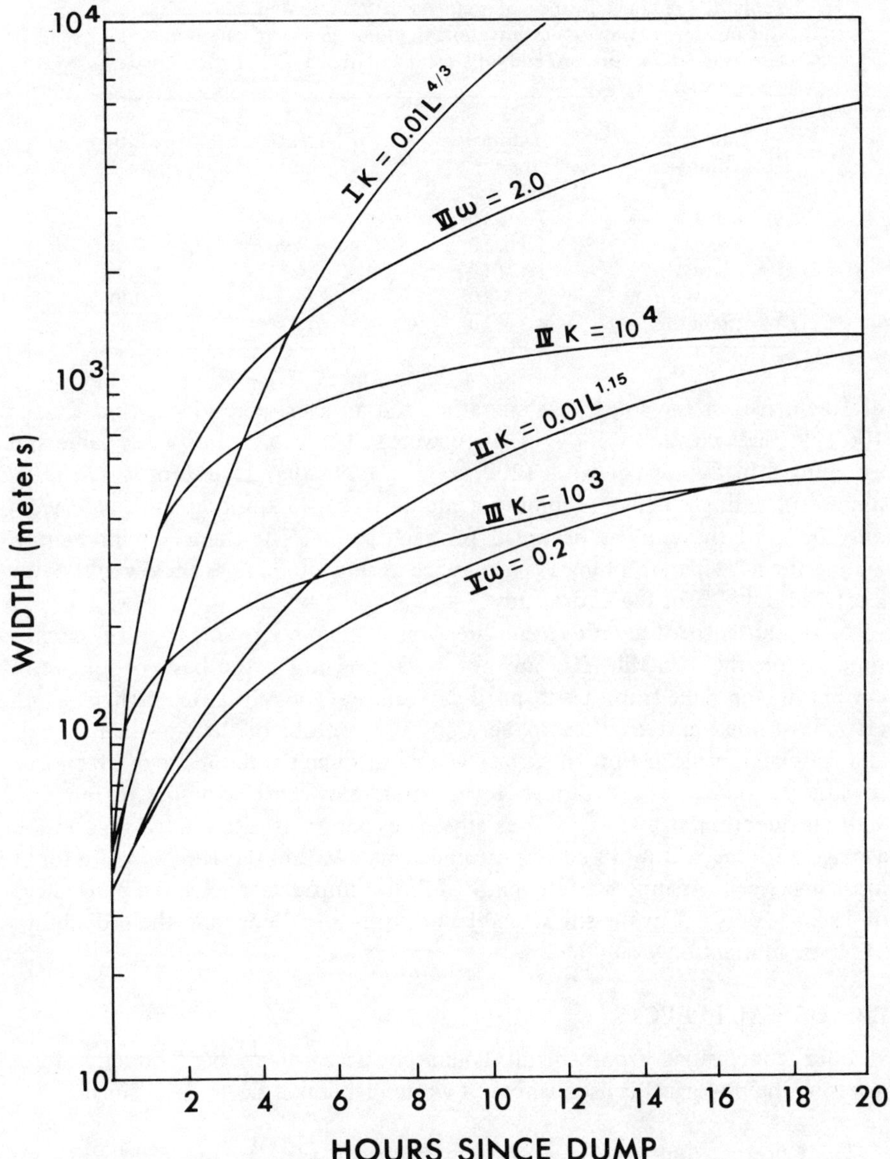

Figure 11. Plume widths as calculated for Models I-VI.

of up to about 50 cm/sec (1 kt). A dump of DuPont-Edge Moor waste, which was followed during 1977, moved about 60 km during the 26 hours it was observed and was about 30 km east of the 106-Mile Site at the end of the experiment (Kester *et al.*, 1981).

The rate at which eddies move toward the southwest is variable, and a single eddy need not move at a uniform speed. Bisagni (1976) found an average speed

Table 1. Predicted history of plume width, minimum dilution factor, and maximum waste concentration if plume grows in one horizontal direction with a constant eddy diffusion coefficient, K = 10^3 cm^2/sec.

Time since dump	Dilution factor	Concentration (ppm)	Width (m)
1 day	7.8×10^4	13	550
2 days	1.1×10^5	9	770
1 week	2.1×10^5	5	1,400
2 weeks	2.9×10^5	3	2,000
1 month	4.3×10^5	2	2,900

of 10 cm/sec for the southwest migration and an average residence time for the 106-Mile Site within an eddy of three weeks. Using the three-week value and assuming four dumps per week (200/yr), it follows that 12 dumps would leave plumes in a single eddy. Assuming a plume to be 45 km long and 1 km wide, only 0.6% of the eddy is occupied for each plume. The chance of interaction among the 12 separate plumes is low, since as separate plumes they would cover a total of only 7% of the entire eddy.

In the absence of an eddy, the direction of plume movement and residence time within the 106-Mile Site are not well-known. On the basis of long-term current meter data from a site northeast of the 106-Mile Site, Ingham *et al.* (1977) assumed current flow to be along the contour of the continental shelf and estimated a mean flow of surface waters through the dumpsite of 10 cm/sec toward the southwest. Although local wind may cause a plume to move in another direction at a faster rate at any given period, it is estimated that on the average a plume will move southwest and remain within the 106-Mile Site for at most one week. Assuming a frequency of four dumps per week and a plume area of 45 km^2, only 3% of the site would be occupied per dump, and the probability of plume interaction would be low.

BIOLOGICAL EFFECTS

Four conclusions from chemical and physical studies bear directly upon assessing the biological consequences of waste disposal at deep ocean sites:

1. At present dumping rates, concentrations of waste are reduced to 50 ppm or less within about two hours of a dump. They may remain near these levels for more than one day. The maximum duration of these levels is unknown.
2. Aside from being diluted, wastes presented to the biological community may be altered from their original condition by precipitate formation, neutralization, or abiotic transformation of organic compounds.
3. Waste plumes are relatively small (approximately 50 km^2) and are restricted in depth to the mixed layer.

Table 2. Average effective concentrations (EC) found to decrease cell counts of *Skeletonema costatum* by 50% during 96-hour exposures (from industry-supplied data).

Waste source	96-hour - EC_{50} (ppm)
DuPont-Edge Moor	2,010[1]
DuPont-Grasselli	970[2]
American Cyanamid	260[3]

[1] Average of results obtained with individual barge samples from May through November 1977 (n = 7; range = 1,100-2,690 ppm).

[2] Average of results obtained with individual barge samples from February through September 1977, excluding value of 8,600 ppm from August (n = 7; range = 330-3,100 ppm).

[3] Average of monthly results from 1976 (n = 12; range = 19-750 ppm).

4. At the present frequency of dumping, each plume is a separate incident affecting a small portion of ocean; about 200 such incidents occur per year.

Components of the marine ecosystem that are most likely to suffer from the practice of deep ocean dumping are those organisms that occupy surface layers and that move passively with and remain in given parcels of water. Free-swimming organisms may encounter plumes, but, unless they are attracted to plumes, their exposure should be brief. Biological studies, therefore, have centered on planktonic bacteria, phytoplankton, and zooplankton.

A measure of the toxicity of wastes on the diatom, *Skeletonema costatum,* is given in Table 2. These data imply little, if any, effect on phytoplankton within waste plumes generated by two of the three dumpers at the 106-Mile Site. However, toxicity results based on single organisms may be misleading. In tests of DuPont-Grasselli waste on growth of *S. costatum,* Murphy *et al.* (1981) found results to vary with the source of the organisms (Figure 12). The most sensitive of the tested strains of *S. costatum* was isolated from open ocean water, while coastal or estuarine isolates of the same species were generally more resistant. Similar results were found for the coccolithophorid, *Emiliania huxleyi* (Figure 12), and for the diatom, *Thalassiosira pseudonana.* In no case were the tested phytoplankton affected by DuPont-Grasselli waste at concentrations as low as those observed within the plumes. Similar tests with other wastes are in progress. While there is some overlap in sensitivity between coastal and oceanic clones of the same species, the general conclusion is that oceanic phytoplankton are less resistant to wastes than those from coastal or estuarine regions. As Murphy *et al.* (1981) pointed out, the relative response among the coastal clones is related to the extent of contamination presumed to be endemic to the

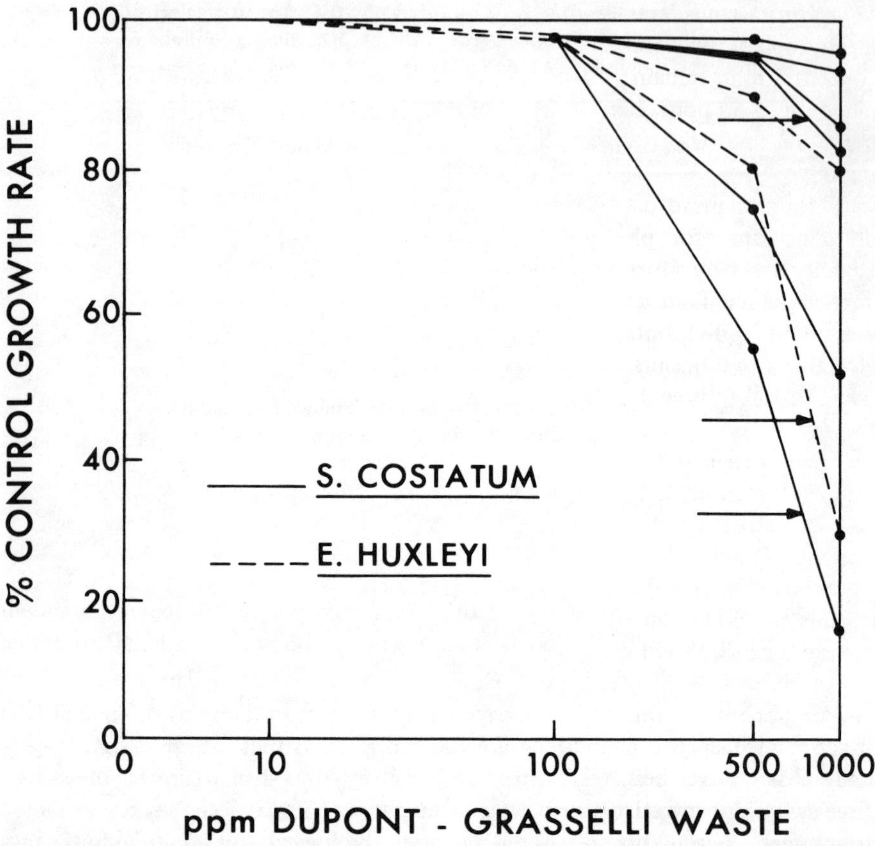

Figure 12. Effect of DuPont-Grasselli waste on growth of *Skeletonema costatum* (six clones) and *Emiliania huxleyi* (two clones). Each point is the average of three determinations with decreases in growth among determinations varying by 0.05% or less. Arrows indicate oceanic clones; other clones were isolated from coastal areas (Murphy *et al.*, 1981).

area from which each clone was isolated. Continuous exposure to contamination may yield clones with a general resistance to anthropogenic inputs (also see Murphy *et al.*, this volume).

Extrapolation of laboratory results to nature is difficult because it is not possible to test all wastes against all possible planktonic species (Hulbert and Jones, 1977; Sherman *et al.*, 1977). Since effects upon growth of phytoplankton vary with species and among subgroups of the same species, it is possible that within waste plumes the species composition of the phytoplankton community is altered. Such an effect has been postulated by Fisher (1975) as a result of chlorinated hydrocarbon additions to marine systems. Thomas and Seibert (1977) observed a shift in community structure toward smaller species when copper was added to controlled ecosystems moored in coastal waters. O'Connors *et al.* (this volume) discuss similar alterations as the result of PCB additions to estuarine systems.

A crucial consideration in the hypothesis that waste dumps result in changes in phytoplankton community structure over a small scale is the time over which concentrations remain effective. If, for instance, waste concentrations in the range of 5-50 ppm can induce community changes, the time required for induction may be greater than the period during which wastes remain at effective levels. Community changes due to copper additions were not evident until four or more days after the additions (Thomas and Seibert, 1977). The doubling time for phytoplankton grown under optimum conditions during toxicity tests by Murphy *et al.* (1981) was about 12 hours. If that time scale applies under field conditions and five doublings are needed to produce a measurable redistribution of species, then concentrations would have to remain effective for 60 hours. In tropical waters the doubling time for phytoplankton may be only three hours (Sheldon and Sutcliffe, 1978), allowing for faster responses.

There are numerous natural controls on phytoplankton growth, such as light, temperature, nutrient concentrations, and grazing pressure, that bear upon the size and structure of communities. It is not known yet if waste dumping, as presently practiced, can exert a significant effect relative to natural variations. The consequence of changes in species composition of phytoplankton is that organisms that are unacceptable as food to zooplankton may dominate; this may disrupt localized food webs over at least short periods.

An extention of the hypothesis that phytoplankton communities within plumes may be altered is that the alteration could be erased as the plume grows. If, as seems to be the case, each plume is a separate incident with the surrounding water being pristine, then the plume increases in size and decreases in waste concentration by entraining seawater with a normal planktonic assemblage. Given that at some plume size and waste concentration the phytoplankton are no longer affected, then as normal phytoplankton growth and dynamics proceed, a temporarily altered population that grew in a smaller and more concentrated plume should vanish since artificially dominant organisms no longer have a competitive advantage.

Capuzzo and Lancaster (1981) found that DuPont-Grasselli waste at concentrations in the 10-50 ppm range caused some mortalities upon the copepod, *Centropages typicus* (Figure 13), and decreased the feeding rates in both that organism and the copepod, *Pseudocalanus* sp. (Figure 14). Since the generation times of zooplankton are longer than those of phytoplankton, it is not clear that small effects would lead to measurable changes in zooplankton populations, though they may contribute to ecological alterations at the phytoplankton level by modifying grazing pressures.

Among organisms higher in the food chain than phytoplankton, the effects of entrainment into contaminant plumes probably are manifested in the same organisms rather than in offspring. Two important questions remain unanswered: do organisms suffer any damage that is not abated as the plume grows and becomes less concentrated? and do organisms assimilate waste components? The latter concern relates to biological dispersion of waste and, if assimilated

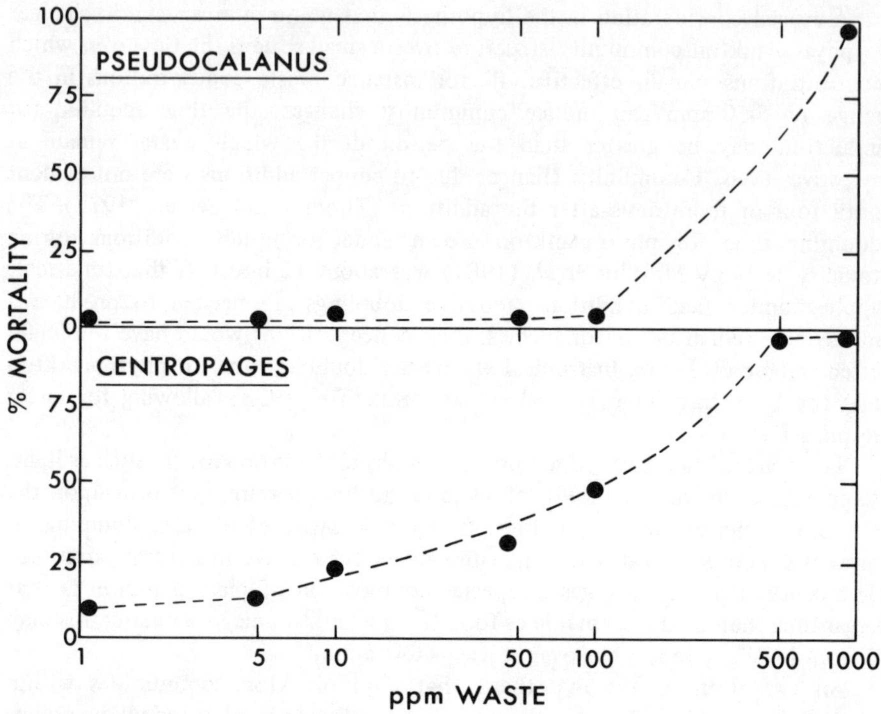

Figure 13. Effect of DuPont-Grasselli waste on mortality of two species of zooplankton. Each point is the mean of six determinations (Capuzzo and Lancaster, 1981).

material is not depurated when concentrations in surrounding water decrease, to possible food chain transfer of disposed chemicals. These aspects of the ocean dumping problem are being addressed through histopathological and chemical studies.

FUTURE CONSIDERATIONS

Some ongoing or planned studies already have been indicated. These include field and laboratory studies of the interaction of wastes with phytoplankton species and populations, laboratory studies with zooplankton, and field sampling for histopathological and chemical analysis. The long-term consequences of ocean dumping depend on the rate of ocean mixing, the advective paths of wastes, and the relative input rates of clean seawater and waste to the dumpsite. These aspects are being addressed through physical measurements of mixing processes, long-term waste tracking experiments, sampling for waste-derived organic compounds at stations away from the dumpsite and identifiable plumes, and radio-tracking from shore of drogues initially placed at the dumpsite. Finally, the rate of decay of waste-derived organic chemicals in the marine environment will be addressed.

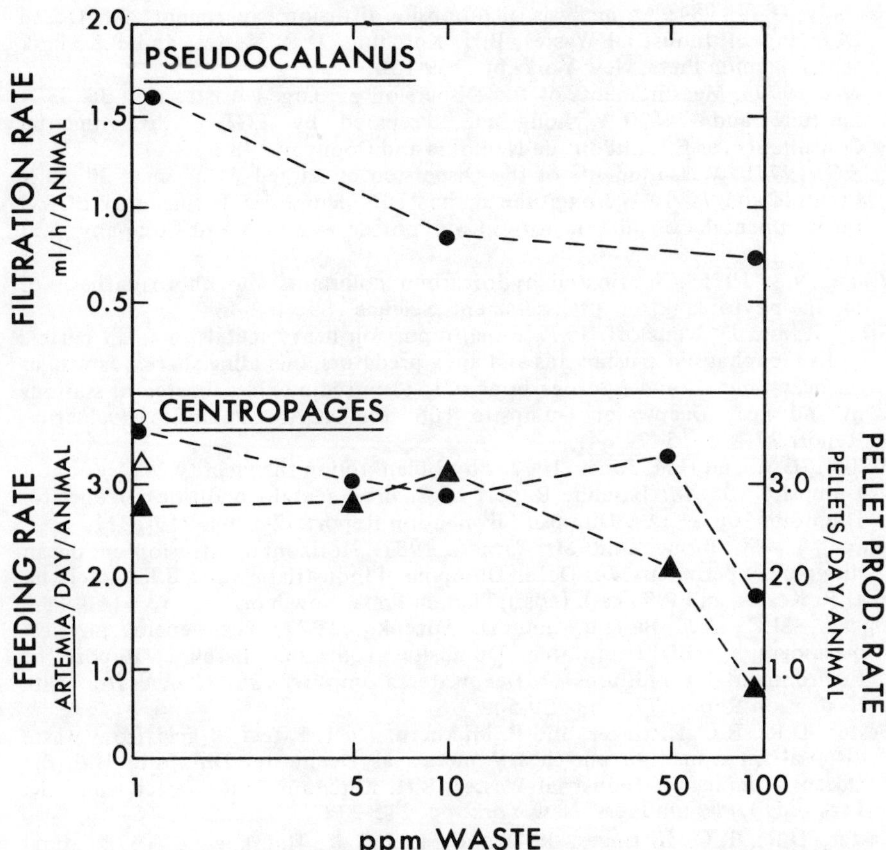

Figure 14. Filtration rates of *Pseudocalanus* and feeding rates and fecal pellet production rates of *Centropages* exposed to DuPont-Grasselli waste for 96 hours. For filtration rates, each point is the mean of 24 determinations; for feeding rates (circles) and fecal pellet production rates (triangles), points represent means of six determinations. Open symbols are control values (Capuzzo and Lancaster, 1981).

REFERENCES

Anderson, P.W.. and R.T. Dewling. 1981. Industrial ocean dumping in EPA Region II–regulatory aspects. *In:* Ocean Dumping of Industrial Wastes, B.H. Ketchum, D.R. Kester, and P.K. Park (eds.), Plenum Press, New York. pp. 25-38.

Bisagni, J.J. 1976. Passage of anticyclonic Gulf Stream eddies through Deepwater Dumpsite 106 during 1974 and 1975. NOAA Dumpsite Evaluation Report 76-1. 39 pp.

Capuzzo, J.M. and B.A. Lancaster. 1981. The effects of pollutants on marine zooplankton at Deepwater Dumpsite 106–preliminary findings. *In:* Ocean Dumping of Industrial Wastes, B.H. Ketchum, D.R. Kester, and P.K. Park (eds.), Plenum Press, New York. pp. 411-420.

Csanady, G.T. 1973. Turbulent Diffusion in the Environment. D. Reidel, Publ., Boston. 248 pp.

Csanady, G.T. 1981. An analysis of dumpsite diffusion experiments. *In:* Ocean Dumping of Industrial Wastes, B.H. Ketchum, D.R. Kester, and P.K. Park (eds.), Plenum Press, New York. pp. 109-130.

EG&G. 1977a. Measurements of the Dispersion of Barged Wastes near 38°33'N Latitude and 74°20'W Longitude. Prepared by EG&G Environmental Consultants for E.I. duPont de Nemours and Company. 78 pp.

EG&G. 1977b. Measurements of the Dispersion of Barged Wastes near 38°50'N Latitude and 72°15'W Longitude at the "106" Dumpsite. Prepared by EG&G Environmental Consultants for E.I. duPont de Nemours and Company. 120 pp.

Fisher, N.S. 1975. Chlorinated hydrocarbon pollutants and photosynthesis of marine phytoplankton; a reassessment. Science 189: 463-464.

Greig, R. and D. Wenzloff. 1977. Final report on heavy metals in small pelagic finfish, euphausiid crustaceans and apex predators, including sharks, as well as on heavy metals and hydrocarbons (C15+) in sediments collected at stations in and near Deepwater Dumpsite 106. *In:* NOAA Dumpsite Evaluation Report 77-1. pp. 547-564.

Hulbert, E.M. and C.M. Jones. 1977. Phytoplankton in the vicinity of Deepwater Dumpsite 106. *In:* Baseline Report of Environmental Conditions Deepwater Dumpsite 106, NOAA Dumpsite Evaluation Report 77-1. pp. 219-231.

Ichiye, T., M. Inoue, and M. Carnes. 1981. Horizontal diffusion in ocean dumping experiments. *In:* Ocean Dumping of Industrial Wastes, B.K. Ketchum, D.R. Kester, and P.K. Park (eds.), Plenum Press, New York. pp. 131-160.

Ingham, M.C., J.J. Bisagni, and D. Mizenko. 1977. The general physical oceanography of Deepwater Dumpsite 106. *In:* Baseline Report of Environmental Conditions at Deepwater Dumpsite 106, NOAA Dumpsite Evaluation Report 77-1. pp. 29-54.

Kester, D.R., R.C. Hittinger, and P. Mukherji. 1981. Effect of acid from waste disposal on transition and heavy metals at Deepwater Dumpsite 106. *In:* Ocean Dumping of Industrial Wastes, B.H. Ketchum, D.R. Kester, and P.K. Park (eds.), Plenum Press, New York. pp. 215-232.

Kester, D.R., R.C. Hittinger, R. Mukherji, and K. Hausknecht. 1978. Metal Studies at Deepwater Dumpsite 106. Final Report, NOAA Grant No. 04-7-158-44027/44064. 49 pp.

Mizenko, D. and J.L. Chamberlain. In press. Gulf Stream anticyclonic eddies and shelf water at Deep Water Dumpsite 106 during 1977. NOAA Dumpsite Evaluation Report 79-1.

Mukherji, P. and D.R. Kester. 1980. Comparison of summer-winter test dumps of acid iron at Deep Water Dumpsite-106. Univ. Rhode Island Mar. Tech. Rep. 80-3.

Murphy, L.S., R.R.L. Guillard, and J. Gavis. This volume. Evolution of resistant phytoplankton strains through exposure to marine pollutants. pp. 401-412.

Murphy, L.S., P.R. Hoar, and R.A. Belastock. 1981. The effects of industrial waste on marine phytoplankton. *In:* Ocean Dumping of Industrial Wastes, B.H. Ketchum, D.R. Kester, and P.K. Park (eds.), Plenum Press, New York. pp. 399-410.

Okubo, A. 1971. Oceanic diffusion diagrams. Deep-Sea Res. 18: 789-802.

O'Connors, H.B., J., C.D. Powers, C.F. Wurster, K.D. Wyman, G.M. Nau-Ritter, and R.G. Rowland. This volume. Fates and biological effects on plankton of particle-sorbed PCB's in coastal waters. pp. 423-449.

Orr, M.H., L. Baxter, and F.R. Hess. 1980. Remote acoustic sensing of the particulate phase of industrial chemical wastes and sewage sludge: report on the seasonal variability of the dispersion of the particulate phase as observed from three cruises: July 1977, January-February 1978, and April 1978. Woods Hole Oceanogr. Inst. Tech. Rep. WHOI 79-38. 153 pp.

Orr, M.H. and F.R. Hess. 1978. Acoustic monitoring of industrial chemical waste released at Deep Water Dumpsite 106. J. Geophys. Res. 83: 6145-6154.

Sheldon, R.W. and W.H. Sutcliffe. 1978. Generation times of 3 h for Sargasso Sea microplankton determined by ATP analysis. Limnol. Oceanogr. 23: 1051-1055.

Sherman, K., D. Busch, and D. Bearse. 1977. Deepwater Dumpsite 106: zooplankton studies. *In:* Baseline Report of Environmental Conditions at Deep Water Dumpsite 106, NOAA Dumpsite Evaluation Report 77-1. pp. 233-308.

Thomas, W.H. and D.L.R. Seibert. 1977. Effects of copper on the dominance and the diversity of algae: Controlled Ecosystems Pollution Experiment. Bull. Mar. Sci. 27: 23-33.

SYMPOSIUM PARTICIPANTS

Robert Adamski
Division of Water Pollution Control
New York City Department of
Environmental Protection
2358 Municipal Building
New York, NY 10007

Jack W. Anderson
Battelle
Pacific Northwest Laboratories
Marine Research Laboratory
Sequim, WA 98382

Peter W. Anderson
U.S. Environmental Protection Agency
Region II, Marine and Wetlands
Protection Branch
26 Federal Plaza
New York, NY 10278

Rudolf H. Bieri
Virginia Institute of Marine Science
Gloucester Point, VA 23062

Donald F. Boesch
Louisiana Universities Marine Consortium
Star Route
Box 541
Chauvin, LA 70344

William F. Bohlen
Marine Sciences Institute
University of Connecticut
Avery Point
Groton, CT 06340

Henry J. Bokuniewicz
Marine Sciences Research Center
State University of New York
Stony Brook, NY 11794

Anthony Calabrese
NOAA National Marine Fisheries Service
Northeast Fisheries Center
Milford Laboratory
Milford, CT 06460

Melbourne R. Carriker
College of Marine Studies
University of Delaware
Lewes, DE 19958

John Ciancia
U.S. Environmental Protection Agency
Region II, Surveillance and Analysis
Division
Edison, NJ 08817

Norman H. Cutshall
Oak Ridge National Laboratory
Building 3037
Oak Ridge, TN 37830

Will P. Davis
U.S. Environmental Protection Agency
Gulf Breeze Laboratory
Sabine Island
Gulf Breeze, FL 32561

Richard Delgado
Division of Water Resources
New Jersey Department of Environmental
Protection
Trenton, NJ 08625

Maurice de Picciotto
Marine Terminals Department
The Port Authority of New York and
New Jersey
1 World Trade Center
New York, NY 10048

Richard T. Dewling
U.S. Environmental Protection Agency
Region II
26 Federal Plaza
New York, NY 10007

Billy Edge
Dames and Moore
7101 Wisconsin Avenue
Washington, D.C. 20014

Stephen C. Esser
New Jersey Marine Sciences Consortium
Building 22
Fort Hancock, NJ 07732

Richard Field
U.S. Environmental Protection Agency
Storm and Combined Sewer Section
Building 10
Edison, NJ 08817

David R. Franz
Biology Department
Brooklyn College
City University of New York
Brooklyn, NY 11210

Robert B. Gordon
Department of Geology and Geophysics
Yale University
New Haven, CT 06520

Edith Gould
NOAA National Marine Fisheries Service
Northeast Fisheries Center
Milford Laboratory
Milford, CT 06460

Charles G. Gunnerson
International Bank for Reconstruction and Development
Transportation, Water, and Telecommunications Department
1818 H Street, NW
Washington, DC 20433

Evelyn Haines
University of Georgia Marine Institute
Sapelo Island, GA 31327

Fred C. Hart
Fred C. Hart Associates, Inc.
527 Madison Avenue
New York, NY 10022

Jaret Johnson
J.B.F. Scientific Corp.
2 Jewel Drive
Wilmington, MA 01887

DeAnne S. Julius
The World Bank
1818 H Street, NW
Washington, DC 20433

Kenneth S. Kamlet
National Wildlife Federation
1412 16th Street, NW
Washington, DC 20036

Gary S. Kleppell
Department of Biological Sciences
Allan Hancock Foundation
University of Southern California
University Park
Los Angeles, CA 90007

Theodore J. Kneip
New York University Medical Center
Institute of Environmental Medicine
A.J. Lanza Laboratories
Long Meadow Road
Tuxedo, NY 10987

Donald W. Lear
U.S. Environmental Protection Agency
Central Regional Laboratory
839 Bestgate Road
Annapolis, MD 21401

John J. Lee
Department of Biology
City College of New York
Convent Avenue at 138th Street
New York, NY 10031

Richard Lee
Skidaway Institute of Oceanography
P.O. Box 13687
Savannah, GA 31406

Albert Levinson
Department of Economics
Queens College
Kissena Boulevard
Flushing, NY 11367

Paul J. Lioy
New York University Medical Center
Institute of Environmental Medicine
A.J. Lanza Laboratories
Long Meadow Road
Tuxedo, NY 10987

Carol D. Litchfield
E.I. du Pont de Nemours and Company
Haskell Laboratory
Wilmington, DE 19898

Robert J. Livingston
Department of Biological Science
Florida State University
Tallahasee, FL 32306

Arlene C. Longwell
NOAA National Marine Fisheries Service
Northeast Fisheries Center
Milford Laboratory
Milford, CT 06460

Bruce B. McCain
NOAA National Marine Fisheries Service
Northwest and Alaska Fisheries Center
2725 Montlake Boulevard East
Seattle, WA 98112

Richard A. McGrath
Taxon, Inc.
50 Grove Street
Salem, MA 01970

J.L. McHugh
Marine Sciences Research Center
State University of New York
Stony Brook, NY 11794

Thomas F. McKinney
Dames and Moore
6 Commerce Drive
Cranford, NJ 07016

John J.A. McLaughlin
Louis Calder Conservation and
Ecology Study Center
Armonk, NY 10504

Thomas C. Malone
Oceanographic Sciences Division
Brookhaven National Laboratory
Upton, NY 11973

John Mancini
Environmental Engineering and
Science Program
Manhattan College
Manhattan College Parkway
Riverdale, NY 10471

Garry F. Mayer
NOAA Office of Marine Pollution
Assessment
Northeast Office
State University of New York
Stony Brook, NY 11794

Alan J. Mearns
NOAA Office of Marine Pollution
Assessment
Pacific Office
7600 Sand Point Way, NE
Seattle, WA 98115

Allan D. Michael
Taxon, Inc.
50 Grove Street
Salem, MA 01970

Joseph T. Miller
Bureau of Water Pollution Control
New York City Department of
Environmental Protection
40 Worth Street
New York, NY 10013

Raymond P. Morgan II
Appalachian Environmental
Laboratory–UMCEES
Gunter Hall
Frostburg State College Campus
Frostburg, MD 21532

James A. Mueller
Environmental Engineering and
Science Program
Manhattan College
Manhattan College Parkway
Riverdale, NY 10471

Robert A. Murchelano
NOAA National Marine Fisheries Service
Northeast Fisheries Center
Oxford Laboratory
Oxford, MD 21654

Lynda S. Murphy
Bigelow Laboratory for Ocean Sciences
West Boothbay Harbor, ME 04575

Edward P. Myers
NOAA Office of Minerals and Energy
2001 Wisconsin Avenue, NW
Washington, DC 20235

Alan I. Mytelka
Interstate Sanitation Commission
10 Columbus Circle
New York, NY 10019

Richard Newman
New York State Department of Environmental Conservation
2 World Trade Center
New York, NY 10047

DelWayne R. Nimmo
Environmental Research and Technology, Inc.
1716 Heath Parkway
Fort Collins, CO 80522

Joel S. O'Connor
NOAA Office of Marine Pollution Assessment
Northeast Office
State University of New York
Stony Brook, NY 11794

Joseph M. O'Connor
New York University Medical Center
Institute of Environmental Medicine
A.J. Lanza Laboratories
Long Meadow Road
Tuxedo, NY 10987

Thomas P. O'Connor
NOAA Office of Marine Pollution Assessment
Ocean Dumping Program
Rockville, MD 20852

Harold B. O'Connors
Marine Sciences Research Center
State University of New York
Stony Brook, NY 11794

John B. Pearce
NOAA National Marine Fisheries Service
Northeast Fisheries Center
Sandy Hook Laboratory
Highlands, NJ 07732

Lois S. Peters
Center for Science and Technology Policy
New York University
50 West 4th Street
New York, NY 10013

Michael Pilson
Graduate School of Oceanography
University of Rhode Island
Kingston, RI 02881

Jan C. Prager
U.S. Environmental Protection Agency
Environmental Research Laboratories
South Ferry Road
Narragansett, RI 02882

William Pressman
New York City Department of Water Resources
Municipal Building
40 Worth Street
New York, NY 10007

E.A. Regna
Allied Chemical Corporation
Industrial Chemical Division
P.O. Box 1139R
Morristown, NJ 07960

John J. Roswell
Water Quality Development
New York City Department of Environmental Protection
40 Worth Street
New York, NY 10013

Marwan Sadat
Office of Industrial Pretreatment and Sludge Management
Division of Water Resources
New Jersey Department of Environmental Protection
Trenton, NJ 08625

Thomas K. Sawyer
NOAA National Marine Fisheries Service
Northeast Fisheries Center
Oxford Laboratory
Oxford, MD 21654

Stephen Schroeder
National Resources Defense Council
917 15th Street, NW
Washington, DC 20005

Jerry Schubel
Marine Sciences Research Center
State University of New York
Stony Brook, NY 11794

Douglas A. Segar
SEAMOcean
Box 1627
Wheaton, MD 20902

Jonathan H. Sharp
College of Marine Studies
University of Delaware
Lewes, DE 19958

Marjorie J. Sherwood
Southern California Coastal Water
Research Project
646 West Pacific Coast Highway
Long Beach, CA 90806

Lowell V. Sick
NOAA National Marine Fisheries Service
Southeast Fisheries Center
P.O. Box 12607
Charleston, SC 29412

Carl J. Sindermann
NOAA National Marine Fisheries Service
Northeast Fisheries Center
Sandy Hook Laboratory
Highlands, NJ 07732

Christopher Sobotowski
Division of Water Resources
New Jersey Department of Environmental
Protection
Box CN029
Trenton, NJ 08625

Paul R. Spitzer
New Zealand Wildlife Service
Box 501
Invercargill
New Zealand

Claire Stern
Claire Stern Associates
1466 Broadway
New York, NY 10036

Dennis J. Suszkowski
U.S. Army Corps of Engineers
New York District
Operations Division
26 Federal Plaza
New York, NY 10007

R. Lawrence Swanson
NOAA Office of Marine Pollution
Assessment
Rockville, MD 20852

John Szeligowski
Tippetts-Abbett-McCarthy-Stratton
Engineers and Architects
655 Third Avenue
New York, NY 10017

John H. Tietjen
Department of Biology
City College of New York
Convent Avenue at 138th Street
New York, NY 10031

John F. Timoney
Department of Microbiology
New York State Veterinary College
Cornell University
Ithaca, NY 14850

James T.B. Tripp
Environmental Defense Fund
475 Park Avenue South
New York, NY 10016

John J. Walsh
Oceanographic Sciences Division
Brookhaven National Laboratory
Upton, NY 11973

Mitchell Wendell
Interstate Sanitation Commission
10 Columbus Circle
New York, NY 10019

Douglas A. Wolfe
NOAA Office of Marine Pollution
Assessment
325 S. Broadway
Boulder, CO 80303

Jack Q. Word
College of Fisheries
University of Washington
Seattle, WA 98195

Thomas D. Wright
U.S. Army Engineer Waterways
Experimental Station
P.O. Box 631
Vicksburg, MS 39180

Kevin D. Wyman
Oceanographic Sciences Division
Brookhaven National Laboratory
Upton, NY 11973

David K. Young
Naval Ocean Research and Development
Activity
Oceanography Division
NSTL Station, MS 39529

David R. Young
Marine Services Group
Dames and Moore
1100 Glendon Avenue
Los Angeles, CA 90024

Martha W. Young
U.S. Fish and Wildlife Service
National Coastal Ecosystems Team
NASA/Slidell Computer Complex
Slidell, LA 70458

INDEX